The IMA Volumes in Mathematics and its Applications

Volume 144

Series Editors
Douglas N. Arnold   Arnd Scheel

Institute for Mathematics and its Applications (IMA)

The **Institute for Mathematics and its Applications** was established by a grant from the National Science Foundation to the University of Minnesota in 1982. The primary mission of the IMA is to foster research of a truly interdisciplinary nature, establishing links between mathematics of the highest caliber and important scientific and technological problems from other disciplines and industries. To this end, the IMA organizes a wide variety of programs, ranging from short intense workshops in areas of exceptional interest and opportunity to extensive thematic programs lasting a year. IMA Volumes are used to communicate results of these programs that we believe are of particular value to the broader scientific community.

The full list of IMA books can be found at the Web site of the Institute for Mathematics and its Applications:

http://www.ima.umn.edu/springer/volumes.html

Presentation materials from the IMA talks are available at

http://www.ima.umn.edu/talks/

Douglas N. Arnold, Director of the IMA

* * * * * * * * * *

IMA ANNUAL PROGRAMS

1982–1983	Statistical and Continuum Approaches to Phase Transition
1983–1984	Mathematical Models for the Economics of Decentralized Resource Allocation
1984–1985	Continuum Physics and Partial Differential Equations
1985–1986	Stochastic Differential Equations and Their Applications
1986–1987	Scientific Computation
1987–1988	Applied Combinatorics
1988–1989	Nonlinear Waves
1989–1990	Dynamical Systems and Their Applications
1990–1991	Phase Transitions and Free Boundaries
1991–1992	Applied Linear Algebra
1992–1993	Control Theory and its Applications
1993–1994	Emerging Applications of Probability
1994–1995	Waves and Scattering
1995–1996	Mathematical Methods in Material Science
1996–1997	Mathematics of High Performance Computing

Continued at the back

Michael Eastwood   Willard Miller, Jr.
Editors

Symmetries and Overdetermined Systems of Partial Differential Equations

Michael Eastwood
School of Mathematical Sciences
The University of Adelaide
SA 5005, Australia
http://www.maths.adelaide.edu.au/pure/staff/meastwood.html

Willard Miller, Jr.
School of Mathematics
University of Minnesota
127 Vincent Hall, 206 Church St. SE
Minneapolis, MN 55455
http://www.ima.umn.edu/~miller/

Series Editors
Douglas N. Arnold
Arnd Scheel
Institute for Mathematics and its Applications
University of Minnesota
Minneapolis, MN 55455
USA

ISBN 978-1-4419-2533-6 e-ISBN 978-0-387-73831-4

Mathematics Subject Classification (2000): 58J70, 35-06, 53-06, 70-06, 81-06

Camera-ready copy provided by the IMA.

Printed on acid-free paper.

9 8 7 6 5 4 3 2 1

springer.com

FOREWORD

This IMA Volume in Mathematics and its Applications

Symmetries and Overdetermined Systems of Partial Differential Equations

contains expository and research papers that were presented at the IMA Summer Program which was held July 17–August 4, 2006. This summer program was dedicated to the memory of Thomas P. Branson, who played a leading role in its conception and organization, but did not live to see its realization. We would like to thank Michael Eastwood (University of Adelaide) and Willard Miller, Jr. (University of Minnesota) for their superb role as workshop organizers and editors of the proceedings.

We take this opportunity to thank the National Science Foundation for its support of the IMA.

Series Editors

Douglas N. Arnold, Director of the IMA

Arnd Scheel, Deputy Director of the IMA

IN MEMORY OF PROFESSOR THOMAS P. BRANSON
OCTOBER 10, 1953 – MARCH 11, 2006

PREFACE

The idea for this Summer Program at the Institute for Mathematics and its Applications came from Thomas Branson. In March 2006 preparations were in full swing when Tom suddenly passed away. It was a great shock to all who knew him. These Proceedings are dedicated to his memory.

In 1979 Thomas Branson gained his PhD from MIT under the supervision of Irving Segal. After holding various positions elsewhere, he worked at the University of Iowa since 1985. He has published over 70 substantial scientific papers with motivation coming from geometry, physics, and symmetry. His work is distinguished as a highly original blend of these three elements. Tom's influence can be clearly seen in these Proceedings and the impact of his work will be always felt. His untimely death is deeply saddening and we shall all miss him sorely.

The symmetries that were studied in the Summer Program naturally arise in several different ways. Firstly, there are the symmetries of a differential geometric structure. By definition, these are the vector fields that preserve the structure in question–the Killing fields of Riemannian differential geometry, for example. Secondly, the symmetries can be those of another differential operator. For example, the Riemannian Killing equation itself is projectively invariant whilst the ordinary Euclidean Laplacian gives rise to conformal symmetries. In addition, there are higher symmetries defined by higher order operators. Physics provides other natural sources of symmetries, especially through string theory and twistor theory.

These symmetries are usually highly constrained–viewed as differential operators, they themselves are overdetermined or have symbols that are subject to overdetermined differential equations. As a typical example, the symbol of a symmetry of the Laplacian must be a conformal Killing field (or a conformal Killing tensor for a higher order symmetry). The Summer Program considered the consequences of overdeterminacy and partial differential equations of finite type.

The question of what it means to be able to solve explicitly a classical or quantum mechanical system, or to solve it in multiple ways, is the subject matter of the integrability theory and superintegrability theory of Hamiltonian systems. Closely related is the theory of exactly solvable and quasi-exactly solvable systems. All of these approaches are associated with the structure of the spaces of higher dimensional symmetries of these systems.

Symmetries of classical equations are intimately connected with special coordinate systems, separation of variables, conservation laws, and integrability. But only the simplest equations are currently understood from these points of view. The Summer Program provided an opportunity for comparison and consolidation, especially in relation to the Dirac equation and massless fields of higher helicity.

Parabolic differential geometry provides a synthesis and generalization of various classical geometries including conformal, projective, and CR. It also provides a very rich geometrical source of overdetermined partial differential operators. Even in the flat model G/P, for G a semisimple Lie group with P a parabolic subgroup, there is much to be gleaned from the representation theory of G. In particular, the Bernstein-Gelfand-Gelfand (BGG) complex is a series of G-invariant differential operators, the first of which is overdetermined. Conformal Killing tensors, for example, may be viewed in this way. Exterior differential systems provide the classical approach to such overdetermined operators. But there are also tools from representation theory and especially the cohomology of Lie algebras that can be used. The Summer Program explored these various approaches.

There are many areas of application. In particular, there are direct links with physics and especially conformal field theory. The AdS/CFT correspondence in physics (or Fefferman-Graham ambient metric construction in mathematics) provides an especially natural route to conformal symmetry operators. There are direct links with string theory and twistor theory. Also, there are numerical schemes based on finite element methods via the BGG complex, moving frames, and other symmetry based methods. The BGG complex arises in many areas of mathematics, both pure and applied. When it is recognized as such, there are immediate consequences.

There are close connections and even overlapping work being done in several areas of current research related to the topics above. The main idea of this Summer Program was to bring together relevant research groups for the purpose of intense discussion, interaction, and fruitful collaboration.

In summary, the topics considered in the Summer Program included:

- Symmetries of geometric structures and differential operators,
- Overdetermined systems of partial differential equations,
- Separation of variables and conserved quantities,
- Integrability, superintegrability and solvable systems,
- Parabolic geometry and the Bernstein-Gelfand-Gelfand complex,
- Interaction with representation theory,
- Exterior differential systems,
- Finite element schemes, discrete symmetries, moving frames, and numerical analysis,
- Interaction with string theory and twistor theory.

The first week was devoted to expository/overview sessions. The articles derived from these lectures form the first half of this collection. We recommend these articles as particularly valuable for people new to the subject seeking an overall understanding of overdetermined systems and their applications.

The following two weeks focussed on more specialized research talks with one or two themes each day. The second half of this volume consists of articles derived from these talks.

It remains to thank the IMA for sponsoring and hosting this event. We believe that the participants found it extremely valuable and hope that these Proceedings will preserve and convey some of the spirit of the meeting itself.

Michael Eastwood
School of Mathematical Sciences
University of Adelaide

Willard Miller, Jr.
School of Mathematics
University of Minnesota

CONTENTS

Part I: Expository Articles

OVERDETERMINED SYSTEMS, CONFORMAL DIFFERENTIAL GEOMETRY, AND THE BGG COMPLEX*

ANDREAS ČAP†

Abstract. This is an expanded version of a series of two lectures given at the IMA summer program "Symmetries and overdetermined systems of partial differential equations". The main part of the article describes the Riemannian version of the prolongation procedure for certain overdetermined systems obtained recently in joint work with T.P. Branson, M.G. Eastwood, and A.R. Gover. First a simple special case is discussed, then the (Riemannian) procedure is described in general.

The prolongation procedure was derived from a simplification of the construction of Bernstein–Gelfand–Gelfand (BGG) sequences of invariant differential operators for certain geometric structures. The version of this construction for conformal structures is described next. Finally, we discuss generalizations of both the prolongation procedure and the construction of invariant operators to other geometric structures.

Key words. Overdetermined system, prolongation, invariant differential operator, conformal geometry, parabolic geometry.

AMS(MOS) subject classifications. 35N10, 53A30, 53A40, 53C15, 58J10, 58J70.

1. Introduction. The plan for this article is as follows. I'll start by describing a simple example of the Riemannian version of the prolongation procedure of [4]. Next, I will explain how the direct observations used in this example can be replaced by tools from representation theory to make the procedure work in general. The whole procedure is based on an inclusion of the group $O(n)$ into $O(n+1,1)$. Interpreting this inclusion geometrically leads to a relation to conformal geometry, that I will discuss next. Via the conformal Cartan connection, the ideas used in the prolongation procedure lead to a construction of conformally invariant differential operators from a twisted de–Rham sequence. On manifolds which are locally conformally flat, this leads to resolutions of certain locally constant sheaves, which are equivalent to the (generalized) Bernstein–Gelfand–Gelfand (BGG) resolutions from representation theory. In the end, I will outline generalizations to other geometric structures.

It should be pointed out right at the beginning, that this presentation basically reverses the historical development. The BGG resolutions in representation theory were originally introduced in [3] and [15] in the 1970's. The constructions were purely algebraic and combinatorial, based

*Supported by project P15747–N05 of the "Fonds zur Förderung der wissenschaftlichen Forschung" (FWF) and by the Insitute for Mathematics and its Applications (IMA).

†Fakultät für Mathematik, Universität Wien, Nordbergstraße 15, A–1090 Wien, Austria; and International Erwin Schrödinger Institute for Mathematical Physics, Boltzmanngasse 9, A-1090 Wien, Austria (andreas.cap@esi.ac.at).

on the classification of homomorphisms of Verma modules. It was known to the experts that there is a relation to invariant differential operators on homogeneous spaces, with conformally invariant operators on the sphere as a special case. However it took some time until the relevance of ideas and techniques from representation theory in conformal geometry was more widely appreciated. An important step in this direction was the work on the curved translation principle in [12]. In the sequel, there were some attempts to construct invariant differential operators via a geometric version of the generalized BGG resolutions for conformal and related structures, in particular in [2].

This problem was completely solved in the general setting of parabolic geometries in [9], and the construction was significantly simplified in [5]. In these constructions, the operators occur in patterns, and the first operators in each pattern form an overdetermined system. For each of these systems, existence of solutions is an interesting geometric condition. In [4] it was shown that weakening the requirement on invariance (for example forgetting conformal aspects and just thinking about Riemannian metrics) the construction of a BGG sequence can be simplified. Moreover, it can be used to rewrite the overdetermined system given by the first operator(s) in the sequence as a first order closed system, and this continues to work if one adds arbitrary lower order terms.

I am emphasizing these aspects because I hope that this will clarify two points which would otherwise remain rather mysterious. On the one hand, we will not start with some overdetermined system and try to rewrite this in closed form. Rather than that, our starting point is an auxiliary first order system of certain type which is rewritten equivalently in two different ways, once as a higher order system and once in closed form. Only in the end, it will follow from representation theory, which systems are covered by the procedure.

On the other hand, if one starts the procedure in a purely Riemannian setting, there are some choices which seem unmotivated. These choices are often dictated if one requires conformal invariance.

2. An example of the prolongation procedure.

2.1. The setup. The basics of Riemannian geometry are closely related to representation theory of the orthogonal group $O(n)$. Any representation of $O(n)$ gives rise to a natural vector bundle on n–dimensional Riemannian manifolds and any $O(n)$–equivariant map between two such representation induces a natural vector bundle map. This can be proved formally using associated bundles to the orthonormal frame bundle.

Informally, it suffices to know (at least for tensor bundles) that the standard representation corresponds to the tangent or cotangent bundle, and the correspondence is natural with respect to direct sums and tensor products. A linear map between two representations of $O(n)$ can be expressed in terms of a basis induced from an orthonormal basis in the standard representation. Starting from a local orthonormal frame of the (co)tangent bundle, one may locally use the same formula in induced frames on any Riemannian manifold. Equivariancy under the group $O(n)$ means that the result is independent of the initial choice of a local orthonormal frame. Hence one obtains a global, well defined bundle map.

The basic strategy for our prolongation procedure is to embed $O(n)$ into a larger Lie group G, and then look how representations of G behave when viewed as representations of the subgroup $O(n)$. In this way, representation theory is used as a way to organize symmetries. A well known inclusion of this type is $O(n) \hookrightarrow O(n+1)$, which is related to viewing the sphere S^n as a homogeneous Riemannian manifold. We use a similar, but slightly more involved inclusion.

Consider $\mathbb{V} := \mathbb{R}^{n+2}$ with coordinates numbered from 0 to $n+1$ and the inner product defined by

$$\langle (x_0, \dots, x_{n+1}), (y_0, \dots, y_{n+1}) \rangle := x_0 y_{n+1} + x_{n+1} y_0 + \sum_{i=1}^{n} x_i y_i.$$

For this choice of inner product, the basis vectors $e_1, \dots, e_n$ span a subspace $\mathbb{V}_1$ which is a standard Euclidean $\mathbb{R}^n$, while the two additional coordinates are what physicists call light cone coordinates, i.e. they define a signature $(1,1)$ inner product on $\mathbb{R}^2$. Hence the whole form has signature $(n+1,1)$ and we consider its orthogonal group $G = O(\mathbb{V}) \cong O(n+1,1)$. There is an evident inclusion $O(n) \hookrightarrow G$ given by letting $A \in O(n)$ act on $\mathbb{V}_1$ and leaving the orthocomplement of $\mathbb{V}_1$ fixed.

In terms of matrices, this inclusion maps $A \in O(n)$ to the block diagonal matrix $\begin{pmatrix} 1 & 0 & 0 \\ 0 & A & 0 \\ 0 & 0 & 1 \end{pmatrix}$ with blocks of sizes 1, n, and 1. The geometric meaning of this inclusion will be discussed later.

The representation of A as a block matrix shows that, as a representation of $O(n)$, $\mathbb{V} = \mathbb{V}_0 \oplus \mathbb{V}_1 \oplus V_2$, where $\mathbb{V}_0$ and $\mathbb{V}_2$ are trivial representations spanned by e_{n+1} and e_0, respectively. We will often denote elements of $\mathbb{V}$ by column vectors with three rows, with the bottom row corresponding to $\mathbb{V}_0$.

If we think of $\mathbb{V}$ as representing a bundle, then differential forms with values in that bundle correspond to the representations $\Lambda^k\mathbb{R}^n \otimes \mathbb{V}$ for $k = 0, \dots, n$. Of course, for each k, this representation decomposes as $\oplus_{i=0}^{2}(\Lambda^k\mathbb{R}^n \otimes \mathbb{V}_i)$, but for the middle component $\Lambda^k\mathbb{R}^n \otimes \mathbb{V}_1$, there is a finer decomposition. For example, if $k=1$, then $\mathbb{R}^n \otimes \mathbb{R}^n$ decomposes as

$$\mathbb{R} \oplus S^2_0\mathbb{R}^n \oplus \Lambda^2\mathbb{R}^n$$

into trace–part, tracefree symmetric part and skew part. We actually need only $k = 0, 1, 2$, where we get the picture

$$
\begin{array}{ccc}
\mathbb{R} & \mathbb{R}^n & \Lambda^2\mathbb{R}^n \\
\mathbb{R}^n & \mathbb{R}\oplus S^2_0\mathbb{R}^n\oplus\Lambda^2\mathbb{R}^n & \mathbb{R}^n\oplus W_2\oplus\Lambda^3\mathbb{R}^n \\
\mathbb{R} & \mathbb{R}^n & \Lambda^2\mathbb{R}^n
\end{array}
\tag{2.1}
$$

and we have indicated some components which are isomorphic as representations of $O(n)$. Observe that assigning homogeneity $k+i$ to elements of $\Lambda^k\mathbb{R}^n\otimes\mathbb{V}_i$, we have chosen to identify components of the same homogeneity.

We will make these identifications explicit in the language of bundles immediately, but let us first state how we will use them. For the left column, we will on the one hand define $\partial:\mathbb{V}\to\mathbb{R}^n\otimes\mathbb{V}$, which vanishes on $\mathbb{V}_0$ and is injective on $\mathbb{V}_1\oplus\mathbb{V}_2$. On the other hand, we will define $\delta^*:\mathbb{R}^n\otimes\mathbb{V}\to\mathbb{V}$ by using inverse identifications. For the right hand column, we will only use the identifications from right to left to define $\delta^*:\Lambda^2\mathbb{R}^n\otimes\mathbb{V}\to\mathbb{R}^n\otimes\mathbb{V}$. Evidently, this map has values in the kernel of $\delta^*:\mathbb{R}^n\otimes\mathbb{V}\to\mathbb{V}$, so $\delta^*\circ\delta^*=0$. By constructions, all these maps preserve homogeneity. We also observe that $\ker(\delta^*)=S^2_0\mathbb{R}^n\oplus\operatorname{im}(\delta^*)\subset\mathbb{R}^n\otimes\mathbb{V}$.

Now we can carry all this over to any Riemannian manifold of dimension n. Sections of the bundle V corresponding to $\mathbb{V}$ can be viewed as triples consisting of two functions and a one–form. Since the representation $\mathbb{R}^n$ corresponds to T^*M, the bundle corresponding to $\Lambda^k\mathbb{R}^n\otimes\mathbb{V}$ is $\Lambda^kT^*M\otimes V$. Sections of this bundle are triples consisting of two k–forms and one T^*M–valued k–form. If there is no danger of confusion with abstract indices, we will use subscripts $i=0,1,2$ to denote the component of a section in $\Lambda^kT^*M\otimes V_i$. To specify our maps, we use abstract index notation and define $\partial:V\to T^*M\otimes V$, $\delta^*:T^*M\otimes V\to V$ and $\delta^*:\Lambda^2T^*M\otimes V\to T^*M\otimes V$ by

$$
\partial\begin{pmatrix}h\\ \varphi_b\\ f\end{pmatrix}:=\begin{pmatrix}0\\ hg_{ab}\\ -\varphi_a\end{pmatrix},\quad
\delta^*\begin{pmatrix}h_b\\ \varphi_{bc}\\ f_b\end{pmatrix}:=\begin{pmatrix}\frac{1}{n}\varphi_c{}^c\\ -f_b\\ 0\end{pmatrix},\quad
\delta^*\begin{pmatrix}h_{ab}\\ \varphi_{abc}\\ f_{ab}\end{pmatrix}:=\begin{pmatrix}\frac{-1}{n-1}\varphi_{ac}{}^c\\ \frac{1}{2}f_{ab}\\ 0\end{pmatrix}.
$$

The numerical factors are chosen in such a way that our example fits into the general framework developed in Section 3.

We can differentiate sections of V using the component–wise Levi–Civita connection, which we denote by ∇. Note that this raises homogeneity by one. The core of the method is to mix this differential term with an algebraic one. We consider the operation $\Gamma(V)\to\Omega^1(M,V)$ defined by $\Sigma\mapsto\nabla\Sigma+\partial\Sigma$. Since ∂ is tensorial and linear, this defines a linear connection $\tilde{\nabla}$ on the vector bundle V.

We are ready to define the class of systems that we will look at. Choose a bundle map (not necessarily linear) $A:V_0\oplus V_1\to S^2_0T^*M$, and view it as $A:V\to T^*M\otimes V$. Notice that this implies that A increases homogeneities. Then consider the system

$$
\tilde{\nabla}\Sigma+A(\Sigma)=\delta^*\psi\qquad\text{for some }\psi\in\Omega^2(M,V).
\tag{2.2}
$$

We will show that on the one hand, this is equivalent to a second order system on the V_0–component Σ_0 of Σ and on the other hand, it is equivalent to a first order system on Σ in closed form.

2.2. The first splitting operator. Since A by definition has values in $\ker(\delta^*)$ and $\delta^* \circ \delta^* = 0$, the system (2.2) implies $\delta^*(\tilde{\nabla}\Sigma) = 0$. Hence we first have to analyze the operator $\delta^* \circ \tilde{\nabla} : \Gamma(V) \to \Gamma(V)$. Using abstract indices and denoting the Levi–Civita connection by ∇_a we obtain

$$\Sigma = \begin{pmatrix} h \\ \varphi_b \\ f \end{pmatrix} \overset{\tilde{\nabla}_a}{\mapsto} \begin{pmatrix} \nabla_a h \\ \nabla_a \varphi_b + h g_{ab} \\ \nabla_a f - \varphi_a \end{pmatrix} \overset{\delta^*}{\mapsto} \begin{pmatrix} \frac{1}{n}\nabla^b \varphi_b + h \\ -\nabla_a f + \varphi_a \\ 0 \end{pmatrix},$$

From this we can read off the set of all solutions of $\delta^*\tilde{\nabla}\Sigma = 0$. We can arbitrarily choose f. Vanishing of the middle row then forces $\varphi_a = \nabla_a f$, and inserting this, vanishing of the top row is equivalent to $h = -\frac{1}{n}\nabla^b\nabla_b f = -\frac{1}{n}\Delta f$, where Δ denotes the Laplacian. Hence we get

PROPOSITION 2.2. *For any $f \in C^\infty(M, \mathbb{R})$, there is a unique $\Sigma \in \Gamma(V)$ such that $\Sigma_0 = f$ and $\delta^*(\nabla\Sigma + \delta\Sigma) = 0$. Mapping f to this unique Σ defines a second order linear differential operator $L : \Gamma(V_0) \to \Gamma(V)$, which is explicitly given by*

$$L(f) = \begin{pmatrix} -\frac{1}{n}\Delta f \\ \nabla_a f \\ f \end{pmatrix} = \sum_{i=0}^{2} (-1)^i (\delta^* \nabla)^i \begin{pmatrix} 0 \\ 0 \\ f \end{pmatrix}.$$

The natural interpretation of this result is that V_0 is viewed as a quotient bundle of V, so we have the tensorial projection $\Sigma \mapsto \Sigma_0$. The operator L constructed provides a differential splitting of this tensorial projection, which is characterized by the simple property that its values are in the kernel of $\delta^*\tilde{\nabla}$. Therefore, L and its generalizations are called *splitting operators*.

2.3. Rewriting as a higher order system. We have just seen that the system $\tilde{\nabla}\Sigma + A(\Sigma) = \delta^*\psi$ from 2.1 implies that $\Sigma = L(f)$, where $f = \Sigma_0$. Now by Proposition 2.2, the components of $L(f)$ in V_0 and V_1 are f and ∇f, respectively. Hence $f \mapsto A(L(f))$ is a first order differential operator $\Gamma(V_0) \to \Gamma(S^2_0 T^*M)$. Conversely, any first order operator $D_1 : \Gamma(V_0) \to \Gamma(S^2_0 T^*M) \subset \Omega^1(M, V)$ can be written as $D_1(f) = A(L(f))$ for some $A : V \to T^*M \otimes V$ as in 2.1.

Next, for $f \in \Gamma(V_0)$ we compute

$$\tilde{\nabla}L(f) = \tilde{\nabla}_a \begin{pmatrix} -\frac{1}{n}\Delta f \\ \nabla_b f \\ f \end{pmatrix} = \begin{pmatrix} -\frac{1}{n}\nabla_a \Delta f \\ \nabla_a \nabla_b f - \frac{1}{n} g_{ab}\Delta f \\ 0 \end{pmatrix}.$$

The middle component of this expression is the tracefree part $\nabla_{(a}\nabla_{b)_0} f$ of $\nabla^2 f$.

PROPOSITION 2.3. *For any operator $D_1 : C^\infty(M,\mathbb{R}) \to \Gamma(S^2_0T^*M)$ of first order, there is a bundle map $A : V \to T^*M \otimes V$ such that $f \mapsto L(f)$ and $\Sigma \mapsto \Sigma_0$ induce inverse bijections between the sets of solutions of*

$$\nabla_{(a}\nabla_{b)_0} f + D_1(f) = 0 \tag{2.3}$$

and of the basic system (2.2).

Proof. We can choose $A : V_0 \oplus V_1 \to S^2_0T^*M \subset T^*M \otimes V$ in such a way that $D_1(f) = A(L(f))$ for all $f \in \Gamma(V_0)$. From above we see that $\tilde{\nabla}L(f) + A(L(f))$ has vanishing bottom component and middle component equal to $\nabla_{(a}\nabla_{b)_0} f + D_1(f)$. From (2.1) we see that sections of $\operatorname{im}(\delta^*) \subset T^*M \otimes V$ are characterized by the facts that the bottom component vanishes, while the middle one is skew symmetric. Hence $L(f)$ solves (2.2) if and only if f solves (2.3). Conversely, we know from 2.2 that any solution Σ of (2.2) satisfies $\Sigma = L(\Sigma_0)$, and the result follows. □

Notice that in this result we do not require D_1 to be linear. In technical terms, an operator can be written in the form $f \mapsto \nabla_{(a}\nabla_{b)_0} f + D_1(f)$ for a first order operator D_1, if and only if it is of second order, quasi–linear and its principal symbol is the projection $S^2T^*M \to S^2_0T^*M$ onto the tracefree part.

2.4. Rewriting in closed form. Suppose that Σ is a solution of (2.2), i.e. $\tilde{\nabla}\Sigma + A(\Sigma) = \delta^*\psi$ for some ψ. Then the discussion in 2.3 shows that the two bottom components of $\tilde{\nabla}\Sigma + A(\Sigma)$ actually have to vanish. Denoting the components of Σ as before, there must be a one–form τ_a such that

$$\begin{pmatrix} \nabla_a h + \tau_a \\ \nabla_a \varphi_b + h g_{ab} + A_{ab}(f,\varphi) \\ \nabla_a f - \varphi_a \end{pmatrix} = 0. \tag{2.4}$$

Apart from the occurrence of τ_a, this is a first order system in closed form, so it remains to compute this one–form.

To do this, we use the *covariant exterior derivative* $d^{\tilde{\nabla}} : \Omega^1(M,V) \to \Omega^2(M,V)$ associated to $\tilde{\nabla}$. This is obtained by coupling the exterior derivative to the connection $\tilde{\nabla}$, so in particular on one–forms we obtain

$$d^{\tilde{\nabla}}\omega(\xi,\eta) = \tilde{\nabla}_\xi(\omega(\eta)) - \tilde{\nabla}_\eta(\omega(\xi)) - \omega([\xi,\eta]).$$

Explicitly, on $\Omega^1(M,V)$ the operator $d^{\tilde{\nabla}}$ is given by

$$\begin{pmatrix} h_b \\ \varphi_{bc} \\ f_b \end{pmatrix} \mapsto 2\begin{pmatrix} \nabla_{[a}h_{b]} \\ \nabla_{[a}\varphi_{b]c} - h_{[a}g_{b]c} \\ \nabla_{[a}f_{b]} + \varphi_{[ab]} \end{pmatrix}, \tag{2.5}$$

where square brackets indicate an alternation of abstract indices.

Now almost by definition, $d^{\tilde{\nabla}}\tilde{\nabla}\Sigma$ is given by the action of the curvature of $\tilde{\nabla}$ on Σ. One easily computes directly that this coincides with the component–wise action of the Riemann curvature. In particular, this is only non–trivial on the middle component. On the other hand, since $A(\Sigma) = A_{ab}(f,\varphi)$ is symmetric, we see that $d^{\tilde\nabla}(A(\Sigma))$ is concentrated in the middle component, and it certainly can be written as $\Phi_{abc}(f,\nabla f,\varphi,\nabla\varphi)$ for an appropriate bundle map Φ. Together with the explicit formula, this shows that applying the covariant exterior derivative to (2.4) we obtain

$$\begin{pmatrix} 2\nabla_{[a}\tau_{b]} \\ -R_{ab}{}^{d}{}_{c}\varphi_d + \Phi_{abc}(f,\nabla f,\varphi,\nabla\varphi) - 2\tau_{[a}g_{b]c} \\ 0 \end{pmatrix} = 0.$$

Applying δ^*, we obtain an element with the bottom two rows equal to zero and top row given by

$$\tfrac{1}{n-1}\big(R_a{}^{cd}{}_{c}\varphi_d - \Phi_a{}^{c}{}_{c}(f,\nabla f,\varphi,\nabla\varphi)\big) + \tau_a,$$

which gives a formula for τ_a. Finally, we define a bundle map $C : V \to T^*M\otimes V$ by

$$C\begin{pmatrix} h \\ \varphi_b \\ f \end{pmatrix} := \begin{pmatrix} \frac{-1}{n-1}\big(R_a{}^{cd}{}_{c}\varphi_d - \Phi_a{}^{c}{}_{c}(f,\varphi,\varphi,-hg-A(f,\varphi))\big) \\ A_{ab}(f,\varphi) \\ 0 \end{pmatrix}$$

to obtain

THEOREM 2.4. *Let $D : C^\infty(M,\mathbb{R}) \to \Gamma(S^2_0T^*M)$ be a quasi–linear differential operator of second order whose principal symbol is the projection $S^2T^*M \to S^2_0T^*M$ onto the tracefree part. Then there is a bundle map $C : V \to T^*M\otimes V$ which has the property that $f \mapsto L(f)$ and $\Sigma \mapsto \Sigma_0$ induce inverse bijections between the sets of solutions of $D(f) = 0$ and of $\tilde\nabla\Sigma + C(\Sigma) = 0$. If D is linear, then C can be chosen to be a vector bundle map.*

Since for any bundle map C, a solution of $\tilde\nabla\Sigma + C(\Sigma) = 0$ is determined by its value in a single point, we conclude that any solution of $D(f) = 0$ is uniquely determined by the values of f, ∇f and Δf in one point. Moreover, if D is linear, then the dimension of the space of solutions is always $\leq n+2$. In this case, $\tilde\nabla + C$ defines a linear connection on the bundle V, and the maximal dimension can be only attained if this connection is flat.

Let us make the last step explicit for $D(f) = \nabla_{(a}\nabla_{b)_0}f + A_{ab}f$ with some fixed section $A_{ab} \in \Gamma(S^2_0T^*M)$. From formula (2.5) we conclude that

$$\Phi_{abc}(f,\nabla f,\varphi,\nabla\varphi) = 2f\nabla_{[a}A_{b]c} + 2A_{c[b}\nabla_{a]}f,$$

and inserting we obtain the closed system

$$\begin{cases} \nabla_a h - \frac{1}{n-1}\big(R_a{}^{cd}{}_{c}\varphi_d + f\nabla^c A_{ac} + \varphi^c A_{ac}\big) = 0 \\ \nabla_a\varphi_b + hg_{ab} + fA_{ab} = 0 \\ \nabla_a f - \varphi_a = 0 \end{cases}$$

which is equivalent to $\nabla_{(a}\nabla_{b)_0}f + A_{ab}f = 0$.

2.5. Remark. As a slight detour (which however is very useful for the purpose of motivation) let me explain why the equation $\nabla_{(a}\nabla_{b)_0}f + A_{ab}f = 0$ is of geometric interest. Let us suppose that f is a nonzero function. Then we can use it to conformally rescale the metric g to $\hat{g} := \frac{1}{f^2}g$. Now one can compute how a conformal rescaling affects various quantities, for example the Levi–Civita connection. In particular, we can look at the conformal behavior of the Riemannian curvature tensor. Recall that the Riemann curvature can be decomposed into various components according to the decomposition of $S^2(\Lambda^2\mathbb{R}^n)$ as a representation of $O(n)$. The highest weight part is the *Weyl curvature*, which is independent of conformal rescalings.

Contracting the Riemann curvature via $\mathrm{Ric}_{ab} := R_{ca}{}^c{}_b$, one obtains the *Ricci curvature*, which is a symmetric two tensor. This can be further decomposed into the *scalar curvature* $R := \mathrm{Ric}^a{}_a$ and the tracefree part $\mathrm{Ric}^0_{ab} = \mathrm{Ric}_{ab} - \frac{1}{n}Rg_{ab}$. Recall that a Riemannian metric is called an *Einstein metric* if the Ricci curvature is proportional to the metric, i.e. if $\mathrm{Ric}^0_{ab} = 0$.

The behavior of the tracefree part of the Ricci curvature under a conformal change $\hat{g} := \frac{1}{f^2}g$ is easily determined explicitly, see [1]. In particular, $\hat{g}$ is Einstein if and only if

$$\nabla_{(a}\nabla_{b)_0}f + A_{ij}f = 0$$

for an appropriately chosen $A_{ij} \in \Gamma(S^2_0T^*M)$. Hence existence of a nowhere vanishing solution to this equation is equivalent to the possibility of rescaling g conformally to an Einstein metric.

From above we know that for a general non–trivial solution f of this system and each $x \in M$, at least one of $f(x)$, $\nabla f(x)$, and $\Delta f(x)$ must be nonzero. Hence $\{x : f(x) \neq 0\}$ is a dense open subset of M, and one obtains a conformal rescaling to an Einstein metric on this subset.

3. The general procedure. The procedure carried out in an example in Section 2 can be vastly generalized by replacing the standard representation by an arbitrary irreducible representation of $G \cong O(n+1,1)$. (Things also work for spinor representations, if one uses $Spin(n+1,1)$ instead.) However, one has to replace direct observations by tools from representation theory, and we discuss in this section, how this is done.

3.1. The Lie algebra $\mathfrak{o}(n+1,1)$. We first have to look at the Lie algebra $\mathfrak{g} \cong \mathfrak{o}(n+1,1)$ of $G = O(\mathbb{V})$. For the choice of inner product used in 2.1 this has the form

$$\mathfrak{g} = \left\{ \begin{pmatrix} a & Z & 0 \\ X & A & -Z^t \\ 0 & -X^t & -a \end{pmatrix} : A \in \mathfrak{o}(n), a \in \mathbb{R}, X \in \mathbb{R}^n, Z \in \mathbb{R}^{n*}, \right\}.$$

The central block formed by A represents the subgroup $O(n)$. The element $E := \begin{pmatrix} 1 & 0 & 0 \\ 0 & 0 & 0 \\ 0 & 0 & -1 \end{pmatrix}$ is called the *grading element*. Forming the commutator with

E is a diagonalizable map $\mathfrak{g} \to \mathfrak{g}$ with eigenvalues -1, 0, and 1, and we denote by $\mathfrak{g}_i$ the eigenspace for the eigenvalue i. Hence $\mathfrak{g}_{-1}$ corresponds to X, $\mathfrak{g}_1$ to Z and $\mathfrak{g}_0$ to A and a. Moreover, the Jacobi identity immediately implies that $[\mathfrak{g}_i, \mathfrak{g}_j] \subset \mathfrak{g}_{i+j}$ with the convention that $\mathfrak{g}_{i+j} = \{0\}$ unless $i + j \in \{-1, 0, 1\}$. Such a decomposition is called a $|1|$*-grading* of $\mathfrak{g}$. In particular, restricting the adjoint action to $\mathfrak{o}(n)$, one obtains actions on $\mathfrak{g}_{-1}$ and $\mathfrak{g}_1$, which are the standard representation respectively its dual (and hence isomorphic to the standard representation).

Since the grading element E acts diagonalizably under the adjoint representation, it also acts diagonalizably on any finite dimensional irreducible representation $\mathbb{W}$ of $\mathfrak{g}$. If $w \in \mathbb{W}$ is an eigenvector for the eigenvalue j, and $Y \in \mathfrak{g}_i$, then $E \cdot Y \cdot w = Y \cdot E \cdot w + [E, Y] \cdot w$ shows that $Y \cdot w$ is an eigenvector with eigenvalue $i + j$. From irreducibility it follows easily that denoting by j_0 the lowest eigenvalue, the set of eigenvalues is $\{j_0, j_0 + 1, \dots, j_0 + N\}$ for some $N \geq 1$. Correspondingly, we obtain a decomposition $\mathbb{W} = \mathbb{W}_0 \oplus \dots \oplus \mathbb{W}_N$ such that $\mathfrak{g}_i \cdot \mathbb{W}_j \subset \mathbb{W}_{i+j}$. In particular, each of the subspaces $\mathbb{W}_j$ is invariant under the action of $\mathfrak{g}_0$ and hence in particular under the action of $\mathfrak{o}(n)$. Notice that the decomposition $\mathbb{V} = \mathbb{V}_0 \oplus \mathbb{V}_1 \oplus \mathbb{V}_2$ used in Section 2 is obtained in this way.

One can find a Cartan subalgebra of (the complexification of) $\mathfrak{g}$ which is spanned by E and a Cartan subalgebra of (the complexification of) $\mathfrak{o}(n)$. The theorem of the highest weight then leads to a bijective correspondence between finite dimensional irreducible representations $\mathbb{W}$ of $\mathfrak{g}$ and pairs $(\mathbb{W}_0, r)$, where $\mathbb{W}_0$ is a finite dimensional irreducible representation of $\mathfrak{o}(n)$ and $r \geq 1$ is an integer. Basically, the highest weight of $\mathbb{W}_0$ is the restriction to the Cartan subalgebra of $\mathfrak{o}(n)$ of the highest weight of $\mathbb{W}$, while r is related to the value of the highest weight on E. As the notation suggests, we can arrange things in such a way that $\mathbb{W}_0$ is the lowest eigenspace for the action of E on $\mathbb{W}$. For example, the standard representation $\mathbb{V}$ in this notation corresponds to $(\mathbb{R}, 2)$. The explicit version of this correspondence is not too important here, it is described in terms of highest weights in [4] and in terms of Young diagrams in [10]. It turns out that, given $\mathbb{W}_0$ and r, the number N which describes the length of the grading can be easily computed.

3.2. Kostant's version of the Bott–Borel–Weil theorem. Suppose that $\mathbb{W}$ is a finite dimensional irreducible representation of $\mathfrak{g}$, decomposed as $\mathbb{W}_0 \oplus \dots \oplus \mathbb{W}_N$ as above. Then we can view $\Lambda^k \mathbb{R}^n \otimes \mathbb{W}$ as $\Lambda^k \mathfrak{g}_1 \otimes \mathbb{W}$, which leads to two natural families of $O(n)$–equivariant maps. First we define $\partial^* : \Lambda^k \mathfrak{g}_1 \otimes \mathbb{W} \to \Lambda^{k-1} \mathfrak{g}_1 \otimes \mathbb{W}$ by

$$\partial^*(Z_1 \wedge \dots \wedge Z_k \otimes w) := \textstyle\sum_{i=1}^k (-1)^i Z_1 \wedge \dots \wedge \widehat{Z_i} \wedge \dots \wedge Z_k \otimes Z_i \cdot w,$$

where the hat denotes omission. Note that if $w \in \mathbb{W}_j$, then $Z_i \cdot w \in W_{j+1}$, so this operation preserves homogeneity. On the other hand, we have

$Z_i \cdot Z_j \cdot w - Z_j \cdot Z_i \cdot w = [Z_i, Z_j] \cdot w = 0$, since $\mathfrak{g}_1$ is a commutative subalgebra. This easily implies that $\partial^* \circ \partial^* = 0$.

Next, there is an evident duality between $\mathfrak{g}_{-1}$ and $\mathfrak{g}_1$, which is compatible with Lie theoretic methods since it is induced by the Killing form of $\mathfrak{g}$. Using this, we can identify $\Lambda^k \mathfrak{g}_1 \otimes \mathbb{W}$ with the space of k–linear alternating maps $\mathfrak{g}_{-1}^k \to \mathbb{W}$. This gives rise to a natural map $\partial = \partial_k : \Lambda^k \mathfrak{g}_1 \otimes \mathbb{W} \to \Lambda^{k+1} \mathfrak{g}_1 \otimes \mathbb{W}$ defined by

$$\partial\alpha(X_0, \dots, X_k) := \sum_{i=0}^k (-1)^i X_i \cdot \alpha(X_0, \dots, \widehat{X_i}, \dots, X_k).$$

In this picture, homogeneity boils down to the usual notion for multilinear maps, i.e. $\alpha : (\mathfrak{g}_{-1})^k \to \mathbb{W}$ is homogeneous of degree ℓ if it has values in $\mathbb{W}_{\ell - k}$. From this it follows immediately that ∂ preserves homogeneities, and $\partial \circ \partial = 0$ since $\mathfrak{g}_{-1}$ is commutative.

As a first step towards the proof of his version of the Bott–Borel–Weil–theorem (see [14]), B. Kostant proved the following result:

LEMMA 3.2. *The maps ∂ and ∂^* are adjoint with respect to an inner product of Lie theoretic origin. For each degree k, one obtains an algebraic Hodge decomposition*

$$\Lambda^k \mathfrak{g}_1 \otimes \mathbb{W} = \operatorname{im}(\partial) \oplus (\ker(\partial) \cap \ker(\partial^*)) \oplus \operatorname{im}(\partial^*),$$

with the first two summands adding up to $\ker(\partial)$ *and the last two summands adding up to* $\ker(\partial^*)$.

In particular, the restrictions of the canonical projections to the subspace $\mathbb{H}_k := \ker(\partial) \cap \ker(\partial^*)$ *induce isomorphisms* $\mathbb{H}_k \cong \ker(\partial)/\operatorname{im}(\partial)$ *and* $\mathbb{H}_k \cong \ker(\partial^*)/\operatorname{im}(\partial^*)$.

Since ∂ and ∂^* are $\mathfrak{g}_0$–equivariant, all spaces in the lemma are naturally representations of $\mathfrak{g}_0$ and all statements include the $\mathfrak{g}_0$–module structure. Looking at the Hodge decomposition more closely, we see that for each k, the map ∂ induces an isomorphism

$$\Lambda^k \mathfrak{g}_1 \otimes \mathbb{W} \supset \operatorname{im}(\partial^*) \to \operatorname{im}(\partial) \subset \Lambda^{k+1} \mathfrak{g}_1 \otimes \mathbb{W},$$

while ∂^* induces an isomorphism in the opposite direction. In general, these two map are not inverse to each other, so we replace ∂^* by the map δ^* which vanishes on $\ker(\partial^*)$ and is inverse to ∂ on $\operatorname{im}(\partial)$. Of course, $\delta^* \circ \delta^* = 0$ and it computes the same cohomology as ∂^*.

Kostant's version of the BBW–theorem computes (in a more general setting to be discussed below) the representations $\mathbb{H}_k$ in an explicit and algorithmic way. We only need the cases $k = 0$ and $k = 1$ here, but to formulate the result for $k = 1$ we need a bit of background. Suppose that $\mathbb{E}$ and $\mathbb{F}$ are finite dimensional representations of a semisimple Lie algebra. Then the tensor product $\mathbb{E} \otimes \mathbb{F}$ contains a unique irreducible component whose highest weight is the sum of the highest weights of $\mathbb{E}$ and $\mathbb{F}$. This component is called the *Cartan product* of $\mathbb{E}$ and $\mathbb{F}$ and denoted by $\mathbb{E} \circledcirc \mathbb{F}$.

Moreover, there is a nonzero equivariant map $\mathbb{E} \otimes \mathbb{F} \to \mathbb{E} \circledcirc \mathbb{F}$, which is unique up to multiples. This equivariant map is also referred to as the *Cartan product.*

The part of Kostant's version of the BBW–theorem that we need (proved in [4] in this form) reads as follows,

THEOREM 3.2. *Let $\mathbb{W} = \mathbb{W}_0 \oplus \cdots \oplus \mathbb{W}_N$ be the irreducible representation of $\mathfrak{g}$ corresponding to the pair $(\mathbb{W}_0, r)$.*
(i) In degree zero, $\operatorname{im}(\partial^*) = \mathbb{W}_1 \oplus \cdots \oplus \mathbb{W}_N$ and $\mathbb{H}_0 = \ker(\partial) = \mathbb{W}_0$.
(ii) The subspace $\mathbb{H}_1 \subset \mathfrak{g}_1 \otimes \mathbb{W}$ is isomorphic to $S^r_0\mathfrak{g}_1 \circledcirc \mathbb{W}_0$. It is contained in $\mathfrak{g}_1 \otimes \mathbb{W}_{r-1}$ and it is the only irreducible component of $\Lambda^*\mathfrak{g}_1 \otimes \mathbb{W}$ of this isomorphism type.

3.3. Some more algebra. Using Theorem 3.2 we can now deduce the key algebraic ingredient for the procedure. For each $i \geq 1$ we have $\partial : \mathbb{W}_i \to \mathfrak{g}_1 \otimes \mathbb{W}_{i-1}$. Next, we consider $(\mathrm{id} \otimes \partial) \circ \partial : \mathbb{W}_i \to \otimes^2 \mathfrak{g}_1 \otimes \mathbb{W}_{i-2}$, and so on, to obtain $\mathfrak{g}_0$–equivariant maps

$$\varphi_i := (\mathrm{id} \otimes \cdots \otimes \mathrm{id} \otimes \partial) \circ \ldots \circ (\mathrm{id} \otimes \partial) \otimes \partial : \mathbb{W}_i \to \otimes^i \mathfrak{g}_1 \otimes \mathbb{W}_0$$

for $i = 1, \ldots, N$, and we put $\varphi_0 = \mathrm{id}_{\mathbb{W}_0}$.

PROPOSITION 3.3. *Let $\mathbb{W} = \mathbb{W}_0 \oplus \cdots \oplus \mathbb{W}_N$ correspond to $(\mathbb{W}_0, r)$ and let $\mathbb{K} \subset S^r\mathfrak{g}_1 \otimes \mathbb{W}_0$ be the kernel of the Cartan product. Then we have*
(i) For each i, the map $\varphi_i : \mathbb{W}_i \to \otimes^i \mathfrak{g}_1 \otimes \mathbb{W}_0$ is injective and hence an isomorphism onto its image. This image is given by

$$\operatorname{im}(\varphi_i) = \begin{cases} S^i\mathfrak{g}_1 \otimes \mathbb{W}_0 & i < r \\ (S^i\mathfrak{g}_1 \otimes \mathbb{W}_0) \cap (S^{i-r}\mathfrak{g}_1 \otimes \mathbb{K}) & i \geq r. \end{cases}$$

(ii) For each $i < r$, the restriction of the map $\delta^* \otimes \varphi_{i-1}^{-1}$ to $S^i\mathfrak{g}_1 \otimes \mathbb{W}_0 \subset \mathfrak{g}_1 \otimes S^{i-1}\mathfrak{g}_1 \otimes \mathbb{W}_0$ coincides with φ_i^{-1}.

Proof. (sketch) (i) Part (i) of Theorem 3.2 shows that $\partial : \mathbb{W}_i \to \mathfrak{g}_1 \otimes \mathbb{W}_{i-1}$ is injective for each $i \geq 1$, so injectivity of the φ_i follows. Moreover, for $i \neq r$, the image of this map coincides with the kernel of $\partial_1 : \mathfrak{g}_1 \otimes \mathbb{W}_{i-1} \to \Lambda^2\mathfrak{g}_1 \otimes \mathbb{W}_{i-2}$, while for $i = r$ this kernel in addition contains a complementary subspace isomorphic to $S^r\mathfrak{g}_1 \circledcirc \mathbb{W}_0$. A moment of thought shows that ∂_1 can be written as $2\mathrm{Alt} \circ (\mathrm{id} \otimes \partial_0)$, where Alt denotes the alternation. This immediately implies that the φ_i all have values in $S^i\mathfrak{g}_1 \otimes \mathbb{W}_0$ as well as the claim about the image for $i < r$.

It further implies that $\mathrm{id} \otimes \varphi_{r-1}$ restricts to isomorphisms

$$\begin{array}{ccc} \mathfrak{g}_1 \otimes \mathbb{W}_{r-1} & \longrightarrow & \mathfrak{g}_1 \otimes S^{r-1}\mathfrak{g}_1 \otimes \mathbb{W}_0 \\ \uparrow & & \uparrow \\ \ker(\partial) & \longrightarrow & S^r\mathfrak{g}_1 \otimes \mathbb{W}_0 \\ \uparrow & & \uparrow \\ \operatorname{im}(\partial) & \longrightarrow & \mathbb{K} \end{array}$$

which proves the claim on the image for $i = r$. For $i > r$ the claim then follows easily as above.
(ii) This follows immediately from the fact that $\delta^*|_{\mathrm{im}(\partial)}$ inverts $\partial|_{\mathrm{im}(\delta^*)}$. □

3.4. Step one of the prolongation procedure. The developments in 3.1–3.3 carry over to an arbitrary Riemannian manifold (M,g) of dimension n. The representation $\mathbb{W}$ corresponds to a vector bundle $W = \oplus_{i=0}^N W_i$. Likewise, $\mathbb{H}_1$ corresponds to a direct summand $H_1 \subset T^*M \otimes W_{r-1}$ which is isomorphic to $S_0^r T^*M \circledcirc W_0$. The maps ∂, ∂^*, and δ^* induce vector bundle maps on the bundles $\Lambda^k T^*M \otimes W$ of W–valued differential forms, and for $i = 0,\dots,N$, the map φ_i induces a vector bundle map $W_i \to S^i T^*M \otimes W_0$. We will denote all these maps by the same symbols as their algebraic counterparts. Finally, the Cartan product gives rise to a vector bundle map $S^r T^*M \otimes W_0 \to H_1$, which is unique up to multiples.

We have the component–wise Levi–Civita connection ∇ on W. We will denote a typical section of W by Σ. The subscript i will indicate the component in $\Lambda^k T^*M \otimes W_i$. Now we define a linear connection $\tilde{\nabla}$ on W by $\tilde{\nabla}\Sigma := \nabla\Sigma + \partial(\Sigma)$, i.e. $(\tilde{\nabla}\Sigma)_i = \nabla\Sigma_i + \partial(\Sigma_{i+1})$. Next, we choose a bundle map $A : W_0 \oplus \cdots \oplus W_{r-1} \to H_1$, view it as $A : W \to T^*M \otimes W$ and consider the system

$$\tilde{\nabla}\Sigma + A(\Sigma) = \delta^*\psi \quad \text{for some } \psi \in \Omega^2(M,W). \tag{3.1}$$

Since A has values in $\ker(\delta^*)$ and $\delta^* \circ \delta^* = 0$, any solution Σ of this system has the property that $\delta^*\tilde{\nabla}\Sigma = 0$.

To rewrite the system equivalently as a higher order system, we define a linear differential operator $L : \Gamma(W_0) \to \Gamma(W)$ by

$$L(f) := \textstyle\sum_{i=0}^N (-1)^i (\delta^* \circ \nabla)^i f.$$

PROPOSITION 3.4. *(i) For $f \in \Gamma(W_0)$ we have $L(f)_0 = f$ and $\delta^*\tilde{\nabla}L(f) = 0$, and $L(f)$ is uniquely determined by these two properties.*
(ii) For $\ell = 0,\dots,N$ the component $L(f)_\ell$ depends only on the ℓ–jet of f. More precisely, denoting by $J^\ell W_0$ the ℓth jet prolongation of the bundle W_0, the operator L induces vector bundle maps $J^\ell W_0 \to W_0 \oplus \cdots \oplus W_\ell$, which are isomorphisms for all $\ell < r$.

Proof. (i) Putting $\Sigma = L(f)$ it is evident that $\Sigma_0 = f$ and $\Sigma_{i+1} = -\delta^*\nabla\Sigma_i$ for all $i \geq 0$. Therefore,

$$(\tilde{\nabla}\Sigma)_i = \nabla\Sigma_i + \partial(\Sigma_{i+1}) = \nabla\Sigma_i - \partial\delta^*\nabla\Sigma_i$$

for all i. Since $\delta^*\partial$ is the identity on $\mathrm{im}(\delta^*)$, we get $\delta^*\tilde{\nabla}\Sigma = 0$.

Conversely, expanding the equation $0 = \delta^*\tilde{\nabla}\Sigma$ in components we obtain

$$\Sigma_{i+1} = \delta^*\partial\Sigma_{i+1} = -\delta^*\nabla\Sigma_i,$$

which inductively implies $\Sigma = L(\Sigma_0)$.

(ii) By definition, $L(f)_\ell$ depends only on ℓ derivatives of f. Again by definition, $L(f)_1 = \delta^*\nabla f$, and if $r > 1$, this equals $\varphi_1^{-1}(\nabla f)$. Naturality of δ^* implies that

$$L(f)_2 = \delta^*\nabla\delta^*\nabla f = \delta^* \circ (\mathrm{id}\otimes\delta^*)(\nabla^2 f) = \delta^* \circ (\mathrm{id}\otimes\varphi_1^{-1})(\nabla^2 f).$$

Replacing ∇^2 by its symmetrization changes the expression by a term of order zero, so we see that, if $r > 2$ and up to lower order terms, $L(f)_2$ is obtained by applying φ_2^{-1} to the symmetrization of $\nabla^2 f$. Using part (ii) of Proposition 3.3 and induction, we conclude that for $\ell < r$ and up to lower order terms $L(f)_\ell$ is obtained by applying φ_ℓ^{-1} to the symmetrized ℓth covariant derivative of ℓ, and the claim follows. □

Note that part (ii) immediately implies that for a bundle map A as defined above, $f \mapsto A(L(f))$ is a differential operator $\Gamma(W_0) \to \Gamma(H_1)$ of order at most $r-1$ and any such operator is obtained in this way.

3.5. The second step of the procedure. For a section $f \in \Gamma(W_0)$ we next define $D^{\mathrm{W}}(f) \in \Gamma(H_1)$ to be the component of $\tilde{\nabla}L(f)$ in $\Gamma(H_1) \subset \Omega^1(M,W)$. We know that $(\tilde{\nabla}L(f))_{r-1} = \nabla L(f)_{r-1} + \partial L(f)_r$, and the second summand does not contribute to the H_1–component. Moreover, from the proof of Proposition 3.4 we know that, up to lower order terms, $L(f)_{r-1}$ is obtained by applying φ_{r-1}^{-1} to the symmetrized $(r-1)$–fold covariant derivative of f. Hence up to lower order terms, $\nabla L(f)_{r-1}$ is obtained by applying $\mathrm{id}\otimes\varphi_{r-1}^{-1}$ to the symmetrized r–fold covariant derivative of f. Using the proof of Proposition 3.3 this easily implies that the principal symbol of D^{W} is (a nonzero multiple of) the Cartan product $S^rT^*M \otimes W_0 \to S^r_0T^*M \circledcirc W_0 = H_1$.

PROPOSITION 3.5. *Let $D : \Gamma(W_0) \to \Gamma(H_1)$ be a quasi-linear differential operator of order r whose principal symbol is given by the Cartan product $S^rT^*M \otimes W_0 \to S^r_0T^*M \circledcirc W_0$. Then there is a bundle map $A : W \to T^*M \otimes W$ as in 3.4 such that $\Sigma \mapsto \Sigma_0$ and $f \mapsto L(f)$ induce inverse bijections between the sets of solutions of $D(f) = 0$ and of the basic system* (3.1).

Proof. This is completely parallel to the proof of Proposition 2.3: The conditions on D exaclty mean that it can be written in the form $D(f) = D^{\mathrm{W}}(f) + A(L(f))$ for an appropriate choice of A as above. Then $\tilde{\nabla}L(f) + A(L(f))$ is a section of the subbundle $\ker(\delta^*)$ and the component in H_1 of this section equals $D(f)$. Of course, being a section of $\mathrm{im}(\delta^*)$ is equivalent to vanishing of the H_1–component.

Conversely, Proposition 3.4 shows that any solution Σ of (3.1) is of the form $\Sigma = L(\Sigma_0)$. □

3.6. The last step of the procedure. To rewrite the basic system (3.1) in first order closed form, we use the covariant exterior derivative $d^{\tilde{\nabla}}$. Suppose that $\alpha \in \Omega^1(M,W)$ has the property that its components α_i vanish for $i = 0,\dots,\ell$. Then one immediately verifies that $(d^{\tilde{\nabla}}\alpha)_i = 0$

for $i = 0, \dots, \ell - 1$ and $(d^{\tilde{\nabla}}\alpha)_\ell = \partial(\alpha_{\ell+1})$, so $(\delta^* d^{\tilde{\nabla}}\alpha)_i$ vanishes for $i \le \ell$ and equals $\delta^*\partial\alpha_{\ell+1}$ for $i = \ell + 1$. If we in addition assume that α is a section of the subbundle $\operatorname{im}(\delta^*)$, then the same is true for $\alpha_{\ell+1}$ and hence $\delta^*\partial\alpha_{\ell+1} = \alpha_{\ell+1}$.

Suppose that Σ solves the basic system (3.1). Then applying $\delta^* d^{\tilde{\nabla}}$, we obtain

$$\delta^*(R \bullet \Sigma + d^{\tilde{\nabla}}(A(\Sigma))) = \delta^* d^{\tilde{\nabla}} \delta^* \psi,$$

where we have used that, as in 2.4, $d^{\tilde{\nabla}}\tilde{\nabla}\Sigma$ is given by the action of the Riemann curvature R. From above we see that we can compute the lowest nonzero homogeneous component of $\delta^*\psi$ from this equation. We can then move this to the other side in (3.1) to obtain an equivalent system whose right hand side starts one homogeneity higher. The lowest nonzero homogeneous component of the right hand side can then be computed in the same way, and iterating this we conclude that (3.1) can be equivalently written as

$$\tilde{\nabla}\Sigma + B(\Sigma) = 0 \tag{3.2}$$

for a certain differential operator $B : \Gamma(W) \to \Omega^1(M, W)$.

While B is a higher order differential operator in general, it is crucial that the construction gives us a precise control on the order of the individual components of B. From the construction it follows that $B(\Sigma)_i \in \Omega^1(M, V_i)$ depends only on $\Sigma_0, \dots, \Sigma_i$, and the dependence is tensorial in Σ_i, first order in Σ_{i-1} and so on up to ith order in Σ_0.

In particular, the component of (3.2) in $\Omega^1(M, W_0)$ has the form $\nabla\Sigma_0 = C_0(\Sigma_0, \Sigma_1)$. Next, the component in $\Omega^1(M, W_1)$ has the form $\nabla\Sigma_1 = \tilde{C}_1(\Sigma_0, \Sigma_1, \Sigma_2, \nabla\Sigma_0)$, and we define

$$C_1(\Sigma_0, \Sigma_1, \Sigma_2) := \tilde{C}_1(\Sigma_0, \Sigma_1, \Sigma_2, -C_0(\Sigma_0, \Sigma_1)).$$

Hence the two lowest components of (3.2) are equivalent to

$$\begin{cases} \nabla\Sigma_1 = C_1(\Sigma_0, \Sigma_1, \Sigma_2) \\ \nabla\Sigma_0 = C_0(\Sigma_0, \Sigma_1). \end{cases}$$

Differentiating the lower row and inserting for $\nabla\Sigma_0$ and $\nabla\Sigma_1$ we get an expression for $\nabla^2\Sigma_0$ in terms of $\Sigma_0, \Sigma_1, \Sigma_2$. Continuing in this way, one proves

THEOREM 3.6. *Let $D : \Gamma(W_0) \to \Gamma(H_1)$ be a quasi-linear differential operator of order r with principal symbol the Cartan product $S^rT^*M \otimes W_0 \to S^rT^*M_0 \circledcirc W_0$. Then there is a bundle map $C : W \to T^*M \otimes W$ such that $\Sigma \mapsto \Sigma_0$ and $f \mapsto L(f)$ induce inverse bijections between the sets of solutions of $D(f) = 0$ and of $\tilde{\nabla}\Sigma + C(\Sigma) = 0$. If D is linear, then C can be chosen to be a vector bundle map.*

This in particular shows that any solution of $D(f) = 0$ is determined by the value of $L(f)$ in one point, end hence by the N–jet of f in one point.

For linear D, the dimension of the space of solutions is bounded by $\dim(\mathbb{W})$ and equality can be only attained if the linear connection $\tilde{\nabla} + C$ on W is flat. A crucial point here is of course that $\mathbb{W}$, and hence $\dim(\mathbb{W})$ and N can be immediately computed from $\mathbb{W}_0$ and r, so all this information is available in advance, without going through the procedure. As we shall see later, both the bound on the order and the bound on the dimension are sharp.

To get a feeling for what is going on, let us consider some examples. If we look at operators on smooth functions, we have $\mathbb{W}_0 = \mathbb{R}$. The representation associated to $(\mathbb{R}, r)$ is $S_0^{r-1}\mathbb{V}$, the tracefree part of the $(r-1)$st symmetric power of the standard representation $\mathbb{V}$. A moment of thought shows that the eigenvalues of the grading element E on this representation range from $-r+1$ to $r-1$, so $N = 2(r-1)$. On the other hand, for $r \geq 3$ we have

$$\dim(S_0^{r-1}\mathbb{V}) = \dim(S^{r-1}\mathbb{V}) - \dim(S^{r-3}\mathbb{V}) = (n+2r-2)\frac{(n+2r-2)!}{n!(r-1)!},$$

and this is the maximal dimension of the space of solutions of any system with principal part $f \mapsto \nabla_{(a_1}\nabla_{a_2} \cdots \nabla_{a_r)_0} f$ for $f \in C^\infty(M, \mathbb{R})$.

As an extreme example let us consider the conformal Killing equation on tracefree symmetric tensors. Here $W_0 = S_0^k TM$ for some k and $r = 1$. The principal part in this case is simply

$$f^{a_1 \dots a_k} \mapsto \nabla^{(a} f^{a_1 \dots a_k)_0}.$$

The relevant representation $\mathbb{W}$ in this case turns out to be $\circledcirc^k \mathfrak{g}$, i.e. the highest weight subspace in $S^k\mathfrak{g}$. In particular $N = 2k$ in this case, so even though we consider first order systems, many derivatives are needed to pin down a solution. The expression for $\dim(\mathbb{W})$ is already reasonably complicated in this case, namely (see [11])

$$\dim(\mathbb{W}) = \frac{(n+k-3)!(n+k-2)!(n+2k)!}{k!(k+1)!(n-2)!n!(n+2k-3)!}.$$

The conformal Killing equation $\nabla^{(a} f^{a_1 \dots a_k)_0} = 0$ plays an important role in the description of symmetries of the Laplacian on a Riemannian manifold, see [11].

4. Conformally invariant differential operators. We now move to the method for constructing conformally invariant differential operators, which gave rise to the prolongation procedure discussed in the last two sections.

4.1. Conformal geometry. Let M be a smooth manifold of dimension $n \geq 3$. As already indicated in 2.4, two Riemannian metrics g and $\hat{g}$ on M are called *conformally equivalent* if and only if there is a positive smooth function φ on M such that $\hat{g} = \varphi^2 g$. A *conformal structure* on M

is a conformal equivalence class $[g]$ of metrics, and then $(M,[g])$ is called a conformal manifold. A *conformal isometry* between conformal manifolds $(M,[g])$ and $(\tilde{M},[\tilde{g}])$ is a local diffeomorphism which pulls back one (or equivalently any) metric from the class $[\tilde{g}]$ to a metric in $[g]$.

A Riemannian metric on M can be viewed as a reduction of structure group of the frame bundle to $O(n) \subset GL(n,\mathbb{R})$. In the same way, a conformal structure is a reduction of structure group to $CO(n) \subset GL(n,\mathbb{R})$, the subgroup generated by $O(n)$ and multiples of the identity.

We want to clarify how the inclusion $O(n) \hookrightarrow G \cong O(n+1,1)$ which was the basis for our prolongation procedure is related to conformal geometry. For the basis $\{e_0,\dots,e_{n+1}\}$ used in 2.1, this inclusion was simply given by $A \mapsto \left(\begin{smallmatrix} 1 & 0 & 0 \\ 0 & A & 0 \\ 0 & 0 & 1 \end{smallmatrix}\right)$. In 3.1 we met the decomposition $\mathfrak{g} = \mathfrak{g}_{-1} \oplus \mathfrak{g}_0 \oplus \mathfrak{g}_1$ of the Lie algebra $\mathfrak{g}$ of G. We observed that this decomposition is preserved by $O(n) \subset G$ and in that way $\mathfrak{g}_{\pm 1}$ is identified with the standard representation. But there is a larger subgroup with these properties. Namely, for elements of

$$G_0 := \left\{ \left(\begin{smallmatrix} a & 0 & 0 \\ 0 & A & 0 \\ 0 & 0 & a^{-1} \end{smallmatrix}\right) : a \in \mathbb{R}\setminus 0, A \in O(n) \right\} \subset G,$$

the adjoint action preserves the grading, and maps $X \in \mathfrak{g}_{-1}$ to $a^{-1}AX$, so $G_0 \cong CO(\mathfrak{g}_{-1})$. Note that $G_0 \subset G$ corresponds to the Lie subalgebra $\mathfrak{g}_0 \subset \mathfrak{g}$.

Now there is a more conceptual way to understand this. Consider the subalgebra $\mathfrak{p} := \mathfrak{g}_0 \oplus \mathfrak{g}_1 \subset \mathfrak{g}$ and let $P \subset G$ be the corresponding Lie subgroup. Then P is the subgroup of matrices which are block–upper–triangular with blocks of sizes 1, n, and 1. Equivalently, P is the stabilizer in G of the isotropic line spanned by the basis vector e_0. The group G acts transitively on the space of all isotropic lines in $\mathbb{V}$, so one may identify this space with the homogeneous space G/P.

Taking coordinates z_i with respect to an orthonormal basis of $\mathbb{V}$ for which the first $n+1$ vectors are positive and the last one is negative, a vector is isotropic if and only if $\sum_{i=0}^n z_i^2 = z_{n+1}^2$. Hence for a nonzero isotropic vector the last coordinate is nonzero and any isotropic line contains a unique vector whose last coordinate equals 1. But this shows that the space of isotropic lines in $\mathbb{V}$ is an n–sphere, so $G/P \cong S^n$.

Given a point $x \in G/P$, choosing a point v in the corresponding line gives rise to an identification $T_xS^n \cong v^\perp/\mathbb{R}v$ and that space carries a positive definite inner product induced by $\langle\ ,\ \rangle$. Passing from v to λv, this inner product gets scaled by λ^2, so we get a canonical conformal class of inner products on each tangent space, i.e. a conformal structure on S^n. This conformal structure contains the round metric of S^n.

The action ℓ_g of $g \in G$ on the space of null lines by construction preserves this conformal structure, so G acts by conformal isometries. It turns out, that this identifies $G/\{\pm \mathrm{id}\}$ with the group of all conformal isometries of S^n. For the base point $o = eP \in G/P$, the tangent space

$T_o(G/P)$ is naturally identified with $\mathfrak{g}/\mathfrak{p} \cong \mathfrak{g}_{-1}$. Let $P_+ \subset P$ be the subgroup of those $g \in P$ for which $T_o\ell_g = \text{id}$. Then one easily shows that $P/P_+ \cong G_0$ and the isomorphism $G_0 \cong CO(\mathfrak{g}_{-1})$ is induced by $g \mapsto T_o\ell_g$. Moreover, P_+ has Lie algebra $\mathfrak{g}_1$ and $\exp : \mathfrak{g}_1 \to P_+$ is a diffeomorphism.

4.2. Conformally invariant differential operators. Let $(M, [g])$ be a conformal manifold. Choosing a metric g from the conformal class, we get the Levi–Civita connection ∇ on each Riemannian natural bundle as well as the Riemann curvature tensor R. Using ∇, g, its inverse, and R, we can write down differential operators, and see how they change if g is replaced by a conformally equivalent metric $\hat{g}$. Operators obtained in that way, which do not change at all under conformal rescalings are called *conformally invariant.* In order to do this successfully one either has to allow density bundles or deal with conformal weights, but I will not go into these details here. The best known example of such an operator is the conformal Laplacian or Yamabe operator which is obtained by adding an appropriate amount of scalar curvature to the standard Laplacian.

The definition of conformally invariant operators immediately suggests a naive approach to their construction. First choose a principal part for the operator. Then see how this behaves under conformal rescalings and try to compensate the changes by adding lower order terms involving curvature quantities. This approach (together with a bit of representation theory) easily leads to a complete classification of conformally invariant first order operators, see [13]. Passing to higher orders, the direct methods get surprisingly quickly out of hand.

The basis for more invariant approaches is provided by a classical result of Elie Cartan, which interprets general conformal structures as analogs of the homogeneous space $S^n \cong G/P$ from 4.1. As we have noted above, a conformal structure $[g]$ on M can be interpreted as a reduction of structure group. This means that a conformal manifold $(M, [g])$ naturally carries a principal bundle with structure group $CO(n)$, the *conformal frame bundle.* Recall from 4.1 that the conformal group $G_0 = CO(\mathfrak{g}_{-1}) \cong CO(n)$ can be naturally viewed as a quotient of the group P. Cartan's result says that the conformal frame bundle can be canonically extended to a principal fiber bundle $\mathcal{G} \to M$ with structure group P, and $\mathcal{G}$ can be endowed with a canonical Cartan connection $\omega \in \Omega^1(\mathcal{G}, \mathfrak{g})$. The form ω has similar formal properties as the Maurer–Cartan form on G, i.e. it defines a trivialization of the tangent bundle $T\mathcal{G}$, which is P–equivariant and reproduces the generators of fundamental vector fields.

While the canonical Cartan connection is conformally invariant, it is not immediately clear how to use it to construct differential operators. The problem is that, unlike principal connections, Cartan connections do not induce linear connections on associated vector bundles.

4.3. The setup for the conformal BGG machinery. Let us see how the basic developments from 3.1–3.3 comply with our new point of view. First of all, for $g \in P$, the adjoint action does not preserve the grading $\mathfrak{g} = \mathfrak{g}_{-1} \oplus \mathfrak{g}_0 \oplus \mathfrak{g}_1$, but it preserves the subalgebras $\mathfrak{p} = \mathfrak{g}_0 \oplus \mathfrak{g}_1$, and $\mathfrak{g}_1$. More generally, if $\mathbb{W} = \mathbb{W}_0 \oplus \cdots \oplus \mathbb{W}_N$ is an irreducible representation of $\mathfrak{g}$ decomposed according to eigenspaces of the grading element E, then each of the subspaces $\mathbb{W}_i \oplus \cdots \oplus \mathbb{W}_N$ is P–invariant. Since P naturally acts on $\mathfrak{g}_1$ and on $\mathbb{W}$, we get induced actions on $\Lambda^k \mathfrak{g}_1 \otimes \mathbb{W}$ for all k. The formula for $\partial^* : \Lambda^k \mathfrak{g}_1 \otimes \mathbb{W} \to \Lambda^{k-1}\mathfrak{g}_1 \otimes \mathbb{W}$ uses only the action of $\mathfrak{g}_1$ on $\mathbb{W}$, so ∂^* is P–equivariant.

In contrast to this, the only way to make P act on $\mathfrak{g}_{-1}$ is via the identification with $\mathfrak{g}/\mathfrak{p}$. However, the action of $\mathfrak{g}_{-1}$ on $\mathbb{W}$ has no natural interpretation in this identification, and $\partial : \Lambda^k \mathfrak{g}_1 \otimes \mathbb{W} \to \Lambda^{k+1}\mathfrak{g}_1 \otimes \mathbb{W}$ is *not* P–equivariant.

Anyway, given a conformal manifold $(M, [g])$ we can now do the following. Rather than viewing $\mathbb{W}$ just as sum of representations of $G_0 \cong CO(n)$, we can view it as a representation of P, and form the associated bundle $\mathcal{W} := \mathcal{G} \times_P \mathbb{W} \to M$. Bundles obtained in this way are called *tractor bundles.* I want to emphasize at this point that the bundle $\mathcal{W}$ is of completely different nature than the bundle W used in Section 3. To see this, recall that elements of the subgroup $P_+ \subset P$ act on G/P by diffeomorphisms which fix the base point $o = eP$ to first order. Therefore, the action of such a diffeomorphism on the fiber over o of any tensor bundle is the identity. On the other hand, it is easy to see that on the fiber over o of any tractor bundle, this action is always non–trivial. Hence tractor bundles are unusual geometric objects.

Examples of tractor bundles have already been introduced as an alternative to Cartan's approach in the 1920's and 30's, in particular in the work of Tracy Thomas, see [17]. Their key feature is that the canonical Cartan connection ω induces a canonical linear connection, called the *normal tractor connection* on each tractor bundle. This is due to the fact that these bundles do not correspond to general representations of P, but only to representations which extend to the big group G. We will denote the normal tractor connection on $\mathcal{W}$ by $\nabla^{\mathcal{W}}$. These connections automatically combine algebraic and differential parts.

The duality between $\mathfrak{g}_1$ and $\mathfrak{g}_{-1}$ induced by the Killing form, is more naturally viewed as a duality between $\mathfrak{g}_1$ and $\mathfrak{g}/\mathfrak{p}$. Via the Cartan connection ω, the associated bundle $\mathcal{G} \times_P (\mathfrak{g}/\mathfrak{p})$ is isomorphic to the tangent bundle TM. Thus, the bundle $\mathcal{G} \times_P (\Lambda^k \mathfrak{g}_1 \otimes \mathbb{W})$ is again the bundle $\Lambda^k T^*M \otimes \mathcal{W}$ of $\mathcal{W}$–valued forms. Now it turns out that in a well defined sense (which however is rather awkward to express), the lowest nonzero homogeneous component of $\nabla^{\mathcal{W}}$ is of degree zero, it is tensorial and induced by the Lie algebra differential ∂.

Equivariancy of ∂^* implies that it defines bundle maps

$$\partial^* : \Lambda^k T^*M \otimes \mathcal{W} \to \Lambda^{k-1}T^*M \otimes \mathcal{W}$$

for each k. In particular, $\mathrm{im}(\partial^*) \subset \ker(\partial^*) \subset \Lambda^k T^*M \otimes \mathcal{W}$ are natural subbundles, and we can form the subquotient $H_k := \ker(\partial^*)/\operatorname{im}(\partial^*)$. It turns out that these bundles are always naturally associated to the conformal frame bundle, so they are usual geometric objects like tensor bundles. The explicit form of the bundles H_k can be computed algorithmically using Kostant's version of the BBW–theorem.

4.4. The conformal BGG machinery. The normal tractor connection $\nabla^{\mathcal{W}}$ extends to the covariant exterior derivative, which we denote by $d^{\mathcal{W}} : \Omega^k(M, \mathcal{W}) \to \Omega^{k+1}(M, \mathcal{W})$. The lowest nonzero homogeneous component of $d^{\mathcal{W}}$ is of degree zero, tensorial, and induced by ∂.

Now for each k, the operator $\partial^* d^{\mathcal{W}}$ on $\Omega^k(M, V)$ is conformally invariant and its lowest nonzero homogeneous component is the tensorial map induced by $\partial^*\partial$. By Theorem 3.2, $\partial^*\partial$ acts invertibly on $\operatorname{im}(\partial^*)$. Hence we can find a (non–natural) bundle map β on $\operatorname{im}(\partial^*)$ such that $\beta\partial^* d^{\mathcal{W}}$ reproduces the lowest nonzero homogeneous component of sections of $\operatorname{im}(\partial^*)$. Therefore, the operator $\operatorname{id} - \beta\partial^* d^{\mathcal{W}}$ is (at most N–step) nilpotent on $\Gamma(\operatorname{im}(\partial^*))$, which easily implies that

$$\left(\sum_{i=0}^{N} (\operatorname{id} - \beta\partial^* d^{\mathcal{W}})^i\right) \beta$$

defines a differential operator Q on $\Gamma(\operatorname{im}(\partial^*))$ which is inverse to $\partial^* d^{\mathcal{W}}$ and therefore conformally invariant.

Next, we have a canonical bundle map

$$\pi_H : \ker(\partial^*) \to \ker(\partial^*)/\operatorname{im}(\partial^*) = H_k,$$

and we denote by the same symbol the induced tensorial projection on sections. Given $f \in \Gamma(H_k)$ we can choose $\varphi \in \Omega^k(M, \mathcal{W})$ such that $\partial^*\varphi = 0$ and $\pi_H(\varphi) = f$, and consider $\varphi - Q\partial^* d^{\mathcal{W}}\varphi$. By construction, φ is uniquely determined up to adding sections of $\operatorname{im}(\partial^*)$. Since these are reproduced by $Q\partial^* d^{\mathcal{W}}$, the above element is independent of the choice of φ and hence defines $L(f) \in \Omega^k(M, \mathcal{W})$. Since Q has values in $\Gamma(\operatorname{im}(\partial^*))$ we see that $\pi_H(L(f)) = f$, and since $\partial^* d^{\mathcal{W}} Q$ is the identity on $\Gamma(\operatorname{im}(\partial^*))$ we get $\partial^* d^{\mathcal{W}} L(f) = 0$. If φ satisfies $\pi_H(\varphi) = f$ and $\partial^* d^{\mathcal{W}}\varphi = 0$, then

$$L(f) = \varphi - Q\partial^* d^{\mathcal{W}}\varphi = \varphi,$$

so $L(f)$ is uniquely determined by these two properties.

By construction, the operator $L : \Gamma(H_k) \to \Omega^k(M, \mathcal{W})$ is conformally invariant. Moreover, $d^{\mathcal{W}} L(f)$ is a section of $\ker(\partial^*)$, so we can finally define the BGG–operators $D^{\mathcal{W}} : \Gamma(H_k) \to \Gamma(H_{k+1})$ by $D^{\mathcal{W}}(f) := \pi_H(d^{\mathcal{W}} L(f))$. They are conformally invariant by construction.

To obtain additional information, we have to look at structures which are locally conformally flat or equivalently locally conformally isometric to the sphere S^n. It is a classical result that local conformal flatness is equivalent to vanishing of the curvature of the canonical Cartan connection.

PROPOSITION 4.4. *On locally conformally flat manifolds, the BGG operators form a complex* $(\Gamma(H_*), D^{\mathcal{W}})$*, which is a fine resolution of the constant sheaf* $\mathbb{W}$.

Proof. The curvature of any tractor connection is induced by the Cartan curvature (see [7]), so on locally conformally flat structures, all tractor connections are flat. This implies that the covariant exterior derivative satisfies $d^{\mathcal{W}} \circ d^{\mathcal{W}} = 0$. Thus $(\Omega^*(M, \mathcal{W}), d^{\mathcal{W}})$ is a fine resolution of the constant sheaf $\mathbb{W}$.

For $f \in \Gamma(H_k)$ consider $d^{\mathcal{W}} L(f)$. By construction, this lies in the kernel of ∂^* and since $d^{\mathcal{W}} \circ d^{\mathcal{W}} = 0$, it also lies in the kernel of $\partial^* d^{\mathcal{W}}$. From above we know that this implies that

$$d^{\mathcal{W}} L(f) = L(\pi_H(d^{\mathcal{W}} L(f))) = L(D^{\mathcal{W}}(f)).$$

This shows that $L \circ D^{\mathcal{W}} \circ D^{\mathcal{W}} = d^{\mathcal{W}} \circ d^{\mathcal{W}} \circ L = 0$ and hence $D^{\mathcal{W}} \circ D^{\mathcal{W}} = 0$, so $(\Gamma(H_*), D^{\mathcal{W}})$ is a complex. The operators L define a chain map from this complex to $(\Omega^*(M, \mathcal{W}), d^{\mathcal{W}})$, and we claim that this chain map induces an isomorphism in cohomology.

First, take $\varphi \in \Omega^k(M, \mathcal{W})$ such that $d^{\mathcal{W}} \varphi = 0$. Then

$$\tilde{\varphi} := \varphi - d^{\mathcal{W}} Q \partial^* \varphi$$

is cohomologous to φ and satisfies $\partial^* \tilde{\varphi} = 0$. Moreover, $d^{\mathcal{W}} \tilde{\varphi} = d^{\mathcal{W}} \varphi = 0$, so $\partial^* d^{\mathcal{W}} \varphi = 0$. Hence $\tilde{\varphi} = L(\pi_H(\tilde{\varphi}))$ and $D^{\mathcal{W}}(\pi_H(\tilde{\varphi})) = 0$, so the induced map in cohomology is surjective.

Conversely, assume that $f \in \Gamma(H_k)$ satisfies $D^{\mathcal{W}}(f) = 0$ and that $L(f) = d^{\mathcal{W}} \varphi$ for some $\varphi \in \Omega^{k-1}(M, \mathcal{W})$. As before, replacing φ by $\varphi - d^{\mathcal{W}} Q \partial^* \varphi$ we may assume that $\partial^* \varphi = 0$. But together with $\partial^* L(f) = 0$ this implies $\varphi = L(\pi_H(\varphi))$ and thus $f = \pi_H(L(f)) = D^{\mathcal{W}}(\pi_H(\varphi))$. Hence the induced map in cohomology is injective, too. Since this holds both locally and globally, the proof is complete. □

Via a duality between invariant differential operators and homomorphisms of generalized Verma modules, this reproduces the original BGG resolutions as constructed in [15]. Via the classification of such homomorphisms, one also concludes that this construction produces a large subclass of all those conformally invariant operators which are non–trivial on locally conformally flat structures.

Local exactness of the BGG sequence implies that all the operators $D^{\mathcal{W}}$ are nonzero on locally conformally flat manifolds. Passing to general conformal structures does not change the principal symbol of the operator $D^{\mathcal{W}}$, so we always get non–trivial operators.

On the other hand, we can also conclude that the bounds obtained from Theorem 3.6 are sharp. From Theorem 3.2 we conclude that any choice of metric in the conformal class identifies $H_0 = \mathcal{W} / \operatorname{im}(\partial^*)$ with the bundle W_0 and H_1 with its counterpart from Section 3, and we consider the operator $D^{\mathcal{W}} : \Gamma(H_0) \to \Gamma(H_1)$. By conformal invariance, the system

$D^{\mathcal{W}}(f) = 0$ must be among the systems covered by Theorem 3.6, and the above procedure identifies its solutions with parallel sections of $\mathcal{W}$. Since $\nabla^{\mathcal{W}}$ is flat in the locally conformally flat case, the space of parallel sections has dimension $\dim(\mathbb{W})$. Moreover, two solutions of the system coincide if and only if their images under L have the same value in one point.

5. Generalizations. In this last part, we briefly sketch how the developments of sections 3 and 4 can be carried over to larger classes of geometric structures.

5.1. The prolongation procedure for general $|1|$–graded Lie algebras. The algebraic developments in 3.1–3.3 generalize without problems to a semisimple Lie algebra $\mathfrak{g}$ endowed with a $|1|$–grading, i.e. a grading of the form $\mathfrak{g} = \mathfrak{g}_{-1} \oplus \mathfrak{g}_0 \oplus \mathfrak{g}_1$. Given such a grading it is easy to see that it is the eigenspace decomposition of $\operatorname{ad}(E)$ for a uniquely determined element $E \in \mathfrak{g}_0$. The Lie subalgebra $\mathfrak{g}_0$ is automatically the direct sum of a semisimple part $\mathfrak{g}_0'$ and a one–dimensional center spanned by E. This gives rise to E–eigenspace decompositions for irreducible representations. Again irreducible representations of $\mathfrak{g}$ may be parametrized by pairs consisting of an irreducible representation of $\mathfrak{g}_0'$ and an integer ≥ 1. Then all the developments of 3.1–3.3 work without changes.

Next choose a Lie group G with Lie algebra $\mathfrak{g}$ and let $G_0 \subset G$ be the subgroup consisting of those elements whose adjoint action preserves the grading of $\mathfrak{g}$. Then this action defines an infinitesimally effective homomorphism $G_0 \to GL(\mathfrak{g}_{-1})$. In particular, the semisimple part G_0' of G_0 is a (covering of a) subgroup of $GL(\mathfrak{g}_{-1})$, so this defines a type of geometric structure on manifolds of dimension $\dim(\mathfrak{g}_{-1})$. This structure is linked to representation theory of G_0' in the same way as Riemannian geometry is linked to representation theory of $O(n)$.

For manifolds endowed with a structure of this type, there is an analog of the prolongation procedure described in 3.4–3.6 with closely parallel proofs, see [4]. The only change is that instead of the Levi–Civita connection one uses any linear connection on TM which is compatible with the reduction of structure group. There are some minor changes if this connection has torsion. The systems that this procedure applies to are the following. One chooses an irreducible representation $\mathbb{W}_0$ of G_0' and an integer $r \geq 1$. Denoting by W_0 the bundle corresponding to $\mathbb{W}_0$, one then can handle systems whose principal symbol is (a multiple of) the projection from $S^rTM \otimes W_0$ to the subbundle corresponding to the irreducible component of maximal highest weight in $S^r\mathfrak{g}_1 \otimes \mathbb{W}_0$.

The simplest example of this situation is $\mathfrak{g} = \mathfrak{sl}(n+1, \mathbb{R})$ endowed with the grading $\begin{pmatrix} \mathfrak{g}_0 & \mathfrak{g}_1 \\ \mathfrak{g}_{-1} & \mathfrak{g}_0 \end{pmatrix}$ with blocks of sizes 1 and n. Then $\mathfrak{g}_{-1}$ has dimension n and $\mathfrak{g}_0 \cong \mathfrak{gl}(n, \mathbb{R})$. For the right choice of group, one obtains $G_0' = SL(n, \mathbb{R})$, so the structure is just a volume form on an n–manifold.

There is a complete description of $|1|$–gradings of semisimple Lie algebras in terms of structure theory and hence a complete list of the other geometries for which the procedure works. One of these is related to almost quaternionic structures, the others can be described in terms of identifications of the tangent bundle with a symmetric or skew symmetric square of an auxiliary bundle or with a tensor product of two auxiliary bundles.

5.2. Invariant differential operators for AHS–structures. For a group G with Lie algebra $\mathfrak{g} = \mathfrak{g}_{-1} \oplus \mathfrak{g}_0 \oplus \mathfrak{g}_1$ as in 5.1, one defines $P \subset G$ as the subgroup of those elements whose adjoint action preserves the subalgebra $\mathfrak{g}_0 \oplus \mathfrak{g}_1 =: \mathfrak{p}$. It turns out that $\mathfrak{p}$ is the Lie algebra of P and $G_0 \subset P$ can also be naturally be viewed as a quotient of P.

On manifolds of dimension $\dim(\mathfrak{g}_{-1})$ we may consider reductions of structure group to the group G_0. The passage from G_0' as discussed in 5.1 to G_0 is like the passage from Riemannian to conformal structures. As in 4.2, one may look at extensions of the principal G_0–bundle defining the structure to a principal P–bundle $\mathcal{G}$ endowed with a normal Cartan connection $\omega \in \Omega^1(\mathcal{G}, \mathfrak{g})$. In the example $\mathfrak{g} = \mathfrak{sl}(n+1, \mathbb{R})$ with $\mathfrak{g}_0 = \mathfrak{gl}(n, \mathbb{R})$ from 5.1, the principal G_0–bundle is the full frame bundle, so it contains no information. One shows that such an extension is equivalent to the choice of a projective equivalence class of torsion free connections on TM. In all other cases (more precisely, one has to require that no simple summand has this form) Cartan's result on conformal structures can be generalized to show that such an extension is uniquely possible for each given G_0–structure, see e.g. [8].

The structures equivalent to such Cartan connections are called AHS–structures in the literature. Apart from conformal and projective structures, they also contain almost quaternionic and almost Grassmannian structures as well as some more exotic examples, see [8]. For all these structures, the procedure from section 4 can be carried out without changes to construct differential operators which are intrinsic to the geometry.

5.3. More general geometries. The construction of invariant differential operators from section 4 applies to a much larger class of geometric structures. Let $\mathfrak{g}$ be a semisimple Lie algebra endowed with a $|k|$–grading, i.e. a grading of the form $\mathfrak{g} = \mathfrak{g}_{-k} \oplus \dots \oplus \mathfrak{g}_k$ for some $k \geq 1$, such that $[\mathfrak{g}_i, \mathfrak{g}_j] \subset \mathfrak{g}_{i+j}$ and such that the Lie subalgebra $\mathfrak{g}_- := \mathfrak{g}_{-k} \oplus \dots \oplus \mathfrak{g}_{-1}$ is generated by $\mathfrak{g}_{-1}$. For any such grading, the subalgebra $\mathfrak{p} := \mathfrak{g}_0 \oplus \dots \oplus \mathfrak{g}_k \subset \mathfrak{g}$ is a *parabolic* subalgebra in the sense of representation theory. Conversely, any parabolic subalgebra in a semisimple Lie algebra gives rise to a $|k|$–grading. Therefore, $|k|$–gradings are well understood and can be completely classified in terms of the structure theory of semisimple Lie algebras.

Given a Lie group G with Lie algebra $\mathfrak{g}$ one always finds a closed subgroup $P \subset G$ corresponding to the Lie algebra $\mathfrak{p}$. The homogeneous space G/P is a so–called *generalized flag variety.* Given a smooth manifold M of the same dimension as G/P, a *parabolic geometry* of type (G, P) on

M is given by a principal P–bundle $p : \mathcal{G} \to M$ and a Cartan connection $\omega \in \Omega^1(\mathcal{G}, \mathfrak{g})$.

In pioneering work culminating in [16], N. Tanaka has shown that assuming the conditions of regularity and normality on the curvature of the Cartan connection, such a parabolic geometry is equivalent to an underlying geometric structure. These underlying structures are very diverse, but during the last years a uniform description has been established, see the overview article [6]. Examples of these underlying structures include partially integrable almost CR structures of hypersurface type, path geometries, as well as generic distributions of rank two in dimension five, rank three in dimension six, and rank four in dimension seven. For all these geometries, the problem of constructing differential operators which are intrinsic to the structure is very difficult.

The BGG machinery developed in [9] and [5] offers a uniform approach for this construction, but compared to the procedure of Section 4 some changes have to be made. One again has a grading element E which leads to an eigenspace decomposition $\mathbb{W} = \mathbb{W}_0 \oplus \cdots \oplus \mathbb{W}_N$ of any finite dimensional irreducible representation of $\mathfrak{g}$. As before, we have $\mathfrak{g}_i \cdot \mathbb{W}_j \subset \mathbb{W}_{i+j}$. Correspondingly, this decomposition is only invariant under a subgroup $G_0 \subset P$ with Lie algebra $\mathfrak{g}_0$, but each of the subspaces $\mathbb{W}_i \oplus \cdots \oplus \mathbb{W}_N$ is P–invariant. The theory of tractor bundles and tractor connections works in this more general setting without changes, see [7].

Via the Cartan connection ω, the tangent bundle TM can be identified with $\mathcal{G} \times_P (\mathfrak{g}/\mathfrak{p})$ and therefore $T^*M \cong \mathcal{G} \times_P (\mathfrak{g}/\mathfrak{p})^*$. Now the annihilator of $\mathfrak{p}$ under the Killing form of $\mathfrak{g}$ is the subalgebra $\mathfrak{p}_+ := \mathfrak{g}_1 \oplus \cdots \oplus \mathfrak{g}_k$. For a tractor bundle $\mathcal{W} = \mathcal{G} \times_P \mathbb{W}$, the bundles of $\mathcal{W}$–valued forms are therefore associated to the representations $\Lambda^k \mathfrak{p}_+ \otimes \mathbb{W}$.

Since we are now working with the nilpotent Lie algebra $\mathfrak{p}_+$ rather than with an Abelian one, we have to adapt the definition of ∂^*. In order to obtain a differential, we have to add terms which involve the Lie bracket on $\mathfrak{p}_+$. The resulting map ∂^* is P–equivariant, and the quotients $\ker(\partial^*)/\operatorname{im}(\partial^*)$ can be computed as representations of $\mathfrak{g}_0$ using Kostant's theorem. As far as ∂ is concerned, we have to identify $\mathfrak{g}/\mathfrak{p}$ with the nilpotent subalgebra $\mathfrak{g}_- := \mathfrak{g}_{-k} \oplus \cdots \oplus \mathfrak{g}_{-1}$. Then we can add terms involving the Lie bracket on $\mathfrak{g}_-$ to obtain a map ∂ which is a differential. As the identification of $\mathfrak{g}/\mathfrak{p}$ with $\mathfrak{g}_-$, the map ∂ is not equivariant for the P–action but only for the action of a subgroup G_0 of P with Lie algebra $\mathfrak{g}_0$.

The P–equivariant map ∂^* again induces vector bundle homomorphisms on the bundles of $\mathcal{W}$–valued differential forms. We can extend the normal tractor connection to the covariant exterior derivative $d^{\mathcal{W}}$. As in 4.4, the lowest homogeneous component of $d^{\mathcal{W}}$ is tensorial and induced by ∂ which is all that is needed to get the procedure outlined in 4.4 going. Also the results for structures which are locally isomorphic to G/P discussed in 4.4 extend to general parabolic geometries.

The question of analogs of the prolongation procedure from section 3 for arbitrary parabolic geometries has not been completely answered yet. It is clear that some parts generalize without problems. For other parts, some modifications will be necessary. In particular, the presence of non-trivial filtrations of the tangent bundle makes it necessary to use the concept of weighted order rather than the usual concept of order of a differential operator and so on. Research in this direction is in progress.

REFERENCES

[1] T.N. Bailey, M.G. Eastwood, and A.R. Gover, *Thomas's structure bundle for conformal, projective and related structures*, Rocky Mountain J. (1994), **24**: 1191–1217.

[2] R.J. Baston, *Almost Hermitian symmetric manifolds, I: Local twistor theory; II: Differential invariants*, Duke Math. J., (1991), **63**: 81–111, 113–138.

[3] I.N. Bernstein, I.M. Gelfand, and S.I. Gelfand, *Differential operators on the base affine space and a study of* $\mathfrak{g}$*–modules*, in "Lie Groups and their Representations" (ed. I.M. Gelfand) Adam Hilger 1975, 21–64.

[4] T. Branson, A. Čap, M.G. Eastwood, and A.R. Gover, *Prolongation of geometric overdetermined systems*, Internat. J. Math. (2006), **17**(6): 641–664, available online as `math.DG/0402100`.

[5] D.M.J.Calderbank and T.Diemer, *Differential invariants and curved Bernstein-Gelfand-Gelfand sequences*, J. Reine Angew. Math. (2001), **537**: 67–103.

[6] A. Čap, *Two constructions with parabolic geometries*, Rend. Circ. Mat. Palermo Suppl. (2006), **79**: 11–37; available online as `math.DG/0504389`.

[7] A. Čap and A.R. Gover, *Tractor Calculi for Parabolic Geometries*, Trans. Amer. Math. Soc. (2002), **354**: 1511-1548.

[8] A. Čap, J. Slovák, and V. Souček, *Invariant operators on manifolds with almost Hermitian symmetric structures, II. Normal Cartan connections*, Acta Math. Univ. Commenianae, (1997), **66**: 203–220.

[9] A. Čap, J. Slovák, and V. Souček, *Bernstein–Gelfand–Gelfand sequences*, Ann. of Math. (2001), **154**(1): 97–113.

[10] M.G. Eastwood, *Prolongations of linear overdetermined systems on affine and Riemannian manifolds*, Rend. Circ. Mat. Palermo Suppl. (2005), **75**: 89–108.

[11] M.G. Eastwood, *Higher symmetries of the Laplacian*, Ann. of Math. (2005), **161**(3): 1645–1665.

[12] M.G. Eastwood and J.W. Rice, *Conformally invariant differential operators on Minkowski space and their curved analogues*, Commun. Math. Phys. (1987), **109**: 207–228.

[13] H.D. Fegan, *Conformally invariant first order differential operators*, Quart. J. Math. (1976), **27**: 371–378.

[14] B. Kostant, *Lie algebra cohomology and the generalized Borel–Weil theorem*, Ann. of Math. (1961), **74**(2): 329–387.

[15] J. Lepowsky,*A generalization of the Bernstein–Gelfand–Gelfand resolution*, J. of Algebra (1977), **49**: 496–511.

[16] N. Tanaka, *On the equivalence problem associated with simple graded Lie algebras*, Hokkaido Math. J. (1979), **8**: 23–84.

[17] T.Y. Thomas, *On conformal geometry*, Proc. N.A.S. (1926), **12**: 352–359; *Conformal tensors*, Proc. N.A.S. (1931), **18**: 103–189.

GENERALIZED WILCZYNSKI INVARIANTS FOR NON-LINEAR ORDINARY DIFFERENTIAL EQUATIONS

BORIS DOUBROV*

Abstract. We show that classical Wilczynski–Se-ashi invariants of linear systems of ordinary differential equations are generalized in a natural way to contact invariants of non-linear ODEs. We explore geometric structures associated with equations that have vanishing generalized Wilczynski invariants and establish relationship of such equations with deformation theory of rational curves on complex algebraic surfaces.

Key words. Differential invariants, projective curves, non-linear equations, twistor, symmetries of differential equations.

AMS(MOS) subject classifications. 34A26, 53B15.

1. Introduction. This paper is devoted to a very important class of contact invariants of (non-linear) systems of ordinary differential equations. They can be considered as a direct generalization of classical Wilczynski invariants of linear differential equations, which are, in its turn, closely related to projective invariants of non-parametrized curves.

The construction of these invariants is based on the fundamental idea of approximating the non-linear objects by linear. In case of differential equations we can consider linearization of a given non-linear along each solution. Roughly speaking, this linearization describes the tangent space to the solution set of the given equation. Note that in general this set is not Hausdorff, but we can still speak about linearization of the equation without any loss of generality.

Unlike the class of all non-linear equations of a fixed order, which is stable with respect contact transformations, the class of linear equations forms a category with a much smaller set of morphisms. This makes possible to describe the set of all invariants of linear equations explicitly. The main goal of this paper is to show the general geometric procedure that extends these invariants to the class of non-linear equations via the notion of the linearization.

In short, the major result of this paper is that *invariants of the linearization of a given non-linear equation are contact invariants of this equation.*

In fact, these invariants have been known for ordinary differential equations of low order: third order equations [3, 13], fourth order ODEs [1, 8, 6], systems of second order [7, 9], but it was not understood up to now that these invariants come from the linearization of non-linear equations.

In the theory of linear differential equations Wilczynski invariants generate all invariants and, in particular, are responsible for the trivialization

*Belarussian State University, Skoriny 4, 220050, Minsk, Belarus (doubrov@islc.org).

of the equation. In non-linear case the generalized Wilczynski invariants form only part of the generators in the algebra of all contact invariants. However, the equations with vanishing generalized Wilczynski invariants have a remarkable property, that its solution space carries a natural geometric structure (see [1] and [6]). We discuss this structure in more detail in Section 5 for a single ODE. Finally, in Section 6 we show that all results of this paper can be extended to the case of systems of ordinary differential equations.

2. Naive approach. As an example, let us outline this idea for the case of a single non-linear ODE of order ≥ 3. Note that all equations of smaller order are contact equivalent to each other and, thus, do not have any non-trivial invariants.

The complete set of invariants for a single linear ODE was described in the classical work of Wilczynski [17]. Namely, consider the class of linear homogeneous differential equations

$$y^{(n+1)} + p_n(x)y^{(n)} + \cdots + p_0(x)y = 0, \tag{2.1}$$

viewed up to all invertible transformations of the form:

$$(x, y) \mapsto (\lambda(x), \mu(x)y). \tag{2.2}$$

In fact, these are the most general transformations preserving the class of linear equations.

A function I of the coefficients $p_i(x)$ and their derivatives is called *a relative invariant of weight l* if it is transformed by the rule $I \mapsto (\lambda')^l I$ under the change of variables (2.2). In particular, if such relative invariant vanishes identically for the initial equation, it will also vanish for the transformed one. Relative invariants of weight 0 are called *(absolute) invariants of the linear equation.*

E. Wilczynski [17] gave the complete description of all relative (and, thus, absolute) invariants of linear ODEs of any order. It is well-known that each equation (2.1) can be brought by the transformation (2.2) to the so-called Laguerre–Forsyth canonical form:

$$y^{(n+1)} + q_{n-2}(x)y^{(n-2)} + \cdots + q_0(x)y = 0.$$

The set of transformations preserving this canonical form is already a finite-dimensional Lie group:

$$(x, y) \mapsto \left(\frac{ax+b}{cx+d}, \frac{ey}{(cx+d)^{n+1}}\right).$$

This group acts on coefficients $q_0, \dots, q_{n-2}$ of the canonical form, and the relative invariants of this action are identified with the relative invariants of the general linear equation. The simplest $n-1$ relative invariants $\theta_3, \dots, \theta_{n+1}$ linear in $q_i^{(j)}$ have the form:

$$\theta_k = \sum_{j=1}^{k-2} (-1)^j \frac{(2k-j-1)!(n-k+j)!}{(k-j)!(j-1)!} q_{n-k+j}^{(j-1)} \qquad k = 3, \ldots, n+1. \quad (2.3)$$

Each invariant θ_k has weight k. Wilczynski proved that all other invariants can be expressed in terms of the invariants θ_k and their derivatives.

Although these formulas express invariants in terms of coefficients of the canonical form, they can be written explicitly in terms of the initial coefficients of the general equation. Moreover, it can be shown that each of these invariants θ_i is polynomial in terms of functions $p_i(x)$ and their derivatives.

Equivalence theory of linear ODEs is intimately related with the projective theory of non-parametrized curves. Namely, let $\{y_0(x), \ldots, y_n(x)\}$ be a fundamental set of solutions of a linear equation $\mathcal{E}$ given by (2.1). Consider the curve

$$L_{\mathcal{E}} = \{[y_0(x) : y_1(x) : \cdots : y_n(x)] \mid x \in \mathbb{R}\}$$

in the n-dimensional projective space $\mathbb{R}\mathrm{P}^n$. Since the solutions do not vanish simultaneously for any $x \in \mathbb{R}$, this curve is well-defined. Moreover, since the fundamental set of solutions is linearly independent, this curve is *non-degenerate*, i.e., it is not contained in any hyperplane. Finally, since the set of fundamental solutions is defined up to any non-degenerate linear transformation, we see that the curve $L_{\mathcal{E}}$ is defined up to projective transformations.

Consider now what happens with this curve, if we apply the transformations (2.2) to the equation $\mathcal{E}$. The transformations $(x, y) \mapsto (x, \mu(x)y)$ do not change the curve, since they just multiply each solution by $\mu(x)$. The transformations $(x, y) \mapsto (\lambda(x), y)$ are equivalent to reparametrizations of $L_{\mathcal{E}}$.

Thus, we see that to each linear equation $\mathcal{E}$ we can assign the set of projectively-equivalent non-degenerate curves in $\mathbb{R}\mathrm{P}^n$. It is easy to see that this correspondence is one-to-one. Indeed, having a non-degenerate curve in $\mathbb{R}\mathrm{P}^n$ we can fix a parameter x on it and write it explicitly as $[y_0(x) : y_1(x) : \cdots : y_n(x)]$, where the coordinates $y_i(x)$ are defined modulo a non-zero multiplier $\mu(x)$. Since the curve is non-degenerate, the set of functions $\{y_i(x)\}_{i=0,\ldots,n}$ is linearly independent and defines a unique linear equation $\mathcal{E}$ having these functions as the set of fundamental solutions.

Each relative invariant I of weight k can be naturally interpreted as a section of the line bundle $S^k(TL)$ invariant with respect to projective transformations. In particular, relative invariants of weight 0 are just projective differential invariant of non-parametrized curves in the projective space. First examples of such invariants were constructed by Sophus Lie [11] for curves on the projective plane and then generalized by Halphen [10] to the case of projective spaces of higher dimensions. Note that they can be constructed via the standard Cartan moving frame method.

E. Wilczynski also proved the following result characterizing the equations with vanishing invariants:

THEOREM 1. *Let $\mathcal{E}$ be a linear ODE given by* (2.1). *The following conditions are equivalent:*

1. *invariants $\theta_3, \dots, \theta_{n+1}$ vanish identically.*
2. *the equation $\mathcal{E}$ is equivalent to the trivial equation $y^{(n+1)} = 0$.*
3. *the curve $L_{\mathcal{E}}$ is an open part of the normal rational curve in $\mathbb{R}P^n$.*
4. *the symmetry algebra of $L_{\mathcal{E}}$ is isomorphic to the subalgebra $\mathfrak{sl}(2, \mathbb{R}) \subset \mathfrak{sl}(n+1, \mathbb{R})$ acting irreducibly on $\mathbb{R}^{n+1}$.*

Let us show how to extend Wilczynski invariants to arbitrary non-linear ordinary differential equations via the notion of linearization. Indeed, consider now an arbitrary non-linear ODE solved with respect to the highest derivative:

$$y^{(n+1)} = f(x, y, y', \dots, y^{(n)}). \tag{2.4}$$

Let $y_0(x)$ be any solution of this equation. Then we can consider the linearization of (2.4) along this solution. It is a linear equation

$$h^{(n+1)} = \frac{\partial f}{\partial y^{(n)}} h^{(n)} + \dots + \frac{\partial f}{\partial y} h, \tag{2.5}$$

where all coefficients are evaluated at the solution $y_0(x)$. It describes all deformations $y_\epsilon(x) = y_0(x) + \epsilon h(x)$ of the solution $y_0(x)$, which satisfy the equation (2.4) modulo $o(\epsilon)$.

Consider now Wilczynski invariants $\theta_3, \dots, \theta_{n+1}$ of the linearization (2.5). They are polynomial in terms of the coefficients $\frac{\partial f}{\partial y^{(i)}}$ and their derivatives (evaluated at the solution $y_0(x)$). In general, let $F(x, y, y', \dots, y^{(n)})$ be any function of x, $y(x)$ and its derivatives evaluated at any solution of the equation (2.4). Then, differentiating it by x is equivalent to applying the operator of total derivative:

$$D = \frac{\partial}{\partial x} + y' \frac{\partial}{\partial y} + \dots + y^{(n)} \frac{\partial}{\partial y^{(n-1)}} + f \frac{\partial}{\partial y^{(n)}}.$$

Thus, analytically, Wilczynski invariants of (2.5) can be expressed as polynomials in terms of $D^j\left(\frac{\partial f}{\partial y^{(i)}}\right)$ $(0 \le i \le n,\ j \ge 0)$ and are independent of the solution $y_0(x)$ we started with. More precisely, there are $(n-1)$ well-defined expressions $W_3, \dots, W_{n+1}$ polynomial in terms of $D^j\left(\frac{\partial f}{\partial y^{(i)}}\right)$, which, being evaluated at any solution $y_0(x)$ of the equation (2.4), give Wilczynski invariants of its linearization along $y_0(x)$. We call these expressions $W_3, \dots, W_{n+1}$ the *generalized Wilczynski invariants of non-linear ordinary differential equations.*

To understand, what geometric objects correspond to generalized Wilczynski invariants we need to consider jet interpretation of differential equations. Let $J^i = J^i(\mathbb{R}^2)$ be the space of all jets of order i of (non-parametrized) curves on the plane. Then for each curve L on the plane we can define its lift $L^{(i)}$ to the jet space J^i of order i. Then the equation (2.4) can be considered as a submanifold $\mathcal{E} \subset J^{n+1}$ of codimension 1, and, by the existence and uniqueness theorem for ODEs, all lifts of its solutions form a one-dimensional foliation on $\mathcal{E}$. Let us denote its tangent one-dimensional distribution by E. Let (x, y) be any local coordinate system on the plane. Then it naturally defines a local coordinate system $(x, y_0, y_1, \dots, y_i)$ on J^i such that the lift of the graph $(x, y(x))$ is equal to $(x, y(x), y'(x), \dots, y^{(i)})$. In these coordinates the submanifold $\mathcal{E}$ is given by equation $y_{n+1} = f(x, y_0, y_1, \dots, y_n)$, functions $(x, y_0, y_1, \dots, y_n)$ form a local coordinate system on $\mathcal{E}$ and the distribution E is generated by the vector field

$$D = \frac{\partial}{\partial x} + y_1 \frac{\partial}{\partial y_0} + \dots + y_n \frac{\partial}{\partial y_{n-1}} + f \frac{\partial}{\partial y_n}.$$

The expressions $W_3, \dots, W_{n+1}$ can be naturally interpreted as functions on the equation manifold $\mathcal{E}$. In fact, each invariant W_i, being a relative invariant of weight i of all linearizations, defines a section $\mathcal{W}_i$ of the line bundle $S^i E^*$ by the formula:

$$\mathcal{W}_i(D, D, \dots, D) = W_i.$$

The largest set of invertible transformations preserving the class of ODEs of fixed order i consists of so-called contact transformations, which are the most general transformations of J^i preserving the class of lifts of plane curves. The main result of this paper can be formulated as follows:

THEOREM 2. *Sections $\mathcal{W}_i$, $i = 3, \dots, n+1$ of line bundles $S^i E^*$ are invariant with respect to contact transformations of J^{n+1}.*

3. Algebraic model of Wilczynski invariants.

In this section we give an alternative algebraic description of Wilczynski invariants, which is due to Se-ashi [14, 15]. There are two main reasons for providing an algebraic picture behind Wilczynski invariants. First, it can be easily generalized to invariants of systems of ODEs or even to more general classes of linear finite type equations (see [14]). Wilczynski himself described only invariants for a single ODE of arbitrary order and for linear systems of second order ODEs. It seems that analytic methods become too elaborate to proceed with systems of higher order.

Second, Se-ashi construction gives alternative analytic formulas for computing Wilczynski invariants, which are independent of Laguerre-Forsyth canonical form. In particular, this explains why Wilczynski invariants are polynomial in the initial coefficients and their derivatives, while the coefficients of the canonical form are not.

Denote by V_n the set $S^n(\mathbb{R}^2)$ of all homogeneous polynomials of degree n in two variables v_1, v_2. The standard $GL(2, \mathbb{R})$-action on $\mathbb{R}^2$ is naturally extended to V_n and turns it into an irreducible $GL(2,\mathbb{R})$-module. Denote by $\rho_n\colon GL(2,\mathbb{R}) \to GL(V_n)$ the corresponding representation mapping.

Consider the corresponding action of $\mathfrak{gl}(2,\mathbb{R})$ on $V_n = S^n(\mathbb{R}^2)$. Denote by X, Y, H, Z the following basis in $\mathfrak{gl}(2,\mathbb{R})$:

$$X = \begin{pmatrix} 0 & 1 \\ 0 & 0 \end{pmatrix}, \quad H = \begin{pmatrix} 1 & 0 \\ 0 & -1 \end{pmatrix}, \quad Y = \begin{pmatrix} 0 & 0 \\ 1 & 0 \end{pmatrix}, \quad Z = \begin{pmatrix} 1 & 0 \\ 0 & 1 \end{pmatrix}.$$

Then the action of these basis elements on V_n is equivalent to the action of the following vector fields on $S^n(\mathbb{R}^2)$:

$$X = v_1 \frac{\partial}{\partial v_2}, \quad H = v_1 \frac{\partial}{\partial v_1} - v_2 \frac{\partial}{\partial v_2}, \quad Y = v_2 \frac{\partial}{\partial v_1}, \quad Z = v_1 \frac{\partial}{\partial v_1} + v_2 \frac{\partial}{\partial v_2}.$$

In the sequel we shall identify $\mathfrak{gl}(2,\mathbb{R})$ with its image in $\mathfrak{gl}(V_n)$ defined by this action.

We define a gradation on V_n such that the polynomials $E_0 = v_1^n$, $E_1 = v_1^{n-1} v_2, \dots, E_n = v_2^n$ have degrees $-n-1, -n, \dots, -1$ respectively. Then the elements X, H, Y, Z define operators of degrees -1, 0, 1 and 0 respectively. Denote also by $V_n^{(i)}$ the set of all elements in V_n of degree $\leq i$. These subspaces define a filtration on V_n:

$$\{0\} = V_n^{(-n-2)} \subset V_n^{(-n-1)} \subset \dots \subset V_n^{(-1)} = V_n.$$

Let E be a one-dimensional vector bundle over a one-dimensional manifold M with a local coordinate x. Denote by $J^n(E)$ the n-th order jet bundle of E, which is a $(n+1)$-dimensional vector bundle over M. Then any $(n+1)$-th order linear homogeneous ODE can be considered as a connection on $J^n(E)$ such that all its solutions, being lifted to $J^n(E)$, are horizontal. Let $\mathcal{F}(J^n(E))$ be the frame bundle of $J^n(E)$. Since $\dim V_n = n+1$, we can identify $\mathcal{F}(J^n(E))$ as a set of all isomorphisms $\phi_x\colon V_n \to J^n_x(E)$. This turns $\mathcal{F}(J^n(E))$ into a principle $GL(V_n)$-bundle over M.

Denote by ω the corresponding connection form on $\mathcal{F}(J^n(E))$. In brief, the main idea of Se-ashi works is that Wilczynski invariants can be interpreted in terms of a natural reduction of the $GL(V_n)$-bundle $\mathcal{F}(J^n(E))$ to some G-subbunlde P characterized by the following conditions:

1. G is the image of the lower-triangular matrices in $GL(2,\mathbb{R})$ under the representation $\rho_n\colon GL(2,\mathbb{R}) \to GL(V_n)$;
2. $\omega|_P$ takes values in the subspace $\langle X, H, Z, Y, Y^2, \dots, Y^n \rangle$.

The form $\omega|_P$ can be decomposed into the sum of $\omega_{\mathfrak{gl}}$ with values in $\mathfrak{gl}(2,\mathbb{R}) \subset \mathfrak{gl}(V_n)$ and $\sum_{i=2}^n \omega_i Y^i$. Then $\omega_{\mathfrak{gl}}$ defines a flat projective structure on the manifold M, while the forms ω_i (or, more precisely, their values on the vector field $\frac{\partial}{\partial x}$) coincide up to the constant with Wilczynski invariants θ_{i+1}.

Let us describe this reduction in more detail. Denote by $G^{(0)}$ the subgroup of $GL(V_n)$ consisting of all elements preserving the filtration $V_n^{(i)}$ on

V_n introduced above. These are exactly all elements of $GL(V_n)$ represented by lower-triangular matrices in the basis $\{E_0, \dots, E_n\}$. For $k \geq 1$ denote by $\mathfrak{gl}^{(k)}(V_n)$ the following subalgebra in $\mathfrak{gl}(V_n)$:

$$\mathfrak{gl}^{(k)}(V_n) = \{\phi \in \mathfrak{gl}(V_n) \mid \phi(V_n^{(i)}) \subset V_n^{(i+k)}\}.$$

Let $GL^{(k)}(V_n)$ be the corresponding unipotent subgroup in $GL(V_n)$. Define the subgroups $G^{(k)} \subset G^{(0)}$ as the products $GL^{(0)}(2, \mathbb{R})GL_k(V_n)$ for each $k \geq 0$, where $GL^{(0)}(2, \mathbb{R})$ is the intersection of $G^{(0)}$ with $\rho_n(GL(2, \mathbb{R}))$. Denote also by W the subspace in $\mathfrak{gl}(V_n)$ spanned by the endomorphisms $Y^2, \dots, Y^n$ (here $Y \subset \mathfrak{gl}(2, \mathbb{R})$ is identified with the corresponding element in $\mathfrak{gl}(V_n)$).

The reduction $P \subset \mathcal{F}(J^n(E))$ is constructed via series of reductions $P_{k+1} \subset P_k$, where P_k is a principal $G^{(k)}$-bundle characterized by the following conditions:

a) P_0 consists of all frames $\phi_x \colon V_n \to J^n_x(E)$, which map the filtration of V_n into the filtration on each fiber $J^n_x(E)$;
b) for $k \geq 1$ the form $\omega_k = \omega|_{P_k}$ takes values in the subspace $W_k = W + \mathfrak{gl}^{(k-1)}(V_n) + \mathfrak{gl}(2, \mathbb{R})$.

At the end of this procedure we arrive at the principal bundle $P = P_{n+1}$ with the structure group $G = G_{n+1}$ and the 1-from $\omega = \omega_{n+1}$ with values in $W + \mathfrak{gl}(2, \mathbb{R})$.

In fact, the second condition can be considered as a definition of P_k for $k \geq 1$. Indeed, let $(x, y_0, y_1, \dots, y_n)$ be a local coordinate system on $J^n(E)$, such that the n-jet of the section $y(x)$ of $J(E)$ is given by $y_i(x) = y^{(i)}(x)$. Then the connection on $J^n(E)$ corresponding to the linear homogeneous equation (2.1) is defined as an annihilator of the forms:

$$\theta_i = dy_i - y_{i+1}dx,$$
$$\theta_n = dy_n + \left(\sum_{i=0}^{n} p_i(x)y_i\right) dx.$$

If $s \colon x \mapsto (y_0(x), \dots, y_n(x))$ is any section of $J^n(E)$, then the covariant derivative of s along the vector field ∂_x has the form:

$$\nabla_{\partial_x} s = \left(y_0'(x) - y_1(x), y_1'(x) - y_2(x), \dots, y_n'(x) + \sum_{i=0}^{n} p_i(x)y_i(x)\right).$$

Let s_i, $i = 0, \dots, n$, be the standard sections defined by $y_j(x) = \delta_{ij}$. Then $s = \{s_0, \dots, s_n\}$ is a local section of the frame bundle P_0. Let ω be the connection form. The pull-back $s^*\omega$ can be written in this coordinate system as:

$$s^*\omega = \begin{pmatrix} 0 & -1 & 0 & \dots & 0 \\ 0 & 0 & -1 & \dots & 0 \\ \vdots & \vdots & \vdots & \ddots & \vdots \\ 0 & 0 & 0 & \dots & -1 \\ p_0(x) & p_1(x) & p_2(x) & \dots & p_n(x) \end{pmatrix} dx.$$

Then Se-ashi reduction theorem says that there exists such gauge transformation $C\colon M \to G^{(0)}$ that

$$C^{-1}(s^*\omega)C + C^{-1}dC = \Big(-X + \alpha H + \beta Z + \gamma Y + \sum_{i=2}^{n} \bar\theta_{i+1} Y^i \Big)\, dx,$$

and the set of such transformations forms a principal G-bundle. As shown in [14], the coefficients $\bar\theta_{i+1}$ coincide with classical Wilczynski invariants θ_{i+1} defined by (2.3) up to the constant and some polynomial expression of invariants of lower weight:

$$\bar\theta_{i+1} = c\theta_{i+1} + P_{i+1}(\theta_3, \dots, \theta_i),$$

where P_{i+1} is some fixed polynomial without free term. In particular, Theorem 1 remains true if we substitute invariants θ_i with their modified versions $\bar\theta_i$.

We shall not repeat the computations from Se-ashi work [14], just mentioning that it is based on the following simple technical fact. Namely, for any $k \geq 1$ consider the subspace $\mathfrak{gl}_k(V_n)$ of all operators of degree k and the mapping $\operatorname{ad}_k(X)\colon \mathfrak{gl}_k(V_n) \to \mathfrak{gl}_{k-1}(V_n)$, $A \mapsto [X, A]$. Then we have the decomposition $\mathfrak{gl}_k(V_n) = \langle Y^k \rangle \oplus \operatorname{Im} \operatorname{ad}_{k+1} X$ for all $k \geq 1$, which allows to carry effectively the reduction from P_k to P_{k+1}. On each step such reduction involves only the operation of solving linear equations with constant coefficients and differentiation. In particular, this proves, that Wilczynski invariants are polynomial in terms of the coefficients $p_0(x), \dots, p_n(x)$ of the initial equation (2.1).

In general, we can not make further reductions without assumption that some of these invariants do not vanish. And if all these invariants vanish, then our equation is equivalent to the trivial equation $y^{(n+1)} = 0$.

4. Generalization of Wilczynski invariants to non-linear ODEs. Let $\mathcal{E}$ be now an arbitrary non-linear ODE of order $(n+1)$. As above, we identify it with a submanifold of codimension 1 in the jet space J^{n+1}. Then in local coordinates $(x, y_0, \dots, y_{n+1})$ the equation $\mathcal{E}$ is given by $y_{n+1} = f(x, y_0, \dots, y_n)$, and the functions $(x, y_0, \dots, y_n)$ form a local coordinate system on $\mathcal{E}$.

Let us recall that there is a canonical contact C^k distribution defined on each jet space $J^k(\mathbb{R}^2)$, which is generated by tangent lines to all lifts of curves from the plane. In local coordinates it is generated by two vector fields:

$$C^k = \left\langle \frac{\partial}{\partial y_k}, \frac{\partial}{\partial x} + y_1 \frac{\partial}{\partial y_0} + \dots y_k \frac{\partial}{\partial y_{k-1}} \right\rangle.$$

We also have natural projections $\pi_{k,l}\colon J^k(\mathbb{R}^2) \to J^l(\mathbb{R}^2)$ for all $l < k$.

The contact distribution C^{n+1} defines a line bundle on the equation $\mathcal{E}$ as follows $E_p = T_p\mathcal{E} \cap C_p^{n+1}$ for all $p \in \mathcal{E}$. Integral curves of this bundle are precisely lifts of the solutions of the given ODE. In local coordinates this line bundle is generated by the vector field:

$$D = \frac{\partial}{\partial x} + y_1 \frac{\partial}{\partial y_0} + \dots + y_n \frac{\partial}{\partial y_{n-1}} + f \frac{\partial}{\partial y_n},$$

which defines also the operator of total derivative.

We are interested in the frames on the normal bundle to E, that is the bundle $N(\mathcal{E}) = T\mathcal{E}/E$. We have a natural filtration of this bundle defined by means of the projections $\pi_{n+1,i}\colon \mathcal{E} \to J^i(\mathbb{R}^2)$ for $i < n$. Namely, we define N^i as the intersection of $\pi_{n+1,i}^* C^i$ with $T\mathcal{E}$ modulo the line bundle E for all $i = 0, \dots, n$. Then it is easy to see that the sequence

$$N = N_0 \supset N_1 \supset \dots N_n \supset 0,$$

is strictly decreasing and $\dim N_i = n + 1 - i$. In local coordinates we have

$$N_i = \left\langle \frac{\partial}{\partial y_i}, \dots, \frac{\partial}{\partial y_n} \right\rangle + E.$$

The vector field D defines a first order operator on N, which is compatible with this filtration, i.e. $D(N_i) \subset N_{i-1}$ for all $i = 1, \dots, n$.

As in case of linear equations, we can define the frame bundle, consisting of all maps $\phi_p\colon V_n \to N_p$ from the $GL(2, R)$-module V_n into N_p, which preserve the filtrations. This defines a $G^{(0)}$-bundle over $\mathcal{E}$.

To proceed further with similar reductions, we need to have an analog of the connection form. In general, we don't have any natural connection on the normal bundle $N(\mathcal{E})$. However, we can define the covariant derivative along all vectors lying in the line bundle E generated by D. This gives us a so-called partial connection with the connection form $\omega\colon E \to \mathfrak{gl}(V_n)$. This connection form it is enough for our purposes. We can form the similar set of reductions, which, being restricted to any solution L, coincides with the principle bundle constructed from the linearization of $\mathcal{E}$ along this solution. As a result, we get the set of well-defined generalized Wilczynski invariants $\mathcal{W}_i$, where each $\mathcal{W}_i$ is a global section of the bundle $S^i E^*$.

Since the bundle $N(\mathcal{E})$ and the connection form ω are defined totally in terms of the contact geometry of the jet space J^{n+1} and the reduction procedure does not depend on any external data, we see that generalized Wilczynski invariants are contact invariants of the original equation. It can be formulated rigorously in the following way:

THEOREM 3. *Assume that $n \geq 2$. Let $\phi\colon J^{n+1} \to J^{n+1}$ be a local contact transformation establishing the local equivalence of two equations $\mathcal{E}$ and $\overline{\mathcal{E}}$. Let E ($\overline{E}$) be the line bundle on $\mathcal{E}$ (resp. $\overline{\mathcal{E}}$) defining the solution foliation. Let $\mathcal{W}_i \in \Gamma(S^i E^*)$ (resp. $\overline{\mathcal{W}}_i \in \Gamma(S^i \overline{E}^*)$ be generalized Wilczynski invariants of $\mathcal{E}$ (resp. $\overline{\mathcal{E}}$). Then we have $\phi_*(E) = \overline{E}$ and $\phi^*\overline{\mathcal{W}}^i = \mathcal{W}_i$ for all $i = 3, \dots, n+1$.*

This theorem was first proven in the work [5] via direct computation of the canonical Cartan connection associated with any ordinary differential equation of order ≥ 4 (see [16, 4] for more details).

5. Equations with vanishing Wilczynski invariants.

5.1. Structure of the solution space. Suppose now that all generalized Wilczynski invariants vanish identically for a given equation $\mathcal{E}$. For linear equations this would mean that the given equation is trivializable, and the associated curve $L_{\mathcal{E}}$ is an open part of the rational normal curve in $\mathbb{R}\mathrm{P}^n$.

Even though the curve $L_{\mathcal{E}}$ is defined up to contact transformations, we can define geometric structures on the solution space itself, which will be independent of the choice of fundamental solutions. Indeed, let $V(\mathcal{E})$ be the solution space of the linear equation (2.1). Then at each point x_0 on the line we can define a subspace $V_{x_0}(\mathcal{E})$ in $V(\mathcal{E})$ consisting of all solutions vanishing at x_0. Thus, assigning the line $V_{x_0}^{\perp}$ in the dual space $V(\mathcal{E})^*$ to each point x_0 we get a well-defined curve in the projective space $\mathrm{P}V_{\mathcal{E}}^{\perp}$.

Using the duality principle, we can also construct the curve in the projective space $\mathrm{P}V_{\mathcal{E}}$. To do this, we can consider the line $l_{x_0}(\mathcal{E})$ in $V(\mathcal{E})$ consisting of all solutions vanishing at x_0 with their derivatives up to order $n-1$. The mapping $x \mapsto l_{x_0}(\mathcal{E})$ defines a non-degenerate curve in the projective space $\mathrm{P}V(\mathcal{E})$.

Suppose now, that all Wilczynski invariants of the linear equation $\mathcal{E}$ vanish identically and consider the symmetry algebra $\mathfrak{g} \subset \mathfrak{sl}(V(\mathcal{E}))$ of the constructed curve. Fixing any fundamental set of solutions and applying Theorem 1, we can show that this Lie algebra is isomorphic to $\mathfrak{sl}(2, \mathbb{R})$ and acts irreducibly on $V(\mathcal{E})$. Let G be the corresponding subgroup in $SL(V(\mathcal{E})$ and let $\bar{G}$ be its product with the central subgroup in $GL(V(\mathcal{E}))$. Then $\bar{G}$ is naturally isomorphic to $GL(2, \mathbb{R})$, and $\bar{G}$-module $V(\mathcal{E})$ is equivalent to the $GL(2, \mathbb{R})$-module $S^n(\mathbb{R}^2)$. Thus, any equation with vanishing Wilczynski invariants defines a $GL(2, \mathbb{R})$-structure on its solution space.

For a non-linear equation vanishing of generalized Wilczynski invariants is not sufficient for being trivializable. The paper [5] describes the extra set of invariants that should vanish to guarantee the trivializability of the equation:

THEOREM 4 ([5]). *The equation* (2.4) *is contact equivalent to the trivial equation $y^{(n+1)} = 0$ if and only if its generalized Wilczynski invariants vanish identically and in addition:*

for $n = 2$: $f_{2222} = 0$;

for $n = 3$*:* $f_{333} = 6f_{233} + f_{33}^2 = 0$;
for $n = 4$*:* $f_{44} = 6f_{234} - 4f_{333} - 3f_{34}^2 = 0$;
for $n = 5$*:* $f_{55} = f_{45} = 0$;
for $n \geq 6$*:* $f_{n,n} = f_{n,n-1} = f_{n-1,n-1} = 0$.

Yet, even if the equation is not trivializable, its linearization at each solution is trivializable, and we can construct a family of associated rational normal curves. It will define a $GL(2,\mathbb{R})$-structure on the solution space $\mathcal{S}$, if it is a Hausdorff manifold.

Let $\mathcal{E}$ be an arbitrary non-linear ODE. Suppose that its solution set $\mathcal{S}$ is Hausdorff (i.e., there exist a factor of $\mathcal{E}$ by the foliation formed by all solutions). Let $y_0(x)$ is any solution of the non-linear ODE, which is just a point in the manifold $\mathcal{S}$. The tangent space $T_{y_0}\mathcal{S}$ to $\mathcal{S}$ at $y_0(x)$ can be naturally identified with a solution space of the linearization of $\mathcal{E}$ along $y_0(x)$. By above, we can naturally construct a curve l in the projectivization of $T_{y_0}\mathcal{S}$, which means that we have a well-defined two-dimensional cone C_{y_0} in each tangent space $T_{y_0}\mathcal{S}$.

Suppose now that all generalized Wilczynski invariants of $\mathcal{E}$ vanish identically. Then linearizations of $\mathcal{E}$ along all solutions are trivializable, and all cones $C_{y_0} \subset T_{y_0}\mathcal{S}$ are locally equivalent to the cone in $\mathbb{R}^{n+1}$ corresponding to the normal curve in $\mathbb{RP}^n$.

Thus, we arrive at the following result.

THEOREM 5. *Let $\mathcal{E}$ be an arbitrary (non-linear) ODE with vanishing generalized Wilczynski invariants. Suppose that its solution space $\mathcal{S}$ is Hausdorff and, hence, is a smooth manifold. Then there exists a natural irreducible $GL(2,\mathbb{R})$-structure on $\mathcal{S}$.*

This structure was constructed in [6] in a slightly different way and is called a paraconformal structure.

5.2. Examples from twistor theory. In general, explicit examples of equations with vanishing Wilczynski are very difficult to construct. However, there is a large class of examples coming from twistor theory.

The whole theory above is also valid in complex analytic category. Consider an arbitrary complex surface S with a rational curve L on it. Suppose that the normal bundle of L has Grothendieck type $\mathcal{O}(n)$. Then Kodaira theory states that this rational curve L is included into a complete $(n+1)$-parameter family $\mathcal{L} = \{L_a\}$ of all deformations of L. Clearly, deformations do not change the topologic type of L. So, the whole family will consist of rational curves.

The family $\mathcal{L}$ uniquely defines an ordinary differential equation of order $(n+1)$ such that all these curves are its solutions. In more detail, we have to consider the lifts of the curves L_a to the jet space $J^{n+1}(S)$, where they will form a submanifold of codimension 1. This defines an ordinary differential equation $\mathcal{E}$.

THEOREM 6. *Let $\mathcal{E}$ be an ordinary differential equation defining the complete set of deformations of a rational curve L on a complex surface S.*

Then all generalized Wilczynski invariant of $\mathcal{E}$ vanish identically.

Proof. Each generalized Wilczynski invariant $\mathcal{W}_i$, being restricted to the lift of any solution L_a, defines a global section of the line bundle S^iTL_a, which has Grothendieck type $\mathcal{O}(-i)$. Since the only global section of the line bundle of this type is zero, we see that all generalized Wilczynski invariants will vanish identically. □

Consider a number of explicit examples.

EXAMPLE 1. Let $S = \mathbb{C}\mathrm{P}^2$ and let L be a quadric in S. Then the normal bundle of L has type $\mathcal{O}(4)$, and the complete family of deformations is 5-dimensional. Clearly, all quadrics in $\mathbb{C}\mathrm{P}^2$ belong to this family and depend on exactly 5 parameters.

So, we see that the family $\mathcal{L}$ in this case is a family of all quadrics on the complex projective plane. The differential equation of all quadrics is well-known and has the form:

$$9(y'')^2y^{(5)} - 45y''y'''y^{(4)} + 40(y''')^3 = 0.$$

By above, we have $\mathcal{W}_3 = \mathcal{W}_4 = \mathcal{W}_5 = 0$ for this equation. However, this equation is not trivializable, since, for example, its symmetry algebra is only 8-dimensional, while the trivial equation $y^{(5)} = 0$ has a 9-dimensional symmetry algebra.

EXAMPLE 2. Let S be a rational surface $S_k = \mathrm{P}(\mathcal{O}(0) + \mathcal{O}(k))$ viewed as a projective bundle over $\mathbb{C}\mathrm{P}^1$. Then all sections of this bundle with a fixed intersection number l with fibers form a complete family of deformations depending on $k + l + 1$ parameter. In the appropriate coordinate system on S_k they can be explicitly written as:

$$y(x) = \frac{a_0 + a_1x + \cdots + a_kx^k}{b_0 + b_1x + \cdots + b_lx^l}.$$

Assume that $k \geq l$. (Otherwise we can substitute y with $1/y$.) This set of rational curves defines the following differential equation:

$$F_{k,l} = \begin{vmatrix} z_{k-l+1} & z_{k-l+2} & \cdots & z_{k+1} \\ z_{k-l+2} & z_{k-l+3} & \cdots & z_{k+2} \\ \vdots & \vdots & \ddots & \vdots \\ z_{k+1} & z_{k+2} & \cdots & z_{k+l+1} \end{vmatrix} = 0, \qquad \text{where } z_i = \frac{(-1)^i}{i!}y^{(i)}.$$

For example, in the simplest case $l = 0$ we get just the trivial equation $y^{(k+1)} = 0$, while for $l = 1$ we get the following equation:

$$(k+1)y^{(k)}y^{(k+2)} - (k+2)(y^{(k+1)})^2 = 0.$$

For $k \geq 2$ the symmetry algebra of this equation is $k+2$-dimensional, and, hence, it is not trivializable.

For $l \geq 2$ the equation $F_{k,l} = 0$ is never trivializable, since, for example, being solved with respect to the highest derivative $y^{(k+l+1}$, it is not linear in term $y^{(k+l)}$.

Since all solutions of the equation $F_{k,l} = 0$ are by construction rational curves, all its generalized Wilczynski invariants vanish identically.

EXAMPLE 3. Consider the standard symplectic form σ on the 4-dimensional vector space $\mathbb{C}^4$. Then it defines the contact structure on $\mathbb{C}P^3$ invariant with respect to the induced action of $SP(4, \mathbb{C})$. It is easy to see that the set of all rational normal curves in $\mathbb{C}P^3$, which are at the same time integral curves of the contact distribution, depends on 7 parameters and forms the complete family of deformations of any of such curves.

The equation describing this set is given by:

$$10(y''')^3 y^{(7)} - 70(y''')^2 y^{(4)} y^{(6)} - 49(y''')^2 (y^{(5)})^2 + 280 y''' (y^{(4)})^2 y^{(5)} - 175 (y^{(4)})^4 = 0.$$

Again, since all solutions of this equation are rational curves, all its generalized Wilczynski invariants vanish identically. Yet, this equation is not trivializable, since its symmetry algebra is 10-dimensional (in fact, it coincides with $\mathfrak{sp}(4, \mathbb{C})$), while the trivial equation $y^{(7)} = 0$ has 11-dimensional symmetry algebra.

All equations from the examples above have the common property that they have the symmetry algebra of submaximal dimension [12].

6. Wilczynski invariants for systems of ODEs. Since all results of this paper are based on two ideas, namely the notion of linearization and the invariants and structures associated with linear equations, they are directly generalized to systems of ordinary differential equations and even to equations of finite type.

The generalization of Wilczynski invariants to the systems of ODEs was obtained by Se-ashi [15]. Consider an arbitrary system of linear ordinary differential equations:

$$y^{(n+1)} + P_n(x) y^{(n)} + \cdots + P_0(x) y(x) = 0,$$

where $y(x)$ is an $\mathbb{R}^m$-valued vector function. The canonical Laguerre-Forsyth form of these equations is defined by conditions $P_n = 0$ and $\operatorname{tr} P_{n-1} = 0$. Then, as in the case of a singe ODE, the following expressions:

$$\Theta_k = \sum_{j=1}^{k-1} (-1)^j \frac{(2k-j-1)!(n-k+j)!}{(k-j)!(j-1)!} P_{n-k+j}^{(j-1)}, \quad k = 2, \ldots, n+1 \quad (6.1)$$

are the $\operatorname{End}(\mathbb{R}^m)$-valued relative invariants, where each invariant θ_i has weight i. Note that unlike the case of a single ODE, the first non-trivial Wilczynski invariant has weight 2.

We can also consider a characteristic curve associated with any linear system, which will take values in the Grassmann manifold $\mathrm{Gr}_m(\mathbb{R}^{(n+1)m})$. Se-ashi also proved that the system is trivializable if and only if all these invariants vanish identically, or, equivalently, if the characteristic curve is an open part of the curve defined by a trivial equation. The symmetry group of this curve is isomorphic to the direct product of $SL(2,\mathbb{R})$ and $GL(m,\mathbb{R})$ with an action on $\mathbb{R}^{(n+1)m}$ equivalent to the natural action on $S^{(n+1)}(\mathbb{R}^2)\otimes\mathbb{R}^m$.

All that leads us immediately to the following generalization of Theorems 2 and 5.

THEOREM 7. *Let $\mathcal{E}$ be an arbitrary (non-linear) system of m ordinary differential equations of order $(n+1)$. Then there exist generalized Wilczynski invariants $\mathcal{W}_2, \dots, \mathcal{W}_{n+1}$ that, being restricted to each solution, coincide with Wilczynski invariants of the linearization along this solution.*

Suppose that all these invariants vanish identically and the solution space $\mathcal{S}$ is a smooth manifold. Then there exists a natural irreducible $SL(2,\mathbb{R})\times GL(m,\mathbb{R})$ structure on $\mathcal{S}$.

REMARK. Let us note that in the case of systems of ODEs the invariant $\mathcal{W}_i$ is a section the vector bundle $S^iE^*\otimes\mathrm{End}(V_n)$, where V_n is a subspace in the normal bundle $N(\mathcal{E}) = T\mathcal{E}/E$ defined as $\pi^*_{n+1,n}C^n\cap T\mathcal{E}$ modulo E.

EXAMPLE 4. Consider the simplest non-trivial case of second order systems of ODEs. The linear homogeneous system can be written as:

$$y'' = A(x)y' + B(x)y, \tag{6.2}$$

where $y(x)$ is the unknown $\mathbb{R}^m$-valued function and $A(x), B(x)\in\mathrm{End}(\mathbb{R}^m)$. The set of all transformations preserving the class of such systems has the form $(x,y)\mapsto(\lambda(x),\mu(x)y)$, where $\lambda(x)$ is a local line reparametrization and $\mu(x)\in GL(m,\mathbb{R})$.

We can bring the equation (6.2) into the semi-canonical form with vanishing coefficient $A(x)$ by means of a certain gauge transformation $(x,y)\mapsto(x,\mu(x)y)$. Then, applying a reparametrization $(x,y)\mapsto(\lambda(x),y)$ and an appropriate gauge transformation to bring the equation back to the semi-canonical form, we can make trace of $B(x)$ vanish.

The traceless part of the coefficient $B(x)$ in the semicanonical form gives us the Wilczynski invariant Θ_2 for linear systems of second order. Explicitly, it can be written as:

$$\Theta_2 = \Phi(x) - 1/m\,\mathrm{tr}\,\Phi(x), \quad \text{where } \Phi(x) = B(x) - \frac{1}{2}A'(x) + \frac{1}{4}A(x)^2. \tag{6.3}$$

According to Se-ashi [15], this is the only Wilczynski invariant available for systems of second order, and the equation (6.2) is trivializable if and only if the invariant Θ_2 vanishes identically.

Consider now an arbitrary non-linear system of second order:

$$y_i'' = f_i(x, y_j, y_k'), \qquad i = 1,\dots,m.$$

Then its generalized Wilczynski invariant can be obtained from (6.3) by substituting $A(x)$ with the matrix $\left(\frac{\partial f_i}{\partial y'_k}\right)$, $B(x)$ with the matrix $\left(\frac{\partial f_i}{\partial y_j}\right)$, and the usual derivative d/dx with operator of total derivative:

$$D = \frac{\partial}{\partial x} + \sum_{i=1}^{m} y'_i \frac{\partial}{\partial y_i} + \sum_{j=1}^{m} f_j \frac{\partial}{\partial y'_j}.$$

Denote by W_2 the invariant we get in this way. It coincides (up to the constant multiplier) with the invariant constructed by M. Fels [7].

If the invariant W_2 vanishes identically, then we get the $GL(2,\mathbb{R}) \times GL(m,R)$ structure on the solution space. This structure is also called Segre structure and was first constructed for solution space of second order ODEs with vanishing generalized Wilczynski invariant by D. Grossman [9].

Acknowledgments. This paper is based in large on the lectures given at the IMA workshop "Symmetries and overdetermined systems of partial differential equations". I would like to thank organizers of this workshop for inviting me to give these lectures. I also would like to thank Eugene Ferapontov and Igor Zelenko for valuable discussions on the topic of this paper.

REFERENCES

[1] R. Bryant, *Two exotic holonomies in dimension four, path geometries, and twistor theory*, Proc. Symp. Pure Math. (1991), **53**: 33–88.

[2] E. Cartan, *Sur les variétés à connexion projective*, Bull. Soc. Math. France (1924), **52**: 205–241.

[3] S.-S. Chern, *The geometry of the differential equation $y''' = F(x, y, y', y'')$*, Sci. Rep. Nat. Tsing Hua Univ. (1950), **4**: 97–111.

[4] B. Doubrov, B. Komrakov, and T. Morimoto, *Equivalence of holonomic differential equations*, Lobachevskij Journal of Mathematics (1999), **3**: 39–71.

[5] B. Doubrov, *Contact trivialization of ordinary differential equations*, Differential Geometry and Its applications, 2001, pp. 73–84.

[6] M. Dunajski and P. Todd, *Paraconformal geometry of n-th order ODEs, and exotic holonomy in dimension four*, J. Geom. Phys. (2006), **56**: 1790-1809.

[7] M. Fels, *The equivalence problem for systems of second order ordinary differential equations*, Proc. London Math. Soc. (1995), **71**(1): 221–240.

[8] M. Fels, The inverse problem of the calculus of variations for scalar fourth-order ordinary differential equations, Trans. Amer. Math. Soc. (1996), **348**: 5007–5029.

[9] D. Grossman, *Torsion-free path geometries and integrable second order ODE systems*, Sel. Math., New Ser. (2000), **6**: 399-442.

[10] G.H. Halphen, *Sur les invariants differentiels des courbes ganches*, Journ. de l'Ecole Polytechnique (1880), **28**: 1–25.

[11] S. Lie, *Klassifikation und Integration von gewönlichen Differentialgleichungen zwischen x, y, die eine Gruppe von Transformationen gestatten, I-IV*, Gesamelte Abhandlungen, Vol. 5, Leipzig–Teubner, 1924, S. 240–310, 362–427, 432–448.

[12] P.J. Olver, *Symmetry, invariants, and equivalence*, New York, Springer Verlag, 1995.

[13] H. Sato and A.Y. Yoshikawa, *Third order ordinary differential equations and Legendre connections*, J. Math. Soc. Japan (1998), **50**: 993–1013.
[14] Yu. Se-ashi, *On differential invariants of integrable finite type linear differential equations*, Hokkaido Math. J. (1988), **17**: 151–195.
[15] Yu. Se-ashi, *A geometric construction of Laguerre-Forsyth's canonical forms of linear ordinary differential equations*, Adv. Studies in Pure Math. (1993), **22**: 265–297.
[16] N. Tanaka, *Geometric theory of ordinary differential equations*, Report of Grant-in-Aid for Scientific Research MESC Japan, 1989.
[17] E.J. Wilczynski, *Projective differential geometry of curves and ruled surfaces*, Leipzig, Teubner, 1905.

NOTES ON PROJECTIVE DIFFERENTIAL GEOMETRY*

MICHAEL EASTWOOD†

(In memory of Thomas Branson)

AMS(MOS) subject classifications. Primary 53A20; Secondary 53B20, 58J70.

1. Introduction. Projective differential geometry was initiated in the 1920s, especially by Élie Cartan and Tracey Thomas. Nowadays, the subject is not so well-known. These notes aim to remedy this deficit and present several reasons why this should be done at this time. The deeper underlying reason is that projective differential geometry provides the most basic application of what has come to be known as the 'Bernstein-Gelfand-Gelfand machinery'. As such, it is completely parallel to conformal differential geometry. On the other hand, there are direct applications within Riemannian differential geometry. We shall soon see, for example, a good geometric reason why the symmetries of the Riemann curvature tensor constitute an irreducible representation of $\mathrm{SL}(n,\mathbb{R})$ (rather than $\mathrm{SO}(n)$ as one might naïvely expect). Projective differential geometry also provides the simplest setting in which overdetermined systems of partial differential equations naturally arise.

These notes are in no way meant to be comprehensive, neither in treatment nor in references to the extensive literature. Rather, their aim is quickly to arrive at some selected topics, as indicated above, to discuss projectively invariant linear differential operators, and especially to describe the Bernstein-Gelfand-Gelfand sequence in fairly concrete terms.

As is often done in differential geometry, we shall adorn tensors with upper or lower indices in correspondence with the tangent or cotangent bundle. We shall also use the Einstein summation convention to denote the natural pairing of vectors and covectors. Thus, X^a denotes a vector or a vector field, ω_a denotes a covector or a 1-form, and $X^a\omega_a$ denotes the pairing between them. For any tensor ϕ_{abc}, we shall denote its skew part by $\phi_{[abc]}$ and its symmetric part by $\phi_{(abc)}$. For example, if ω_{ab} is a 2 form, then

$$\nabla_{[a}\omega_{bc]} \quad \text{and} \quad X^a\nabla_a\omega_{bc} - 2(\nabla_{[b}X^a)\omega_{c]a},$$

*This article is based on two introductory lectures given at the 2006 Summer Program at the Institute for Mathematics and its Applications at the University of Minnesota. The author would like to thank the IMA for hospitality during this time and the referee of this article for helpful suggestions. The author is supported by the Australian Research Council.

†School of Mathematical Sciences, University of Adelaide, SA 5005, Australia (meastwoo@member.ams.org).

for any torsion-free connection ∇_a, are the exterior derivative and the Lie derivative in the direction of the vector field X^a, respectively. Such formulae are not meant to imply any choice of local coördinates. (More precisely, this is Penrose's abstract index notation [24]. It formalises the conventions used by many classical authors—see, for example, the discussion of projective differential geometry by Schouten [25].)

2. Definitions and an example of projective invariance. Let M be a smooth real manifold of dimension $n \geq 2$. There are two ways to define a projective differential geometry on M. One is geometric and intuitive. The other is more operational and useful in practice. Their equivalence is the subject of the following proposition.

PROPOSITION 2.1. *Two torsion-free connections ∇_a and $\widehat{\nabla}_a$ on M have the same geodesics as unparameterised curves if and only if*

$$\widehat{\nabla}_a\omega_b = \nabla_a\omega_b - \Upsilon_a\omega_b - \Upsilon_b\omega_a \tag{2.1}$$

for some 1-form Υ_a.

Proof. Let $\pi : TM \to M$ denote the tangent bundle to M and let V denote the vertical subbundle of $T(TM)$ so that we have the exact sequence

$$0 \to V \to T(TM) \to \pi^*TM \to 0 \tag{2.2}$$

of vector bundles on TM. The connection ∇_a may be viewed as defining a splitting of this exact sequence, in other words defining a horizontal subbundle complementary to V. Each element $X^a \in T_pM$ then pulls back to a unique horizontal vector. We obtain a vector field $\mathcal{X}$ on TM whose integral curves define the geodesics spray of ∇_a. To say that the connection $\widehat{\nabla}_a$ has the same unparameterised geodesics as ∇_a is to say that the corresponding vector field $\widehat{\mathcal{X}}$ differs from $\mathcal{X}$ by a multiple of the Euler field along the fibres of V.

Any two torsion-free connections are related by

$$\widehat{\nabla}_a\omega_b = \nabla_a\omega_b - \Gamma_{ab}{}^c\omega_c$$

for some tensor $\Gamma_{ab}{}^c = \Gamma_{(ab)}{}^c$. This tensor defines the corresponding change of splitting of (2.2). Specifically, at $X^a \in T_pM$ the change is given by the tensor $X^a\Gamma_{ab}{}^c$, regarded as a homomorphism from T_pM to $T_pM = V_p$. It follows that ∇_a and $\widehat{\nabla}_a$ have the same unparameterised geodesics if and only if $X^aX^b\Gamma_{ab}{}^c$ is a multiple of X^c for all X^a. But it is a matter of linear algebra to check that

$$X^aX^b\Gamma_{ab}{}^{[c}X^{d]}=0 \text{ for all } X^a \quad \text{if and only if } \Gamma_{(ab)}{}^c=\Upsilon_a\delta_b{}^c+\Upsilon_b\delta_a{}^c \text{ for some } \Upsilon_a.$$

This completes the proof. □

DEFINITION 2.1. *We shall say that two torsion-free connections ∇_a and $\widehat{\nabla}_a$ on M are projectively equivalent if and only if they have the same* *geodesics as unparameterised curves. A projective structure on M is a projective equivalence class of torsion-free connections on M.*

Proposition 2.1 gives an alternative, more operational, definition of projective equivalence according to (2.1). From this point of view, is it also clear that a projective structure is really a local notion (that can be patched together with a partition of unity).

To proceed further, let us now consider the consequences of (2.1) for other tensor fields. For X^a a vector field we have, dual to (2.1),

$$\widehat{\nabla}_a X^b = \nabla_a X^b + \Upsilon_a X^b + \Upsilon_c X^c \delta_a{}^b.$$

If ω_{ab} is a 2-form (a covariant tensor $\omega_{ab} = \omega_{[ab]}$), then

$$\begin{aligned}\widehat{\nabla}_a \omega_{bc} &= \nabla_a \omega_{bc} - 2\Upsilon_a \omega_{bc} - \Upsilon_b \omega_{ac} - \Upsilon_c \omega_{ba} \\ &= \nabla_a \omega_{bc} - 3\Upsilon_a \omega_{bc} + \Upsilon_a \omega_{bc} + \Upsilon_b \omega_{ca} + \Upsilon_c \omega_{ab} \\ &= \nabla_a \omega_{bc} - 3\Upsilon_a \omega_{bc} + 3\Upsilon_{[a} \omega_{bc]}\end{aligned}$$

and, more generally, for an p-form $\omega_{bc\cdots d}$,

$$\widehat{\nabla}_a \omega_{bc\cdots d} = \nabla_a \omega_{bc\cdots d} - (p+1)\Upsilon_a \omega_{bc\cdots d} + (p+1)\Upsilon_{[a} \omega_{bc\cdots d]}.$$

In particular, for an n-form $\omega_{bc\cdots de}$ we find that

$$\widehat{\nabla}_a \omega_{bc\cdots de} = \nabla_a \omega_{bc\cdots de} - (n+1)\Upsilon_a \omega_{bc\cdots de}$$

or, more succinctly

$$\widehat{\nabla}_a \sigma = \nabla_a \sigma - (n+1)\Upsilon_a \sigma$$

for a volume form σ (for simplicity let us suppose that M is oriented and write Λ^n for the bundle of volume forms). If we introduce the terminology projective density of weight w for sections of the line bundle $(\Lambda^n)^{-w/(n+1)}$, then we see that

$$\widehat{\nabla}_a \sigma = \nabla_a \sigma + w\Upsilon_a \sigma$$

when σ is such a density. Let us write $\mathcal{E}(w)$ for the bundle of projective densities of weight w (and also for its sheaf of smooth sections). Let us also write $\mathcal{E}_a$ for the bundle of 1-forms and $\mathcal{E}_a(w)$ for the bundle of 1-forms of weight w obtained by tensoring the 1-forms with $\mathcal{E}(w)$. For σ_a such a projectively weighted 1-form we see that

$$\widehat{\nabla}_a \sigma_b = \nabla_a \sigma_b + (w-1)\Upsilon_a \sigma_b - \Upsilon_b \sigma_a.$$

In particular, when $w = 2$ we conclude that

$$\widehat{\nabla}_a \sigma_b = \nabla_a \sigma_b + \Upsilon_a \sigma_b - \Upsilon_b \sigma_a \quad \text{whence} \quad \widehat{\nabla}_{(a} \sigma_{b)} = \nabla_{(a} \sigma_{b)}.$$

In other words

$$\mathcal{E}_a(2) \ni \omega_a \longmapsto \nabla_{(a}\sigma_{b)} \in \mathcal{E}_{(ab)}(2) \tag{2.3}$$

is <u>projectively invariant</u>. Similarly

$$\mathcal{E}_a \ni \omega_a \longmapsto \nabla_{[a}\sigma_{b]} \in \mathcal{E}_{[ab]}$$

is projectively invariant: this is the familiar exterior derivative $d : \Lambda^1 \to \Lambda^2$.

3. A link with Riemannian geometry. Now let us suppose that the projective structure on M arises from a Riemannian metric g_{ab}. A vector field X^a on M said to be a <u>Killing field</u> if and only if $\mathcal{L}_X g_{ab} = 0$ where $\mathcal{L}_X$ is the Lie derivative along X^a. Equivalently,

$$\nabla_{(a}X_{b)} = 0, \tag{3.1}$$

where ∇_a is the Levi Civita connection associated to g_{ab} and the vector field X^a is identified with the 1-form X_a by means of the metric. Geometrically, the Killing fields are the infinitesimal isometries of M. In the presence of the metric g_{ab}, the bundle of volume forms Λ^n is canonically trivialised and so we may regard X_a as having projective weight 2 if we so wish. In this sense, we have shown in the previous section:

PROPOSITION 3.1. *The Killing operator* $X^a \mapsto \nabla_{(a}X_{b)}$ *is projectively invariant.*

This observation may seem a little contrived but it acquires more significance when it is realised that (2.3) is the first in a natural sequence of projectively invariant differential operators. In order the describe this sequence we need firstly to develop a little more basic projective differential geometry and we start with some curvature conventions.

For any torsion-free connection ∇_a on TM, define its curvature tensor $R_{ab}{}^c{}_d$ by

$$(\nabla_a\nabla_b - \nabla_b\nabla_a)X^c = R_{ab}{}^c{}_d X^d$$

or, equivalently,

$$(\nabla_a\nabla_b - \nabla_b\nabla_a)\omega_d = -R_{ab}{}^c{}_d\omega_c. \tag{3.2}$$

It satisfies the Bianchi symmetry $R_{[ab}{}^c{}_{d]} = 0$ and may be uniquely and conveniently written as

$$R_{ab}{}^c{}_d = W_{ab}{}^c{}_d + 2\delta_{[a}{}^c\mathrm{P}_{b]d} + \beta_{ab}\delta^c{}_d, \tag{3.3}$$

where

$$W_{[ab}{}^c{}_{d]} = 0, \quad W_{ab}{}^c{}_d \text{ is totally trace-free}, \quad \beta_{ab} = -2\mathrm{P}_{[ab]}.$$

If we replace the connection ∇_a by $\widehat{\nabla}_a$ is accordance with (2.1), then we find that

$$\widehat{W}_{ab}{}^c{}_d = W_{ab}{}^c{}_d, \quad \widehat{\mathrm{P}}_{ab} = \mathrm{P}_{ab} - \nabla_a \Upsilon_b + \Upsilon_a \Upsilon_b, \quad \widehat{\beta}_{ab} = \beta_{ab} + 2\nabla_{[a}\Upsilon_{b]}. \tag{3.4}$$

In particular, the Weyl curvature $W_{ab}{}^c{}_d$ is projectively invariant. The Bianchi identity $\nabla_{[a}R_{bc]}{}^d{}_e = 0$ may be rewritten as

$$\begin{aligned} 4\nabla_{[a}\mathrm{P}_{b][c}\delta_{d]}{}^e - \nabla_a W_{cd}{}^e{}_b + \nabla_b W_{cd}{}^e{}_a \\ = 4\nabla_{[c}\mathrm{P}_{d][a}\delta_{b]}{}^e - \nabla_c W_{ab}{}^e{}_d + \nabla_d W_{ab}{}^e{}_c. \end{aligned} \tag{3.5}$$

It has the following consequences

$$\nabla_c W_{ab}{}^c{}_d = 2(n-2)\nabla_{[a}\mathrm{P}_{b]d} \quad \text{and} \quad \nabla_{[a}\beta_{bc]} = 0. \tag{3.6}$$

Therefore, the cohomology class $[\beta] \in H^2(M, \mathbb{R})$ is a global invariant of the projective structure and the obstruction to choosing a connection in the projective class with symmetric Schouten tensor P_{ab}. The tensor β_{ab} also finds a geometric interpretation as the curvature on densities. Specifically,

$$(\nabla_a\nabla_b - \nabla_b\nabla_a)\sigma = w\beta_{ab}\sigma \quad \text{for } \sigma \in \mathcal{E}(w). \tag{3.7}$$

If there is a connection in the projective class with $\beta_{ab} = 0$ (and this is always the case locally), then we may work exclusively with such connections to obtain a more restricted notion of equivalence. Specifically, we allow only closed 1-forms in (2.1). The resulting structure is called special projective or equi-projective. The analogy with the conformal case is stronger and many of the formulae below are simpler for special projective structures (this approach is adopted in [2]).

We are now in a position to construct the claimed next operator in the sequence. There is a general theory to be explained later. Here, we shall construct it 'by hand'. It will be a second order operator acting on $\mathcal{E}_{(ab)}(2)$ so let us consider now $\nabla_a\nabla_c h_{bd}$ for a symmetric covariant tensor h_{bd} of projective weight 2. Under a projective change of connection, we find

$$\begin{aligned} \widehat{\nabla}_c h_{bd} &= \nabla_c h_{bd} - \Upsilon_b h_{cd} - \Upsilon_d h_{bc} \\ \Longrightarrow \widehat{\nabla}_a \widehat{\nabla}_c h_{bd} &= \nabla_a(\nabla_c h_{bd} - \Upsilon_b h_{cd} - \Upsilon_d h_{bc}) \\ &\quad -\Upsilon_a(\nabla_c h_{bd} - \Upsilon_b h_{cd} - \Upsilon_d h_{bc}) \\ &\quad -\Upsilon_c(\nabla_a h_{bd} - \Upsilon_b h_{ad} - \Upsilon_d h_{ba}) \\ &\quad -\Upsilon_b(\nabla_c h_{ad} - \Upsilon_a h_{cd} - \Upsilon_d h_{ac}) \\ &\quad -\Upsilon_d(\nabla_c h_{ba} - \Upsilon_b h_{ca} - \Upsilon_a h_{bc}) \\ &= \nabla_a\nabla_c h_{bd} - 2\Upsilon_b\nabla_{(a}h_{c)d} \\ &\quad -2\Upsilon_d\nabla_{(a}h_{c)b} - 2\Upsilon_{(a}\nabla_{c)}h_{bd} \\ &\quad -(\nabla_a\Upsilon_b - \Upsilon_a\Upsilon_b)h_{cd} - (\nabla_a\Upsilon_d - \Upsilon_a\Upsilon_d)h_{bc} \\ &\quad +2\Upsilon_b\Upsilon_{(a}h_{c)d} + 2\Upsilon_d\Upsilon_{(a}h_{c)b} + 2\Upsilon_b\Upsilon_d h_{ac}, \end{aligned}$$

which we may rewrite using (3.4) as

$$\begin{aligned}\widehat{\nabla}_a\widehat{\nabla}_c h_{bd} - 2\widehat{\mathrm{P}}_{a(b}h_{d)c} &= \nabla_a\nabla_c h_{bd} - 2\mathrm{P}_{a(b}h_{d)c}\\ &\quad -2\Upsilon_b\nabla_{(a}h_{c)d} - 2\Upsilon_d\nabla_{(a}h_{c)b} - 2\Upsilon_{(a}\nabla_{c)}h_{bd}\\ &\quad +2\Upsilon_b\Upsilon_{(a}h_{c)d} + 2\Upsilon_d\Upsilon_{(a}h_{c)b} + 2\Upsilon_b\Upsilon_d h_{ac}.\end{aligned}$$

It follows straightforwardly that

$$\begin{array}{c}(\nabla_{(a}\nabla_{c)} + \mathrm{P}_{(ac)})h_{bd} - (\nabla_{(b}\nabla_{c)} + \mathrm{P}_{(bc)})h_{ad}\\ -(\nabla_{(a}\nabla_{d)} + \mathrm{P}_{(ad)})h_{bc} + (\nabla_{(b}\nabla_{d)} + \mathrm{P}_{(bd)})h_{ac}\end{array}$$

is projectively invariant. Notice that the resulting tensor r_{abcd} has the symmetries of the usual Riemann tensor:

$$r_{abcd} = -r_{bacd}, \quad r_{abcd} = -r_{abdc}, \quad r_{abcd} + r_{bcad} + r_{cabd} = 0.$$

For a flat metric, the tensor P_{ab} is zero and the operator is

$$h_{ab} \longmapsto r_{abcd} \equiv 2(\nabla_c\nabla_{[a}h_{b]d} - \nabla_c\nabla_{[d}h_{b]a}) \tag{3.8}$$

Next, it is easily verified that if r_{abcd} has Riemann tensor symmetries and is of projective weight 2, then

$$r_{abcd} \longmapsto B_{abcde} \equiv \nabla_{[a}r_{bc]de}$$

is projectively invariant (cf. Bianchi identity). We pause to introduce some terminology to accommodate the more complicated tensors that are now naturally arising. Rather than writing $\mathcal{E}_a$ for the bundle of 1-forms let us write a single box like this: $\square$. More generally, let us write a Young tableau to denote the bundle of covariant tensors enjoying the corresponding symmetries. For example,

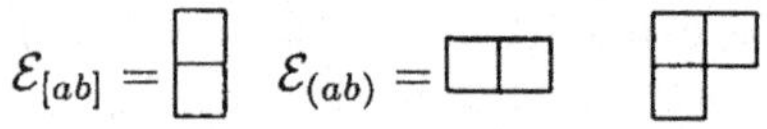

$$\mathcal{E}_{[ab]} = \ydiagram{1,1} \quad \mathcal{E}_{(ab)} = \ydiagram{2} \quad \ydiagram{2,1} = \{\phi_{abc} \text{ s.t. } \phi_{abc} = \phi_{[ab]c} \text{ and } \phi_{[abc]} = 0\}.$$

As discussed in [24], there is a 1-1 correspondence between irreducible tensors and Young tableau. As before, let us incorporate a conformal weight w into the notation by appending (w). Then the sequence of differential operators we have just constructed may be written in the flat case and continues, schematically, as follows:

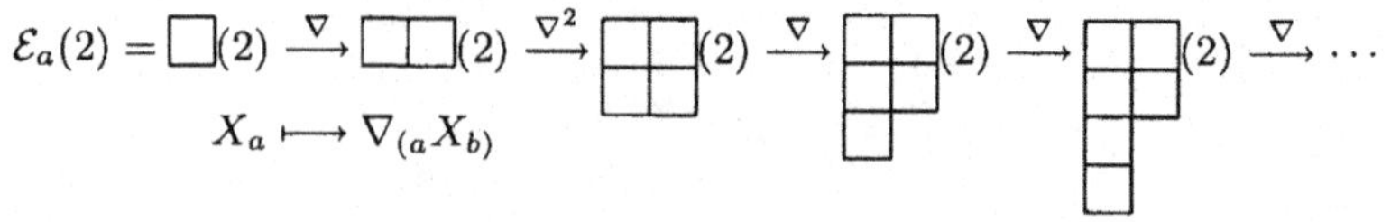

$$\begin{array}{l}\mathcal{E}_a(2) = \square(2) \xrightarrow{\nabla} \ydiagram{2}(2) \xrightarrow{\nabla^2} \ydiagram{2,2}(2) \xrightarrow{\nabla} \ydiagram{2,2,1}(2) \xrightarrow{\nabla} \ydiagram{2,2,1,1}(2) \xrightarrow{\nabla} \cdots\\ \qquad X_a \longmapsto \nabla_{(a}X_{b)}\end{array} \tag{3.9}$$

where ∇^k is the operator that differentiates k times and then imposes the appropriate symmetries. It is readily verified that this is a complex (the composition of any two successive operators is zero) as is done in [6] where Calabi introduced the same complex on projective space for similar reasons. It terminates when the number of rows exceeds n. As such, it is highly reminiscent of the de Rham complex

$$\mathcal{E} \xrightarrow{\nabla} \begin{array}{|c|}\hline \ \\ \hline\end{array} \xrightarrow{\nabla} \begin{array}{|c|}\hline \ \\ \hline \ \\ \hline\end{array} \xrightarrow{\nabla} \begin{array}{|c|}\hline \ \\ \hline \ \\ \hline \ \\ \hline\end{array} \xrightarrow{\nabla} \begin{array}{|c|}\hline \ \\ \hline \ \\ \hline \ \\ \hline \ \\ \hline\end{array} \xrightarrow{\nabla} \cdots$$

$$\omega_a \mapsto \nabla_{[a}\omega_{b]}$$

and we shall see later that there is a good reason for this.

There is a good geometric interpretation of the complex (3.9), not only in the flat case but also in the projectively flat case where it remains a complex owing to its projective invariance. It is the natural deformation complex in Riemannian geometry. In particular, the differential operator $h_{ab} \mapsto r_{abcd}$ is the linearisation at a projectively flat metric of the mapping that associates to a general metric its Riemann curvature tensor. In fact, the only projectively flat metrics are those of constant curvature (see, for example, [17]). Here, we shall not pursue this deformation complex further but note that it is a common feature in parabolic geometry [9].

4. Tractors. In contrast to Riemannian geometry, a projective structure does not give rise to a canonically defined connection on the tangent bundle. There is, however, a canonically defined connection on an auxiliary bundle, which forms the basis of an invariantly defined calculus on manifolds with projective structure. In this approach, the bundle and its connection are due to Tracey Thomas [26, 27]. For this reason, and also to follow the general pattern of vector, tensor, spinor, twistor, ..., Andrew Hodges suggested the terminology 'tractor' for constructions of this type.

We define a canonical rank $n+1$ vector bundle $\mathcal{E}_A$ on M as follows. For each choice of connection in the projective class $\mathcal{E}_A$ is identified as the direct sum

$$\mathcal{E}_A = \mathcal{E}(1) \oplus \mathcal{E}_a(1).$$

Under change of connection (2.1), however, this splitting changes according to

$$\widehat{\begin{pmatrix} \sigma \\ \mu_a \end{pmatrix}} = \begin{pmatrix} \sigma \\ \mu_a + \Upsilon_a\sigma \end{pmatrix}.$$

Notice that there is a canonical exact sequence

$$0 \to \mathcal{E}_a(1) \to \mathcal{E}_A \to \mathcal{E}(1) \to 0. \tag{4.1}$$

The bundle $\mathcal{E}_A$ is called a tractor bundle and comes equipped with an invariantly defined connection. For a particular connection ∇_a in the projective class define

$$\nabla_a \begin{pmatrix} \sigma \\ \mu_b \end{pmatrix} = \begin{pmatrix} \nabla_a\sigma - \mu_a \\ \nabla_a\mu_b + \mathrm{P}_{ab}\sigma \end{pmatrix}.$$

It is straightforward to check that this definition is projectively invariant:

$$\begin{aligned}
\widehat{\nabla}_a \widehat{\begin{pmatrix} \sigma \\ \mu_a \end{pmatrix}} &= \widehat{\nabla}_a \begin{pmatrix} \sigma \\ \mu_b + \Upsilon_b \sigma \end{pmatrix} \\
&= \begin{pmatrix} \widehat{\nabla}_a \sigma - (\mu_a + \Upsilon_a \sigma) \\ \widehat{\nabla}_a(\mu_b + \Upsilon_b \sigma) + \widehat{\mathrm{P}}_{ab}\sigma \end{pmatrix} \\
&= \begin{pmatrix} \nabla_a \sigma + \Upsilon_a \sigma - (\mu_a + \Upsilon_a \sigma) \\ \nabla_a(\mu_b + \Upsilon_b \sigma) - \Upsilon_b(\mu_a + \Upsilon_a \sigma) + (\mathrm{P}_{ab} - \nabla_a \Upsilon_b + \Upsilon_a \Upsilon_b)\sigma \end{pmatrix} \\
&= \begin{pmatrix} \nabla_a \sigma - \mu_a \\ \nabla_a \mu_b + \Upsilon_b \nabla_a \sigma - \Upsilon_b \mu_a + \mathrm{P}_{ab}\sigma \end{pmatrix} \\
&= \begin{pmatrix} \nabla_a \sigma - \mu_a \\ \nabla_a \mu_b + \mathrm{P}_{ab}\sigma + \Upsilon_b(\nabla_a \sigma - \mu_a) \end{pmatrix} \\
&= \widehat{\begin{pmatrix} \nabla_a \sigma - \mu_a \\ \nabla_a \mu_b + \mathrm{P}_{ab}\sigma \end{pmatrix}} = \nabla_a \widehat{\begin{pmatrix} \sigma \\ \mu_b \end{pmatrix}}.
\end{aligned}$$

The curvature of the tractor connection is easily calculated:

$$\begin{aligned}
\nabla_a \nabla_b \begin{pmatrix} \sigma \\ \mu_c \end{pmatrix} &= \nabla_a \begin{pmatrix} \nabla_b \sigma - \mu_b \\ \nabla_b \mu_c + \mathrm{P}_{bc}\sigma \end{pmatrix} \\
&= \begin{pmatrix} \nabla_a(\nabla_b \sigma - \mu_b) - (\nabla_b \mu_a + \mathrm{P}_{ba}\sigma) \\ \nabla_a(\nabla_b \mu_c + \mathrm{P}_{bc}\sigma) + \mathrm{P}_{ac}(\nabla_b \sigma - \mu_b) \end{pmatrix}
\end{aligned}$$

$$\Longrightarrow (\nabla_a \nabla_b - \nabla_b \nabla_a) \begin{pmatrix} \sigma \\ \mu_c \end{pmatrix}$$

$$\begin{aligned}
&= \begin{pmatrix} (\nabla_a \nabla_b - \nabla_b \nabla_a)\sigma - \mathrm{P}_{ba}\sigma + \mathrm{P}_{ab}\sigma \\ (\nabla_a \nabla_b - \nabla_b \nabla_a)\mu_c + 2(\nabla_{[a}\mathrm{P}_{b]c})\sigma - 2\mu_{[b}\mathrm{P}_{a]c} \end{pmatrix} \\
&= \begin{pmatrix} \beta_{ab}\sigma + 2\mathrm{P}_{[ab]}\sigma \\ -R_{ab}{}^d{}_c \mu_d + \beta_{ab}\mu_c + 2(\nabla_{[a}\mathrm{P}_{b]c})\sigma - 2\mu_{[b}\mathrm{P}_{a]c} \end{pmatrix}.
\end{aligned}$$

However, from (3.3) we obtain

$$R_{ab}{}^d{}_c \mu_d = (W_{ab}{}^d{}_c + 2\delta_{[a}{}^d \mathrm{P}_{b]c} + \beta_{ab}\delta^d{}_c)\mu_d = W_{ab}{}^d{}_c \mu_d + 2\mu_{[a}\mathrm{P}_{b]c} + \beta_{ab}\mu_c.$$

Also, recall that $\beta_{ab} = -2\mathrm{P}_{ab}$. Therefore,

$$(\nabla_a \nabla_b - \nabla_b \nabla_a) \begin{pmatrix} \sigma \\ \mu_c \end{pmatrix} = \begin{pmatrix} 0 \\ -W_{ab}{}^d{}_c \mu_d + 2(\nabla_{[a}\mathrm{P}_{b]c})\sigma \end{pmatrix}.$$

Of course, this curvature must be projectively invariant: it is easily verified that

$$\widehat{\nabla}_{[a}\widehat{\mathrm{P}}_{b]c} = \nabla_{[a}\mathrm{P}_{b]c} + \tfrac{1}{2}W_{ab}{}^d{}_c \Upsilon_d. \tag{4.2}$$

THEOREM 4.1. *The tractor connection is flat if and only if*

$$\begin{aligned} W_{ab}{}^c{}_d &= 0 \quad \text{if } n \geq 3 \\ \nabla_{[a}\mathrm{P}_{b]c} &= 0 \quad \text{if } n = 2. \end{aligned}$$

Proof. If $n \geq 3$ and $W_{ab}{}^c{}_d = 0$, then (3.6) implies that $\nabla_{[a}\mathrm{P}_{b]c}$ also vanishes. When $n = 2$, however, the Weyl curvature automatically vanishes by symmetry considerations and (4.2) says that the Cotton-York tensor $\nabla_{[a}\mathrm{P}_{b]c}$ is projectively invariant. □

An alternative and more well-known approach to projective differential geometry is due to Élie Cartan [13]. In this approach, firstly one constructs a principal P-bundle over the manifold with projective structure, where P is a certain subgroup of $\mathrm{SL}(n+1,\mathbb{R})$ as discussed in §5. This construction is essentially equivalent to the exact sequence (4.1). One can take, as principal P-bundle, the bundle of frames for $\mathcal{E}_A$ adapted to the exact sequence (4.1). Conversely, the standard representation of $\mathrm{SL}(n+1,\mathbb{R})$ on $\mathbb{R}^{n+1}$ gives $\mathcal{E}_A$ as the associated vector bundle. Next, one constructs on this Cartan bundle an $\mathfrak{sl}(n+1,\mathbb{R})$-valued 1-form, which behaves very much like a connection on a principal bundle (except that it does not have values in the Lie algebra of P). It is called a Cartan connection and has the property that, for any representation of $\mathrm{SL}(n+1,\mathbb{R})$, the associated vector bundle is thereby equipped with a connection. The tractor connection on $\mathcal{E}_A$ arises in this way (and, conversely, characterises the Cartan connection).

More generally, any bundle arising from an irreducible representation of $\mathrm{SL}(n+1,\mathbb{R})$ will also be called a tractor bundle whilst $\mathcal{E}_A$ will be called the standard tractor bundle. Any tractor bundle is thereby equipped with a tractor connection. For example,

$$\Lambda^2\mathcal{E}_A = \mathcal{E}_{[AB]} = \mathcal{E}_a(2) + \mathcal{E}_{[ab]}(2)$$

transforms by

$$\widehat{\begin{pmatrix} \sigma_a \\ \mu_{ab} \end{pmatrix}} = \begin{pmatrix} \sigma_a \\ \mu_{ab} + 2\Upsilon_{[a}\sigma_{b]} \end{pmatrix}$$

and inherits a canonical tractor connection defined by

$$\nabla_a \begin{pmatrix} \sigma_b \\ \mu_{bc} \end{pmatrix} = \begin{pmatrix} \nabla_a\sigma_b - \mu_{ab} \\ \nabla_a\mu_{bc} + 2\mathrm{P}_{a[b}\sigma_{c]} \end{pmatrix}$$

with curvature given by

$$(\nabla_a\nabla_b - \nabla_b\nabla_a)\begin{pmatrix} \sigma_c \\ \mu_{cd} \end{pmatrix} = \begin{pmatrix} -W_{ab}{}^d{}_c\sigma_d \\ 2W_{ab}{}^e{}_{[c}\mu_{d]e} + 4(\nabla_{[a}\mathrm{P}_{b][c})\sigma_{d]} \end{pmatrix}. \qquad (4.3)$$

We may prolong the Killing equation (3.1) as follows. To compare with tractors, let us write σ_a instead of X_a. Then (3.1) may be written as

$$\nabla_a\sigma_b = \mu_{ab}$$

for some skew tensor μ_{ab}. Since σ_a has weight 2, from (3.2) and (3.7) we find

$$(\nabla_a\nabla_b - \nabla_b\nabla_a)\sigma_d = -R_{ab}{}^c{}_d\sigma_c + 2\beta_{ab}\sigma_d$$

whence

$$\nabla_{[a}\mu_{bc]} = \nabla_{[a}\nabla_b\sigma_{c]} = \beta_{[ab}\sigma_{c]},$$

which we may rewrite as

$$\begin{aligned}
\nabla_a\mu_{bc} &= \nabla_c\mu_{ba} - \nabla_b\mu_{ca} + 3\beta_{[ab}\sigma_{c]} \\
&= \nabla_c\nabla_b\sigma_a - \nabla_b\nabla_c\sigma_a + 3\beta_{[ab}\sigma_{c]} \\
&= R_{bc}{}^d{}_a\sigma_d - 2\beta_{bc}\sigma_a + 3\beta_{[ab}\sigma_{c]} \\
&= R_{bc}{}^d{}_a\sigma_d - \beta_{bc}\sigma_a + 2\beta_{a[b}\sigma_{c]}.
\end{aligned}$$

However, from (3.3) we find that

$$\begin{aligned}
R_{bc}{}^d{}_a\sigma_d &= (W_{bc}{}^d{}_a + 2\delta_{[b}{}^d\mathrm{P}_{c]a} + \beta_{bc}\delta^d{}_a)\sigma_d \\
&= W_{bc}{}^d{}_a\sigma_d + 2\sigma_{[b}\mathrm{P}_{c]a} + \beta_{bc}\sigma_a \\
&= W_{bc}{}^d{}_a\sigma_d - 2\mathrm{P}_{a[b}\sigma_{c]} - 2\beta_{a[b}\sigma_{c]} + \beta_{bc}\sigma_a.
\end{aligned}$$

Therefore,

$$\nabla_a\mu_{bc} = W_{bc}{}^d{}_a\sigma_d - 2\mathrm{P}_{a[b}\sigma_{c]}$$

and we conclude that Killing fields are equivalent to parallel sections of the connection

$$D_a\begin{pmatrix}\sigma_b\\ \mu_{bc}\end{pmatrix} = \begin{pmatrix}\nabla_a\sigma_b - \mu_{ab}\\ \nabla_a\mu_{bc} + 2\mathrm{P}_{a[b}\sigma_{c]} - W_{bc}{}^d{}_a\sigma_d\end{pmatrix}.$$

Somewhat unexpectedly (but compare with [9, §3.2] and [17, Theorem 5.1]), this is not the tractor connection on $\mathcal{E}_{[BC]}$. Specifically,

$$D_a\begin{pmatrix}\sigma_b\\ \mu_{bc}\end{pmatrix} = \nabla_a\begin{pmatrix}\sigma_b\\ \mu_{bc}\end{pmatrix} - \begin{pmatrix}0\\ W_{bc}{}^d{}_a\sigma_d\end{pmatrix}.$$

The curvature of the connection D_a is computed as follows.

$$\begin{aligned}
&D_aD_b\begin{pmatrix}\sigma_c\\ \mu_{cd}\end{pmatrix}\\
&= D_a\nabla_b\begin{pmatrix}\sigma_c\\ \mu_{cd}\end{pmatrix} - D_a\begin{pmatrix}0\\ W_{cd}{}^e{}_b\sigma_e\end{pmatrix}\\
&= \nabla_a\nabla_b\begin{pmatrix}\sigma_c\\ \mu_{cd}\end{pmatrix} - \begin{pmatrix}0\\ W_{cd}{}^e{}_a(\nabla_b\sigma_e - \mu_{be})\end{pmatrix} - \nabla_a\begin{pmatrix}0\\ W_{cd}{}^e{}_b\sigma_e\end{pmatrix}\\
&= \nabla_a\nabla_b\begin{pmatrix}\sigma_c\\ \mu_{cd}\end{pmatrix} - \begin{pmatrix}0\\ W_{cd}{}^e{}_a(\nabla_b\sigma_e - \mu_{be})\end{pmatrix} - \begin{pmatrix}-W_{ac}{}^e{}_b\sigma_e\\ \nabla_a(W_{cd}{}^e{}_b\sigma_e)\end{pmatrix}\\
&= \nabla_a\nabla_b\begin{pmatrix}\sigma_c\\ \mu_{cd}\end{pmatrix} - \begin{pmatrix}-W_{ac}{}^e{}_b\sigma_e\\ 2W_{cd}{}^e{}_{(a}\nabla_{b)}\sigma_e - W_{cd}{}^e{}_a\mu_{be} + (\nabla_aW_{cd}{}^e{}_b)\sigma_e\end{pmatrix}.
\end{aligned}$$

Therefore, bearing (4.3) and some Bianchi symmetry in mind,

$$(D_aD_b - D_bD_a)\begin{pmatrix}\sigma_c\\ \mu_{cd}\end{pmatrix} = \begin{pmatrix} 0 \\ 2W_{ab}{}^e{}_{[c}\mu_{d]e} + 2W_{cd}{}^e{}_{[a}\mu_{b]e} \\ +4(\nabla_{[a}\mathrm{P}_{b][c})\sigma_{d]} - (\nabla_a W_{cd}{}^e{}_b)\sigma_e + (\nabla_b W_{cd}{}^e{}_a)\sigma_e \end{pmatrix}.$$

Notice that the Bianchi identity (3.5) implies that the tensor

$$\begin{aligned} r_{abcd} \equiv\ & 2W_{ab}{}^e{}_{[c}\mu_{d]e} + 2W_{cd}{}^e{}_{[a}\mu_{b]e} + 4(\nabla_{[a}\mathrm{P}_{b][c})\sigma_{d]} \\ & -(\nabla_a W_{cd}{}^e{}_b)\sigma_e + (\nabla_b W_{cd}{}^e{}_a)\sigma_e \end{aligned}$$

has the symmetries of the usual Riemann tensor.

5. The flat model of projective differential geometry. Central projection from the upper hemisphere

$$\{x = (x_0, x_1, x_2 \ldots, x_n) \in \mathbb{R}^{n+1} \text{ s.t. } \|x\| = 1 \text{ and } x_0 > 0\}$$

to the hyperplane

$$\{(1, x_1, x_2 \ldots, x_n) \in \mathbb{R}^{n+1}\} \cong \mathbb{R}^n$$

sends great circles on the sphere to straight lines in $\mathbb{R}^n$. Hence, the projective differential geometry of $\mathbb{R}^n$ is locally the same as that of the n-sphere with its round metric. If we further quotient the n-sphere using antipodal identification then we obtain real projective n-space and $\mathbb{R}^n \hookrightarrow \mathbb{RP}_n$ as a standard affine coördinate patch. As far as local projective differential goes, $\mathbb{RP}_n$ with its usual notion of lines is just as good as $\mathbb{R}^n$ with its straight lines. In addition to its being compact, $\mathbb{RP}_n$ also has the advantage that any local diffeomorphism defined on a connected open set and preserving the projective structure (i.e., preserving colinearity or, equivalently by Proposition 2.1, pulling back the round connection to a projectively equivalent one) extends to a global projective automorphism. Furthermore, it is clear that any linear transformation of $\mathbb{R}^{n+1}$ induces a projective transformation of $\mathbb{RP}_n$ and, conversely, it is easy to check that all global projective automorphisms of $\mathbb{RP}_n$ are so induced. Without loss of generality, we may restrict attention to linear transformations of unit determinant. Hence, we shall take as the flat model of projective differential geometry, $\mathbb{RP}_n$ as a homogenous space

$$\mathbb{RP}_n = G/P = \mathrm{SL}(n+1,\mathbb{R}) \Big/ \left\{ \begin{bmatrix} * & * & \cdots & * \\ 0 & * & \cdots & * \\ \vdots & \vdots & \ddots & \vdots \\ 0 & * & \cdots & * \end{bmatrix} \right\}. \tag{5.1}$$

The group $G = \mathrm{SL}(n+1,\mathbb{R})$ is a simple Lie group and P is a parabolic subgroup. For this reason, projective differential geometry is perhaps the simplest case of what is now called parabolic differential geometry [9, 10].

In any case, it is clear that a first step towards understanding projective differential geometry is to understand the flat model. To study projectively invariant differential operators, for example, a first task is to study the G-invariant operators on $\mathbb{RP}_n$. These operators should act between homogeneous vector bundles [4] on $\mathbb{RP}_n$. If we restrict attention to the irreducible bundles (i.e., those induced an irreducible representation of P), then we may classify the linear G-invariant operators between them. This is discussed in the next section. Already, the basic homogeneous line bundles on $\mathbb{RP}_n$ have arisen in this exposition. The bundle $\mathcal{E}(w)$ is the usual line bundle whose sections correspond to homogeneous functions of degree w.

The flat model serves as a guide for the 'curved' geometry of a general projective structure. The tractor bundle $\mathcal{E}_A$ on $\mathbb{RP}_n$ is induced by the standard representation of P on $\mathbb{R}^{n+1}$ (strictly speaking, its dual). As a homogeneous vector bundle this is non-trivial but, solely as a vector bundle, it is a product $\mathbb{RP}_n \times \mathbb{R}^{n+1}$. Better said, the tractor connection is flat on $\mathbb{RP}_n$ and, in general, the curvature of the tractor connection measures its deviation from $\mathbb{RP}_n$. The exact sequence (4.1) for the flat model is the familiar Euler sequence. In the next section is described the classification of $\mathrm{SL}(n+1,\mathbb{R})$-invariant linear differential operators between irreducible homogeneous vector bundles on $\mathbb{RP}_n$. A good example, is the second order operator (3.8) written in the flat metric and acting on $\mathcal{E}_{(ab)}(2)$. As is clear from its derivation, this particular operator has a curved analogue in the sense that there is a projectively invariant operator in general that differs from (3.8) only in the addition of lower order curvature correction terms. This is another way in which the flat model makes its presence felt in the curved setting.

The diligent reader should be warned that it is, strictly speaking, better to take as the flat model of projective geometry, the n-sphere with its projective structure induced by the round metric. Equivalently, we should take the projective space of oriented lines through the origin in $\mathbb{R}^{n+1}$ (as is done in [15]). The difference is that the subgroup P now consists of matrices of the form

$$\begin{bmatrix} \lambda & * & \cdots & * \\ 0 & * & \cdots & * \\ \vdots & \vdots & \ddots & \vdots \\ 0 & * & \cdots & * \end{bmatrix} \qquad \text{for } \lambda > 0$$

and the bundle $\mathcal{E}(w)$ is induced by the character λ^{-w}, which makes sense for any real w, not necessarily integral.

6. Invariant operators and the BGG sequence. As discussed in the previous section, we should firstly try to understand the invariant operators on the flat model and then investigate the consequences in curved projective differential geometry. Certainly, an invariant operator in the curved setting will be invariant under the projective automorphisms on the flat model. This means we should consider the G-invariant (some authors say 'G-equivariant') differential operators on $\mathbb{RP}_n$ as a homogeneous space (5.1) under $G = \mathrm{SL}(n+1, \mathbb{R})$.

As one might imagine, this is an algebraic task in Lie theory. Rather than go into detail here, suffice it to say that the appropriate theory exists and is quite simple in this particular case. The appropriate tool is the theory of Verma modules and a detailed exposition for the two cases

$$\mathrm{SL}(2,\mathbb{R})\Big/\left\{\begin{bmatrix} * & * \\ 0 & * \end{bmatrix}\right\} \quad \text{and} \quad \mathrm{SL}(4,\mathbb{C})\Big/\left\{\begin{bmatrix} * & * & * & * \\ * & * & * & * \\ 0 & 0 & * & * \\ 0 & 0 & * & * \end{bmatrix}\right\}$$

occurs in [14] and [18], respectively. The G-invariant linear differential operators acting between irreducible homogeneous vector bundles on $\mathbb{RP}_n$ may be classified. The result is as follows.

Firstly, there is the exterior derivative and, indeed, the de Rham complex:

$$0 \to \mathbb{R} \to \Lambda^0 \xrightarrow{d} \Lambda^1 \xrightarrow{d} \Lambda^2 \xrightarrow{d} \cdots \xrightarrow{d} \Lambda^{n-1} \xrightarrow{d} \Lambda^n \to 0.$$

These operators are projectively invariant. Better, they are diffeomorphism invariant (and are characterised by this property [21]). All the G-invariant operators on $\mathbb{RP}_n$ fall into similar patterns and the only stumbling block is having a suitable notation in which to write these patterns and appreciate the operators within them. Let us write these patterns using the more general notation of [3] and then rephrase the results using the Young tableau introduced towards the end of §3. For simplicity, let us describe the results in case $n = 5$. It is clear how the patterns extend to general n. There are three good reasons for using the Dynkin diagram notation. Firstly, it is more compact. Secondly, it naturally generalises to cover arbitrary homogeneous spaces G/P for G semisimple and P parabolic. Thirdly, it is well suited to the standard conventions of Lie theory in which the irreducible representations of semisimple and reductive Lie groups are specified by their highest weights, usually written as a linear combination of fundamental weights. There follows an introduction to this notation. Details may be found in [3]. A precise correspondence between the Dynkin diagram notation and the Young tableau notation will be given shortly: see (6.4) and (6.5). Those readers not familiar with Dynkin diagrams and who wish to know only the classification of G-invariant operators on $\mathbb{RP}_n$ in terms of Young tableau can safely skip to (6.4) and omit (6.5).

The Lie group $\mathrm{SL}(6, \mathbb{R})$ or, more precisely, its complexified Lie algebra $\mathfrak{sl}(6, \mathbb{C})$ will be denoted by the corresponding Dynkin diagram

•—•—•—•—•

where each node represents a copy of $\mathfrak{sl}(2,\mathbb{C})$ contained within $\mathfrak{sl}(6,\mathbb{C})$ as a subalgebra. The second node, for example, corresponds to the subalgebra generated by

$$h_2 = \begin{bmatrix} 0&0&0&0&0&0\\ 0&1&0&0&0&0\\ 0&0&-1&0&0&0\\ 0&0&0&0&0&0\\ 0&0&0&0&0&0\\ 0&0&0&0&0&0 \end{bmatrix} \qquad x_2 = \begin{bmatrix} 0&0&0&0&0&0\\ 0&0&1&0&0&0\\ 0&0&0&0&0&0\\ 0&0&0&0&0&0\\ 0&0&0&0&0&0\\ 0&0&0&0&0&0 \end{bmatrix} \qquad y_2 = \begin{bmatrix} 0&0&0&0&0&0\\ 0&0&0&0&0&0\\ 0&1&0&0&0&0\\ 0&0&0&0&0&0\\ 0&0&0&0&0&0\\ 0&0&0&0&0&0 \end{bmatrix}$$

satisfying the usual commutation relations

$$[h_2, x_2] = 2x_2 \qquad [h_2, y_2] = -2y_2 \qquad [x_2, y_2] = h_2.$$

The edges in the Dynkin diagram specify how these subalgebras interact in order to generate the full $\mathfrak{sl}(6,\mathbb{C})$. Details may be found in any standard text on representation theory (see for example [19] and especially Serre's Theorem). This is the simple Lie algebra A_5 and, more generally, $\mathfrak{sl}(n+1,\mathbb{C})$ is A_n. An irreducible representation of any simple Lie algebra has a unique highest weight vector up to scale and the representation is characterised up to isomorphism by the weight of this vector, which is necessarily dominant and integral. Here is not the place to explain what this means. Again, details may be found in any standard text such as [19]. The upshot is that an irreducible representation may be denoted by assigning non-negative integers to the nodes of the Dynkin diagram. In fact, following [3], it is more convenient to use lowest weights whence

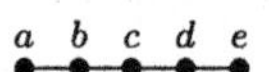

means that $-[a, b, c, d, e]$ is the lowest weight of this representation with respect to the standard basis of fundamental weights. The parabolic subalgebras of any complex simple Lie algebra may be thrown into standard form by conjugation and this standard form may also be recorded on the Dynkin diagram, replacing some of the nodes by crosses. The parabolic subalgebra so denoted is generated by all of the elements h_i and x_i but only those y_i corresponding to the uncrossed nodes. For example

$$\bullet\!\!-\!\!\times\!\!-\!\!\times\!\!-\!\!\bullet\!\!-\!\!\bullet = \left\{ \begin{bmatrix} *&*&*&*&*&*\\ *&*&*&*&*&*\\ 0&0&*&*&*&*\\ 0&0&0&*&*&*\\ 0&0&0&*&*&*\\ 0&0&0&*&*&* \end{bmatrix} \right\}$$

and

$$\times\!\!-\!\!\bullet\!\!-\!\!\bullet\!\!-\!\!\bullet\!\!-\!\!\bullet = \left\{ \begin{bmatrix} *&*&*&*&*&*\\ 0&*&*&*&*&*\\ 0&*&*&*&*&*\\ 0&*&*&*&*&*\\ 0&*&*&*&*&*\\ 0&*&*&*&*&* \end{bmatrix} \right\}.$$

In fact, we shall further abuse this notation and write a crossed Dynkin diagram to represent not only a parabolic subalgebra but also the corresponding parabolic subgroup of a complex simple Lie group G and also for the homogeneous space G/P. We shall even maintain the same notation for a real form of this homogeneous space (should P have a real form). In particular, we shall write

$$\mathbb{RP}_5 = \times\!-\!\bullet\!-\!\bullet\!-\!\bullet\!-\!\bullet.$$

Finally, this diagram is further abused by assigning numbers to its nodes in order to denote an irreducible representation of P and thus a homogeneous bundle on G/P or its real form. These numbers should be non-negative for the uncrossed nodes and (for the case of $\mathbb{RP}_n$) arbitrary real numbers for the crossed nodes. As before, this specifies an irreducible representation in terms of its lowest weight but now it is a representation of P carried by its reductive part. Some simple examples of this notation for bundles on $\mathbb{RP}_5$ are as follows

$$\overset{1}{\times}\!-\!\overset{0}{\bullet}\!-\!\overset{0}{\bullet}\!-\!\overset{0}{\bullet}\!-\!\overset{1}{\bullet} = \text{the tangent bundle} \qquad \overset{-2}{\times}\!-\!\overset{1}{\bullet}\!-\!\overset{0}{\bullet}\!-\!\overset{0}{\bullet}\!-\!\overset{0}{\bullet} = \text{the cotangent bundle}$$

$$\overset{w-4}{\times}\!-\!\overset{2}{\bullet}\!-\!\overset{0}{\bullet}\!-\!\overset{0}{\bullet}\!-\!\overset{0}{\bullet} = \mathcal{E}_{(ab)}(w) \qquad \overset{-3}{\times}\!-\!\overset{0}{\bullet}\!-\!\overset{1}{\bullet}\!-\!\overset{0}{\bullet}\!-\!\overset{0}{\bullet} = \Lambda^2 \qquad \overset{w-6}{\times}\!-\!\overset{0}{\bullet}\!-\!\overset{2}{\bullet}\!-\!\overset{0}{\bullet}\!-\!\overset{0}{\bullet} = \boxplus(w).$$

Details may be found in [3]. The convenience of this notation is perhaps not appreciated until some non-trivial interaction with Lie theory is encountered. For example, readers familiar with Lie theory theory will probably notice that the weights in the following theorem are related by the affine action of the Weyl group ($\lambda \mapsto w(\lambda + \rho) - \rho$ where ρ is half the sum of the positive roots) and this is inevitable by Harish-Chandra's Theorem on central character (again see [19]).

THEOREM 6.1. *The general G-invariant linear differential operator on $\mathbb{RP}_5$ occurs in precisely once in*

$$0 \to \overset{a}{\bullet}\!-\!\overset{b}{\bullet}\!-\!\overset{c}{\bullet}\!-\!\overset{d}{\bullet}\!-\!\overset{e}{\bullet} \to \overset{a}{\times}\!-\!\overset{b}{\bullet}\!-\!\overset{c}{\bullet}\!-\!\overset{d}{\bullet}\!-\!\overset{e}{\bullet} \xrightarrow{\nabla^{a+1}} \overset{-a-2}{\times}\!-\!\overset{a+b+1}{\bullet}\!-\!\overset{c}{\bullet}\!-\!\overset{d}{\bullet}\!-\!\overset{e}{\bullet} \xrightarrow{\nabla^{b+1}} \overset{-a-b-3}{\times}\!-\!\overset{a}{\bullet}\!-\!\overset{b+c+1}{\bullet}\!-\!\overset{d}{\bullet}\!-\!\overset{e}{\bullet}$$

$$\xrightarrow{\nabla^{c+1}} \overset{-a-b-c-4}{\times}\!-\!\overset{a}{\bullet}\!-\!\overset{b}{\bullet}\!-\!\overset{c+d+1}{\bullet}\!-\!\overset{e}{\bullet} \xrightarrow{\nabla^{d+1}} \overset{-a-b-c-d-5}{\times}\!-\!\overset{a}{\bullet}\!-\!\overset{b}{\bullet}\!-\!\overset{c}{\bullet}\!-\!\overset{d+e+1}{\bullet} \xrightarrow{\nabla^{e+1}} \overset{-a-b-c-d-e-6}{\times}\!-\!\overset{a}{\bullet}\!-\!\overset{b}{\bullet}\!-\!\overset{c}{\bullet}\!-\!\overset{d}{\bullet} \to 0$$

for some non-negative integers a, b, c, d, e. Here $\overset{a}{\bullet}\!-\!\overset{b}{\bullet}\!-\!\overset{c}{\bullet}\!-\!\overset{d}{\bullet}\!-\!\overset{e}{\bullet}$ is a finite-dimensional real representation of G viewed as a locally constant sheaf on $\mathbb{RP}_n$. These sequences are exact on the sheaf level and referred to as Bernstein-Gelfand-Gelfand resolutions (cf. [22]).

For example, the representation $\overset{0}{\bullet}\!-\!\overset{0}{\bullet}\!-\!\overset{0}{\bullet}\!-\!\overset{0}{\bullet}\!-\!\overset{0}{\bullet}$ is trivial and the resolution is the de Rham complex whilst the case

$$0 \to \underbrace{\overset{0}{\bullet}\!-\!\overset{1}{\bullet}\!-\!\overset{0}{\bullet}\!-\!\overset{0}{\bullet}\!-\!\overset{0}{\bullet}}_{\text{the representation } \Lambda^2\mathbb{R}^6 \text{ of } \mathrm{SL}(6,\mathbb{R})} \to \overset{0}{\times}\!-\!\overset{1}{\bullet}\!-\!\overset{0}{\bullet}\!-\!\overset{0}{\bullet}\!-\!\overset{0}{\bullet} \to \overset{-2}{\times}\!-\!\overset{2}{\bullet}\!-\!\overset{0}{\bullet}\!-\!\overset{0}{\bullet}\!-\!\overset{0}{\bullet} \to \overset{-4}{\times}\!-\!\overset{0}{\bullet}\!-\!\overset{2}{\bullet}\!-\!\overset{0}{\bullet}\!-\!\overset{0}{\bullet} \to \overset{-5}{\times}\!-\!\overset{0}{\bullet}\!-\!\overset{1}{\bullet}\!-\!\overset{1}{\bullet}\!-\!\overset{0}{\bullet}$$

$$\to \overset{-6}{\times}\!-\!\overset{0}{\bullet}\!-\!\overset{1}{\bullet}\!-\!\overset{0}{\bullet}\!-\!\overset{1}{\bullet} \to \overset{-7}{\times}\!-\!\overset{0}{\bullet}\!-\!\overset{1}{\bullet}\!-\!\overset{0}{\bullet}\!-\!\overset{0}{\bullet} \to 0$$

is the deformation complex (3.9) constructed 'by hand' in §3. There is a systematic way of manufacturing these resolutions from the de Rham complex using tractors. In the flat case, this is Zuckerman's translation functor but the method extends to the curved setting [18] and eventually to quite sophisticated machinery [7, 8, 12] the upshot of which is that all the G-invariant operators (as implicitly listed above) admit curved analogues. In the curved setting, however, when the operators are fitted into patterns, as above, the resulting sequences are no longer exact on the sheaf level. In fact, they are not even complexes. Typically, the composition of two successive operators is an operator whose symbol entails the Weyl curvature in a non-trivial fashion.

Even in the (projectively) flat case, the construction of these complexes from the basic de Rham complex is an interesting manœuvre. An introductory discussion appears in [15] and the Riemannian deformation complex in 3 dimensions, namely

$$0 \to \overset{0}{\bullet}{-}\overset{1}{\bullet}{-}\overset{0}{\bullet} \to \overset{0}{\times}{-}\overset{1}{\bullet}{-}\overset{0}{\bullet} \xrightarrow{\nabla} \overset{-2}{\times}{-}\overset{2}{\bullet}{-}\overset{0}{\bullet} \xrightarrow{\nabla^2} \overset{-4}{\times}{-}\overset{0}{\bullet}{-}\overset{2}{\bullet} \xrightarrow{\nabla} \overset{-5}{\times}{-}\overset{0}{\bullet}{-}\overset{1}{\bullet} \to 0, \qquad (6.1)$$

is constructed in [16] by manœuvring with tractors, roughly as follows. The Euler sequence (4.1) on $\mathbb{RP}_3$ gives rise to

$$0 \to \mathcal{E}_{[ab]}(2) \to \mathcal{E}_{[AB]} \to \mathcal{E}_a(2) \to 0. \qquad (6.2)$$

In the notation of [3], this is

$$0 \to \overset{-1}{\times}{-}\overset{0}{\bullet}{-}\overset{1}{\bullet} \to \overset{0}{\bullet}{-}\overset{1}{\bullet}{-}\overset{0}{\bullet} \to \overset{0}{\times}{-}\overset{1}{\bullet}{-}\overset{0}{\bullet} \to 0 \qquad (6.3)$$

and we now consider the de Rham sequence on $\mathbb{RP}_3$ with values in $\overset{0}{\bullet}{-}\overset{1}{\bullet}{-}\overset{0}{\bullet}$:

$$\begin{array}{ccccccccccc}
 & & & & 0 & & 0 & & 0 & & 0 \\
 & & & & \downarrow & & \downarrow & & \downarrow & & \downarrow \\
 & & & & \overset{-1}{\times}{-}\overset{0}{\bullet}{-}\overset{1}{\bullet} & & \overset{-3}{\times}{-}\overset{1}{\bullet}{-}\overset{1}{\bullet} \oplus \overset{-2}{\times}{-}\overset{0}{\bullet}{-}\overset{0}{\bullet} & & \overset{-4}{\times}{-}\overset{0}{\bullet}{-}\overset{2}{\bullet} \oplus \overset{-4}{\times}{-}\overset{1}{\bullet}{-}\overset{0}{\bullet} & & \overset{-5}{\times}{-}\overset{0}{\bullet}{-}\overset{1}{\bullet} \\
 & & & & \downarrow & & \downarrow & & \downarrow & & \downarrow \\
0 & \to & \overset{0}{\bullet}{-}\overset{1}{\bullet}{-}\overset{0}{\bullet} & \to & \Lambda^0 \otimes \overset{0}{\bullet}{-}\overset{1}{\bullet}{-}\overset{0}{\bullet} & \to & \Lambda^1 \otimes \overset{0}{\bullet}{-}\overset{1}{\bullet}{-}\overset{0}{\bullet} & \to & \Lambda^2 \otimes \overset{0}{\bullet}{-}\overset{1}{\bullet}{-}\overset{0}{\bullet} & \to & \Lambda^3 \otimes \overset{0}{\bullet}{-}\overset{1}{\bullet}{-}\overset{0}{\bullet} \to 0 \\
 & & & & \downarrow & & \downarrow & & \downarrow & & \downarrow \\
 & & & & \overset{0}{\times}{-}\overset{1}{\bullet}{-}\overset{0}{\bullet} & & \overset{-2}{\times}{-}\overset{2}{\bullet}{-}\overset{0}{\bullet} \oplus \overset{-1}{\times}{-}\overset{0}{\bullet}{-}\overset{1}{\bullet} & & \overset{-3}{\times}{-}\overset{1}{\bullet}{-}\overset{1}{\bullet} \oplus \overset{-2}{\times}{-}\overset{0}{\bullet}{-}\overset{0}{\bullet} & & \overset{-4}{\times}{-}\overset{1}{\bullet}{-}\overset{0}{\bullet} \\
 & & & & \downarrow & & \| & & \downarrow & & \downarrow \\
 & & & & 0 & & 0 & & 0 & & 0
\end{array}$$

where the columns are induced by (6.3) and consequently are exact. Chasing now effects the cancellation of $\overset{-1}{\times}{-}\overset{0}{\bullet}{-}\overset{1}{\bullet}$, $\overset{-3}{\times}{-}\overset{1}{\bullet}{-}\overset{1}{\bullet}$, $\overset{-2}{\times}{-}\overset{0}{\bullet}{-}\overset{0}{\bullet}$, and $\overset{-4}{\times}{-}\overset{1}{\bullet}{-}\overset{0}{\bullet}$ from this diagram, as detailed in [16], leaving the resolution (6.1). As a halfway manœuvre, if one chases just $\overset{-3}{\times}{-}\overset{1}{\bullet}{-}\overset{1}{\bullet}$ and $\overset{-2}{\times}{-}\overset{0}{\bullet}{-}\overset{0}{\bullet}$ from this diagram, then we obtain

$$0 \to \overset{0}{\bullet}{-}\overset{1}{\bullet}{-}\overset{0}{\bullet} \to \begin{array}{c}\overset{-1}{\times}{-}\overset{0}{\bullet}{-}\overset{1}{\bullet}\\ \oplus \\ \overset{0}{\times}{-}\overset{1}{\bullet}{-}\overset{0}{\bullet}\end{array} \to \begin{array}{c}\overset{-1}{\times}{-}\overset{0}{\bullet}{-}\overset{1}{\bullet}\\ \oplus \\ \overset{-2}{\times}{-}\overset{2}{\bullet}{-}\overset{0}{\bullet}\end{array} \to \begin{array}{c}\overset{-4}{\times}{-}\overset{0}{\bullet}{-}\overset{2}{\bullet}\\ \oplus \\ \overset{-4}{\times}{-}\overset{1}{\bullet}{-}\overset{0}{\bullet}\end{array} \to \begin{array}{c}\overset{-5}{\times}{-}\overset{0}{\bullet}{-}\overset{1}{\bullet}\\ \oplus \\ \overset{-4}{\times}{-}\overset{1}{\bullet}{-}\overset{0}{\bullet}\end{array} \to 0.$$

This is exactly the Arnold-Falk-Winther elasticity complex [1, page 127] formulated to give stable finite element schemes for the equations of linear elasticity through their finite element exterior calculus. More precisely, the complex (6.2) is written in [1] as

$$0 \to \mathbb{V} \to \mathbb{W} \to \mathbb{K} \to 0$$

and the elasticity complex as

$$0 \to \Lambda^2\mathbb{W} \to \Lambda^0(\mathbb{W}) \to \Lambda^1(\mathbb{K}) \to \Lambda^2(\mathbb{V}) \to \Lambda^3(\mathbb{W}) \to 0.$$

The G-invariant differential operators on $\mathbb{RP}_5$ are listed only implicitly in Theorem 6.1. It is an elementary matter to rewrite this list more explicitly and usefully as follows. We ask for the G-invariant differential operators emanating from $\overset{u}{\times}\!-\!\overset{a}{\bullet}\!-\!\overset{b}{\bullet}\!-\!\overset{c}{\bullet}\!-\!\overset{d}{\bullet}$ and find that there are five cases depending on where this particular homogeneous vector bundle can be fitted into a BGG (Bernstein-Gelfand-Gelfand) sequence. For fixed a, b, c, d, its position in a BGG sequence is determined by u:

CASE FIVE | CASE FOUR | CASE THREE | CASE TWO | CASE ONE

$-a-b-c-d-5$ | $-a-b-c-4$ | $-a-b-3$ | $-a-2$ | -1

For each of these cases there is, up to scale, precisely one projectively invariant linear differential operator as follows.

CASE ONE
let $k=u+1$

$$\overset{u}{\times}\!-\!\overset{a}{\bullet}\!-\!\overset{b}{\bullet}\!-\!\overset{c}{\bullet}\!-\!\overset{d}{\bullet} \xrightarrow{\nabla^k} \overset{-u-2}{\times}\!-\!\overset{a+u+1}{\bullet}\!-\!\overset{b}{\bullet}\!-\!\overset{c}{\bullet}\!-\!\overset{d}{\bullet}$$

CASE TWO
let $k=a+u+2$

$$\overset{u}{\times}\!-\!\overset{a}{\bullet}\!-\!\overset{b}{\bullet}\!-\!\overset{c}{\bullet}\!-\!\overset{d}{\bullet} \xrightarrow{\nabla^k} \overset{-a-2}{\times}\!-\!\overset{-u-2}{\bullet}\!-\!\overset{a+b+u+2}{\bullet}\!-\!\overset{c}{\bullet}\!-\!\overset{d}{\bullet}$$

CASE THREE
let $k=a+b+u+3$

$$\overset{u}{\times}\!-\!\overset{a}{\bullet}\!-\!\overset{b}{\bullet}\!-\!\overset{c}{\bullet}\!-\!\overset{d}{\bullet} \xrightarrow{\nabla^k} \overset{-a-b-3}{\times}\!-\!\overset{a}{\bullet}\!-\!\overset{-a-u-3}{\bullet}\!-\!\overset{a+b+c+u+3}{\bullet}\!-\!\overset{d}{\bullet}$$

CASE FOUR
let $k=a+b+c+u+4$

$$\overset{u}{\times}\!-\!\overset{a}{\bullet}\!-\!\overset{b}{\bullet}\!-\!\overset{c}{\bullet}\!-\!\overset{d}{\bullet} \xrightarrow{\nabla^k} \overset{-a-b-c-4}{\times}\!-\!\overset{a}{\bullet}\!-\!\overset{b}{\bullet}\!-\!\overset{-a-b-u-4}{\bullet}\!-\!\overset{a+b+c+d+u+4}{\bullet}$$

CASE FIVE
let $k=a+b+c+d+u+5$

$$\overset{u}{\times}\!-\!\overset{a}{\bullet}\!-\!\overset{b}{\bullet}\!-\!\overset{c}{\bullet}\!-\!\overset{d}{\bullet} \xrightarrow{\nabla^k} \overset{-a-b-c-d-5}{\times}\!-\!\overset{a}{\bullet}\!-\!\overset{b}{\bullet}\!-\!\overset{c}{\bullet}\!-\!\overset{-a-b-c-u-5}{\bullet}$$

It is useful to express this classification in the more congenial Young tableau notation introduced in §3. There is, however, an ambiguity due to the bundle of projective densities of weight -1 being a 6th root of the bundle of 5-forms. In other words, the bundle

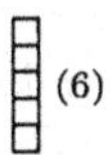

is canonically trivial. Apart from this slight inconvenience, all the irreducible bundles may be represented in this fashion. For the purpose of listing the projectively invariant differential operators, it is convenient to consider the bundle

$$B = \text{[Young tableau with rows of lengths } t, s, r, q, p\text{]} \quad (w+v-5) \tag{6.4}$$

where $v = p + q + r + s + t$ is the valency of the tensor. The ambiguity is

$$(p, q, r, s, t, w) \equiv (p + m, q + m, r + m, s + m, t + m, w + m) \quad \forall m \in \mathbb{Z}.$$

In the Dynkin diagram notation

$$B = \overset{w-t-5}{\times}\!\!-\!\!\overset{t-s}{\bullet}\!\!-\!\!\overset{s-r}{\bullet}\!\!-\!\!\overset{r-q}{\bullet}\!\!-\!\!\overset{q-p}{\bullet}\,. \tag{6.5}$$

Now, the five cases depend on w:

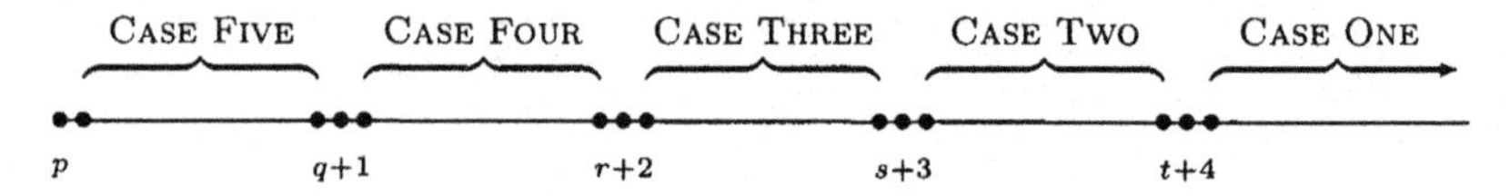

and Theorem 6.1 says that, for each of these cases there is, up to scale, precisely one projectively invariant linear differential operator as follows.

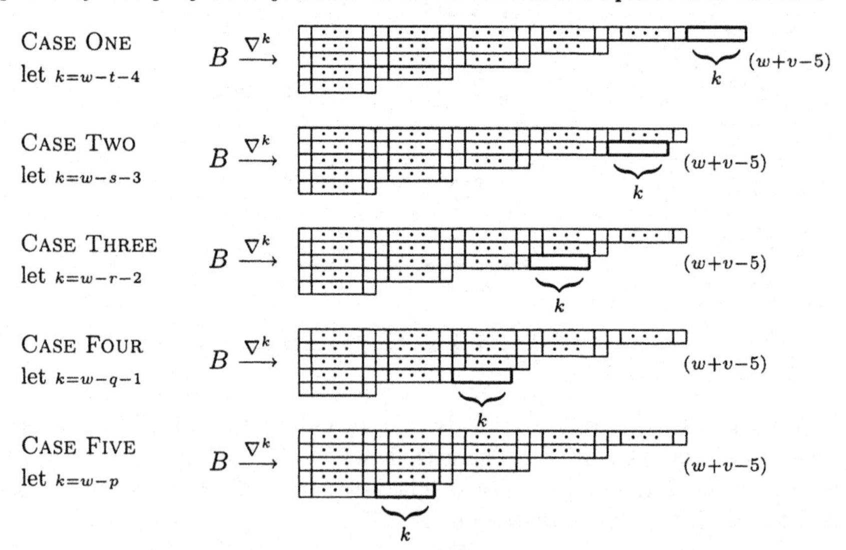

In each case, ∇^k indicates the differential operator of order k and obtained by applying the flat connection k times $B \to \bigodot^k \Lambda^1 \otimes B$ and then projecting onto the appropriate irreducible bundle, which occurs with multiplicity one in $\bigodot^k \Lambda^1 \otimes B$.

When $n = 2$, a classification of projectively invariant first and second order linear differential operators was obtained by Gunning [20]. Specifically,

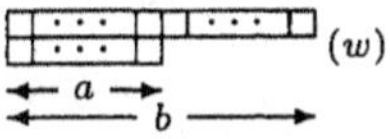

$$\;_{(w)} = T^{w-3a}_{b-a}$$

$\leftarrow a \rightarrow$
$\leftarrow b \rightarrow$

and the operators of [20, Theorem 1]

$$\nabla' : T^{2q}_{q} \to T^{2q}_{q+1} \quad \text{and} \quad \nabla'' : T^{q-1}_{q} \to T^{q-4}_{q-1}$$

become, in our notation,

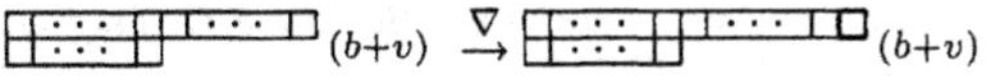

$$_{(b+v)} \xrightarrow{\nabla} \;_{(b+v)}$$

and $$_{(a+v-1)} \xrightarrow{\nabla} \;_{(a+v-1)}$$

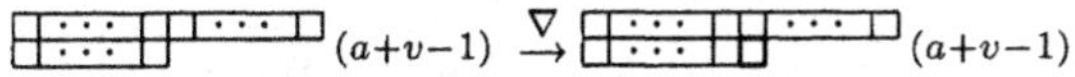

whilst those operators of [20, Theorem 2]

$$\nabla'_2 : T^{2q+1}_{q} \to T^{2q+1}_{q+2} \quad \text{and} \quad \nabla''_2 : T^{q}_{q} \to T^{q-6}_{q-2}$$

become

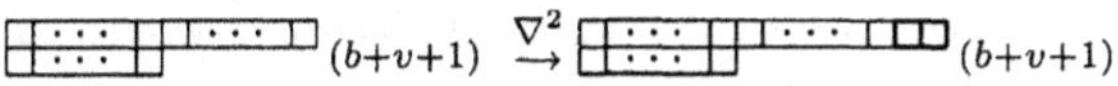

$$_{(b+v+1)} \xrightarrow{\nabla^2} \;_{(b+v+1)}$$

and $$_{(a+v)} \xrightarrow{\nabla^2} \;_{(a+v)}$$

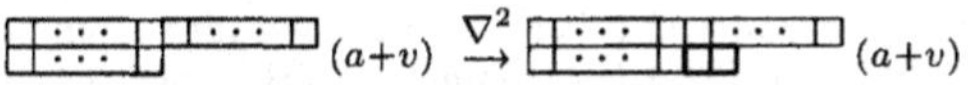

The projective BGG-complex on Euclidean space was also constructed by Olver [23] without employing projective invariance. His construction and proof of exactness uses Schur functors and is especially direct.

Finally, let us notice that the operators of CASE ONE are special in that their kernels, even locally, are finite-dimensional. This is part of Theorem 6.1 but is proved directly in [5] using Lie algebra cohomology (an instance of the 'BGG machinery') to construct explicit finite-type prolongations.

REFERENCES

[1] D.N. ARNOLD, R.S. FALK, AND R. WINTHER, *Finite element exterior calculus, homological techniques, and applications*, Acta Numerica **15** (2006), 1–155.

[2] T.N. BAILEY, M.G. EASTWOOD, AND A.R. GOVER, *Thomas's structure bundle for conformal, projective, and related structures*, Rocky Mtn. Jour. Math. **24** (1994), 1191–1217.

[3] R.J. BASTON AND M.G. EASTWOOD, *The Penrose Transform: its Interaction with Representation Theory*, Oxford University Press 1989.

[4] R. BOTT, *Homogeneous vector bundles*, Ann. Math. **60** (1957), 203–248.

[5] T.P. BRANSON, A. ČAP, M.G. EASTWOOD, AND A.R. GOVER, *Prolongations of geometric overdetermined systems*, Int. Jour. Math. **17** (2006), 641-664.

[6] E. CALABI, *On compact Riemannian manifolds with constant curvature I*, Differential Geometry, Proc. Symp. Pure Math. Vol. III, Amer. Math. Soc. 1961, pp. 155-180.
[7] D.M.J. CALDERBANK AND T. DIEMER, *Differential invariants and curved Bernstein-Gelfand-Gelfand sequences*, Jour. Reine Angew. Math. **537** (2001), 67–103.
[8] D.M.J. CALDERBANK, T. DIEMER, AND V. SOUČEK, *Ricci corrected derivatives and invariant differential operators*, Diff. Geom. Appl. **23** (2005), 149–175.
[9] A. ČAP, *Infinitesimal automorphisms and deformations of parabolic geometries*, preprint ESI 1684 (2005), Erwin Schrödinger Institute, available at http://www.esi.ac.at.
[10] A. ČAP AND H. SCHICHL, *Parabolic geometries and canonical Cartan connections*, Hokkaido Math. Jour. **29** (2000), 453–505.
[11] A. ČAP AND J. SLOVÁK, *Parabolic Geometries, Background and General Theory*, Math. Surveys Monogr., Amer. Math. Soc., to appear.
[12] A. ČAP, J. SLOVÁK, AND V. SOUČEK, *Bernstein-Gelfand-Gelfand sequences*, Ann. Math. **154** (2001), 97–113.
[13] E. CARTAN, *Sur les variétés à connexion projective*, Bull. Soc. Math. France **52** (1924), 205–241.
[14] M.G. EASTWOOD, *Notes on conformal differential geometry*, Proc. 15th Czech Winter School on Geometry and Physics, Srní, Suppl. Rendi. Circ. Mat. Palermo **43** (1996), 57–76.
[15] M.G. EASTWOOD, *Variations on the de Rham complex*, Notices Amer. Math. Soc. **46** (1999), 1368–1376.
[16] M.G. EASTWOOD, *A complex from linear elasticity*, Proc. 19th Czech Winter School on Geometry and Physics, Srní, Suppl. Rendi. Circ. Mat. Palermo **63** (2000), 23–29.
[17] M.G. EASTWOOD AND V.S. MATVEEV, *Metric connections in projective differential geometry*, this volume.
[18] M.G. EASTWOOD AND J.W. RICE, *Conformally invariant differential operators on Minkowski space and their curved analogues*, Commun. Math. Phys. **109** (1987), 207–228. Erratum, Commun. Math. Phys. **144** (1992), 213.
[19] J.E. HUMPHREYS, *Introduction to Lie Algebras and Representation Theory*, Grad. Texts in Math. Vol. 9, Springer Verlag 1972.
[20] R.C. GUNNING, *On projective covariant differentiation*, E.B. Christoffel, the Influence of his Work on Mathematics and the Physical Sciences (eds. P.L. Butzer and F. Fehér), Birkhäuser 1981, pp. 584–591.
[21] I. KOLÁŘ, P.W. MICHOR, AND J. SLOVÁK, *Natural Operators in Differential Geometry*, Springer Verlag 1993.
[22] J. LEPOWSKY, *A generalization of the Bernstein-Gelfand-Gelfand resolution*, Jour. Alg. **49** (1977), 496–511.
[23] P.J. OLVER, *Differential hyperforms I*, Mathematics Report 82-101, University of Minnesota 1982, http://www.math.umn.edu/~olver/a_/hyper.pdf.
[24] R. PENROSE AND W. RINDLER, *Spinors and Space-time, Vol. 1*, Cambridge University Press 1984.
[25] J.A. SCHOUTEN, *Ricci-Calculus*, Springer Verlag 1954.
[26] T.Y. THOMAS, *Announcement of a projective theory of affinely connected manifolds*, Proc. Nat. Acad. Sci. **11** (1925), 588–589.
[27] T.Y. THOMAS, *The Differential Invariants of Generalized Spaces*, reprint of 1934 original, AMS Chelsea Publishing 1991.

AMBIENT METRIC CONSTRUCTION OF CR INVARIANT DIFFERENTIAL OPERATORS

KENGO HIRACHI*

(Dedicated to the memory of Tom Branson)

1. Introduction. These notes are based on my lectures at IMA, in which I tried to explain basic ideas of the ambient metric construction by studying the Szegö kernel of the sphere. The ambient metric was introduced in Fefferman [F] in his program of describing the boundary asymptotic expansion of the Bergman kernel of strictly pseudoconvex domain. This can be seen as an analogy of the description of the heat kernel asymptotic in terms of local Riemannian invariants. The counterpart of the Riemannian invariants for the Bergman kernel is invariants of the CR structure of the boundary. Thus the program consists of two parts:

(1) Construct local invariants of CR structures;

(2) Prove that (1) gives all invariants by using the invariant theory.

In the case of the Szegö kernel, (1) is replaced by the construction of local invariants of the Levi form that are invariant under scaling by CR pluriharmonic functions. We formulate the class of invariants in Sections 2 and 3. To simplify the presentation, we confine ourself to the case of the sphere in $\mathbb{C}^n$. It is the model case of the ambient metric construction and the basic tools already appears in this setting. We construct invariants (formulated as CR invariant differential operators) by using the ambient space in Section 4 and then explain, in Section 5, how to prove that we have got all.

The CR invariant operators studied in these notes look similar to the invariants of CR densities studied in [EG1] and [GG]. However, there is a crucial difference. We here deal with operators acting on a quotient of the space of CR densities. They naturally arise in the description of the jets of geometric structures. For example, the jets of CR structures are described by Moser's normal form, which is defined as a slice of the quotient space of CR densities. In Section 6, we explain that the quotient space can be realized as cohomology of a subcomplex of Bernstein-Gelfand-Gelfand (BGG) complex. In particular, Moser's normal form corresponds to the cohomology of the deformation complex, a subcomplex of the BGG complex for the adjoint representation. The space we use for the Szegö kernel comes from the Rumin complex, the BGG complex for the trivial representation.

General theory of the ambient metric construction in CR case has been developed in [F, BEG] and [H2]; see [H3] for a survey and more

*Graduate School of Mathematical Sciences, The University of Tokyo, 3-8-1 Komaba, Meguro, Tokyo 153-8914, JAPAN.

comprehensive references. The proofs of the results in these notes (without citations) are given in my forthcoming paper, which will also deal with the general strictly pseudoconvex manifolds. See also [H1] for the case of 3-dimensions.

2. Transformation rule of the Szegö kernel. To motivate the problem, we first recall the definition of the Szegö kernel and its transformation rule under scaling of the volume form. (One can skip this section and jump to the formulation of the problem in the next section.)

Let $\Omega \subset \mathbb{C}^n$ be a bounded domain with smooth boundary $\partial\Omega$ and $\mathcal{O}(\Omega)$ be the space of holomorphic functions on Ω. We fix a volume form $d\sigma$ on $\partial\Omega$ and define a pre-Hilbert space $CH(\Omega) = C^\infty(\overline{\Omega}) \cap \mathcal{O}(\Omega)$ with inner product $(f_1, f_2) = \int_{\partial\Omega} f_1 \overline{f_2} d\sigma$. The completion of $CH(\Omega)$ is called the Hardy space and denoted by $H^2(\Omega, d\sigma)$. As a set, $H^2(\Omega, d\sigma)$ is independent of the choice of $d\sigma$ but the inner product does depend on the choice. Each element of $H^2(\Omega, d\sigma)$ is realized as a holomorphic function on Ω that has L^2 boundary value; thus we can identify $H^2(\Omega, d\sigma)$ with a subspace of $\mathcal{O}(\Omega)$.

Choose a complete orthonormal system $\{\varphi_j(z)\}_{j=1}^\infty$ of $H^2(\Omega, d\sigma)$ and make a series

$$K_{d\sigma}(z) = \sum_{j=1}^{\infty} |\varphi_j(z)|^2.$$

It converges for $z \in \Omega$ and define a smooth function on Ω, which is independent of the choice of $\{\varphi_j(z)\}_{j=1}^\infty$. The function $K_{d\sigma}(z)$ is called the Szegö kernel of $H^2(\Omega, d\sigma)$.

For the unit ball $\Omega_0 = \{z \in \mathbb{C}^n : |z|^2 = |z_1|^2 + \cdots + |z_n|^2 < 1\}$ with the standard ($U(n)$-invariant) volume form $d\sigma_0$ on the boundary S^m, $m = 2n - 1$, we can compute the Szegö kernel by choosing the monomials of $z_1, z_2, \ldots, z_n$ as a complete orthogonal system. Summing up $|z^\alpha|^2$ after normalization gives

$$K_{d\sigma_0}(z) = c_n(1 - |z|^2)^{-n},$$

where c_n is the inverse of the volume of the sphere S^m. This computation utilizes the symmetry of S^m and can be applied only to very limited cases. Even if one just perturbs the volume form on S^m, it becomes very hard to compute the Szegö kernel.

For small perturbation of the ball and the volume form, we can give the form of the boundary asymptotics of the Szegö kernel.

THEOREM 2.1 (Fefferman, Boutet de Monvel–Sjöstrand). *If Ω is strictly pseudoconvex domain with a smooth defining function ρ, $\Omega = \{\rho > 0\}$, and $d\sigma$ be a smooth volume form on $\partial\Omega$, then*

$$K_{d\sigma}(z) = \varphi(z)\rho(z)^{-n} + \psi(z)\log\rho(z), \quad \varphi, \psi \in C^\infty(\overline{\Omega}).$$

The proof is based on the theory of singular integral operators or Fourier integral operators, and it is not practical to use these calculus to compute the expansion explicitly. We here try to apply the invariant theory to write down the coefficients φ and ψ in terms of the "curvature" of $(\partial\Omega, d\sigma)$. For this purpose, we derive the transformation rule of the Szegö kernel under biholomorphic maps and the scaling of the volume form.

Let $F\colon \Omega \to \Omega'$ be a biholomorphic map between strictly pseudoconvex domains with smooth boundaries. Then F can be extended smoothly to the boundary by Fefferman's theorem. Suppose that we are given volume forms $d\sigma$ on $\partial\Omega$ and $d\sigma'$ on $\partial\Omega'$ such that $F^*(d\sigma') = e^f d\sigma$ for a CR-pluriharmonic function f (i.e., f can be extended to a pluriharmonic function $\widetilde{f}$ on Ω). If $K_{d\sigma}$ and $K_{d\sigma'}$ are respectively the Szegö kernels of $(\Omega, d\sigma)$ and $(\Omega', d\sigma')$, then

$$K_{d\sigma'}(F(z)) = e^{-\widetilde{f}(z)} K_{d\sigma}(z), \quad z \in \Omega.$$

In particular, the coefficients of the logarithmic terms ψ and ψ' of $K_{d\sigma}$ and $K_{d\sigma'}$ satisfy

$$\psi'(F(z)) = e^{-\widetilde{f}(z)}\psi(z) \mod O(\rho^\infty).$$

On the boundary, we may multiply each side by the corresponding volume form and eliminate the scaling factor:

$$F^*(\psi' d\sigma') = \psi d\sigma \quad \text{on } \partial\Omega. \tag{2.1}$$

Our aim here is to classify the differential forms that satisfy this transformation rule. (We also want to study higher order term in ψ and φ, but it requires more geometric tools; see [H2].)

3. Formulation of the problem. We now restrict the domain to the ball Ω_0 and express the transformation rule (2.1) in terms of the automorphism group of Ω_0.

The automorphisms of Ω_0 are given by the action of $G = SU(1,n)$ as rational maps. The action is defined by using the embedding of Ω_0 into the projective space

$$\Omega_0 \ni z \mapsto [1 : z] = [\zeta_0 : \zeta_1 : \cdots : \zeta_n] \in \mathbb{CP}^n.$$

The ball is then defined by

$$L(\zeta) = |\zeta_0|^2 - |\zeta_1|^2 - \cdots - |\zeta_n|^2 > 0$$

and the complex linear transformations that preserve the hermitian form L give automorphisms of Ω_0. Let $\widetilde{e}_0 = {}^t(1, 1, 0, \ldots, 0) \in \mathbb{C}^{n+1}$ and $P = \{h \in$

$G : h\,\widetilde{e}_0 = \lambda\widetilde{e}_0\}$. Then the action of P fixes $e_0 = (1, 0, \ldots, 0) \in S^m$ and we may write $S^m = G/P$. Note that P is a parabolic subgroup of G and the representation theory of P plays an essential role in the characterization of invariant differential operators. For $p, q \in \mathbb{Z}$, we denote by $\sigma_{p,q}$ the (complex) one-dimensional representation of P, where $h \in P$ is represented by $\lambda^{-p}\overline{\lambda}^{-q}$. In case $p = q$, we regard $\sigma_{p,p}$ as a real representation.

The action of G on $\mathbb{C}^{n+1} \setminus \{0\}$ can be also seen as isometries of the Lorentz-hermitian metric

$$\widetilde{g} = d\zeta_0 d\overline{\zeta}_0 - d\zeta_1 d\overline{\zeta}_1 - \cdots - d\zeta_n d\overline{\zeta}_n.$$

The Lorentz-hermitian manifold $(\mathbb{C}^{n+1} \setminus \{0\}, \widetilde{g})$ is called *the ambient space* of S^m. The ambient space admits $\mathbb{C}^*$ action $\zeta \mapsto \lambda\zeta$ and we may define homogeneous functions on it: we say that $\widetilde{f}(\zeta) \in C^\infty(\mathbb{C}^{n+1} \setminus \{0\}, \mathbb{C})$ is homogeneous of degree $(p, q) \in \mathbb{Z}^2$ if $\widetilde{f}(\lambda\zeta) = \lambda^p\overline{\lambda}^q$ for any $\lambda \in \mathbb{C}^*$. We can also define homogeneous functions on the real hypersurface $\mathcal{N} = \{L(\zeta) = 0\} \subset \mathbb{C}^{n+1} \setminus \{0\}$. Let

$$\mathcal{E}(p) = \{f(\zeta) \in C^\infty(\mathcal{N}, \mathbb{R}) : f(\lambda\zeta) = |\lambda|^{2p} f(\zeta),\ \lambda \in \mathbb{C}^*\},$$

which is a G-submodule of $C^\infty(\mathcal{N}, \mathbb{R})$. Since $\mathcal{N}$ is a $\mathbb{C}^*$-bundle over $S^m = G/P$, we can identify $\mathcal{E}(p)$ with the space of the sections of the real line bundle induced from the representation $\sigma_{p,p}$. In particular, $\mathcal{E}(0)$ can be identified with $C^\infty(S^m, \mathbb{R})$, and $\mathcal{E}(-n)$ is identified with the space of volume forms $C^\infty(S^m, \wedge^m S^m)$. The correspondence of the latter is given by

$$f(\zeta_0, \zeta_1, \ldots, \zeta_n) \mapsto f(1, z)\,d\sigma_0.$$

To simplify the notation we write $\mathcal{E} = \mathcal{E}(0)$. Since the space of CR pluriharmonic functions is preserved by the G-action, we may define the submodule

$$\mathcal{P} = \{f \in \mathcal{E} : f \text{ is CR pluriharmonic}\}.$$

When we study the jets of $\mathcal{E}(p)$ at e_0, it is useful to have coordinates centred at e_0. Let

$$\xi_0 = \zeta_0 + \zeta_n,\ \xi_1 = \sqrt{2}\zeta_1, \ldots, \xi_{n-1} = \sqrt{2}\zeta_{n-1}, \xi_n = \zeta_0 - \zeta_n.$$

Then in the coordinates $(\xi_0, \ldots, \xi_n)$, we have $\widetilde{e}_0 = {}^t(1, 0, \ldots, 0)$ and

$$2L(\xi) = \xi_0\overline{\xi}_n + \xi_n\overline{\xi}_0 - \sum_{j=1}^{n-1} |\xi_j|^2.$$

Hence setting $w_j = \xi_j/\xi_0,\ j = 1, \ldots, n$, we may realize Ω_0 as the Siegel domain $2\operatorname{Re} w_n > \sum_{j=1}^{n-1} |w_j|^2$ with $e_0 = 0$.

Now we return to the Szegö kernel. On the sphere, each volume form is written as $d\sigma = e^f d\sigma_0$ for $f \in \mathcal{E}$. Hence the Szegö kernel can be seen as a functional of f. Taking the defining function $1 - |z|^2$, we write

$$K_{d\sigma} = \varphi_f(1 - z^2)^{-n} + \psi_f \log(1 - |z|^2).$$

Here we put the subscript f on φ and ψ to emphasis that these coefficients are determined by f. In particular, we can define a map

$$\Psi\colon \mathcal{E} \to \mathcal{E}(-n), \quad \Psi(f) = \psi_f d\sigma.$$

The proof of Theorem 2.1 also implies that Ψ is a differential operator, which polynomially depends on the jets of f. The transformation rule (2.1) can be now reformulated as follows:

(i) Ψ is G-equivariant;

(ii) $\Psi(f + h) = \Psi(f)$ for any $h \in \mathcal{P}$.

The property (i) follows from the fact that any automorphism F satisfies $F^*(d\sigma_0) = e^f d\sigma_0$ for an $f \in \mathcal{P}$; (ii) is the special case of (2.1) in which F is the identity map. Since the sphere is the model of CR manifolds, we will use the terminology "CR invariant" instead of G-equivariant. Then our problem can be stated as follows.

PROBLEM 1. Write down all CR invariant differential operators from $\mathcal{E}$ to $\mathcal{E}(-n)$ that are invariant under the additions of $\mathcal{P}$.

To be more precise, we should assume that the operator polynomially depends on the jets of f, that is, the value of the operator at e_0 is given by a polynomial in the Taylor coefficients of f at $w = 0$. We will always assume this property in the rest of these notes.

4. Construction of CR invariant differential operators. We first study linear CR invariant differential operators. Such operators define homomorphisms of G-modules and we may apply the results of the representation theory. Taking the jets at e_0, we may reduce the study of invariant differential operators to the one for the homomorphisms between generalized Verma modules. These homomorphisms can be classified by using the affine action of the Weyl group on the highest weight. In particular, we obtain the following

THEOREM 4.1. *There is a unique, up to a constant multiple, linear CR invariant differential operator* $\mathcal{E} \to \mathcal{E}(-n)$.

This theorem ensures that we have a unique linear CR invariant differential operator, but does not provide its explicit formula. We construct the linear operator by using the Laplacian for the ambient metric

$$\Delta = \partial_{\zeta_0}\partial_{\overline{\zeta}_0} - \partial_{\zeta_1}\partial_{\overline{\zeta}_1} - \cdots - \partial_{\zeta_n}\partial_{\overline{\zeta}_n}.$$

This is an analogy of the construction of conformally invariant differential operators in [GJMS].

THEOREM 4.2. *For $f \in \mathcal{E}$, take an $\widetilde{f} \in C^\infty(\mathbb{C}^{n+1} \setminus \{0\})$ homogeneous of degree $(0,0)$ such that $\widetilde{f}|_\mathcal{N} = f$. Then*

$$(\Delta^n \widetilde{f})|_\mathcal{N}$$

depends only on f and defines a differential operator $\mathcal{E} \to \mathcal{E}(-n)$.

Proof. Since Δ^n is linear, it suffices to show that $\widetilde{f}|_\mathcal{N} = 0$ implies $(\Delta^n \widetilde{f})|_\mathcal{N} = 0$. Write $\widetilde{f} = Lh$ with a function h of homogeneous degree $(-1,-1)$. Then we have, for a positive integer k, that

$$\begin{aligned} \Delta^k(Lh) &= [\Delta^k, L]h + L\Delta^k h \\ &= k(Z + \overline{Z} + n + k)\Delta^{k-1}h + O(L) \\ &= k(n-k)\Delta^{k-1}h + O(L). \end{aligned}$$

Here $Z = \sum_{j=0}^n \zeta_j \partial_{\zeta_j}$ and we have used the fact that $\Delta^{k-1}h$ is homogeneous of degree $(-k,-k)$. In particular, substituting $k = n$, we get $\Delta^n(Lh) = O(L)$. □

We denote the linear CR invariant differential operator defined above by $Q(f)$, as it agrees with the CR Q-curvature of the volume form $d\sigma = e^f d\sigma_0$ or the contact form θ satisfying $d\sigma = \theta \wedge (d\theta)^{n-1}$. See [FH] for more discussion about the ambient metric construction of the Q-curvature. Since $h \in \mathcal{P}$ can be extended to a pluriharmonic function on the ambient space, we have $Q(h) = 0$ for $h \in \mathcal{P}$; thus Q satisfies the condition of Problem 1.

We can also use the ambient Laplacian Δ to construct non-linear CR invariant differential operators. In this case, we need to specify the ambient extension $\widetilde{f}$ more precisely.

LEMMA 4.1. *Each $f \in \mathcal{E}$ can be extended to a smooth homogeneous function $\widetilde{f}$ on $\mathbb{C}^{n+1} \setminus \{0\}$ satisfying*

$$\Delta \widetilde{f} = O(L^{n-1}). \tag{4.1}$$

Such an $\widetilde{f}$ is unique modulo $O(L^n)$.

Proof. We construct $\widetilde{f}$ by induction. Suppose that we have a function $\widetilde{f}_k$ homogeneous of degree $(0,0)$ such that $\widetilde{f}_k = f + O(L)$ and $\Delta \widetilde{f}_k = L^{k-1}h$ for an h homogeneous of degree $(-k,-k)$. (The second condition is vacuous if $k = 1$ and we have $\widetilde{f}_1$.) We set $\widetilde{f}_{k+1} = \widetilde{f}_k + L^k g$ for g homogeneous of degree $(-k,-k)$ and try to achieve $\Delta \widetilde{f}_{k+1} = O(L^k)$. Since

$$[\Delta, L^k] = k\, L^{k-1}(Z + \overline{Z} + n + k),$$

we have

$$\begin{aligned} \Delta \widetilde{f}_{k+1} &= \Delta \widetilde{f}_k + [\Delta, L^k]g + O(L^k) \\ &= L^{k-1}h + k(n-k)L^{k-1}g + O(L^k). \end{aligned}$$

Thus, as long as $k \neq n$, we may set $g = -h/(k(n-k))$ and get $\widetilde{f}_{k+1}$. The inductive step starting with $\widetilde{f}_1$ gives $\widetilde{f}_n$. The uniqueness of $\widetilde{f}$ is also clear from the construction. □

From this proof, one can also observe that $Q(f)$ is the only obstruction to the existence of an exact formal solution to $\Delta\widetilde{f} = 0$. In fact, applying Δ^{n-1} to $\Delta\widetilde{f} = L^{n-1}h$, one gets $\Delta^n\widetilde{f} = \text{const.}\, h + O(L)$. Thus $Q(f) = 0$ implies $h = O(L)$ and so $\Delta\widetilde{f} = O(L^n)$. We can then apply the inductive step to get $\widetilde{f}_k$ for any $k > n$.

The metric $\widetilde{g}$ defines a flat hermitian connection $\widetilde{\nabla} = \widetilde{\nabla}^{1,0} + \widetilde{\nabla}^{0,1}$. For iterated derivatives $\widetilde{\nabla}^{p+q}$, we denote by $\widetilde{\nabla}^{p,q}$ the part of type (p,q). Then we can define scalar valued nonlinear differential operators for f by making complete contractions:

$$W(\widetilde{f}) = \text{contr}(\widetilde{\nabla}^{p_1,q_1}\widetilde{f} \otimes \cdots \otimes \widetilde{\nabla}^{p_r,q_r}\widetilde{f})|_{\mathcal{N}}. \tag{4.2}$$

Here $r \geq 2$, $p_j, q_j \geq 1$, $p_1 + \cdots + p_r = q_1 + \cdots + q_r = n$ and the contraction is taken with respect to $\widetilde{g}$ for some paring of holomorphic and anti-holomorphic indices. In view of the ambiguity of $\widetilde{f}$, we can show that $W(\widetilde{f})$ depends only on f and define a CR invariant differential operator $W\colon \mathcal{E} \to \mathcal{E}(-n)$. Since $\widetilde{\nabla}^{1,1}$ kills pluriharmonic functions, we also see that $W(f+h) = W(f)$ for any $h \in \mathcal{P}$. We now give a solution to Problem 1.

THEOREM 4.3. *Let $S : \mathcal{E} \to \mathcal{E}(-n)$ be a CR invariant differential operator that satisfies $S(f+h) = S(f)$ for $h \in \mathcal{P}$. Then S is a linear combination of $Q(f)$ and complete contractions of the form* (4.2).

In case $n = 2$, the theorem implies

$$\Psi(f) = c_2\, Q(f) + c_2'\, \|\widetilde{\nabla}^{1,1}\widetilde{f}\|^2,$$

where $\|\widetilde{\nabla}^{p,q}\widetilde{f}\|^2 = \text{contr}(\widetilde{\nabla}^{p,q}\widetilde{f} \otimes \widetilde{\nabla}^{q,p}\widetilde{f})$. However, we will see in the next section that $\|\widetilde{\nabla}^{1,1}\widetilde{f}\|^2 = 0$ by using the fact that the rank of $T^{1,0}S^m$ is one. Hence $\Psi(f) = c_2 Q(f)$. The constant $c_2 = 1/(24\pi^2)$ is identified in [H1]; see also [FH].

In case $n = 3$, we have

$$\Psi(f) = c_3\, Q(f) + c_3'\, \|\widetilde{\nabla}^{2,1}\widetilde{f}\|^2 + c_3''\, \text{contr}(\widetilde{\nabla}^{1,1}\widetilde{f} \otimes \widetilde{\nabla}^{1,1}\widetilde{f} \otimes \widetilde{\nabla}^{1,1}\widetilde{f}),$$

where each term is nontrivial. We know $c_3 \neq 0$ but haven't computed other constants.

5. Jet isomorphism theorem and invariant theory. We explain how we reduce Theorem 4.3 to a purely algebraic theorem of representation theory. Let $J\mathcal{E}$ denote the space of ∞-jet of smooth functions at $e_0 \in S^m$. If one fixes coordinates of S^m around e_0, $J\mathcal{E}$ is identified with the space of formal power series centered at e_0. We denote by $J\mathcal{P}$ the subspace of $J\mathcal{E}$ consisting of jets of CR-pluriharmonic functions in a neighborhood of e_0.

It is clear that a CR invariant (or G-equivariant) differential operator defines a P-equivariant map

$$I\colon J\mathcal{E}/J\mathcal{P} \to \sigma_{-n,-n}.$$

Conversely, by Frobenius reciprocity, we can extends I uniquely to a $(\mathfrak{g}, P)$-equivariant map $\widetilde{I}\colon J\mathcal{E}/J\mathcal{P} \to J\mathcal{E}(-n)$ and it extends to a G-equivariant differential operator on $\mathcal{E}/\mathcal{P} \to \mathcal{E}(-n)$. Thus Problem 1 is reduced to

PROBLEM 2. *Write down all P-equivariant map $J\mathcal{E}/J\mathcal{P} \to \sigma_{-n,-n}$.*

Let us call "P-equivariant map" simply "CR invariant." When we study the CR invariants of the quotient module $J\mathcal{E}/J\mathcal{P}$, it is useful to have a slice of the coset. We here follow Moser's argument [CM] of making normal form for real hypersurfaces in $\mathbb{C}^n$.

In the coordinates $(w', w_n) \in \mathbb{C}^{n-1} \times \mathbb{C}$, the sphere S^m is given by $2\operatorname{Re} w_n = |w'|^2$ and (w', v), $v = \operatorname{Im} w_n$, give local coordinates of S^m with $e_0 = 0$. Thus we may identify $J\mathcal{E}$ with the space of real formal power series of $(w', \overline{w}', v)$. Each element of $J\mathcal{P}$ is given by the real part of a formal power series of ξ at $\widetilde{e}_0 = (1, 0, \ldots, 0)$ homogeneous of degree 0. Hence substituting $\xi_0 = 1, \xi' = w', \xi_n = |w'|^2/2 + iv$, we see that $J\mathcal{P}$ consists of formal power series of the form

$$\operatorname{Re} \sum_{|\alpha|\geq 0, l\geq 0} A^l_\alpha w'^\alpha (|w'|^2 + 2iv)^l, \quad \text{where } A^l_\alpha \in \mathbb{C}.$$

To define a complementary space of $J\mathcal{P}$, we write $f \in J\mathcal{E}$ as

$$f(w', \overline{w}', v) = \sum_{p,q\geq 0} A_{p,q}(v), \tag{5.1}$$

where

$$A_{p,q}(v) = \sum_{\substack{|\alpha| = p, |\beta| = q \\ l \geq 0}} A^l_{\alpha\overline{\beta}} w'^\alpha \overline{w}'^\beta v^l.$$

Let $\mathcal{N}_0$ be the subspace (not a submodule) of $\mathcal{E}$ defined by the equations

$$A_{p,q} = 0 \text{ if } \min(p, q) = 0, \quad \Delta' A_{1,1} = 0,$$

where $\Delta' = \sum_{j=1}^{n-1} \partial_{w_j} \partial_{\overline{w}_j}$. Then we have

LEMMA 5.1. *As a vector space $J\mathcal{E} = \mathcal{N}_0 \oplus J\mathcal{P}$.*

We have $J\mathcal{E}/J\mathcal{P} \cong \mathcal{N}_0$ as vector spaces and may define P-action on $\mathcal{N}_0$ via this isomorphism. We identify $\mathcal{N}_0$ with the space of lists $(A^l_{\alpha\overline{\beta}})$; so P acts on the list $(A^l_{\alpha\overline{\beta}})$. Then CR invariants of weight $(-n, -n)$ are P-equivariant polynomial of $(A^l_{\alpha\overline{\beta}})$ that takes value in $\sigma_{-n,-n}$. We define the weight of $A^l_{\alpha\overline{\beta}}$ to be $|\alpha|+|\beta|+2l$ and extend the weight to the monomials of $(A^l_{\alpha\overline{\beta}})$ by summing up the weight of each variable. It then turns out that a CR invariant of weight $(-n,-n)$ is a polynomial of $(A^l_{\alpha\overline{\beta}})$ homogeneous of weight $2n$ (examine the action of dilation $(w', w_n) \mapsto (\lambda w', |\lambda|^2 w_n)$). Since each nontrivial $A^l_{\alpha\overline{\beta}}$ has weight ≥ 2, we see that a nonlinear monomial of homogeneous weight $2n$ depend only on the variables $(A^l_{\alpha\overline{\beta}})$ of weight $\leq 2n-2$. Thus we may restrict our attention to the polynomials on

$$\mathcal{N}_0^{[2n]} = \{(A^l_{\alpha\overline{\beta}})_{|\alpha|+|\beta|+2l<2n} : (A^l_{\alpha\overline{\beta}}) \in \mathcal{N}_0\}.$$

Note that the kernel of the projection $\mathcal{N}_0 \to \mathcal{N}_0^{[2n]}$ is a P-submodule and so $\mathcal{N}_0^{[2n]}$ has a structure of quotient P-module. As we already know the linear CR invariants, we may reduce Problem 2 to

PROBLEM 3. *Write down all CR invariant* $\mathcal{N}_0^{[2n]} \to \sigma_{-n,-n}$.

In case $n=2$, we can determine CR invariants of weight $(-n,-n)$ just by counting the weight of the variables $(A^l_{\alpha\overline{\beta}})$. Observe that the condition $\Delta' A_{1,1} = 0$ implies $A_{1,1} = 0$. Thus $f \in \mathcal{N}_0$ is of the form

$$f = A_{2,1} + A_{1,2} + \sum_{p+q\geq 4, p,q\geq 1} A_{p,q}.$$

It follows that each nontrivial $A^l_{\alpha\overline{\beta}}$ has weight ≥ 3. Thus nonlinear monomial should have weight ≥ 6 and it cannot appear in an invariant of weight $(-2,-2)$. Since $A_{2,2}(0)$ is the only term of weight 4, we see that a CR invariant of weight $(-2,-2)$ must be a multiple of $A_{2,2}(0)$. (We know the existence of a CR invariant of weight $(-2,-2)$; so $A_{2,2}(0)$ must be a CR invariant.)

We next examine the dependence of $\widetilde{\nabla}^{1,1}\widetilde{f}$ on $\mathcal{N}_0$. We use the coordinates $\xi_0, \ldots, \xi_n$ to define the components of

$$\widetilde{\nabla}^{p,q}\widetilde{f}(\widetilde{e}_0) = (T_{I_1\cdots I_p\overline{J}_1\cdots\overline{J}_q}(\widetilde{f})).$$

The action of P on $\widetilde{f}$ induces an action on the tensors given by

$$\mathbb{T}_0^{p,q} = \bigodot^p V^* \otimes \bigodot^q \overline{V}^* \otimes \sigma_{-p,-q},$$

where $V = \mathbb{C}^{n+1}$ on which P acts by right multiplication on column vectors. We next introduce a weight on each component of $\mathbb{T}_0^{p,q}$. Set $\|0\| = 0$, $\|j\| = 1$, $j = 1, 2, \ldots, n-1$, $\|n\| = 2$ and extend it to the list of indices by

$$\|I_1 \cdots I_p \overline{J}_1 \cdots \overline{J}_q\| = \|I_1\| + \cdots + \|I_p\| + \|J_1\| + \cdots + \|J_q\|.$$

Since $\widetilde{f}$ is unique modulo $O(L^n)$, we have the following estimate of the ambiguity.

LEMMA 5.2. *If* $\|\mathcal{I}\overline{\mathcal{J}}\| < 2n$, *then* $T_{\mathcal{I}\overline{\mathcal{J}}}(\tilde{f})$ *depends only on* $f \in J\mathcal{E}$ *modulo* $J\mathcal{P}$, *where* $\mathcal{I} = I_1 \cdots I_p$ *and* $\overline{\mathcal{J}} = \overline{J}_1 \cdots \overline{J}_q$ *with* $p, q \geq 1$.

Let $\mathbb{T}_0 = \prod_{p,q\geq 1} \mathbb{T}_0^{p,q}$ and set

$$\mathbb{T}_0^{[2n]} = \{(T_{\mathcal{I}\overline{\mathcal{J}}})_{\|\mathcal{I}\overline{\mathcal{J}}\|<2n} : (T_{\mathcal{I}\overline{\mathcal{J}}}) \in \mathbb{T}_0\}.$$

Since the kernel of the projection $\mathbb{T}_0 \to \mathbb{T}_0^{[2n]}$ is shown to be a submodule, we may define P action on $\mathbb{T}_0^{[2n]}$ as the quotient. The lemma above ensures that the map

$$T \colon J\mathcal{E}/J\mathcal{P} \to \mathbb{T}_0^{[2n]}, \quad T(f) = (T_{\mathcal{I}\overline{\mathcal{J}}}(f))_{\|\mathcal{I}\overline{\mathcal{J}}\|<2n},$$

is a well-defined P-equivariant map. It is easy to verify that the image of T satisfies the equations

$$\overline{T_{\mathcal{I}\overline{\mathcal{J}}}} = T_{\mathcal{J}\overline{\mathcal{I}}}, \quad \tilde{g}^{K\overline{L}} T_{\mathcal{I}K\overline{L}\overline{\mathcal{J}}} = 0, \quad T_{\mathcal{I}0\overline{\mathcal{J}}} = -|\mathcal{I}| T_{\mathcal{I}\overline{\mathcal{J}}}, \tag{5.2}$$

where $|I_1 \cdots I_p| = p$ and we also consider the case $p = 0$. These respectively follow from the facts that $\tilde{f}$ is real, Δ-harmonic and homogenous of degree $(0,0)$. Let us denote by $\mathcal{H}_0$ the submodule of $\mathbb{T}_0$ defined by these equations and define $\mathcal{H}_0^{[2n]}$ to be its projection to $\mathbb{T}_0^{[2n]}$. It is clear that $T(f)$ depends only on finite jets of $J\mathcal{E}/J\mathcal{P} \cong \mathcal{N}_0$. More precisely, we can show that $T_{\mathcal{I}\overline{\mathcal{J}}}(f)$ of weight w is a homogeneous polynomial of $(A^l{}_{\alpha\overline{\beta}})$ of weight w. It follows that $T \colon \mathcal{N}_0^{[2n]} \to \mathcal{H}_0^{[2n]}$ is well-defined.

THEOREM 5.1 (Jet Isomorphism Theorem). *The map* $T \colon \mathcal{N}_0^{[2n]} \to \mathcal{H}_0^{[2n]}$ *is an isomorphism of* P*-modules.*

The isomorphism T reduces the problem of determining invariants on $\mathcal{N}_0^{[2n]}$ to the one on $\mathcal{H}_0^{[2n]}$. It is clear that an invariant polynomial on $\mathcal{H}_0^{[2n]}$ can be also seen as an invariant polynomial on $\mathcal{H}_0$. The converse is also true because an invariant taking values in $\sigma_{-n,-n}$ depends only on the components in $\mathcal{H}_0^{[2n]}$. Hence Problem 3 is reduced to

PROBLEM 4. *Write down all CR invariant* $\mathcal{H}_0 \to \sigma_{-n,-n}$.

Fortunately, this problem has been completely solved in more general setting.

THEOREM 5.2 ([BEG]). *Let* $I \colon \mathcal{H}_0 \to \sigma_{q,q}$ *be a CR invariant for some integer* q. *Then* I *is a linear combination of complete contractions of the form*

$$\mathrm{contr}(T^{p_1,q_1} \otimes \cdots \otimes T^{p_k,q_k}), \quad T^{p,q} \in \mathbb{T}_0^{p,q}.$$

Note that $T^{p,q}$ is trace-free and there is no linear CR invariant defined on $\mathcal{H}_0$. On the other hand, the linear CR invariant operator $Q(f)$

is the complete contraction of $T^{n,n}(\widetilde{f})$, which is not covered by the jet isomorphism theorem. Thus there is no hope to generalize Theorem 5.1 to $T\colon \mathcal{N}_0^{[w]} \to \mathcal{H}_0^{[w]}$, $w > 2n$. To get some relation between $\mathcal{N}_0$ and $\mathcal{H}_0$ for higher jet, one needs to introduce additional parameter to resolve the ambiguity of the harmonic extension. This approach is discussed in [H2].

6. Bernstein-Gelfand-Gelfand resolution. We explain why it is natural to realize the quotient space $J\mathcal{E}/J\mathcal{P}$ as tensor space in $\mathbb{T}_0$ from the point of view of the BGG resolution. We here only consider some explicit examples of (the dual of) BGG resolutions; more complete treatment of BGG should be contained in other articles in this volume. We first assume that $n > 2$. The case $n = 2$ will be discussed at the end of this section.

Recall that the complexified tangent space of S^m admits a natural subbundle $T_b^{1,0} = \mathbb{C}TS^m \cap T^{1,0}\mathbb{C}^n$ and its conjugate $T_b^{0,1}$. Hence the restrictions of the exterior derivative of a function give $\partial_b f = df|_{T^{1,0}}$ and $\overline{\partial}_b f = df|_{T^{0,1}}$. Let $\mathcal{E}^{1,0}$ and $\mathcal{E}^{0,1}$ denote the space of the sections of $(T_b^{1,0})^*$ and $(T_b^{0,1})^*$ respectively. For $p + q \leq n - 1$, we also define $\mathcal{E}^{p,q}$ to be the space of the sections of $\bigwedge^p(T_b^{1,0})^* \otimes_\circ \bigwedge^q(T_b^{0,1})^*$, where $\otimes_\circ$ means the trace-free part of the bundle with respect to the Levi from. Note that these bundles are induced from irreducible representations of P. We fix a subbundle N such that

$$\mathbb{C}TS^m = N \oplus T^{1,0} \oplus T^{0,1}$$

and identify $f \in \mathcal{E}^{p,q}$ with a $(p+q)$-form. If $p + q < n$, we can define

$$\mathcal{E}^{p,q} \begin{array}{l} \xrightarrow{\ \partial_b\ } \mathcal{E}^{p+1,q} \\ \xrightarrow[\ \overline{\partial}_b\ ]{} \mathcal{E}^{p,q+1} \end{array}$$

by applying exterior derivative to the $(p+q)$-form and projecting to appropriate bundles; these maps are independent of the choice of N.

The decompositions of the exterior derivatives on functions and one-forms give a complex of G-modules ($\mathbb{C}$ is regarded as the trivial representation)

$$0 \to \mathbb{C} \to \mathcal{E}^{\mathbb{C}} \begin{array}{l} \nearrow \mathcal{E}^{0,1} \\ \searrow \mathcal{E}^{1,0} \end{array} \begin{array}{l} \nearrow \mathcal{E}^{0,2} \\ \searrow\!\!\!\!\nearrow \mathcal{E}^{1,1} \\ \searrow \mathcal{E}^{2,0} \end{array} \tag{6.1}$$

Here $\mathcal{E}^{\mathbb{C}}$ is the space of complex valued smooth functions. By composition, we can also define a G-equivariant map

$$R\colon \mathcal{E}^{\mathbb{C}} \to \mathcal{E}^{1,1}, \quad R(f) = \text{trace-free part of } i\,\overline{\partial}_b\partial_b f.$$

Restricting R to $\mathcal{E}$ and then taking the jets at e_0, we obtain a homomorphism of P-modules

$$R : J\mathcal{E} \to J\mathcal{E}^{1,1}.$$

The kernel is given by $J\mathcal{P}$ (see e.g. [L]) and we get an injection

$$J\mathcal{E}/J\mathcal{P} \to J\mathcal{E}^{1,1}.$$

The map $T\colon J\mathcal{E}/J\mathcal{P} \to \mathbb{T}_0^{[n]}$ is given by taking the jet at $\widetilde{e}_0$ of $\widetilde{\nabla}^{1,1}\widetilde{f}$ and $\widetilde{\nabla}^{1,1}$ can be seen as the lift of R to the ambient space. In terms of the slice $\mathcal{N}_0$, the value of $R(f)$ at e_0 is given by $A_{1,1}(0)$, the leading term of the normal form. Since $\Delta' A_{1,1} = 0$, it defines a trace-free tensor.

We next realize $J\mathcal{E}/J\mathcal{P}$ as a cohomology of a subcomplex of (6.1). Since $J\mathcal{E}^{\mathbb{C}} \to J\mathcal{E}^{0,1} \to J\mathcal{E}^{0,2}$ is exact (see e.g. [B]), it gives no local invariants. However, by replacing $\mathcal{E}^{\mathbb{C}}$ by $\mathcal{E}$, we obtain a complex

$$0 \to \mathbb{R} \to J\mathcal{E} \to J\mathcal{E}^{0,1} \to J\mathcal{E}^{0,2}$$

which has nontrivial first cohomology

$$H^1 = \frac{\ker(\overline{\partial}_b : J\mathcal{E}^{0,1} \to J\mathcal{E}^{0,2})}{\operatorname{image}(\overline{\partial}_b : J\mathcal{E} \to J\mathcal{E}^{0,1})}.$$

We will show that this is isomorphic to $J\mathcal{E}/J\mathcal{P}$.

PROPOSITION 6.1. *The map $J\mathcal{E} \to J\mathcal{E}^{0,1}$ induces an isomorphism $J\mathcal{E}/J\mathcal{P} \cong H^1$.*

Proof. By the exactness of $J\mathcal{E}^{\mathbb{C}} \to J\mathcal{E}^{0,1} \to J\mathcal{E}^{0,2}$, we may find for each $f \in \ker(\overline{\partial}_b\colon J\mathcal{E}^{0,1} \to J\mathcal{E}^{0,2})$ a jet of function $u + iv \in J\mathcal{E}^{\mathbb{C}}$ such that $f = \overline{\partial}_b(u+iv) \equiv i\overline{\partial}_b v \mod \overline{\partial}_b(J\mathcal{E})$. Thus $i\overline{\partial}_b\colon \mathcal{E} \to H^1$ is surjective. It suffices to compute the kernel. If $i\overline{\partial}_b v \in \overline{\partial}_b J\mathcal{E}$, we may take $u \in J\mathcal{E}$ so that $i\overline{\partial}_b v = \overline{\partial}_b u$, or equivalently, $\overline{\partial}_b(u - iv) = 0$. Thus $v \in J\mathcal{P}$. □

We next apply the construction of the quotient module to the BGG resolution of the adjoint representation $\mathfrak{su}(n,1)$. We first complexify the representation and give the BGG for $\mathfrak{sl}(n+1,\mathbb{C})$:

$$0 \to \mathfrak{sl}(n+1,\mathbb{C}) \to \mathcal{E}^{\mathbb{C}}(1) \begin{smallmatrix} \nearrow & \mathcal{E}^{(1,1)}(1) & \begin{smallmatrix}\nearrow\\ \searrow\end{smallmatrix} & \begin{smallmatrix}\mathcal{E}^{(2,1)}(1)\\ \\ \mathcal{E}^{(1,1)(1,1)}(1)\end{smallmatrix} \\ \searrow & \overline{\mathcal{E}^{(1,1)}(1)} & \begin{smallmatrix}\nearrow\\ \searrow\end{smallmatrix} & \overline{\mathcal{E}^{(2,1)}(1)} \end{smallmatrix}$$

Here $\mathcal{E}^{(p,q)}(1)$ denotes the space of the sections of $\bigwedge^p (T^{0,1})^* \otimes \bigwedge^q (T^{0,1})^* \otimes \sigma_{1,1}$ satisfying $f_{[\alpha_1\cdots\alpha_p\beta_1]\beta_2\cdots\beta_q} = 0$, namely, $f_{\alpha_1\ldots\beta_q}$ has the symmetry corresponds to the Young diagram with two columns of hight p and q. $\mathcal{E}^{(1,1)(1,1)}(1)$ is the space of the sections of $\bigodot^2 (T^{1,0})^* \otimes_\circ \bigodot^2 (T^{0,1})^* \otimes \sigma_{1,1}$.

Composing the homomorphisms in the complex, we obtain a map

$$R\colon \mathcal{E}^{\mathbb{C}}(1) \to \mathcal{E}^{(1,1)(1,1)}(1).$$

It has order 4 and is given by the projection to $\mathcal{E}^{(1,1)(1,1)}(1)$ of $\nabla^4 f$, where ∇ is the Tanaka-Webster connection. The kernel of the map

$$R\colon J\mathcal{E}(1) \to J\mathcal{E}^{(1,1)(1,1)},$$

denoted by $J\mathcal{P}_1$, is given by the jets of the form

$$\operatorname{Re} \sum_{j=0}^{n} \overline{\xi}_j f_j(\xi),$$

where f_j is a formal power series of ξ at $\widetilde{e}_0$ of homogeneous degree 1. In this setting, we can say that the formal theory of Moser's normal form gives a slice of the coset of $J\mathcal{E}(1)/J\mathcal{P}_1$. Moser's normal form is a slice of equivalence classes of the real hypersurfaces of the form

$$2\operatorname{Re} w_n = |w'|^2 + f(w', \overline{w}', v)$$

modulo holomorphic coordinates changes that fix 0. $J\mathcal{P}_1$ describes the linear part of the effect by coordinates changes. By setting $\xi_0 = 1$, we identify $J\mathcal{E}(1)$ with the space of real formal power series $f(w', \overline{w}', v)$ of the form (5.1). Then the space of Moser's normal form $\mathcal{N}_1$ is given by the conditions:

$$\begin{aligned} A_{p,q} &= 0 \quad \text{if } \min(p,q) \le 1, \\ (\Delta')^{j+k+1} A_{2+j,2+k} &= 0 \quad \text{for } 0 \le j,k \le 1. \end{aligned}$$

PROPOSITION 6.2 ([CM]). *As a vector space* $J\mathcal{E}(1) = \mathcal{N}_1 \oplus J\mathcal{P}_1$.

If we identify $J\mathcal{E}(1)/J\mathcal{P}_1 \cong \mathcal{N}_1$, the value of $R(f)$ at e_0 is given by $A_{2,2}(0)$, which can be idenfitied with a trace-free tensor because of the normalization condition $\Delta' A_{2,2}(0) = 0$.

We next realize $J\mathcal{E}(1)/J\mathcal{P}_1$ as a cohomology. If we start with the adjoint representation, we obtain

$$0 \to \mathfrak{su}(1,n) \to \mathcal{E}(1) \to \mathcal{E}^{(1,1)}(1) \begin{array}{l} \nearrow \ \mathcal{E}^{(2,1)}(1) \\ \searrow \ H\mathcal{E}_\circ^{(1,1)(1,1)}(1) \end{array}$$

where $H\mathcal{E}_\circ^{(1,1)(1,1)}(1)$ is the subspace of $\mathcal{E}_\circ^{(1,1)(1,1)}(1)$ consisting of tensors with hermitian symmetry. The subcomplex

$$0 \to \mathcal{E}(1) \xrightarrow{D_0} \mathcal{E}^{(1,1)}(1) \xrightarrow{D_1} \mathcal{E}^{(2,1)}(1) \to \cdots \to \mathcal{E}^{(n-1,1)}(1) \to 0$$

is called the deformation complex (see[C]): $\ker D_1$ parametrizes infinitesimal integrable deformation of CR structure of S^m, and image D_0 corresponds to trivial deformations. In analogy with Proposition 6.1 we have

$$J\mathcal{E}(1)/J\mathcal{P}_1 \cong \frac{\ker(D_1 : J\mathcal{E}^{(1,1)}(1) \to J\mathcal{E}^{(2,1)}(1))}{\mathrm{image}(D_0 : J\mathcal{E}(1) \to J\mathcal{E}^{(1,1)}(1))}.$$

We finally consider the case $n = 2$. For the trivial representation, we have the following complex of CR invariant differential operators

$$0 \to \mathbb{C} \to \mathcal{E}^{\mathbb{C}} \to \begin{matrix} \mathcal{E}^{0,1} \\ \oplus \\ \mathcal{E}^{1,0} \end{matrix} \to \begin{matrix} \mathcal{E}^{0,1}(-1) \\ \oplus \\ \mathcal{E}^{1,0}(-1) \end{matrix} \to \mathcal{E}^{\mathbb{C}}(-2) \to 0,$$

which is called the Rumin complex. The operator in the middle has order 2 and the composition

$$R \colon \mathcal{E} \to \mathcal{E}^{0,1}(-1)$$

has order 3. $R(f) = 0$ characterizes CR pluriharmonic functions and $J\mathcal{E}/J\mathcal{P}$ can be embedded into $\mathcal{E}^{0,1}(-1)$. Under the identification $J\mathcal{E}/J\mathcal{P} \cong \mathcal{N}_0$, the value of $R(f)$ at e_0 is given by $A_{1,2}(0)$, which is the third order term of f. It seems rather strange that the one form valued operator R lifted to the ambient space as a $(1,1)$-tensor valued operator. This type of change also appears in the ambient metric construction of 3-dimensional conformal structures, where Cotton tensor lifted to the ambient curvature tensor.

For the adjoint representation, we have

$$0 \to \mathfrak{sl}(3,\mathbb{C}) \to \mathcal{E}^{\mathbb{C}}(1) \to \begin{matrix} \mathcal{E}^{(1,1)}(1) \\ \oplus \\ \overline{\mathcal{E}^{(1,1)}(1)} \end{matrix} \to \begin{matrix} \mathcal{E}^{(1,1)}(-1) \\ \oplus \\ \overline{\mathcal{E}^{(1,1)}(-1)} \end{matrix} \to \mathcal{E}^{\mathbb{C}}(-3) \to 0.$$

The operator in the middle has order 4 and the composition $\mathcal{E}(1) \to \overline{\mathcal{E}^{(1,1)}(-1)}$ has order 6. Under the identification $J\mathcal{E}(1)/J\mathcal{P}_1 \cong \mathcal{N}_1$, the value of $R(f)$ at e_0 is given by $A_{2,4}(0)$. This is the leading term of Moser's normal form and has order 6. The ambient lift of R is $\widetilde{\nabla}^{2,2}$; it again change the type of the tensor.

7. Concluding remarks. The arguments in Section 5 can be applied to CR invariants $J\mathcal{E}/J\mathcal{P} \to \sigma_{-p,-p}$ for $p = 0, 1, 2, \ldots, n$. These invariants are given by complete contractions on the ambient space. To study the case $p > n$, we need to generalize the jet isomorphism theorem. One way of doing this is to allow logarithmic term in the solution to $\Delta\widetilde{f} = 0$. Since the solution is not unique, we need to introduce a parameter space to formulate the jet isomorphism theorem. This causes a new difficulty in the construction of CR invariants and the classification problem in this case is still open [H2].

To deal with the general strictly pseudoconvex real hypersurfaces in a complex manifold X, we need to start with the construction of the ambient space. The ambient metric is defined on the canonical bundle of X (with the zero section removed) and is constructed by solving a complex Monge-Ampère equation. The ambient metric construction in Section 4 also holds for this curved ambient metric. The details will be given in my forthcoming paper.

REFERENCES

[BEG] T.N. Bailey, M.G. Eastwood, and C.R. Graham, *Invariant theory for conformal and CR geometry*, Ann. of Math. **139** (1994), 491–552.

[B] A. Boggess, CR manifolds and the tangential Cauchy-Riemann complex. Studies in Advanced Math., CRC Press, 1991.

[C] A. Čap, *Infinitesimal Automorphisms and Deformations of Parabolic Geometries*, `arXiv:math.DG/0508535`

[CM] S.S. Chern and J.K. Moser, *Real hypersurfaces in complex manifolds*, Acta Math. **133** (1974), 219–271.

[EG1] M.G. Eastwood and C.R. Graham, *Invariants of CR densities*, in "Several complex variables and complex geometry", Part 2 (Santa Cruz, CA, 1989), Proc. Sympos. Pure Math. **52**, 117–133, A.M.S., 1991.

[EG2] M.G. Eastwood and C.R. Graham, *Invariants of conformal densities*, Duke Math. J. **63** (1991), 633–671.

[F] C. Fefferman, *Parabolic invariant theory in complex analysis*, Adv. in Math. **31** (1979), 131–262.

[FH] C. Fefferman and K. Hirachi, *Ambient metric construction of Q-curvature in conformal and CR geometries,* Math. Res. Lett. **10** (2003), 819–832.

[GG] A.R. Gover and C.R. Graham, *CR invariant powers of the sub-Laplacian*, J. Reine Angew. Math. **583** (2005), 1–27.

[GJMS] C.R. Graham, R. Jenne, L.J. Mason, and G.A.J. Sparling, *Conformally invariant powers of the Laplacian. I, Existence*, J. London Math. Soc. **46** (1992), 557–565.

[H1] K. Hirachi, *Scalar pseudo-hermitian invariants and the Szegö kernel on three-dimensional CR manifolds*, in "Complex Geometry," Lect. Notes in Pure and Appl. Math. **143**, 67–76, Dekker, 1992.

[H2] K. Hirachi, *Construction of boundary invariants and the logarithmic singularity of the Bergman kernel*, Ann. of Math. **151** (2000), 151–191.

[H3] K. Hirachi, *Invariant theory of the Bergman kernel of strictly pseudoconvex domains*, Sugaku Expositions **17** (2004), 151–169.

[L] J.M. Lee, *Pseudo-Einstein structures on CR manifolds*, Amer. J. Math. **110** (1988), 157–178.

FINE STRUCTURE FOR SECOND ORDER SUPERINTEGRABLE SYSTEMS

ERNIE G. KALNINS*, JONATHAN M. KRESS†, AND WILLARD MILLER, JR.‡

Abstract. A classical (or quantum) superintegrable system is an integrable n-dimensional Hamiltonian system with potential that admits $2n-1$ functionally independent constants of the motion polynomial in the momenta, the maximum possible. If the constants are all quadratic the system is second order superintegrable. Such systems have remarkable properties: multi-integrability and multi-separability, a quadratic algebra of symmetries whose representation theory yields spectral information about the Schrödinger operator, deep connections with special functions and with QES systems. For n=2 we have worked out the structure and classified the possible spaces and potentials, and for n=3 on conformally flat spaces with nondegenerate potentials we determined the structure theory and made major progress on the classification.

The quadratic algebra closes at order 6 and there is a 1-1 classical-quantum relationship. All such systems are Stäckel transforms of systems on complex Euclidean space or the complex 3-sphere. We survey these results and announce a series of new results concerning the structure of superintegrable systems with degenerate potentials. In several cases the classification theory for such systems reduces to the study of polynomial ideals on which the symmetry group of the correspond manifold acts.

1. Introduction and examples. In this paper we report on recent work concerning the structure of second order superintegrable systems, both classical and quantum mechanical. We concentrate on the basic ideas; the details of the proofs can be (or will be) found elsewhere. The results on the quadratic algebra structure of 2D and 3D conformally flat systems with nondegenerate potential have appeared recently, but the results on fine structure of 3D superintegrable systems with degenerate potentials, and the relation of superintegrable systems to polynomial ideals are announced here.

Here we consider only superintegrable systems on complex conformally flat spaces. This is no restriction at all in two dimensions. An n-dimensional complex Riemannian space is conformally flat if and only if it admits a set of local coordinates $x_1, \cdots, x_n$ such that the contravariant metric tensor takes the form $g^{ij} = \delta^{ij}/\lambda(\mathbf{x})$. Thus the metric is $ds^2 = \lambda(\mathbf{x})(\sum_{i=1}^n dx_i^2)$. A classical superintegrable system $\mathcal{H} = \sum_{ij} g^{ij} p_i p_j + V(\mathbf{x})$ on the phase space of this manifold is one that admits $2n-1$ functionally independent generalized symmetries (or constants of the motion) $\mathcal{S}_k$, $k = 1, \cdots, 2n-1$ with $\mathcal{S}_1 = \mathcal{H}$ where the $\mathcal{S}_k$ are polynomials in the momenta p_j. That is, $\{\mathcal{H}, \mathcal{S}_k\} = 0$ where

*Department of Mathematics, University of Waikato, Hamilton, New Zealand (math0236@math.waikato.ac.nz).

†School of Mathematics, The University of New South Wales, Sydney NSW 2052, Australia. (j.kress@unsw.edu.au).

‡School of Mathematics, University of Minnesota, Minneapolis, MN 55455, U.S.A. (miller@ima.umn.edu).

$$\{f,g\} = \sum_{j=1}^{n} (\partial_{x_j} f \partial_{p_j} g - \partial_{p_j} f \partial_{x_j} g)$$

is the Poisson bracket for functions $f(\mathbf{x},\mathbf{p}), g(\mathbf{x},\mathbf{p})$ on phase space [1–8]. It is easy to see that $2n-1$ is the maximum possible number of functionally independent symmetries and, locally, such (in general nonpolynomial) symmetries always exist.

Superintegrable systems can lay claim to be the most symmetric Hamiltonian systems though many such systems admit no group symmetry. Generically, every trajectory $\mathbf{p}(t), \mathbf{x}(t)$ in phase space, i.e., solution of the Hamilton equations of motion for the sytem, is obtained as the common intersection of the (constants of the motion) hypersurfaces

$$\mathcal{S}_k(\mathbf{p},\mathbf{x}) = c_k, \quad k = 0, \cdots, 2n-2.$$

The trajectories can be found without solving the equations of motion. This is better than integrability. The superintegrabilty of the Kepler-Coulomb two-body problem is the reason that Kepler was able to determine the planetary orbits before the invention of calculus.

A system is second order superintegrable if the $2n-1$ functionally independent symmetries can be chosen to be quadratic in the momenta. Usually a superintegrable system is also required to be integrable, i.e., it is assumed that n of the constants of the motion are in involution, although we do not make that assumption here. Sophisticated tools such as R-matrix theory can be applied to the general study of superintegrable systems, e.g., [9–11]. However, the most detailed and complete results are known for second order superintegrable systems because separation of variables methods for the associated Hamilton-Jacobi equations can be applied. Standard orthogonal separation of variables techniques are associated with second-order symmetries, e.g., [12–17] and multiseparable Hamiltonian systems provide numerous examples of superintegrability. Here we concentrate on such systems, i.e., on those in which the symmetries take the form $\mathcal{S} = \sum a^{ij}(\mathbf{x}) p_i p_j + W(\mathbf{x})$, quadratic in the momenta. However, the ultimate goal is to develop tools that will enable us to study the structure of superintegrable systems of all orders and to develop a classification theory.

There is an analogous definition for second-order quantum superintegrable systems with Schrödinger operator

$$H = \Delta + V(\mathbf{x}), \quad \Delta = \frac{1}{\sqrt{g}} \sum_{ij} \partial_{x_i} (\sqrt{g} g^{ij}) \partial_{x_j},$$

the Laplace-Beltrami operator plus a potential function, [12]. Here there are $2n-1$ second-order symmetry operators

$$S_k = \frac{1}{\sqrt{g}} \sum_{ij} \partial_{x_i} (\sqrt{g} a^{ij}_{(k)}) \partial_{x_j} + W^{(k)}(\mathbf{x}), \quad k = 1, \cdots, 2n-1$$

with $S_1 = H$ and $[H, S_k] \equiv HS_k - S_kH = 0$. Again multiseparable systems yield many examples of superintegrability, though not all multiseparable systems are superintegrable and not all second-order superintegrable systems are multiseparable.

A basic motivation for studying these systems is that they can be solved explicitly and in multiple ways. It is the information gleaned from comparing the distinct solutions and expressing one solution set in terms of another that is a primary reason for their interest.

Two dimensional second order superintegrable systems have been studied and classified and the structure of three dimensional systems with nondegenerate potentials has been worked out in a recent series of papers [18–22]. We survey these results and announce a series of new results concerning the structure of superintegrable systems with degenerate potentials. In several cases the classification theory for such systems reduces to the study of polynomial ideals on which the symmetry group of the correspond manifold acts.

We start with some simple 3D examples. (To make clearer the connection with quantum theory and Hilbert space methods we shall, for these examples alone, adopt standard physical normalizations, such as using the factor $-\frac{1}{2}$ in front of the free Hamiltonian.) Consider the Schrödinger equation $H\Psi = E\Psi$ or ($\hbar = m = 1, x_1 = x, x_2 = y, x_3 = z$)

$$H\Psi = -\frac{1}{2}\left(\frac{\partial^2}{\partial x^2} + \frac{\partial^2}{\partial y^2} + \frac{\partial^2}{\partial z^2}\right)\Psi + V(x, y, z)\Psi = E\Psi.$$

The generalized anisotropic oscillator corresponds to the 4-parameter potential

$$V(x, y, z) = \frac{\omega^2}{2}\left(x^2 + y^2 + 4z^2\right) + \frac{1}{2}\left[\frac{\alpha}{x^2} + \frac{\beta}{y^2}\right] + \gamma z.$$

(This potential is "nondegenerate" in a precise sense that we will explain later.) The corresponding Schrödinger equation has separable solutions in five coordinate systems: Cartesian, cylindrical polar, cylindrical elliptic, cylindrical parabolic and parabolic coordinates. A basis for the second order symmetry operators is

$$M_1 = \partial_x^2 - \omega^2 x^2 + \frac{\alpha}{x^2}, \qquad M_2 = \partial_y^2 - \omega^2 y^2 - \frac{\beta}{y^2},$$

$$P = \partial_z^2 - 4\omega^2(z+\rho)^2, \qquad L = L_{12}^2 - \alpha\frac{y^2}{x^2} - \beta\frac{x^2}{y^2} - \frac{1}{2},$$

$$S_1 = -\frac{1}{2}(\partial_x L_{13} + L_{13}\partial_x) + \rho\partial_x^2 + (z+\rho)\left(\omega^2 x^2 - \frac{\alpha}{x^2}\right),$$

$$S_2 = -\frac{1}{2}(\partial_y L_{23} + L_{23}\partial_y) + \rho\partial_y^2 + (z+\rho)\left(\omega^2 y^2 - \frac{\beta}{y^2}\right),$$

where $L_{ij} = x_i\partial_{x_j} - x_j\partial_{x_i}$. It can be verified that these symmetries generate a "quadratic algebra" that closes at level six. Indeed, the nonzero commutators of the above generators are

$$[M_1, L] = [L, M_2] = Q, \qquad [L, S_1] = [S_2, L] = B,$$
$$[M_i, S_i] = A_i, \qquad [P, S_i] = -A_i.$$

Nonzero commutators of the basis symmetries with Q (4th order symmetries) are expressible in terms of the second order symmetries:

$$[M_i, Q] = [Q, M_2] = 4\{M_1, M_2\} + 16\omega^2 L, \qquad [S_1, Q] = [Q, S_2] = 4\{M_1, M_2\},$$
$$[L, Q] = 4\{M_1, L\} - 4\{M_2, L\} + 16\left(\frac{3}{4} - \alpha\right)M_1 - 16\left(\frac{3}{4} - \beta\right)M_2.$$

There are similar expressions for commutators with B and the A_i. Also the squares of Q, B, A_i and products such as $\{Q, B\}$, (all 6th order symmetries) are all expressible in terms of 2nd order symmetries. Indeed

$$\begin{aligned}
Q^2 &= \frac{8}{3}\{L, M_1, M_2\} + 8\omega^2\{L, L\} - 16\left(\frac{3}{4} - \alpha\right)M_1^2 - 16\left(\frac{3}{4} - \beta\right)M_2^2 \\
&\quad + \frac{64}{3}\{M_1, M_2\} - \frac{128}{3}\omega^2 L - 128\omega^2\left(\frac{3}{4} - \alpha\right)\left(\frac{3}{4} - \beta\right), \\
\{Q, B\} &= -\frac{8}{3}\{M_2, L, S_1\} - \frac{8}{3}\{M_1, L, S_2\} + 16\left(\frac{3}{4} - \alpha\right)\{M_2, S_2\} \\
&\quad + 16\left(\frac{3}{4} - \beta\right)\{M_1, S_1\} - \frac{64}{3}\{M_1, S_2\} - \frac{64}{3}\{M_2, S_1\}.
\end{aligned}$$

Here $\{C_1, \cdots, C_j\}$ is the completely symmetrized product of operators $C_1, \cdots, C_j$. (For details see [23].) The point is that the algebra generated by products and commutators of the 2nd order symmetries closes at order 6. This is a remarkable fact, and ordinarily not the case for an integrable system. The algebra provides information about the spectra of the generators.

A counterexample to the existence of a quadratic algebra in Euclidean space is given by the Schrödinger equation with 3-parameter extended Kepler-Coulomb potential:

$$\left(\frac{\partial^2\Psi}{\partial x^2} + \frac{\partial^2\Psi}{\partial y^2} + \frac{\partial^2\Psi}{\partial z^2}\right) + \left[2E + \frac{2\alpha}{\sqrt{x^2+y^2+z^2}} - \left(\frac{\beta}{x^2} + \frac{\gamma}{y^2}\right)\right]\Psi = 0. \quad (1.1)$$

This equation admits separable solutions in four coordinate systems: spherical, sphero-conical, prolate spheroidal and parabolic coordinates. Again the bound states are degenerate and important special function identities arise by expanding one basis of separable eigenfunctions in terms of another. However, the space of second order symmetries is only 5 dimensional and, although there are useful identities among the generators and commutators that enable one to derive spectral properties algebraically, there is no finite quadratic algebra structure. The key difference with our first example is, as we shall show later, that the 3-parameter Kepler-Coulomb potential is degenerate and cannot be extended to a 4-parameter potential.

In [20, 21] there are examples of superintegrable systems on the 3-sphere that admit a quadratic algebra structure. A more general set of examples arises from a space with metric

$$ds^2 = \lambda(A, B, C, D, E, \mathbf{x})(dx^2 + dy^2 + dz^2)$$

and classical potential $V = \lambda(\alpha, \beta, \gamma, \delta, \eta, \mathbf{x})/\lambda(A, B, C, D, E, \mathbf{x})$, where

$$\begin{aligned}\lambda = A(x+iy) &+ B\left(\frac{3}{4}(x+iy)^2 + \frac{z}{4}\right)\\ &+ C\left((x+iy)^3 + \frac{1}{16}(x-iy) + \frac{3z}{4}(x+iy)\right)\\ &+ D\left(\frac{5}{16}(x+iy)^4 + \frac{z^2}{16} + \frac{1}{16}(x^2+y^2) + \frac{3z}{8}(x+iy)^2\right) + E.\end{aligned}$$

If $A = B = C = D = 0$ this is a nondegenerate metric on complex Euclidean space. The quadratic algebra always closes, and for general values of A, B, C, D the space is not of constant curvature. As will be explained later, this is an example of a superintegrable system that is Stäckel equivalent to a system on complex Euclidean space.

Observed common features of superintegrable systems are that they are usually multiseparable and that the eigenfunctions of one separable system can be expanded in terms of the eigenfunctions of another. This is the source of nontrivial special function expansion theorems [24]. The symmetry operators are in formal self-adjoint form and suitable for spectral analysis. The representation theory of the abstract quadratic algebra can be used to derive spectral properties of the second order generators in a manner analogous to the use of Lie algebra representation theory to derive spectral properties of quantum systems that admit Lie symmetry algebras, [24–27].

Another common feature of quantum superintegrable systems is that they can be modified by a gauge transformation so that the Schrödinger and symmetry operators are acting on a space of polynomials, [28]. This is closely related to the theory of exactly and quasi-exactly solvable systems, [29, 30]. The characterization of ODE quasi-exactly solvable systems as embedded in PDE superintegrable systems provides considerable insight into the nature of these phenomena [31].

The classical analogs of the above examples are obtained by the replacements $\partial_{x_i} \rightarrow p_{x_i}$ and modification of the potential by curvature terms. Commutators go over to Poisson brackets. The operator symmetries become second order constants of the motion. Symmetrized operators become products of functions. The quadratic algebra relations simplify: the highest order terms agree with the operator case but there are fewer nonzero lower order terms.

Many examples of 3D superintegrable systems are known, although they have not been classified, [32–37]. Here, we employ theoretical methods based on integrability conditions to derive structure common to all such systems, with a view to complete classification. In each case we work out the classical problem first and then quantize. Finally, we exhibit a deep connection between 3D second order superintegrable systems with nondegenerate potential and a polynomial ideal in 10 variables.

2. 2D classical structure theory. For any complex 2D Riemannian manifold we can always find local coordinates x, y such that the classical Hamiltonian takes the form

$$\mathcal{H} = \frac{1}{\lambda(x,y)}(p_1^2 + p_2^2) + V(x,y), \qquad (x,y) = (x_1, x_2),$$

i.e., the complex metric is $ds^2 = \lambda(x,y)(dx^2 + dy^2)$. Necessary and sufficient conditions that $\mathcal{S} = \sum a^{ji}(x,y)p_j p_i + W(x,y)$ be a symmetry of $\mathcal{H}$ are the Killing equations

$$\begin{aligned} a_i^{ii} &= -\frac{\lambda_1}{\lambda}a^{i1} - \frac{\lambda_2}{\lambda}a^{i2}, \quad i = 1,2 \\ 2a_i^{ij} + a_j^{ii} &= -\frac{\lambda_1}{\lambda}a^{j1} - \frac{\lambda_2}{\lambda}a^{j2}, \quad i,j = 1,2, \ i \neq j, \end{aligned} \tag{2.1}$$

and the Bertrand-Darboux (B-D) conditions on the potential $\partial_i W_j = \partial_j W_i$ or

$$\begin{aligned} &(V_{22} - V_{11})a^{12} + V_{12}(a^{11} - a^{22}) \\ &\quad = \left[\frac{(\lambda a^{12})_1 - (\lambda a^{11})_2}{\lambda}\right]V_1 + \left[\frac{(\lambda a^{22})_1 - (\lambda a^{12})_2}{\lambda}\right]V_2. \end{aligned}$$

There are similar but more complicated conditions for the higher order symmetries. From the 3 second order constants of the motion we get 3 B-D equations and can solve them to obtain fundamental PDEs for the potential of the form

$$V_{22} - V_{11} = A^{(22}(\mathbf{x})V_1 + B^{22}(\mathbf{x})V_2, \qquad V_{12} = A^{12}(\mathbf{x})V_1 + B^{12}(\mathbf{x})V_2. \tag{2.2}$$

If the B-D equations provide no further conditions on the potential and if the integrability conditions for these PDEs are satisfied identically, we say that the potential is **nondegenerate**. That means, at each regular point $\mathbf{x}_0$ where the A^{ij}, B^{ij} are defined and analytic, we can prescribe the values of V, V_1, V_2 and V_{11} arbitrarily and there will exist a unique potential $V(\mathbf{x})$ with these values at $\mathbf{x}_0$. Nondegenerate potentials depend on 3 parameters, in addition to the trivial additive parameter. Degenerate potentials depend on < 3 parameters.

To study the possible quadratic algebra structure it is important to compute the dimensions of the spaces of symmetries of these nondegenerate systems that are of orders 2,3,4 and 6. These symmetries are necessarily of a special type.

- The highest order terms in the momenta are independent of the parameters in the potential.
- The terms of order 2 less in the momenta less are linear in these parameters.
- Those of order 4 less are quadratic and those of order 6 less are cubic.

The system is **2nd order superintegrable** with **nondegenerate potential** if

- it admits 3 functionally independent second-order symmetries (here $2N - 1 = 3$)
- the potential is 3-parameter (in addition to the usual additive parameter) in the sense that it satisfies equations (2.2) and their integrability conditions.

We say that n functionally independent symmetries are **functionally linearly independent** if at each regular point $\mathbf{x}_0$ the n matrices

$$a^{ij}_{(1)}(\mathbf{x}_0),\ a^{ij}_{(2)}(\mathbf{x}_0), \cdots\ a^{ij}_{(n)}(\mathbf{x}_0)$$

are linearly independent. This functional linear independence criterion splits superintegrable systems of all orders into two classes with different properties. In 2D there is essentially only one functionally linearly dependent superintegrable system, namely $\mathcal{H} = p_z p_{\bar{z}} + V(z)$, where $V(z)$ is an arbitrary function of z alone. This system separates in only one set of coordinates $z, \bar{z}$. For functionally linearly independent 2D systems the theory is much more interesting.

THEOREM 2.1. *Let $\mathcal{H}$ be the Hamiltonian of a 2D superintegrable (functionally linearly independent) system with nondegenerate potential.*

- *The space of second order constants of the motion is 3-dimensional.*
- *The space of third order constants of the motion is 1-dimensional.*
- *The space of fourth order constants of the motion is 6-dimensional.*
- *The space of sixth order constants is 10-dimensional.*

THEOREM 2.2. *Let $\mathcal{K}$ be a third order constant of the motion for a superintegrable system with nondegenerate potential V:*

$$\mathcal{K} = \sum_{k,j,i=1}^{2} a^{kji}(x,y) p_k p_j p_i + \sum_{\ell=1}^{2} b^{\ell}(x,y) p_\ell .$$

Then $b^\ell(x,y) = \sum_{j=1}^{2} f^{\ell,j}(x,y) \frac{\partial V}{\partial x_j}(x,y)$ with $f^{\ell,j} + f^{j,\ell} = 0, \quad 1 \le \ell, j \le 2$. The a^{ijk}, b^ℓ are uniquely determined by the number $f^{1,2}(x_0, y_0)$ at some regular point (x_0, y_0) of V.

This result enables us to choose standard bases for second and higher order symmetries. Indeed, given any 2×2 symmetric matrix $\mathcal{A}_0$, and any regular point (x_0, y_0) there exists one and only one second order symmetry (or constant of the motion) $\mathcal{S}$ such that $\{a^{kj}_{(i)}(x_0, y_0)\} = \mathcal{A}_0$ and $W(x_0, y_0) = 0$. Further, if

$$\mathcal{S}_1 = \sum a^{kj}_{(1)} p_k p_j + W_{(1)}, \quad \mathcal{S}_2 = \sum a^{kj}_{(2)} p_k p_j + W_{(2)}$$

are second order constants of the the motion and $\mathcal{A}_{(i)}(x, y) = \{a^{kj}_{(i)}(x, y)\}$, $i = 1, 2$ are 2×2 symmetric matrix functions, then the Poisson bracket of these symmetries is given by

$$\{\mathcal{S}_1, \mathcal{S}_2\} = \sum_{k,j,i=1}^{2} a^{kji}(x, y) p_k p_j p_i + b^{\ell}(x, y) p_\ell$$

where

$$f^{k,\ell} = 2\lambda \sum_j (a^{kj}_{(2)} a^{j\ell}_{(1)} - a^{kj}_{(1)} a^{j\ell}_{(2)}).$$

Thus $\{\mathcal{S}_1, \mathcal{S}_2\}$ is uniquely determined by the skew-symmetric matrix

$$[\mathcal{A}_{(2)}, \mathcal{A}_{(1)}] \equiv \mathcal{A}_{(2)}\mathcal{A}_{(1)} - \mathcal{A}_{(1)}\mathcal{A}_{(2)},$$

hence by the constant matrix $[\mathcal{A}_{(2)}(x_0, y_0), \mathcal{A}_{(1)}(x_0, y_0)]$ evaluated at a regular point. There is a standard structure that allows the identification of the space of second order constants of the motion with the space of 2×2 symmetric matrices and identification of the space of third order constants of the motion with the space of 2×2 skew-symmetric matrices.

Let $\mathcal{E}^{ij}$ be the 2×2 matrix with a 1 in row i, column j and 0 for every other matrix element. Then the symmetric matrices

$$\mathcal{A}^{(ij)} = \frac{1}{2}(\mathcal{E}^{ij} + \mathcal{E}^{ji}) = \mathcal{A}^{(ji)}, \quad i, j = 1, 2$$

form a basis for the 3-dimensional space of symmetric matrices. Moreover,

$$[\mathcal{A}^{(ij)}, \mathcal{A}^{(k\ell)}] = \frac{1}{2}\left(\delta_{jk}\mathcal{B}^{(i\ell)} + \delta_{j\ell}\mathcal{B}^{(ik)} + \delta_{ik}\mathcal{B}^{(j\ell)} + \delta_{i\ell}\mathcal{B}^{(jk)}\right)$$

where

$$\mathcal{B}^{(ij)} = \frac{1}{2}(\mathcal{E}^{ij} - \mathcal{E}^{ji}) = -\mathcal{B}^{(ji)}, \quad i, j = 1, 2.$$

Here $\mathcal{B}^{(ii)} = 0$ and $\mathcal{B}^{(12)}$ forms a basis for the space of skew-symmetric matrices. This gives the commutation relations for the second order symmetries.

We define a standard set of basis symmetries $\mathcal{S}_{(k\ell)} = \sum a^{ij}(\mathbf{x})p_ip_j + W_{(k\ell)}(\mathbf{x})$ corresponding to a regular point $\mathbf{x}_0$ by

$$\begin{pmatrix} f_1^1 & f_2^1 \\ f_1^2 & f_2^2 \end{pmatrix}_{\mathbf{x}_0} = \lambda(\mathbf{x}_0)\begin{pmatrix} a^{11} & a^{12} \\ a^{21} & a^{22} \end{pmatrix}_{\mathbf{x}_0} = \lambda(\mathbf{x}_0)\mathcal{A}^{(k\ell)}, \quad W_{(k\ell)}(\mathbf{x}_0) = 0.$$

These structure results and standard separation of variables theory [15] give us the tools to prove multiseparabilty of 2D systems.

COROLLARY 2.1. *Let V be a superintegrable nondegenerate potential and L be a second order constant of the motion with matrix function $\mathcal{A}(\mathbf{x})$. If at some regular point $\mathbf{x}_0$ the matrix $\mathcal{A}(\mathbf{x}_0)$ has 2 distinct eigenvalues, then H, L characterize an orthogonal separable coordinate system.*
Since a generic 2×2 symmetric matrix has distinct roots, it follows that any superintegrable nondegenerate potential is multiseparable.

We prove the existence of a quadratic algebra structure by demonstrating that polynomials in the basis symmetries span the space of fourth and sixth order symmetries:

THEOREM 2.3. *The 6 distinct monomials*

$$(\mathcal{S}_{(11)})^2,\ (\mathcal{S}_{(22)})^2,\ (\mathcal{S}_{(12)})^2,\ \mathcal{S}_{(11)}\mathcal{S}_{(22)},\ \mathcal{S}_{(11)}\mathcal{S}_{(12)},\ \mathcal{S}_{(12)}\mathcal{S}_{(22)},$$

form a basis for the space of fourth order symmetries.

THEOREM 2.4. *The 10 distinct monomials*

$$(\mathcal{S}_{(ii)})^3,\ (\mathcal{S}_{(ij)})^3,\ (\mathcal{S}_{(ii)})^2\mathcal{S}_{(jj)},\ (\mathcal{S}_{(ii)})^2\mathcal{S}_{(ij)},\ (\mathcal{S}_{(ij)})^2\mathcal{S}_{(ii)},$$

$\mathcal{S}_{(11)}\mathcal{S}_{(12)}\mathcal{S}_{(22)}$, *for $i, j = 1, 2,\ i \neq j$ form a basis for the space of sixth order symmetries.*
The analogous results for 5th order symmetries follow directly from the Jacobi equality.

All 2D nondegenerate superintegrable systems in Euclidean space and on the 2-sphere have been classified [38, 39, 33]. There are 12 families of nondegenerate potentials in flat space (4 in real Euclidean space) and 6 families on the complex 2-sphere (2 on the real sphere). The principal tool in the classification of such systems on general Riemannian manifolds is the Stäckel transform. Suppose we have a (classical or quantum) superintegrable system

$$\mathcal{H} = \frac{1}{\lambda(x,y)}(p_1^2 + p_2^2) + V(x,y), \quad H = \frac{1}{\lambda(x,y)}(\partial_{11} + \partial_{22}) + V(x,y)$$

in local orthogonal coordinates, with nondegenerate potential $V(x, y)$ and suppose $U(x, y)$ is a particular case of the 3-parameter potential V, nonzero in an open set. Then the transformed systems

$$\tilde{\mathcal{H}} = \frac{1}{\tilde{\lambda}(x,y)}(p_1^2 + p_2^2) + \tilde{V}(x,y), \quad \tilde{H} = \frac{1}{\tilde{\lambda}(x,y)}(\partial_{11} + \partial_{22}) + \tilde{V}(x,y)$$

are also superintegrable, where $\tilde{\lambda} = \lambda U, \quad \tilde{V} = V/U$.

THEOREM 2.5.

$$\{\tilde{\mathcal{H}}, \tilde{\mathcal{S}}\} = 0 \iff \{\mathcal{H}, \mathcal{S}\} = 0,$$

$$\tilde{\mathcal{S}} = \sum_{ij} \frac{1}{\lambda U} p_i \left(\left(a^{ij} + \delta^{ij} \frac{1 - W_U}{\lambda U} \right) \lambda U \right) p_j + \left(W - \frac{W_U V}{U} + \frac{V}{U} \right),$$

$$[\tilde{H}, \tilde{S}] = 0 \iff [H, S] = 0,$$

$$\tilde{S} = \sum_{ij} \frac{1}{\lambda U} \partial_i \left(\left(a^{ij} + \delta^{ij} \frac{1 - W_U}{\lambda U} \right) \lambda U \right) \partial_j + \left(W - \frac{W_U V}{U} + \frac{V}{U} \right).$$

COROLLARY 2.2. *If $\mathcal{S}^{(1)}, \mathcal{S}^{(2)}$ are second order constants of the motion for $\mathcal{H}$, then*

$$\{\tilde{\mathcal{S}}^{(1)}, \tilde{\mathcal{S}}^{(2)}\} = 0 \iff \{\mathcal{S}^{(1)}, \mathcal{S}^{(2)}\} = 0.$$

If $S^{(1)}, S^{(2)}$ are second order symmetry operators for H, then

$$[\tilde{S}^{(1)}, \tilde{S}^{(2)}] = 0 \iff [S^{(1)}, S^{(2)}] = 0.$$

This transform of one (classical or quantum) superintegrable system into another on a different manifold is called the **Stäckel transform.** Two such systems related by a Stäckel transform are called **Stäckel equivalent**.

THEOREM 2.6. *Every nondegenerate second-order classical or quantum superintegrable system in two variables is Stäckel equivalent to a superintegrable system on a constant curvature space, i.e., flat space or the sphere.*

There is a fundamental duality between Killing tensors and the B-D equation (only in the 2D case) that enables us to get a complete classification of the possible manifolds.

THEOREM 2.7. *If $ds^2 = \lambda(dx^2 + dy^2)$ is the metric of a nondegenerate superintegrable system (expressed in coordinates x, y such that $\lambda_{12} = 0$) then $\lambda = \mu$ is a solution of the system*

$$\mu_{12} = 0, \qquad \mu_{22} - \mu_{11} = 3\mu_1(\ln a^{12})_1 - 3\mu_2(\ln a^{12})_2 + \left(\frac{a^{12}_{11} - a^{12}_{22}}{a^{12}} \right) \mu,$$

where either

I) $a^{12} = X(x)Y(y), \quad X'' = \alpha^2 X, \quad Y'' = -\alpha^2 Y,$

or

II) $a^{12} = \dfrac{2X'(x)Y'(y)}{C(X(x) + Y(y))^2},$

$$(X')^2 = F(X), \quad X'' = \frac{1}{2}F'(X), \quad (Y')^2 = G(Y), \quad Y'' = \frac{1}{2}G'(Y)$$

where

$$F(X) = \frac{\alpha}{24}X^4 + \frac{\gamma_1}{6}X^3 + \frac{\gamma_2}{2}X^2 + \gamma_3 X + \gamma_4,$$
$$G(Y) = -\frac{\alpha}{24}Y^4 + \frac{\gamma_1}{6}Y^3 - \frac{\gamma_2}{2}Y^2 + \gamma_3 Y - \gamma_4.$$

Conversely, every solution λ of one of these systems of equations defines a nondegenerate superintegrable system. If λ is a solution then the remaining solutions μ are exactly the nondegenerate superintegrable systems that are Stäckel equivalent to λ.

In a tour de force, Koenigs [40] has classified all 2D manifolds (i.e., zero potential) that admit exactly 3 second-order Killing tensors and listed them in two tables: Tableau VI and Tableau VII. Our methods show that these are exactly the spaces that admit superintegrable systems with non-degenerate potentials. (An alternate derivation can be found in [41].)

TABLEAU VI

$$[1]\ ds^2 = \left[\frac{c_1\cos x + c_2}{\sin^2 x} + \frac{c_3\cos y + c_4}{\sin^2 y}\right](dx^2 - dy^2)$$
$$[2]\ ds^2 = \left[\frac{c_1\cosh x + c_2}{\sinh^2 x} + \frac{c_3 e^y + c_4}{e^{2y}}\right](dx^2 - dy^2)$$
$$[3]\ ds^2 = \left[\frac{c_1 e^x + c_2}{e^{2x}} + \frac{c_3 e^y + c_4}{e^{2y}}\right](dx^2 - dy^2)$$
$$[4]\ ds^2 = \left[c_1(x^2 - y^2) + \frac{c_2}{x^2} + \frac{c_3}{y^2} + c_4\right](dx^2 - dy^2)$$
$$[5]\ ds^2 = \left[c_1(x^2 - y^2) + \frac{c_2}{x^2} + c_3 y + c_4\right](dx^2 - dy^2)$$
$$[6]\ ds^2 = \left[c_1(x^2 - y^2) + c_2 x + c_3 y + c_4\right](dx^2 - dy^2)$$

TABLEAU VII

$$[1]\ ds^2 = \left[c_1\left(\frac{1}{\mathrm{sn}^2(x,k)}\right) - \frac{1}{\mathrm{sn}^2(y,k)} + c_2\left(\frac{1}{\mathrm{cn}^2(x,k)} - \frac{1}{\mathrm{cn}^2(y,k)}\right)\right.$$
$$\left. + c_3\left(\frac{1}{\mathrm{dn}^2(x,k)} - \frac{1}{\mathrm{dn}^2(y,k)}\right) + c_4(\mathrm{sn}^2(x,k) - \mathrm{sn}^2(y,k)\right]$$
$$\times (dx^2 - dy^2)$$
$$[2]\ ds^2 = \left[c_1\left(\frac{1}{\sin^2 x} - \frac{1}{\sin^2 y}\right) + c_2\left(\frac{1}{\cos^2 x} - \frac{1}{\cos^2 y}\right)\right.$$
$$\left. + c_3(\cos 2x - \cos 2y)c_4(\cos 4x - \cos 4y)\right](dx^2 - dy^2)$$

$$[3]\ ds^2 = \Big[c_1(\sin 4x - \sin 4y) + c_2(\cos 4x - \cos 4y)$$
$$+ c_3(\sin 2x - \sin 2y c_4(\cos 2x - \cos 2y)\Big](dx^2 - dy^2)$$

$$[4]\ ds^2 = \Big[c_1(\frac{1}{x^2} - \frac{1}{y^2}) + c_2(x^2 - y^2) + c_3(x^4 - y^4)$$
$$+ c_4(x^6 - y^6)\Big](dx^2 - dy^2)$$

$$[5]\ ds^2 = \Big[c_1(x - y) + c_2(x^2 - y^2) + c_3(x^3 - y^3) + c_4(x^4 - y^4)\Big]$$
$$\times\ (dx^2 - dy^2).$$

The quantization of our 2D results is relatively straightforward. For a manifold with metric $ds^2 = \lambda(x, y)(dx^2 + dy^2)$ the Hamiltonian system

$$\mathcal{H} = \frac{p_1^2 + p_2^2}{\lambda(x, y)} + V(x, y)$$

is replaced by the Hamiltonian (Schrödinger) operator with potential

$$H = \frac{1}{\lambda(x, y)}(\partial_{11} + \partial_{22}) + V(x, y).$$

A second-order symmetry of the Hamiltonian system

$$\mathcal{S} = \sum_{k,j=1}^{2} a^{kj}(x, y)p_k p_j + W(x, y),$$

with $a^{kj} = a^{jk}$, corresponds to the formally self-adjoint operator

$$S = \frac{1}{\lambda(x, y)} \sum_{k,j=1}^{2} \partial_k(a^{kj}(x, y)\lambda(x, y)\partial_j) + W(x, y), \quad a^{kj} = a^{jk}.$$

LEMMA 2.1.

$$\{\mathcal{H}, \mathcal{S}\} = 0 \iff [H, S] = 0.$$

This result is not generally true for higher dimensional manifolds.

Through use of the self-adjointness of even order operator symmetries and the skew adjointness of odd order symmetries, it follows that the main classical results for symmetries corresponding to a nondegenerate potential can be taken over with little change [22]. In particular there is closure of the quadratic algebra for 2D quantum superintegrable potentials: All fourth order and sixth order symmetry operators can be expressed as symmetric polynomials in the second order symmetry operators.

3. Conformally flat spaces in three dimensions. The 3D results are considerably more complicated, but essential features are preserved. We limit ourselves to conformally flat spaces. For each such space there always exists a local coordinate system x, y, z and a nonzero function $\lambda(x, y, z) = \exp G(x, y, z)$ such that the Hamiltonian is

$$\mathcal{H} = (p_1^2 + p_2^2 + p_3^2)/\lambda + V(x, y, z).$$

A quadratic constant of the motion (or generalized symmetry)

$$\mathcal{S} = \sum_{k,j=1}^{3} a^{kj}(x, y, z) p_k p_j + W(x, y, z) \equiv \mathcal{L} + W, \quad a^{jk} = a^{kj}$$

must satisfy $\{\mathcal{H}, \mathcal{S}\} = 0$. i.e.,

$$\begin{aligned} a_i^{ii} &= -G_1 a^{1i} - G_2 a^{2i} - G_3 a^{3i} \\ 2a_i^{ij} + a_j^{ii} &= -G_1 a^{1j} - G_2 a^{2j} - G_3 a^{3j}, \quad i \neq j \\ a_k^{ij} + a_j^{ki} + a_i^{jk} &= 0, \quad i, j, k \text{ distinct} \end{aligned}$$

and

$$W_k = \lambda \sum_{s=1}^{3} a^{sk} V_s, \quad k = 1, 2, 3. \tag{3.1}$$

(Here a subscript j denotes differentiation with respect to x_j.) The requirement that $\partial_{x_\ell} W_j = \partial_{x_j} W_\ell$, $\ell \neq j$ leads from (3.1) to the second order B-D partial differential equations for the potential.

$$\sum_{s=1}^{3} \left[V_{sj} \lambda a^{s\ell} - V_{s\ell} \lambda a^{sj} + V_s \left((\lambda a^{s\ell})_j - (\lambda a^{sj})_\ell \right) \right] = 0. \tag{3.2}$$

For second order superintegrabilty in 3D there must be five functionally independent constants of the motion (including the Hamiltonian itself). Thus the Hamilton-Jacobi equation admits four additional constants of the motion:

$$\mathcal{S}_h = \sum_{j,k=1}^{3} a_{(h)}^{jk} p_k p_j + W_{(h)} = \mathcal{L}_h + W_{(h)}, \qquad h = 1, \cdots, 4.$$

We assume that the four functions $\mathcal{S}_h$ together with $\mathcal{H}$ are functionally linearly independent in the six-dimensional phase space. In [20] it is shown that the matrix of the 15 B-D equations for the potential has rank at least 5, hence we can solve for the second derivatives of the potential in the form

$$\begin{aligned} V_{22} &= V_{11} + A^{22} V_1 + B^{22} V_2 + C^{22} V_3, \\ V_{33} &= V_{11} + A^{33} V_1 + B^{33} V_2 + C^{33} V_3, \\ V_{ij} &= A^{ij} V_1 + B^{ij} V_2 + C^{ij} V_3, \end{aligned} \tag{3.3}$$

where $1 \leq i < j \leq 3$. If the matrix has rank > 5 then there will be additional conditions on the potential and it will depend on fewer parameters. $D^1_{(s)}V_1 + D^2_{(s)}V_2 + D^3_{(s)}V_3 = 0$. Here the $A^{ij}, B^{ij}, C^{ij}, D^i_{(s)}$ are functions of x, symmetric in the superscripts, that can be calculated explicitly. Suppose now that the superintegrable system is such that the rank is exactly 5 so that the relations are only (3.3). Further, suppose the integrability conditions for system (3.3) are satisfied identically. In this case we say that the potential is *nondegenerate*. Otherwise the potential is *degenerate*. If V is nondegenerate then at any point $\mathbf{x}_0$, where the A^{ij}, B^{ij}, C^{ij} are defined and analytic, there is a unique solution $V(\mathbf{x})$ with arbitrarily prescribed values of $V_1(\mathbf{x}_0), V_2(\mathbf{x}_0), V_3(\mathbf{x}_0), V_{11}(\mathbf{x}_0)$ (as well as the value of $V(\mathbf{x}_0)$ itself.) The points $\mathbf{x}_0$ are called *regular*. The points of singularity for the A^{ij}, B^{ij}, C^{ij} form a manifold of dimension < 3. Degenerate potentials depend on fewer parameters. For example, it may be that the rank of the B-D equations is exactly 5 but the integrability conditions are not satisfied identically. This occurs for the generalized Kepler-Coulomb potential.

Assuming that V is nondegenerate, we substitute the requirement for a nondegenerate potential (3.3) into the B-D equations (3.2) and obtain three equations for the derivatives a_i^{jk}, the first of which is

$$\begin{aligned}
&(a_3^{11} - a_1^{31})V_1 + (a_3^{12} - a_1^{32})V_2 + (a_3^{13} - a_1^{33})V_3 \\
&+ a^{12}(A^{23}V_1 + B^{23}V_2 + C^{23}V_3) \\
&- (a^{33} - a^{11})(A^{13}V_1 + B^{13}V_2 + C^{13}V_3) \\
&- a^{23}(A^{12}V_1 + B^{12}V_2 + C^{12}V_3) + a^{13}(A^{33}V_1 + B^{33}V_2 + C^{33}V_3) \\
&= (-G_3a^{11} + G_1a^{13})V_1 + (-G_3a^{12} + G_1a^{23})V_2 + (-G_3a^{13} + G_1a^{33})V_3,
\end{aligned} \tag{3.4}$$

and the other two are obtained in a similar fashion.

Since V is a nondegenerate potential we can equate coefficients of V_1, V_2, V_3, V_{11} on each side of the conditions $\partial_1 V_{23} = \partial_2 V_{13} = \partial_3 V_{12}$, $\partial_3 V_{23} = \partial_2 V_{33}$, etc., to obtain integrability conditions, the simplest of which are

$$\begin{aligned}
A^{23} = B^{13} = C^{12}, \qquad & B^{12} - A^{22} = C^{13} - A^{33}, \\
B^{23} = A^{31} + C^{22}, \qquad & C^{23} = A^{12} + B^{33}.
\end{aligned} \tag{3.5}$$

In general, the integrability conditions satisfied by the potential equations can be expressed as follows. We introduce the vector $\mathbf{w} = (V_1, V_2, V_3, V_{11})^{\mathrm{T}}$, and the matrices $\mathbf{A}^{(j)}$, $j = 1, 2, 3$, such that

$$\partial_{x_j}\mathbf{w} = \mathbf{A}^{(j)}\mathbf{w} \qquad j = 1, 2, 3. \tag{3.6}$$

The integrability conditions for this system are

$$A_i^{(j)} - A_j^{(i)} = A^{(i)}A^{(j)} - A^{(j)}A^{(i)} \equiv [A^{(i)}, A^{(j)}]. \tag{3.7}$$

The integrability conditions (3.5) and (3.7) are analytic expressions in x_1, x_2, x_3 and must hold identically. Then the system has a solution V depending on 4 parameters (plus an arbitrary additive parameter).

Using the nondegenerate potential condition and the B-D equations we can solve for all of the first partial derivatives a_i^{jk} of a quadratic symmetry to obtain

$$\begin{aligned}
a_1^{11} &= -G_1a^{11} - G_2a^{12} - G_3a^{13},\\
a_2^{22} &= -G_1a^{12} - G_2a^{22} - G_3a^{23},\\
a_3^{33} &= -G_1a^{13} - G_2a^{23} - G_3a^{33},\\
3a_1^{12} &= a^{12}A^{22} - (a^{22} - a^{11})A^{12} - a^{23}A^{13} + a^{13}A^{23}\\
&\quad + G_2a^{11} - 2G_1a^{12} - G_2a^{22} - G_3a^{23},\\
3a_2^{11} &= -2a^{12}A^{22} + 2(a^{22} - a^{11})A^{12} + 2a^{23}A^{13} - 2a^{13}A^{23}\\
&\quad - 2G_2a^{11} + G_1a^{12} - G_2a^{22} - G_3a^{23},\\
3a_3^{13} &= -a^{12}C^{23} + (a^{33} - a^{11})C^{13} + a^{23}C^{12} - a^{13}C^{33}\\
&\quad - G_1a^{11} - G_2a^{12} - 2G_3a^{13} + G_1a^{33},\\
3a_1^{33} &= 2a^{12}C^{23} - 2(a^{33} - a^{11})C^{13} - 2a^{23}C^{12} + 2a^{13}C^{33}\\
&\quad - G_1a^{11} - G_2a^{12} + G_3a^{13} - 2G_1a^{33},\\
3a_2^{23} &= a^{23}(B^{33} - B^{22}) - (a^{33} - a^{22})B^{23} - a^{13}B^{12} + a^{12}B^{13}\\
&\quad - G_1a^{13} - 2G_2a^{23} - G_3a^{33} + G_3a^{22},\\
3a_3^{22} &= -2a^{23}(B^{33} - B^{22}) + 2(a^{33} - a^{22})B^{23} + 2a^{13}B^{12} - 2a^{12}B^{13}\\
&\quad - G_1a^{13} + G_2a^{23} - G_3a^{33} - 2G_3a^{22},\\
3a_1^{13} &= -a^{23}A^{12} + (a^{11} - a^{33})A^{13} + a^{13}A^{33} + a^{12}A^{23}\\
&\quad - 2G_1a^{13} - G_2a^{23} - G_3a^{33} + G_3a^{11},\\
3a_3^{11} &= 2a^{23}A^{12} + 2(a^{33} - a^{11})A^{13} - 2a^{13}A^{33} - 2a^{12}A^{23}\\
&\quad + G_1a^{13} - G_2a^{23} - G_3a^{33} - 2G_3a^{11},\\
3a_2^{33} &= -2a^{13}C^{12} + 2(a^{22} - a^{33})C^{23} + 2a^{12}C^{13} - 2a^{23}(C^{22} - C^{33})\\
&\quad - G_1a^{12} - G_2a^{22} + G_3a^{23} - 2G_2a^{33},\\
3a_3^{23} &= a^{13}C^{12} - (a^{22} - a^{33})C^{23} - a^{12}C^{13} - a^{23}(C^{33} - C^{22})\\
&\quad - G_1a^{12} - G_2a^{22} - 2G_3a^{23} + G_2a^{33},\\
3a_2^{12} &= -a^{13}B^{23} + (a^{22} - a^{11})B^{12} - a^{12}B^{22} + a^{23}B^{13}\\
&\quad - G_1a^{11} - 2G_2a^{12} - G_3a^{13} + G_1a^{22},\\
3a_1^{22} &= 2a^{13}B^{23} - 2(a^{22} - a^{11})B^{12} + 2a^{12}B^{22} - 2a^{23}B^{13}\\
&\quad - G_1a^{11} + G_2a^{12} - G_3a^{13} - 2G_1a^{22},\\
3a_1^{23} &= a^{12}(B^{23} + C^{22}) + a^{11}(B^{13} + C^{12}) - a^{22}C^{12} - a^{33}B^{13}\\
&\quad + a^{13}(B^{33} + C^{23}) - a^{23}(C^{13} + B^{12})\\
&\quad - 2G_1a^{23} + G_2a^{13} + G_3a^{12},
\end{aligned} \tag{3.8}$$

$$
\begin{aligned}
3a_3^{12} &= a^{12}(-2B^{23}+C^{22}) + a^{11}(C^{12}-2B^{13}) - a^{22}C^{12} + 2a^{33}B^{13} \\
&\quad + a^{13}(-2B^{33}+C^{23}) + a^{23}(-C^{13}+2B^{12}) \\
&\quad - 2G_3a^{12} + G_2a^{13} + G_1a^{23}, \\
3a_2^{13} &= a^{12}(B^{23}-2C^{22}) + a^{11}(B^{13}-2C^{12}) + 2a^{22}C^{12} - a^{33}B^{13} \\
&\quad + a^{13}(B^{33}-2C^{23}) + a^{23}(2C^{13}-B^{12}) \\
&\quad - 2G_2a^{13} + G_1a^{23} + G_3a^{12},
\end{aligned}
$$

plus the linear relations (3.5). Using the linear relations we can express $C^{12}, C^{13}, C^{22}, C^{23}$ and B^{13} in terms of the remaining 10 functions.

Since the above system of first order partial differential equations is involutive the general solution for the 6 functions a^{jk} can depend on at most 6 parameters, the values $a^{jk}(\mathbf{x}_0)$ at a fixed regular point $\mathbf{x}_0$. For the integrability conditions we define the vector-valued function

$$
\mathbf{h}(x,y,z) = \left(a^{11}, a^{12}, a^{13}, a^{22}, a^{23}, a^{33}\right)^{\mathrm{T}}
$$

and directly compute the 6×6 matrix functions $\mathcal{A}^{(j)}$ to get the first-order system

$$
\partial_{x_j}\mathbf{h} = \mathcal{A}^{(j)}\mathbf{h}, \qquad j = 1,2,3.
$$

The integrability conditions for this system are

$$
\mathcal{A}_i^{(j)}\mathbf{h} - \mathcal{A}_j^{(i)}\mathbf{h} = \mathcal{A}^{(i)}\mathcal{A}^{(j)}\mathbf{h} - \mathcal{A}^{(j)}\mathcal{A}^{(i)}\mathbf{h} \equiv [\mathcal{A}^{(i)}, \mathcal{A}^{(j)}]\mathbf{h}. \tag{3.9}
$$

By assumption we have 5 functionally linearly independent symmetries, so at each regular point the solutions sweep out a 5 dimensional subspace of the 6 dimensional space of symmetric matrices. However, from the conditions derived above there seems to be no obstruction to construction of a 6 dimensional space of solutions. Indeed in [20] we show that this construction can always be done.

THEOREM 3.1. ($5 \Longrightarrow 6$) *Let V be a nondegenerate potential corresponding to a conformally flat space in 3 dimensions that is superintegrable, i.e., suppose V satisfies the equations (3.3) whose integrability conditions hold identically, and there are 5 functionally independent constants of the motion. Then the space of second order symmetries for the Hamiltonian $\mathcal{H} = (p_x^2+p_y^2+p_z^2)/\lambda(x,y,z) + V(x,y,z)$ (excluding multiplication by a constant) is of dimension $D = 6$.*

COROLLARY 3.1. *If $\mathcal{H}+V$ is a superintegrable conformally flat system with nondegenerate potential, then the dimension of the space of 2nd order symmetries*

$$
\mathcal{S} = \sum_{k,j=1}^{3} a^{kj}(x,y,z)p_kp_j + W(x,y,z)
$$

is 6. At any regular point (x_0, y_0, z_0), and given constants $\alpha^{kj} = \alpha^{jk}$, there is exactly one symmetry $\mathcal{S}$ (up to an additive constant) such that $a^{kj}(x_0, y_0, z_0) = \alpha^{kj}$. Given a set of 5 functionally independent 2nd order symmetries $\mathcal{L} = \{\mathcal{S}_\ell : \ell = 1, \cdots 5\}$ associated with the potential, there is always a 6th second order symmetry $\mathcal{S}_6$ that is functionally dependent on $\mathcal{L}$, but linearly independent.

It appears that the functional relationship between these 6 symmetries is always expressible in terms of a polynomial of order 8 in the momenta, but we do not yet have a proof.

As in the 2D case, the key to understanding the structure of the space of constants of the motion for 3D superintegrable systems with nondegenerate potential is an investigation of third order constants of the motion. We have

$$\mathcal{K} = \sum_{k,j,i=1}^{3} a^{kji}(x, y, z) p_k p_j p_i + b^\ell(x, y, z) p_\ell, \tag{3.10}$$

which must satisfy $\{\mathcal{H}, \mathcal{K}\} = 0$. Here the third order Killing tensor a^{kji} is symmetric in the indices k, j, i. We are interested in such third order symmetries that could possibly arise as commutators of second order symmetries. Thus we require that the Killing tensor terms be independent of the four independent parameters in V. However, the b^ℓ must depend on these parameters. We set

$$b^\ell(x, y, z) = \sum_{j=1}^{3} f^{\ell,j}(x, y, z) V_j(x, y, z). \tag{3.11}$$

In [20] the following result is obtained.

THEOREM 3.2. *Let $\mathcal{K}$ be a third order constant of the motion for a conformally flat superintegrable system with nondegenerate potential V. Then $f^{\ell,j} + f^{j,\ell} = 0, \quad 1 \le \ell, j \le 3$. The a^{ijk}, b^ℓ are uniquely determined by the four numbers $f^{1,2},\ f^{1,3},\ f^{2,3},\ f_3^{1,2}$ at any regular point (x_0, y_0, z_0) of V.*

Let

$$\mathcal{S}_1 = \sum a^{kj}_{(1)} p_k p_j + W_{(1)}, \quad \mathcal{S}_2 = \sum a^{kj}_{(2)} p_k p_j + W_{(2)}$$

be second order constants of the the motion for a superintegrable system with nondegenerate potential and let $\mathcal{A}_{(i)}(x, y, z) = \{a^{kj}_{(i)}(x, y, z)\}$, $i = 1, 2$ be 3×3 matrix functions. Then the Poisson bracket of these symmetries is given by

$$\{\mathcal{S}_1, \mathcal{S}_2\} = \sum_{k,j,i=1}^{3} a^{kji}(x, y, z) p_k p_j p_i + b^\ell(x, y, z) p_\ell$$

where

$$f^{k,\ell} = 2\lambda \sum_j (a^{kj}_{(2)} a^{j\ell}_{(1)} - a^{kj}_{(1)} a^{j\ell}_{(2)}).$$

Differentiating, we find

$$\begin{aligned} f^{k,\ell}_i = 2\lambda \sum_j \Big(\partial_i a^{kj}_{(2)} a^{j\ell}_{(1)} + a^{kj}_{(2)} \partial_i a^{j\ell}_{(1)} \\ - \partial_i a^{kj}_{(1)} a^{j\ell}_{(2)} - a^{kj}_{(1)} \partial_i a^{j\ell}_{(2)} \Big) + G_i f^{k,\ell}. \end{aligned} \tag{3.12}$$

Thus, there is a standard structure allowing the identification of the space of second order constants of the motion with the space S_3 of 3×3 symmetric matrices, as well as identification of the space of third order constants of the motion with a subspace of the space $K_3 \times F$ of 3×3 skew-symmetric matrices K_3. crossed with the line. $F = \{\mathcal{F}(\mathbf{x}_0)\}$. A consequence of these results is [20].

COROLLARY 3.2. *Let V be a superintegrable nondegenerate potential on a conformally flat space. Then the space of third order constants of the motion is 4-dimensional and is spanned by Poisson brackets of the second order constants of the motion.*

Let $\mathcal{E}^{ij}$ be the 3×3 matrix with a 1 in row i, column j and 0 for every other matrix element. Then the symmetric matrices

$$\mathcal{A}^{(ij)} = \frac{1}{2}(\mathcal{E}^{ij} + \mathcal{E}^{ji}) = \mathcal{A}^{(ji)}, \qquad i,j = 1,2,3 \tag{3.13}$$

form a basis for the 6-dimensional space of symmetric matrices.

COROLLARY 3.3. *We can define a standard set of 6 second order basis symmetries*

$$\mathcal{S}^{(jk)} = \sum a^{hs}_{(jk)}(\mathbf{x}) p_h p_s + W^{(jk)}(\mathbf{x})$$

corresponding to a regular point $\mathbf{x}_0$ by $(a_{(jk)})(\mathbf{x}_0) = \mathcal{A}^{(jk)}$, $W^{(jk)}(\mathbf{x}_0) = 0$.

In [20] we proved the following.

THEOREM 3.3. *The dimension of the space of fourth order symmetries for a nondegenerate 3D potential is* 21. *The dimension of the space of sixth order symmetries is* 56.

THEOREM 3.4. *The 21 distinct standard monomials $\mathcal{S}^{(ij)}\mathcal{S}^{(jk)}$, defined with respect to a regular point $\mathbf{x}_0$, form a basis for the space of fourth order symmetries.*

THEOREM 3.5. *The 56 distinct standard monomials $\mathcal{S}^{(hi)}\mathcal{S}^{(jk)}\mathcal{S}^{(\ell m)}$, defined with respect to a regular $\mathbf{x}_0$, form a basis for the space of sixth order symmetries.*

We conclude that the quadratic algebra closes.

From the general theory of variable separation for Hamilton-Jacobi equations [16, 17] we know that second order symmetries $\mathcal{S}_1, \mathcal{S}_2$ define a separable coordinate system for the equation

$$H = \frac{p_x^2 + p_y^2 + p_z^2}{\lambda(x,y,z)} + V(x,y,z) = E$$

if and only if

1. The symmetries $\mathcal{H}, \mathcal{S}_1, \mathcal{S}_2$ form a linearly independent set as quadratic forms.
2. $\{\mathcal{S}_1, \mathcal{S}_2\} = 0$.
3. The three quadratic forms have a common eigenbasis of differential forms.

This last requirement means that, expressed in coordinates x, y, z, at least one of the matrices $\mathcal{A}_{(j)}(\mathbf{x})$ can be diagonalized by conjugacy transforms in a neighborhood of a regular point and that $[\mathcal{A}_{(2)}(\mathbf{x}), \mathcal{A}_{(1)}(\mathbf{x})] = 0$. However, for nondegenerate superintegrable potentials in a conformally flat space we see that

$$\{\mathcal{S}_1, \mathcal{S}_2\} = 0 \Longleftrightarrow [\mathcal{A}_{(2)}(\mathbf{x}_0), \mathcal{A}_{(1)}(\mathbf{x}_0)] = 0, \quad \text{and} \quad \mathcal{F}(\mathbf{x}_0) = 0$$

so that the intrinsic conditions for the existence of a separable coordinate system are simplified.

THEOREM 3.6. *Let V be a superintegrable nondegenerate potential in a 3D conformally flat space. Then V defines a multiseparable system.*

The Stäckel transform for 3D systems [42], or coupling constant metamorphosis [43], can be constructed in exact analogy with the 2D case.

THEOREM 3.7. *Every superintegrable system with nondegenerate potential on a 3D conformally flat space is Stäckel equivalent to a superintegrable system on either 3D flat space or the 3-sphere.*

See [21] for the details of the proofs. In this same reference we exhibited 9 superintegrable systems in Euclidean space and another on the sphere that was not Stäckel equivalent to any of these. Eight of these systems are "generic" in the sense that they correspond to generic Jacobi separable coordinates and are uniquely determined by this correspondence.

The quantization of 3D results is carried out in [22]. For a manifold with metric $ds^2 = \lambda(x,y,z)(dx^2 + dy^2 + dz^2)$ the Hamiltonian system $\mathcal{H} = (p_1^2 + p_2^2 + p_3^2)/\lambda(x,y,z) + V(x,y,z)$ is replaced by the Hamiltonian (Schrödinger) operator with potential

$$H = \sum_{i=1}^{3} \frac{1}{\lambda^{\frac{3}{2}}} \partial_i(\lambda^{\frac{1}{2}} \partial_i) - \frac{1}{8} R + V,$$

where R is the scalar curvature. Similarly a second order constant of the motion is replaced by the formally self-adjoint symmetry operator

$$S = \sum_{i,j=1}^{3} \left(\frac{1}{\lambda^{\frac{3}{2}}} \partial_i(a^{ij} \lambda^{\frac{3}{2}} \partial_j) + a^{ij}(\mathcal{R}_{ij} + 5\mathcal{R}_i\mathcal{R}_j) + a_i^{ij}\mathcal{R}_j \right) + W,$$

where $\mathcal{R} = \frac{1}{4}\ln\lambda$. Both of these expressions can be written in covariant form.

Through use of the self-adjointness of even order operator symmetries and the skew adjointness of odd order symmetries, it follows again that the main classical results for symmetries corresponding to a nondegenerate potential can be taken over with relatively little change, except that the potential is modified by bits of the metric. In particular there is closure of the quadratic algebra for 3D quantum superintegrable potentials: All fourth order and sixth order symmetry operators can be expressed as symmetric polynomials in the second order symmetry operators.

4. Fine structure of superintegrable systems. For fine structure of superintegrable systems we drop the requirement of nondegeneracy and study the various possibilities for systems with potentials depending on fewer parameters. For 2D systems the structure is very simple.

THEOREM 4.1. *Every 2D system with a two-parameter potential and 3 functionally linearly independent second-order symmetries is the restriction of some nondegenerate (three-parameter) potential. Every 2D system with a one-parameter potential and 3 functionally linearly independent second-order symmetries is the restriction of some nondegenerate to a single parameter, such that the restricted potential is annihilated by some Killing vector of the underlying space.*

For 3D systems the results are much more complicated and have not yet been fully determined. We announce some new results here whose detailed proofs will appear subsequently. We first consider those systems that just fail to be nondegenerate in the sense that the four functions $\mathcal{S}_h$ together with $\mathcal{H}$ are functionally linearly independent in the six-dimensional phase space but that the associated potential functions V span only a 3 dimensional subspace of the 4 dimensional space of solutions of equations (3.3), ignoring the trivial added constant. In particular, we stipulate that we can arbitrarily prescribe V_1, V_2, V_3 at a regular point, but not V_{11} independently of these. This circumstance can occur in only two ways: either the potential is a 3-parameter restriction of a nondegenerate potential, or the integrability conditions for the system (3.3) are not satisfied identically and an additional condition is imposed. In either case equations (3.3) are replaced by the 6 equations

$$V_{ij} = \tilde{A}^{ij}V_1 + \tilde{B}^{ij}V_2 + \tilde{C}^{ij}V_3, \qquad i \leq j, \tag{4.1}$$

whose integrability conditions are satisfied identically. Equations (3.3) still hold, but with the identifications

$$D^{ij} = \tilde{D}^{ij},\ 1 \leq i < j \leq 3, \qquad D^{kk} = \tilde{D}^{kk} - \tilde{D}^{11},\ k = 2, 3,$$

where $D = A, B, C$. For short, we will call the solutions of (4.1) **3-parameter potentials**. In analogy to the nondegenerate potential case we can compute the full set of integrability conditions satisfied by the potential, and we can use the 10 second order Killing tensor equations and the $3 \times 3 = 9$ conditions for the derivatives $a^{\ell m}_h$ that result from substituting relations (4.1) into the 3 B-D equations and equating coefficients of V_1, V_2, V_3, respectively. Thus there are 19 conditions for the 18 derivatives $a^{\ell m}_h$. We get exactly equations (3.8) and the remaining condition

$$\begin{aligned}&a^{11}(\tilde{C}^{12}-\tilde{B}^{13})+a^{22}(\tilde{A}^{23}-\tilde{C}^{12})+a^{33}(\tilde{B}^{13}-\tilde{A}^{23})\\&+a^{12}(\tilde{A}^{13}+\tilde{C}^{22}-\tilde{C}^{11}-\tilde{B}^{23})+a^{13}(\tilde{C}^{23}+\tilde{B}^{11}-\tilde{B}^{33}-\tilde{A}^{12})\\&+a^{23}(\tilde{B}^{12}+\tilde{A}^{33}-\tilde{A}^{22}-\tilde{C}^{13})=0,\end{aligned} \tag{4.2}$$

which we can regard as an obstruction to extending the assumed 5 dimensional space of second order symmetries to the full 6 dimensional space. Note that the analogous obstruction equation appears for the nondegenerate potential case, but there the linear integrability conditions (3.5) for the nondegenerate potential cause the obstruction to vanish identically. By exploitation of the integrability conditions for the potential and for equations (3.8) we have obtained the following results:

THEOREM 4.2. *A 3D 3-parameter potential is a restriction of a nondegenerate potential if and only if the obstruction (4.2) vanishes identically. If the obstruction doesn't vanish then the space of second order symmetries is 5 dimensional and the system is uniquely determined by the values of* $\tilde{D}^{ij}, i \le j$, $D = A, B, C$ *at a single regular point.*

The extended Kepler-Coulomb system (1.1) is an example of a 3-parameter potential with obstruction, as are two other real Euclidean space potentials in Evans' list [2]. Another example is defined by the potential

$$V = \frac{\alpha}{\sqrt{x^2+y^2+z^2}} + \frac{\beta}{(x+iy)^2} + \frac{\gamma(x-iy)}{(x+iy)^3}.$$

These are true 3-parameter potentials in the sense that they cannot be extended to nondegenerate potentials.

Third order constants of the motion for a true 3-parameter potential superintegrable system again take the form (3.10), (3.11) where $f^{\ell,j}+f^{j,\ell}=0, \quad 1 \le \ell, j \le 3$.

THEOREM 4.3. *For a true 3-parameter system the* a^{ijk}, b^ℓ *are uniquely determined by the three numbers* $f^{1,2}$, $f^{1,3}$, $f^{2,3}$, *at any regular point* (x_0, y_0, z_0) *of* V.

COROLLARY 4.1. *Let* V *be a superintegrable true 3-parameter potential on a conformally flat space. Then the space of third order constants of the motion is 3-dimensional and is spanned by Poisson brackets of the second order constants of the motion. The Poisson bracket of two second order constants of the motion is uniquely determined by the matrix commutator of the second order constants at a regular point.*

THEOREM 4.4. *Let* V *be a superintegrable true 3-parameter potential in a 3D conformally flat space. Then* V *defines a multiseparable system.*

THEOREM 4.5. *Every superintegrable system with true 3-parameter potential on a 3D conformally flat space is Stäckel equivalent to a superintegrable system on either 3D flat space or the 3-sphere.*

Although the spaces of higher order symmetries for true 3-parameter systems have an interesting structure, the quadratic algebra doesn't close.

THEOREM 4.6. *For a superintegrable system with true 3-parameter potential on a 3D conformally flat space there exist two second order constants of the motion $\mathcal{S}_1, \mathcal{S}_2$ such that $\{\mathcal{S}_1, \mathcal{S}_2\}^2$ is not expressible as a cubic polynomial in the second order constants of the motion.*

5. Polynomial ideals. In this section we introduce a very different way of studying and classifying superintegrable systems, through polynomial ideals. Here we confine our analysis to 3D Euclidean superintegrable systems with nondegenerate potentials, though the approach is also effective in 2D and for spheres. The equations for the second order symmetries in this case are just (3.8) with $G \equiv 0$. Due to the linear conditions (3.5) all of the functions A^{ij}, B^{ij}, C^{ij} can be expressed in terms of the 10 basic terms

$$(A^{12}, A^{13}, A^{22}, A^{23}, A^{33}, B^{12}, B^{22}, B^{23}, B^{33}, C^{33}). \tag{5.1}$$

Since the equations (3.8) admit 6 linearly independent solutions a^{hk} the integrability conditions $\partial_i a_\ell^{hk} = \partial_\ell a_i^{hk}$ for these equations must be satisfied identically. As follows from [20], these conditions plus the integrability conditions (3.7) for the potential allow us to compute the 30 derivatives $\partial_\ell D^{ij}$ of the 10 basic terms. Each is a quadratic polynomial in the 10 terms. In addition there are 5 quadratic conditions remaining [20]:

$$\begin{aligned}
&a) \quad -A^{23}B^{33} - A^{12}A^{23} + A^{13}B^{12} + B^{22}A^{23} + B^{23}A^{33} - A^{22}B^{23} &= 0,\\
&b) \quad \left(A^{33}\right)^2 + B^{12}A^{33} - A^{33}A^{22} - B^{33}A^{12} - C^{33}A^{13} + B^{22}A^{12}&\\
&\qquad -B^{12}A^{22} + A^{13}B^{23} - \left(A^{12}\right)^2 + &\\
&c) \quad -\left(B^{33}\right)^2 - B^{33}A^{12} + B^{33}B^{22} + B^{12}A^{33} + B^{23}C^{33} - \left(B^{23}\right)^2&\\
&\qquad + \left(B^{12}\right)^2 &= 0,\\
&d) \quad -B^{12}A^{23} - A^{33}A^{23} + A^{13}B^{33} + A^{12}B^{23} &= 0,\\
&e) \quad A^{12}B^{12} + C^{33}A^{23} - A^{23}B^{23} + B^{33}A^{22} - B^{33}A^{33} &= 0.
\end{aligned}$$

These 5 polynomials determine an ideal Σ'. Already we see that the values of the 10 terms at a fixed regular point must uniquely determine a superintegrable system. However, choosing those values such that the 5 conditions (5.2) are satisfied will not guarantee the existence of a solution, because the conditions may be violated for values of (x, y, z) away from the chosen regular point. To test this we compute the derivatives $\partial_i \Sigma'$ and obtain a single new condition, the square of the quadratic expression

$$\begin{aligned}
f) \quad & A^{13}C^{33} + 2A^{13}B^{23} + B^{22}B^{33} - (B^{33})^2\\
& + A^{33}A^{22} - (A^{33})^2 + 2A^{12}B^{22} + (A^{12})^2\\
& - 2B^{12}A^{22}(B^{12})^2 + B^{23}C^{33} - (B^{23})^2 - 3(A^{23})^2 = 0.
\end{aligned} \tag{5.2}$$

The polynomial (5.2) extends the ideal. Let Σ be the ideal generated by the 6 quadratic polynomials. It can be verified that $\partial_i\Sigma \subseteq \Sigma$, so that the system is closed under differentiation! This leads us to a fundamental result.

THEOREM 5.1. *Choose the 10-tuple (5.1) at a regular point, such that the 6 polynomial identities (5.2), (5.2) are satisfied. Then there exists one and only one Euclidean superintegrable system with nondegenerate potential that takes on these values at a point.*

We see that all possible nondegenerate 3D Euclidean superintegrable systems are encoded into the 6 quadratic polynomial identities. These identities define an algebraic variety that generically has dimension 6, though there are singular points, such as the origin $(0, \cdots, 0)$, where the dimension of the tangent space is greater. This result gives us the means to classify all superintegrable systems.

An issue is that many different 10-tuples correspond to the same superintegrable system. How do we sort this out? The key is that the Euclidean group E(3,C) acts as a transformation group on the variety and gives rise to a foliation. The action of the translation subgroup is determined by the derivatives $\partial_k D^{ij}$ that we have already determined. The action of the rotation subgroup on the D^{ij} can be determined from the behavior of the canonical equations (3.3) under rotations. The local action on a 10-tuple is then given by 6 Lie derivatives that are a basis for the Euclidean Lie algebra $e(3, C)$. For "most" 10-tuples $\mathbf{D}_0$ on the 6 dimensional variety the action of the Euclidean group is locally transitive with isotropy subgroup only the identity element. Thus the group action on such points sweeps out a solution surface homeomorphic to the 6 parameter $E(3, C)$ itself. This occurs for the generic Jacobi elliptic system with potential

$$V = \alpha(x^2 + y^2 + z^2) + \frac{\beta}{x^2} + \frac{\gamma}{y^2} + \frac{\delta}{z^2}.$$

At the other extreme the isotropy subgroup of the origin $(0, \cdots, 0)$ is $E(3, C)$ itself, i.e., the point is fixed under the group action. This corresponds to the isotropic oscillator with potential

$$V = \alpha(x^2 + y^2 + z^2) + \beta x + \gamma y + \delta z.$$

More generally, the isotropy subgroup at $\mathbf{D}_0$ will be H and the Euclidean group action will sweep out a solution surface homeomorphic to the homogeneous space $E(3, C)/H$ and define a unique superintegrable system. For example, the isotropy subalgebra formed by the translation and rotation generators $\{P_1, P_2, P_3, J_1 + iJ_2\}$ determines a system with potential

$$\alpha\left((x - iy)^3 + 6(x^2 + y^2 + z^2)\right) + \beta\left((x - iy)^2 + 2(x + iy)\right) + \gamma(x - iy) + \delta z.$$

Indeed, each class of Stäckel equivalent Euclidean superintegrable systems is associated with a unique isotropy subalgebra of $e(3, C)$, although not all subalgebras occur. (Indeed, there is no isotropy subalgebra conjugate to $\{P_1, P_2, P_3\}$.) Thus to find all superintegrable systems we need to determine a list of all subalgebras of $e(3, C)$, defined up to conjugacy, and then for each subalgebra to determine if it occurs as an isotropy subalgebra. Further we must resolve the degeneracy problem in which more than one superintegrable system may correspond to a single isotropy subalgebra.

6. Outlook. We have given an overview of some of the tools used and results obtained in the study of second order superintegrable systems. The basic problems for 2D systems have been solved, and the extension of these methods to complete the fine structure analysis for 3D systems appears relatively straightforward. The 3D fine structure analysis can be extended to analyze 2 parameter and 1 parameter potentials with 5 functionally linearly independent second order symmetries. Here first order PDEs for the potential appear as well as second order, and Killing vectors may occur. The other class of 3D superintegrable systems is that for which the 5 functionally independent symmetries are functionally linearly dependent. This class is related to the Calogero potential [44–46] and necessarily leads to first order PDEs for the potential, as well as second order [22]. However, the integrability methods discussed here should be able to handle this class with no special difficulties. On a deeper level, we think that the algebraic geometry methods of the last section can be extended to classify the possible superintegrable systems in all these cases.

Whereas 2D superintegrable systems are very special, the 3D systems seem to be good guides to the structure of general nD systems, and we intend to proceed with this analysis. The ultimate aim is to understand the structure of superintegrable systems in general. We have started with second order systems because of their historical connection to the Kepler-Coulomb problem and to separation of variables. However, most of the methods that we have developed make use of integrability conditions alone, not separation of variables (a purely second order phenomenon) and show promise of being extendable to higher order superintegrable systems.

Finally, the algebraic geometry related results that we have sketched in the last section suggest strongly that there is an underlying geometric structure to superintegrable systems that is not apparent from the usual presentations of these systems.

REFERENCES

[1] WOJCIECHOWSKI S., Superintegrability of the Calogero-Moser System. *Phys. Lett.*, 1983, A **95**: 279–281.

[2] EVANS N.W., Superintegrability in Classical Mechanics; *Phys. Rev.* 1990, A **41**, 5666–5676; Group Theory of the Smorodinsky-Winternitz System; *J. Math. Phys.* 1991, **32**: 3369.

[3] EVANS N.W., Super-Integrability of the Winternitz System; *Phys. Lett.* 1990, A **147**: 483–486.

[4] FRIŠ J., MANDROSOV V., SMORODINSKY YA.A, UHLÍR M., AND WINTERNITZ P., On Higher Symmetries in Quantum Mechanics; *Phys. Lett.* 1965, **16**: 354–356.

[5] FRIŠ J., SMORODINSKII YA.A., UHLÍR M., AND WINTERNITZ P., Symmetry Groups in Classical and Quantum Mechanics; *Sov. J. Nucl. Phys.* 1967, **4**: 444–450.

[6] MAKAROV A.A., SMORODINSKY YA.A., VALIEV KH., AND WINTERNITZ P., A Systematic Search for Nonrelativistic Systems with Dynamical Symmetries. *Nuovo Cimento*, 1967, **52**: 1061–1084.

[7] CALOGERO F., Solution of a Three-Body Problem in One Dimension. *J. Math. Phys.* 1969, **10**: 2191–2196.

[8] CISNEROS A. AND MCINTOSH H.V., Symmetry of the Two-Dimensional Hydrogen Atom. *J. Math. Phys.* 1969, **10**: 277–286.

[9] SKLYANIN E.K., Separation of variables in the Gaudin model. *J. Sov. Math.* 1989, **47**: 2473–2488.

[10] FADDEEV L.D. AND TAKHTAJAN L.A., Hamiltonian Methods in the Theory of Solitons *Springer*, Berlin 1987.

[11] HARNAD J., Loop groups, R-matrices and separation of variables. In "Integrable Systems: From Classical to Quantum" J. Harnad, G. Sabidussi and P. Winternitz eds. CRM Proceedings and Lecture Notes, **26**: 21–54, 2000.

[12] EISENHART L.P., *Riemannian Geometry.* Princeton University Press, 2nd printing, 1949.

[13] MILLER W.JR., Symmetry and Separation of Variables. *Addison-Wesley Publishing Company*, Providence, Rhode Island, 1977.

[14] KALNINS E.G. AND MILLER W.JR., Killing tensors and variable separation for Hamilton-Jacobi and Helmholtz equations. *SIAM J. Math. Anal.*, 1980, **11**: 1011–1026.

[15] MILLER W., The technique of variable separation for partial differential equations. Proceedings of School and Workshop on Nonlinear Phenomena, Oaxtepec, Mexico, November 29 - December 17, 1982, Lecture Notes in Physics, **189**: Springer-Verlag, New York, 1983.

[16] KALNINS E.G., *Separation of Variables for Riemannian Spaces of Constant Curvature*, Pitman, Monographs and Surveys in Pure and Applied Mathematics, **28**: 184–208, Longman, Essex, England, 1986.

[17] MILLER W.JR., Mechanisms for variable separation in partial differential equations and their relationship to group theory. In *Symmetries and Non-linear Phenomena* pp. 188–221, World Scientific, 1988.

[18] KALNINS E.G., KRESS J.M, AND MILLER W.JR., Second order superintegrable systems in conformally flat spaces. I: 2D classical structure theory. *J. Math. Phys.*, 2005, **46**: 053509.

[19] KALNINS E.G., KRESS J.M, AND MILLER W.JR., Second order superintegrable systems in conformally flat spaces. II: The classical 2D Stäckel transform. *J. Math. Phys.*, 2005, **46**: 053510.

[20] KALNINS E.G., KRESS J.M, AND MILLER W.JR., Second order superintegrable systems in conformally flat spaces. III: 3D classical structure theory. *J. Math. Phys.*, 2005, **46**: 103507.

[21] KALNINS E.G., KRESS J.M, AND MILLER W.JR., Second order superintegrable systems in conformally flat spaces. IV: The classical 3D Stäckel transform and 3D classification theory. *J. Math. Phys.*, 2006, **47**: 043514.

[22] KALNINS E.G., KRESS J.M, AND MILLER W.JR., Second order superintegrable systems in conformally flat spaces. V: 2D and 3D quantum systems. *J. Math. Phys.*, 2006, **47**: 093501.

[23] KALNINS E.G., MILLER W.JR. AND POGOSYAN G.S., Superintegrability in three dimensional Euclidean space. *J. Math. Phys.*, 1999, **40**: 708–725.

[24] KALNINS E.G., MILLER W.JR., AND POGOSYAN G.S., Superintegrability and associated polynomial solutions. Euclidean space and the sphere in two dimensions. *J.Math.Phys.*, 1996, **37**: 6439–6467.

[25] BONATOS D., DASKALOYANNIS C., AND KOKKOTAS K., Deformed Oscillator Algebras for Two-Dimensional Quantum Superintegrable Systems; *Phys. Rev.*, 1994, A **50**: 3700–3709.

[26] DASKALOYANNIS C., Quadratic Poisson algebras of two-dimensional classical superintegrable systems and quadratic associate algebras of quantum superintegrable systems. *J. Math. Phys.*, 2001, **42**: 1100–1119.

[27] SMITH S.P., A class of algebras similar to the enveloping algebra of $sl(2)$. *Trans. Amer. Math. Soc.*, 1990, **322**, 285–314.

[28] KALNINS E.G., MILLER W., AND TRATNIK M.V., Families of orthogonal and biorthogonal polynomials on the n-sphere. *SIAM J. Math. Anal.*, 1991, **22**: 272–294.

[29] USHVERIDZE A.G., *Quasi-Exactly solvable models in quantum mechanics.* Institute of Physics, Bristol, 1993.

[30] LETOURNEAU P. AND VINET L., Superintegrable systems: Polynomial Algebras and Quasi-Exactly Solvable Hamiltonians. *Ann. Phys.*, 1995, **243**: 144–168.

[31] KALNINS E.G., MILLER W.JR., AND POGOSYAN G.S., Exact and quasi-exact solvability of second order superintegrable systems. I. Euclidean space preliminaries (submitted).

[32] GROSCHE C., POGOSYAN G.S., AND SISSAKIAN A.N., Path Integral Discussion for Smorodinsky-Winternitz Potentials: I. Two- and Three Dimensional Euclidean Space. *Fortschritte der Physik*, 1995, **43**: 453–521.

[33] KALNINS E.G., KRESS J.M., MILLER W.JR., AND POGOSYAN G.S., *Completeness of superintegrability in two-dimensional constant curvature spaces.* *J. Phys. A: Math. Gen.*, 2001, **34**: 4705–4720.

[34] KALNINS E.G., KRESS J.MN., AND WINTERNITZ P., *Superintegrability in a two-dimensional space of non-constant curvature.* *J. Math. Phys.*, 2002, **43**: 970–983.

[35] KALNINS E.G., KRESS J.M., MILLER, W.JR., AND WINTERNITZ P., *Superintegrable systems in Darboux spaces.* *J. Math. Phys.*, 2003, **44**: 5811–5848.

[36] RAÑADA M.F., Superintegrable n=2 systems, quadratic constants of motion, and potentials of Drach. *J. Math. Phys.*, 1997, **38**: 4165–4178.

[37] KALNINS E.G., MILLER W.JR., WILLIAMS G.C., AND POGOSYAN G.S., On superintegrable symmetry-breaking potentials in n-dimensional Euclidean space. *J. Phys. A: Math. Gen.*, 2002, **35**: 4655–4720.

[38] KALNINS E.G., MILLER W.JR., AND POGOSYAN G.S., Completeness of multiseparable superintegrability in $E_{2,C}$. *J. Phys. A: Math. Gen.*, 2000, **33**: 4105.

[39] KALNINS E.G., MILLER W.JR., AND POGOSYAN G.S., Completeness of multiseparable superintegrability on the complex 2-sphere. *J. Phys. A: Math. Gen.*, 2000, **33**: 6791–6806.

[40] KOENIGS G., Sur les géodésiques a intégrales quadratiques. A note appearing in "Lecons sur la théorie générale des surfaces". G. Darboux., **4**: 368–404, *Chelsea Publishing*, 1972.

[41] DASKALOYANNIS C. AND YPSILANTIS K., Unified treatment and classification of superintegrable systems with integrals quadratic in momenta on a two dimensional manifold. (Preprint) (2005).

[42] BOYER C.P., KALNINS E.G., AND MILLER W., *Stäckel - equivalent integrable Hamiltonian systems.* *SIAM J. Math. Anal.*, 1986, **17**: 778–797.

[43] HIETARINTA J., GRAMMATICOS B., DORIZZI B., AND RAMANI A., *Coupling-constant metamorphosis and duality between integrable Hamiltonian systems.* *Phys. Rev. Lett.*, 1984, **53**: 1707–1710.

[44] CALOGERO F., Solution to the one-dimensional N-body problems with quadratic and/or inversely quadratic pair potentials. *J. Math. Phys.*, 1971, **12**: 419–436.

[45] Rauch-Wojciechowski S. and Waksjö C., What an effective criterion of separability says about the Calogero type systems. *J. Nonlinear Math. Phys.*, 2005, **12**(1): 535–547.
[46] Horwood J.T., McLenaghan R.G., and Smirnov R.G., Invariant classification of orthogonally separable Hamiltonian systems in Euclidean space. *Comm. Math. Phys.*, 2005, **259**: 679–709.

DIFFERENTIAL GEOMETRY OF SUBMANIFOLDS OF PROJECTIVE SPACE*

JOSEPH M. LANDSBERG†

Abstract. These are lecture notes on the rigidity of submanifolds of projective space "resembling" compact Hermitian symmetric spaces in their homogeneous embeddings. The results of [16, 20, 29, 18, 19, 10, 31] are surveyed, along with their classical predecessors. The notes include an introduction to moving frames in projective geometry, an exposition of the Hwang-Yamaguchi ridgidity theorem and a new variant of the Hwang-Yamaguchi theorem.

1. Overview.

- Introduction to the local differential geometry of submanifolds of projective space.
- Introduction to moving frames for projective geometry.
- How much must a submanifold $X \subset \mathbb{P}^N$ resemble a given submanifold $Z \subset \mathbb{P}^M$ infinitesimally before we can conclude $X \simeq Z$?
- To what order must a line field on a submanifold $X \subset \mathbb{P}^N$ have contact with X before we can conclude the lines are contained in X?
- Applications to algebraic geometry.
- A new variant of the Hwang-Yamaguchi rigidity theorem.
- An exposition of the Hwang-Yamaguchi rigidity theorem in the language of moving frames.

Representation theory and algebraic geometry are natural tools for studying submanifolds of projective space. Recently there has also been progress the other way, using projective differential geometry to prove results in algebraic geometry and representation theory. These talks will focus on the basics of submanifolds of projective space, and give a few applications to algebraic geometry. For further applications to algebraic geometry the reader is invited to consult chapter 3 of [11] and the references therein.

Due to constraints of time and space, applications to representation theory will not be given here, but the interested reader can consult [23] for an overview. Entertaining applications include new proofs of the classification of compact Hermitian symmetric spaces, and of complex simple Lie algebras, based on the geometry of rational homogeneous varieties (instead of root systems), see [22]. The applications are not limited to classical representation theory. There are applications to Deligne's conjectured categorical generalization of the exceptional series [25], to Vogel's proposed

*Supported by NSF grant DMS-0305829.

†Department of Mathematics, Texas A & M University, College Station, TX 77843 (jml@math.tamu.edu).

Universal Lie algebra [27], and to the study of the intermediate Lie algebra $\mathfrak{e}_{7\frac{1}{2}}$ [26].

Notations, conventions. I mostly work over the complex numbers in the complex analytic category, although most of the results are valid in the C^∞ category and over other fields, even characteristic p, as long as the usual precautions are taken. When working over $\mathbb{R}$, some results become more complicated as there are more possible normal forms. I use notations and the ordering of roots as in [2] and label maximal parabolic subgroups accordingly, e.g., P_k refers to the maximal parabolic obtained by omitting the spaces corresponding to the simple root α_k. $\langle v_1, ..., v_k\rangle$ denotes the linear span of the vectors $v_1, ..., v_k$. If $X \subset \mathbb{P}V$ is a subset, $\hat{X} \subset V$ denotes the corresponding cone in $V\backslash 0$, the inverse image of X under the projection $V\backslash 0 \to \mathbb{P}V$. and $\overline{X} \subset \mathbb{P}V$ denotes the Zariski closure of X, the zero set of all the homogeneous polynomials vanishing on X. When we write X^n, we mean $\dim(X) = n$. We often use Id to denote the identity matrix or identity map. Repeated indicies are to be summed over.

Acknowledgements. These notes are based on lectures given at Seoul National University in June 2006, the IMA workshop *Symmetries and overdetermined systems of partial differential equations*, July 2006 and at CIMAT (Guanajuato) August 2006. It is a pleasure to thank Professors Han, Eastwood and Hernandez for inviting me to give these respective lecture series. I would also like to thank Professor Yamaguchi for carefully explaining his results with Hwang to me at the workshop and Professor Robles for reading a draft of this article and providing corrections and suggestions for improvement.

2. Submanifolds of projective space.

2.1. Projective geometry. Let V be a vector space and let $\mathbb{P}V$ denote the associated projective space. We think of $\mathbb{P}V$ as the quotient of $GL(V)$, the general linear group of invertible endomorphisms of V, by the subgroup P_1 preserving a line. For example if we take the line

$$\langle \begin{pmatrix} 1 \\ 0 \\ \vdots \\ 0 \end{pmatrix} \rangle$$

then

$$P_1 = \begin{pmatrix} * & * & \dots & * \\ 0 & * & \dots & * \\ \vdots & * & \dots & * \\ 0 & * & \dots & * \end{pmatrix}$$

where, if $\dim V = N+1$ the blocking is $(1, N) \times (1, N)$, so P_1 is the group of invertible matrices with zeros in the lower left hand block.

In the spirit of Klein, we consider two submanifolds $M_1, M_2 \subset \mathbb{P}V$ to be *equivalent* if there exists some $g \in GL(V)$ such that $g.M_1 = M_2$ and define the corresponding notion of local equivalence.

Just as in the geometry of submanifolds of Euclidean space, we will look for *differential invariants* that will enable us to determine if a neighborhood (germ) of a point of M_1 is equivalent to a neighborhood (germ) of a point of M_2. These invariants will be obtained by taking derivatives at a point in a geometrically meaningful way. Recall that second derivatives furnish a complete set of differential invariants for surfaces in Euclidean three space- the vector-bundle valued Euclidean first and second fundamental forms, and two surfaces are locally equivalent iff there exists a local diffeomorphism $f : M_1 \to M_2$ preserving the first and second fundamental forms.

The group of admissible motions in projective space is larger than the corresponding Euclidean group so we expect to have to take more derivatives to determine equivalence in the projective case than the Euclidean. For example, it was long known that for hypersurfaces $X^n \subset \mathbb{P}^{n+1}$ that one needs at least three derivatives, and Jensen and Musso [12] showed that for most hypersurfaces, when $n > 1$, three derivatives are sufficient. For curves in the plane, by a classical result of Monge, one needs six derivatives!

In order to take derivatives in a way that will facilitate extracting geometric information from them, we will use the *moving frame*. Before developing the moving frame in §3, we discuss a few coarse invariants without machinery and state several rigidity results.

2.2. Asymptotic directions. Fix $x \in X^n \subset \mathbb{P}V$. After taking one derivative, we have the tangent space $T_xX \subset T_x\mathbb{P}V$, which is the set of tangent directions to lines in $\mathbb{P}V$ having contact with X at x to order at least one. Since we are discussing directions, it is better to consider $\mathbb{P}T_xX \subset \mathbb{P}T_x\mathbb{P}V$. Inside $\mathbb{P}T_xX$ is $\mathcal{C}_{2,X,x} \subset \mathbb{P}T_xX$, the set of tangent directions to lines having contact at least two with X at x, these are called the *asymptotic directions* in Euclidean geometry, and we continue to use the same terminology in the projective setting. Continuing, we define $\mathcal{C}_{k,X,x}$ for all k, and finally, $\mathcal{C}_{\infty,X,x}$, which, in the analytic category, equals $\mathcal{C}_{X,x}$, the lines on (the completion of) X through x. When X is understood we sometimes write $\mathcal{C}_{k,x}$ for $\mathcal{C}_{k,X,x}$.

What does $\mathcal{C}_{2,X,x}$, or more generally $\mathcal{C}_{k,X,x}$ tell us about the geometry of X?

That is, what can we learn of the macroscopic geometry of X from the microscopic geometry at a point? To increase the chances of getting meaningful information, from now on, when we are in the analytic or algebraic category, we will work at a *general point.* Loosely speaking, after taking k

derivatives there will be both discrete and continuous invariants. A general point is one where all the discrete invariants are locally constant.

(To be more precise, if one is in the analytic category, one should really speak of k-general points (those that are general to order k), to insure there is just a finite number of discrete invariants. In everything that follows we will be taking just a finite number of derivatives and we should say we are working at a k-general point where k is larger than the number of derivatives we are taking.)

When we are in the C^∞ category, we will work in open subsets and require whatever property we are studying at a point holds at all points in the open subset.

For example, if X^n is a hypersurface, then $\mathcal{C}_{2,X,x}$ is a degree two hypersurface in $\mathbb{P}T_xX$ (we will prove this below), and thus its only invariant is its rank r. In particular, if X is a smooth algebraic variety and $x \in X_{general}$ the rank is n (see e.g., [6, 11]) and thus we do not get much information. (In contrast, if $r < n$, then the Gauss map of X is degenerate and X is (locally) ruled by $\mathbb{P}^{n-r}$'s.)

More generally, if $X^n \subset \mathbb{P}^{n+a}$, then $\mathcal{C}_{2,X,x}$ is the intersection of at most min $(a, \binom{n+1}{2})$ quadric hypersurfaces, and one generally expects that equality holds. In particular, if the codimension is sufficiently large we expect $\mathcal{C}_{2,x}$ to be empty and otherwise it should have codimension a. When this fails to happen, there are often interesting consequences for the macroscopic geometry of X.

2.3. The Segre variety and Griffiths-Harris conjecture. Let A, B be vector spaces and let $V = A \otimes B$. Let

$$X = \mathbb{P}(\text{rank one tensors}) \subset \mathbb{P}V.$$

Recall that every rank one matrix (i.e., rank one tensor expressed in terms of bases) is the matrix product of a column vector with a row vector, and that this representation is unique up to a choice of scale, so when we projectivize (and thus introduce another choice of scale) we obtain

$$X \simeq \mathbb{P}A \times \mathbb{P}B.$$

X is called the *Segre variety* and is often written $X = Seg(\mathbb{P}A \times \mathbb{P}B) \subset \mathbb{P}(A \otimes B)$.

We calculate $\mathcal{C}_{2,x}$ for the Segre. We first must calculate $T_xX \subset T_x\mathbb{P}V$. We identify $T_x\mathbb{P}V$ with V mod $\hat{x}$ and locate T_xX as a subspace of V mod $\hat{x}$.

Let $x = [a_0 \otimes b_0] \in Seg(\mathbb{P}A \times \mathbb{P}B)$. A curve $x(t)$ in X with $x(0) = x$ is given by curves $a(t) \subset A$, $b(t) \subset B$, with $a(0) = a_0, b(0) = b_0$ by taking $x(t) = [a(t) \otimes b(t)]$.

$$\frac{d}{dt}|_{t=0} a_t \otimes b_t = a_0' \otimes b_0 + a_0 \otimes b_0'$$

and thus

$$T_xX = (A/a_0) \otimes b_0 \oplus a_0 \otimes B/b_0 \text{ mod } a_0 \otimes b_0.$$

Write $A' = (A/a_0) \otimes b_0$, $B' = a_0 \otimes (B/b_0)$ so

$$T_xX \simeq A' \oplus B'.$$

We now take second derivatives modulo the tangent space to see which tangent directions have lines osculating to order two (these will be the derivatives that are zero modulo the tangent space).

$$\frac{d^2}{(dt)^2}|_{t=0} a_t \otimes b_t = a_0'' \otimes b_0 + a_0' \otimes b_0' + a_0 \otimes b_0'' \text{ mod } \hat{x} \tag{2.1}$$

$$\equiv a_0' \otimes b_0' \text{ mod } \hat{T}_xX. \tag{2.2}$$

Thus we get zero iff either $a_0' = 0$ or $b_0' = 0$, i.e.,

$$C_{2,X,x} = \mathbb{P}A' \sqcup \mathbb{P}B' \subset \mathbb{P}(A' \oplus B')$$

i.e., $C_{2,X,x}$ is the disjoint union of two linear spaces, of dimensions $\dim A - 2, \dim B - 2$. Note that $\dim C_{2,x}$ is much larger than expected.

For example, consider the case $Seg(\mathbb{P}^2 \times \mathbb{P}^2) \subset \mathbb{P}^8$. Here $C_{2,x}$ is defined by four quadratic polynomials on $\mathbb{P}^3 = \mathbb{P}(T_xX)$, so one would have expected $C_{2,x}$ to be empty. This rather extreme pathology led Griffiths and Harris to conjecture:

CONJECTURE 2.1 (Griffiths-Harris, 1979 [6]). *Let $Y^4 \subset \mathbb{P}^8$ be a variety not contained in a hyperplane and let $y \in Y_{general}$. If $C_{2,Y,y} = \mathbb{P}^1 \sqcup \mathbb{P}^1 \subset \mathbb{P}^3 = \mathbb{P}(T_yY)$, then Y is isomorphic to $Seg(\mathbb{P}^2 \times \mathbb{P}^2)$.*

(The original statement of the conjecture was in terms of the projective second fundamental form defined below.) Twenty years later, in [16] I showed the conjecture was true, and moreover in [16, 20] I showed:

THEOREM 2.1. *Let $X^n = G/P \subset \mathbb{P}V$ be a rank two compact Hermitian symmetric space (CHSS) in its minimal homogeneous embedding, other than a quadric hypersurface. Let $Y^n \subset \mathbb{P}V$ be a variety not contained in a hyperplane and let $y \in Y_{general}$. If $C_{2,Y,y} \simeq C_{2,X,x}$ then Y is projectively isomorphic to X.*

An analogous result is true in the C^∞ category, namely

THEOREM 2.2. *Let $X^n = G/P \subset \mathbb{P}V$ be a rank two compact Hermitian symmetric space (CHSS) in its minimal homogeneous embedding, other than a quadric hypersurface. Let $Y^n \subset \mathbb{P}W$ be a smooth submanifold not contained in a hyperplane. If $C_{2,Y,y} \simeq C_{2,X,x}$ for all $y \in Y$, then Y is projectively isomorphic to an open subset of X.*

The situation of the quadric hypersurface is explained below (Fubini's theorem) - to characterize it, one must have $C_{3,Y,y} = C_{3,Q,x}$.

The rank two CHSS are $Seg(\mathbb{P}A \times \mathbb{P}B)$, the Grassmanians of two-planes $G(2, V)$, the quadric hypersurfaces, the complexified Cayley plane

$\mathbb{OP}^2 = E_6/P_6$, and the spinor variety D_5/P_5 (essentially the isotropic 5-planes through the origin in $\mathbb{C}^{10}$ equipped with a quadratic form - the set of such planes is disconnected and the spinor variety is one (of the two) isormorphic components. The minimal homogenous embedding is also in a smaller linear space than the Plucker embedding of the Grassmannian.) The only rank one CHSS is projective space $\mathbb{P}^n$. The rank two CHSS and projective space are examples of *rational homogeneous varieties.*

2.4. Homogeneous varieties. Let $G \subset GL(V)$ be a reductive group acting irreducibly on a vector space V. Then there exists a unique closed orbit $X = G/P \subset \mathbb{P}V$, which is called a *rational homogeneous variety.* (Equivalently, X may be characterized as the orbit of a highest weight line, or as the minimal orbit.)

Note that if $X = G/P \subset \mathbb{P}V$ is homogenous, T_xX inherits additional structure beyond that of a vector space. Namely, consider $x = [Id]$ as the class of the identity element for the projection $G \to G/P$. Then P acts on T_xX and, as a P-module $T_xX \simeq \mathfrak{g}/\mathfrak{p}$. For example, in the case of the Segre, T_xX was the direct sum of two vector spaces.

A homogeneous variety $X = G/P$ is a *compact Hermitian symmetric space*, or CHSS for short, if P acts irreducibly on T_xX. The *rank* of a CHSS is the number of its last nonzero fundamental form in its minimal homogeneous embedding. (This definition agress with the standard one.) Fundamental forms are defined in §3.4.

EXERCISE 2.1. *The Grassmannian of k-planes through the origin in V, which we denote $G(k, V)$, is homogenous for $GL(V)$ (we have already seen the special case $G(1, V) = \mathbb{P}V$).*

Determine the group $P_k \subset GL(V)$ that stabilizes a point. Show that $T_EG(k, V) \simeq E^* \otimes V/E$ in two different ways - by an argument as in the Segre case above and by determining the structure of $\mathfrak{g}/\mathfrak{p}$.

While all homogeneous varieties have many special properties, the rank at most two CHSS (other than the quadric hypersurface) are distinguished by the following property:

PROPOSITION 2.1. *[21] Theorem 2.1 is sharp in the sense that no other homogeneous variety is completely determined by its asymptotic directions at a general point other than a linearly embedded projective space.*

Nevertheless, there are significant generalizations of Theorem 2.1 due to Hwang-Yamaguchi and Robles discussed below in §3.5. To state these results we will need definitions of the fundamental forms and Fubini cubic forms, which are given in the next section.

However, with an additional hypothesis - namely that the unknown variety has the correct codimension, we obtain the following result (which appears here for the first time):

THEOREM 2.3. *Let $X^n \subset \mathbb{CP}^{n+a}$ be a complex submanifold not contained in a hyperplane. Let $x \in X$ be a general point. Let $Z^n \subset \mathbb{P}^{n+a}$ be an irreducible compact Hermitian symmetric space in its minimal homo-*

geneous embedding, other than a quadric hypersurface. If $C_{2,X,x} = C_{2,Z,z}$ *then* $\overline{X} = Z$.

REMARK 2.1. *The Segre variety has* $C_{2,x} = C_x = \mathbb{P}A' \sqcup \mathbb{P}B' \subset \mathbb{P}(A' \oplus B') = \mathbb{P}T_xX$. *To see this, note that a matrix has rank one iff all its* 2×2 *minors are zero, and these minors provide defining equations for the Segre. In general, if a variety is defined by equations of degree at most d, then any line having contact to order d at any point must be contained in the variety. In fact, by an unpublished result of Kostant, all homogeneously embedded rational homogeneous varieties* G/P *are cut out by quadratic equations so* $C_{2,G/P,x} = C_{G/P,x}$.

3. Moving frames and differential invariants. For more details regarding this section, see chapter 3 of [11].

Once and for all fix index ranges $1 \leq \alpha, \beta, \gamma \leq n$, $n+1 \leq \mu, \nu \leq n+a$, $0 \leq A, B, C \leq n+a = N$.

3.1. The Maurer-Cartan form of $GL(V)$. Let $\dim V = N+1$, denote an element $f \in GL(V)$ by $f = (e_0, ..., e_N)$ where we may think of the e_A as column vectors providing a basis of V. (Once a reference basis of V is fixed, $GL(V)$ is isomorphic to the space of all bases of V.) Each e_A is a V-valued function on $GL(V)$, $e_A : GL(V) \to V$. For any differentiable map between manifolds we can compute the induced differential

$$de_A|_f : T_fGL(V) \to T_{e_A}V$$

but now since V is a vector space, we may identify $T_{e_A}V \simeq V$ and consider

$$de_A : T_fGL(V) \to V$$

i.e., de_A is a V-valued one-form on $GL(V)$. As such, we may express it as

$$de_A = e_0\omega^0_A + e_1\omega^1_A + \cdots + e_N\omega^N_A$$

where $\omega^A_B \in \Omega^1(GL(V))$ are ordinary one-forms (This is because $e_0, ..., e_N$ is a basis of V so any V-valued one form is a linear combination of these with scalar valued one forms as coefficients.) Collect the forms ω^A_B into a matrix $\Omega = (\omega^A_B)$. Write $df = (de_0, ..., de_N)$, so $df = f\Omega$ or

$$\Omega = f^{-1}df$$

Ω is called the *Maurer-Cartan form* for $GL(V)$. Note that ω^A_B measures the infinitesimal motion of e_B towards e_A.

Amazing fact: we can compute the exterior derivative of Ω algebraically! We have $d\Omega = d(f^{-1}) \wedge df$ so we need to calculate $d(f^{-1})$. Here is where an extremely useful fact comes in:

The derivative of a constant function is zero.

We calculate $0 = d(Id) = d(f^{-1}f) = d(f^{-1})f + f^{-1}df$, and thus $d\Omega = -f^{-1}dff^{-1} \wedge df$ but we can move the scalar valued-matrix f^{-1} across the wedge product to conclude

$$d\Omega = -\Omega \wedge \Omega$$

which is called the *Maurer-Cartan equation.* The notation is such that $(\Omega \wedge \Omega)^A_B = \omega^A_C \wedge \omega^C_B$.

3.2. Moving frames for $X \subset \mathbb{P}V$. Now let $X^n \subset \mathbb{P}^{n+a} = \mathbb{P}V$ be a submanifold. We are ready to take derivatives. Were we working in coordinates, to take derivatives at $x \in X$, we might want to choose coordinates such that x is the origin. We will make the analoguous adaptation using moving frames, but the advantage of moving frames is that all points will be as if they were the origin of a coordinate system. To do this, let $\pi : \mathcal{F}^0_X := GL(V)|_X \to X$ be the restriction of $\pi : GL(V) \to \mathbb{P}V$.

Similarly, we might want to choose local coordinates $(x^1, ..., x^{n+a})$ about $x = (0, ..., 0)$ such that T_xX is spanned by $\frac{\partial}{\partial x^1}, , ..., \frac{\partial}{\partial x^n}$. Again, using moving frames the effect will be as if we had chosen such coordinates about each point simultaneously. To do this, let $\pi : \mathcal{F}^1 \to X$ denote the sub-bundle of $\mathcal{F}^0_X$ preserving the flag

$$\hat{x} \subset \hat{T}_xX \subset V.$$

Recall $\hat{x} \subset V$ denotes the line corresponding to x and $\hat{T}_xX$ denotes the affine tangent space $T_v\hat{X} \subset V$, where $[v] = x$. Let $(e_0, ..., e_{n+a})$ be a basis of V with dual basis $(e^0, ..., e^{n+a})$ adapted such that $e_0 \in \hat{x}$ and $\{e_0, e_\alpha\}$ span $\hat{T}_xX$. Write $T = T_xX$ and $N = N_xX = T_x\mathbb{P}V/T_xX$.

REMARK 3.1. *(Aside for the experts) I am slightly abusing notation in this section by identifying $\hat{T}_xX/\hat{x}$ with $T_xX := (\hat{T}_xX/\hat{x}) \otimes \hat{x}^*$ and similarly for N_xX.*

The fiber of $\pi : \mathcal{F}^1 \to X$ over a point is isomorphic to the group

$$G_1 = \left\{ g = \begin{pmatrix} g^0_0 & g^0_\beta & g^0_\nu \\ 0 & g^\alpha_\beta & g^\alpha_\nu \\ 0 & 0 & g^\mu_\nu \end{pmatrix} \Big| \, g \in GL(V) \right\}.$$

While $\mathcal{F}^1$ is not in general a Lie group, since $\mathcal{F}^1 \subset GL(V)$, we may pull back the Maurer-Cartan from on $GL(V)$ to $\mathcal{F}^1$. Write the pullback of the Maurer-Cartan form to $\mathcal{F}^1$ as

$$\omega = \begin{pmatrix} \omega^0_0 & \omega^0_\beta & \omega^0_\nu \\ \omega^\alpha_0 & \omega^\alpha_\beta & \omega^\alpha_\nu \\ \omega^\mu_0 & \omega^\mu_\beta & \omega^\mu_\nu \end{pmatrix}.$$

The definition of $\mathcal{F}^1$ implies that $\omega^\mu_0 = 0$ because

$$de_0 = \omega^0_0 e_0 + \cdots + \omega^n_0 e_n + \omega^{n+1}_0 e_{n+1} + \cdots + \omega^{n+a}_0 e_{n+1}$$

but we have required that e_0 only move towards $e_1, ..., e_n$ to first order. Similarly, because $\dim X = n$, the adaptation implies that the forms ω_0^α are all linearly independent.

At this point you should know what to do - seeing something equal to zero, we differentiate it. Thanks to the Maurer-Cartan equation, we may calculate the derivative algebraically. We obtain

$$0 = d(\omega_0^\mu) = -\omega_\alpha^\mu \wedge \omega_0^\alpha \ \forall \mu.$$

Since the one-forms ω_0^α are all linearly independent, it is clear that the ω_α^μ must be linear combinations of the ω_0^β, and in fact the *Cartan lemma* (see e.g., [11], p 314) implies that the dependence is symmetric. More precisely (exercise!) there exist functions

$$q_{\alpha\beta}^\mu : \mathcal{F}^1 \to \mathbb{C}$$

with $\omega_\alpha^\mu = q_{\alpha\beta}^\mu \omega_0^\beta$ and moreover $q_{\alpha\beta}^\mu = q_{\beta\alpha}^\mu$. One way to understand the equation $\omega_\alpha^\mu = q_{\alpha\beta}^\mu \omega_0^\beta$ is that the infinitesimal motion of the embedded tangent space (the infinitesimal motion of the e_α's in the direction of the e_μ's) is determined by the motion of e_0 towards the e_α's and the coeffiencients $q_{\alpha\beta}^\mu$ encode this dependence.

Now $\pi : \mathcal{F}^1 \to X$ was defined geometrically (i.e., without making any arbitrary choices) so any function on $\mathcal{F}^1$ invariant under the action of G_1 descends to be a well defined function on X, and will be a *differential invariant.* Our functions $q_{\alpha\beta}^\mu$ are not invariant under the action of G_1, but we can form a tensor from them that is invariant, which will lead to a *vector-bundle valued differential invariant* for X (the same phenomenon happens in the Euclidean geometry of submanifolds).

Consider

$$\tilde{II}_f = F_{2,f} := \omega_0^\alpha \omega_\alpha^\mu \otimes (e_\mu \bmod \hat{T}_x X) = q_{\alpha\beta}^\mu \omega_0^\alpha \omega_0^\beta \otimes (e_\mu \bmod \hat{T}_x X)$$

$\tilde{II} \in \Gamma(\mathcal{F}^1, \pi^*(S^2T^*X \otimes NX))$ is constant on the fiber and induces a tensor $II \in \Gamma(X, S^2T^*X \otimes NX)$. called the *projective second fundamental form.*

Thinking of $II_x : N_x^*X \to S^2T_x^*X$, we may now properly define the asymptotic directions by

$$C_{2,x} := \mathbb{P}(Zeros(II_x(N_x^*X)) \subset \mathbb{P}T_xX.$$

3.3. Higher order differential invariants: the Fubini forms. We continue differentiating constant functions:

$$0 = d(\omega_\alpha^\mu - q_{\alpha\beta}^\mu \omega_0^\beta)$$

yields functions $r_{\alpha\beta\gamma}^\mu : \mathcal{F}^1 \to \mathbb{C}$, symmetric in their lower indices, that induce a tensor $F_3 \in \Gamma(\mathcal{F}^1, \pi^*(S^3T^*X \otimes NX))$ called the *Fubini cubic form.*

Unlike the second fundamental form, it does *not* descend to be a tensor over X because it varies in the fiber. We discuss this variation in §6. Such tensors provide *relative differential invariants* and by sucessive differentiations, one obtains a series of invariants $F_k \in \Gamma(\mathcal{F}^1, \pi^*(S^kT^* \otimes N))$. For example,

$$F_3 = r^\mu_{\alpha\beta\gamma}\omega^\alpha_0\omega^\beta_0\omega^\gamma_0 \otimes e_\mu$$
$$F_4 = r^\mu_{\alpha\beta\gamma\delta}\omega^\alpha_0\omega^\beta_0\omega^\gamma_0\omega^\delta_0 \otimes e_\mu$$

where the functions $r^\mu_{\alpha\beta\gamma}, r^\mu_{\alpha\beta\gamma\delta}$ are given by

$$r^\mu_{\alpha\beta\gamma}\omega^\gamma_0 = -dq^\mu_{\alpha\beta} - q^\mu_{\alpha\beta}\omega^0_0 - q^\nu_{\alpha\beta}\omega^\mu_\nu + q^\mu_{\alpha\delta}\omega^\delta_\beta + q^\mu_{\beta\delta}\omega^\delta_\alpha \tag{3.1}$$

$$\begin{aligned} r^\mu_{\alpha\beta\gamma\delta}\omega^\delta_0 &= -dr^\mu_{\alpha\beta\gamma} - 2r^\mu_{\alpha\beta\gamma}\omega^0_0 - r^\nu_{\alpha\beta\gamma}\omega^\mu_\nu \\ &\quad + \mathfrak{S}_{\alpha\beta\gamma}(r^\mu_{\alpha\beta\epsilon}\omega^\epsilon_\gamma + q^\mu_{\alpha\beta}\omega^0_\gamma - q^\mu_{\alpha\epsilon}q^\nu_{\beta\gamma}\omega^\epsilon_\nu). \end{aligned} \tag{3.2}$$

We define $\mathcal{C}_{k,x} := Zeros(F_{2,f}, \ldots, F_{k,f}) \subset \mathbb{P}T_xX$, which is independent of our choice of $f \in \pi^{-1}(x)$.

If one chooses local affine coordinates $(x^1, \ldots, x^{n+a})$ such that $x = (0, \ldots, 0)$ and $T_xX = \langle \frac{\partial}{\partial x^\alpha} \rangle$, and writes X as a graph

$$x^\mu = q^\mu_{\alpha\beta}x^\alpha x^\beta - r^\mu_{\alpha\beta\gamma}x^\alpha x^\beta x^\gamma + r^\mu_{\alpha\beta\gamma\delta}x^\alpha x^\beta x^\gamma x^\delta + \cdots$$

then there exists a local section of $\mathcal{F}^1$ such that

$$F_2|_x = q^\mu_{\alpha\beta}dx^\alpha dx^\beta \otimes \frac{\partial}{\partial x^\mu}$$
$$F_3|_x = r^\mu_{\alpha\beta\gamma}dx^\alpha dx^\beta dx^\gamma \otimes \frac{\partial}{\partial x^\mu}$$
$$F_4|_x = r^\mu_{\alpha\beta\gamma\delta}dx^\alpha dx^\beta dx^\gamma dx^\delta \otimes \frac{\partial}{\partial x^\mu}$$

and similarly for higher orders.

(3.1) is a system of $a\binom{n+1}{2}$ equations with one-forms as coefficients for the $a\binom{n+2}{3}$ coefficients of F_3 and is overdetermined if we assume $dq^\mu_{\alpha\beta} = 0$, as we do in the rigidity problems. One can calculate directly in the Segre $Seg(\mathbb{P}^m \times \mathbb{P}^r)$, $m, r > 1$ case that the only possible solutions are normalizable to zero by a fiber motion as described in (6.4). The situation is the same for F_4, F_5 in this case.

In general, once $F_k, \ldots, F_{2k-1}$ are normalized to zero at a general point, it is automatic that all higher F_j are zero, see [15]. Thus one has the entire Taylor series and has completely identified the variety. This was the method of proof used in [16], as the rank two CHSS have all F_k normalizable to zero when $k > 2$.

Another perspective, for those familiar with G-structures, is that one obtains rigidity by reducing $\mathcal{F}^1$ to a smaller bundle which is isomorphic to G, where the homogeneous model is G/P.

Yet another perspective, for those familiar with exterior differential systems, is that after three prolongations, the EDS defined by $I = \{\omega_0^\mu, \omega_\alpha^\mu - q_{\alpha\beta}^\mu \omega_0^\beta\}$ on $GL(V)$ becomes involutive, in fact Frobenius.

3.4. The higher fundamental forms. A component of F_3 does descend to a well defined tensor on X. Namely, considering $F_3 : N^* \to S^3T^*$, if we restrict $F_3|_{\ker F_2}$, we obtain a tensor $\mathbb{F}_3 = III \in S^3T^* \otimes N_3$ where $N_3 = T_x\mathbb{P}V/\{T_xX + II(S^2T_xX)\}$. One continues in this manner to get a series of tensors $\mathbb{F}_k$ called the *fundamental forms.*

Geometrically, II measures how X is leaving its embedded tangent space at x to first order, III measures how X is leaving its second osculating space at x to first order while F_3 mod III measures how X is moving away from its embedded tangent space to second order.

3.5. More rigidity theorems. Now that we have defined fundamental forms, we may state:

THEOREM 3.1 (Hwang-Yamaguchi). *[10] Let $X^n \subset \mathbb{CP}^{n+a}$ be a complex submanifold. Let $x \in X$ be a general point. Let Z be an irreducible rank r compact Hermitian symmetric space in its natural embedding, other than a quadric hypersurface. If there exists linear maps $f : T_xX \to T_zZ$, $g_k : N_{k,x}X \to N_{k,z}Z$ such that the induced maps $S^kT_x^*X \otimes N_xX \to S^kT_z^*Z \otimes N_zZ$ take $\mathbb{F}_{k,X,x}$ to $\mathbb{F}_{k,Z,z}$ for $2 \le k \le r$, then $\overline{X} = Z$.*

In [22] we calculated the differential invariants of the *adjoint varieties,* the closed orbits in the projectivization of the adjoint representation of a simple Lie algebra. (These are the homogeneous complex contact manifolds in their natural homogeneous embedding.) The adjoint varieties have $III = 0$, but, in all cases but $v_2(\mathbb{P}^{2n-1}) = C_n/P_1 \subset \mathbb{P}(\mathfrak{c}_n) = \mathbb{P}(S^2\mathbb{C}^{2n})$ which we exclude from discussion in the remainder of this paragraph, the invariants F_3, F_4 are not normalizable to zero, even though $C_{3,x} = C_{4,x} = C_{2,x} = C_x$. In a normalized frame C_x is contained in a hyperplane H and F_4 is the equation of the *tangential variety* of C_x in H, where the tangential variety $\tau(X) \subset \mathbb{P}V$ of an algebraic manifold $X \subset \mathbb{P}V$ is the union of the points on the embedded tangent lines ($\mathbb{P}^1$'s) to the manifold. In this case the tangential variety is a hypersurface in H, except for $\mathfrak{g} = \mathfrak{a}_n = \mathfrak{sl}_{n+1}$ which is discussed below. Moreover, F_3 consists of the defining equations for the singular locus of $\tau(C_x)$. In [22] we speculated that the varieties X_{ad}, with the exception of $v_2(\mathbb{P}^{2n-1})$ (which is rigid to order three, see [16]) would be rigid to order four, but not three, due to the nonvanishing of F_4. Thus the following result came as a suprise to us:

THEOREM 3.2 (Robles, [31]). *Let $X^{2(m-2)} \subset \mathbb{CP}^{m^2-2}$ be a complex submanifold. Let $x \in X$ be a general point. Let $Z \subset \mathbb{P}\mathfrak{sl}_m$ be the adjoint variety. If there exist linear maps $f : T_xX \to T_zZ$, $g : N_xX \to N_zZ$ such that the induced maps $S^kT_x^*X \otimes N_xX \to S^kT_z^*Z \otimes N_zZ$ take $F_{k,X,x}$ to $F_{k,Z,z}$ for $k = 2, 3$, then $\overline{X} = Z$.*

Again, the corresponding result holds in the C^∞ category.

The adjoint variety of $\mathfrak{sl}_m = \mathfrak{sl}(W)$ has the geometric interpretation of the variety of flags of lines inside hyperplanes inside W, or equivalently as the traceless, rank one matrices. It has $C_{2,z}$ the union of two disjoint linear spaces in a hyperplane in $\mathbb{P}T_zZ$. The quartic F_4 is the square of a quadratic equation (whose zero set contains the two linear spaces), and the cubics in F_3 are the derivatives of this quartic, see [22], §6.

3.6. The prolongation property and proof of Theorem 2.3. The precise restrictions II places on the F_k in general is not known at this time. However, there is a strong restriction II places on the higher fundamental forms that dates back to Cartan. We recall a definition from exterior differential systems:

Let U, W be vector spaces. Given a linear subspace $A \subset S^kU^* \otimes W$, define the *j-th prolongation* of A to be $A^{(j)} := (A \otimes S^jU^*) \cap (S^{k+j}U^* \otimes W)$. Thinking of A as a collection of W-valued homogeneous polynomials on U, the j-th prolongation of A is the set of all homogeneous W-valued polynomials of degree $k+j$ on U with the property that all their j-th order partial derivatives lie in A.

PROPOSITION 3.1 (Cartan [3] p 377). *Let $X^n \subset \mathbb{P}^{n+a}$, and let $x \in X$ be a general point. Then $\mathbb{F}_{k,x}(N_k^*) \subseteq \mathbb{F}_{2,x}(N_2^*)^{(k-2)}$. (Here W is taken to be the trivial vector space $\mathbb{C}$ and $U = T_xX$.)*

PROPOSITION 3.2. *[21] Let $X = G/P \subset \mathbb{P}V$ be a CHSS in its minimal homogeneous embedding. Then $\mathbb{F}_{k,x}(N_k^*) = \mathbb{F}_{2,x}(N_2^*)^{(k-2)}$. Moreover, the only nonzero components of the F_k are the fundamental forms.*

The only homogeneous varieties having the property that the only nonzero components of the F_k are the fundamental forms are the CHSS.

Proof. [Proof of Theorem 2.3] The strict prolongation property for CHSS in their minimal homogenous embedding implies that any variety with the same second fundamental form at a general point as a CHSS in its minimal homogeneous embedding can have codimension at most that of the corresponding CHSS, and equality holds iff all the other fundamental forms are the prolongations of the second. □

4. Bertini type theorems and applications. The results discussed so far dealt with homogeneous varieties. We now broaden our study to various pathologies of the $C_{k,x}$.

Let T be a vector space. The classical *Bertini theorem* implies that for a linear subspace $A \subset S^2T^*$, if $q \in A$ is such that $\text{rank}\,(q) \geq \text{rank}\,(q')$ for all $q' \in A$, then $u \in q_{sing} := \{v \in T \mid q(v,w) = 0\ \forall w \in T\}$ implies $u \in \text{Zeros}\,(A) := \{v \in T \mid Q(v,v) = 0\ \forall Q \in A\}$.

THEOREM 4.1 (Mobile Bertini). *Let $X^n \subset \mathbb{P}V$ be a complex manifold and let $x \in X$ be a general point. Let $q \in II(N_x^*X)$ be a generic quadric. Then q_{sing} is tangent to a linear space on $\overline{X}$.*

For generalizations and variations, see [20].

The result holds in the C^∞ category if one replaces a general point by all points and that the linear space is contained in X as long as X continues (e.g., it is contained in X if X is complete).

Proof. Assume $v = e_1 \in q_{sing}$ and $q = q^{n+1}_{\alpha\beta}\omega^\alpha_0\omega^\beta_0$. Our hypotheses imply $q^{n+1}_{1\beta} = 0$ for all β. Formula (3.1) reduces to

$$r^{n+1}_{11\beta}\omega^\beta_0 = -q^\mu_{11}\omega^{n+1}_\mu.$$

If q is generic we are working on a reduction of $\mathcal{F}^1$ where the ω^μ_ν are independent of each other and independent of the semi-basic forms (although the ω^α_β will no longer be independent of the $\omega^\alpha_0, \omega^\mu_\nu$); thus the coefficients on both sides of the equality are zero, proving both the classical Bertini theorem and $v \in r_{sing}$ where r is a generic cubic in $F_3(N^*)$. Then using the formula for F_4 one obtains $v \in Zeros(F_3)$ and $v \in s_{sing}$ where s is a generic element of $F_4(N^*)$. One then concludes by induction. □

REMARK 4.1. *The mobile Bertini theorem essentially dates back to B. Segre [35], and was rediscovered in various forms in [6, 4]. Its primarary use is in the study of varieties $X^n \subset \mathbb{P}V$ with defective dual varieties $X^* \subset \mathbb{P}V^*$, where the dual variety of a smooth variety is the set of tangent hyperplanes to X, which is usually a hypersurface. The point is that a generic quadric in $II(N^*_{x,X})$ (with $x \in X_{general}$) is singular of rank r iff $codim\, X^* = n - r + 1$.*

EXAMPLE 1. *Taking $X = Seg(\mathbb{P}A \times \mathbb{P}B)$ and keeping the notations of above, let Y have the same second fundamental form of X at a general point $y \in Y$ so we inherit an identification $T_yY \simeq A' \oplus B'$. If $dim\, B = b > a = dim\, A$, then mobile Bertini implies that $\mathbb{P}B'$ is actually tangent to a linear space on Y, because the maximum rank of a quadric is $a-1$. So at any point $[b] \in \mathbb{P}B'$ there is even a generic quadric singular at $[b]$. (Of course the directions of $\mathbb{P}A'$ are also tangent to lines on Y because the Segre is rigid.)*

While in [16] I did not calculate the rigidity of $Seg(\mathbb{P}^1 \times \mathbb{P}^n) \subset \mathbb{P}(\mathbb{C}^2 \otimes \mathbb{C}^{n+1})$, the rigidity follows from the same calculations, however one must take additional derivatives to get the appropriate vanishing of the Fubini forms. However there is also an elementary proof of rigidity in this case using the mobile Bertini theorem [20]. Given a variety $Y^{n+1} \subset \mathbb{P}^N$ such that at a general point $y \in Y$, $C_{2,y}$ contains a $\mathbb{P}^{n-1}$, by 4.1 the resulting n-plane field on TY is integrable and thus Y is ruled by $\mathbb{P}^n$'s. Such a variety arises necessarily from a curve in the Grassmannian $G(n+1, N+1)$ (as the union of the points on the $\mathbb{P}^n$'s in the curve). But in order to also have the $\mathbb{P}^0$ factor in $C_{2,y}$, such a curve must be a line and thus Y must be the Segre.

The mobile Bertini theorem describes consequences of $C_{2,x}$ being pathological. Here are some results when $C_{k,x}$ is pathological for $k > 2$.

THEOREM 4.2 (Darboux). *Let $X^2 \subset \mathbb{P}^{2+a}$ be an analytic submanifold and let $x \in X_{general}$. If there exists a line l having contact to order three with X at x, then $l \subset X$. In other words, for surfaces in projective space,*

$$C_{3,x} = C_x \ \forall x \in X_{general}.$$

The C^∞ analogue holds replacing general points by all points and lines by line segments contained in X.

There are several generalizations of this result in [18, 19]. Here is one of them:

THEOREM 4.3. *[19] Let $X^n \subset \mathbb{P}^{n+1}$ be an analytic submanifold and let $x \in X_{general}$. If $\Sigma \subseteq C_{k,x}$ is an irreducible component with $\dim \Sigma > n-k$, then $\Sigma \subset C_x$.*

The C^∞ analog holds with the by now obvious modifications.

The proof is similar to that of the mobile Bertini theorem.

EXERCISE 4.1. *One of my favorite problems to put on an undergraduate differential geometry exam is: Prove that a surface in Euclidean three space that has more than two lines passing through each point is a plane (i.e., has an infinite number of lines passing through). In [30], Mezzetti and Portelli showed that a 3-fold having more than six lines passing through a general point must have an infinite number. Show that an n-fold having more than n! lines passing through a general point must have an infinite number passing through each point. (See [19] if you need help.)*

The rigidity of the quadric hypersurface is a classical result:

THEOREM 4.4 (Fubini). *Let $X^n \subset \mathbb{P}^{n+1}$ be an analytic submanifold or algebraic variety and let $x \in X_{general}$. Say $C_{3,x} = C_{2,x}$. Let $r = rank C_{2,x}$. Then*

- *If $r > 1$, then X is a quadric hypersurface of rank r.*
- *If $r = 1$, then X has a one-dimensional Gauss image. In particular, it is ruled by $\mathbb{P}^{n-1}$'s.*

In all situations, the dimension of the Gauss image of X is r, see [11], §3.4.

One way to prove Fubini's theorem (assuming $r > 1$) is to first use mobile Bertini to see that X contains large linear spaces, then to note that degree is invariant under linear section, so one can reduce to the case of a surface. But then it is elementary to show that the only analytic surface that is doubly ruled by lines is the quadric surface.

Another way to prove Fubini's theorem (assuming $r > 1$) is to use moving frames to reduce the frame bundle to $O(n+2)$.

What can we say in higher codimension? Consider codimension two. What are the varieties $X^n \subset \mathbb{P}^{n+2}$ such that for general $x \in X$ we have $C_{3,x} = C_{2,x}$?

Note that there are two principal difficulties in codimension two. First, in codimension one, having $C_{3,x} = C_{2,x}$ implies that F_3 is normalizable to zero - this is no longer true in codimension greater than one. Second, in codimension one, there is just one quadratic form in II, so its only invariant is its rank. In larger codimension there are moduli spaces, although for pencils at least there are normal forms, convienently given in [9].

For examples, we have inherited from Fubini's theorem:

0. $\mathbb{P}^n \subset \mathbb{P}^{n+2}$ as a linear subspace

0'. $Q^n \subset \mathbb{P}^{n+1} \subset \mathbb{P}^{n+2}$ a quadric

0". A variety with a one-dimensional Gauss image.

To these it is easy to see the following are also possible:

1. The (local) product of a curve with a variety with a one dimensional Gauss image.

2. The intersection of two quadric hypersurfaces.

3. A (local) product of a curve with a quadric hypersurface.

There is one more example we have already seen several times in these lectures:

4. The Segre $Seg(\mathbb{P}^1 \times \mathbb{P}^2) \subset \mathbb{P}^5$ or a cone over it.

THEOREM 4.5 (Codimension two Fubini). *[29] Let $X^n \subset \mathbb{P}^{n+2}$ be an analytic submanifold and let $x \in X_{general}$. If $C_{3,x} = C_{2,x}$ then X is (an open subset of) one of $0, 0', 0'', 1-4$ above.*

Here if one were to work over $\mathbb{R}$, the corresponding result would be more complicated as there are more normal forms for pencils of quadrics.

5. Applications to algebraic geometry. One nice aspect of algebraic geometry is that spaces parametrizing algebraic varieties tend to also be algebraic varieties (or at least *stacks*, which is the algebraic geometer's version of an orbifold). For example, let $Z^n \subset \mathbb{P}^{n+1}$ be a hypersurface. The study of the images of holomorphic maps $f : \mathbb{P}^1 \to Z$, called *rational curves on Z* is of interest to algebraic geometers and physicists. One can break this into a series of problems based on the degree of $f(\mathbb{P}^1)$ (that is, the number of points in the intersection $f(\mathbb{P}^1) \cap H$ where $H = \mathbb{P}^n$ is a general hyperplane). When the degree is one, these are just the lines on Z, and already here there are many open questions. Let $\mathbb{F}(Z) \subset \mathbb{G}(\mathbb{P}^1, \mathbb{P}^{n+1}) = G(2, \mathbb{C}^{n+2})$ denote the variety of lines (i.e. linear $\mathbb{P}^1$'s) on Z.

Note that if the lines are distributed evenly on Z and $z \in Z_{general}$, then $\dim \mathbb{F}(Z) = \dim C_{z,Z} + n - 1$. We always have $\dim \mathbb{F}(Z) \geq \dim C_{z,Z} + n - 1$ (exercise!), so the microscopic geometry bounds the macroscopic geometry.

EXAMPLE 2. *Let $Z = \mathbb{P}^n$, then $\mathbb{F}(Z) = G(2, n+1)$. In particular, $\dim \mathbb{F}(Z) = 2n - 2$ and this is the largest possible dimension, and $\mathbb{P}^n$ is the only variety with $\dim \mathbb{F}(Z) = 2n - 2$.*

Which are the varieties $X^n \subset \mathbb{P}V$ with $\dim \mathbb{F}(X) = 2n - 3$? A classical theorem states that in this case X must be a quadric hypersurface.

Rogora, in [32], classified all $X^n \subset \mathbb{P}^{n+a}$ with $\dim \mathbb{F}(X) = 2n - 4$, with the extra hypothesis codim $X > 2$. The only "new" example is the Grassmannian $G(2, 5)$. (A classification in codimension one would be quite difficult.) A corollary of the codimension two Fubini theorem is that Rogora's theorem in codimension two is nearly proved - nearly and not completely because one needs to add the extra hypothesis that $C_{2,x}$ has only one component or that $C_{3,x} = C_{2,x}$.

If one has extra information about X, one can say more. Say X is a hypersurface of degree d. Then it is easy to show that $\dim \mathbb{F}(X) \geq 2n - 1 - d$.

CONJECTURE 5.1 (Debarre, de Jong). *If $X^n \subset \mathbb{P}^{n+1}$ is smooth and $n > d = deg(X)$, then $dim \mathbb{F}(X) = 2n - 1 - d$.*

Without loss of generality it would be sufficient to prove the conjecture when $n = d$ (slice by linear sections to reduce the dimension). The conjecture is easy to show when $n = 2$, it was proven by Collino when $n = 3$, by Debarre when $n = 4$, and the proof of the $n = 5$ case was was the PhD thesis of R. Beheshti [1]. Beheshti's thesis had three ingredients, a general lemma (that $\mathbb{F}(X)$ could not be uniruled by rational curves), Theorem 4.3 above, and a case by case argument. As a corollary of the codimension two Fubini theorem one obtains a new proof of Beheshti's theorem eliminating the case by case argument (but there is a different case by case argument buried in the proof of the codimension two Fubini theorem). More importantly, the techniques should be useful in either proving the theorem, or pointing to where one should look for potential counter-examples for $n > 5$.

6. Moving frames proof of the Hwang-Yamaguchi theorem. The principle of calculation in [20] was to use mobile Bertini theorems and the decomposition of the spaces $S^dT^* \otimes N$ into irreducible R-modules, where $R \subset GL(T) \times GL(N)$ is the subgroup preserving $II \in S^2T^* \otimes N$. One can isolate where each F_k can "live" as the intersection of two vector spaces (one of which is $S^kT^* \otimes N$, the other is $(\mathfrak{g}^{\perp})_{k-3}$ defined below). Then, since R acts on fibers, we can decompose $S^kT^* \otimes N$ and $(\mathfrak{g}^{\perp})_{k-3}$ into R-modules and in order for a module to appear, it most be in both the vector spaces. This combined with mobile Bertini theorems reduces the calculations to almost nothing.

Hwang and Yamaguchi use representation theory in a more sophisticated way via a theory developed by Se-ashi [34]. What follows is a proof of their result in the language of moving frames.

Let $Z = G/P \subset \mathbb{P}W$ be a CHSS in its minimal homogeneous embedding. For the moment we restrict to the case where $III_Z = 0$ (i.e., rank $Z = 2$). Let $X \subset \mathbb{P}V$ be an analytic submanifold, let $x \in X$ be a general point and assume $II_{X,x} \simeq II_{Z,z}$. We determine sufficient conditions that imply X is projectively isomorphic to Z.

We have a filtration of V, $V_0 = \hat{x} \subset V_1 = \hat{T}_xX \subset V = V_2$. Write $L = V_0, T = V_1/V_0, N = V/V_1$. We have an induced grading of $\mathfrak{gl}(V)$ where

$$\begin{aligned}
\mathfrak{gl}(V)_0 &= \mathfrak{gl}(L) \oplus \mathfrak{gl}(T) \oplus \mathfrak{gl}(N),\\
\mathfrak{gl}(V)_{-1} &= L^* \otimes T \oplus T^* \otimes N,\\
\mathfrak{gl}(V)_1 &= L \otimes T^* \oplus T \otimes N^*,\\
\mathfrak{gl}(V)_{-2} &= L^* \otimes N,\\
\mathfrak{gl}(V)_2 &= L \otimes N^*.
\end{aligned}$$

In what follows we can no longer ignore the twist in defining II, that is, we have $II \in S^2T^* \otimes N \otimes L$.

Let $\mathfrak{g}_{-1} \subset \mathfrak{gl}(V)_{-1}$ denote the image of

$$T \to L^* \otimes T + T^* \otimes N$$
$$e_\alpha \mapsto e^0 \otimes e_\alpha + q^\mu_{\alpha\beta} e^\beta \otimes e_\mu.$$

Let $\mathfrak{g}_0 \subset \mathfrak{gl}(V)_0$ denote the subalgebra annhilating II. More precisely

$$u = \begin{pmatrix} x^0_0 & & \\ & x^\alpha_\beta & \\ & & x^\mu_\nu \end{pmatrix} \in \mathfrak{gl}(V)_0$$

is in $\mathfrak{g}_0$ iff

$$u.II := (-x^\mu_\nu q^\nu_{\alpha\beta} + x^\gamma_\alpha q^\mu_{\gamma\beta} + x^\gamma_\beta q^\mu_{\alpha\gamma} - x^0_0 q^\mu_{\alpha\beta}) e^\alpha \circ e^\beta \otimes e_\mu \otimes e_0 = 0. \quad (6.1)$$

Let $\mathfrak{g}_1 \subset \mathfrak{gl}(V)_1$ denote the maximal subspace such that $[\mathfrak{g}_1, \mathfrak{g}_{-1}] = \mathfrak{g}_0$. Note that $\mathfrak{g} = \mathfrak{g}_{-1} \oplus \mathfrak{g}_0 \oplus \mathfrak{g}_1$ coincides with the $\mathbb{Z}$-graded semi-simple Lie algebra giving rise to $Z = G/P$ and that the inclusion $\mathfrak{g} \subset \mathfrak{gl}(V)$ coincides with the embedding $\mathfrak{g} \to \mathfrak{gl}(W)$. The grading on $\mathfrak{g}$ induced from the grading of $\mathfrak{gl}(V)$ agrees with the grading induced by P. In particular, $\mathfrak{g}_{\pm 2} = 0$.

We let $\mathfrak{g}^\perp = \mathfrak{gl}(V)/\mathfrak{g}$ and note that $\mathfrak{g}^\perp$ is naturally a $\mathfrak{g}$-module. Alternatively, one can work with $\mathfrak{sl}(V)$ instead of $\mathfrak{gl}(V)$ and define $\mathfrak{g}^\perp$ as the Killing-orthogonal complement to $\mathfrak{g}_0$ in $\mathfrak{sl}(V)$ (Working with $\mathfrak{sl}(V)$-frames does not effect projective geometry since the actual group of projective transformations is $PGL(V)$).

Recall that F_3 arises by applying the Cartan lemma to the equations

$$0 = -\omega^\mu_\alpha \wedge \omega^\alpha_\beta - \omega^\mu_\nu \wedge \omega^\nu_\beta + q^\mu_{\alpha\beta}(\omega^\alpha_0 \wedge \omega^0_0 + \omega^\alpha_\gamma \wedge \omega^\gamma_0)\ \forall \mu, \beta \quad (6.2)$$

i.e., the tensor

$$(-q^\nu_{\beta\gamma}\omega^\mu_\nu + q^\mu_{\alpha\gamma}\omega^\alpha_\beta + q^\mu_{\alpha\beta}\omega^\alpha_\gamma - q^\mu_{\gamma\beta}\omega^0_0) \wedge \omega^\gamma_0 \otimes e^\beta \otimes e_\mu \quad (6.3)$$

must vanish. All forms appearing in the term in parenthesis in (6.3) are $\mathfrak{gl}(V)_0$-valued. Comparing with (6.1), we see that the $\mathfrak{g}_0$-valued part will be zero and $(\mathfrak{g}^\perp)_0$ bijects to the image in parenthesis.

Thus we may think of obtaining the coefficients of F_3 at x in two stages, first we write the $(\mathfrak{g}^\perp)_0$ component of the Maurer-Cartan form of $GL(V)$ as an arbitrary linear combination of semi-basic forms, i.e., we choose a map $T \to (\mathfrak{g}^\perp)_0$. Once we have chosen such a map, substituting the image into (6.3) yields a $(\Lambda^2 T^* \otimes \mathfrak{gl}(V)_{-1})$-valued tensor. But by the definition of $\mathfrak{g}$, it is actually $(\Lambda^2 T^* \otimes (\mathfrak{g}^\perp)_{-1})$-valued. Then we require moreover that that this tensor is zero. In other words, pointwise we have a map

$$\partial^{1,1} : T^* \otimes (\mathfrak{g}^\perp)_0 \to \Lambda^2 T^* \otimes (\mathfrak{g}^\perp)_{-1}$$

and the (at this stage) admissible coefficients of F_3 are determined by a choice of map $T \to (\mathfrak{g}^\perp)_0$ which is in the kernel of $\partial^{1,1}$.

Now the variation of F_3 as one moves in the fiber is given by a map $T \to (\mathfrak{g}^\perp)_0$ induced from the action of $(\mathfrak{g}^\perp)_1$ on $(\mathfrak{g}^\perp)_0$. We may express it as:

$$\begin{aligned} x^0_\beta e_0 \otimes e^\beta + x^\alpha_\nu e_\alpha \otimes e^\nu \mapsto x^0_\beta \omega^\beta_0 e_0 \otimes e^0 + (x^0_\beta \omega^\alpha_0 \\ + x^\alpha_\nu \omega^\nu_\beta) \otimes e^\beta \otimes e_\alpha \\ + x^\alpha_\nu \omega^\mu_\alpha \otimes e^\nu \otimes e_\mu \end{aligned} \tag{6.4}$$

where

$$\begin{pmatrix} 0 & x^0_\beta & 0 \\ 0 & 0 & x^\gamma_\nu \\ 0 & 0 & 0 \end{pmatrix}$$

is a general element of $\mathfrak{gl}(V)_1$. The kernel of this map is $\mathfrak{g}_1$ so it induces a linear map with source $(\mathfrak{g}^\perp)_1$. Similarly, the image automatically takes values in $T^* \otimes (\mathfrak{g}^\perp)_0$. In summary, (6.4) may be expressed as a map

$$\partial^{2,0} : (\mathfrak{g}^\perp)_1 \to T^* \otimes (\mathfrak{g}^\perp)_0.$$

Let $C^{p,q} := \Lambda^q T^* \otimes (\mathfrak{g}^\perp)_{p-1}$. Considering $\mathfrak{g}^\perp$ as a T-module (via the embedding $T \to \mathfrak{g}$) we have the Lie algebra cohomology groups

$$H^{p,1}(T, \mathfrak{g}^\perp) := \frac{\ker \partial^{p,1} : C^{p,1} \to C^{p-1,2}}{\text{Image}\, \partial^{p+1,0} : C^{p+1,0} \to C^{p,1}}.$$

We summarize the above discussion:

PROPOSITION 6.1. *Let $X \subset \mathbb{P}V$ be an analytic submanifold, let $x \in X_{general}$ and suppose $II_{X,x} \simeq II_{Z,z}$ where $Z \subset \mathbb{P}W$ is a rank two CHSS in its minimal homogeneous embedding. The choices of $F_{3,X,x}$ imposed by* (6.2), *modulo motions in the fiber, is isomorphic to the Lie algebra cohomology group $H^{1,1}(T, \mathfrak{g}^\perp)$.*

Now if F_3 is normalizable and normalized to zero, differentiating again, we obtain the equation

$$(q^\mu_{\alpha\beta}\omega^0_\gamma + q^\mu_{\alpha\epsilon} q^\nu_{\beta\gamma}\omega^\epsilon_\nu) \wedge \omega^\gamma_0 \otimes e_\mu \otimes e^\alpha \circ e^\beta = 0$$

which determines the possible coefficients of F_4.

We conclude any choice of F_4 must be in the kernel of the map

$$\partial^{2,1} : T^* \otimes (\mathfrak{g}^\perp)_1 \to \Lambda^2 T^* \otimes (\mathfrak{g}^\perp)_0.$$

The variation of of F_4 in the fiber of $\mathcal{F}^1$ is given by the image of the map

$$x^0_\nu e_0 \otimes e^\nu \mapsto x^0_\nu \omega^\epsilon_0 e_\epsilon \otimes e^\nu$$

as $v \wedge w$ ranges over the decomposable elements of $\Lambda^2 T$. Without indicies, the variation of F_4 is the image of the map

$$\partial^{3,0} : (\mathfrak{g}^{\perp})_2 \to T^* \otimes (\mathfrak{g}^{\perp})_1.$$

We conclude that if F_3 has been normalized to zero, then F_4 is normalizable to zero if $H^{2,1}(T, \mathfrak{g}^{\perp}) = 0$.

Finally, if F_3, F_4 are normalized to zero, then the coefficients of F_5 are given by

$$\ker \partial^{3,1} : T^* \otimes (\mathfrak{g}^{\perp})_2 \to \Lambda^2 T^* \otimes (\mathfrak{g}^{\perp})_1,$$

and there is nothing to quotient by in this case because $\mathfrak{gl}(V)_3 = 0$. In summary

PROPOSITION 6.2. *A sufficient condition for second order rigidity to hold for a rank two CHSS in its minimal homogeneous embedding $Z = G/P \subset \mathbb{P}W$ is that $H^{1,1}(T, \mathfrak{g}^{\perp}) = 0, H^{2,1}(T, \mathfrak{g}^{\perp}) = 0, H^{3,1}(T, \mathfrak{g}^{\perp}) = 0$.*

Assume for simplicity that G is simple and $P = P_{\alpha_{i_0}}$ is the maximal parabolic subgroup obtained by deleting all root spaces whose roots have a negative coefficient on the simple root α_{i_0}. By Kostant's results [13], the $\mathfrak{g}_0$-module $H^{*,1}(T, \Gamma)$ for *any* irreducible $\mathfrak{g}$-module Γ of highest weight λ is the irreducible $\mathfrak{g}_0$-module with highest weight $\sigma_{\alpha_{i_0}}(\lambda + \rho) - \rho$, where $\sigma_{\alpha_{i_0}}$ is the simple reflection in the Weyl group corresponding to α_{i_0} and ρ is half the sum of the simple roots.

But now as long as $G/P_{\alpha_{i_0}}$ is not projective space or a quadric hypersurface, Hwang and Yamaguchi, following [33], observe that any non-trivial $\mathfrak{g}$-module Γ yields a $\mathfrak{g}_0$-module in $H^{*,1}$ that has a non-positive grading, so one concludes that the above groups are *a priori* zero. Thus one only need show that the trivial representation is not a submodule of $\mathfrak{g}^{\perp}$.

REMARK 6.1. *We note that the condition in Proposition 6.2 is not necessary for second order rigidity. It holds in all rigid rank two cases but one, $Seg(\mathbb{P}^1 \times \mathbb{P}^n)$, with $n > 1$, which we saw, in §1, is indeed rigid to order two. Note that in that case the naïve moving frames approach is significantly more difficult as one must prolong several times before obtaining the vanishing of the normalized F_3.*

To prove the general case of the Hwang-Yamaguchi theorem, say the last nonzero fundamental form is the k-th. Then one must show $H^{1,1}, ..., H^{k+1,1}$ are all zero. But again, Kostant's theory applies to show all groups $H^{p,1}$ are zero for $p > 0$. Note that in this case, $H^{1,1}$ governs the vanishing $F_{3,2}, ..., F_{k,k-1}$, where $F_{k,l}$ denotes the component of F_k in $S^k T^* \otimes N_l$. In general, $H^{l,1}$ governs $F_{2+l,2}, ..., F_{k+l,k-1}$.

REFERENCES

[1] R. BEHESHTI, *Lines on projective hypersurfaces*, J. Reine Angew. Math. (2006), **592**: 1–21.

[2] N. BOURBAKI, *Groupes et algèbres de Lie*, Hermann, Paris, 1968, MR0682756.

[3] E. CARTAN, *Sur les variétés de courbure constante d'un espace euclidien ou non euclidien*, Bull. Soc. Math France (1919), **47**: 125–160, and (1920), **48**: 132–208; see also pp. 321–432 in Oeuvres Complètes Part III, Gauthier-Villars, 1955.

[4] L. EIN, *Varieties with small dual varieties, I*, Inventiones Math. (1986), **86**: 63–74.

[5] G. FUBINI, *Studi relativi all'elemento lineare proiettivo di una ipersuperficie*, Rend. Acad. Naz. dei Lincei (1918), 99–106.

[6] P.A. GRIFFITHS AND J. HARRIS, *Algebraic Geometry and Local Differential Geometry*, Ann. scient. Ec. Norm. Sup. (1979), **12**: 355–432, MR0559347.

[7] J.-M. HWANG, *Geometry of minimal rational curves on Fano manoflds*, ICTP lecture notes, www.ictp.trieste.it./~pub_off/services.

[8] J.-M. HWANG AND N. MOK, *Uniruled projective manifolds with irreducible reductive G-structures*, J. Reine Angew. Math. (1997), **490**: 55–64, MR1468924.

[9] W.V.D. HODGE AND D. PEDOE, *Methods of algebraic geometry*, Vol. II, Cambridge University Press, Cambridge (1994), p. 394+.

[10] J.-M. HWANG AND K. YAMAGUCHI, *Characterization of Hermitian symmetric spaces by fundamental forms*, Duke Math. J. (2003), **120**(3): 621–634.

[11] T. IVEY AND J.M. LANDSBERG, *Cartan for beginners: differential geometry via moving frames and exterior differential systems*, Graduate Studies in Mathematics, **61**, American Mathematical Society, Providence, RI, 2003, MR2003610.

[12] G. JENSEN AND E. MUSSO, *Rigidity of hypersurfaces in complex projective space*, Ann. scient. Ec. Norm. (1994), **27**: 227–248.

[13] B. KOSTANT, *Lie algebra cohomology and the generalized Borel-Weil theorem*. Ann. of Math. (1961), **74**(2): 329–387. MR0142696

[14] J.M. LANDSBERG, *On second fundamental forms of projective varieties*, Inventiones Math. (1994), **117**: 303–315, MR1273267.

[15] ——, *Differential-geometric characterizations of complete intersections*, J. Differential Geom. (1996), **44**: 32–73, MR1420349.

[16] ——, *On the infinitesimal rigidity of homogeneous varieties*, Compositio Math. (1999), **118**: 189–201, MR1713310.

[17] ——, *Algebraic geometry and projective differential geometry*, Seoul National University concentrated lecture series 1997, Seoul National University Press, 1999, MR1712467.

[18] ——, *Is a linear space contained in a submanifold? — On the number of derivatives needed to tell*, J. reine angew. Math. (1999), **508**: 53–60.

[19] ——, *Lines on projective varieties*, J. Reine Angew. Math. (2003), **562**: 1–3, MR2011327

[20] ——, *Griffiths-Harris rigidity of compact Hermitian symmetric spaces*, J. Differential Geometry (2006), **74**: 395–405.

[21] J.M. LANDSBERG AND L. MANIVEL, *On the projective geometry of rational homogeneous varieties*, Comment. Math. Helv. (2003), **78**(1): 65–100, MR1966752.

[22] ——, *Classification of simple Lie algebras via projective geometry*, Selecta Mathematica (2002), **8**: 137–159, MR1890196.

[23] ——, *Representation theory and projective geometry*, Algebraic Transformation Groups and Algebraic Varieties, Ed. V.L. Popov, Encyclopaedia of Mathematical Sciences, Springer 2004, **132**: 71–122.

[24] ——, *Series of Lie groups*, Michigan Math. J. (2004), **52**(2): 453–479, MR2069810.

[25] ——, *Triality, exceptional Lie algebras, and Deligne dimension formulas*, Adv. Math. (2002), **171**: 59–85.

[26] ———, *The sextonions and* $E_{7\frac{1}{2}}$, Adv. Math. (2006), **201**(1): 143–179, MR2204753

[27] ———, *A universal dimension formula for complex simple Lie algebras*, Adv. Math. (2006), **201**(2): 379–407, MR2211533

[28] ———, *Legendrian varieties*, math.AG/0407279. To appear in Asian Math. J.

[29] J.M. LANDSBERG AND C. ROBLES, *Fubini's theorem in codimension two*, preprint math.AG/0509227.

[30] E. MEZZETTI AND D. PORTELLI, *On threefolds covered by lines*, Abh. Math. Sem. Univ. Hamburg (2000), **70**: 211–238, MR1809546.

[31] C. ROBLES, *Rigidity of the adjoint variety of* $\mathfrak{sl}_n$, preprint math.DG/0608471.

[32] E. ROGORA, *Varieties with many lines*, Manuscripta Math. (1994), **82**(2): 207–226.

[33] T. SASAKI, K. YAMAGUCHI, AND M. YOSHIDA, *On the rigidity of differential systems modelled on Hermitian symmetric spaces and disproofs of a conjecture concerning modular interpretations of configuration spaces*, in CR-geometry and overdetermined systems (Osaka, 1994), 318–354, Adv. Stud. Pure Math., **25**, Math. Soc. Japan, Tokyo, 1997.

[34] Y. SE-ASHI, *On differential invariants of integrable finite type linear differential equations*, Hokkaido Math. J. (1988), **17**(2): 151–195. MR0945853

[35] B. SEGRE, *Bertini forms and Hessian matrices*, J. London Math. Soc. (1951), **26**: 164–176.

PSEUDO–GROUPS, MOVING FRAMES, AND DIFFERENTIAL INVARIANTS

PETER J. OLVER* AND JUHA POHJANPELTO†

Abstract. We survey recent developments in the method of moving frames for infinite-dimensional Lie pseudo-groups. These include a new, direct approach to the construction of invariant Maurer–Cartan forms and the Cartan structure equations for pseudo-groups, and new algorithms, based on constructive commutative algebra, for establishing the structure of their differential invariant algebras.

1. Introduction.

Lie pseudo-groups are the infinite-dimensional counterparts of local Lie groups of transformations. In Lie's day, abstract Lie groups were as yet unknown, and, as a result, no significant distinction was drawn between finite-dimensional and infinite-dimensional theory. However, since then the two subjects have traveled along radically different paths. The finite-dimensional theory has been rigorously formalized, and is a well-established and widely used mathematical tool. In contrast, the theory of infinite-dimensional pseudo-groups remains surprisingly primitive in its current overall state of development. Since there is still no generally accepted abstract objects to play the role of infinite-dimensional Lie groups, Lie pseudo-groups only arise through their concrete action on a space. This makes the classification problems and analytical foundations of the subject thorny, particularly in the intransitive situation. We refer the reader to the original papers of Lie, Medolaghi, Tresse and Vessiot, [38, 49, 73, 75], for the classical theory of pseudo-groups, to Cartan, [13], for their reformulation in terms of exterior differential systems, and [20, 30, 31, 36, 37, 40, 39, 65, 66, 70, 72] for a variety of modern approaches. Various nonconstructive approaches to the classification of differential invariants of Lie pseudo-groups are studied in [34, 35, 52, 73].

Lie pseudo-groups appear in many fundamental physical and geometrical contexts, including gauge symmetries, [5], Hamiltonian mechanics and symplectic and Poisson geometry, [54], conformal geometry of surfaces and conformal field theory, [19, 21], the geometry of real hypersurfaces, [16], symmetry groups of both linear and nonlinear partial differential equations, such as the Navier-Stokes and Kadomtsev–Petviashvili (KP) equations appearing in fluid and plasma mechanics, [4, 18, 54], Vessiot's group splitting method for producing explicit solutions to nonlinear partial differential equations, [46, 53, 64, 75], mathematical morphology and computer vision, [69, 76], and geometric numerical integration, [47]. Pseudogroups

*School of Mathematics, University of Minnesota, Minneapolis, MN 55455 (olver@math.umn.edu, http://www.math.umn.edu/~olver). Supported in part by NSF Grants DMS 05-05293.

†Department of Mathematics, Oregon State University, Corvallis, OR 97331 (juha@math.oregonstate.edu, http://oregonstate.edu/~pohjanpp). Supported in part by NSF Grants DMS 04-53304 and OCE 06-21134.

also appear as foliation-preserving groups of transformations, with the associated characteristic classes defined by certain invariant forms, cf. [24]. Also, keep in mind that all (sufficiently regular) local Lie group actions can be regarded as Lie pseudo-groups.

In a series of collaborative papers, starting with [22, 23], the first author has successfully reformulated the classical theory of moving frames, [11, 25], in a general, algorithmic, and equivariant framework that can be readily applied to a wide range of finite-dimensional Lie group actions. Applications have included complete classifications of differential invariants and their syzygies, [59], equivalence and symmetry properties of submanifolds, rigidity theorems, invariant signatures in computer vision, [2, 6, 8, 10, 26, 57], joint invariants and joint differential invariants, [7, 57], rational and algebraic invariants of algebraic group actions, [28, 29], invariant numerical algorithms, [32, 58, 76], classical invariant theory, [3, 56], Poisson geometry and solitons, [43–45], Killing tensors arising in separation of variables and general relativity, [48, 71], and the calculus of variations, [33]. New applications of these methods to computation of symmetry groups and classification of partial differential equations can be found in [42, 50, 51]. MAPLE software implementing the moving frame algorithms, written by E. Hubert, can be found at [27]

Our main goal in this contribution is to survey the extension of the moving frame theory to general Lie pseudo-groups recently put forth by the authors in [60–63], and in [14, 15] in collaboration with J. Cheh. Following [33], we develop the theory in the context of two different variational bicomplexes — the first over the infinite jet bundle $\mathcal{D}^{(\infty)} \subset \mathrm{J}^\infty(M, M)$ of local diffeomorphisms of M, and the second over the infinite jet bundle $\mathrm{J}^\infty(M, p)$ of p-dimensional submanifolds $N \subset M$, [1, 33, 74]. The interplay between these two bicomplexes underlies our moving frame constructions. Importantly, the invariant contact forms on the diffeomorphism jet bundle $\mathcal{D}^{(\infty)}$ will play the role of Maurer–Cartan forms for the diffeomorphism pseudo-group. This identification enables us to explicitly formulate the diffeomorphism structure equations in the form of a simple formal power series identity. Restricting the diffeomorphism-invariant forms to the pseudo-group subbundle $\mathcal{G}^{(\infty)} \subset \mathcal{D}^{(\infty)}$ yields a complete system of Maurer–Cartan forms for the pseudo-group. The remarkable fact is that the Maurer–Cartan forms satisfy an "invariantized" version of the linear infinitesimal determining equations for the pseudo-group, and, as a result, we can immediately produce an explicit form of the pseudo-group structure equations. Application of these results to the design of a practical computational algorithm for directly determining the structure of symmetry (pseudo-)groups of partial differential equations can be found in [4, 14, 15, 51].

Assuming freeness (as defined below) of the prolonged pseudo-group action at sufficiently high order, the explicit construction of the moving frame is founded on the Cartan normalization procedure associated with a choice of local cross-section to the pseudo-group orbits in $\mathrm{J}^\infty(M, p)$. The moving frame induces an invariantization process that projects general differential functions and differential forms on $J^\infty(M,p)$ to invariant counterparts. In particular, invariantization of the standard jet coordinates results in a complete local system of normalized differential invariants, while invariantization of the horizontal and contact one-forms yields an invariant coframe. The corresponding dual invariant total derivative operators will map invariants to invariants of higher order. The structure of the algebra of differential invariants, including the specification of a finite generating set as well as a finite generating system for their syzygies (differential relations), will then follow from the recurrence formulae that relate the differentiated and normalized differential invariants. Remarkably, this final step requires only linear algebra and differentiation based on the infinitesimal determining equations of the pseudo-group action, and not the explicit formulae for either the differential invariants, the invariant differential operators, or the moving frame. Except possibly for some low order complications, the underlying structure of the differential invariant algebra is then entirely governed by two commutative algebraic modules: the symbol module of the infinitesimal determining system of the pseudo-group and a new module, named the "prolonged symbol module", built up from the symbols of the prolonged infinitesimal generators.

The paper begins with a discussion of the most basic example — the diffeomorphism pseudo-group of a manifold. The usual variational bicomplex structure on the diffeomorphism jets is employed to construct the Maurer–Cartan forms as invariant contact forms, and write out the complete system of structure equations. Section 3 shows how the structure equations of a Lie pseudo-group are obtained by restricting the diffeomorphism structure equations to the solution space to the infinitesimal determining equations. In Section 4, we develop the moving frame constructions for the prolonged action on submanifold jets, and explain how to determine a complete system of differential invariants. In Section 5, we explicitly derive the recurrence formulae for the differentiated invariants, demonstrating, in particular,that the differential invariants of any transitive pseudo-group form a non-commutative rational differential algebra. Finally, in Section 6 we present a constructive version of the Basis Theorem that provides a finite system of generating differential invariants for a large class of pseudo-group actions and the generators of their differential syzygies. Lack of space precludes us from including any serious examples, and, for this, we refer the reader to [14, 15, 61–63].

2. The Diffeomorphism Pseudo–Group. Let M be a smooth m-dimensional manifold. Let $\mathcal{D} = \mathcal{D}(M)$ denote the pseudo-group of all local diffeomorphisms[1] $\varphi\colon M \to M$. For each $0 \le n \le \infty$, let $\mathcal{D}^{(n)} = \mathcal{D}^{(n)}(M) \subset J^n(M,M)$ denote the nth order diffeomorphism jet groupoid, [41], with

[1]Our notation for maps and functions allows the possibility that their domain be an open subset of the source space, so in this case $\operatorname{dom}\varphi \subset M$.

source map $\sigma^{(n)}(\mathrm{j}_n\varphi|_z) = z$ and target map $\tau^{(n)}(\mathrm{j}_n\varphi|_z) = \varphi(z) = Z$. The groupoid multiplication is induced by composition of diffeomorphisms. Following Cartan, [12, 13], we will consistently use lower case letters, $z, x, u, \ldots$ for the source coordinates and the corresponding upper case letters $Z, X, U, \ldots$ for the target coordinates of our diffeomorphisms $Z = \varphi(z)$. Given local coordinates $(z, Z) = (z^1, \ldots, z^m, Z^1, \ldots, Z^m)$ on an open subset of $M \times M$, the induced local coordinates of $\mathrm{j}_n\varphi|_z \in \mathcal{D}^{(n)}$ are denoted $g^{(n)} = (z, Z^{(n)})$, where the components Z^a_B of $Z^{(n)}$, for $a = 1, \ldots, m$, $\#B \leq n$, represent the partial derivatives $\partial^B \varphi^a / \partial z^B$ of φ at the source point $z = \sigma^{(n)}(g^{(n)})$.

Since $\mathcal{D}^{(\infty)} \subset \mathrm{J}^\infty(M, M)$, the inherited variational bicomplex structure, [1, 74], provides a natural splitting of the cotangent bundle $T^*\mathcal{D}^{(\infty)}$ into horizontal and vertical (contact) components, [1, 55], and we use $d = d_M + d_G$ to denote the induced splitting of the differential. In terms of local coordinates $g^{(\infty)} = (z, Z^{(\infty)})$, the horizontal subbundle of $T^*\mathcal{D}^{(\infty)}$ is spanned by the one-forms $dz^a = d_M\, z^a$, $a = 1, \ldots, m$, while the vertical subbundle is spanned by the basic *contact forms*

$$\Upsilon^a_B = d_G\, Z^a_B = dZ^a_B - \sum_{c=1}^m Z^a_{B,c}\, dz^c, \qquad a = 1, \ldots, m, \qquad \#B \geq 0. \quad (2.1)$$

Composition of local diffeomorphisms induces an action of $\psi \in \mathcal{D}$ by right multiplication on diffeomorphism jets: $\mathrm{R}_\psi(\mathrm{j}_n\varphi|_z) = \mathrm{j}_n(\varphi \circ \psi^{-1})|_{\psi(z)}$. A differential form μ on $\mathcal{D}^{(n)}$ is *right-invariant* if $\mathrm{R}^*_\psi\, \mu = \mu$, where defined, for every $\psi \in \mathcal{D}$. Since the splitting of forms on $\mathcal{D}^{(\infty)}$ is invariant under this action, if μ is any right-invariant differential form, so are $d_M\, \mu$ and $d_G\, \mu$. The target coordinate functions $Z^a \colon \mathcal{D}^{(0)} \to \mathbb{R}$ are obviously right-invariant, and hence their horizontal differentials

$$\sigma^a = d_M\, Z^a = \sum_{b=1}^m Z^a_b\, dz^b, \qquad a = 1, \ldots, m, \quad (2.2)$$

form an invariant horizontal coframe, while their vertical differentials

$$\mu^a = d_G\, Z^a = \Upsilon^a = dZ^a - \sum_{b=1}^m Z^a_b\, dz^b, \qquad a = 1, \ldots, m, \quad (2.3)$$

are the zeroth order invariant contact forms. Let $\mathbb{D}_{Z^1}, \ldots, \mathbb{D}_{Z^m}$ be the total derivative operators dual to the horizontal forms (2.2), so that

$$d_M\, F = \sum_{a=1}^m \mathbb{D}_{z^a} F\; dz^a \qquad \text{for any} \qquad F \colon \mathcal{D}^{(\infty)} \to \mathbb{R}. \quad (2.4)$$

Then the higher-order invariant contact forms are obtained by successively Lie differentiating the invariant contact forms (2.3):

$$\mu^a_B = \mathbb{D}^B_Z \mu^a = \mathbb{D}^B_Z \Upsilon^a, \qquad a = 1, \ldots, m, \qquad k = \#B \geq 0, \quad (2.5)$$

where $\mathbb{D}_Z^B = \mathbb{D}_{Z^{b_1}} \cdots \mathbb{D}_{Z^{b_k}}$. As explained in [61], the right-invariant contact forms $\mu^{(\infty)} = (\ \ldots\ \mu_B^a\ \ldots\ )$ are to be viewed as the *Maurer–Cartan forms* for the diffeomorphism pseudo-group.

The next step in our program is to establish the structure equations for the diffeomorphism groupoid $\mathcal{D}^{(\infty)}$. Let $\mu[\![\,H\,]\!]$ denote the column vector whose components are the invariant contact form-valued formal power series

$$\mu^a[\![\,H\,]\!] = \sum_{\#B\geq 0} \frac{1}{B!}\,\mu_B^a\,H^B, \qquad a = 1,\ldots,m, \tag{2.6}$$

depending on the formal parameters $H = (H^1,\ldots,H^m)$. Further, let $dZ = \mu[\![\,0\,]\!] + \sigma$ denote column vectors of one-forms whose entries are $dZ^a = \mu^a + \sigma^a$ for $a = 1,\ldots,m$.

THEOREM 2.1. *The complete structure equations for the diffeomorphism pseudo-group are obtained by equating coefficients in the power series identities*

$$\begin{aligned} d\mu[\![\,H\,]\!] &= \nabla_H \mu[\![\,H\,]\!] \wedge \big(\,\mu[\![\,H\,]\!] - dZ\,\big), \\ d\,\sigma &= -\,d\mu[\![\,0\,]\!] = \nabla_H \mu[\![\,0\,]\!] \wedge \sigma. \end{aligned} \tag{2.7}$$

Here $\nabla_H \mu[\![\,H\,]\!] = \left(\dfrac{\partial \mu^a}{\partial H^b}\,[\![\,H\,]\!] \right)$ *denotes the* $m \times m$ *formal power series Jacobian matrix.*

3. Lie Pseudo–Groups. The literature contains several variants of the precise technical definition of a Lie pseudo-group. Ours is:

DEFINITION 3.1. A sub-pseudo-group $\mathcal{G} \subset \mathcal{D}$ will be called a *Lie pseudo-group* if there exists $n_0 \geq 1$ such that for all finite $n \geq n_0$:

a) the corresponding sub-groupoid $\mathcal{G}^{(n)} \subset \mathcal{D}^{(n)}$ forms a smooth, embedded subbundle,
b) every smooth local solution $Z = \varphi(z)$ to the determining system $\mathcal{G}^{(n)}$ belongs to $\mathcal{G}$,
c) $\mathcal{G}^{(n)} = \mathrm{pr}^{(n-n_0)}\,\mathcal{G}^{(n_0)}$ is obtained by prolongation.

The minimal value of n_0 is called the *order* of the pseudo-group.

Thus, on account of conditions (a) and (c), for $n \geq n_0$, the pseudo-group jet subgroupoid $\mathcal{G}^{(n)} \subset \mathcal{D}^{(n)}$ is defined in local coordinates by a formally integrable system of nth order nonlinear partial differential equations

$$F^{(n)}(z, Z^{(n)}) = 0, \tag{3.1}$$

known as the *determining equations* for the pseudo-group. Condition (b) says that the set of local solutions $Z = \varphi(z)$ to the determining equations coincides with the set of pseudo-group transformations.

The key to analyzing pseudo-group actions is to work infinitesimally, using the generating Lie algebra[2] of vector fields. Let $\mathcal{X} = \mathcal{X}(M)$ denote the space of locally defined vector fields on M, which we write in local coordinates as

$$\mathbf{v} = \sum_{a=1}^{m} \zeta^a(z) \frac{\partial}{\partial z^a} \,. \tag{3.2}$$

Let $\mathrm{J}^n TM$, for $0 \leq n \leq \infty$, denote the tangent n-jet bundle. Local coordinates on $\mathrm{J}^n TM$ are indicated by $(z, \zeta^{(n)}) = (\ \ldots\ z^a\ \ldots\ \zeta^a_B\ \ldots\ )$, $a = 1, \ldots, m$, $\#B \leq n$, where the fiber coordinate ζ^a_B represents the partial derivative $\partial^B \zeta^a / \partial z^B$.

Let $\mathfrak{g} \subset \mathcal{X}$ denote the space of infinitesimal generators of the pseudo-group, i.e., the set of locally defined vector fields (3.2) whose flows belong to $\mathcal{G}$. In local coordinates, we can view $\mathrm{J}^n \mathfrak{g} \subset \mathrm{J}^n TM$ as defining a formally integrable linear system of partial differential equations

$$L^{(n)}(z, \zeta^{(n)}) = 0 \tag{3.3}$$

for the vector field coefficients (3.2), called the *linearized* or *infinitesimal determining equations* for the pseudo-group. They can be obtained by linearizing the nth order determining equations (3.1) at the identity jet. If $\mathcal{G}$ is the symmetry group of a system of differential equations, then the linearized determining equations (3.3) are (the involutive completion of) the usual determining equations for its infinitesimal generators obtained via Lie's algorithm, [54].

As with finite-dimensional Lie groups, the structure of a pseudo-group can be described by its invariant Maurer–Cartan forms. A complete system of right-invariant one-forms on $\mathcal{G}^{(\infty)} \subset \mathcal{D}^{(\infty)}$ is obtained by restricting (or pulling back) the Maurer–Cartan forms (2.2–2.5). For simplicity, we continue to denote these forms by σ^a, μ^a_B. The restricted Maurer–Cartan forms are, of course, no longer linearly independent, but are subject to certain constraints prescribed by the pseudo-group. Remarkably, these constraints can be explicitly characterized by an invariant version of the linearized determining equations (3.3), which is formally obtained by replacing source coordinates z^a by the corresponding target coordinates Z^a and vector field jet coordinates ζ^a_B by the corresponding Maurer–Cartan form μ^a_B.

THEOREM 3.1. *The linear system*

$$L^{(n)}(Z, \mu^{(n)}) = 0 \tag{3.4}$$

serves to define the complete set of dependencies among the right-invariant Maurer–Cartan forms $\mu^{(n)}$ on $\mathcal{G}^{(n)}$. Consequently, the structure equations

[2]Here, we are using the term "Lie algebra" loosely, since, technically, the vector fields may only be locally defined, and so their Lie brackets only make sense on their common domains of definition.

for $\mathcal{G}$ are obtained by restriction of the diffeomorphism structure equations (2.7) *to the kernel of the linearized involutive system* (3.4).

In this way, we effectively and efficiently bypass Cartan's more complicated prolongation procedure, [9, 13], for accessing the pseudo-group structure equations. Examples of this procedure can be found in [14, 61]; see also [51] for a comparison with other approaches.

EXAMPLE 3.1. Let us consider the particular pseudo-group

$$\begin{gathered} X = f(x), \qquad\qquad Y = e(x,y) \equiv f'(x)\,y + g(x), \\ U = u + \frac{e_x(x,y)}{f'(x)} = u + \frac{f''(x)\,y + g'(x)}{f'(x)}, \end{gathered} \tag{3.5}$$

acting on $M = \mathbf{R}^3$, with local coordaintes (x, y, u). Here $f(x) \in \mathcal{D}(\mathbf{R})$, while $g(x) \in \mathrm{C}^\infty(\mathbf{R})$. The determining equations are the first order involutive system

$$X_y = X_u = 0, \quad Y_y = X_x \neq 0, \quad Y_u = 0, \quad Y_x = (U-u)X_x, \quad U_u = 1. \tag{3.6}$$

The infinitesimal generators of the pseudo-group have the form

$$\begin{aligned} \mathbf{v} &= \xi\,\frac{\partial}{\partial x} + \eta\,\frac{\partial}{\partial y} + \varphi\,\frac{\partial}{\partial u} \\ &= a(x)\,\frac{\partial}{\partial x} + \big[\,a'(x)\,y + b(x)\,\big]\,\frac{\partial}{\partial y} + \big[\,a''(x)\,y + b'(x)\,\big]\,\frac{\partial}{\partial u}, \end{aligned} \tag{3.7}$$

where $a(x), b(x)$ are arbitrary smooth functions. The infinitesimal generators (3.7) form the general solution to the first order involutive infinitesimal determining system

$$\xi_x = \eta_y, \qquad \xi_y = \xi_u = \eta_u = \varphi_u = 0, \qquad \eta_x = \varphi, \tag{3.8}$$

obtained by linearizing (3.6) at the identity.

The Maurer–Cartan forms are obtained by repeatedly differentiating $\mu = d_G X$, $\widetilde{\mu} = d_G Y$, $\nu = d_G U$, so that $\mu_{j,k,l} = \mathbf{D}_X^j \mathbf{D}_Y^k \mathbf{D}_U^l \mu$, etc. According to Theorem 3.1, they are subject to the linear relations

$$\mu_X = \widetilde{\mu}_Y, \qquad \mu_Y = \mu_U = \widetilde{\mu}_U = \nu_U = 0, \qquad \widetilde{\mu}_X = \nu, \tag{3.9}$$

along with all their "differential" consequences. Writing out (2.7), we are led to the following structure equations

$$\begin{aligned} d\mu_n &= \sigma \wedge \mu_{n+1} - \sum_{j=1}^{[(n+1)/2]} \frac{n-2j+1}{n+1} \binom{n+1}{j}\, \mu_j \wedge \mu_{n+1-j}, \\ d\widetilde{\mu}_n &= \sigma \wedge \widetilde{\mu}_{n+1} + \widetilde{\sigma} \wedge \mu_{n+1} - \sum_{j=0}^{n-1} \frac{n-2j-1}{n+1} \binom{n+1}{j+1}\, \widetilde{\mu}_{j+1} \wedge \mu_{n-j}, \\ d\sigma &= -\,d\mu = -\,\sigma \wedge \mu_X, \\ d\widetilde{\sigma} &= -\,d\widetilde{\mu} = -\,\sigma \wedge \widetilde{\mu}_X - \widetilde{\sigma} \wedge \mu_X, \\ d\tau &= -\,d\nu = -\,d\widetilde{\mu}_X = -\,\sigma \wedge \widetilde{\mu}_{XX} - \widetilde{\sigma} \wedge \mu_{XX}, \end{aligned} \tag{3.10}$$

in which $\sigma = d_M X, \widetilde{\sigma} = d_M Y, \tau = d_M U$, and $\mu_n = \mu_{n,0,0}$, $\widetilde{\mu}_n = \widetilde{\mu}_{n,0,0}$, for $n = 0, 1, 2, \ldots$, form a basis for the Maurer–Cartan forms of the pseudo-group. See [61] for full details.

4. Pseudo–Group Actions on Submainfolds. Our primary focus is to study the induced action of pseudo-groups on submanifolds. For $0 \leq n \leq \infty$, let $J^n = J^n(M,p)$ denote the nth order (extended) jet bundle consisting of equivalence classes of p-dimensional submanifolds $S \subset M$ under the equivalence relation of nth order contact, cf. [55]. We employ the standard local coordinates

$$z^{(n)} = (x, u^{(n)}) = (\ \ldots\ x^i\ \ldots\ u^\alpha_J\ \ldots\ ) \tag{4.1}$$

on J^n induced by a splitting of the local coordinates

$$z = (x,u) = (x^1, \ldots, x^p, u^1, \ldots, u^q)$$

on M into p independent and $q = m - p$ dependent variables, [54, 55]. The choice of independent and dependent variables induces the variational bicomplex structure on J^∞, [1, 74]. The basis *horizontal forms* are the differentials $dx^1, \ldots, dx^p$ of the independent variables, while the basis *contact forms* are denoted by

$$\theta^\alpha_J = du^\alpha_J - \sum_{i=1}^p u^\alpha_{J,i}\, dx^i, \qquad \alpha = 1, \ldots, q, \qquad \#J \geq 0. \tag{4.2}$$

This decomposition splits the differential $d = d_H + d_V$ on J^∞ into horizontal and vertical (or contact) components, and endows the space of differential forms with the structure of a variational bicomplex, [1, 33, 74].

Local diffeomorphisms $\varphi \in \mathcal{D}$ preserve the contact equivalence relation between submanifolds, and thus induce an action on the jet bundle $J^n = J^n(M,p)$, known as the nth prolonged action, which, by the chain rule, factors through the diffeomorphism jet groupoid $\mathcal{D}^{(n)}$. Let $\mathcal{H}^{(n)}$ denote the groupoid obtained by pulling back the pseudo-group jet groupoid $\mathcal{G}^{(n)} \to M$ via the projection $\widetilde{\pi}^n_0 \colon J^n \to M$. Local coordinates on $\mathcal{H}^{(n)}$ are written $(x, u^{(n)}, g^{(n)})$, where $(x, u^{(n)})$ are the submanifold jet coordinates on J^n, while the fiber coordinates $g^{(n)}$ serve to parametrize the pseudo-group jets.

DEFINITION 4.1. A *moving frame* $\rho^{(n)}$ of *order* n is a $\mathcal{G}^{(n)}$ equivariant local section of the bundle $\mathcal{H}^{(n)} \to J^n$.

Thus, in local coordinates, the moving frame section has the form

$$\rho^{(n)}(x, u^{(n)}) = (x, u^{(n)}, \gamma^{(n)}(x, u^{(n)})), \quad \text{where} \quad g^{(n)} = \gamma^{(n)}(x, u^{(n)}) \tag{4.3}$$

defines a right equivariant map to the pseudo-group jets. A moving frame $\rho^{(k)} \colon J^k \to \mathcal{H}^{(k)}$ of order $k > n$ is *compatible* with $\rho^{(n)}$ provided $\widehat{\pi}^k_n \circ \rho^{(k)} = \rho^{(n)} \circ \widetilde{\pi}^k_n$ where defined, with $\widehat{\pi}^k_n \colon \mathcal{H}^{(k)} \to \mathcal{H}^{(n)}$ and $\widetilde{\pi}^k_n \colon J^k \to J^n$ denoting the evident projections. A *complete moving frame* is provided by a mutually compatible collection of moving frames of all orders $k \geq n$.

As in the finite-dimensional construction, [23], the (local) existence of a moving frame requires that the prolonged pseudo-group action be free and regular.

DEFINITION 4.2. The pseudo-group $\mathcal{G}$ acts *freely* at $z^{(n)} \in \mathrm{J}^n$ if its *isotropy subgroup* is trivial,

$$\mathcal{G}^{(n)}_{z^{(n)}} = \left\{ \mathbf{g}^{(n)} \in \mathcal{G}^{(n)} \;\middle|\; \mathbf{g}^{(n)} \cdot z^{(n)} = z^{(n)} \right\} = \{ \mathbb{1}^{(n)}_z \}, \qquad (4.4)$$

and *locally freely* if $\mathcal{G}^{(n)}_{z^{(n)}}$ is a discrete subgroup.

Warning: According to the standard definition, [23], any (locally) free action of a finite-dimensional Lie group satisfies the (local) freeness condition of Definition 4.2, but *not* necessarily conversely.

The pseudo-group acts locally freely at $z^{(n)}$ if and only if the prolonged pseudo-group orbit through $z^{(n)}$ has dimension $r_n = \dim \mathcal{G}^{(n)}|_z$. Thus, freeness of the pseudo-group at order n requires, at the very least, that

$$r_n = \dim \mathcal{G}^{(n)}|_z \leq \dim \mathrm{J}^n = p + (m-p)\binom{p+n}{p}. \qquad (4.5)$$

Freeness thus provides an alternative and simpler means of quantifying the Spencer cohomological growth conditions imposed on the pseudo-group in [34, 35]. Pseudo-groups having too large a fiber dimension r_n will, typically, act transitively on (a dense open subset of) J^n, and thus possess no non-constant differential invariants. A key result of [63], generalizing the finite-dimensional case, is the persistence of local freeness.

THEOREM 4.1. *Let $\mathcal{G}$ be a Lie pseudo-group acting on an m-dimensional manifold M. If $\mathcal{G}$ acts locally freely at $z^{(n)} \in \mathrm{J}^n$ for some $n > 0$, then it acts locally freely at any $z^{(k)} \in \mathrm{J}^k$ with $\widetilde{\pi}^k_n(z^{(k)}) = z^{(n)}$, for $k \geq n$.*

As in the finite-dimensional version, [23], moving frames are constructed through a normalization procedure based on a choice of *cross-section* to the pseudo-group orbits, i.e., a transverse submanifold of the complementary dimension.

THEOREM 4.2. *Suppose $\mathcal{G}^{(n)}$ acts freely on an open subset $\mathcal{V}^n \subset \mathrm{J}^n$, with its orbits forming a regular foliation. Let $K^n \subset \mathcal{V}^n$ be a (local) cross-section to the pseudo-group orbits. Given $z^{(n)} \in \mathcal{V}^n$, define $\rho^{(n)}(z^{(n)}) \in \mathcal{H}^{(n)}$ to be the unique pseudo-group jet such that $\widetilde{\boldsymbol{\sigma}}^{(n)}(\rho^{(n)}(z^{(n)})) = z^{(n)}$ and $\widetilde{\boldsymbol{\tau}}^{(n)}(\rho^{(n)}(z^{(n)})) \in K^n$ (when such exists). Then $\rho^{(n)} \colon \mathrm{J}^n \to \mathcal{H}^{(n)}$ is a moving frame for $\mathcal{G}$ defined on an open subset of $\mathcal{V}^n$ containing K^n.*

Usually — and, to simplify the development, from here on — we select a coordinate cross-section of minimal order, defined by fixing the values of

r_n of the individual submanifold jet coordinates $(x, u^{(n)})$. We write out the explicit formulae $(X, U^{(n)}) = F^{(n)}(x, u^{(n)}, g^{(n)})$ for the prolonged pseudo-group action in terms of a convenient system of pseudo-group parameters $g^{(n)} = (g_1, \ldots, g_{r_n})$. The r_n components corresponding to our choice of cross-section variables serve to define the *normalization equations*

$$F_1(x, u^{(n)}, g^{(n)}) = c_1, \qquad \ldots \qquad F_{r_n}(x, u^{(n)}, g^{(n)}) = c_{r_n}, \tag{4.6}$$

which, when solved for the pseudo-group parameters $g^{(n)} = \gamma^{(n)}(x, u^{(n)})$, produces the moving frame section (4.3).

With the moving frame in place, the general invariantization procedure introduced in [33] in the finite-dimensional case adapts straightforwardly. To compute the invariantization of a function, differential form, differential operator, etc., one writes out how it explicitly transforms under the pseudo-group, and then replaces the pseudo-group parameters by their moving frame expressions (4.3). Invariantization defines a morphism that projects the exterior algebra differential functions and forms onto the algebra of invariant differential functions and forms. In particular, invariantizing the coordinate functions on J^∞ leads to the *normalized differential invariants*

$$H^i = \iota(x^i), \quad i = 1, \ldots, p, \qquad I^\alpha_J = \iota(u^\alpha_J), \quad \alpha = 1, \ldots, q, \quad \#J \geq 0, \tag{4.7}$$

collectively denoted by $(H, I^{(n)}) = \iota(x, u^{(n)})$. The normalized differential invariants naturally split into two subspecies: those appearing in the normalization equations (4.6) will be constant, and are known as the *phantom differential invariants*. The remaining $s_n = \dim \mathrm{J}^n - r_n$ components, called the *basic differential invariants*, form a complete system of functionally independent differential invariants of order $\leq n$ for the prolonged pseudo-group action on p-dimensional submanifolds.

Secondly, invariantization of the basis horizontal one-forms leads to the invariant one-forms

$$\varpi^i = \iota(dx^i) = \omega^i + \kappa^i, \qquad i = 1, \ldots, p, \tag{4.8}$$

where ω^i, κ^i denote, respectively, the horizontal and vertical (contact) components. If the pseudo-group acts projectably, then the contact components vanish: $\kappa^i = 0$. The horizontal forms $\omega^1, \ldots, \omega^p$ provide, in the language of [55], a contact-invariant coframe on J^∞. The dual invariant differential operators $\mathcal{D}_1, \ldots, \mathcal{D}_p$ are uniquely defined by the formula

$$dF = \sum_{i=1}^{p} \mathcal{D}_i F \, \varpi^i + \cdots, \tag{4.9}$$

valid for any differential function F, where the dots indicate contact components which are not needed here, but do play an important role in the study of invariant variational problems, cf. [33]. The invariant differential operators $\mathcal{D}_i$ map differential invariants to differential invariants. In general, they do not commute, but are subject to linear commutation relations of the form

$$[\mathcal{D}_i, \mathcal{D}_j] = \sum_{k=1}^{p} Y_{ij}^k \, \mathcal{D}_k, \qquad i,j = 1, \ldots, p, \tag{4.10}$$

where the coefficients Y_{ij}^k are certain differential invariants. Finally, invariantizing the basis contact one-forms

$$\vartheta_K^\alpha = \iota(\theta_K^\alpha), \qquad \alpha = 1, \ldots, q, \qquad \#K \geq 0, \tag{4.11}$$

provide a complete system of invariant contact one-forms. The invariant coframe serves to define the invariant variational complex for the pseudo-group, [33].

The *Basis Theorem* for differential invariants states that, assuming freeness of the sufficiently high order prolonged pseudo-group action, then locally, there exist a finite number of *generating differential invariants* $I_1, \ldots, I_\ell$, with the property that every differential invariant can be locally expressed as a function of the generating invariants and their invariant derivatives:

$$\mathcal{D}_J I_\kappa = \mathcal{D}_{j_1} \mathcal{D}_{j_2} \cdots \mathcal{D}_{j_k} I_\kappa .$$

The differentiated invariants are not necessarily independent, but may be subject to certain functional relations or *differential syzygies* of the form

$$H(\ \ldots \ \mathcal{D}_J I_\kappa \ \ldots \ ) \equiv 0. \tag{4.12}$$

A consequence of our moving frame methods is a constructive algorithm for producing a (not necessarily minimal) system of generating differential invariants, as well as a complete, finite system of generating syzygies, meaning that any other syzygy is a differential consequence thereof.

EXAMPLE 4.1. Consider the action of the pseudo-group (3.5) on surfaces $u = h(x,y)$. Under the pseudo-group transformations, the basis horizontal forms dx, dy are mapped to the one-forms

$$d_H X = f_x \, dx, \qquad d_H Y = e_x \, dx + f_x \, dy. \tag{4.13}$$

The prolonged pseudo-group transformations are found by applying the dual implicit differentiations

$$\mathrm{D}_X = \frac{1}{f_x} \, \mathrm{D}_x - \frac{e_x}{f_x^2} \, \mathrm{D}_y, \qquad \mathrm{D}_Y = \frac{1}{f_x} \, \mathrm{D}_y,$$

successively to $U = u + e_x / f_x$, so that

$$
\begin{aligned}
U_X &= \frac{u_x}{f_x} + \frac{e_{xx} - e_x\,u_y}{f_x^2} - 2\,\frac{f_{xx}\,e_x}{f_x^3}, \qquad U_Y = \frac{u_y}{f_x} + \frac{f_{xx}}{f_x^2}, \\
U_{XX} &= \frac{u_{xx}}{f_x^2} + \frac{e_{xxx} - e_{xx}\,u_y - 2\,e_x\,u_{xy} - f_{xx}\,u_x}{f_x^3} \\
&\quad + \frac{e_x^2\,u_{yy} + 3\,e_x f_{xx}\,u_y - 4\,e_{xx}\,f_{xx} - 3\,e_x\,f_{xxx}}{f_x^4} + 8\,\frac{e_x\,f_{xx}^2}{f_x^5}, \\
U_{XY} &= \frac{u_{xy}}{f_x^2} + \frac{f_{xxx} - f_{xx}\,u_y - e_x\,u_{yy}}{f_x^3} - 2\,\frac{f_{xx}^2}{f_x^4}, \qquad U_{YY} = \frac{u_{yy}}{f_x^2},
\end{aligned}
\tag{4.14}
$$

and so on. In these formulae, the jet coordinates $f, f_x, f_{xx}, \ldots, e, e_x, e_{xx}, \ldots$ are to be regarded as the independent pseudo-group parameters. The pseudo-group cannot act freely on J^1 since $r_1 = \dim \mathcal{G}^{(1)}|_z = 6 > \dim \mathrm{J}^1 = 5$. On the other hand, $r_2 = \dim \mathcal{G}^{(2)}|_z = 8 = \dim \mathrm{J}^2$, and the action on J^2 is, in fact, locally free and transitive on the sets $\mathcal{V}^2{}_+ = \mathrm{J}^2 \cap \{u_{yy} > 0\}$ and $\mathcal{V}^2{}_- = \mathrm{J}^2 \cap \{u_{yy} < 0\}$. Moreover, as predicted by Theorem 4.1, $\mathcal{G}^{(n)}$ acts locally freely on the corresponding open subsets of J^n for any $n \geq 2$.

To construct the moving frame, we successively solve the following coordinate cross-section equations for the pseudo-group parameters:

$$
\begin{aligned}
X &= 0, & f &= 0, \\
Y &= 0, & e &= 0, \\
U &= 0, & e_x &= -u\,f_x, \\
U_Y &= 0, & f_{xx} &= -u_y\,f_x, \\
U_X &= 0, & e_{xx} &= (u\,u_y - u_x)\,f_x, \\
U_{YY} &= 1, & f_x &= \sqrt{u_{yy}}\,, \\
U_{XY} &= 0, & f_{xxx} &= -\sqrt{u_{yy}}\,\big(u_{xy} + uu_{yy} - u_y^2\big), \\
U_{XX} &= 0, & e_{xxx} &= -\sqrt{u_{yy}}\,\big(u_{xx} - uu_{xy} - 2\,u^2u_{yy} - 2u_xu_y + uu_y^2\big).
\end{aligned}
$$

At this stage, we can construct the first two fundamental differential invariants:

$$
J_1 = \iota(u_{xyy}) = \frac{u_{xyy} + uu_{yyy} + 2\,u_y u_{yy}}{u_{yy}^{3/2}}, \qquad J_2 = \iota(u_{yyy}) = \frac{u_{yyy}}{u_{yy}^{3/2}}. \tag{4.15}
$$

Higher order differential invariants are found by continuing this procedure, or by employing the more powerful Taylor series method developed in [62]. Further, substituting the pseudo-group normalizations into (4.13) fixes the invariant horizontal coframe

$$
\omega^1 = \iota(dx) = \sqrt{u_{yy}}\;dx, \qquad \omega^2 = \iota(dy) = \sqrt{u_{yy}}\,(dy - u\,dx). \tag{4.16}
$$

The dual invariant total derivative operators are

$$
\mathcal{D}_1 = \frac{1}{\sqrt{u_{yy}}}\,(\mathrm{D}_x + u\,\mathrm{D}_y), \qquad \mathcal{D}_2 = \frac{1}{\sqrt{u_{yy}}}\,\mathrm{D}_y. \tag{4.17}
$$

The higher-order differential invariants can be generated by successively applying these differential operators to the pair of basic differential invariants (4.15). The commutation relation is

$$[\mathcal{D}_1, \mathcal{D}_2] = -\tfrac{1}{2} J_2 \mathcal{D}_1 + \tfrac{1}{2} J_1 \mathcal{D}_2. \tag{4.18}$$

Finally, there is a single generating syzygy among the differentiated invariants:

$$\mathcal{D}_1 J_2 - \mathcal{D}_2 J_1 = 2, \tag{4.19}$$

from which all others can be deduced by invariant differentiation.

5. Recurrence Formulae. Since the basic differential invariants arising from invariantization of the jet coordinates form a complete system, any other differential invariant, e.g., one obtained by application of the invariant differential operators, can be locally written as a function thereof. The recurrence formulae, cf. [23, 33], connect the differentiated invariants and invariant differential forms with their normalized counterparts. These formulae are fundamental, since they prescribe the structure of the algebra of (local) differential invariants, underly a full classification of generating differential invariants and their differential syzygies, as well as the structure of invariant variational problems and, indeed, the entire invariant variational bicomplex. As in the finite-dimensional version, the recurrence formulae are established, using only linear algebra and differentiation, using only the formulas for the prolonged infinitesimal generators and the cross-section. In particular, they do *not* require the explicit formulae for either the moving frame, or the Maurer–Cartan forms, or the normalized differential invariants and invariant forms, or even the invariant differential operators!

Let $\nu^{(\infty)} = (\rho^{(\infty)})^* \mu^{(\infty)}$ denote the one-forms on J^∞ obtained by pulling back the Maurer–Cartan forms on $\mathcal{H}^{(\infty)}$ via the complete moving frame section $\rho^{(\infty)} \colon \mathrm{J}^\infty \to \mathcal{H}^{(\infty)}$, with individual components

$$\nu_A^b = (\rho^{(\infty)})^* (\mu_A^b) = \sum_{i=1}^{p} S_{A,i}^b \, \omega^i + \sum_{\alpha,K} T_{A,\alpha}^{b,K} \vartheta_K^\alpha, \qquad \begin{array}{l} b = 1, \ldots, m, \\ \# A \geq 0. \end{array} \tag{5.1}$$

The coefficients $S_{A,i}^b, T_{A,\alpha}^{b,K}$ will be called the *Maurer–Cartan invariants.* Their explicit formulas will be a direct consequence of the recurrence relations for the phantom differential invariants. In view of Theorem 3.1, the pulled-back Maurer–Cartan forms are subject to the linear relations

$$L^{(n)}(H, I, \nu^{(n)}) = \iota\big[\, L^{(n)}(z, \zeta^{(n)}) \,\big] = 0, \qquad n \geq 0, \tag{5.2}$$

obtained by invariantizing the original linear determining equations (3.3), where we set $\iota(\zeta_A^b) = \nu_A^b$, and where $(H, I) = \iota(x, u) = \iota(z)$ are the zeroth

order differential invariants in (4.7). In particular, if $\mathcal{G}$ acts transitively on M, then, since we are using a minimal order moving frame, (H, I) are constant phantom invariants.

Given a locally defined vector field

$$\mathbf{v} = \sum_{a=1}^{m} \zeta^a(z) \frac{\partial}{\partial z^a} = \sum_{i=1}^{p} \xi^i(x,u) \frac{\partial}{\partial x^i} + \sum_{\alpha=1}^{q} \varphi^\alpha(x,u) \frac{\partial}{\partial u^\alpha} \in \mathcal{X}(M), \quad (5.3)$$

let

$$\mathbf{v}^{(\infty)} = \sum_{i=1}^{p} \xi^i(x,u) \frac{\partial}{\partial x^i} + \sum_{\alpha=1}^{q} \sum_{k=\#J\geq 0} \widehat{\varphi}^\alpha_J(x,u^{(k)}) \frac{\partial}{\partial u^\alpha_J} \in \mathcal{X}(\mathrm{J}^\infty(M,p)) \quad (5.4)$$

denote its infinite prolongation. The coefficients are computed via the usual prolongation formula,

$$\widehat{\varphi}^\alpha_J = D_J Q^\alpha + \sum_{i=1}^{p} u^\alpha_{J,i} \xi^i, \quad (5.5)$$

where

$$Q^\alpha = \varphi^\alpha - \sum_{i=1}^{p} u^\alpha_i \xi^i, \qquad \alpha = 1, \ldots, q, \quad (5.6)$$

are the components of the *characteristic* of $\mathbf{v}$; cf. [54, 55]. Consequently, each prolonged vector field coefficient

$$\widehat{\varphi}^\alpha_J = \Phi^\alpha_J(u^{(n)}, \zeta^{(n)}) \quad (5.7)$$

is a certain universal linear combination of the vector field jet coordinates, whose coefficients are polynomials in the submanifold jet coordinates u^β_K for $1 \leq \#K \leq n$. Let

$$\eta^i = \iota(\xi^i) = \nu^i, \qquad \widehat{\psi}^\alpha_J = \iota(\widehat{\varphi}^\alpha_J) = \Phi^\alpha_J(I^{(n)}, \nu^{(n)}), \quad (5.8)$$

denote their invariantizations, which are certain linear combinations of the pulled-back Maurer–Cartan forms ν^b_A, whose coefficients are polynomials in the normalized differential invariants I^β_K for $1 \leq \#K \leq \#J$.

With all these in hand, we can formulate the *universal recurrence formula*, from which all other recurrence formulae follow.

THEOREM 5.1. *If Ω is any differential form on J^∞, then*

$$d\,\iota(\Omega) = \iota\big[\, d\,\Omega + \mathbf{v}^{(\infty)}(\Omega) \,\big], \quad (5.9)$$

where $\mathbf{v}^{(\infty)}(\Omega)$ denotes the Lie derivative of Ω with respect to the prolonged vector field (5.4), *and we use* (5.8) *and its analogs for the partial derivatives of the prolonged vector field coefficients when invariantizing the result.*

Specializing Ω in (5.9) to be one of the coordinate functions x^i, u^α_J yields recurrence formulae for the normalized differential invariants (4.7),

$$\begin{aligned} dH^i &= \iota\big(dx^i + \xi^i \big) = \varpi^i + \eta^i, \\ dI^\alpha_J &= \iota\big(du^\alpha_J + \widehat{\varphi}^\alpha_J \big) = \iota\left(\sum_{i=1}^p u^\alpha_{J,i}\, dx^i + \theta^\alpha_J + \widehat{\varphi}^\alpha_J \right) \\ &= \sum_{i=1}^p I^\alpha_{J,i}\, \varpi^i + \vartheta^\alpha_J + \widehat{\psi}^\alpha_J, \end{aligned} \tag{5.10}$$

where, as in (5.8), each $\widehat{\psi}^\alpha_J$ is written in terms of the pulled-back Maurer–Cartan forms ν^b_A, which are subject to the linear constraints (5.2). Each phantom differential invariant is, by definition, normalized to a constant value, and hence has zero differential. Consequently, the phantom recurrence formulae in (5.10) form a system of linear algebraic equations which can, as a result of the transversality of the cross-section, be uniquely solved for the pulled-back Maurer–Cartan forms.

THEOREM 5.2. *If the pseudo-group acts locally freely on $\mathcal{V}^n \subset \mathrm{J}^n$, then the nth order phantom recurrence formulae can be uniquely solved to express the pulled-back Maurer–Cartan forms ν^b_A of order $\#A \leq n$ as invariant linear combinations of the invariant horizontal and contact one-forms $\varpi^i, \vartheta^\alpha_J$.*

Substituting the resulting expressions (5.1) into the remaining, non-phantom recurrence formulae in (5.10) leads to a complete system of recurrence relations, for both the vertical and horizontal differentials of all the normalized differential invariants. In particular, equating the coefficients of the forms ω^i leads to individual recurrence formulae for the normalized differential invariants:

$$\mathcal{D}_i H^j = \delta^j_i + M^j_i, \qquad\qquad \mathcal{D}_i I^\alpha_J = I^\alpha_{J,i} + M^\alpha_{J,i}, \tag{5.11}$$

where δ^j_i is the Kronecker delta, and the *correction terms* $M^j_i, M^\alpha_{J,i}$ are certain invariant linear combinations of the Maurer–Cartan invariants $S^b_{A,i}$. One complication, to be dealt with in the following section, is that the correction term $M^\alpha_{J,i}$ can have the same order as the initial differential invariant $I^\alpha_{J,i}$.

It is worth pointing out that, since the prolonged vector field coefficients $\widehat{\varphi}^\alpha_J$ are polynomials in the jet coordinates u^β_K of order $\#K \geq 1$, their invariantizations are polynomial functions of the differential invariants I^β_K for $\#K \geq 1$. Since the correction terms are constructed by solving a linear system for the invariantized Maurer–Cartan forms (5.1), the Maurer–Cartan invariants depend rationally on these differential invariants. Thus, in most cases (including the majority of applications), the resulting differential invariant algebra is endowed with an entirely rational algebraic recurrence structure.

THEOREM 5.3. *If $\mathcal{G}$ acts transitively on M, or, more generally, its infinitesimal generators depend rationally on the coordinates $z = (x, u) \in M$, then the correction terms $M_i^j, M_{J,i}^\alpha$ in the recurrence formulas (5.10) are rational functions of the basic differential invariants.*

6. The Symbol Modules. While the devil is in the details, the most important properties (Cartan characters, ellipticity, finite or infinite type, etc.) of a system of partial differential equations are fixed by the algebraic properties of its symbol module. For the action of a pseudo-group on submanifolds, there are, in fact, two interrelated submodules that prescribe the key structural features of the pseudo-group and its induced differential invariant algebra: the symbol module of its infinitesimal determining equations and the related prolonged symbol module governing its prolonged infinitesimal generators on the submanifold jet space. In particular, except for some low order complications, the generators of the differential invariant algebra and the associated differential syzygies can be identified with the algebraic generators and algebraic syzygies of an invariantized version of the prolonged symbol module. In this manner, constructive Gröbner basis techniques from commutative algebra can be applied to pin down the non-commutative differential algebraic structure of the pseudo-group's differential invariants.

To avoid technical complications, we will work in the analytic category. Let $\mathcal{G}$ be a pseudo-group, and let (3.3) be the formally integrable completion of its linearized determining equations. At each $z \in M$, we let $\mathcal{I}|_z$ denote the *symbol module* of the determining equations, which, by involutivity, forms a submodule of the $\mathbf{R}[t]$ module

$$\mathcal{T} = \left\{ \eta(t, T) = \sum_{a=1}^{m} \eta_a(t)\, T^a \right\} \simeq \mathbf{R}[t] \otimes \mathbf{R}^m \tag{6.1}$$

consisting of real polynomials in $t = (t_1, \ldots, t_m)$ and $T = (T^1, \ldots, T^m)$ that are linear in the T's. Assuming regularity, the symbol module's *Hilbert polynomial*, [17],

$$H(n) = \sum_{i=0}^{d} b_i \binom{n}{d-i}, \tag{6.2}$$

where $b_0, b_1, \ldots, b_d \in \mathbf{Z}$, does not depend on $z \in M$. The integer $0 \leq d \leq m$ is the *dimension*, while $b = b_0$, its *degree*, is strictly positive unless $\mathcal{I}|_z = \mathcal{T}$, in which case $\widetilde{H}(n) \equiv 0$ and the pseudo-group is purely discrete. Assuming solvability by the Cartan–Kähler Theorem, [9, 55], the general solution to the determining equations — that is, the general pseudo-group transformation — can be written in terms of b arbitrary functions of d variables. In particular, the system is of finite type — and hence $\mathcal{G}$ is, in fact, a b-dimensional Lie group action — if and only if the symbol module has dimension $d = 0$. See Seiler, [67, 68], for additional details.

The prolonged infinitesimal generators of the pseudo-group on the submanifold jet bundle have an analogous *prolonged symbol module.* Let

$$\widehat{\mathcal{S}} = \left\{ \widehat{\sigma}(s,S) = \sum_{\alpha=1}^{q} \widehat{\sigma}_\alpha(s)\, S^\alpha \right\} \simeq \mathbf{R}[s] \otimes \mathbf{R}^q \tag{6.3}$$

be the $\mathbf{R}[s]$ module consisting of polynomials in $s = (s_1, \ldots, s_p)$, $S = (S^1, \ldots, S^q)$, which are linear in the S's. At each submanifold 1-jet $z^{(1)} = (x, u^{(1)}) = (\ldots x^i \ldots u^\alpha \ldots u_i^\alpha \ldots) \in \mathrm{J}^1(M,p)$, we define a linear map

$$\beta|_{z^{(1)}} : \mathbf{R}^m \times \mathbf{R}^m \to \mathbf{R}^m$$

by the formulas

$$\begin{aligned} s_i &= \beta_i(z^{(1)};t) = t_i + \sum_{\alpha=1}^{q} u_i^\alpha\, t_{p+\alpha}, & i &= 1,\ldots,p, \\ S^\alpha &= B^\alpha(z^{(1)};T) = T^{p+\alpha} - \sum_{i=1}^{p} u_i^\alpha\, T^i, & \alpha &= 1,\ldots,q. \end{aligned} \tag{6.4}$$

The induced pull-back map

$$\begin{aligned} &(\beta|_{z^{(1)}})^* \left[\widehat{\sigma}(s_1,\ldots,s_p,S_1,\ldots,S_q) \right] \\ &\qquad = \widehat{\sigma}\big(\beta_1(z^{(1)};t),\ldots,\beta_p(z^{(1)};t), B^1(z^{(1)};T)\ldots, B^q(z^{(1)};T) \big) \end{aligned} \tag{6.5}$$

defines an injection $(\beta|_{z^{(1)}})^* : \widehat{\mathcal{S}} \to \mathcal{T}$.

DEFINITION 6.1. The *prolonged symbol submodule* at $z^{(1)} \in \mathrm{J}^1|_z$ is the inverse image of the symbol module under this pull-back map:

$$\mathcal{J}|_{z^{(1)}} = ((\beta|_{z^{(1)}})^*)^{-1}(\mathcal{I}|_z) = \left\{ \sigma(s,S) \;\middle|\; (\beta|_{z^{(1)}})^*(\sigma) \in \mathcal{I}|_z \right\} \subset \widehat{\mathcal{S}}. \tag{6.6}$$

It can be proved that, as long as $n > n^\star$, the module $\mathcal{J}|_{z^{(1)}}$ coincides with the symbol module associated with the prolonged infinitesimal generators (5.4); see [63] for precise details.

To relate this construction to the differential invariant algebra, we need to invariantize the modules using our moving frame. In general, the invariantization of a prolonged symbol polynomial

$$\sigma(x,u^{(1)};s,S) = \sum_{\alpha,J} h_\alpha^J(x,u^{(1)})\; s_J S^\alpha \in \mathcal{J}|_{z^{(1)}}, \quad \text{where} \quad z^{(1)} = (x,u^{(1)}),$$

is given by

$$\widetilde{\sigma}(H,I^{(1)};s,S) = \iota\left[\sigma(x,u^{(1)};s,S) \right] = \sum_{\alpha,J} h_\alpha^J(H,I^{(1)})\; s_J S^\alpha. \tag{6.7}$$

Let $\widetilde{\mathcal{J}}|_{(H,I^{(1)})} = \iota(\mathcal{J}|_{z^{(1)}})$ denote the resulting *invariantized prolonged symbol submodule.* We identify each parametrized symbol polynomial (6.7) with the differential invariant

$$I_{\widetilde{\sigma}} = \sum_{\alpha,J} h^J_\alpha(H, I^{(1)})\, I^\alpha_J. \tag{6.8}$$

If $\mathcal{G}$ acts transitively on an open subset of J^1, then $\widetilde{\mathcal{J}} = \widetilde{\mathcal{J}}|_{(H,I^{(1)})}$ is a fixed module, independent of the submanifold jet coordinates, and (6.8) is a linear, constant coefficient combination of the normalized differential invariants.

The recurrence formulae for these differential invariants take the form

$$\mathcal{D}_i I_{\widetilde{\sigma}} = I_{s_i\,\widetilde{\sigma}} + M_{\widetilde{\sigma},i}, \tag{6.9}$$

in which, as long as $n = \deg\widetilde{\sigma} > n^\star$, the *leading term* $I_{s_i\,\widetilde{\sigma}}$ is a differential invariant of order $= n+1$, while, unlike in (5.11), the *correction term* $M_{\widetilde{\sigma},i}$ is of lower order $\leq n$. Iteration leads to the higher order recurrences

$$\mathcal{D}_J I_{\widetilde{\sigma}} = I_{s_J\widetilde{\sigma}} + M_{\widetilde{\sigma},J}, \tag{6.10}$$

where $J = (j_1, \ldots, j_k)$ is an *ordered* multi-index of order k, and, assuming order $I_{\widetilde{\sigma}} = \deg\widetilde{\sigma} = n > n^\star$, the correction term $M_{\widetilde{\sigma},J}$ has order $< k + n = \deg\big[\, s_J\,\widetilde{\sigma}(s,S)\,\big]$.

With this in hand, we are able to state a Constructive Basis Theorem for the differential invariant algebra of an eventually locally freely acting pseudo-group.

THEOREM 6.1. *Let $\mathcal{G}$ be a Lie pseudo-group that acts locally freely on an open subset of the submanifold jet bundle at order $n^\star$. Then the following constitute a finite generating system for its algebra of local differential invariants:*

a) the differential invariants $I_\nu = I_{\sigma_\nu}$, whose polynomials $\sigma_1, \ldots, \sigma_l$ form a Gröbner basis for the invariantized prolonged symbol submodule, and, possibly,

b) a finite number of additional differential invariants of order $\leq n^\star$.

As noted above, the listed differential invariants do not typically form a minimal generating system, and the characterization of minimal generators remains a challenging open problem.

We are also able to exhibit a finite generating system of differential invariant syzygies — again not necessarily minimal. First, owing to the non-commutative nature of the the invariant differential operators, (4.10), we have the *commutator syzygies*

$$\mathcal{D}_J I_{\widetilde{\sigma}} - \mathcal{D}_{\widetilde{J}} I_{\widetilde{\sigma}} = M_{\widetilde{\sigma},J} - M_{\widetilde{\sigma},\widetilde{J}} \equiv N_{J,\widetilde{J},\widetilde{\sigma}}, \quad \text{whenever} \quad \widetilde{J} = \pi(J) \tag{6.11}$$

for some permutation π of the multi-index J. Provided $\deg \widetilde{\sigma} > n^\star$, the right hand side $N_{J,\widetilde{J},\widetilde{\sigma}}$ is a differential invariant of lower order than those on the left hand side.

In addition, any commutative algebraic syzygy satisfied by polynomials in the prolonged symbol module $\widetilde{\mathcal{J}}|_{(H,I^{(1)})}$ provides an additional "essential" syzygy amongst the differentiated invariants. In detail, to each invariantly parametrized polynomial

$$q(H, I^{(1)}; s) = \sum_J q_J(H, I^{(1)}) s_J \in \mathbf{R}[s] \tag{6.12}$$

we associate an invariant differential operator

$$q(H, I^{(1)}; \mathcal{D}) = \sum_J q_J(H, I^{(1)}) \mathcal{D}_J. \tag{6.13}$$

Our convention is that the sums range over non-decreasing multi-indices $1 \leq j_1 \leq j_2 \leq \cdots \leq j_k \leq p$, for $k = \#J$, and where, for specificity, we adopt the normal ordering when writing $\mathcal{D}_J = \mathcal{D}_{j_1} \mathcal{D}_{j_2} \cdots \mathcal{D}_{j_k}$. In view of (6.10), whenever $\widetilde{\sigma}(H, I^{(1)}; s, S) \in \widetilde{\mathcal{J}}|_{(H,I^{(1)})}$, we can write

$$q(H, I^{(1)}; \mathcal{D})\, I_{\widetilde{\sigma}(H,I^{(1)};s,S)} = I_{q(H,I^{(1)};s)\,\widetilde{\sigma}(H,I^{(1)};s,S)} + R_{q,\widetilde{\sigma}}, \tag{6.14}$$

where $R_{q,\widetilde{\sigma}}$ has order $< \deg q + \deg \widetilde{\sigma}$. In particular, any algebraic syzygy

$$\sum_{\nu=1}^{l} q_\nu(H, I^{(1)}, s)\, \sigma_\nu(H, I^{(1)}; s, S) = 0 \tag{6.15}$$

among the Gröbner basis polynomials of the invariantized prolonged symbol module induces a syzygy among the generating differential invariants:

$$\sum_{\nu=1}^{l} q_\nu(H, I^{(1)}, \mathcal{D})\, I_{\widetilde{\sigma}_\nu}(H, I^{(1)}; s, S) = R, \tag{6.16}$$

where $\operatorname{order} R < \max\{\deg q_\nu + \deg \widetilde{\sigma}_\nu\}$.

THEOREM 6.2. *Every differential syzygy among the generating differential invariants is a combination of the following:*

a) the syzygies among the differential invariants of order $\leq n^\star$,
b) the commutator syzygies,
c) syzygies coming from an algebraic syzygy among the Gröbner basis polynomials.

In this manner, we deduce a finite system of generating differential syzygies for the differential invariant algebra of our pseudo-group.

Further details, and applications of these results can be found in our papers listed in the references.

REFERENCES

[1] ANDERSON I.M., *The Variational Bicomplex*, Utah State Technical Report, 1989, `http://math.usu.edu/~fg_mp`.

[2] BAZIN P.-L., and BOUTIN M., Structure from motion: theoretical foundations of a novel approach using custom built invariants, *SIAM J. Appl. Math.* **64** (2004), 1156–1174.

[3] BERCHENKO I.A., and OLVER P.J., Symmetries of polynomials, *J. Symb. Comp.* **29** (2000), 485–514.

[4] BÍLĂ N., MANSFIELD E.L., and CLARKSON P.A., Symmetry group analysis of the shallow water and semi-geostrophic equations, *Quart. J. Mech. Appl. Math.* **59** (2006), 95–123.

[5] BLEECKER D., *Gauge Theory and Variational Principles*, Addison–Wesley Publ. Co., Reading, Mass., 1981.

[6] BOUTIN M., Numerically invariant signature curves, *Int. J. Computer Vision* **40** (2000), 235–248.

[7] BOUTIN M., On orbit dimensions under a simultaneous Lie group action on n copies of a manifold, *J. Lie Theory* **12** (2002), 191–203.

[8] BOUTIN M., Polygon recognition and symmetry detection, *Found. Comput. Math.* **3** (2003), 227–271.

[9] BRYANT R.L., CHERN S.-S., GARDNER R.B., GOLDSCHMIDT H.L., and GRIFFITHS P.A., *Exterior Differential Systems*, Math. Sci. Res. Inst. Publ., Vol. 18, Springer–Verlag, New York, 1991.

[10] CALABI E., OLVER P.J., SHAKIBAN C., TANNENBAUM A., and HAKER S., Differential and numerically invariant signature curves applied to object recognition, *Int. J. Computer Vision* **26** (1998), 107–135.

[11] CARTAN É., *La Méthode du Repère Mobile, la Théorie des Groupes Continus, et les Espaces Généralisés*, Exposés de Géométrie, no. 5, Hermann, Paris, 1935.

[12] CARTAN É., Sur la structure des groupes infinis de transformations, *in*: *Oeuvres Complètes*, Part. II, Vol. 2, Gauthier–Villars, Paris, 1953, pp. 571–714.

[13] CARTAN É., La structure des groupes infinis, *in*: *Oeuvres Complètes*, part. II, Vol. 2, Gauthier–Villars, Paris, 1953, pp. 1335–1384.

[14] CHEH J., OLVER P.J., and POHJANPELTO J., Maurer–Cartan equations for Lie symmetry pseudo-groups of differential equations, *J. Math. Phys.* **46** (2005), 023504.

[15] CHEH J., OLVER P.J., and POHJANPELTO J., Algorithms for differential invariants of symmetry groups of differential equations, *Found. Comput. Math.*, to appear.

[16] CHERN S.S., and MOSER J.K., Real hypersurfaces in complex manifolds, *Acta Math.* **133** (1974), 219–271; also *Selected Papers*, Vol. 3, Springer–Verlag, New York, 1989, pp. 209–262.

[17] COX D., LITTLE J., and O'SHEA D., *Ideals, Varieties, and Algorithms*, 2nd ed., Springer–Verlag, New York, 1996.

[18] DAVID D., KAMRAN N., LEVI D., and WINTERNITZ P., Subalgebras of loop algebras and symmetries of the Kadomtsev-Petviashivili equation, *Phys. Rev. Lett.* **55** (1985), 2111–2113.

[19] DI FRANCESCO P., MATHIEU P., and SÉNÉCHAL D., *Conformal Field Theory*, Springer–Verlag, New York, 1997.

[20] EHRESMANN C., Introduction à la théorie des structures infinitésimales et des pseudo-groupes de Lie, *in*: *Géometrie Différentielle*, Colloq. Inter. du Centre Nat. de la Rech. Sci., Strasbourg, 1953, pp. 97–110.

[21] FEFFERMAN C., and GRAHAM C.R., Conformal invariants, *in*: *Élie Cartan et les Mathématiques d'aujourd'hui*, Astérisque, hors série, Soc. Math. France, Paris, 1985, pp. 95–116.

[22] FELS M., and OLVER P.J., Moving coframes. I. A practical algorithm, *Acta Appl. Math.* **51** (1998), 161–213.

[23] FELS M., and OLVER P.J., Moving coframes. II. Regularization and theoretical foundations, *Acta Appl. Math.* **55** (1999), 127–208.

[24] FUCHS D.B., GABRIELOV A.M., and GEL'FAND I.M., The Gauss–Bonnet theorem and Atiyah–Patodi–Singer functionals for the characteristic classes of foliations, *Topology* **15** (1976), 165–188.

[25] GUGGENHEIMER H.W., *Differential Geometry*, McGraw–Hill, New York, 1963.

[26] HANN C.E., and HICKMAN M.S., Projective curvature and integral invariants, *Acta Appl. Math.* **74** (2002), 177–193.

[27] Hubert, E., The AIDA Maple package, `http://www.inria.fr/cafe/Evelyne.Hubert/aida`, 2006..

[28] HUBERT E., and KOGAN I.A., Rational invariants of an algebraic group action. Construction and rewriting, *J. Symb. Comp.*, to appear.

[29] HUBERT E., and KOGAN I.A., Smooth and algebraic invariants of a group action. Local and global constructions, *Found. Comput. Math.* , to appear.

[30] KAMRAN N., Contributions to the study of the equivalence problem of Elie Cartan and its applications to partial and ordinary differential equations, *Mém. Cl. Sci. Acad. Roy. Belg.* **45** (1989), Fac. 7.

[31] KAMRAN N., and ROBART T., A manifold structure for analytic isotropy Lie pseudogroups of infinite type, *J. Lie Theory* **11** (2001), 57–80.

[32] KIM P., *Invariantization of Numerical Schemes for Differential Equations Using Moving Frames*, Ph.D. Thesis, University of Minnesota, Minneapolis, 2006.

[33] KOGAN I.A., and OLVER P.J., Invariant Euler-Lagrange equations and the invariant variational bicomplex, *Acta Appl. Math.* **76** (2003), 137–193.

[34] KRUGLIKOV B., and LYCHAGIN V., Invariants of pseudogroup actions: homological methods and finiteness theorem, preprint, arXiv: math.DG/0511711, 2005.

[35] KUMPERA A., Invariants différentiels d'un pseudogroupe de Lie, *J. Diff. Geom.* **10** (1975), 289–416.

[36] KURANISHI M., On the local theory of continuous infinite pseudo groups I, *Nagoya Math. J.* **15** (1959), 225–260.

[37] KURANISHI M., On the local theory of continuous infinite pseudo groups II, *Nagoya Math. J.* **19** (1961), 55–91.

[38] LIE S., Die Grundlagen für die Theorie der unendlichen kontinuierlichen Transformationsgruppen, *Leipzig. Ber.* **43** (1891), 316–393; also *Gesammelte Abhandlungen*, Vol. 6, B.G. Teubner, Leipzig, 1927, pp. 300–364.

[39] LISLE I.G., and REID G.J., Geometry and structure of Lie pseudogroups from infinitesimal defining systems, *J. Symb. Comp.* **26** (1998), 355–379.

[40] LISLE I.G., and REID G.J., Cartan structure of infinite Lie pseudogroups, *in*: *Geometric Approaches to Differential Equations*, P.J. Vassiliou and I.G. Lisle, eds., Austral. Math. Soc. Lect. Ser., 15, Cambridge Univ. Press, Cambridge, 2000, pp. 116–145.

[41] MACKENZIE K., *Lie Groupoids and Lie Algebroids in Differential Geometry*, London Math. Soc. Lecture Notes, Vol. 124, Cambridge University Press, Cambridge, 1987.

[42] MANSFIELD E.L., Algorithms for symmetric differential systems, *Found. Comput. Math.* **1** (2001), 335–383.

[43] MARÍ BEFFA G., Relative and absolute differential invariants for conformal curves, *J. Lie Theory* **13** (2003), 213–245.

[44] MARÍ BEFFA G., Poisson geometry of differential invariants of curves in some nonsemisimple homogeneous spaces, *Proc. Amer. Math. Soc.* **134** (2006), 779–791.

[45] MARÍ BEFFA G., and OLVER P.J., Differential invariants for parametrized projective surfaces, *Commun. Anal. Geom.* **7** (1999), 807–839.

[46] MARTINA L., SHEFTEL M.B., and WINTERNITZ P., Group foliation and non-invariant solutions of the heavenly equation, *J. Phys. A* **34** (2001), 9243–9263.

[47] MCLACHLAN R.I., and QUISPEL G.R.W., What kinds of dynamics are there? Lie pseudogroups, dynamical systems and geometric integration, *Nonlinearity* **14** (2001), 1689–1705.

[48] McLenaghan R.G., Smirnov R.G., and The D., An extension of the classical theory of algebraic invariants to pseudo-Riemannian geometry and Hamiltonian mechanics, *J. Math. Phys.* **45** (2004), 1079–1120.
[49] Medolaghi P., Classificazione delle equazioni alle derivate parziali del secondo ordine, che ammettono un gruppo infinito di trasformazioni puntuali, *Ann. Mat. Pura Appl.* 1 (3) (1898), 229–263.
[50] Morozov O., Moving coframes and symmetries of differential equations, *J. Phys. A* **35** (2002), 2965–2977.
[51] Morozov O.I., Structure of symmetry groups via Cartan's method: survey of four approaches, *SIGMA: Symmetry Integrability Geom. Methods Appl.* **1** (2005), paper 006.
[52] Muñoz J., Muriel F.J., and Rodríguez J., On the finiteness of differential invariants, *J. Math. Anal. Appl.* **284** (2003), 266–282.
[53] Nutku Y., and Sheftel M.B., Differential invariants and group foliation for the complex Monge–Ampère equation, *J. Phys. A* **34** (2001), 137–156.
[54] Olver P.J., *Applications of Lie Groups to Differential Equations*, Second Edition, Graduate Texts in Mathematics, Vol. 107, Springer–Verlag, New York, 1993.
[55] Olver P.J., *Equivalence, Invariants, and Symmetry*, Cambridge University Press, Cambridge, 1995.
[56] Olver P.J., *Classical Invariant Theory*, London Math. Soc. Student Texts, Vol. 44, Cambridge University Press, Cambridge, 1999.
[57] Olver P.J., Joint invariant signatures, *Found. Comput. Math.* **1** (2001), 3–67.
[58] Olver P.J., Geometric foundations of numerical algorithms and symmetry, *Appl. Alg. Engin. Commun. Comput.* **11** (2001), 417–436.
[59] Olver P.J., Generating differential invariants, *J. Math. Anal. Appl.*, to appear.
[60] Olver P.J., and Pohjanpelto J., Regularity of pseudogroup orbits, *in*: *Symmetry and Perturbation Theory*, G. Gaeta, B. Prinari, S. Rauch-Wojciechowski, S. Terracini, eds., World Scientific, Singapore, 2005, pp. 244–254.
[61] Olver P.J., and Pohjanpelto J., Maurer–Cartan forms and the structure of Lie pseudo-groups, *Selecta Math.* **11** (2005), 99–126.
[62] Olver P.J., and Pohjanpelto J., Moving frames for Lie pseudo–groups, *Canadian J. Math.*, to appear.
[63] Olver P.J., and Pohjanpelto J., On the algebra of differential invariants of a Lie pseudo-group, preprint, University of Minnesota, 2007.
[64] Ovsiannikov L.V., *Group Analysis of Differential Equations*, Academic Press, New York, 1982.
[65] Pommaret J.F., *Systems of Partial Differential Equations and Lie Pseudogroups*, Gordon and Breach, New York, 1978.
[66] Robart T., and Kamran N., Sur la théorie locale des pseudogroupes de transformations continus infinis I, *Math. Ann.* **308** (1997), 593–613.
[67] Seiler W., On the arbitrariness of the solution to a general partial differential equation, *J. Math. Phys.* **35** (1994), 486–498.
[68] Seiler W., *Involution*, in preparation.
[69] Serra J., *Image Analysis and Mathematical Morphology*, Academic Press, London 1982.
[70] Singer I.M., and Sternberg S., The infinite groups of Lie and Cartan. Part I (the transitive groups), *J. Analyse Math.* **15** (1965), 1–114.
[71] Smirnov R., and Yue J., Covariants, joint invariants and the problem of equivalence in the invariant theory of Killing tensors defined in pseudo-Riemannian spaces of constant curvature, *J. Math. Phys.* **45** (2004), 4141–4163.
[72] Stormark O., *Lie's Structural Approach to PDE Systems*, Cambridge University Press, Cambridge, 2000.
[73] Tresse A., Sur les invariants différentiels des groupes continus de transformations, *Acta Math.* **18** (1894), 1–88.
[74] Tsujishita T., On variational bicomplexes associated to differential equations, *Osaka J. Math.* **19** (1982), 311–363.

[75] VESSIOT E., Sur l'intégration des systèmes différentiels qui admettent des groupes continues de transformations, *Acta. Math.* **28** (1904), 307–349.

[76] WELK M., KIM P., and OLVER P.J., Numerical invariantization for morphological PDE schemes, *in*: *Scale Space and Variational Methods in Computer Vision*, F. Sgallari, A. Murli and N. Paragios, eds., Lecture Notes in Computer Science, Springer–Verlag, New York, 2007.

GEOMETRY OF LINEAR DIFFERENTIAL SYSTEMS TOWARDS CONTACT GEOMETRY OF SECOND ORDER

KEIZO YAMAGUCHI*

0. Introduction. This is a lecture note on the geometry of **linear differential systems**. By a (linear) differential system (or Pfaffian system) (M, D), we mean a subbundle D of the tangent bundle $T(M)$ of a manifold M. Locally D is defined by 1-forms $\omega_1, \dots, \omega_s$ such that $\omega_1 \wedge \cdots \wedge \omega_s \neq 0$ at each point , where r is the rank of D and $r+s = \dim M$;

$$D = \{\,\omega_1 = \cdots = \omega_s = 0\,\}.$$

(M, D) is called **completely integrable**, if there exist coordinate systems $(x_1, \dots, x_d)$ around each point of M such that $D = \{\,dx_1 = \cdots = dx_s = 0\,\}$ locally.

As important examples of non-integrable differential systems, we will start with the discussion of the canonical (or contact) systems on jet spaces. In fact we will start with the space $J(M,n)$ of n-dimensional **contact elements** to a manifold M and the **canonical system** C on $J(M,n)$ in §1. Moreover we state the **Bäcklund Theorem** for $(J(M,n), C)$ and the **Darboux Theorem** for contact manifolds. We also prepare the basic notions (**Derived Systems** and **Cauchy Characteristic Systems**) for (non-integrable) linear differential systems. To familiarize the reader with the notion of Cauchy characteristic systems, we will give a proof of the Darboux theorem in the end of §1.

We will discuss the **Tanaka Theory** of linear differential systems in §2. A differential system (M, D) is called regular if the k-th weak derived system is a subbundle for each integer k. As a first invariant for non-integrable differential systems, we define the **symbol algebra** $\mathfrak{m}(x)$ of a regular differential system (M, D) at $x \in M$. $\mathfrak{m}(x)$ is endowed with the bracket product so that $\mathfrak{m}(x)$ becomes a nilpotent graded Lie algebra satisfying the generating condition (for the precise definition, see §2). Conversely, starting from a nilpotent graded Lie algebra $\mathfrak{m} = \bigoplus_{p<0} \mathfrak{g}_p$ over $\mathbb{R}$ satisfying the generating condition, we will construct the model differential system $(M(\mathfrak{m}), D_{\mathfrak{m}})$, called the **Standard Differential System of Type** $\mathfrak{m}$, group-theoretically. Moreover we introduce the notion of the (algebraic) **Prolongation** $\mathfrak{g}(\mathfrak{m}) = \bigoplus_{p\in\mathbb{Z}} \mathfrak{g}_p(\mathfrak{m})$ of $\mathfrak{m}$, which describe the structure of the Lie algebra of infinitesimal automorphisms of $(M(\mathfrak{m}), D_{\mathfrak{m}})$, where $\mathfrak{g}_p(\mathfrak{m}) = \mathfrak{g}_p$ for $p < 0$ and $\mathfrak{g}_0(\mathfrak{m})$ is the Lie algebra consisting of gradation preserving derivations of $\mathfrak{m}$. Given a subalgebra $\mathfrak{g}_0$ of $\mathfrak{g}_0(\mathfrak{m})$, we also

*Department of Mathematics, Faculty of Science, Hokkaido University, Sapporo 060-0810, Japan (yamaguch@math.sci.hokudai.ac.jp).

define the prolongation $\mathfrak{g}(\mathfrak{m}, \mathfrak{g}_0)$ of $(\mathfrak{m}, \mathfrak{g}_0)$ as the subalgebra of $\mathfrak{g}(\mathfrak{m})$. As an application of the notion of symbol algebra, we will give a proof of the Bäcklund theorem in the end of §2.

We will formulate the submanifold theory of the second order jet space for a scalar function, as the geometry of **PD-manifolds** in §3. Utilizing the **Realization Lemma**, which characterizes submanifolds of $J(M, n)$, we will characterize submanifolds R of $L(J)$, such that $p : R \to J$ is a submersion, in terms of the pair of differential systems D^1 and D^2, where D^1 and D^2 are the restrictions to R of the first and second canonical systems on $L(J)$ (for the precise argument, see §3) and form the notion of **PD-manifolds $(R; D^1, D^2)$ of the Second Order**. We also discuss the **Reduction Theorem** for PD-manifolds, which will be used in §5. In the rest of §3, we will discuss the geometric construction of higher order jet spaces and discuss **Drapeau Theorem** as related subjects.

In §4, we will discuss *when* $\mathfrak{g}(\mathfrak{m})$ *or* $\mathfrak{g}(\mathfrak{m}, \mathfrak{g}_0)$ *becomes finite dimensional and simple*, utilizing the structure theory of simple Lie algebras over $\mathbb{C}$. For a simple Lie algebra $\mathfrak{g}$ over $\mathbb{C}$, we first classify gradations $\mathfrak{g} = \bigoplus_{p \in \mathbb{Z}} \mathfrak{g}_p$ of $\mathfrak{g}$ satisfying the generating condition; $\mathfrak{g}_{p-1} = [\mathfrak{g}_p, \mathfrak{g}_{-1}]$ for $p < 0$ on their negative parts $\mathfrak{m} = \bigoplus_{p<0} \mathfrak{g}_p$, which turns out to be equivalent to classify parabolic subalgebras $\mathfrak{p} = \bigoplus_{p \geqq 0} \mathfrak{g}_p$ of $\mathfrak{g}$. We will describe gradations of $\mathfrak{g}$ in terms of the root space decomposition and also explicitly in terms of the matrix representations in the classical cases. Then we answer in **Theorem 4.2** when $\mathfrak{g}$ is the prolongation of $\mathfrak{m}$ or $(\mathfrak{m}, \mathfrak{g}_0)$. In view of the **Tanaka Theory of Normal Cartan Connection** [T4], in case $\mathfrak{g}$ is the prolongation of $(\mathfrak{m}, \mathfrak{g}_0)$, each gradation $\mathfrak{g} = \bigoplus_{p \in \mathbb{Z}} \mathfrak{g}_p$ represents the symbol of the corresponding **(Parabolic) Geometry**.

Following a short passage from Cartan's **Five Variables Paper** [C1], we will discuss the local classification of regular differential systems (M, D) of lower dimension ($\dim M \leqq 5$) in §5 and see how the **Exceptional Simple Lie Algebra G_2** shows up as a Lie algebra of infinitesimal automorphisms of a regular differential system of type $\mathfrak{m}_5$, where $\mathfrak{m}_5 = \mathfrak{g}_{-3} \oplus \mathfrak{g}_{-2} \oplus \mathfrak{g}_{-1}$, $\dim \mathfrak{g}_{-3} = \dim \mathfrak{g}_{-1} = 2$ and $\dim \mathfrak{g}_{-2} = 1$. We will explicitly describe the **Standard Differential System of Type $\mathfrak{m}_5$**, given by E.Cartan. Moreover, utilizing the Reduction Theorem in §3.3, we will show explicitly how to construct the model overdetermined involutive system of second order, whose symmetry algebra of contact transformations is isomorphic to G_2, from the Standard Differential System of Type $\mathfrak{m}_5$.

Finally in §6, to find important subclasses of parabolic geometries given in Theorem 4.2, we will discuss the **Se-ashi's Principle** to form good classes of linear differential equations of finite type from the view point of contact geometry and pseudo-product structures. In this course, we will find classes of parabolic geometries (**Theorem 6.1**) among the lists of Theorem 4.2, which are associated with differential equations of finite type, and which generalize the cases of second and third order ODE for scalar functions. Namely we will exhibit classes of **differential equations**

of finite type, for which the model equations admit nonlinear contact transformations and their symmetry algebras become finite dimensional and simple.

This lecture note is originally prepared for the IMA Summer Program: Symmetries and Overdetermined Systems of Partial Differential Equations, July 17- August 4, 2006. Meanwhile the author had the opportunity to give 5 lectures on the same title at Workshop on Differential Geometry and Application at La Trobe University, Australia, June 13 - 16, 2006. The contents of sections 1 through 5 were lectured at La Trobe. At IMA, the first lecture covered most of the materials in sections 1 and 2 together with the second half of section 3. The materials of the second lecture consisted of sections 4 and 6. The author is grateful to the participants of both lectures for their intimate interest and concern.

1. Geometry of jet spaces. Let us start with the geometric construction of jet spaces and prepare the basic notions for linear differential systems.

1.1. Spaces of contact elements (Grassmann Bundle). The notion of contact manifolds originates from the following space $J(M,n)$ of contact elements: Let M be a (real or complex) manifold of dimension $m+n$. Fixing the number n, we consider the space of n-dimensional **contact elements** to M, i.e., the **Grassmannian bundle** over M consisting of all n-dimensional contact elements to M;

$$J(M,n) = \bigcup_{x \in M} J_x \xrightarrow{\pi} M,$$

where $J_x = \mathrm{Gr}(T_x(M),n)$ is the Grassmann manifold of all n-dimensional subspaces of the tangent space $T_x(M)$ to M at x. Each element $u \in J(M,n)$ is a linear subspace of $T_x(M)$ of codimension m, where $x = \pi(u)$. Hence we have a differential system C of codimension m on $J(M,n)$ by putting:

$$C(u) = \pi_*^{-1}(u) \subset T_u(J(M,n)) \xrightarrow{\pi_*} T_x(M)$$

for each $u \in J(M,n)$. C is called the **Canonical System** on $J(M,n)$. We can introduce the *inhomogeneous Grassmann coordinate* of $J(M,n)$ around $u_o \in J(M,n)$ as folllows; Take a coordinate system $U'; (x_1, \cdots, x_n, z^1, \cdots, z^m)$ of M around $x_o = \pi(u_o)$ such that $dx_1 \wedge \cdots \wedge dx_n \mid_{u_o} \neq 0$. Then we have the coordinate system $(x_1, \cdots, x_n, z^1, \cdots, z^m, p_1^1, \cdots, p_n^m)$ on the neighborhood

$$U = \{u \in \pi^{-1}(U') \mid \pi(u) = x \in U' \quad \text{and} \quad dx_1 \wedge \cdots \wedge dx_n \mid_u \neq 0\};$$

of u_o by defining functions $p_i^\alpha(u)$ on U as follows;

$$dz^\alpha \mid_u = \sum_{i=1}^{n} p_i^\alpha(u)\; dx_i \mid_u .$$

On a canonical coordinate system$(x_1, \cdots, x_n, z^1, \cdots, z^m, p_1^1, \cdots, p_n^m)$, C is clearly defined by

$$C = \{\varpi^1 = \cdots = \varpi^m = 0\},$$

where

$$\varpi^\alpha = dz^\alpha - \sum_{i=1}^{n} p_i^\alpha \, dx_i, \qquad (\alpha = 1, \cdots, m).$$

$(J(M,n), C)$ is the (geometric) 1-jet space and especially, in case $m = 1$, is the so-called contact manifold. Let M, $\hat{M}$ be manifolds (of dimension $m+n$) and $\varphi : M \to \hat{M}$ be a diffeomorphism between them. Then φ induces the **isomorphism** $\varphi_* : (J(M,n), C) \to (J(\hat{M}, n), \hat{C})$, i.e., the differential map $\varphi_* : J(M,n) \to J(\hat{M}, n)$ is a diffeomorphism sending C onto $\hat{C}$. The reason why the case $m = 1$ is special is explained by the following theorem of Bäcklund.

THEOREM 1.1 (**Bäcklund**). *Let M and $\hat{M}$ be manifolds of dimension $m+n$. Assume $m \geqq 2$. Then, for an isomorphism $\Phi : (J(M,n), C) \to (J(\hat{M}, n), \hat{C})$, there exists a diffeomorphism $\varphi : M \to \hat{M}$ such that $\Phi = \varphi_*$.*

We will give a proof of this theorem in §2.4. as an application of the notion of the symbol algebra of $(J(M,n), C)$, which will be introdued in §2.1.

1.2. Contact manifolds. Let J be a manifold and C be a (linear) differential system on J of codimension 1. Namely C is a subundle of $T(J)$ of codimension 1. Thus, locally at each point u of J, there exists a 1-form ϖ defined around $u \in J$ such that

$$C = \{\varpi = 0\}.$$

Then (J, C) is called a **contact manifold** if $\varpi \wedge (d\varpi)^n$ forms a volume element of J. This condition is equivalent to the following conditions (1) or (2);

(1) The restriction $d\varpi \mid_C$ of $d\varpi$ to $C(u)$ is non-degenerate at each point $u \in J$.

(2) There exists a coframe $\{\varpi, \omega_1, \ldots, \omega_n, \pi_1, \ldots, \pi_n\}$ defined around $u \in J$ such that the following holds

$$d\varpi \equiv \omega_1 \wedge \pi_1 + \cdots + \omega_n \wedge \pi_n \qquad (\text{mod} \quad \varpi).$$

A contact manifold (J, C) of dimension $2n+1$ can be regarded locally as a space of 1-jets for one unknown function by the following theorem of Darboux.

THEOREM 1.2 (**Darboux**). *At each point of a* **contact manifold** *(J, C), there exists a canonical coordinate system $(x_1, \ldots, x_n, z, p_1, \ldots, p_n)$ such that*

$$C = \{dz - \sum_{i=1}^{n} p_i\, dx_i = 0\}$$

We will give a proof of this theorem in §1.4.

Starting from a contact manifold (J, C), we can construct the geometric second order jet space $(L(J), E)$ as follows: We consider the **Lagrange-Grassmann bundle** $L(J)$ over J consisting of all n-dimensional integral elements of (J, C);

$$L(J) = \bigcup_{u \in J} L_u,$$

where L_u is the Grassmann manifolds of all lagrangian (or **legendrian**) subspaces of the symplectic vector space $(C(u), d\varpi)$. Here ϖ is a local contact form on J. Let π be the projection of $L(J)$ onto J. Then the canonical system E on $L(J)$ is defined by

$$E(v) = \pi_*^{-1}(v) \subset T_v(L(J)) \xrightarrow{\pi_*} T_u(J), \qquad \text{for} \quad v \in L(J).$$

Let us fix a point $v_o \in L(J)$. Starting from a canonical coordinate system $(x_1, \cdots, x_n, z, p_1, \cdots, p_n)$ defined on a neiborhood U' of the contact manifold (J, C) around $u_o = \pi(v_o)$ such that $dx_1 \wedge \cdots \wedge dx_n \mid_{v_o} \neq 0$, we can introduce a coordinate system (x_i, z, p_i, p_{ij}) $(1 \leqq i \leqq j \leqq n)$ on

$$U = \{v \in \pi^{-1}(U') \mid \pi(v) = u \in U' \quad \text{and} \quad dx_1 \wedge \cdots \wedge dx_n \mid_v \neq 0\} \subset L(J)$$

by defining functions $p_{ij}(v)$ on U as follows;

$$dp_i \mid_v = \sum_{i-!}^{n} p_{ij}(v) dx_j \mid_v .$$

Then, since $v \in C(u)$, we have $dz \mid_v = \sum_{i-1}^{n} p_i(u) dx \mid_v$ and , since $d\varpi \mid_v = 0$, we get $p_{ij} = p_{ji}$ from

$$d\varpi \mid_v = \sum_{i=1}^{n} dx_i \mid_v \wedge dp_i \mid_v = \sum_{i,j=1}^{n} p_{ij}(v) dx_i \mid_v \wedge dx_j \mid_v = 0.$$

Thus E is defined on this canonical coordinate system by

$$E = \{\varpi = \varpi_1 = \cdots = \varpi_n = 0\},$$

where

$$\varpi = dz - \sum_{i=1}^{n} p_i\, dx_i, \quad \text{and} \quad \varpi_i = dp_i - \sum_{j=1}^{n} p_{ij}\, dx_j \qquad \text{for} \quad i = 1, \cdots, n.$$

Let (J, C), $(\hat{J}, \hat{C})$ be contact manifolds of dimension $2n+1$ and $\varphi : (J, C) \to (\hat{J}, \hat{C})$ be a contact diffeomorphism between them. Then φ induces an isomorphism $\varphi_* : (L(J), E) \to (L(\hat{J}), \hat{E})$. Conversely we have (cf. Theorem 3.2 [Y1])

THEOREM 1.3. *Let (J, C) and $(\hat{J}, \hat{C})$ be contact manifolds of dimension $2n + 1$. Then, for an isomorphism $\Phi : (L(J), E) \to (L(\hat{J}), \hat{E})$, there exists a contact diffeomorphism $\varphi : (J, C) \to (\hat{J}, \hat{C})$ such that $\Phi = \varphi_*$.*

Our first aim is to formulate the submanifold theory for $(L(J), E)$, which will be given in §3.

1.3. Derived sytems and Cauchy characteristic systems. Now we prepare basic notions for **linear differential systems** (or Pfaffian systems). By a (linear) differential system (M, D), we mean a subbundle D of the tangent bundle $T(M)$ of a manifold M of dimension d. Locally D is defined by 1-forms $\omega_1, \ldots, \omega_{d-r}$ such that $\omega_1 \wedge \cdots \wedge \omega_{d-r} \neq 0$ at each point, where r is the rank of D;

$$D = \{\, \omega_1 = \cdots = \omega_{d-r} = 0 \,\}.$$

For two differential systems (M, D) and $(\hat{M}, \hat{D})$, a diffeomorphism φ of M onto $\hat{M}$ is called an **isomorphism** of (M, D) onto $(\hat{M}, \hat{D})$ if the differential map φ_* of φ sends D onto $\hat{D}$.

By the **Frobenius Theorem**, we know that D is completely integrable if and only if

$$d\omega_i \equiv 0 \pmod{\omega_1, \ldots, \omega_s} \qquad \text{for } i = 1, \ldots, s,$$

or equivalently, if and only if

$$[\mathcal{D}, \mathcal{D}] \subset \mathcal{D}$$

where $s = d - r$ and $\mathcal{D} = \Gamma(D)$ denotes the space of sections of D.

Thus, for a non-integrable differential system D, we are led to consider the **Derived System** ∂D of D, which is defined, in terms of sections, by

$$\partial\mathcal{D} = \mathcal{D} + [\mathcal{D}, \mathcal{D}].$$

Furthermore the **Cauchy Characteristic System** $\mathrm{Ch}\,(D)$ of (M, D) is defined at each point $x \in M$ by

$$\mathrm{Ch}\,(D)(x) = \{X \in D(x) \mid X \rfloor d\omega_i \equiv 0 \pmod{\omega_1, \ldots, \omega_s} \quad \text{for } i = 1, \ldots, s\,\},$$

where $\rfloor$ denotes the interior multiplication, i.e., $X \rfloor d\omega(Y) = d\omega(X, Y)$. When $\mathrm{Ch}\,(D)$ is a differential system (i.e., has constant rank), it is always completely integrable (see §1.4.).

Moreover **Higher Derived Systems** $\partial^k D$ are usually defined successively (cf. $[BCG_3]$) by

$$\partial^k D = \partial(\partial^{k-1} D),$$

where we put $\partial^0 D = D$ for convention.

On the other hand we define the k-th **Weak Derived System** $\partial^{(k)} D$ of D inductively by

$$\partial^{(k)}\mathcal{D} = \partial^{(k-1)}\mathcal{D} + [\mathcal{D}, \partial^{(k-1)}\mathcal{D}],$$

where $\partial^{(0)} D = D$ and $\partial^{(k)}\mathcal{D}$ denotes the space of sections of $\partial^{(k)} D$. This notion is one of the key point in the **Tanaka Theory** ([T1]).

1.4. Proof of the Darboux theorem. First of all, we will show that, for a differential system (M, D), the Cauchy characteristic system $\mathrm{Ch}\,(D)$ is completely integrable if $\mathrm{Ch}\,(D)$ is of constant rank, i.e., if $\mathrm{Ch}\,(D)$ is a subbundle of $T(M)$, where we assume D is locally defined by

$$D = \{\, \omega_1 = \cdots = \omega_s = 0 \,\}.$$

We will show that $[X, Y] \in \Gamma(\mathrm{Ch}\,(D)) = \mathrm{Ch}\,(\mathcal{D})$ for $X, Y \in \mathrm{Ch}\,(\mathcal{D})$. From

$$d\omega_\alpha(X, Y) = X(\omega_\alpha(Y)) - Y(\omega_\alpha(X)) - \omega_\alpha([X, Y]),$$

it follows that

$$\omega_\alpha([X, Y]) = -d\omega_\alpha(X, Y) = -(X \rfloor d\omega_\alpha)(Y) = 0$$

for $X, Y \in \mathrm{Ch}\,(\mathcal{D})$. Hence $[X, Y] \in \mathcal{D}$. Moreover, from $[L_X, i_Y] = i_{[X,Y]}$, where i_Y denotes the interior multiplication by Y, we calculate

$$\begin{aligned} [X, Y] \rfloor d\omega_\alpha &= [L_X, i_Y](d\omega_\alpha) = L_X i_Y d\omega_\alpha - i_Y L_X d\omega_\alpha \\ &= L_X(Y \rfloor d\omega_\alpha) - Y \rfloor d(L_X \omega_\alpha) \end{aligned}$$

The first term of the last equality vanishes because

$$L_X\omega_\beta = d(i_X\omega_\beta) + i_X d\omega_\beta = X \rfloor d\omega_\beta \equiv 0 \pmod{\omega_1, \ldots, \omega_s}$$

for $X \in \mathrm{Ch}\,(\mathcal{D})$. As for the second term, writing $L_X\omega_\alpha = \sum A^\alpha_\beta \omega_\beta$, we get

$$Y \rfloor d(L_X\omega_\alpha) = \sum dA^\alpha_\beta(Y)\, \omega_\beta + \sum A^\alpha_\beta\, (Y \rfloor d\omega_\beta) \equiv 0 \pmod{\omega_1, \ldots, \omega_s}$$

from $d(L_X\omega_\alpha) = \sum dA^\alpha_\beta \wedge \omega_\beta + \sum A^\alpha_\beta d\omega_\beta$. Thus we obtain $[X, Y] \rfloor d\omega_\alpha \equiv 0 \pmod{\omega_1, \ldots, \omega_s}$. This implies $[X, Y] \in \mathrm{Ch}\,(\mathcal{D})$.

Now let (J, C) be a contact manifold of dimension $2n + 1$. Let us fix a point u_o of J. Then there exists a coframe $\{\, \varpi, \omega_1, \ldots, \omega_n, \pi_1, \ldots, \pi_n \}$ defined around $u_o \in J$ such that the following holds

$$d\varpi \equiv \omega_1 \wedge \pi_1 + \cdots + \omega_n \wedge \pi_n \qquad \pmod{\varpi}.$$

Then, from the definition of $\mathrm{Ch}\,(C)$, it follows

$$\mathrm{Ch}\,(C) = \{\, \varpi = \omega_1 = \cdots = \omega_n = \pi_1 = \cdots = \pi_n = 0 \,\} = \{0\}.$$

In fact, (J, C) is a contact manifold if and only if $\mathrm{Ch}\,(C)$ is trivial.

Let us take a function x_1 defined aroud u_o such that $\varpi \wedge dx_1 \wedge \omega_2 \wedge \cdots \wedge \omega_n \wedge \pi_1 \wedge \cdots \wedge \pi_n \neq 0$ around u_o and consider the differential system C^1 defined by

$$C^1 = \{\, \varpi = dx_1 = 0 \,\}.$$

We can write

$$\omega_1 \equiv \sum_{i=2}^{n} a_i \omega_i + \sum_{i=1}^{n} b_i \pi_i \pmod{\varpi, dx_1}.$$

Then we calculate

$$\begin{aligned} d\varpi &\equiv \left(\sum_{i=2}^{n} a_i\omega_i + \sum_{i=1}^{n} b_i\pi_i \right) \wedge \pi_1 + \omega_2 \wedge \pi_2 + \cdots + \omega_n \wedge \pi_n \pmod{\varpi, dx_1} \\ &= \left(\sum_{i=2}^{n} b_i\pi_i \right) \wedge \pi_1 + \omega_2 \wedge (\pi_2 + a_2\pi_1) + \cdots + \omega_n \wedge (\pi_n + a_n\pi_1) \\ &= (\omega_2 - b_2\pi_1) \wedge (\pi_2 + a_2\pi_1) + \cdots + (\omega_n - b_n\pi_1) \wedge (\pi_n + a_n\pi_1). \end{aligned}$$

Thus, putting $\hat{\omega}_i = \omega_i - b_i\pi_1$, $\hat{\pi}_i = \pi_i + a_i\pi_1$ for $2 \leqq i \leqq n$, we get

$$d\varpi \equiv \hat{\omega}_2 \wedge \hat{\pi}_2 + \cdots + \hat{\omega}_n \wedge \hat{\pi}_n \pmod{\varpi, dx_1}.$$

Hence we obtain

$$\begin{aligned} \mathrm{Ch}\,(C^1) &= \{X \in C^1(u) \mid X \rfloor d\varpi \equiv 0 \pmod{\varpi, dx_1}\} \\ &= \{\varpi = dx_1 = \hat{\omega}_2 = \cdots = \hat{\omega}_n = \hat{\pi}_2 = \cdots = \hat{\pi}_n = 0\}. \end{aligned}$$

Moreover we have

$$\{0\} = \mathrm{Ch}\,(C) \subset \mathrm{Ch}\,(C^1) \subset C^1 \subset C.$$

Now let us take a first integral x_2 of $\mathrm{Ch}\,(C^1)$ such that $\varpi \wedge dx_1 \wedge dx_2 \wedge \hat{\omega}_3 \wedge \cdots \wedge \hat{\omega}_n \wedge \hat{\pi}_2 \wedge \cdots \wedge \hat{\pi}_n \neq 0$ around u_o and consider the differential system C^2 defined by

$$C^2 = \{\, \varpi = dx_1 = dx_2 = 0 \,\}.$$

Thus

$$\mathrm{Ch}\,(C^1) = \{\varpi = dx_1 = dx_2 = \hat{\omega}_3 = \cdots = \hat{\omega}_n = \hat{\pi}_2 = \cdots = \hat{\pi}_n = 0\},$$

so that we can write

$$\hat{\omega}_2 \equiv \sum_{i=3}^{n} \hat{a}_i \hat{\omega}_i + \sum_{i=2}^{n} \hat{b}_i \hat{\pi}_i \pmod{\varpi, dx_1, dx_2}.$$

Then, as in the above calculation, we get

$$d\varpi \equiv \tilde{\omega}_3 \wedge \tilde{\pi}_3 + \cdots + \tilde{\omega}_n \wedge \tilde{\pi}_n \quad (\mathrm{mod}\ \varpi, dx_1, dx_2)$$

where we put $\tilde{\omega}_i = \hat{\omega}_i - \hat{b}_i \hat{\pi}_2$, $\tilde{\pi}_i = \hat{\pi}_i + \hat{a}_i \hat{\pi}_2$ for $3 \leqq i \leqq n$. Thus we obtain

$$\begin{aligned}\mathrm{Ch}\,(C^2) &= \{X \in C^2(u) \mid X \rfloor d\varpi \equiv 0 \ (\mathrm{mod}\ \varpi, dx_1, dx_2)\} \\ &= \{\varpi = dx_1 = dx_2 = \tilde{\omega}_3 = \cdots = \tilde{\omega}_n = \tilde{\pi}_3 = \cdots = \tilde{\pi}_n = 0\}.\end{aligned}$$

Moreover we have

$$\{0\} = \mathrm{Ch}\,(C) \subset \mathrm{Ch}\,(C^1) \subset \mathrm{Ch}\,(C^2) \subset C^2 \subset C^1 \subset C.$$

If we repeat this procedure n times, we obtain first integrals x_i of $\mathrm{Ch}\,(C^{i-1})$ defined around u_o for $i = 2, \ldots, n$ such that $\varpi \wedge dx_1 \wedge \cdots \wedge dx_n \neq 0$ around u_o, and that

$$C^i = \{\varpi = dx_1 = \cdots = dx_i\} \qquad \text{for} \ \ i = 1, \ldots, n.$$

Moreover we have

$$d\varpi \equiv 0 \qquad (\mathrm{mod}\ \varpi, dx_1, \ldots, dx_n),$$

i.e., $C^n = \mathrm{Ch}\,(C^n)$ is completely integrable.

Finally let us take a first integral z of C^n such that $dz \wedge dx_1 \wedge \cdots \wedge dx_n \neq 0$ around u_o. Then we have

$$C^n = \{dz = dx_1 = \cdots = dx_n\},$$

so that

$$\varpi = a\left(dz - \sum_{i=1}^{n} p_i dx_i\right),$$

for some functions $a, p_1, \ldots, p_n$ defined around u_o such that $a(u_o) \neq 0$. Hence we obtain

$$C = \left\{dz - \sum_{i=1}^{n} p_i dx_i = 0\right\}.$$

Then, from $X \rfloor d\hat{\varpi} = \sum_{i=1}^{n} (dx_i(X) dp_i - dp_i(X) dx_i)$ for $\hat{\varpi} = dz - \sum_{i=1}^{n} p_i dx_i$, we get

$$\begin{aligned}&\{\hat{\varpi} = dx_1 = \cdots = dx_n = dp_1 = \cdots = dp_n = 0\} \\ &\quad = \{dz = dx_1 = \cdots = dx_n = dp_1 = \cdots = dp_n = 0\} \subset \mathrm{Ch}\,(C) = \{0\}\end{aligned}$$

which implies that $dz \wedge dx_1 \wedge \cdots \wedge dx_n \wedge dp_1 \wedge \cdots \wedge dp_n \neq 0$ aroud $u_o \in J$. This completes the proof of the Darboux theorem.

2. Tanaka theory of linear differential sytems. We will define **symbol algebras** of regular differential systems as a first invariant of non-integrable differential systems, which form nilpotent graded Lie algebras satisfying the generating condition (FGLA in short) at each point. Conversely, starting from a FGLA $\mathfrak{m}$, we will construct a model differential system $(M(\mathfrak{m}), D_{\mathfrak{m}})$(Standard Differential System of type $\mathfrak{m}$) and discuss the important notion of (algebraic) **prolongation** $\mathfrak{g}(\mathfrak{m})$ of $\mathfrak{m}$.

2.1. Symbol algebras of (M, D). A differential system (M, D) is called **regular**, if $D^{-(k+1)} = \partial^{(k)}D$ are subbundles of $T(M)$ for every integer $k \geqq 1$. For a regular differential system (M, D), we have ([T2], Proposition 1.1)

($S1$) *There exists a unique integer $\mu > 0$ such that, for all $k \geqq \mu$,*

$$D^{-k} = \cdots = D^{-\mu} \supsetneqq D^{-\mu+1} \supsetneqq \cdots \supsetneqq D^{-2} \supsetneqq D^{-1} = D,$$

($S2$) $[\mathcal{D}^p, \mathcal{D}^q] \subset \mathcal{D}^{p+q}$ *for all* $p, q < 0$

where $\mathcal{D}^p$ denotes the space of sections of D^p. ($S2$) can be checked easily by induction on q. Thus $D^{-\mu}$ is the smallest completely integrable differential system, which contains $D = D^{-1}$.

Let (M, D) be a regular differential system such that $T(M) = D^{-\mu}$. As a first invariant for non-integrable differential systems, we now define the **symbol algebra** $\mathfrak{m}(x)$ *associated with a differential system* (M, D) at $x \in M$, which was introduced by N. Tanaka [T2].

We put $\mathfrak{g}_{-1}(x) = D^{-1}(x)$, $\mathfrak{g}_p(x) = D^p(x)/D^{p+1}(x)$ $(p < -1)$ and

$$\mathfrak{m}(x) = \bigoplus_{p=-1}^{-\mu} \mathfrak{g}_p(x).$$

Let ϖ_p be the projection of $D^p(x)$ onto $\mathfrak{g}_p(x)$. Then, for $X \in \mathfrak{g}_p(x)$ and $Y \in \mathfrak{g}_q(x)$, the bracket product $[X, Y] \in \mathfrak{g}_{p+q}(x)$ is defined by

$$[X, Y] = \varpi_{p+q}([\tilde{X}, \tilde{Y}]_x),$$

where $\tilde{X}$ and $\tilde{Y}$ are any element of $\mathcal{D}^p$ and $\mathcal{D}^q$ respectively such that $\varpi_p(\tilde{X}_x) = X$ and $\varpi_q(\tilde{Y}_x) = Y$.

Endowed with this bracket operation, by ($S2$) above, $\mathfrak{m}(x)$ becomes a nilpotent graded Lie algebra such that $\dim \mathfrak{m}(x) = \dim M$ and satisfies

$$\mathfrak{g}_p(x) = [\mathfrak{g}_{p+1}(x), \mathfrak{g}_{-1}(x)] \qquad \text{for } p < -1.$$

We call $\mathfrak{m}(x)$ the **symbol algebra of** (M, D) at $x \in M$ for short.

Furthermore, let $\mathfrak{m}$ be a FGLA (fundamental graded Lie algebra) of μ-th kind, that is,

$$\mathfrak{m} = \bigoplus_{p=-1}^{-\mu} \mathfrak{g}_p$$

is a nilpotent graded Lie algebra such that

$$\mathfrak{g}_p = [\mathfrak{g}_{p+1}, \mathfrak{g}_{-1}] \qquad \text{for } p < -1.$$

Then (M, D) is called of type $\mathfrak{m}$ if the symbol algebra $\mathfrak{m}(x)$ is isomorphic with $\mathfrak{m}$ at each $x \in M$.

2.2. Standard differential system $(M(\mathfrak{m}), D_{\mathfrak{m}})$ of type $\mathfrak{m}$. Conversely, given a FGLA $\mathfrak{m} = \bigoplus_{p=-1}^{-\mu} \mathfrak{g}_p$, we can construct a model differential system of type $\mathfrak{m}$ as follows: Let $M(\mathfrak{m})$ be the simply connected Lie group with Lie algebra $\mathfrak{m}$. Identifying $\mathfrak{m}$ with the Lie algebra of left invariant vector fields on $M(\mathfrak{m})$, $\mathfrak{g}_{-1}$ defines a left invariant subbundle $D_{\mathfrak{m}}$ of $T(M(\mathfrak{m}))$. By definition of symbol algebras, it is easy to see that $(M(\mathfrak{m}), D_{\mathfrak{m}})$ is a regular differential system of type $\mathfrak{m}$. $(M(\mathfrak{m}), D_{\mathfrak{m}})$ is called the standard differential system of type $\mathfrak{m}$. The Lie algebra $\mathfrak{g}(\mathfrak{m})$ of all infinitesimal automorphisms of $(M(\mathfrak{m}), D_{\mathfrak{m}})$ can be calculated algebraically as the prolongation of $\mathfrak{m}$ ([T1], cf. [Y5]). We will discuss in §4 the question: *when does $\mathfrak{g}(\mathfrak{m})$ become finite dimensional and simple*?

As an example to calculate symbol algebras, let us show that $(L(J), E)$ is a regular differential system of type $\mathfrak{c}^2(n)$:

$$\mathfrak{c}^2(n) = \mathfrak{c}_{-3} \oplus \mathfrak{c}_{-2} \oplus \mathfrak{c}_{-1},$$

where $\mathfrak{c}_{-3} = \mathbb{R}$, $\mathfrak{c}_{-2} = V^*$ and $\mathfrak{c}_{-1} = V \oplus S^2(V^*)$. Here V is a vector space of dimension n and the bracket product of $\mathfrak{c}^2(n)$ is defined accordingly through the pairing between V and V^* such that V and $S^2(V^*)$ are both abelian subspaces of $\mathfrak{c}_{-1}$. This fact can be checked as follows: Let us take a canonical coordinate system $U; (x_i, z, p_i, p_{ij})$ $(1 \leqq i \leqq j \leqq n)$ of $(L(J), E)$. Then we have a coframe $\{\varpi, \varpi_i, dx_i, dp_{ij}\}$ $(1 \leqq i \leqq j \leqq n)$ at each point in U, where $\varpi = dz - \sum_{i=1}^n p_i\, dx_i$, $\varpi_i = dp_i - \sum_{j=1}^n p_{ij}\, dx_j$ $(i = 1, \cdots, n)$. Now take the dual frame $\{\frac{\partial}{\partial z}, \frac{\partial}{\partial p_i}, \frac{d}{dx_i}, \frac{\partial}{\partial p_{ij}}\}$, of this coframe, where

$$\frac{d}{dx_i} = \frac{\partial}{\partial x_i} + p_i \frac{\partial}{\partial z} + \sum_{j=1}^n p_{ij} \frac{\partial}{\partial p_j}$$

is the classical notation. Notice that $\{\frac{d}{dx_i}, \frac{\partial}{\partial p_{ij}}\}$ $(i = 1, \cdots, n)$ forms a free basis of $\Gamma(E)$. Then an easy calculation shows the above fact. Moreover we see that the derived system ∂E of E satisfies the following :

$$\partial E = \{\varpi = 0\} = \pi_*^{-1} C, \qquad \mathrm{Ch}\,(\partial E) = \mathrm{Ker}\, \pi_*.$$

These facts provide the proof of Theorem 1.3 (cf. Theorem 3.2 [Y1]).

Similarly we see that $(J(M, n), C)$ is a regular differential system of type $\mathfrak{c}^1(n, m)$:

$$\mathfrak{c}^1(n, m) = \mathfrak{c}_{-2} \oplus \mathfrak{c}_{-1},$$

where $\mathfrak{c}_{-2} = W$ and $\mathfrak{c}_{-1} = V \oplus W \otimes V^*$ for vector spaces V and W of dimension n and m respectively, and the bracket product of $\mathfrak{c}^1(n,m)$ is defined accordingly through the pairing between V and V^* such that V and $W \otimes V^*$ are both abelian subspaces of $\mathfrak{c}_{-1}$.

2.3. Prolongation $\mathfrak{g}(\mathfrak{m})$ of symbol algebras $\mathfrak{m}$. Let $\mathfrak{m} = \bigoplus_{p<0} \mathfrak{g}_p$ be a fundamental graded Lie algebra of μ-th kind defined over a field K. Here K denotes the field of real numbers $\mathbb{R}$ or that of complex numbers $\mathbb{C}$. We put

$$\mathfrak{g}(\mathfrak{m}) = \bigoplus_{p \in \mathbb{Z}} \mathfrak{g}_p(\mathfrak{m}),$$

where $\mathfrak{g}_p(\mathfrak{m}) = \mathfrak{g}_p$ for $p < 0$, $\mathfrak{g}_0(\mathfrak{m})$ is the Lie algebra of all (gradation preserving) derivations of graded Lie algebra $\mathfrak{m}$ and $\mathfrak{g}_k(\mathfrak{m})$ is defined inductively by the following for $k \geqq 1$

$$\mathfrak{g}_k(\mathfrak{m}) = \left\{ u \in \bigoplus_{p<0} \mathfrak{g}_{p+k} \otimes \mathfrak{g}_p^* \mid u([Y,Z]) = [u(Y),Z] - [u(Z),Y] \right\}.$$

Thus, as a vector space over K, $\mathfrak{g}_k(\mathfrak{m})$ is a linear subspace of $\operatorname{End}(\mathfrak{m}, \mathfrak{m}^k) = \mathfrak{m}^k \otimes \mathfrak{m}^*$, where $\mathfrak{m}^k = \mathfrak{m} \oplus \mathfrak{g}_0(\mathfrak{m}) \oplus \cdots \oplus \mathfrak{g}_{k-1}(\mathfrak{m})$. The bracket operation of $\mathfrak{g}(\mathfrak{m})$ is given as follows: First, since $\mathfrak{g}_0(\mathfrak{m})$ is the (gradation preserving) derivation algebra of graded Lie algebra $\mathfrak{m}$, we see that $\bigoplus_{p \leqq 0} \mathfrak{g}_p(\mathfrak{m})$ becomes a graded Lie algebra by putting

$$[u, X] = -[X, u] = u(X) \qquad \text{for } u \in \mathfrak{g}_0(\mathfrak{m}) \text{ and } X \in \mathfrak{m}.$$

Similarly, for $u \in \mathfrak{g}_k(\mathfrak{m}) \subset \mathfrak{m}^k \otimes \mathfrak{m}^*$ $(k > 0)$ and $X \in \mathfrak{m}$, we put $[u, X] = -[X, u] = u(X)$. Now, for $u \in \mathfrak{g}_k(\mathfrak{m})$ and $v \in \mathfrak{g}_\ell(\mathfrak{m})$ $(k, \ell \geqq 0)$, by induction on the integer $k + \ell \geqq 0$, we define $[u, v] \in \mathfrak{m}^{k+\ell} \otimes \mathfrak{m}^*$ by

$$[u, v](X) = [[u, X], v] + [u, [v, X]] \qquad \text{for } X \in \mathfrak{m}.$$

Here we note that, as the first case $k = \ell = 0$, this definition begins with that of the bracket product in $\mathfrak{g}_0(\mathfrak{m})$. It follows easily that $[u, v] \in \mathfrak{g}_{k+\ell}(\mathfrak{m})$. With this bracket product, $\mathfrak{g}(\mathfrak{m})$ becomes a graded Lie algebra. In fact the Jacobi identity

$$[[u, v], w] + [[v, w], u] + [[w, u], v] = 0,$$

for $u \in \mathfrak{g}_p(\mathfrak{m})$, $v \in \mathfrak{g}_q(\mathfrak{m})$ and $w \in \mathfrak{g}_r(\mathfrak{m})$, follows by definition when one of p, q or r is negative, and can be shown by induction on the integer $p+q+r \geqq 0$, when all of p, q and r are non-negative. The structure of the Lie algebra $\mathcal{A}(M(\mathfrak{m}), D_\mathfrak{m})$ of all infinitesimal automorphisms of $(M(\mathfrak{m}), D_\mathfrak{m})$ can be described by $\mathfrak{g}(\mathfrak{m})$. Especially $\mathcal{A}(M(\mathfrak{m}), D_\mathfrak{m})$ is isomorphic with $\mathfrak{g}(\mathfrak{m})$, when $\mathfrak{g}(\mathfrak{m})$ is finite dimensional ([T1], cf. [Y5]).

Let $\mathfrak{g}_0$ be a subalgebra of $\mathfrak{g}_0(\mathfrak{m})$. We define a subspace $\mathfrak{g}_k$ of $\mathfrak{g}_k(\mathfrak{m})$ for $k \geqq 1$ inductively by

$$\mathfrak{g}_k = \{ u \in \mathfrak{g}_k(\mathfrak{m}) \mid [u, \mathfrak{g}_{-1}] \subset \mathfrak{g}_{k-1} \}.$$

Then, putting

$$\mathfrak{g}(\mathfrak{m}, \mathfrak{g}_0) = \mathfrak{m} \oplus \bigoplus_{k \geqq 0} \mathfrak{g}_k,$$

we see, with the generating condition of $\mathfrak{m}$, that $\mathfrak{g}(\mathfrak{m}, \mathfrak{g}_0)$ is a graded subalgebra of $\mathfrak{g}(\mathfrak{m})$. $\mathfrak{g}(\mathfrak{m}, \mathfrak{g}_0)$ is called the prolongation of $(\mathfrak{m}, \mathfrak{g}_0)$.

REMARK 2.1 The notion of the prolongation of $\mathfrak{m}$ or $(\mathfrak{m}, \mathfrak{g}_0)$ plays quite an important role in the equivalence problems for the geometric structures subordinate to regular differential systems of type $\mathfrak{m}$, e.g., CR-structures, pseudo-product structures or Lie contact structures. We could not touch upon the more important geometric aspect of the prolongation theory of these structures. On these subjects, we refer the reader to foundational papers [T2, T3, T4] of N. Tanaka.

2.4. Proof of the Bäcklund theorem (Theorem 1.4 [Y3]). Let $J(M, n)$ be the space of n-dimensional contact elements to M and C be the canonical system on $J(M, n)$. Recall that $(J(M, n), C)$ is a regular differential system of type $\mathfrak{c}^1(n, m) = \mathfrak{c}^1(V, W)$:

$$\mathfrak{c}^1(V, W) = \mathfrak{c}_{-2} \oplus \mathfrak{c}_{-1},$$

where $\mathfrak{c}_{-2} = W$ and $\mathfrak{c}_{-1} = V \oplus W \otimes V^*$ for vector spaces V and W of dimension n and m respectively. Put $\mathfrak{f} = W \otimes V^*$. First we will characterize the abelian subspace $\mathfrak{f}$ of $\mathfrak{c}_{-1}$. Namely we first claim : *If* $\dim W \geqq 2$, *then*

$$\mathfrak{f} = \langle \{ X \in \mathfrak{c}_{-1} \mid \text{rank } \operatorname{ad}(X) \leqq 1 \} \rangle,$$

i.e., $\mathfrak{f}$ is the span of elements $X \in \mathfrak{c}_{-1}$ such that rank $\operatorname{ad}(X) = 1$. In fact, let $X = v_X + f_X$ be any element of $\mathfrak{c}_{-1}(V, W)$, where $v_X \in V$ and $f_X \in \mathfrak{f} = W \otimes V^*$. Then we have

$$\begin{aligned} \operatorname{ad}(X)(v) &= [X, v] = f_X(v) \qquad &&\text{for} \quad v \in V, \\ \operatorname{ad}(X)(f) &= [X, f] = -f(v_X) \qquad &&\text{for} \quad f \in W \otimes V^*. \end{aligned}$$

Thus we see that rank $\operatorname{ad}(X) = \dim W$ if $v_X \neq 0$ and rank $\operatorname{ad}(X) =$ rank f_X if $v_X = 0$. On the other hand it is clear that $\mathfrak{f} = W \otimes V^*$ is spanned by elements of rank 1. Put $E = \langle \{ X \in \mathfrak{c}_{-1} \mid \text{rank } \operatorname{ad}(X) \leqq 1 \} \rangle$. Then it follows that $E = \mathfrak{c}_{-1}(V, W)$ if $\dim W = 1$ and $E = \mathfrak{f}$ otherwise.

To prove Theorem 1.1, assume that $m = \dim W \geqq 2$ and let u be any point of $J(M, n)$. Let $\mathfrak{c}(u)$(resp. $\hat{\mathfrak{c}}(\Phi(u))$) be the symbol algebra of $(J(M, n), C))$ (resp. $(J(\hat{M}, n), \hat{C})$) at u (resp. $\Phi(u)$). Then there exist graded Lie algebra isomorphisms $\nu : \mathfrak{c}^1(V, W) \to \mathfrak{c}(u)$ and $\hat{\nu} : \mathfrak{c}^1(V, W) \to \hat{\mathfrak{c}}(\Phi(u))$ such that $\nu(\mathfrak{f}) = \operatorname{Ker} \pi_*$ and $\hat{\nu}(\mathfrak{f}) = \operatorname{Ker} \hat{\pi}_*$. Then, by the above claim, we get $\Phi_*(\operatorname{Ker} \pi_*) = \operatorname{Ker} \hat{\pi}_*$. Since each fibre of $J(M, n)$ and $J(\hat{M}, n)$ is connected, we see that Φ is fibre-preserving. Hence Φ induces a unique diffeomorphism φ of M onto $\hat{M}$ such that $\hat{\pi} \cdot \Phi = \varphi \cdot \pi$. Put $\hat{\Phi} = \Phi \cdot (\varphi^{-1})_*$. Then $\hat{\Phi}$ is an isomorphism of $(J(M, n), C)$ onto itself such that $\pi \cdot \hat{\Phi} = \pi$ Finally $\hat{\Phi} = id$ easily follows from $\hat{\Phi}_*(C) = C$ and the very definiton of the canonical system on $J(M, n)$, which implies $\Phi = \varphi_*$.

3. ***PD*-manifolds of second order.** We will here formulate the submanifold theory for $(L(J), E)$ as the geometry of PD-manifolds ([Y1]). Moreover we will discuss structures of higher order jet spaces.

3.1. Submanifolds in $L(J)$. Let R be a submanifold of $L(J)$ satisfying the following condition:

$$(R.0) \qquad p : R \to J; \text{ submersion,}$$

where $p = \pi\,|_R$ and $\pi : L(J) \to J$ is the projection. There are two differential systems $C^1 = \partial E$ and $C^2 = E$ on $L(J)$. We denote by D^1 and D^2 those differential systems on R obtained by restricting these differential systems to R. Moreover we denote by the same symbols those 1-forms obtained by restricting the defining 1-forms $\{\varpi, \varpi_1, \cdots, \varpi_n\}$ of the canonical system E to R. Then it follows from $(R.0)$ that these 1-forms are independent at each point on R and that

$$D^1 = \{\varpi = 0\}, \qquad D^2 = \{\varpi = \varpi_1 = \cdots = \varpi_n = 0\}.$$

In fact $(R; D^1, D^2)$ further satisfies the following conditions:

$(R.1)$ *D^1 and D^2 are differential systems of codimension 1 and $n+1$ respectively.*

$(R.2)$ $\partial D^2 \subset D^1$.

$(R.3)$ $\operatorname{Ch}(D^1)$ *is a subbundle of D^2 of codimension n.*

$(R.4)$ $\operatorname{Ch}(D^1)(v) \cap \operatorname{Ch}(D^2)(v) = \{0\}$ *at each $v \in R$.*

The last condition follows easily from the Realization Lemma below.

3.2. Realization lemma. Conversely these four conditions characterize submanifolds in $L(J)$ satisfying $(R.0)$. To see this , we first recall the following **Realization Lemma**, which characterize a submanifold of $(J(M, n), D)$.

Realization Lemma. *Let R and M be manifolds. Assume that the quadruple (R, D, p, M) satisfies the following conditions:*

(1) *p is a map of R into M of constant rank.*

(2) *D is a differential system on R such that $F = Ker\ p_*$ is a subbundle of D of codimension n.*

Then there exists a unique map ψ of R into $J(M,n)$ satisfying $p = \pi \cdot \psi$ and $D = \psi_*^{-1}(C)$, where C is the canonical differential system on $J(M,n)$ and $\pi : J(M,n) \to M$ is the projection. Furthermore, let v be any point of R. Then ψ is in fact defined by

$$\psi(v) = p_*(D(v)) \qquad \text{as a point of } Gr\ (T_{p(v)}(M)),$$

and satisfies

$$Ker\ (\psi_*)_v = F(v) \cap Ch(D)(v)$$

where $Ch(D)$ is the Cauchy Characteristic System of D.

For the proof, see Lemma 1.5 [Y1].

In view of this Lemma, we call the triplet $(R; D^1, D^2)$ of a manifold and two differential systems on it a **PD-manifold** if these satisfy the above four conditions $(R.1)$ to $(R.4)$. We have the (local) Realization Theorem for PD-manifolds as follows: From conditions $(R.1)$ and $(R.3)$, it follows that the codimension of the foliation defined by the completely integrable system $\mathrm{Ch}\,(D^1)$ is $2n+1$. Assume that R is regular with respect to $\mathrm{Ch}\,(D^1)$, i.e., the space $J = R/\mathrm{Ch}\,(D^1)$ of leaves of this foliation is a manifold of dimension $2n+1$. Then D^1 drops down to J. Namely there exists a differential system C on J of codimension 1 such that $D^1 = p_*^{-1}(C)$, where $p : R \to J = R/\mathrm{Ch}\,(D^1)$ is the projection. Obviously (J, C) becomes a contact manifold of dimension $2n+1$. Conditions $(R.1)$ and $(R.2)$ guarantees that the image of the following map ι is a legendrian subspace of (J, C):

$$\iota(v) = p_*(D^2(v)) \subset C(u), \qquad u = p(v).$$

Finally the condition $(R.4)$ shows that $\iota : R \to L(J)$ is an immersion by Realization Lemma for (R, D^2, p, J). Furthermore we have (Corollary 5.4 [Y1])

THEOREM 3.1. *Let $(R; D^1, D^2)$ and $(\hat{R}; \hat{D}^1, \hat{D}^2)$ be PD-manifolds. Assume that R and $\hat{R}$ are regular with respect to $Ch(D^1)$ and $Ch(\hat{D}^1)$ respectively. Let (J, C) and $(\hat{J}, \hat{C})$ be the associated contact manifolds. Then an isomorphism $\Phi : (R; D^1, D^2) \to (\hat{R}; \hat{D}^1, \hat{D}^2)$ induces a contact diffeomorphism $\varphi : (J, C) \to (\hat{J}, \hat{C})$ such that the following commutes;*

$$\begin{array}{ccc} R & \xrightarrow{\ \iota\ } & L(J) \\ {\scriptstyle\Phi}\downarrow & & \downarrow{\scriptstyle\varphi_*} \\ \hat{R} & \xrightarrow{\ \hat{\iota}\ } & L(\hat{J}). \end{array}$$

By this theorem, the submanifold theory for $(L(J), E)$ is reformulated as the geometry of PD-manifolds.

3.3. Reduction theorem. When $D^1 = \partial D^2$ holds for a PD-manifold $(R; D^1, D^2)$, the geometry of $(R; D^1, D^2)$ reduces to that of (R, D^2) and the Tanaka theory is directly applicable to this case. We will treat one of examples of this case in §6. Concerning about this situation, the following theorem is known under the compatibility condition (C) below:

(C) $\quad p^{(1)} : R^{(1)} \to R$ *is onto.*

where $R^{(1)}$ is the first prolongation of $(R; D^1, D^2)$,i.e.,

$$R^{(1)} = \{n\text{-dim. } \textit{integral elements} \text{ of } (R, D^2), \\ \text{transversal to } F = \operatorname{Ker} p_*\} \subset J(R, n),$$

(cf. Proposition 5.11 [Y1]).

THEOREM 3.2. *Let $(R; D^1, D^2)$ be a PD-manifold satisfying the condition (C) above. Then the following equality holds at each point v of R:*

$$\dim D^1(v) - \dim \partial D^2(v) = \dim Ch(D^2)(v).$$

In particular $D^1 = \partial D^2$ holds if and only if $Ch(D^2) = \{0\}$.

When PD-manifold $(R; D^1, D^2)$ admits a non-trivial Cauchy characteristics, i.e.,when $\operatorname{rank} \mathrm{Ch}\,(D^2) > 0$, the geometry of $(R; D^1, D^2)$ is further reducible to the geometry of a single differential system. Here we will be concerned with the local equivalence of $(R; D^1, D^2)$, hence we may assume that R is regular with respect to $\mathrm{Ch}\,(D^2)$, i.e., the leaf space $X = R/\mathrm{Ch}\,(D^2)$ is a manifold such that the projection $\rho : R \to X$ is a submersion and there exists a differential system D on X satisfying $D^2 = \rho_*^{-1}(D)$. Then the local equivalence of $(R; D^1, D^2)$ is further reducible to that of (X, D) as in the following

THEOREM 3.3. *Let (R, D^1, D^2) and $(\hat{R}; \hat{D}^1, \hat{D}^2)$ be PD-manifolds satisfying the condition (C) such that $Ch(D^2)$ and $Ch(\hat{D}^2)$ are subbundles of rank r $(0 < r < n)$. Assume that R and $\hat{R}$ are regular with respect to $Ch(D^2)$ and $Ch(\hat{D}^2)$ respectively. Let (X, D) and $(\hat{X}, \hat{D})$ be the leaf spaces, where $X = R/Ch(D^2)$ and $\hat{X} = \hat{R}/Ch(\hat{D}^2)$. Let us fix points $v_o \in R$ and $\hat{v}_o \in \hat{R}$ and put $x_o = \rho(v_o)$ and $\hat{x}_o = \hat{\rho}(\hat{v}_o)$. Then a local isomorphism $\psi : (R; D^1, D^2) \to (\hat{R}; \hat{D}^1, \hat{D}^2)$ such that $\psi(v_o) = \hat{v}_o$ induces a local isomorphism $\varphi : (X, D) \to (\hat{X}, \hat{D})$ such that $\varphi(x_o) = \hat{x}_o$ and $\varphi_*(\kappa(x_o)) = \hat{\kappa}(\hat{x}_o)$, and vice versa.*

3.4. Higher order jet spaces. The essential part of the Bäcklund's Theorem is to show that $F = \operatorname{Ker} \pi_*$ is the covariant system of $(J(M, n), C)$ for $m \geq 2$. Namely an isomorphism $\Phi : (J(M, n), C) \to (J(\hat{M}, n), \hat{C})$ sends F onto $\hat{F} = \operatorname{Ker} \hat{\pi}_*$ for $m \geq 2$.

In case $m = 1$, it is a well known fact that the group of isomorphisms of $(J(M, n), C)$, i.e., the group of contact transformations, is larger than the group of diffeomorphisms of M. Therefore, when we consider the geometric

2-jet spaces, the situation differs according to whether the number m of dependent variables is 1 or greater.

(1) **Case m = 1.** We should start from a contact manifold (J, C) of dimension $2n+1$, which is locally a space of 1-jet for one dependent variable by Darboux's theorem. Then we can construct the geometric second order jet space $(L(J), E)$ as the Lagrange-Grassmann bundle $L(J)$ over J consisting of all n-dimensional integral elements of (J, C), while E is the restriction to $L(J)$ of the canonical system on $J(L(J), n)$.

Now we put

$$(J^2(M,n), C^2) = (L(J(M,n)), E),$$

where M is a manifold of dimension $n+1$.

(2) **Case m ≥ 2.** Since $F = \operatorname{Ker} \pi_*$ is a covariant system of $(J(M,n), C)$, we define $J^2(M,n) \subset J(J(M,n), n)$ by

$$J^2(M,n) = \{n\text{-dim. } \textit{integral elements} \text{ of } (J(M,n), C), \ \text{transversal to } F\},$$

C^2 is defined as the restriction to $J^2(M,n)$ of the canonical system on $J(J(M,n), n)$.

Now the higher order (geometric) jet spaces $(J^{k+1}(M,n), C^{k+1})$ for $k \geq 2$ are defined (simultaneously for all m) by induction on k. Namely, for $k \geq 2$, we define $J^{k+1}(M,n) \subset J(J^k(M,n), n)$ and C^{k+1} inductively as follows:

$$\begin{aligned} J^{k+1}(M,n) = \{ & n\text{-dim. } \textit{integral elements} \text{ of } (J^k(M,n), C^k), \\ & \text{transversal to Ker } (\pi^k_{k-1})_* \ \}, \end{aligned}$$

where $\pi^k_{k-1} : J^k(M,n) \to J^{k-1}(M,n)$ is the projection. Here we have

$$\text{Ker } (\pi^k_{k-1})_* = \operatorname{Ch}(\partial C^k),$$

and C^{k+1} is defined as the restriction to $J^{k+1}(M,n)$ of the canonical system on $J(J^k(M,n), n)$. Then we have ([Y1],[Y3])

$$\begin{array}{ccccccc} C^k & \subset \cdots \subset & \partial^{k-2} C^k & \subset & \partial^{k-1} C^k & \subset \partial^k C^k = T(J^k(M,n)) \\ \cup & & \cup & & \cup & \\ \{0\} = \operatorname{Ch}(C^k) \subset \operatorname{Ch}(\partial C^k) & \subset \cdots \subset & \operatorname{Ch}(\partial^{k-1} C^k) & \subset & F & \end{array}$$

where $\operatorname{Ch}(\partial^{i+1} C^k)$ is a subbundle of $\partial^i C^k$ of codimension n for $i = 0, \ldots, k-2$ and, when $m \geq 2$, F is a subbundle of $\partial^{k-1} C^k$ of codimension n. The transversality conditions are expressed as

$$C^k \cap F = \operatorname{Ch}(\partial C^k) \ \text{ for } m \geqq 2, \quad C^k \cap \operatorname{Ch}(\partial^{k-1} C^k) = \operatorname{Ch}(\partial C^k) \ \text{ for } m = 1$$

By the above diagram together with the rank condition, Jet spaces $(J^k(M,n), C^k)$ can be characterized as higher order contact manifolds as in [Y1] and [Y3].

Here we observe that, if we drop the transversality condition in our definition of $J^k(M,n)$ and collect all n-dimensional integral elements, we may have some singularities in $J^k(M,n)$ in general. However, since every 2-form vanishes on 1-dimensional subspaces, in case $n=1$, the integrability condition for $v \in J(J^{k-1}(M,1),1)$ reduces to $v \subset C^{k-1}(u)$ for $u = \pi^k_{k-1}(v)$. Hence, in this case, we can safely drop the transversality condition in the above construction as in the following "Rank 1 Prolongation" , which constitutes the key construction for the Drapeau theorem for m-flags (see [SY]).

3.5. Drapeau theorem. We say that (R, D) is an m-flag of length k, if $\partial^i D$ is a subbundle of $T(R)$ for any i and has a derived length k, i.e., $\partial^k D = T(R)$;

$$D \subset \partial D \subset \cdots \subset \partial^{k-2} D \subset \partial^{k-1} D \subset \partial^k D = T(R),$$

such that rank $D = m+1$ and rank $\partial^i D =$ rank $\partial^{i-1} D + m$ for $i = 1, \ldots, k$. In particular $\dim R = (k+1)m + 1$.

Moreover, for a differential system (R, D), the **Rank 1 Prolongation** $(P(R), \widehat{D})$ is defined as follows;

$$P(R) = \bigcup_{x \in R} P_x \subset J(R, 1),$$

where

$$\begin{aligned} P_x &= \{\text{1-dim. integral elements of } (R, D)\} \\ &= \{u \subset D(x) \mid \text{1-dim. subspaces}\} \cong \mathbb{P}^m. \end{aligned}$$

We define the canonical system $\widehat{D}$ on $P(R)$ as the restriction to $P(R)$ of the canonical system on $J(R,1)$. It can be shown that the Rank 1 Prolongation of an m-flag of length k becomes a m-flag of length $k+1$.

Especially (R, D) is called a *Goursat flag* (un drapeau de Goursat) of length k when $m = 1$. Historically, by Engel, Goursat and Cartan, it is known that a Goursat flag (R, D) of length k is locally isomorphic, at a generic point, to the canonical system $(J^k(M,1), C^k)$ on the k-jet spaces of 1 independent and 1 dependent variable. The characterization of the canonical (contact) systems on jet spaces was given by R. Bryant in [B] for the first order systems and in [Y1] and [Y3] for higher order systems for n independent and m dependent variables. However, it was first explicitly exhibited by A.Giaro, A. Kumpera and C. Ruiz in [GKR] that a Goursat flag of length 3 has singuralities and the research of singularities of Goursat flags of length k ($k \geq 3$) began as in [M]. To this situation, R. Montgomery and M. Zhitomirskii constructed the "Monster Goursat manifold" by successive applications of the "Cartan prolongation of rank 2 distributions [BH]" to a surface and showed that every germ of a Goursat flag (R, D) of length k appears in this "Monster Goursat manifold" in [MZ], by first exhibitting the following Sandwich Lemma for (R, D);

$$\begin{array}{ccccccc} D & \subset & \partial D & \subset \cdots \subset & \partial^{k-2} D & \subset \partial^{k-1} D \subset & \partial^k D = T(R) \\ \cup & & \cup & & \cup & & \\ \mathrm{Ch}\,(D) \subset & & \mathrm{Ch}\,(\partial D) \subset \mathrm{Ch}\,(\partial^2 D) \subset & \cdots \subset & \mathrm{Ch}\,(\partial^{k-1} D) & & \end{array}$$

where $\mathrm{Ch}\,(\partial^i D)$ is the Cauchy characteristic system of $\partial^i D$ and $\mathrm{Ch}\,(\partial^i D)$ is a subbundle of $\partial^{i-1} D$ of corank 1 for $i = 1, \ldots, k-1$. Moreover, after [MZ], P.Mormul defined the notion of a *special m- flag* of length k for $m \geq 2$ to characterize those m-flags which are obtained by successive applications of the Rank 1 Prolongations to the space of 1-jets of 1 independent and m dependent variables.

To be precise, starting from a manifold M of dimension $m+1$, we put, for $k \geqq 2$,

$$(P^k(M), C^k) = (P(P^{k-1}(M)), \widehat{C}^{k-1})$$

where $(P^1(M), C^1) = (J(M,1), C)$. When $m = 1$, $(P^k(M), C^k)$ are called **"Monster Goursat Manifolds"** in [MZ].

Then we have (Corollary 5.8. [SY])

THEOREM 3.4. *An m-flag (R, D) of length k for $m \geq 3$ is locally isomorphic to $(P^k(M), C^k)$ if and only if $\partial^{k-1} D$ is of Cartan rank 1, and, moreover for $m \geq 4$, if and only if $\partial^{k-1} D$ is of Engel rank 1.*

Here, the *Cartan rank* of (R, C) is the smallest integer ρ such that there exist 1-forms $\{\pi^1, \ldots, \pi^\rho\}$, which are independent modulo $\{\omega_1, \ldots, \omega_s\}$ and satisfy

$$d\alpha \wedge \pi^1 \wedge \cdots \wedge \pi^\rho \equiv 0 \pmod{\omega_1, \ldots, \omega_s} \qquad \text{for } \forall \alpha \in C^\perp = \Gamma(C^\perp),$$

where $C = \{\omega_1 = \cdots = \omega_s = 0\}$. Furthermore the *Engel (half) rank* of (R, C) is the smallest integer ρ such that

$$(d\alpha)^{\rho+1} \equiv 0 \pmod{\omega_1, \ldots, \omega_s} \qquad \text{for } \forall \alpha \in C^\perp,$$

Moreover we have for an m-flag of length k for $m \geq 2$ (Corollary 6.3. [SY]) ,

THEOREM 3.5. *An m-flag (R, D) of length k is locally isomorphic to $(P^k(M), C^k)$ if and only if there exists a completely integrable subbundle F of $\partial^{k-1} D$ of corank 1.*

4. Differential sytems associated with simple graded lie algebras. Utilizing the structure theory of simple Lie algebras, we will answer to the following question ; *When does* $\mathfrak{g}(\mathfrak{m})$ *or* $\mathfrak{g}(\mathfrak{m}, \mathfrak{g}_0)$ *become finite dimensional and simple*? To do this, for a simple graded Lie algebra $\mathfrak{g}$, we first classify gradations of $\mathfrak{g}$ satisfying the generating conditions on their negative parts, which turns out to be equivalent to classify parabolic subalgebras of $\mathfrak{g}$.

4.1. Gradation of $\mathfrak{g}$ in terms of root space decomposition. Let $\mathfrak{g}$ be a finite dimensional simple Lie algebra over $\mathbb{C}$. Let us fix a Cartan subalgebra $\mathfrak{h}$ of $\mathfrak{g}$ and choose a simple root system $\Delta = \{\alpha_1, \ldots, \alpha_\ell\}$ of the root system Φ of $\mathfrak{g}$ relative to $\mathfrak{h}$. Then every $\alpha \in \Phi$ is an (all non-negative or all non-positive) integer coefficient linear combination of elements of Δ and we have the root space decomposition of $\mathfrak{g}$;

$$\mathfrak{g} = \bigoplus_{\alpha\in\Phi^+} \mathfrak{g}_\alpha \oplus \mathfrak{h} \oplus \bigoplus_{\alpha\in\Phi^+} \mathfrak{g}_{-\alpha},$$

where $\mathfrak{g}_\alpha = \{X \in \mathfrak{g} \mid [h, X] = \alpha(h)X \quad \text{for } h \in \mathfrak{h}\}$ is (1-dimensional) root space (corresponding to $\alpha \in \Phi$) and Φ^+ denotes the set of positive roots.

Now let us take a nonempty subset Δ_1 of Δ. Then Δ_1 defines the partition of Φ^+ as in the following and induces the gradation of $\mathfrak{g} = \bigoplus_{p\in\mathbb{Z}} \mathfrak{g}_p$ as follows:

$$\Phi^+ = \cup_{p\geqq 0}\Phi_p^+, \qquad \Phi_p^+ = \left\{\alpha = \sum_{i=1}^{\ell} n_i\alpha_i \mid \sum_{\alpha_i\in\Delta_1} n_i = p\right\},$$

$$\mathfrak{g}_p = \bigoplus_{\alpha\in\Phi_p^+} \mathfrak{g}_\alpha, \quad \mathfrak{g}_0 = \bigoplus_{\alpha\in\Phi_0^+} \mathfrak{g}_\alpha \oplus \mathfrak{h} \oplus \bigoplus_{\alpha\in\Phi_0^+} \mathfrak{g}_{-\alpha}, \quad \mathfrak{g}_{-p} = \bigoplus_{\alpha\in\Phi_p^+} \mathfrak{g}_{-\alpha},$$

$$[\mathfrak{g}_p, \mathfrak{g}_q] \subset \mathfrak{g}_{p+q} \qquad \text{for} \quad p, q \in \mathbb{Z}.$$

Moreover the negative part $\mathfrak{m} = \bigoplus_{p<0} \mathfrak{g}_p$ satisfies the following generating condition :

$$\mathfrak{g}_p = [\mathfrak{g}_{p+1}, \mathfrak{g}_{-1}] \quad \text{for} \quad p < -1.$$

We denote the SGLA (simple graded Lie algebra) $\mathfrak{g} = \bigoplus_{p=-\mu}^{\mu} \mathfrak{g}_p$ obtained from Δ_1 in this manner by (X_ℓ, Δ_1), when $\mathfrak{g}$ is a simple Lie algebra of type X_ℓ. Here X_ℓ stands for the Dynkin diagram of $\mathfrak{g}$ representing Δ and Δ_1 is a subset of vertices of X_ℓ. Moreover we have

$$\mu = \sum_{\alpha_i\in\Delta_1} n_i(\theta),$$

where $\theta = \sum_{i=1}^{\ell} n_i(\theta)\, \alpha_i$ is the highest root of Φ^+.

Conversely we have (Theorem 3.12 [Y5])

THEOREM 4.1. *Let $\mathfrak{g} = \bigoplus_{p\in\mathbb{Z}} \mathfrak{g}_p$ be a simple graded Lie algebra over $\mathbb{C}$ satisfying the generating condition. Let X_ℓ be the Dynkin diagram of $\mathfrak{g}$. Then $\mathfrak{g} = \bigoplus_{p\in\mathbb{Z}} \mathfrak{g}_p$ is isomorphic with a graded Lie algebra (X_ℓ, Δ_1) for some $\Delta_1 \subset \Delta$. Moreover (X_ℓ, Δ_1) and (X_ℓ, Δ_1') are isomorphic if and only if there exists a diagram automorphism ϕ of X_ℓ such that $\phi(\Delta_1) = \Delta_1'$.*

In the real case, we can utilize the Satake diagram of $\mathfrak{g}$ to describe gradations of $\mathfrak{g}$ (Theorem 3.12 [Y5]).

4.2. Gradation of $\mathfrak{g}$ in terms of matrix representations. Let $\mathfrak{g}$ be a simple Lie algebra over $\mathbb{C}$ of the classical type. We shall describe gradations of $\mathfrak{g}$ in terms of matrices. Here we reproduce the matrices description of the root space decomposition of $\mathfrak{g}$ from §7 of [Tk] (cf. [K-A], [V, Chapter 4.4]), which gives us explicit pictures of $M_{\mathfrak{g}}$.

(1) A_ℓ type ($\ell \geqq 1$). $\mathfrak{g} = \mathfrak{sl}(\ell+1, \mathbb{C})$. We take a Cartan subalgebra $\mathfrak{h}$ consisting of all diagonal elements of $\mathfrak{sl}(\ell+1, \mathbb{C})$, whose member we denote by $\operatorname{diag}(a_1, \ldots, a_{\ell+1})$. Let $\lambda_1, \ldots, \lambda_{\ell+1}$ be the linear form on $\mathfrak{h}$ defined by $\lambda_i : \operatorname{diag}(a_1, \ldots, a_{\ell+1}) \mapsto a_i$. We write E_{ij} $(1 \leqq i, j \leqq \ell+1)$ for the matrix whose (i,j)-component is 1 and all of whose other components are 0. Then we have

$$[H, E_{ij}] = (\lambda_i - \lambda_j)(H)\, E_{ij} \qquad \text{for } H \in \mathfrak{h}.$$

Hence $\Phi = \{\lambda_i - \lambda_j \in \mathfrak{h}^* \ (1 \leqq i, j \leqq \ell+1,\ i \neq j)\}$ and E_{ij} spans the root subspace for $\lambda_i - \lambda_j \in \Phi$. Let us choose a simple root system $\Delta = \{\alpha_1, \ldots, \alpha_\ell\}$ by putting

$$\alpha_i = \lambda_i - \lambda_{i+1}.$$

We have $\lambda_i - \lambda_j = \alpha_i + \cdots + \alpha_{j-1}$ when $i < j$. Hence $\theta = \alpha_1 + \cdots + \alpha_\ell$. Then we see that the gradation of $(A_\ell, \{\alpha_i\})$ is given by $\mathfrak{sl}(\ell+1, \mathbb{C}) = \mathfrak{g}_{-1} \oplus \mathfrak{g}_0 \oplus \mathfrak{g}_1$;

$$\mathfrak{g}_{-1} = \left\{ \begin{pmatrix} 0 & 0 \\ C & 0 \end{pmatrix} \,\middle|\, C \in M(j,i) \right\}, \quad \mathfrak{g}_1 = \left\{ \begin{pmatrix} 0 & D \\ 0 & 0 \end{pmatrix} \,\middle|\, D \in M(i,j) \right\},$$

$$\mathfrak{g}_0 = \left\{ \begin{pmatrix} A & 0 \\ 0 & B \end{pmatrix} \,\middle|\, A \in M(i,i),\ B \in M(j,j) \text{ and } \operatorname{tr}A + \operatorname{tr}B = 0 \right\},$$

where $j = \ell - i + 1$ and $M(p,q)$ denotes the set of $p \times q$ matrices. This decomposition can be described schematically by the following diagram,

	i	j
i	0	1
j	−1	0

where the vertical (resp. horizontal) line stands for the i-th vertical (resp. horizontal) intermediate line of a matrix in $\mathfrak{sl}(\ell+1, \mathbb{C})$. Then, for example, the diagram of $(A_\ell, \{\alpha_i, \alpha_j\})$ $(i < j)$ is obtained by superposing the diagrams of $(A_\ell, \{\alpha_i\})$ and $(A_\ell, \{\alpha_j\})$.

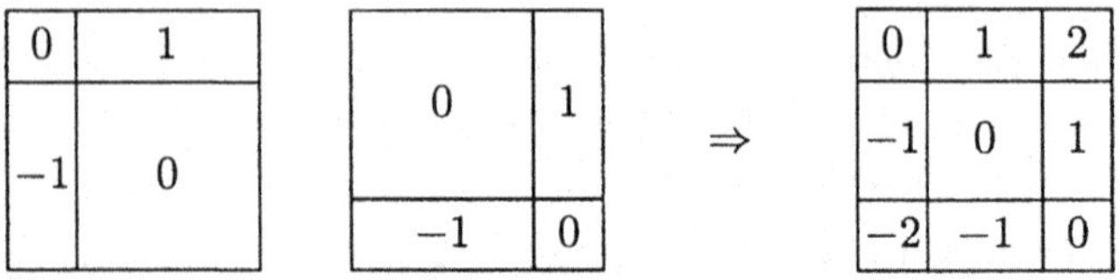

$$\begin{array}{|c|c|}\hline 0 & 1 \\ \hline -1 & 0 \\ \hline \end{array} \qquad \begin{array}{|c|c|}\hline 0 & 1 \\ \hline -1 & 0 \\ \hline \end{array} \qquad \Rightarrow \qquad \begin{array}{|c|c|c|}\hline 0 & 1 & 2 \\ \hline -1 & 0 & 1 \\ \hline -2 & -1 & 0 \\ \hline \end{array}$$

In general the diagram of $(A_\ell, \{\alpha_{i_1}, \ldots, \alpha_{i_k}\})$ is obtained by superposing the k diagrams of $(A_\ell, \{\alpha_{i_1}\})$, ..., $(A_\ell, \{\alpha_{i_k}\})$. Namely the gradation of $(A_\ell, \{\alpha_{i_1}, \ldots, \alpha_{i_k}\})$ is obtained by subdividing matrices by both vertical and horizontal k lines. Here i-th intermediate line corresponds to the simple root α_i.

By this description of gradations, we see that the model space $M_\mathfrak{g}$ of $(A_\ell, \{\alpha_i\})$ is the complex Grassmann manifold $Gr(i, V)$ consisting of all i-dimensional subspaces of $V = \mathbb{C}^{\ell+1}$. Furthermore the model space $M_\mathfrak{g}$ of $(A_\ell, \{\alpha_{i_1}, \ldots, \alpha_{i_k}\})$ $(1 \leqq i_1 < \cdots < i_k \leqq \ell)$ is the flag manifold $F(i_1, \ldots, i_k; V)$ consisting of all flags $\{V_1 \subset \cdots \subset V_k\}$ in V such that $\dim V_j = i_j$ for $j = 1, \ldots, k$ (cf. [Tt]).

(2) C_ℓ type $(\ell \geqq 2)$. Let $(V, \langle\, ,\, \rangle)$ be a symplectic vector space over $\mathbb{C}$ of dimension 2ℓ, that is, $\langle\, ,\, \rangle$ is a non-degenerate skew symmetric bilinear form on V. Then $\mathfrak{g} = \mathfrak{sp}(V)$. Let us take a symlectic basis $\{e_1, \ldots, e_\ell, f_1, \ldots, f_\ell\}$ of V such that $\langle e_i, e_j\rangle = \langle f_i, f_j\rangle = 0$ and $\langle f_i, e_{\ell+1-j}\rangle = \delta_{ij}$ for i, $j = 1, \ldots, \ell$. Thus we have a matrix representation

$$\mathfrak{g} = \{\, X \in \mathfrak{gl}(2\ell, \mathbb{C}) \mid {}^tXJ + JX = 0\,\}, \quad \text{where } J = \begin{pmatrix} 0 & K \\ -K & 0 \end{pmatrix},$$

and K is the $\ell \times \ell$ matrix whose (i,j)-component is $\delta_{i,\ell+1-j}$. We put $A' = KAK$ for $A \in \mathfrak{gl}(\ell, \mathbb{C})$. Namely A' is the "transposed" matrix of A with respect to the anti-diagonal line. Each $X \in \mathfrak{g}$ is expressed as a matrix of the following form;

$$X = \begin{pmatrix} A & B \\ C & -A' \end{pmatrix},$$

where A, B, C are $\ell \times \ell$ matrices such that B and C satisfy $B = B'$ and $C = C'$. Namely both B and C are symmetric with respect to the anti-diagonal line. Thus we see that X is determined by its upper anti-diagonal part. In the following we write $X = (A, B, C)$ in short.

We take a Cartan subalgebra $\mathfrak{h}$ consisting of all diagonal elements of the form $H = (\mathrm{diag}(a_1, \ldots, a_\ell), 0, 0)$. Let $\lambda_1, \ldots, \lambda_\ell$ be the linear form on $\mathfrak{h}$

defined by $\lambda_i : H \mapsto a_i$. We put $F_{ij} = E_{ij} + E'_{ij}$, where $E'_{ij} = E_{\ell+1-j,\ell+1-i}$. Then we have

$$\begin{aligned}
[H, (E_{ij}, 0, 0)] &= (\lambda_i - \lambda_j)(H)(E_{ij}, 0, 0), \\
[H, (0, F_{ij}, 0)] &= (\lambda_i + \lambda_{\ell+1-j})(H)(0, F_{ij}, 0), \\
[H, (0, 0, F_{ij})] &= -(\lambda_{\ell+1-i} + \lambda_j)(H)(0, 0, F_{ij}).
\end{aligned}$$

Hence $\Phi = \{\lambda_i - \lambda_j \ (i \neq j),\ \pm(\lambda_i + \lambda_j)\ (1 \leqq i \leqq j \leqq \ell)\}$ and $(E_{ij}, 0, 0)$, $(0, F_{i,\ell+1-j}, 0)$, $(0, 0, F_{\ell+1-i,j})$ are root vectors for $\lambda_i - \lambda_j$, $\lambda_i + \lambda_j$, $-(\lambda_i + \lambda_j) \in \Phi$ respectively. Let us choose a simple root system $\Delta = \{\alpha_1, \ldots, \alpha_\ell\}$ by putting

$$\begin{cases} \alpha_i = \lambda_i - \lambda_{i+1} & \text{for } i = 1, \ldots, \ell - 1, \\ \alpha_\ell = 2\,\lambda_\ell. \end{cases}$$

We have

$$\begin{cases} \lambda_i - \lambda_j = \alpha_i + \cdots + \alpha_{j-1} & (1 \leqq i < j \leqq \ell), \\ \lambda_i + \lambda_j = (\alpha_i + \cdots + \alpha_{\ell-1}) + (\alpha_j + \cdots + \alpha_\ell) & (1 \leqq i \leqq j \leqq \ell). \end{cases}$$

Hence $\theta = 2\,\alpha_1 + \cdots + 2\,\alpha_{\ell-1} + \alpha_\ell$. Then we see that the gradation of $(C_\ell, \{\alpha_i\})$ is given by the following diagram.

	i		i
i	0	1	2
	−1	0	1
i	−2	−1	0

$(1 \leqq i < \ell)$

0	1
−1	0

$(i = \ell)$

Then the diagram of $(C_\ell, \{\alpha_{i_1}, \ldots, \alpha_{i_k}\})$ is obtained by superposing the k diagrams of $(C_\ell, \{\alpha_{i_1}\})$, ..., $(C_\ell, \{\alpha_{i_k}\})$. Here two intermediate lines (i-th and $(2\ell - i)$-th lines) correspond to the simple root $\{\alpha_i\}$ for $i = 1, \ldots, \ell - 1$ and the center line corresponds to $\{\alpha_\ell\}$.

By this description of gradation, we see that the model space $M_{\mathfrak{g}}$ of $(C_\ell, \{\alpha_i\})$ is the Grassmann manifold $Sp\text{-}Gr(i, V)$ consisting of all i-dimensional isotropic subspaces of $(V, \langle\, ,\, \rangle)$. Furthermore the model space $M_{\mathfrak{g}}$ of $(C_\ell, \{\alpha_{i_1}, \ldots, \alpha_{i_k}\})$ $(1 \leqq i_1 < \cdots < i_k \leqq \ell)$ is the flag manifold $Sp\text{-}F(i_1, \ldots, i_k; V)$ consisting of all flags $\{V_1 \subset \cdots \subset V_k\}$ in V such that V_j is an i_j dimensional isotropic subspace of $(V, \langle\, ,\, \rangle)$ (cf. [Tt]).

(3) B_ℓ $(\ell \geqq 3)$, D_ℓ $(\ell \geqq 4)$ type. Let $(V, (\,|\,))$ be an inner product space over $\mathbb{C}$ of dimension 2ℓ or $2\ell + 1$, that is, $(\,|\,)$ is a non-degenerate symmetric bilinear form on V. Then $\mathfrak{g} = \mathfrak{o}(V)$. Let us take a basis $\{e_1, \ldots, e_\ell, e_{\ell+1}, f_1, \ldots, f_\ell\}$ of V such that $(e_i|e_j) = (e_{\ell+1}|e_i) = (e_{\ell+1}|f_i) =$

$(f_i|f_j) = 0$, $(e_{\ell+1}|e_{\ell+1}) = 1$ and $(e_i|f_{\ell+1-j}) = \delta_{ij}$ for $i, j = 1, \ldots, \ell$. Here we neglect $e_{\ell+1}$, when $\dim V = 2\ell$. Then we have a matrices representation

$$\mathfrak{g} = \{ X \in \mathfrak{gl}(n, \mathbb{C}) \mid {}^tXS + SX = 0 \}, \quad \text{where } S = \begin{pmatrix} 0 & 0 & K \\ 0 & 1 & 0 \\ K & 0 & 0 \end{pmatrix}$$

and $n = 2\ell$ or $2\ell + 1$. Each $X \in \mathfrak{g}$ is expressed as a matrix of the form

$$X = \begin{pmatrix} A & a & B \\ \xi & 0 & -a' \\ C & -\xi' & -A' \end{pmatrix}$$

where A, B, C are $\ell \times \ell$ matrices such that $B = -B'$, $C = -C'$ and a, ξ are column and row ℓ-vector respectively such that a' and ξ' are given by $a' = (a_\ell, \ldots, a_1)$, $\xi' = {}^t(\xi_\ell, \ldots, \xi_1)$ for $a = {}^t(a_1, \ldots, a_\ell)$, $\xi = (\xi_1, \ldots, \xi_\ell)$ respectively. Here the center column and the center row of X should be deleted when $\dim V = 2\ell$. Both B and C are skew symmetric with respect to the anti-diagonal line. In particular all the anti-diagonal components $x_{i,n+1-i}$ of X are 0. Thus X is determined by its upper anti-diagonal part. We write $X = (A, B, C, a, \xi)$, in short.

We take a Cartan subalgebra $\mathfrak{h}$ consisting of all diagonal elements of the form $H = (\mathrm{diag}(a_1, \ldots, a_\ell), 0, 0, 0, 0)$. Let $\lambda_1, \ldots, \lambda_\ell$ be the linear form on $\mathfrak{h}$ defined by $\lambda_i : H \mapsto a_i$. We put $G_{ij} = E_{ij} - E'_{ij}$ and $E_i = (\delta_{1i}, \ldots, \delta_{\ell i}) \in \mathbb{C}^\ell$. Then we have

$$\begin{aligned}
[H, (E_{ij}, 0, 0, 0, 0)] &= (\lambda_i - \lambda_j)(H)(E_{ij}, 0, 0, 0, 0), \\
[H, (0, G_{ij}, 0, 0, 0)] &= (\lambda_i + \lambda_{\ell+1-j})(H)(0, G_{ij}, 0, 0, 0), \\
[H, (0, 0, G_{ij}, 0, 0)] &= -(\lambda_{\ell+1-i} + \lambda_j)(H)(0, 0, G_{ij}, 0, 0), \\
[H, (0, 0, 0, E_i, 0)] &= \lambda_i(H)(0, 0, 0, E_i, 0), \\
[H, (0, 0, 0, 0, E_i)] &= -\lambda_i(H)(0, 0, 0, 0, E_i).
\end{aligned}$$

Hence we have

$$\Phi = \begin{cases} \{\lambda_i - \lambda_j \ (i \neq j), \ \pm(\lambda_i + \lambda_j) \ (1 \leqq i < j \leqq \ell)\} & \text{if } n = 2\ell, \\ \{\pm\lambda_i \ (1 \leqq i \leqq \ell), \ \lambda_i - \lambda_j \ (i \neq j), & \\ \qquad\qquad \pm(\lambda_i + \lambda_j) \ (1 \leqq i < j \leqq \ell)\} & \text{if } n = 2\ell + 1. \end{cases}$$

$(E_{ij}, 0, 0, 0, 0)$, $(0, G_{i,\ell+1-j}, 0, 0, 0)$, $(0, 0, G_{\ell+1-i,j}, 0, 0)$, $(0, 0, 0, E_i, 0)$ and $(0, 0, 0, 0, E_i)$ are root vectors for $\lambda_i - \lambda_j$, $\lambda_i + \lambda_j$, $-(\lambda_i + \lambda_j)$, λ_i and $-\lambda_i \in \Phi$ respectively. Let us choose a simple root system $\Delta = \{\alpha_1, \ldots, \alpha_\ell\}$ by putting

(i) B_ℓ type
$$\begin{cases} \alpha_i = \lambda_i - \lambda_{i+1} & \text{for } i = 1, \ldots, \ell - 1, \\ \alpha_\ell = \lambda_\ell. \end{cases}$$

(ii) D_ℓ type
$$\begin{cases} \alpha_i = \lambda_i - \lambda_{i+1} & \text{for } i = 1, \ldots, \ell - 1, \\ \alpha_\ell = \lambda_{\ell-1} + \lambda_\ell. \end{cases}$$

Then we have

(i) B_ℓ type

$$\begin{cases} \lambda_i - \lambda_j = \alpha_i + \cdots + \alpha_{j-1} & (1 \leqq i < j \leqq \ell), \\ \lambda_i = \alpha_i + \cdots + \alpha_\ell & (1 \leqq i \leqq \ell), \\ \lambda_i + \lambda_j = \alpha_i + \cdots + \alpha_{j-1} + 2\,\alpha_j + \cdots + 2\,\alpha_\ell & (1 \leqq i < j \leqq \ell). \end{cases}$$

Hence $\theta = \alpha_1 + 2\,\alpha_2 + \cdots + 2\,\alpha_\ell$.

(ii) D_ℓ type

$$\begin{cases} \lambda_i - \lambda_j = \alpha_i + \cdots + \alpha_{j-1} & (1 \leqq i < j \leqq \ell), \\ \lambda_i + \lambda_\ell = \alpha_i + \cdots + \alpha_{\ell-2} + \alpha_\ell & (1 \leqq i \leqq \ell - 2), \\ \lambda_{\ell-1} + \lambda_\ell = \alpha_\ell & \\ \lambda_i + \lambda_{\ell-1} = \alpha_i + \cdots + \alpha_{\ell-1} + \alpha_\ell & (1 \leqq i \leqq \ell - 2), \\ \lambda_i + \lambda_j = \alpha_i + \cdots + \alpha_{j-1} + 2\,\alpha_j + \cdots + 2\,\alpha_{\ell-2} + \alpha_{\ell-1} + \alpha_\ell & \\ & (1 \leqq i < j \leqq \ell - 2). \end{cases}$$

Hence $\theta = \alpha_1 + 2\,\alpha_2 + \cdots + 2\,\alpha_{\ell-2} + \alpha_{\ell-1} + \alpha_\ell$.

Then we see that the gradation of $(B_\ell, \{\alpha_i\})$ is given by the following diagram.

1	$n-2$	1
0	1	*
-1	0	1
*	-1	0

$(i = 1)$

i	$n-2i$	i
0	1	2
-1	0	1
-2	-1	0

$(1 < i \leqq \ell)$

The gradation of $(D_\ell, \{\alpha_i\})$ is given by the same diagram as above for $i = 1, \ldots, \ell - 2$ and the above diagram with $i = \ell - 1$ is that of $(D_\ell, \{\alpha_{\ell-1}, \alpha_\ell\})$. Moreover the diagrams of $(D_\ell, \{\alpha_{\ell-1}\})$ and $(D_\ell, \{\alpha_\ell\})$ are given as follows

$\ell-1$	0	1	0	1
1	-1	0	*	0
1	0	*	0	1
$\ell-1$	-1	0	-1	0

$(i = \ell - 1)$

ℓ	0	1
ℓ	-1	0

$(i = \ell)$

Clearly, by interchanging e_ℓ and f_1, matrices representations of $(D_\ell, \{\alpha_{\ell-1}\})$ and $(D_\ell, \{\alpha_\ell\})$ transforms each other, i.e., $(D_\ell, \{\alpha_{\ell-1}\})$ and

$(D_\ell, \{\alpha_\ell\})$ are conjugate. The other gradations of B_ℓ or D_ℓ type can be obtained by the principle of superposition as in the previous cases. Here two intermediate lines (i-th and $(n-i)$-th lines) correspond to the simple root $\{\alpha_i\}$ for $i = 1, \ldots, \ell$ in case of type B_ℓ and for $i = 1, \ldots, \ell - 2$ in case of type D_ℓ. Moreover in case of type D_ℓ, $(\ell-1)$-th and $(\ell+1)$-th intermediate lines correspond to the pair $\{\alpha_{\ell-1}, \alpha_\ell\}$ and the center line corresponds to $\{\alpha_\ell\}$.

By this description of gradations, we see that the Grassmann manifold $O\text{-}Gr(i, V)$ consisting of all i-dimensional isotropic subspaces of $(V, (\,|\,))$ is the model space $M_\mathfrak{g}$ of $(B_\ell, \{\alpha_i\})$ or $(D_\ell, \{\alpha_i\})$ according as $\dim V = 2\ell+1$ or 2ℓ, except for the case when $i = \ell-1$ and $\dim V = 2\ell$. In the latter case $O\text{-}Gr(\ell-1, V)$ is the model space $M_\mathfrak{g}$ of $(D_\ell, \{\alpha_{\ell-1}, \alpha_\ell\})$, where $\dim V = 2\ell$. Thus, for D_ℓ type, we make a following convention for a subset Δ_1 of Δ: If $\alpha_{\ell-1} \in \Delta_1$ and $\alpha_\ell \notin \Delta_1$, we replace $\alpha_{\ell-1}$ by α_ℓ (the conjugacy class of (D_ℓ, Δ_1) does not change by this replacement), and if both $\alpha_{\ell-1}$ and $\alpha_\ell \in \Delta_1$, we write $\alpha^*_{\ell-1} = \{\alpha_{\ell-1}, \alpha_\ell\}$. Under this convention, we see that the model space $M_\mathfrak{g}$ of $(B_\ell, \{\alpha_{i_1}, \ldots, \alpha_{i_k}\})$ or $(D_\ell, \{\alpha_{i_1}, \ldots, \alpha_{i_k}\})$ $(1 \leqq i_1 < \cdots < i_k \leqq \ell)$ is the flag manifold $O\text{-}F(i_1, \ldots, i_k; V)$ consisting of all flags $\{V_1 \subset \cdots \subset V_k\}$ in V such that V_j is an i_j-dimensional isotropic subspace of $(V, (\,|\,))$, according as $\dim V = 2\ell + 1$ or 2ℓ (cf. [Tt]).

4.3. Theorem on prolongations. By Theorem 4.1, the classification of gradations $\mathfrak{g} = \bigoplus_{p\in\mathbb{Z}} \mathfrak{g}_p$ of simple Lie algebras $\mathfrak{g}$ satisfying the generating condition coincides with that of parabolic subalgebras $\mathfrak{g}' = \bigoplus_{p\geqq 0} \mathfrak{g}_p$ of $\mathfrak{g}$. Accordingly, to each SGLA (X_ℓ, Δ_1), there corresponds a unique R-space $M_\mathfrak{g} = G/G'$ (compact simply connected homogeneous complex manifold). Furthermore, when $\mu \geqq 2$, there exists the G-invariant differential system $D_\mathfrak{g}$ on $M_\mathfrak{g}$, which is induced from $\mathfrak{g}_{-1}$, and $(M(\mathfrak{m}), D_\mathfrak{m})$ (Standard differential system of type $\mathfrak{m}$) becomes an open submanifold of $(M_\mathfrak{g}, D_\mathfrak{g})$. For the Lie algebras of all infinitesimal automorphisms of $(M_\mathfrak{g}, D_\mathfrak{g})$, hence of $(M(\mathfrak{m}), D_\mathfrak{m})$, we have the following theorem (Theorem 5.2 [Y5]).

THEOREM 4.2. *Let $\mathfrak{g} = \bigoplus_{p\in\mathbb{Z}} \mathfrak{g}_p$ be a simple graded Lie algebra over $\mathbb{C}$ satisfying the generating condition. Then $\mathfrak{g} = \bigoplus_{p\in\mathbb{Z}} \mathfrak{g}_p$ is the prolongation of $\mathfrak{m} = \bigoplus_{p<0} \mathfrak{g}_p$ except for the following three cases.*

(1) $\mathfrak{g} = \mathfrak{g}_{-1} \oplus \mathfrak{g}_0 \oplus \mathfrak{g}_1$ *is of depth* 1 *(i.e., $\mu = 1$).*

(2) $\mathfrak{g} = \bigoplus_{p=-2}^{2} \mathfrak{g}_p$ *is a (complex) contact gradation.*

(3) $\mathfrak{g} = \bigoplus_{p\in\mathbb{Z}} \mathfrak{g}_p$ *is isomorphic with $(A_\ell, \{\alpha_1, \alpha_i\})$ $(1 < i < \ell)$ or $(C_\ell, \{\alpha_1, \alpha_\ell\})$.*

Furthermore $\mathfrak{g} = \bigoplus_{p\in\mathbb{Z}} \mathfrak{g}_p$ is the prolongation of $(\mathfrak{m}, \mathfrak{g}_0)$ except when $\mathfrak{g} = \bigoplus_{p\in\mathbb{Z}} \mathfrak{g}_p$ is isomorphic with $(A_\ell, \{\alpha_1\})$ or $(C_\ell, \{\alpha_1\})$.

Here R-spaces corresponding to the above exceptions (1), (2) and (3) are as follows: (1) correspond to compact irreducible hermitian symmetric spaces. (2) correspond to contact manifolds of Boothby type (Standard contact manifolds), which exist uniquely for each simple Lie algebra other than $\mathfrak{sl}(2, \mathbb{C})$(see §5.1 below). In case of (3), $(J(\mathbb{P}^\ell, i), C)$ corresponds to

$(A_\ell, \{\alpha_1, \alpha_i\})$ and $(L(\mathbb{P}^{2\ell-1}), E)$ corresponds to $(C_\ell, \{\alpha_1, \alpha_\ell\})$ $(1 < i < \ell)$, where $\mathbb{P}^\ell$ denotes the ℓ-dimensional complex projective space and $\mathbb{P}^{2\ell-1}$ is the Standard contact manifold of type C_ℓ. Here we note that R-spaces corresponding to (2) and (3) are all Jet spaces of the first or second order.

For the real version of this theorem, we refer the reader to Theorem 5.3 [Y5].

4.4. Standard contact manifolds. Each simple Lie algebra $\mathfrak{g}$ over $\mathbb{C}$ has the highest root θ. Let Δ_θ denote the subset of Δ consisting of all vertices which are connected to $-\theta$ in the Extended Dynkin diagram of X_ℓ $(\ell \geqq 2)$. This subset Δ_θ of Δ, by the construction in §4, defines a gradation (or a partition of Φ^+), which distinguishes the highest root θ. Then, this gradation (X_ℓ, Δ_θ) turns out to be a contact gradation, which is unique up to conjugacy.

Moreover we have the adjoint (or equivalently coadjoint) representation, which has θ as the highest weight. The R-space $J_\mathfrak{g}$ corresponding to (X_ℓ, Δ_θ) can be obtained as the projectiviation of the (co-)adjoint orbit of G passing through the root vector of θ. By this construction, $J_\mathfrak{g}$ has the natural contact structure $C_\mathfrak{g}$ induced from the symplectic structure as the coadjoint orbit, which corresponds to the contact gradation (X_ℓ, Δ_θ) (cf. [Y5, §4]). Standard contact manifolds $(J_\mathfrak{g}, C_\mathfrak{g})$ were first found by Boothby ([Bo]) as compact simply connected homogeneous complex contact manifolds.

5. G_2-Geometry of overdetermined systems. This topic has its origin in the following paper of E. Cartan.

[C1] *Les systèmes de Pfaff à cinq variables et les équations aux derivèes partielles du second ordre*, Ann. Ec. Normale (1910), **27**: 109–192.

In this paper, following the tradition of geometric theory of partial differential equations of 19th century, E.Cartan dealt with the equivalence problem of two classes of second order partial differential equations in two independent variables under "contact transformations". One class consists of overdetermined systems, which are involutive, and the other class consists of single equations of Goursat type, i.e., single equations of parabolic type whose Monge characteristic systems are completely integrable. Especially in the course of the investigation, he found out the following facts: the symmetry algebras (i.e., the Lie algebra of infinitesimal contact transformations) of the following overdetermined system (involutive system) (A) and the single Goursat type equation (B) are both isomorphic with the 14-dimensional exceptional simple Lie algebra G_2.

$$\frac{\partial^2 z}{\partial x^2} = \frac{1}{3}\left(\frac{\partial^2 z}{\partial y^2}\right)^3, \quad \frac{\partial^2 z}{\partial x \partial y} = \frac{1}{2}\left(\frac{\partial^2 z}{\partial y^2}\right)^2, \tag{A}$$

$$9r^2 + 12t^2(rt - s^2) + 32s^3 - 36rst = 0, \tag{B}$$

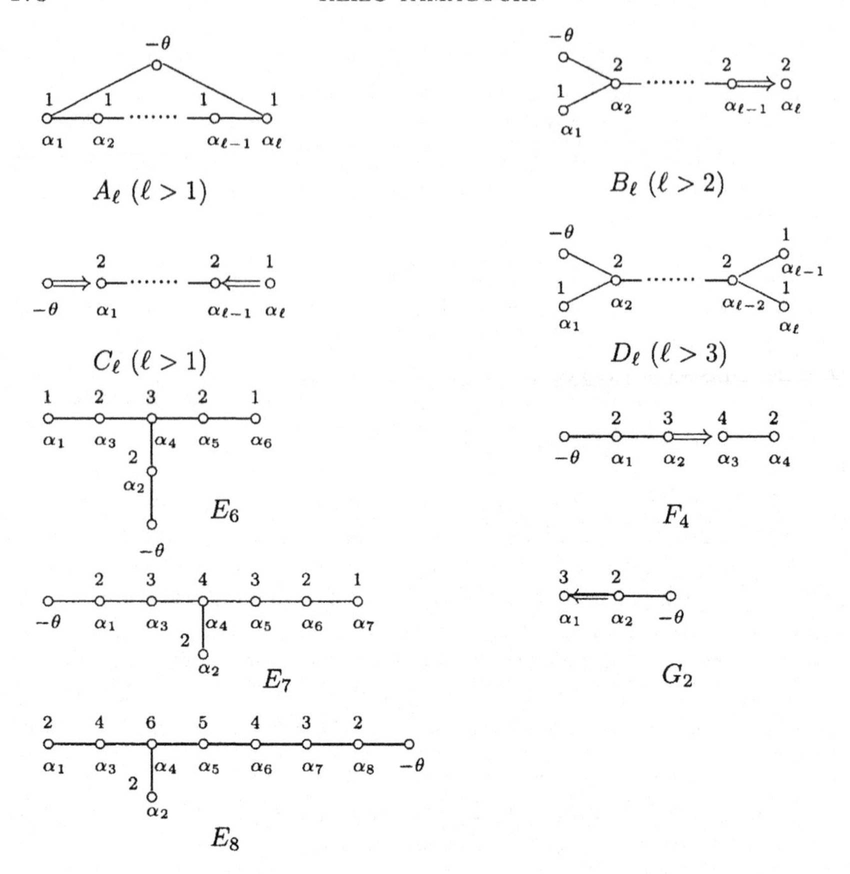

Extended Dynkin Diagrams with the coefficient of Highest Root (cf. [Bu]).

where

$$r = \frac{\partial^2 z}{\partial x^2}, \qquad s = \frac{\partial^2 z}{\partial x \partial y}, \qquad t = \frac{\partial^2 z}{\partial y^2}$$

are the classical terminology.

In the following, we will see how the model equation (A) can be obtained. For the details, including the case of the Goursat type equations and their generalizations to other exceptional simple Lie algebras, we refer the reader to [Y6].

5.1. Gradation of G_2. The Dynkin diagram of G_2 is given by

$$\underset{\alpha_1}{\circledcirc} \Leftarrow \underset{\alpha_2}{\circledcirc}$$

and the set Φ^+ of positive roots consists of six elements (cf. [Bu]):

$$\Phi^+ = \{\alpha_1, \alpha_2, \alpha_1 + \alpha_2, 2\alpha_1 + \alpha_2, 3\alpha_1 + \alpha_2, 3\alpha_1 + 2\alpha_2\}.$$

Here $\theta = 3\alpha_1 + 2\alpha_2$ and we have three choices for $\Delta_1 \subset \Delta = \{\alpha_1, \alpha_2\}$. Namely $\Delta_1 = \{\alpha_1\}$, $\{\alpha_2\}$ or $\{\alpha_1, \alpha_2\}$. Then the structure of each (G_2, Δ_1) is described as follows.

(1) $(G_2, \{\alpha_1\})$. We have $\mu = 3$ and Φ^+ decomposes as follows;

$$\Phi_3^+ = \{3\alpha_1 + \alpha_2, 3\alpha_1 + 2\alpha_2\}, \qquad \Phi_2^+ = \{2\alpha_1 + \alpha_2\},$$
$$\Phi_1^+ = \{\alpha_1, \alpha_1 + \alpha_2\}, \qquad \Phi_0^+ = \{\alpha_2\}.$$

Thus $\dim \mathfrak{g}_{-3} = \dim \mathfrak{g}_{-1} = 2$, $\dim \mathfrak{g}_{-2} = 1$ and $\dim \mathfrak{g}_0 = 4$. In the following section §5.2, we will see how the regular differential system of this type showed up historically.

(2) $(G_2, \{\alpha_2\})$. We have $\mu = 2$ and Φ^+ decomposes as follows;

$$\Phi_2^+ = \{3\alpha_1 + 2\alpha_2\}, \qquad \Phi_0^+ = \{\alpha_1\},$$
$$\Phi_1^+ = \{\alpha_2, \alpha_1 + \alpha_2, 2\alpha_1 + \alpha_2, 3\alpha_1 + \alpha_2\}.$$

Thus $\dim \mathfrak{g}_{-2} = 1$ and $\dim \mathfrak{g}_{-1} = \dim \mathfrak{g}_0 = 4$. Hence this is a contact gradation (cf. §4.4).

(3) $(G_2, \{\alpha_1, \alpha_2\})$. We have $\mu = 5$ and Φ^+ decomposes as follows;

$$\Phi_5^+ = \{3\alpha_1 + 2\alpha_2\}, \qquad \Phi_4^+ = \{3\alpha_1 + \alpha_2\}, \qquad \Phi_3^+ = \{2\alpha_1 + \alpha_2\},$$
$$\Phi_2^+ = \{\alpha_1 + \alpha_2\}, \qquad \Phi_1^+ = \{\alpha_1, \alpha_2\}, \qquad \Phi_0^+ = \emptyset.$$

Namely $(G_2, \{\alpha_1, \alpha_2\})$ is a gradation according to the height of roots and $\mathfrak{g}' = \bigoplus_{p \geqq 0} \mathfrak{g}_p$ is a Borel subalgebra. This case shows up in connection with the Hilbert-Cartan equation([Y5, §1.3]).

5.2. Classification of symbol algebras $\mathfrak{m}$ of lower dimension. In this paragraph, following a short passage from Cartan's paper [C1], let us classify FGLAs $\mathfrak{m} = \bigoplus_{p=-1}^{-\mu} \mathfrak{g}_p$ such that $\dim \mathfrak{m} \leqq 5$, which gives us the first invariants towards the classification of regular differential system (M, D) such that $\dim M \leqq 5$.

In the case $\dim \mathfrak{m} = 1$ or 2, $\mathfrak{m} = \mathfrak{g}_{-1}$ should be abelian. To discuss the case $\dim \mathfrak{m} \geqq 3$, we further assume that $\mathfrak{g}_{-1}$ is nondegenerate, i.e., $[X, \mathfrak{g}_{-1}] = 0$ implies $X = 0$ for $X \in \mathfrak{g}_{-1}$. This condition is equivalent to say $\mathrm{Ch}(D) = \{0\}$ for regular differential system (M, D) of type $\mathfrak{m}$. When $\mathfrak{g}_{-1}$ is degenerate, $\mathrm{Ch}(D)$ is non-trivial, hence at least locally, (M, D) induces a regular differential system (X, D^*) on the lower dimensional space X, where $X = M/\mathrm{Ch}(D)$ is the leaf space of the foliation on M defined by $\mathrm{Ch}(D)$ and D^* is the differential system on X such that $D = p_*^{-1}(D^*)$. Here $p : M \to X = M/\mathrm{Ch}(D)$ is the projection. Moreover, for the following discussion, we first observe that the dimension of $\mathfrak{g}_{-2}$ does not exceed $\binom{m}{2}$, where $m = \dim \mathfrak{g}_{-1}$.

In the case $\dim \mathfrak{m} = 3$, we have $\mu \leqq 2$. When $\mu = 2$, $\mathfrak{m} = \mathfrak{g}_{-2} \oplus \mathfrak{g}_{-1}$ is the contact gradation, i.e., $\dim \mathfrak{g}_{-2} = 1$ and $\mathfrak{g}_{-1}$ is nondegenerate. In the case $\dim \mathfrak{m} = 4$, we see that $\mathfrak{g}_{-1}$ is degenerate when $\mu \leqq 2$. When $\mu = 3$, we have $\dim \mathfrak{g}_{-3} = \dim \mathfrak{g}_{-2} = 1$ and $\dim \mathfrak{g}_{-1} = 2$. Moreover it follows that $\mathfrak{m}$ is isomorphic with $\mathfrak{c}^2(1)$ in this case. In the case $\dim \mathfrak{m} = 5$, we have $\dim \mathfrak{g}_{-1} = 4$, 3 or 2. When $\dim \mathfrak{g}_{-1} = 4$, $\mathfrak{m} = \mathfrak{g}_{-2} \oplus \mathfrak{g}_{-1}$ is the contact gradation. When $\dim \mathfrak{g}_{-1} = 3$, $\mathfrak{g}_{-1}$ is degenerate if $\dim \mathfrak{g}_{-2} = 1$, which implies that $\mu = 2$ and $\dim \mathfrak{g}_{-2} = 2$ in this case. Moreover, when $\mu = 2$, it follows that $\mathfrak{m}$ is isomorphic with $\mathfrak{c}^1(1,2)$. When $\dim \mathfrak{g}_{-1} = 2$, we have $\dim \mathfrak{g}_{-2} = 1$ and $\mu = 3$ or 4. Moreover, when $\mu = 4$, it follows that $\mathfrak{m}$ is isomorphic with $\mathfrak{c}^3(1)$, where $\mathfrak{c}^3(1)$ is the symbol algebra of the canonical system on the third order jet spaces for 1 unknown function (cf. §3 [Y1]).

Summarizing the above discussion, we obtain the following classification of the FGLAs $\mathfrak{m} = \bigoplus_{p=-1}^{-\mu} \mathfrak{g}_p$ such that $\dim \mathfrak{m} \leqq 5$ and $\mathfrak{g}_{-1}$ is nondegenerate.

(1) $\dim \mathfrak{m} = 3 \Longrightarrow \mu = 2$

$$\mathfrak{m} = \mathfrak{g}_{-2} \oplus \mathfrak{g}_{-1} \cong \mathfrak{c}^1(1) : \text{contact gradation}$$

(2) $\dim \mathfrak{m} = 4 \Longrightarrow \mu = 3$

$$\mathfrak{m} = \mathfrak{g}_{-3} \oplus \mathfrak{g}_{-2} \oplus \mathfrak{g}_{-1} \cong \mathfrak{c}^2(1)$$

(3) $\dim \mathfrak{m} = 5$, then $\mu \leqq 4$

(*a*) $\mu = 4$ $\quad \mathfrak{m} = \mathfrak{g}_{-4} \oplus \mathfrak{g}_{-3} \oplus \mathfrak{g}_{-2} \oplus \mathfrak{g}_{-1} \cong \mathfrak{c}^3(1)$

(*b*) $\mu = 3$ $\quad \mathfrak{m} = \mathfrak{g}_{-3} \oplus \mathfrak{g}_{-2} \oplus \mathfrak{g}_{-1}$
such that $\dim \mathfrak{g}_{-3} = \dim \mathfrak{g}_{-1} = 2$ and $\dim \mathfrak{g}_{-2} = 1$

(*c*) $\mu = 2$ $\quad \mathfrak{m} = \mathfrak{g}_{-2} \oplus \mathfrak{g}_{-1} \cong \mathfrak{c}^1(1,2)$

(*d*) $\mu = 2$ $\quad \mathfrak{m} = \mathfrak{g}_{-2} \oplus \mathfrak{g}_{-1} \cong \mathfrak{c}^1(2)$: contact gradation.

A notable and rather misleading fact is that, once the dimensions of $\mathfrak{g}_p$ are fixed, the Lie algebra structure of $\mathfrak{m} = \bigoplus_{p=-1}^{-\mu} \mathfrak{g}_p$ is unique in the above classification list. Moreover, except for the cases (*b*) and (*c*), every regular differential system (M, D) of type $\mathfrak{m}$ in the above list is isomorphic with the standard differential system $(M(\mathfrak{m}), D_{\mathfrak{m}})$ of type $\mathfrak{m}$ by Darboux's theorem (cf. Corollary 6.6 [Y1]). The first non-trivial situation that cannot be analyzed on the basis of Darboux's theorem occurs in the cases (*b*) and (*c*) (see [C1], [St]). Regular differential systems of type (*b*) and (*c*) are mutually closely related to each other (cf. [Y6, §6.3] and [C1]). We encountered with the type (*b*) fundamental graded Lie algebra as the case (1) of §5.1. in connection with the root space decomposition of the exceptional simple Lie algebra G_2.

As for the diferential system of type (*b*) above, the following differential system (X, E) on $X = \mathbb{R}^5$ was constructed by E. Cartan [C1];

$$E = \{\, \omega_1 = \omega_2 = \omega_3 = 0 \,\},$$

where

$$\begin{cases} \omega_1 = dx_1 + (x_3 + \frac{1}{2}x_4x_5)\, dx_4, \\ \omega_2 = dx_2 + (x_3 - \frac{1}{2}x_4x_5)\, dx_5, \\ \omega_3 = dx_3 + \frac{1}{2}(x_4\, dx_5 - x_5\, dx_4), \end{cases}$$

and $(x_1, x_2, x_3, x_4, x_5)$ is a coordinate system of $X = \mathbb{R}^5$. We have

$$\begin{cases} d\omega_1 = \omega_3 \wedge \omega_4, \\ d\omega_2 = \omega_3 \wedge \omega_5, \\ d\omega_3 = \omega_4 \wedge \omega_5, \end{cases} \tag{5.1}$$

where $\omega_4 = dx_4$ and $\omega_5 = dx_5$. In this case we may calculate symbol algebras of (X, E) as follows. We take a dual basis $\{X_1, \ldots, X_5\}$ of vector fields on X to a basis of 1-forms $\{\omega_1, \ldots, \omega_5\}$ given above;

$$\begin{aligned} X_1 &= \frac{\partial}{\partial x_1}, \quad X_2 = \frac{\partial}{\partial x_2}, \quad X_3 = \frac{\partial}{\partial x_3}, \\ X_4 &= \frac{\partial}{\partial x_4} + \frac{1}{2}x_5\frac{\partial}{\partial x_3} - (x_3 + \frac{1}{2}x_4x_5)\frac{\partial}{\partial x_1}, \\ X_5 &= \frac{\partial}{\partial x_5} - \frac{1}{2}x_4\frac{\partial}{\partial x_3} - (x_3 - \frac{1}{2}x_4x_5)\frac{\partial}{\partial x_2}. \end{aligned}$$

Then we calculate, or from (5.1),

$$[X_5, X_4] = X_3, \quad [X_5, X_3] = X_2, \quad [X_4, X_3] = X_1,$$

and $[X_i, X_j] = 0$ otherwise. This implies that $E^{-2} = \{\omega_1 = \omega_2 = 0\}$, $E^{-3} = T(X)$ and that (X, E) is isomorphic with the standard differential system of type $\mathfrak{m}_5$, where

$$\mathfrak{m}_5 = \mathfrak{g}_{-3} \oplus \mathfrak{g}_{-2} \oplus \mathfrak{g}_{-1}$$

is the fundamental graded algebra of third kind, whose Maurer-Cartan equation is given by (5.1). Here we note that the Lie algebra structure of $\mathfrak{m}_5$ is uniquely determined by the requirement that $\mathfrak{m}$ is fundamental, $\dim \mathfrak{g}_{-3} = \dim \mathfrak{g}_{-1} = 2$ and $\dim \mathfrak{g}_{-2} = 1$ (cf. [C1], [T2]). In fact $\mathfrak{m}_5$ is the universal fundamental graded algebra of third kind with $\dim \mathfrak{g}_{-1} = 2$ (see [T2, §3]).

5.3. ***G_2*-Geometry.** Let $(J_{\mathfrak{g}}, C_{\mathfrak{g}})$ be the Standard contact manifold of type G_2, i.e., R-space corresponding to $(G_2, \{\alpha_2\})$. If we lift the action of the exceptional group G_2 to $L(J_{\mathfrak{g}})$, then we have the following orbit decomposition:

$$L(J_{\mathfrak{g}}) = O \cup R_1 \cup R_2,$$

where O is the open orbit and R_i is the orbit of codimension i. Here R_1 and R_2 can be considered as the global model of (B) and (A) respectively. Moreover R_2 is compact and is a R-space corresponding to $(G_2, \{\alpha_1, \alpha_2\})$. From this fact, it becomes possible to describe the PD-manifold $(R; D^1, D^2)$ corresponding to (A) in terms of the R-space corresponding to $(G_2, \{\alpha_1, \alpha_2\})$. In fact R_2 has double fibrations onto $J_{\mathfrak{g}}$ (corresponding to $(G_2, \{\alpha_2\})$) and onto $\tilde{X}$ (corresponding to $(G_2, \{\alpha_1\})$).

Now, utilizing the Reduction Theorem (Theorem 3.3), we will construct the model equation (A) from the standard differential system (X, E) in §5.2, which is the local model corresponding to $(G_2, \{\alpha_1\})$. In fact $(R; D^1, D^2)$ is constructed as follows; $R = R(X)$ is the collection of hyperplanes v in each tangent space $T_x(X)$ at $x \in X$ which contains the fibre $\partial E(x)$ of the derived system ∂E of E.

$$R(X) = \bigcup_{x \in X} R_x \subset J(X, 4),$$

$$R_x = \{v \in \mathrm{Gr}(T_x(X), 4) \mid v \supset \partial E(x)\} \cong \mathbb{P}^1.$$

Moreover D^1 is the canonical system obtained by the Grassmaniann construction and D^2 is the lift of E. Precisely, D^1 and D^2 are given by

$$D^1(v) = \nu_*^{-1}(v) \supset D^2(v) = \nu_*^{-1}(E(x)),$$

for each $v \in R(X)$ and $x = \nu(v)$, where $\nu : R(X) \to X$ is the projection.

We introduce a fibre coordinate λ by $\varpi = \omega_1 + \lambda\omega_2$, where

$$D^1 = \{\varpi = 0\} \qquad \text{and} \qquad \partial E = \{\omega_1 = \omega_2 = 0\}.$$

Here $(x_1, \ldots, x_5, \lambda)$ constitutes a coordinate system on $R(X)$. Then we have

$$d\varpi = \omega_3 \wedge (\omega_4 + \lambda\omega_5) + d\lambda \wedge \omega_2,$$
$$\mathrm{Ch}\,(D^1) = \{\ \varpi = \omega_2 = \omega_3 = \omega_4 + \lambda\,\omega_5 = d\lambda = 0\ \},$$
$$D^2 = \{\ \varpi = \omega_2 = \omega_3 = 0\ \} \quad \text{and} \quad \partial D^2 = \{\ \varpi = \omega_2 = 0\ \}.$$

Hence $(R(X); D^1, D^2)$ is a PD-manifold of second order. Now we calculate

$$\begin{aligned}
\varpi &= \omega_1 + \lambda\omega_2 \\
&= dx_1 + \lambda\, dx_2 + \left(x_3 + \frac{1}{2}x_4x_5\right)dx_4 + \lambda\left(x_3 - \frac{1}{2}x_4x_5\right)dx_5 \\
&= d(x_1 + \lambda\, x_2) - x_2 d\lambda + \left(x_3 + \frac{1}{2}x_4x_5\right)(dx_4 + \lambda\, dx_5) - \lambda\, x_4x_5dx_5 \\
&= d(x_1 + \lambda\, x_2) - \left\{x_2 + x_5\left(x_3 + \frac{1}{2}x_4x_5\right)\right\}d\lambda \\
&\quad + \left(x_3 + \frac{1}{2}x_4x_5\right)d(x_4 + \lambda\, x_5) - \lambda\, x_4x_5dx_5.
\end{aligned}$$

Moreover we have

$$\begin{aligned}\lambda\, x_4 x_5 dx_5 &= \frac{1}{2}\lambda\, x_4 dx_5^2 = \frac{1}{2}\{d(\lambda\, x_4 x_5^2) - x_4 x_5^2 d\lambda - \lambda\, x_5^2 dx_4\,\} \\ &= \frac{1}{2}\{d(\lambda\, x_4 x_5^2) - x_4 x_5^2 d\lambda - \lambda\, x_5^2 d(x_4 + \lambda\, x_5) + \lambda\, x_5^2 d(\lambda\, x_5)\,\} \\ &= \frac{1}{2}\{d(\lambda\, x_4 x_5^2) + (\lambda\, x_5^3 - x_4 x_5^2) d\lambda - \lambda\, x_5^2 d(x_4 + \lambda\, x_5) + \lambda^2\, x_5^2 dx_5\,\} \\ &= \frac{1}{2}\left\{d\left(\lambda\, x_4 x_5^2 + \frac{1}{3}\lambda^2 x_5^3\right) - \left(\frac{2}{3}\lambda\, x_5^3 - \lambda\, x_5^3 + x_4 x_5^2\right) d\lambda\right. \\ &\qquad \left. - \lambda\, x_5^2 d(x_4 + \lambda\, x_5)\right\}.\end{aligned}$$

Thus we obtain

$$\begin{aligned}\varpi =& d\left(x_1 + \lambda\, x_2 - \frac{1}{2}\lambda\, x_4 x_5^2 - \frac{1}{6}\lambda^2\, x_5^3\right) - \left(x_2 + x_3 x_5 + \frac{1}{6}\lambda\, x_5^3\right) d\lambda \\ &+ \left(x_3 + \frac{1}{2} x_4 x_5 + \frac{1}{2}\lambda\, x_5^2\right) d(x_4 + \lambda\, x_5).\end{aligned}$$

We put

$$\begin{cases} z = x_1 + \lambda\, x_2 - \frac{1}{2}\lambda\, x_4 x_5^2 - \frac{1}{6}\lambda^2\, x_5^3, \\ x = \lambda, \\ y = x_4 + \lambda\, x_5, \\ p = x_2 + x_3 x_5 + \frac{1}{6}\lambda\, x_5^3, \\ q = -\left(x_3 + \frac{1}{2} x_4 x_5 + \frac{1}{2}\lambda\, x_5^2\right). \end{cases}$$

Then

$$D^1 = \{\, dz - p\, dx - q\, dy = 0\,\},$$

and (x, y, z, p, q) constitutes a canonical coordinate system on $J = R(X)/\mathrm{Ch}\,(D^1)$. Putting $x_5 = a$, we solve

$$\begin{cases} x_4 = y - xa, \\ x_3 = -q - \frac{1}{2}(y - xa)a - \frac{1}{2}xa^2 = -q - \frac{1}{2}ya, \\ x_2 = p + qa + \frac{1}{2}ya^2 - \frac{1}{6}xa^3, \\ x_1 = z - x\left(p + qa + \frac{1}{2}ya^2 - \frac{1}{6}xa^3\right) + \frac{1}{2}x(y - xa)a^2 + \frac{1}{6}x^2 a^3 \\ \quad\; = z - xp - xqa - \frac{1}{6}x^2 a^3. \end{cases}$$

Then, from

$$\begin{cases} x_4x_5 = ya - xa^2, \\ x_3 - \frac{1}{2}x_4x_5 = -q - \frac{1}{2}ya - \frac{1}{2}(y - xa)a = -q - ya + \frac{1}{2}xa^2, \end{cases}$$

we calculate

$$\begin{aligned} \omega_3 &= -d\left(q + ya - \frac{1}{2}xa^2\right) + (y - xa)da = -dq + \frac{1}{2}a^2dx - ady, \\ \omega_2 &= d\left(p + qa + \frac{1}{2}ya^2 - \frac{1}{6}xa^3\right) - \left(q + ya - \frac{1}{2}xa^2\right)da \\ &= dp + adq + \frac{1}{2}a^2dy - \frac{1}{6}a^3dx \\ &= a\left(dq - \frac{1}{2}a^2dx + ady\right) + dp + \frac{1}{3}a^3dx - \frac{1}{2}a^2dy \\ &= dp + \frac{1}{3}a^3dx - \frac{1}{2}a^2dy - a\omega_3. \end{aligned}$$

Putting $a = -t$, we obtain

$$D^2 = \{\varpi = \hat{\omega}_2 = \hat{\omega}_3 = 0\},$$

where

$$\varpi = dz - pdx - qdy, \quad \hat{\omega}_2 = dp - \frac{1}{3}t^3dx - \frac{1}{2}t^2dy, \quad \hat{\omega}_3 = dq - \frac{1}{2}t^2dx - tdy.$$

This implies

$$R(X) = \left\{ r = \frac{1}{3}t^3, \quad s = \frac{1}{2}t^2 \right\} \subset L(J),$$

in terms of the canonical coordinate (x, y, z, p, q, r, s, t) of $L(J)$.

6. Differential equations of finite type associated with SGLA. We discuss here the Se-ashi's principle to form good classes of linear differential equations of finite type from the view point of contact geometry and pseudo-product structures. In this course, we will find classes of parabolic geometries associated with differential equations of finite type, i.e., geometries associated with simple graded Lie algebras, which generalize the cases of second and third order ODE for scalar functions. Namely we will exhibit classes of differential equations of finite type, for which the model equations admit nonlinear contact transformations and their symmetry algebras become finite dimensional and simple.

6.1. Geometry of ODE and its generalizations. The geometry of ordinary differential equations for scalar function is strongly linked to the Lie algebra $\mathfrak{sl}(2,\mathbb{R}) = \mathfrak{sl}(\hat{V})$, where $\hat{V}$ is a vector space of dimension 2. Associated to the geometry of k-th order ordinary differential equation

$$\frac{d^k y}{dx^k} = F\left(x, y, \frac{dy}{dx}, \dots, \frac{d^{k-1}y}{dx^{k-1}}\right),$$

we have the irreducible representation of $\hat{\mathfrak{l}} = \mathfrak{sl}(\hat{V})$ on $S = S^{k-1}(\hat{V}^*)$, where $S^{k-1}(\hat{V}^*)$ is the space of homogeneopus polynomials of degree $k-1$ in two variables and is the solution space of the model equation $\frac{d^k y}{dx^k} = 0$ on the model space $\mathbb{P}^1(\mathbb{R}) = \mathbb{P}(\hat{V})$. It is known that the Lie algebra $\mathfrak{l} = \mathfrak{gl}(\hat{V})$ is the infinitesimal group of linear automorphisms of the model equation (cf. Proposition 4.4.1 [Se]). Moreover the Lie algebra $\mathfrak{g}^k = \mathfrak{g}^k(1,1)$ of infinitesimal contact transformations of $\frac{d^k y}{dx^k} = 0$ is given as follows; (1) $\mathfrak{g}^2$ is isomorphic to $\mathfrak{sl}(3,\mathbb{R})$. (2) $\mathfrak{g}^3$ is isomorphic to $\mathfrak{sp}(2,\mathbb{R})$. (3) Otherwise, for $k \geqq 4$, $\mathfrak{g}^k = S \oplus \mathfrak{l}$ is a subalgebra of the affine Lie algebra $\mathfrak{A}(S) = S \oplus \mathfrak{gl}(S)$(see below). The Lie algebra $\mathfrak{g}^k$ plays the fundamental role in the contact geometry of k-th order ordinary differential equations.

A little generally, we now consider a system of higher order differential equations of finite type of the following form:

$$\frac{\partial^k y^\alpha}{\partial x_{i_1} \cdots \partial x_{i_k}} = F^\alpha_{i_1 \cdots i_k}(x_1, \dots, x_n, y^1, \cdots, y^m, \dots, p^\beta_i, \dots, p^\beta_{j_1 \cdots j_{k-1}})$$
$$(1 \leqq \alpha \leqq m, 1 \leqq i_1 \leqq \cdots \leqq i_k \leqq n),$$

where $p^\beta_{i_1 \cdots i_l} = \frac{\partial^l y^\beta}{\partial x_{i_1} \cdots \partial x_{i_l}}$. These equations define a submanifold R in k-jets space J^k such that the restriction p to R of the bundle projection $\pi^k_{k-1} : J^k \to J^{k-1}$ gives a diffeomorphism

$$p : R \to J^{k-1}; \quad \text{diffeomorphism.} \tag{6.1}$$

On J^k, we have the contact (differential) system C^k defined by

$$C^k = \{\varpi^\alpha = \varpi^\alpha_i = \cdots = \varpi^\alpha_{i_1 \cdots i_{k-1}} = 0\},$$

where

$$\begin{cases} \varpi^\alpha = dy^\alpha - \sum_{i=1}^n p^\alpha_i dx_i, & (1 \leqq \alpha \leqq m) \\ \varpi^\alpha_i = dp^\alpha_i - \sum_{j=1}^n p^\alpha_{ij} dx_j, & (1 \leqq \alpha \leqq m, 1 \leqq i \leqq n) \\ \cdots\cdots\cdots\cdots, & \\ \varpi^\alpha_{i_1 \cdots i_{k-1}} = dp^\alpha_{i_1 \cdots i_{k-1}} - \displaystyle\sum_{j=1}^n p^\alpha_{i_1 \cdots i_{k-1} j} dx_j & \\ \qquad (1 \leqq \alpha \leqq m, 1 \leqq i_1 \leqq \cdots \leqq i_{k-1} \leqq n). & \end{cases} \tag{6.2}$$

Then C^k gives a foliation on R when R is integrable. Namely the restriction E of C^k to R is completely integrable.

Thus, through the diffeomorphism (6.1), R defines a completely integrable differential system $E' = p_*(E)$ on J^{k-1} such that

$$C^{k-1} = E' \oplus F', \quad F' = \mathrm{Ker}\,(\pi^{k-1}_{k-2})_*$$

where $\pi^{k-1}_{k-2} : J^{k-1} \to J^{k-2}$ is the bundle projection. The triplet $(J^{k-1}; E', F')$ is called the **pseudo-product structure** associated with R.

Corresponding to the splitting $D = E \oplus F = (p^{-1})_*(C^{k-1})$, we have the splitting in the symbol algebra of the regular differential system $(R, D) \cong (J^{k-1}, C^{k-1})$ of type $\mathfrak{c}^{k-1}(n, m)$;

$$\mathfrak{c}_{-1} = \mathfrak{e} \oplus \mathfrak{f},$$

where $\mathfrak{e} = V, \mathfrak{f} = W \otimes S^{k-1}(V^*)$. At each point $x \in R$, $\mathfrak{e}$ corresponds to $E(x)$ (the point in $R^{(1)}$ over x) and $\mathfrak{f}$ corresponds to $\mathrm{Ker}\,(\pi^{k-1}_{k-2})_*(p(x))$. Here we recall (see §1.3[YY1] for detail) that the fundamental graded Lie algebra (FGLA) $\mathfrak{c}^{k-1}(n, m)$ is defined by

$$\mathfrak{c}^{k-1}(n, m) = \mathfrak{c}_{-k} \oplus \cdots \oplus \mathfrak{c}_{-2} \oplus \mathfrak{c}_{-1},$$

where $\mathfrak{c}_{-k} = W, \mathfrak{c}_p = W \otimes S^{k+p}(V^*)$, $\mathfrak{c}_{-1} = V \oplus W \otimes S^{k-1}(V^*)$. Here V and W are vector spaces of dimension n and m respectively and the bracket product of $\mathfrak{c}^{k-1}(n, m) = \mathfrak{c}^{k-1}(V, W)$ is defined accordingly through the pairing between V and V^* such that V and $W \otimes S^{k-1}(V^*)$ are both abelian subspaces of $\mathfrak{c}_{-1}$. Here $S^r(V^*)$ denotes the r-th symmetric product of V^*.

Now we put

$$\check{\mathfrak{g}}_0 = \{X \in \mathfrak{g}_0(\mathfrak{c}^{k-1}(n, m)) \mid [X, \mathfrak{e}] \subset \mathfrak{e}, [X, \mathfrak{f}] \subset \mathfrak{f}\,\}$$

and consider the (algebraic) prolongation $\mathfrak{g}^k(n, m)$ of $(\mathfrak{c}^{k-1}(n, m), \check{\mathfrak{g}}_0)$, which is called the **pseudo-projective GLA of order k of bidegree** (n, m) ([T7]). Here $\mathfrak{g}_0(\mathfrak{c}^{k-1}(n, m))$ denotes the Lie algebra of gradation preserving derivations of $\mathfrak{c}^{k-1}(n, m)$.

Let $\check{G}_0 \subset GL(\mathfrak{c}^{k-1}(n, m))$ be the (gradation preserving) automorphism group of $\mathfrak{c}^{k-1}(n, m)$ which also preserve the splitting $\mathfrak{c}_{-1} = \mathfrak{e} \oplus \mathfrak{f}$. Then $\check{G}_0$ is the Lie subgroup of $GL(\mathfrak{c}^{k-1}(n, m))$ with Lie algebra $\check{\mathfrak{g}}_0$. The pseudo-product structure on a k-th order differential equation R of finite type given above, which is called the **pseudo-projective system of order k of bidegree** (n, m) in [T7], can be formulated as the $\check{G}_0^\sharp$-structure over a regular differential system of type $\mathfrak{c}^{k-1}(n, m)$ ([T2], [T7],[DKM]). Thus the prolongation $\mathfrak{g}^k(n, m)$ of $(\mathfrak{c}^{k-1}(n, m), \check{\mathfrak{g}}_0)$ represents the Lie algebra of infinitesimal automorphisms of the (local) model k-th order differential equation R_o of finite type, where

$$R_o = \left\{ \frac{\partial^k y^\alpha}{\partial x_{i_1} \cdots \partial x_{i_k}} = 0 \quad (1 \leqq \alpha \leqq m, 1 \leqq i_1 \leqq \cdots \leqq i_k \leqq n) \right\}.$$

The isomorphism ϕ of the pseudo-product structure on R preserves the differential system $D = E \oplus F$, which is equivalent to the canonical system C^{k-1} on J^{k-1}. Hence, by Bäcklund's Theorem (see §2.4), ϕ is the lift of a point transformation on J^0 when $m \geqq 2$ and $k \geqq 2$ and is the lift of a contact transformation on J^1 when $m = 1$ and $k \geqq 3$. When $(m, k) = (1, 2)$, ϕ is the lift of the point transformation on J^0, since ϕ preserves both D and $F = \mathrm{Ker}\,(\pi_0^1)_*$. Thus the equivalence of the pseudo-product structure on R is the equivalence of the k-th order equation under point or contact transformations. To settle the equivalence problem for the pseudo-projective systems of order k of bidegree (n, m), N.Tanaka constructed the **normal Cartan connections of type** $\mathfrak{g}^k(n, m)$ ([T4], [T5], [T7]).

It is well known that $\mathfrak{g}^k(n, m)$ $(k \geqq 2)$ has the following structure ([T7],[Y5], [DKM], [YY1]);

(1) $k = 2$: $\mathfrak{g}^2(n, m)$ is isomorphic to $\mathfrak{sl}(m + n + 1, \mathbb{R})$ and has the following gradation:

$$\mathfrak{sl}(m + n + 1, \mathbb{R}) = \mathfrak{g}_{-2} \oplus \mathfrak{g}_{-1} \oplus \mathfrak{g}_0 \oplus \mathfrak{g}_1 \oplus \mathfrak{g}_2,$$

where the gradation is given by subdividing matrices as follows;

$$\mathfrak{g}_{-2} = \left\{ \begin{pmatrix} 0 & 0 & 0 \\ 0 & 0 & 0 \\ \xi & 0 & 0 \end{pmatrix} \middle| \quad \xi \in W \cong \mathbb{R}^m \right\},$$

$$\mathfrak{g}_{-1} = \left\{ \begin{pmatrix} 0 & 0 & 0 \\ x & 0 & 0 \\ 0 & A & 0 \end{pmatrix} \middle| \; x \in V \cong \mathbb{R}^n, \; A \in M(m, n) = W \otimes V^* \right\},$$

$$\mathfrak{g}_0 = \left\{ \begin{pmatrix} a & 0 & 0 \\ 0 & B & 0 \\ 0 & 0 & C \end{pmatrix} \middle| \; \begin{array}{l} a \in \mathbb{R}, \, B \in \mathfrak{gl}(V), \, C \in \mathfrak{gl}(W), \\ a + \mathrm{tr}B + \mathrm{tr}C = 0 \end{array} \right\},$$

$$\mathfrak{g}_1 = \{ {}^tX \mid X \in \mathfrak{g}_{-1} \}, \qquad \mathfrak{g}_2 = \{ {}^tX \mid X \in \mathfrak{g}_{-2} \},$$

where $V = M(n, 1)$, $W = M(m, 1)$ and $M(a, b)$ denotes the set of $a \times b$ matrices. This gradation over $\mathbb{C}$ corresponds to $(A_{m+n}, \{\alpha_1, \alpha_{n+1}\})$ in (3) of Theorem 4.2.

(2) $k = 3$ and $m = 1$: $\mathfrak{g}^3(n, 1)$ is isomorphic to $\mathfrak{sp}(n + 1, \mathbb{R})$ and has the following gradation:

$$\mathfrak{sp}(n + 1, \mathbb{R}) = \mathfrak{g}_{-3} \oplus \mathfrak{g}_{-2} \oplus \mathfrak{g}_{-1} \oplus \mathfrak{g}_0 \oplus \mathfrak{g}_1 \oplus \mathfrak{g}_2 \oplus \mathfrak{g}_3.$$

First we describe

$$\mathfrak{sp}(n + 1, \mathbb{R}) = \{ X \in \mathfrak{gl}(2n + 2, \mathbb{R}) \mid {}^tXJ + JX = 0 \},$$

where

$$J = \begin{pmatrix} 0 & 0 & 0 & 1 \\ 0 & 0 & I_n & 0 \\ 0 & -I_n & 0 & 0 \\ -1 & 0 & 0 & 0 \end{pmatrix} \in \mathfrak{gl}(2n+2, \mathbb{R}), \quad I_n = (\delta_{ij}) \in \mathfrak{gl}(n, \mathbb{R}).$$

Here $I_n \in \mathfrak{gl}(n, \mathbb{R})$ is the unit matrix and the gradation is given again by subdividing matrices as follows;

$$\mathfrak{g}_{-3} = \left\{ \begin{pmatrix} 0 & 0 & 0 & 0 \\ 0 & 0 & 0 & 0 \\ 0 & 0 & 0 & 0 \\ 2a & 0 & 0 & 0 \end{pmatrix} \middle| \quad a \in \mathbb{R} \right\},$$

$$\mathfrak{g}_{-2} = \left\{ \begin{pmatrix} 0 & 0 & 0 & 0 \\ 0 & 0 & 0 & 0 \\ \xi & 0 & 0 & 0 \\ 0 & {}^t\xi & 0 & 0 \end{pmatrix} \middle| \quad \xi \in \mathbb{R}^n \cong V^* \right\},$$

$$\mathfrak{g}_{-1} = \left\{ \begin{pmatrix} 0 & 0 & 0 & 0 \\ x & 0 & 0 & 0 \\ 0 & A & 0 & 0 \\ 0 & 0 & -{}^t x & 0 \end{pmatrix} \middle| \quad x \in \mathbb{R}^n = V, \ A \in Sym(n) \cong S^2(V^*) \right\},$$

$$\mathfrak{g}_0 = \left\{ \begin{pmatrix} b & 0 & 0 & 0 \\ 0 & B & 0 & 0 \\ 0 & 0 & -{}^t B & 0 \\ 0 & 0 & 0 & -b \end{pmatrix} \middle| \quad b \in \mathbb{R}, \ B \in \mathfrak{gl}(V), \right\}$$

$$\mathfrak{g}_k = \{ {}^t X \mid X \in \mathfrak{g}_{-k} \}, (k = 1, 2, 3),$$

where $Sym(n) = \{ A \in \mathfrak{gl}(n, \mathbb{R}) \mid {}^t A = A \}$ is the space of symmetric matrices. This gradation over $\mathbb{C}$ corresponds to $(C_{n+1}, \{\alpha_1, \alpha_{n+1}\})$ in (3) of Theorem 4.2.

(3) otherwise For vector spaces V and W of dimension n and m respectively, $\mathfrak{g}^k(n, m) = \bigoplus_{p \in \mathbb{Z}} \mathfrak{g}_p$ has the following description:

$$\mathfrak{g}_k = \{0\} \quad (k \geqq 2), \qquad \mathfrak{g}_1 = V^*, \qquad \mathfrak{g}_0 = \mathfrak{gl}(V) \oplus \mathfrak{gl}(W),$$
$$\mathfrak{g}_{-1} = V \oplus W \otimes S^{k-1}(V^*), \quad \mathfrak{g}_p = W \otimes S^{k+p}(V^*) \quad (p < -1).$$

Here the bracket product in $\mathfrak{g}^k(n, m)$ is given through the natural tensor operations.

For the structure of $\mathfrak{g}^k(n, m)$ in case (3), we observe the following points. We put

$$\begin{aligned}\mathfrak{l} = V \oplus \mathfrak{g}_0 \oplus \mathfrak{g}_1 &= (V \oplus \mathfrak{gl}(V) \oplus V^*) \oplus \mathfrak{gl}(W) \\ &\cong \mathfrak{sl}(\hat{V}) \oplus \mathfrak{gl}(W), \\ S = W \otimes S^{k-1}(\hat{V}^*), &\qquad \hat{V} = \mathbb{R} \oplus V\end{aligned} \tag{6.3}$$

where the gradation of the first kind; $\mathfrak{sl}(\hat{V}) = V \oplus \mathfrak{gl}(V) \oplus V^*$ is given by subdividing matrices corresponding to the decomposition $\hat{V} = \mathbb{R} \oplus V$. Then

$$S^{k-1}(\hat{V}^*) \cong \bigoplus_{l=0}^{k-1} S^l(V^*),$$

and S is a faithful irreducible $\mathfrak{l}$-module such that $\mathfrak{l} = \mathfrak{l}_{-1} \oplus \mathfrak{l}_0 \oplus \mathfrak{l}_1$ is a reductive graded Lie algebras, where $\mathfrak{l}_{-1} = V$, $\mathfrak{l}_0 = \mathfrak{g}_0$, $\mathfrak{l}_1 = \mathfrak{g}_1$. Moreover $\mathfrak{g}^k(n,m) \cong S \oplus \mathfrak{l}$ is the semi-direct product of $\mathfrak{l}$ by S.

6.2. Se-ashi's Principle: Pseudo-product GLA of type $(\mathfrak{l}, S)$. We will now give the notion of the pseudo-product GLA of type $(\mathfrak{l}, S)$, generalizing the pseudo-projective GLA of order k of bidegree (n,m).

Now, let us recall Se-ashi's procedure to form good classes of linear differential equations of finite type, following [Se] and [YY1]. Se-ashi's procedure starts from a reductive graded Lie algebra (GLA) $\mathfrak{l} = \mathfrak{l}_{-1} \oplus \mathfrak{l}_0 \oplus \mathfrak{l}_1$ and a faithful irreducible $\mathfrak{l}$-module S. Then we form the **pseudo-product GLA** $\mathfrak{g} = \bigoplus_{p\in\mathbb{Z}} \mathfrak{g}_p$ **of type** $(\mathfrak{l}, S)$ as follows: Let $\mathfrak{l} = \mathfrak{l}_{-1} \oplus \mathfrak{l}_0 \oplus \mathfrak{l}_1$ be a finite dimensional **reductive GLA** of the first kind such that

(1) The ideal $\hat{\mathfrak{l}} = \mathfrak{l}_{-1} \oplus [\mathfrak{l}_{-1}, \mathfrak{l}_1] \oplus \mathfrak{l}_1$ of $\mathfrak{l}$ is a simple Lie algebra.
(2) The center $\mathfrak{z}(\mathfrak{l})$ of $\mathfrak{l}$ is contained in $\mathfrak{l}_0$.

Let S be a finite dimensional **faithful irreducible** $\mathfrak{l}$**-module**. We put

$$S_{-1} = \{ s \in S \mid \mathfrak{l}_1 \cdot s = 0 \} \qquad \text{and} \qquad S_p = \mathrm{ad}\,(\mathfrak{l}_{-1})^{-p-1} S_{-1} \quad \text{for } p < 0$$

We form the semi-direct product $\mathfrak{g}$ of $\mathfrak{l}$ by S, and put

$$\mathfrak{g} = S \oplus \mathfrak{l}, \qquad [S,S] = 0 \qquad \mathfrak{g}_k = \mathfrak{l}_k \ (k \geqq 0),$$

$$\mathfrak{g}_{-1} = \mathfrak{l}_{-1} \oplus S_{-1}, \qquad \mathfrak{g}_p = S_p \ (p < -1).$$

Then $\mathfrak{g} = \bigoplus_{p\in\mathbb{Z}} \mathfrak{g}_p$ enjoys the following properties (Lemma 2.1[YY1]);

(1) $S = \bigoplus_{p=-1}^{-\mu} S_p$, *where* $S_{-\mu} = \{ s \in S \mid [\mathfrak{l}_{-1}, s] = 0 \}$.
(2) $\mathfrak{m} = \bigoplus_{p<0} \mathfrak{g}_p$ is generated by $\mathfrak{g}_{-1}$.
(3) S_p *is naturally embedded as a subspace of* $W \otimes S^{\mu+p}(\mathfrak{l}_{-1}{}^*)$ *through the bracket operation in* $\mathfrak{m}$, *where* $W = S_{-\mu}$.

Thus $S = S_{-\mu} \oplus S_{-\mu+1} \oplus \cdots \oplus S_{-1} \subset W \oplus W \otimes V^* \oplus \cdots \oplus W \otimes S^{\mu-1}(V^*)$ defines a symbol of μ-th order differential equations of finite type by putting $S_0 = \{0\} \subset W \otimes S^{\mu}(V^*)$. We can construct the model linear equation R_o of finite type, whose symbol at each point is isomorphic to S (see §4 [Se]). R_o is a μ-th order involutive differential equation of finite type. Then, we see that the symbol algebra of (R_o, D_o) is isomorphic to $\mathfrak{m}$, where D_o is the pullback of the canonical system $C^{\mu-1}$ on the $(\mu-1)$-jet space $J^{\mu-1}$. $\mathfrak{m}$ has the splitting $\mathfrak{g}_{-1} = \mathfrak{l}_{-1} \oplus S_{-1}$, corresponding to the pseudo-product structure on R_o, where $V = \mathfrak{l}_{-1}$ and $W = S_{-\mu}$. In this way, $\mathfrak{m}$ is a symbol algebra of μ-th order differential equation of finite type, which is called the **typical symbol of type** $(\mathfrak{l}, S)$.

This class of higher order (linear) differential equations of finite type first appeared in the work of Y.Se-ashi [Se].

6.3. Prolongation theorem. Let $\mathfrak{G} = (\mathfrak{g}, (\mathfrak{g}_p)_{p\in\mathbb{Z}}, \mathfrak{l}_{-1}, S_{-1})$ be a pseudo-product GLA of type $(\mathfrak{l}, S)$, i.e., $\mathfrak{g} = S \oplus \mathfrak{l}$ is endowed with the gradation $(\mathfrak{g}_p)_{p\in\mathbb{Z}}$, $\mathfrak{g} = \bigoplus_{p=-\mu}^{1} \mathfrak{g}_p$ given in §6.2. $\mathfrak{g}$ has also another gradation $(\mathfrak{b}_p)_{p\in\mathbb{Z}}$, $\mathfrak{g} = \bigoplus_{p=-1}^{0} \mathfrak{b}_p$, given by $\mathfrak{b}_{-1} = S$ and $\mathfrak{b}_0 = \mathfrak{l}$. Thus $\mathfrak{g}$ has a bigradation $(\mathfrak{g}_{p,q})_{p,q\in\mathbb{Z}}$, where $\mathfrak{g}_{p,q} = \mathfrak{g}_p \cap \mathfrak{b}_q$. We have the cohomology group $H^*(\mathfrak{m}, \mathfrak{g})$ associated with the adjoint representation of $\mathfrak{m} = \mathfrak{g}_-$ on $\mathfrak{g}$, that is, the cohomology space of the cochain complex $C^*(\mathfrak{m}, \mathfrak{g}) = \bigoplus C^p(\mathfrak{m}, \mathfrak{g})$ with the coboundary operator $\partial : C^p(\mathfrak{m}, \mathfrak{g}) \longrightarrow C^{p+1}(\mathfrak{m}, \mathfrak{g})$, where $C^p(\mathfrak{m}, \mathfrak{g}) = \mathrm{Hom}(\bigwedge^p \mathfrak{m}, \mathfrak{g})$. We put

$$C^p(\mathfrak{m}, \mathfrak{g})_{r,s} = \{\omega \in C^p(\mathfrak{m}, \mathfrak{g}) \mid$$

$$\omega(\mathfrak{g}_{i_1,j_1} \wedge \ldots \wedge \mathfrak{g}_{i_p,j_p}) \subset \mathfrak{g}_{i_1+\cdots+i_p+r, j_1+\cdots+j_p+s} \text{ for all } i_1, \ldots, i_p, j_1, \ldots, j_p\}.$$

As is easily seen, $C^*(\mathfrak{m}, \mathfrak{g})_{r,s} = \bigoplus_p C^p(\mathfrak{m}, \mathfrak{g})_{r,s}$ is a subcomplex of $C^*(\mathfrak{m}, \mathfrak{g})$. Denoting its cohomology space by $H(\mathfrak{m}, \mathfrak{g})_{r,s} = \bigoplus H^p(\mathfrak{m}, \mathfrak{g})_{r,s}$, we obtain the direct sum decomposition

$$H^*(\mathfrak{m}, \mathfrak{g}) = \bigoplus_{p,r,s} H^p(\mathfrak{m}, \mathfrak{g})_{r,s}.$$

The cohomology space, endowed with this tri-gradation, is called the generalized Spencer cohomology space of the PPGLA $\mathfrak{G}$ of type $(\mathfrak{l}, S)$. Note that $H^1(\mathfrak{m}, \mathfrak{g})_{0,0} = 0$ if and only if $\mathfrak{g}_0$ coincides with the Lie algebra of derivations of $\mathfrak{m}$ such that $D(\mathfrak{g}_p) \subset \mathfrak{g}_p$ $(p < 0)$, $D(\mathfrak{l}_{-1}) \subset \mathfrak{l}_{-1}$ and $D(S_{-1}) \subset S_{-1}$.

From now on, we assume that the ground field is the field $\mathbb{C}$ of complex numbers for the sake of simplicity. For the discussion over $\mathbb{R}$, the corresponding results will be obtained easily through the argument of complexification as in §3.2 in [Y5]. We set $\hat{\mathfrak{l}} = \mathfrak{l}_{-1} \oplus [\mathfrak{l}_{-1}, \mathfrak{l}_1] \oplus \mathfrak{l}_1$ and $\mathfrak{u} = \mathcal{D}(\mathfrak{z}_{\mathfrak{l}}(\hat{\mathfrak{l}}))$; then $\mathfrak{l} = \hat{\mathfrak{l}} \oplus \mathfrak{u} \oplus \mathfrak{z}(\mathfrak{l})$, $\mathcal{D}(\mathfrak{l}) = \hat{\mathfrak{l}} \oplus \mathfrak{u}$ and $\hat{\mathfrak{l}} = \mathfrak{l}_{-1} \oplus \hat{\mathfrak{l}}_0 \oplus \mathfrak{l}_1$, where $\hat{\mathfrak{l}}_0 = [\mathfrak{l}_{-1}, \mathfrak{l}_1]$, is a simple GLA. Let us take a Cartan subalgebra $\mathfrak{h}$ of $\mathfrak{l}$ such that $\mathfrak{h} \subset \mathfrak{l}_0$. Then $\mathfrak{h} \cap \hat{\mathfrak{l}}$ (resp. $\mathfrak{h} \cap \mathfrak{u}$) is a Cartan subalgebra of $\hat{\mathfrak{l}}$ (resp. $\mathfrak{u}$). Let $\Delta = \{\alpha_1, \ldots, \alpha_\ell\}$ (resp. $\Delta' = \{\beta_1, \ldots, \beta_m\}$) be a simple root system of $(\hat{\mathfrak{l}}, \mathfrak{h} \cap \hat{\mathfrak{l}})$ (resp. $(\mathfrak{u}, \mathfrak{h} \cap \mathfrak{u})$) such that $\alpha(Z) \geqq 0$ for all $\alpha \in \Delta$, where Z is the characteristic element of the GLA $\mathfrak{l} = \mathfrak{l}_{-1} \oplus \mathfrak{l}_0 \oplus \mathfrak{l}_1$. We assume that $\hat{\mathfrak{l}}$ is a simple Lie algebra of type X_ℓ. We set $\Delta_1 = \{\alpha \in \Delta \mid \alpha(Z) = 1\}$. It is well known that the pair (X_ℓ, Δ_1) is one of the following type (up to a diagram automorphism) (cf. §3 in [Y5]):

$$(A_\ell, \{\alpha_i\})\ (1 \leqq i \leqq [(\ell+1)/2]),$$

$$(B_\ell, \{\alpha_1\})\ (\ell \geqq 3),$$

$$(C_\ell, \{\alpha_\ell\})\ (\ell \geqq 2),$$

$$(D_\ell, \{\alpha_1\})\ (\ell \geqq 4),$$

$$(D_\ell, \{\alpha_{\ell-1}\})\ (\ell \geqq 5),$$

$$(E_6, \{\alpha_1\}),\ (E_7, \{\alpha_7\}).$$

We denote by $\{\varpi_1, \ldots, \varpi_\ell\}$ (resp. $\{\pi_1, \ldots, \pi_n\}$) the set of fundamental weights relative to Δ (resp. Δ'). Since S is a faithful $\mathfrak{l}$-module, we have $\dim \mathfrak{z}(\mathfrak{l}) \leqq 1$. Assume that $\mathfrak{z}(\mathfrak{l}) \neq \{0\}$. Let σ be the element of $\mathfrak{z}(\mathfrak{l})^*$ such that $\sigma(J) = 1$, where J is the characteristic element of the GLA $\mathfrak{g} = \mathfrak{b}_{-1} \oplus \mathfrak{b}_0$. Namely $J = -id_S \in \mathfrak{z}(\mathfrak{l}) \subset \mathfrak{b}_0 = \mathfrak{l}$ as the element of $\mathfrak{gl}(S)$. There is an irreducible $\hat{\mathfrak{l}}$ -module T (resp. $\mathfrak{z}_{\mathfrak{l}}(\hat{\mathfrak{l}})$ -module U) with highest weight χ (resp. $\eta - \sigma$) such that $S = \mathfrak{b}_{-1}$ is isomorphic to $U \otimes T$ as an $\mathfrak{l}$-module, where η is a weight of $\mathfrak{u}$. Then we have (Lemma 4.5 [YY1]).

LEMMA 6.1. *$H^1(\mathfrak{m}, \mathfrak{g})_{0,0} = 0$ if and only if $\mathfrak{z}_{\mathfrak{l}}(\hat{\mathfrak{l}})$ is isomorphic to $\mathfrak{gl}(U)$ and $\eta = \pi_1$. Especially, when $\mathcal{D}(\mathfrak{l}) = \hat{\mathfrak{l}}$, $H^1(\mathfrak{m}, \mathfrak{g})_{0,0} = 0$ if and only if $\mathfrak{l} = \hat{\mathfrak{l}} \oplus \mathfrak{z}(\mathfrak{l})$, where $\mathfrak{z}(\mathfrak{l}) = \langle J \rangle$.* Thus, when $H^1(\mathfrak{m}, \mathfrak{g})_{0,0} = 0$, the semisimple GLA $\mathcal{D}(\mathfrak{l})$ is of type $(X_\ell \times A_n, \{\alpha_i\})$ and S is an irreducible $\mathcal{D}(\mathfrak{l})$-module with highest weight $\Xi = \chi + \pi_1$ when $\dim U > 1$ and $\mathcal{D}(\mathfrak{l})$ is of type $(X_\ell, \{\alpha_i\})$ and S is an irreducible $\hat{\mathfrak{l}}$-module with highest weight χ, when $\mathcal{D}(\mathfrak{l}) = \hat{\mathfrak{l}}$ (i.e., when $\dim U = 1$).

The following theorem was obtained in Theorem 5.2 [YY1] as the answer to the following question:

When is $\mathfrak{g}$ the prolongation of $\mathfrak{m}$ or $(\mathfrak{m}, \mathfrak{g}_0)$?

In the following theorem (a), the simple graded Lie algebra $\mathfrak{b} = \check{\mathfrak{g}} = \bigoplus_{p\in\mathbb{Z}} \check{\mathfrak{g}}_p$ is described by $(Y_{\ell+n+1}, \Sigma_1)$ such that $\mathfrak{g} = \bigoplus_{p=-\mu}^{1} \mathfrak{g}_p$ is a graded subalgebra of $\check{\mathfrak{g}} = \bigoplus_{p=-\mu}^{\mu} \check{\mathfrak{g}}_p$ satisfying $\mathfrak{g}_p = \check{\mathfrak{g}}_p$ for $p \leqq 0$.

THEOREM 6.1. *Let $\mathfrak{G}$ be a pseudo-product GLA of type $(\mathfrak{l}, S)$ satisfying the condition $H^1(\mathfrak{G})_{0,0} = 0$. Let $\mathfrak{b} = \bigoplus_{p\in\mathbb{Z}} \mathfrak{b}_p$ be the prolongation of $\mathfrak{g} = \mathfrak{b}_{-1} \oplus \mathfrak{b}_0$, where $\mathfrak{b}_{-1} = S$ and $\mathfrak{b}_0 = \mathfrak{l}$. Then $\mathfrak{g} = \bigoplus_{p\in\mathbb{Z}} \mathfrak{g}_p$ is the prolongation of $\mathfrak{m} = \oplus_{p<0}\mathfrak{g}_p$ except for the following three cases.*

(a) $\dim \mathfrak{b} < \infty$ *and* $\mathfrak{b}_1 \neq 0$ $(\mathfrak{b} = \mathfrak{b}_{-1} \oplus \mathfrak{b}_0 \oplus \mathfrak{b}_1$*: simple*$)$

$\mathcal{D}(\mathfrak{l}) = [\mathfrak{l}, \mathfrak{l}]$	Δ_1	$\mathfrak{b}_{-1} = S$	$\check{\mathfrak{g}} = Y_{\ell+n+1}$	Σ_1
$A_\ell \times A_n$	$\{\alpha_i\}$	$\varpi_\ell + \pi_1$	$A_{\ell+n+1}$	$\{\gamma_i, \gamma_{\ell+1}\}$
A_ℓ	$\{\alpha_i\}$	$2\varpi_l$	$C_{\ell+1}$	$\{\gamma_i, \gamma_{\ell+1}\}$
A_ℓ ($\ell \geqq 3$)	$\{\alpha_i\}$	$\varpi_{\ell-1}$	$D_{\ell+1}$	$\{\gamma_i, \gamma_{\ell+1}\}$
B_ℓ ($\ell \geqq 2$)	$\{\alpha_1\}$	ϖ_1	$B_{\ell+1}$	$\{\gamma_2, \gamma_1\}$
D_ℓ ($l \geqq 4$)	$\{\alpha_1\}$	ϖ_1	$D_{\ell+1}$	$\{\gamma_2, \gamma_1\}$
D_ℓ ($\ell \geqq 4$)	$\{\alpha_\ell\}$	ϖ_1	$D_{\ell+1}$	$\{\gamma_{\ell+1}, \gamma_1\}$
D_5	$\{\alpha_1\}$	ϖ_5	E_6	$\{\gamma_1, \gamma_6\}$
D_5	$\{\alpha_5\}$	ϖ_5	E_6	$\{\gamma_3, \gamma_1\}$
D_5	$\{\alpha_4\}$	ϖ_5	E_6	$\{\gamma_2, \gamma_1\}$
E_6	$\{\alpha_6\}$	ϖ_6	E_7	$\{\gamma_6, \gamma_7\}$
E_6	$\{\alpha_1\}$	ϖ_6	E_7	$\{\gamma_1, \gamma_7\}$

Here $(Y_{\ell+n+1}, \Sigma_1)$ *is the prolongation of* $\mathfrak{m}$ *except for* $(A_{\ell+n+1}, \{\gamma_1, \gamma_{\ell+1}\})$ *and* $(C_{\ell+1}, \{\gamma_1, \gamma_{\ell+1}\})$.

Moreover the latter two are the prolongations of $(\mathfrak{m}, \mathfrak{g}_0)$.

(*b*) $\dim \mathfrak{b} = \infty$

$\mathcal{D}(\mathfrak{l})$	Δ_1	$\mathfrak{b}_{-1}$	$\mathfrak{g}(\mathfrak{m}, \mathfrak{g}_0)$
A_ℓ	$\{\alpha_i\}$	ϖ_ℓ	$(A_{\ell+1}, \{\gamma_i, \gamma_{\ell+1}\})$
C_ℓ	$\{\alpha_\ell\}$	ϖ_1	$\mathfrak{g}$

In $(C_\ell, \{\alpha_\ell\})$*-case,* $\mu = 2$

$$S_{-2} = V^*, \quad S_{-1} = V, \quad \mathfrak{l}_{-1} = S^2(V^*),$$
$$\mathfrak{l}_0 = V \otimes V^* \oplus \mathbb{C}, \qquad \mathfrak{l}_1 = S^2(V)$$

(*c*) $\mathfrak{g}$ *is a pseudo-projective GLA, i.e.,* $\mathcal{D}(\mathfrak{l}) = (A_\ell \times A_n, \{\alpha_1\})$, $\Xi = k\varpi_\ell + \pi_1$, $(k \geqq 2, n \geqq 1)$, *or* $\mathcal{D}(\mathfrak{l}) = (A_\ell, \{\alpha_1\})$, $\chi = k\varpi_\ell$, $(k \geqq 3, n = 0)$

$$S_{-\mu} = W, \quad S_p = W \otimes S^{\mu+p}(V^*) \ (-\mu < p < 0),$$
$$\mathfrak{l}_{-1} = V, \quad \mathfrak{l}_0 = \mathfrak{gl}(V) \oplus \mathfrak{gl}(W), \quad \mathfrak{l}_1 = V^*,$$

where $\mu = k+1$, $\dim V = \ell$ *and* $\dim W = n+1$.

In this case $\mathfrak{g}$ *is the prolongation of* $(\mathfrak{m}, \mathfrak{g}_0)$.

By Proposition 4.4.1 in [Se], the Lie algebra of infinitesimal linear automorphisms of the model equation of type $(\mathfrak{l}, S)$ coincides with $\mathfrak{l}$. Hence the cases (a) and (b) of the above theorem exhaust classes of the equations of type $(\mathfrak{l}, S)$, for which the model equations admit non trivial nonlinear automorphisms. These cases correspond to the parabolic geometries associated with differential equations of finite type, which generalize the case of second and third order ordinary differential equations. In the cases of $(A_{\ell+1}, \{\gamma_1, \gamma_i\})$ and $(C_{\ell+1}, \{\gamma_1, \gamma_{\ell+1}\})$, $\mathfrak{m}$ coincides with the symbol algebra of the canonical system of the first or second order jet spaces (see §6.1) and $\mathfrak{g}_0$ determines the splitting of $\mathfrak{g}_{-1}$, hence the parabolic geometries associated with these graded Lie algebras are geometries of the pseudo-product structures on the first or second order jet spaces.

In the other cases of the above theorem (a), $(Y_{\ell+n+1}, \Sigma_1)$ is the prolongation of $\mathfrak{m}$. This fact implies that the parabolic geometries associated with these graded Lie algebras are geometries of regular differential system of type $\mathfrak{m}$, which have the (almost) pseudo-product structure corresponding to $\mathfrak{g}_{-1} = S_{-1} \oplus \mathfrak{l}_{-1}$. Moreover every isomorphism of these regular differential system preserves this pseudo-product structure. Thus the parabolic geometries associated with $(Y_{\ell+n+1}, \Sigma_1)$ have the canonical (almost) pseudo-product structures in the regular differential system of type $\mathfrak{m}$ corresponding to the splitting $\mathfrak{g}_{-1} = S_{-1} \oplus \mathfrak{l}_{-1}$.

More precisely, we can classify our parabolic geometries into the following four groups.

(**A**) The parabolic geometry associated with $(A_{\ell+n+1}, \{\gamma_1, \gamma_{\ell+1}\})$ $(n \geqq 0, \ell \geqq 1)$ is the geometry of the pseudo-pojective systems of second order of bidegree $(\ell, n+1)$, i.e. the geometry of the second order equations of ℓ independent and $n+1$ dependent variables by point transformations. The parabolic geometry associated with $(C_{\ell+1}, \{\gamma_1, \gamma_{\ell+1}\})$ $(\ell \geqq 1)$ is the geometry of the pseudo-projective systems of third order of bidegree $(\ell, 1)$, i.e. the geometry of the third order equations of ℓ independent and 1 dependent variables by contact transformations.

(**B**) The parabolic geometries associated with $(D_{\ell+1}, \{\gamma_{\ell+1}, \gamma_1\})$ $(\ell \geqq 4)$, $(D_{\ell+1}, \{\gamma_1, \gamma_{\ell+1}\})$ $(\ell \geqq 3)$, $(D_{\ell+1}, \{\gamma_\ell, \gamma_{\ell+1}\})$ $(\ell \geqq 3)$, $(A_{\ell+n+1}, \{\gamma_i, \gamma_{\ell+1}\})$ $(2 \leqq i \leqq \ell, n \geqq 0)$ and $(E_6, \{\gamma_1, \gamma_6\})$ are the contact geometries of finite type equations of the first order in the following sense.

In this case $\mu = 2$ and the typical symbol $\mathfrak{m}$ has the following description: $\mathfrak{m} = \mathfrak{g}_{-2} \oplus \mathfrak{g}_{-1} \subset \mathfrak{c}^1(V, W)$, where $W = S_{-2}$ and $V = \mathfrak{l}_{-1}$. Moreover $\mathfrak{g}_{-1} = V \oplus S_{-1}$ and $S_{-1} \subset W \otimes V^*$. Let $J^k(n,m)$ be the space of k-jets of n independent and m dependent variables, where $n = \dim V$ and $m = \dim W$. We consider a submanifold R of $J^1(n,m)$ such that $\pi^1_0 \mid_R$: $R \to J^0(n,m)$ is a submersion. Let D be the restriction to R of the canonical systetem C^1 on $J^1(n,m)$ and $R^{(1)} \subset J^2(n,m)$ be the first prolongation of R (cf. §4.2 [Y1]). We assume that $p^{(1)} : R^{(1)} \to R$ is onto. This assumption is equivalent to say that (R, D) has an (n-dimensional) integral element (transversal to the fibre $\mathrm{Ker}\,(\pi^1_0 \mid_R)_*$) at each point of R. Under this integrability condition, (R, D) is a regular differential system of type $\mathfrak{m}$ if and only if the symbols of this equation R are isomorphic to $S_{-1} \subset W \otimes V^*$ at each point of R (see §2.1 in [SYY] for the precise meaning of the isomorphism of the symbol). In this case, by (1) of Proposition 5.1[YY2], integral elements of (R, D) are unique at each point of R so that $p^{(1)} : R^{(1)} \to R$ is a diffeomorphism. Thus (R, D) has the (almost) pseudo-product structure corresponding to the splitting $\mathfrak{g}_{-1} = V \oplus S_{-1}$. In fact S_{-1} corresponds to the fibre direction $\mathrm{Ker}\,(\pi^1_0 \mid_R)_*$ and V corresponds to the restriction to $R^{(1)}$ of the canonical system C^2 on $J^2(n,m)$. Since $\check{\mathfrak{g}}$ is the prolongation of $\mathfrak{m}$, an isomorphism of (R, D) preserves the pseudo-product structure. In particular an isomorphism of (R, D) preserves the projection $\pi^1_0 \mid_R : R \to J^0(n,m)$. Hence a local isomorphism of (R, D) is the lift of a local point transformation of $J^0(n,m)$.

In this class (**B**), we can discuss about the duality of our pseudo-product structures as in [T7]. For example, $(A_{\ell+n+1}, \{\gamma_{n+1}, \gamma_{n+j+1}\})$ corresponds to the dual of $(A_{\ell+n+1}, \{\gamma_i, \gamma_{\ell+1}\})$, where $i + j = \ell + 1$.

By Theorem 2.7 and 2.9 [T4] and Proposition 5.5 [Y5], we observe that parabolic geometries associated with $(A_{\ell+n+1}, \{\gamma_i, \gamma_{\ell+1}\})$ $(3 \leqq i \leqq \ell - 1, n \geqq 2)$,$(D_{\ell+1}, \{\gamma_\ell, \gamma_{\ell+1}\})$ $(\ell \geqq 3)$ and $(E_6, \{\gamma_1, \gamma_6\})$ have no local invariants. Hence in these cases, (R, D), satisfying the integrability condition, is always locally isomorphic to the model equation.

(**C**) The parabolic geometries associated with $(B_{\ell+1}, \{\gamma_2, \gamma_1\})$ $(\ell \geqq 2)$, $(D_{\ell+1}, \{\gamma_2, \gamma_1\})$ $(\ell \geqq 4)$, $(D_{\ell+1}, \{\gamma_2, \gamma_{\ell+1}\})$ $(\ell \geqq 3)$, $(E_6, \{\gamma_2, \gamma_1\})$ and $(E_7, \{\gamma_1, \gamma_7\})$ are the contact geometries of finite type equations of the second order in the following sense.

In this case $\mu = 3$ and the typical symbol $\mathfrak{m}$ has the following description: $\mathfrak{m} = \mathfrak{g}_{-3} \oplus \mathfrak{g}_{-2} \oplus \mathfrak{g}_{-1} \subset \mathfrak{c}^2(V, W)$, where $W = \mathbb{K}$, $V = \mathfrak{l}_{-1}$ and $\dim V = n$. Moreover we have $\mathfrak{g}_{-2} = V^*, \mathfrak{g}_{-1} = V \oplus S_{-1}$ and $S_{-1} \subset S^2(V^*)$. In this case, we note that the standard differential system $(M_{\mathfrak{g}}, D_{\mathfrak{g}})$ of type $(Y_L, \{\gamma_a\})$ is the standard contact manifold of type Y_L (see §4 in [Y5]).

We consider a submanifold R of $J^2(n,1)$ such that $\pi^2_1 \mid_R : R \to J^1(n,1)$ is a submersion. Let D be the restriction to R of the canonical systetem C^2 on $J^2(n,1)$ and $R^{(1)} \subset J^3(n,1)$ be the first prolongation of R. We assume that $p^{(1)} : R^{(1)} \to R$ is onto. Under this integrability condition, (R, D) is a regular differential system of type $\mathfrak{m}$ if and only if the symbols of this equation R are isomorphic to $S_{-1} \subset S^2(V^*)$ at each point of R. In this case, by (1) of Proposition 5.1[YY2], integral elements of (R, D) are unique at each point of R so that $p^{(1)} : R^{(1)} \to R$ is a diffeomorphism. Thus (R, D) has the (almost) pseudo-product structure corresponding to the splitting $\mathfrak{g}_{-1} = V \oplus S_{-1}$. In fact S_{-1} corresponds to the fibre direction $\mathrm{Ker}\,(\pi^2_1 \mid_R)_*$

and V corresponds to the restriction to $R^{(1)}$ of the canonical system C^3 on $J^3(n,1)$. Since $\check{\mathfrak{g}}$ is the prolongation of $\mathfrak{m}$, an isomorphism of (R,D) preserves the pseudo-product structure. In particular an isomorphism of (R,D) preserves the projection $\pi^2_1 \mid_R : R \to J^1(n,1)$ and $\partial D = (\pi^2_1)^{-1}_*(C^1)$. Hence a local isomorphism of (R,D) is the lift of a local contact transformation of $J^1(n,1)$.

By Theorem 2.7 and 2.9 [T4] and Proposition 5.5 [Y5], we observe that parabolic geometries associated with $(D_{\ell+1}, \{\gamma_2, \gamma_{\ell+1}\})$ $(\ell \geqq 3)$, $(E_6, \{\gamma_2, \gamma_1\})$ and $(E_7, \{\gamma_1, \gamma_7\})$ have no local invariants. Hence in these cases, (R,D), satisfying the integrability condition, is always locally isomorphic to the model equation. The rigidity of the parabolic geometry associated with $(D_{\ell+1}, \{\gamma_2, \gamma_{\ell+1}\})$ $(\ell \geqq 3)$ is discussed in [YY1] in connection with the Plücker embedding equations.

(**D**) The parabolic geometries associated with $(C_{\ell+1}, \{\gamma_i, \gamma_{\ell+1}\})$ $(1 < i \leqq \ell, \ell \geqq 2)$, $(D_{\ell+1}, \{\gamma_i, \gamma_{\ell+1}\})$ $(2 < i < \ell, \ell \geqq 4)$, $(E_6, \{\gamma_3, \gamma_1\})$ and $(E_7, \{\gamma_6, \gamma_7\})$ are the geometries of finite type equations of the first order in the following sense.

In this case $\mu = 3$ and the typical symbol $\mathfrak{m}$ has the following description: $\mathfrak{m} = \mathfrak{g}_{-3} \oplus \mathfrak{g}_{-2} \oplus \mathfrak{g}_{-1} \subset \mathfrak{c}^2(V,W)$, where $W = S_{-3}$ and $V = \mathfrak{l}_{-1}$. Moreover $\mathfrak{g}_{-2} = S_{-2}, \mathfrak{g}_{-1} = V \oplus S_{-1}$, $S_{-2} \subset W \otimes V^*$, $S_{-1} \subset W \otimes S^2(V^*)$ and $\dim S_{-2} = \dim V$. In this case we first consider a submanifold R of $J^1(n,m)$ such that $\pi^1_0 \mid_R : R \to J^0(n,m)$ is a submersion, where $n = \dim V$ and $m = \dim W$. Let D be the restriction to R of the canonical systetem C^1 on $J^1(n,m)$ and $R^{(1)} \subset J^2(n,m)$ be the first prolongation of R. We assume that the symbols of this equation R are isomorphic to $S_{-2} \subset W \otimes V^*$ at each point of R and also assume that $p^{(1)} : R^{(1)} \to R$ is onto. Then (R,D) is a regular differential system of type $\hat{\mathfrak{m}} = \hat{\mathfrak{g}}_{-2} \oplus \hat{\mathfrak{g}}_{-1}$, where $\hat{\mathfrak{g}}_{-2} = W$ and $\hat{\mathfrak{g}}_{-1} = V \oplus S_{-2}$. Here the symbol algebra $\hat{\mathfrak{m}}$ is the negative part of the simple graded Lie algebra of type $(Y_L, \{\gamma_a\})$, i.e., of type $(C_{\ell+1}, \{\gamma_i\})$ $(2 \leqq i \leqq \ell)$, $(D_{\ell+1}, \{\gamma_i\})$ $(2 < i < \ell)$, $(E_6, \{\gamma_3\})$ and $(E_7, \{\gamma_6\})$ respectively. Furthermore, by (2) of Proposition 5.1[YY2], the symbols of this equation $R^{(1)}$ are isomorphic to $\rho(S_{-2}) = S_{-1} \subset W \otimes S^2(V^*)$. Let $D^{(1)}$ be the restriction to $R^{(1)}$ of the canonical system C^2 on $J^2(n,m)$ and $R^{(2)} \subset J^3(n,m)$ be the prolongation of $R^{(1)}$. We further assume that $p^{(2)} : R^{(2)} \to R^{(1)}$ is onto. Under these integrability conditions, $(R^{(1)}, D^{(1)})$ becomes a regular differential system of type $\mathfrak{m}$. Actually the set of n-dimensional integral elements of (R,D) forms a bundle over R, which contains $R^{(1)}$ as an open dense subset such that $D^{(1)}$ coincides with the canonical system induced by this Grassmanian construction (cf. §2 in [Y1], §1 in [Y6]). Moreover, by (1) of Proposition 5.1[YY2], integral elements of $(R^{(1)}, D^{(1)})$ are unique at each point of $R^{(1)}$ so that $p^{(2)} : R^{(2)} \to R^{(1)}$ is a diffeomorphism. Thus $(R^{(1)}, D^{(1)})$ has the (almost) pseudo-product structure corresponding to the splitting $\mathfrak{g}_{-1} = V \oplus S_{-1}$. In fact S_{-1} corresponds to the fibre direction $\operatorname{Ker}(p^{(1)})_*$ and V corresponds to the restriction to $R^{(2)}$ of the canonical system C^3 on $J^3(n,m)$. Since $\check{\mathfrak{g}}$ is the prolongation of $\mathfrak{m}$, an isomorphism of $(R^{(1)}, D^{(1)})$ preserves the pseudo-product structure. In particular an isomorphism of $(R^{(1)}, D^{(1)})$ preserves the projection $p^{(1)} : R^{(1)} \to R$ and $\partial D^{(1)} = (p^{(1)})_*^{-1}(D)$. Thus a local isomorphism of $(R^{(1)}, D^{(1)})$ induces that of (R, D) and coincides with the local lift of this isomorphism of (R, D). Hence the local equivalence of $(R^{(1)}, D^{(1)})$ is reducible to that of (R, D).

By Theorem 2.7 and 2.9 [T4] and Proposition 5.5 [Y5], we observe that parabolic geometries associated with $(C_{\ell+1}, \{\gamma_i, \gamma_{\ell+1}\})$ $(2 < i < \ell)$, $(D_{\ell+1}, \{\gamma_i, \gamma_{\ell+1}\})$ $(2 < i < \ell)$, $(E_6, \{\gamma_3, \gamma_1\})$ and $(E_7, \{\gamma_6, \gamma_7\})$ have no local invariants. Hence in these cases, (R, D), satisfying the integrability conditions, is always locally isomorphic to the model equation.

REMARK 6.1 Among the cases in (**A**) and (**C**), notable omissions are the parabolic geometries asociated with $(A_{\ell+1}, \{\gamma_1, \gamma_{i+1}, \gamma_{\ell+1}\})$ $(0 < i < \ell)$, which do not show up in the exceptional lists in Theorem 6.1, but are associated with differential equations of finite type as follows. In fact the standard contact manifold of type $A_{\ell+1}$ is given by $(A_{\ell+1}, \{\gamma_1, \gamma_{\ell+1}\})$. Hence the typical symbol $\mathfrak{m}$ in this case has the same description as in (**C**). The model equation in this case, as the second order system, is given by

$$\frac{\partial^2 y}{\partial x_{i_1} \partial x_{i_2}} = \frac{\partial^2 y}{\partial x_{j_1} \partial x_{j_2}} = 0$$

$$\text{for} \quad 1 \leqq i_1, i_2 \leqq i, \quad \text{and} \quad i < j_1, j_2 < \ell + 1,$$

where y is dependent variable and $x_1, \ldots, x_\ell$ are independent variables.

6.4. Examples. For these parabolic geometries associated with differential equations of finite type, discussed in §6.3, we can describe explicitly the symbol algebras and the model differential equations of finite type by utilizing the explicit matrices description, given in §4.2 , of the simple graded Lie algebra $\mathfrak{b}$ for the classical cases and by describing the structure of $\mathfrak{m}$ explicitly by use of the Chevalley basis of the exceptional simple Lie algebras. For the details of these explicit descriptions, we refer the reader to [YY2]. Here we will show some examples from classes (**B**) ,(**C**) and (**D**).

(**B**) $[(A_{\ell+1}, \{\gamma_\ell, \gamma_{\ell+1})]$.

$\mathfrak{b} = \mathfrak{b}_{-1} \oplus \mathfrak{b}_0 \oplus \mathfrak{b}_1$ is described by $(A_{\ell+1}, \{\gamma_{\ell+1}\})$ and $\check{\mathfrak{g}} = \bigoplus_{p=-\mu}^{\mu} \check{\mathfrak{g}}_p$ is described by $(A_{\ell+1}, \{\gamma_\ell, \gamma_{\ell+1}\})$. Hence $\mu = 2$ and we obtain the following matrix representation of $\check{\mathfrak{g}} = \mathfrak{b} = \mathfrak{sl}(\ell + 2, \mathbb{K})$:

$$\mathfrak{sl}(\ell + 2, \mathbb{K}) = \mathfrak{g}_{-2} \oplus \mathfrak{g}_{-1} \oplus \mathfrak{g}_0 \oplus \check{\mathfrak{g}}_1 \oplus \check{\mathfrak{g}}_2 = S \oplus \mathfrak{l} \oplus S^*,$$

where the gradation is given by subdividing matrices as follows

$$\mathfrak{g}_{-2} = S_{-2} = \left\{ \begin{pmatrix} 0 & 0 & 0 \\ 0 & 0 & 0 \\ y & 0 & 0 \end{pmatrix} \middle| \quad y \in V = M(1,\ell) \right\},$$

$$\mathfrak{g}_{-1} = S_{-1} \oplus \mathfrak{l}_{-1},$$

$$S_{-1} = \left\{ \begin{pmatrix} 0 & 0 & 0 \\ 0 & 0 & 0 \\ 0 & a & 0 \end{pmatrix} \middle| \, a \in \mathbb{K} \right\},$$

$$\mathfrak{l}_{-1} = \left\{ \begin{pmatrix} 0 & 0 & 0 \\ x & 0 & 0 \\ 0 & 0 & 0 \end{pmatrix} \middle| \quad x \in V = M(1,\ell) \right\},$$

$$\mathfrak{g}_0 = \check{\mathfrak{l}}_0 \oplus \mathfrak{u} = \left\{ \begin{pmatrix} F & 0 & 0 \\ 0 & G & 0 \\ 0 & 0 & H \end{pmatrix} \middle| \begin{array}{l} F \in \mathfrak{gl}(\ell,\mathbb{K}),\ g \in \mathbb{K},\ h \in \mathbb{K}, \\ \mathrm{tr}F + g + h = 0 \end{array} \right\},$$

$$\check{\mathfrak{g}}_1 = \{ {}^tX \mid X \in \mathfrak{g}_{-1} \}, \qquad \check{\mathfrak{g}}_2 = \{ {}^tX \mid X \in \mathfrak{g}_{-2} \}.$$

By a direct calculation, we have $[\hat{a}, \hat{x}] = (\check{ax}) \in S_{-2} = V$, i.e., $y = ax$. Thus S_{-1} is embedded as the 1-dimensional subspace of scalar multiplications of $V \otimes V^* = S_{-2} \otimes (\mathfrak{l}_{-1})^*$ through the bracket operation in $\mathfrak{m}$. This implies that the model equation of our typical symbol $\mathfrak{m} = \mathfrak{g}_{-2} \oplus \mathfrak{g}_{-1} \subset \mathfrak{c}^1(V,V)$ is given by

$$\frac{\partial y_p}{\partial x_q} = \delta_{pq} \frac{\partial y_1}{\partial x_1} \qquad \text{for} \quad 1 \leqq p,q \leqq l \tag{6.4}$$

where $y_1, \ldots, y_l$ are dependent variables and $x_1, \ldots, x_l$ are independent variables. By a direct calculation, we see that the prolongation of the first order system (3.1) is given by

$$\frac{\partial^2 y_p}{\partial x_q \partial x_r} = 0 \qquad \text{for} \quad 1 \leqq p,q,r \leqq l. \tag{6.5}$$

(C) $[(E_6, \{\gamma_2, \gamma_1\})]$.

In this case, we have $\mu = 3$,

$$\mathfrak{m} = \check{\mathfrak{g}}_{-3} \oplus \check{\mathfrak{g}}_{-2} \oplus \check{\mathfrak{g}}_{-1} \qquad \text{and} \qquad \check{\mathfrak{g}}_{-1} = S_{-1} \oplus \mathfrak{l}_{-1},$$

where $\check{\mathfrak{g}}_{-3} = S_{-3}$, $\check{\mathfrak{g}}_{-2} = S_{-2}$, S_{-1} and $\mathfrak{l}_{-1}$ are spanned by the root spaces $\mathfrak{g}_{-\beta}$ for $\beta \in \Phi_3^+, \Phi_2^+, \Psi^1$ and Ψ^2 respectively. Hence $\dim S_{-3} = 1$, $\dim S_{-2} = \dim \mathfrak{l}_{-1} = 10$ and $\dim S_{-1} = 5$.

For $\Phi_3^+, \Phi_2^+, \Psi^1$ and Ψ^2, we observe that $\Phi_3^+ = \{\theta\}$, where θ is the highest root, $\alpha+\beta \notin \Phi$ for $\alpha, \beta \in \Phi_3^+ \cup \Phi_2^+ \cup \Psi^1$, $\zeta - \xi \notin \Phi$ for $\xi \in \Psi^2, \zeta \in \Psi^1$ and that $\eta_i + \xi_i = \theta$, $\eta_i - \xi_i \notin \Phi$, $\xi_i + \xi_j \notin \Phi$ and $\eta_i + \xi_j \notin \Phi$ if $i \neq j$ for

$\eta_i \in \Phi_2^+$ and $\xi_i, \xi_j \in \Psi^2$ $(i,j = 1,\ldots,10)$. This implies that $[y_\alpha, y_\beta] = 0$ for $\alpha, \beta \in \Phi_3^+ \cup \Phi_2^+ \cup \Psi^1$, $[y_\zeta, y_\xi] = \pm y_{\zeta+\xi}$ for $\xi \in \Psi^2, \zeta \in \Psi^1$, if $\zeta + \xi \in \Phi$ and that $[y_{\xi_i}, y_{\xi_j}] = 0$, $[y_{\eta_i}, y_{\xi_j}] = \pm\delta_{ij} y_\theta$ for $\eta_i \in \Phi_2^+$, $\xi_i, \xi_j \in \Psi^2$ $(i,j = 1,\ldots,10)$, by the property of the Chevalley basis. Hence, from Planche V in [Bu], we readily obtain the non-trivial bracket relation among $\check{\mathfrak{g}}_{-1}$ and $[\check{\mathfrak{g}}_{-2}, \mathfrak{l}_{-1}]$ as in (6.6) and (6.7) below up to signs.

We fix the signs of y_β for $\beta \in \Phi_3^+, \Phi_2^+, \Psi^2$ and Ψ^1 as follows: First we choose the orientation of y_{γ_i} for simple roots by fixing the root vectors $y_i = y_{\gamma_i} \in \mathfrak{g}_{-\gamma_i}$. For $\zeta \in \Psi^1$, we fix the orientation by the following order;

$$y_{\zeta_1} = y_1, \quad y_{\zeta_2} = [y_3, y_{\zeta_1}], \quad y_{\zeta_3} = [y_4, y_{\zeta_2}], \quad y_{\zeta_4} = [y_5, y_{\zeta_3}], \quad y_{\zeta_5} = [y_6, y_{\zeta_4}].$$

For $\xi \in \Psi^2$, we fix the orientation by the following order;

$$\begin{aligned} &y_{\xi_1} = y_2, && y_{\xi_2} = [y_{\xi_1}, y_4], && y_{\xi_3} = [y_{\xi_2}, y_3], && y_{\xi_4} = [y_{\xi_2}, y_5],\\ &y_{\xi_5} = [y_{\xi_3}, y_5], && y_{\xi_6} = [y_{\xi_4}, y_6], && y_{\xi_7} = [y_{\xi_5}, y_6], && y_{\xi_8} = [y_{\xi_5}, y_4],\\ &y_{\xi_9} = [y_{\xi_8}, y_6], && y_{\xi_{10}} = [y_{\xi_9}, y_5]. \end{aligned}$$

For $\eta \in \Phi_2^+$, we fix the orientation by the following order;

$$\begin{aligned} &y_{\eta_1} = [y_{\zeta_3}, y_{\xi_{10}}], && y_{\eta_2} = [y_{\zeta_2}, y_{\xi_{10}}], && y_{\eta_3} = [y_{\zeta_1}, y_{\xi_{10}}], && y_{\eta_4} = -[y_{\zeta_2}, y_{\xi_9}],\\ &y_{\eta_5} = -[y_{\zeta_1}, y_{\xi_9}], && y_{\eta_6} = [y_{\zeta_2}, y_{\xi_8}], && y_{\eta_7} = [y_{\zeta_1}, y_{\xi_8}], && y_{\eta_8} = [y_{\zeta_1}, y_{\xi_7}],\\ &y_{\eta_9} = -[y_{\zeta_1}, y_{\xi_5}], && y_{\eta_{10}} = [y_{\zeta_1}, y_{\xi_3}]. \end{aligned}$$

Finally, for $\theta \in \Phi_3^+$, we fix the orientation by the following;

$$y_\theta = [y_{\eta_{10}}, y_{\xi_{10}}].$$

Then we obtain

$$[y_{\eta_i}, y_{\xi_j}] = \delta_{ij} y_\theta, \quad [y_{\eta_i} \cdot y_{\eta_j}] = [y_{\xi_i}, y_{\xi_j}] = 0 \quad \text{for} \quad 1 \leqq i,j \leqq 10. \tag{6.6}$$

$$\begin{aligned} y_\theta &= [[y_{\zeta_1}, y_{\xi_3}], y_{\xi_{10}}] = -[[y_{\zeta_1}, y_{\xi_5}], y_{\xi_9}] = [[y_{\zeta_1}, y_{\xi_7}], y_{\xi_8}],\\ y_\theta &= [[y_{\zeta_2}, y_{\xi_2}], y_{\xi_{10}}] = -[[y_{\zeta_2}, y_{\xi_4}], y_{\xi_9}] = [[y_{\zeta_2}, y_{\xi_6}], y_{\xi_8}],\\ y_\theta &= [[y_{\zeta_3}, y_{\xi_1}], y_{\xi_{10}}] = -[[y_{\zeta_3}, y_{\xi_4}], y_{\xi_7}] = [[y_{\zeta_3}, y_{\xi_5}], y_{\xi_6}],\\ y_\theta &= [[y_{\zeta_4}, y_{\xi_1}], y_{\xi_9}] = -[[y_{\zeta_4}, y_{\xi_2}], y_{\xi_7}] = [[y_{\zeta_4}, y_{\xi_3}], y_{\xi_6}],\\ y_\theta &= [[y_{\zeta_5}, y_{\xi_1}], y_{\xi_8}] = -[[y_{\zeta_2}, y_{\xi_2}], y_{\xi_5}] = [[y_{\zeta_5}, y_{\xi_3}], y_{\xi_4}]. \end{aligned} \tag{6.7}$$

From (6.6), we have $S_{-2} = V^*$, by fixing the base of $S_{-3} \cong \mathbb{K}$ and putting $\mathfrak{l}_{-1} = V$. Moreover, from (6.7), S_{-1} is embedded as the 5-dimensional subspace of $S^2(V^*)$ spanned by the following quadratic forms $f_1, \ldots, f_5$;

$$\begin{aligned} &f_1(X,X) = x_3x_{10} - x_5x_9 + x_7x_8, && f_2(X,X) = x_2x_{10} - x_4x_9 + x_6x_8,\\ &f_3(X,X) = x_1x_{10} - x_4x_7 + x_5x_6, && f_4(X,X) = x_1x_9 - x_2x_7 + x_3x_6,\\ &f_5(X,X) = x_1x_8 - x_2x_5 + x_3x_4, \end{aligned}$$

for $X = \sum_{i=1}^{10} x_i y_{\xi_i} \in \mathfrak{l}_{-1}$. Hence the standard differential system $(M(\mathfrak{m}), D_{\mathfrak{m}})$ of type $\mathfrak{m}$ in this case is given by

$$D_{\mathfrak{m}} = \{\varpi = \varpi_1 = \varpi_2 = \cdots = \varpi_{10} = 0\},$$

where

$$\begin{aligned}
&\varpi = dy - p_1 dx_1 - \ldots - p_{10} dx_{10}, \\
&\varpi_1 = dp_1 + q_5 dx_8 + q_4 dx_9 + q_3 dx_{10}, \quad &&\varpi_2 = dp_2 - q_5 dx_5 - q_4 dx_7 + q_2 dx_{10}, \\
&\varpi_3 = dp_3 + q_5 dx_4 + q_4 dx_6 + q_1 dx_{10}, \quad &&\varpi_4 = dp_4 + q_5 dx_3 - q_3 dx_7 - q_2 dx_9, \\
&\varpi_5 = dp_5 - q_5 dx_2 + q_3 dx_6 - q_1 dx_9, \quad &&\varpi_6 = dp_6 + q_4 dx_3 + q_3 dx_5 + q_2 dx_8, \\
&\varpi_7 = dp_7 - q_4 dx_2 - q_3 dx_4 + q_1 dx_8, \quad &&\varpi_8 = dp_8 + q_5 dx_1 + q_2 dx_6 + q_1 dx_7, \\
&\varpi_9 = dp_9 + q_4 dx_1 - q_2 dx_4 - q_1 dx_5 \quad &&\varpi_{10} = dp_{10} + q_3 dx_1 + q_2 dx_2 + q_1 dx_3.
\end{aligned}$$

Here $(x_1, \ldots, x_{10}, y, p_1, \ldots, p_{10}, q_1, \ldots, q_5)$ is a coordinate system of $M(\mathfrak{m}) \cong \mathbb{K}^{26}$. Thus the model equation of our typical symbol $\mathfrak{m} = \breve{\mathfrak{g}}_{-3} \oplus \breve{\mathfrak{g}}_{-2} \oplus \breve{\mathfrak{g}}_{-1} \subset \mathfrak{c}^2(\mathfrak{l}_{-1}, \mathbb{K})$ is given by

$$\begin{aligned}
&\frac{\partial^2 y}{\partial x_3 \partial x_{10}} = -\frac{\partial^2 y}{\partial x_5 \partial x_9} = \frac{\partial^2 y}{\partial x_7 \partial x_8}, \\
&\frac{\partial^2 y}{\partial x_2 \partial x_{10}} = -\frac{\partial^2 y}{\partial x_4 \partial x_9} = \frac{\partial^2 y}{\partial x_6 \partial x_8}, \\
&\frac{\partial^2 y}{\partial x_1 \partial x_{10}} = -\frac{\partial^2 y}{\partial x_4 \partial x_7} = \frac{\partial^2 y}{\partial x_5 \partial x_6}, \\
&\frac{\partial^2 y}{\partial x_1 \partial x_9} = -\frac{\partial^2 y}{\partial x_2 \partial x_7} = \frac{\partial^2 y}{\partial x_3 \partial x_6}, \\
&\frac{\partial^2 y}{\partial x_1 \partial x_8} = -\frac{\partial^2 y}{\partial x_2 \partial x_5} = \frac{\partial^2 y}{\partial x_3 \partial x_4}, \\
&\frac{\partial^2 y}{\partial x_i \partial x_j} = 0 \quad \text{otherwise},
\end{aligned} \tag{6.8}$$

where y is dependent variable and $x_1, \ldots, x_{10}$ are independent variables. By a direct calculation, we see that the prolongation of the second order system (6.8) is given by

$$\frac{\partial^3 y}{\partial x_i \partial x_j \partial x_k} = 0 \qquad \text{for} \quad 1 \leqq i, j, k \leqq 10. \tag{6.9}$$

(D) $[(C_{\ell+1}, \{\gamma_\ell, \gamma_{\ell+1}\})]$.

$\mathfrak{b} = \mathfrak{b}_{-1} \oplus \mathfrak{b}_0 \oplus \mathfrak{b}_1$ is described by $(C_{\ell+1}, \{\gamma_{\ell+1}\})$ and $\breve{\mathfrak{g}} = \bigoplus_{p=-\mu}^{\mu} \breve{\mathfrak{g}}_p$ is described by $(C_{\ell+1}, \{\gamma_\ell, \gamma_{\ell+1}\})$. Hence $\mu = 3$ and $\breve{\mathfrak{g}} = \mathfrak{b}$ is isomorphic to $\mathfrak{sp}(\ell+1, \mathbb{K})$. First we describe

$$\mathfrak{sp}(\ell+1, \mathbb{K}) = \{X \in \mathfrak{gl}(2\ell+2, \mathbb{K}) \mid {}^tXJ + JX = 0\},$$

where

$$J = \begin{pmatrix} 0 & 0 & 0 & I_\ell \\ 0 & 0 & 1 & 0 \\ 0 & -1 & 0 & 0 \\ -I_\ell & 0 & 0 & 0 \end{pmatrix} \in \mathfrak{gl}(2l+2, \mathbb{K}).$$

Here $I_\ell \in \mathfrak{gl}(\ell, \mathbb{K})$ is the unit matrix and the gradation is given again by subdividing matrices as follows;

$$\mathfrak{g}_{-3} = S_{-3} = \left\{ \begin{pmatrix} 0 & 0 & 0 & 0 \\ 0 & 0 & 0 & 0 \\ 0 & 0 & 0 & 0 \\ Y & 0 & 0 & 0 \end{pmatrix} \middle| \quad Y \in Sym(\ell) \right\},$$

$$\mathfrak{g}_{-2} = S_{-2} = \left\{ \begin{pmatrix} 0 & 0 & 0 & 0 \\ 0 & 0 & 0 & 0 \\ \xi & 0 & 0 & 0 \\ 0 & {}^t\xi & 0 & 0 \end{pmatrix} \middle| \quad \xi \in \mathbb{K}^\ell = M(1, \ell) \right\},$$

$$\mathfrak{g}_{-1} = S_{-1} \oplus \mathfrak{l}_{-1},$$

$$S_{-1} = \left\{ \begin{pmatrix} 0 & 0 & 0 & 0 \\ 0 & 0 & 0 & 0 \\ 0 & a & 0 & 0 \\ 0 & 0 & 0 & 0 \end{pmatrix} \middle| \quad a \in \mathbb{K} \right\},$$

$$\mathfrak{l}_{-1} = \left\{ \begin{pmatrix} 0 & 0 & 0 & 0 \\ x & 0 & 0 & 0 \\ 0 & 0 & 0 & 0 \\ 0 & 0 & -{}^t x & 0 \end{pmatrix} \middle| \quad x \in \mathbb{K}^\ell = M(1, \ell) \right\},$$

$$\mathfrak{g}_0 = \check{\mathfrak{l}}_0 = \left\{ \begin{pmatrix} F & 0 & 0 & 0 \\ 0 & g & 0 & 0 \\ 0 & 0 & -g & 0 \\ 0 & 0 & 0 & -{}^t F \end{pmatrix} \middle| \quad F \in \mathfrak{gl}(\ell, \mathbb{K}),\ g \in \mathbb{K}, \right\}$$

$$\check{\mathfrak{g}}_k = \{\, {}^t X \mid X \in \mathfrak{g}_{-k} \,\}, (k = 1, 2, 3),$$

where $Sym(\ell)$ is the space of symmetric matrices. Thus we have

$$\mathfrak{m} = S_{-3} \oplus S_{-2} \oplus (S_{-1} \oplus \mathfrak{l}_{-1})$$

$$= \left\{ \begin{pmatrix} 0 & 0 & 0 & 0 \\ x & 0 & 0 & 0 \\ \xi & a & 0 & 0 \\ Y & {}^t\xi & -{}^t x & 0 \end{pmatrix} = \hat{Y} + \check{\xi} + \hat{a} + \hat{x} \middle| \begin{array}{l} a \in \mathbb{K}, \\ x, \xi \in \mathbb{K}^\ell = M(1, \ell), \\ Y \in Sym(\ell) \end{array} \right\}.$$

By calculating $[\hat{\xi}, \hat{x}]$ and $[[\hat{a}, \hat{x}], \hat{x}]$, we have

$$y_{pq} (= y_{qp}) = \xi_p x_q + \xi_q x_p = 2 a x_p x_q,$$

where $Y = (y_{pq})$, $\xi = (\xi_1, \ldots, \xi_\ell)$ and $x = (x_1, \ldots, x_\ell)$. From the first equality, we can embed S_{-2} as a subspace of $S_{-3} \otimes (\mathfrak{l}_{-1})^*$ and obtain the following first order system as the model equation whose symbol coincides with this subspace:

$$\frac{\partial y_{pq}}{\partial x_r} = 0 \quad \text{for} \quad r \neq p, q, \qquad \frac{\partial y_{pq}}{\partial x_q} = \frac{1}{2}\frac{\partial y_{pp}}{\partial x_p} \qquad \text{for} \quad p \neq q, \tag{6.10}$$

where $y_{pq} = y_{qp}$ $(1 \leqq p \leqq q \leqq \ell)$ are dependent variables and $x_1, \ldots, x_\ell$ are independent variables. Moreover, by a direct calculation, we see that the prolongation of the first order system (6.10) is given by

$$\begin{aligned} &\frac{\partial^2 y_{pq}}{\partial x_r \partial x_s} = 0 \quad \text{for} \quad \{r, s\} \neq \{p, q\}, \\ &\frac{\partial^2 y_{pq}}{\partial x_p \partial x_q} = \frac{1}{2}\frac{\partial^2 y_{pp}}{\partial^2 x_p} = \frac{1}{2}\frac{\partial^2 y_{qq}}{\partial^2 x_q} \quad \text{for} \quad p \neq q. \end{aligned} \tag{6.11}$$

From the second equality, we observe that the above second order system is the model equation of the 1-dimensional embedded subspace S_{-1} in $S_{-3} \otimes S^2((\mathfrak{l}_{-1})^*)$. Furthermore, by a direct calculation, we see that the prolongation of this second order system (6.11) is given by

$$\frac{\partial^3 y_{pq}}{\partial x_r \partial x_s \partial x_t} = 0 \qquad \text{for} \quad 1 \leqq p, q, r, s, t \leqq \ell. \tag{6.12}$$

REFERENCES

[Bo] W.M. Boothby, *Homogeneous complex contact manifolds*, Proc. Symp. Pure Math., Amer. Math. Soc. (1961), **3**: 144–154.

[Bu] N. Bourbaki, *Groupes et algèbles de Lie, Chapitre 4,5 et 6*, Hermann, Paris (1968).

[B] R. Bryant, *Some aspect of the local and global theory of Pfaffian systems*, Thesis, University of North Carolina, Chapel Hill, 1979.

[BCG3] R. Bryant, S.S. Chern, R.B. Gardner, H. Goldschmidt, and P. Griffiths, *Exterior differential systems*, Springer-Verlag, New-York (1986).

[BH] R. Bryant and L. Hsu, *Rigidity of integral curves of rank 2 distributions*, Invent. Math. (1993), **144**: 435–461.

[C1] E. Cartan, *Les systèmes de Pfaff à cinq variables et les équations aux dérivées partielles du second ordre*, Ann. Ec. Normale (1910), **27**: 109–192.

[C2] ———, *Sur les systèmes en involution d'équations aux dérivées partielles du second ordre à une fonction inconnue de trois variables indépendantes*, Bull. Soc. Math. France (1911), **39**: 352–443.

[DKM] B. Doubrov, B. Komrakov, and T. Morimoto, *Equivalence of holonomic differential equations*, Lobachevskii J. of Math. (1999), **3**: 39–71.

[GKR] A. Giaro, A. Kumpera, and C. Ruiz, *Sur la lecture correcte d'un resultat d'Elie Cartan*, C.R. Acad. Sc. Paris (1978), Sér.A **287**: 241–244.

[Hu] J.E. Humphreys, *Introduction to Lie Algebras and Representation Theory*, Springer-Verlag, New York (1972).

[K-A] S. Kaneyuki and H. Asano, *Graded Lie algebras and generalized Jordan triple systems*, Nagoya Math. J. (1988), **112**: 81–115.

[Ko] B. KOSTANT, *Lie algebra cohomology and generalized Borel-Weil theorem*, Ann. Math. (1961), **74**: 329–387.

[Ku] M. KURANISHI, *Lectures on involutive systems of partial differential equations*, Pub. Soc. Mat., São Paulo (1967).

[MZ] R. MONTGOMERY AND M. ZHITOMIRSKII, *Geometric approach to Goursat flags*, Ann. Inst. H. Poincaré-AN (2001), **18**: 459-493.

[M] P. MORMUL, *Goursat Flags:Classification of codimension-one singularities*, J.of Dynamical and Control Systems (2000), **6**(3): 311–330.

[SYY] T. SASAKI, K. YAMAGUCHI, AND M. YOSHIDA, *On the Rigidity of Differential Systems modeled on Hermitian Symmetric Spaces and Disproofs of a Conjecture concerning Modular Interpretations of Configuration Spaces*, Advanced Studies in Pure Math. (1997), **25**: 318–354.

[Se] Y. SE-ASHI, *On differential invariants of integrable finite type linear differential equations*, Hokkaido Math.J. (1988), **17**: 151–195.

[St] S. STERNBERG, *Lectures on Differential Geometry*, Prentice-Hall, New Jersey (1964).

[SY] K. SHIBUYA AND K. YAMAGUCHI, *Drapeau Theorem for Differential Systems*, to appear.

[Tk] M. TAKEUCHI, *Cell decomposition and Morse equalities on certain symmetric spaces*, J. Fac. Sci. Univ. Tokyo (1965), **12**: 81–192.

[T1] N. TANAKA, *On generalized graded Lie algebras and geometric structures I*, J. Math. Soc. Japan (1967), **19**: 215–254.

[T2] ——, *On differential systems, graded Lie algebras and pseudo-groups*, J. Math. Kyoto Univ. (1970), **10**: 1–82.

[T3] ——, *On non-degenerate real hypersurfaces, graded Lie algebras and Cartan connections*, Japan. J. Math. (1976), **2**: 131–190.

[T4] ——, *On the equivalence problems associated with simple graded Lie algebras*, Hokkaido Math. J. (1979), **8**: 23–84.

[T5] ——, *On geometry and integration of systems of second order ordinary differential equations*, Proc. Symposium on Differential Geometry, 1982, pp. 194–205 (in Japanese).

[T6] ——, *On affine symmetric spaces and the automorphism groups of product manifolds*, Hokkaido Math. J. (1985), **14**: 277–351.

[T7] ——, *Geometric theory of ordinary differential equations*, Report of Grant-in-Aid for Scientific Research MESC Japan (1989).

[Ts] A. TSUCHIYA, *Geometric theory of partial differential equations of second order*, Lecture Note at Nagoya University (in Japanese) (1981).

[Tt] J.TITS, *Espaces homogènes complexes compacts*, Comm. Math. Helv. (1962), **37**: 111–120.

[V] V.S. VARADARAJAN, *Lie Groups, Lie Algebras and Their Representations*, Spriger-Verlag, New York, 1984.

[Y1] K. YAMAGUCHI, *Contact geometry of higher order*, Japanese J. of Math. (1982), **8**: 109–176.

[Y2] ——, *On involutive systems of second order of codimension 2*, Proc. of Japan Acad. (1982), Ser A **58**(7): 302–305.

[Y3] ——, *Geometrization of Jet bundles*, Hokkaido Math. J. (1983), **12**: 27–40.

[Y4] ——, *Typical classes in involutive systems of second order*, Japanese J. Math. (1985), **11**: 265–291.

[Y5] ——, *Differential systems associated with simple graded Lie algebras*, Adv. Studies in Pure Math. (1993), **22**: 413–494.

[Y6] ——, *G_2-Geometry of Overdetermined Systems of Second Order*, Trends in Math. (Analysis and Geometry in Several Complex Variables) (1999), Birkhäuser, Boston, pp. 289–314.

[YY1] K. Yamaguchi and T. Yatsui, *Geometry of Higher Order Differential Equations of Finite Type associated with Symmetric Spaces*, Advanced Studies in Pure Mathematics (2002), **37**: 397–458.

[YY2] K. Yamaguchi and T. Yatsui, *Parabolic Geometries associated with Differential Equations of Finite Type*, Progress in Mathematics, **252** (from Geometry to Quantum Mechanics: In honor of Hideki Omori) (2007), pp. 161–209.

Part II: Research Articles

ON GEOMETRIC PROPERTIES OF JOINT INVARIANTS OF KILLING TENSORS

CAROLINE M. ADLAM*, RAYMOND G. MCLENAGHAN†, AND ROMAN G. SMIRNOV‡

Abstract. We employ the language of Cartan's geometry to present a model for studying vector spaces of Killing two-tensors defined in pseudo-Riemannian spaces of constant curvature under the action of the corresponding isometry group. We also discuss geometric properties of joint invariants of Killing two-tensors defined in the Euclidean plane to formulate and prove an analogue of the Weyl theorem on joint invariants. In addition, it is shown how the joint invariants manifest themselves in the theory of superintegrable Hamiltonian systems.

Key words. Joint invariants, killing tensors, superintegrable systems.

AMS(MOS) subject classifications. 53A55, 70H06.

1. Introduction.

The purpose of this article is three-fold. First, we wish to introduce a general model which forms a suitable framework for describing the invariant theory of Killing tensors from the viewpoint of Cartan's geometry (see [12, 13, 15, 18, 35] for more details), in particular its moving frame method and the theory of fiber bundles. Apparently, the idea to study the vector spaces of second order differential operators defined in the Euclidean plane under the action of the corresponding isometry group goes back to the classical 1965 Lie group theory paper by Winternitz and Friš [43]. In 2002 an analogous idea resurfaced in the problem of studying the isometry group action in a vector space of Killing two-tensors in the paper by McLenaghan, Smirnov and The [26] on classical Hamiltonian systems. A series of recent works on the subject represent a steady development of the invariant theory of Killing tensors [1, 5, 16, 17, 24, 26–31, 36–38, 44, 45].

Our second goal is to extend the study of joint invariants of Killing tensors introduced in Smirnov and Yue [36]. More specifically, we formulate and prove an analogue of the Weyl theorem on joint invariants in classical invariant theory (see Olver [33] and Weyl [42] for more details). Furthermore, we introduce the concept of a *resultant* of Killing tensors, which is also inspired by its analogue in classical invariant theory (see Olver [33] for more details).

*Department of Mathematics and Statistics, Dalhousie University, Halifax, Nova Scotia, Canada B3H 3J5.

†Department of Applied Mathematics, University of Waterloo, Waterloo, Ontario, Canada N2L 3G1. The work of the second author was supported in part by an NSERC Discovery Grant.

‡Department of Mathematics and Statistics, Dalhousie University, Halifax, Nova Scotia, Canada B3H 3J5. The work of the third author was supported in part by an NSERC Discovery Grant.

Finally, we employ the resultants to derive a new characterization of superintegrable systems. Recall that although examples of superintegrable systems have been known since the time of Kepler, a systematic development of a general theory of superintegrable systems originated in the pioneering works by Winternitz and collaborators [14, 25] (see also Smirnov and Winternitz [39] for more references). The theory has been actively developed since then for both quantum and classical Hamiltonian systems. Thus, a general classification and structure theory for superintegrable systems defined in two- and three-dimensional conformally flat spaces has been introduced in a series of recent papers by Kalnins, Kress and Miller (see [19–23] and the relevant references therein). As is well-known, a wide range of methods are employed to study superintegrable systems. We shall demonstrate that the geometric methods stemming from the invariant theory of Killing tensors also prove their worth in this study and present a new prospective. Thus, the main theorem presented in this work is a geometric characterization of the Kepler potential in terms of the existence of a vanishing resultant of the associated Killing two-tensors.

2. The model. In what follows, we shall assume that $(M, \mathbf{g})$ is an m-dimensional (pseudo-) Riemannian manifold of constant curvature. Generalized Killing tensors can be defined as the elements of the vector space of solutions to the following overdetermined system of PDEs, called the *generalized Killing tensor equation*, given by

$$[[\dots[\mathbf{K}, \mathbf{g}], \mathbf{g}], \dots, \mathbf{g}] = 0 \quad (n+1 \text{ brackets}), \tag{2.1}$$

where $[\ ,\ ]$ denotes the Schouten bracket [34] and $\mathbf{K}$ is a symmetric contravariant tensor of valence p. The generalized Killing tensor equation (2.1) is determined by given $m \geq 1$, $n \geq 0$ and $p \geq 0$. Let n, m and p be fixed and $\mathcal{K}_n^p(M)$ denote the corresponding vector space of solutions to the generalized Killing tensor equation (2.1). The *Nikitin-Prylypko-Eastwood (NPE)* formula [8, 9, 32]

$$d = \dim \mathcal{K}_n^p(M) = \frac{n+1}{m}\binom{p+m-1}{m-1}\binom{p+n+m}{m-1}, \tag{2.2}$$

represents the dimension d of the vector space $\mathcal{K}_n^p(M)$. The NPE formula (2.2) generalizes the *Delong-Takeuchi-Thompson (DTT)* formula [7, 40, 41] derived for the case $n = 0$. When $(M, \mathbf{g})$ is not of constant curvature, d given by (2.2) is the least upper bound for the dimension of the space of solutions to (2.1). Of particular importance, due to their relevance in various problems of mathematical physics, are the elements of the vector space $\mathcal{K}_0^2(M) \simeq \mathbb{R}^d$ for a given space $(M, \mathbf{g})$ (see, for example, [2, 7, 10, 17, 22, 23, 26, 38, 39, 43] and the relevant references therein).

The geometric study of the elements of $\mathcal{K}_0^2(M)$ having distinct (and real in the case of $(M, \mathbf{g})$ being pseudo-Riemannian) eigenvalues and hypersurface forming eigenvectors (eigenforms) is pivotal in the Hamilton-Jacobi theory of orthogonal separation of variables (see [2, 10, 17, 22, 23, 26, 38] and the relevant references therein). An element $\mathbf{K} \in \mathcal{K}_0^2(M)$ enjoying these properties generates an *orthogonal coordinate web* (see [10, 2, 17] and the relevant references therein) which consists of m foliations, the leaves of which are $m-1$-dimensional hypersurfaces orthogonal to the eigenvectors of $\mathbf{K}$. In this way the orthogonal coordinate web is adapted to the hypersurface forming eigenvectors (eigenforms) of the corresponding Killing tensor $\mathbf{K}$. The most natural framework for such a study is best described in terms of an appropriate Cartan's (or Klein's) geometry [35, 18, 38], depending on a given (pseudo-) Riemannian manifold $(M, \mathbf{g})$ and its geometry. Indeed, let G be the isometry group of $(M, \mathbf{g})$, and $H \subset G$ - a closed subgroup of the Lie group G. The study of the orthogonal coordinate webs generated by the elements of $\mathcal{K}_0^2(M)$ with the properties prescribed above, is established in the study of the principal fiber H-bundle: $\pi_1 : G \to G/H \simeq M$, where the left coset space G/H is identified with M on which G acts transitively. We also have the natural structure of a vector bundle: $\pi_2 : K_0^2(M) \to G/H \simeq M$. The orthogonal coordinate webs are thus defined in the homogeneous space M and the basic equivalence problem can be formulated as follows: Given two orthogonal coordinate webs $\mathcal{W}_1, \mathcal{W}_2 \in M$ generated by the corresponding elements $\mathbf{K}_1, \mathbf{K}_2 \in \mathcal{K}_0^2(M)$ respectively. The coordinate webs $\mathcal{W}_1, \mathcal{W}_2$ are said to be *equivalent* iff there exists a group element $g \in G$ and $\ell \in \mathbb{R}$ such that $g\mathbf{K}_1 = \mathbf{K}_2 + \ell\mathbf{g}$. Thus, given two orthogonal coordinate webs $\mathcal{W}_i$, $i = 1, 2$, one wishes to know whether or not they are equivalent as defined above. More specifically, the transitive action $G \circlearrowright M$ induces the corresponding *non-transitive* action $G \circlearrowright \mathbf{K}_0^2(M)$, solving the equivalence problem then amounts to the study of the orbit space $K_0^2(M)/G$. This structure leads to the principal G-bundle $\pi_3 : \mathcal{K}_0^2(M) \to \mathcal{K}_0^2(M)/G$, provided the action $G \circlearrowright \mathbf{K}_0^2(M)$ is free (for more details on the fiber bundle theory see, for example, Fatibene and Francaviglia [11]). Finally, one can define a map $f : \mathcal{K}_0^2(M)/G \to G$, so that the following diagram commutes.

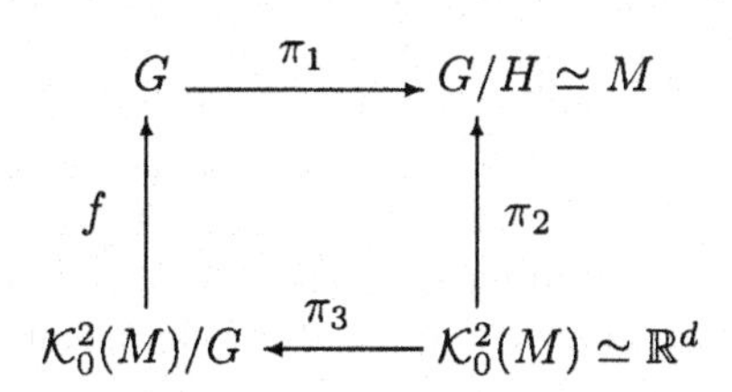

Now let $\mathbf{x} \in G/H$ and an element $\mathbf{K} \in \mathcal{K}_0^2(M)$ has pointwise distinct eigenvalues and hypersurface forming eigenvectors. If $\mathbf{x}$ is not a singular point of the coordinate web (a point where the eigenvalues of $\mathbf{K}$ coincide), we can use the metric $\mathbf{g}$ to orthonormalize the eigenvectors $E_1, \ldots, E_m$ of $\mathbf{K}$ and then use this *frame* of eigenvectors as a basis of $T_{\mathbf{x}}G/H$, which has a group conjugate to H acting on it. Thus, each fiber $\pi_1^{-1}(\mathbf{x})$ can be identified with an orthonormal frame of eigenvectors of $\mathbf{K}$. In the other direction the fiber $\pi_2^{-1}(\mathbf{x})$ can be identified with the corresponding vector space in $\mathcal{K}_0^2(M)$ for $\mathbf{x} \in M$. Furthermore, in this vector space we can fix the parameters so that the resulting element of the vector space is precisely the Killing tensor $\mathbf{K}$ above evaluated at $\mathbf{x} \in M$. It is now a point in the fiber $\pi_2^{-1}(\mathbf{x})$. Next, under the projection π_3 this point is mapped to the corresponding point in the orbit space $\mathcal{K}_0^2(M)/G$, which is nothing, but the orbit generated by $\mathbf{K}$ under the action of $G \circlearrowright \mathcal{K}_0^2(M)$. Finally, choosing a function f lifting the action to G is equivalent to either choosing a *cross-section* K through the orbits of $\mathcal{K}_0^2(M)/G$ or fixing a *frame*. Then (local) *invariants* of the group action $G \circlearrowright \mathcal{K}_0^2(M)$ are the coordinates of the canonical forms obtained as the intersection of the orbits with an appropriately chosen cross-section K [33]. In the latter case one can solve the equivalence problem by employing Cartan's language of the exterior differential forms [15, 35, 4, 18], while in the former case the problem can be solved by using the techniques of the geometric version of the classical moving frames method as developed by Fels and Olver [12, 13, 33] (for more details of how these two approaches interact see [38]). The equivalence of the two approaches is manifested by the fact that the group G acts transitively on the bundle of frames. We note however, that in practice the most effective approach is to use both methods. Thus, the former method can be used to find canonical forms of the orbits of $\mathcal{K}_0^2(M)/G$, while the latter approach comes into play when, for example, one needs to determine the moving frames map for a given representative of an orbit of $\mathcal{K}_0^2(M)/G$. Note also that the composition map $\gamma = \pi_3 \circ f : \mathcal{K}_0^2(M) \to G$ is the *moving frame* [15, 12, 13, 33], corresponding to the cross-section (or the frame of eigenvectors) prescribed by a chosen map $f : \mathcal{K}_0^2(M)/G \to G$. Since "The art of doing mathematics consists in finding that special case which contains all the germs of generality." (D. Hilbert), in the example that follows we shall do just that.

3. Illustrative example. Consider the following example. $M = \mathbb{E}^2$, $K_0^2(M) = \mathcal{K}_0^2(\mathbb{E}^2)$, $G = SE(2)$ and $H = SO(2)$. This example has been extensively studied in the literature from different points of view [4, 24, 26, 36, 38, 43]. For the first time the orbit problem $SE(2) \circlearrowright \mathcal{K}_0^2(\mathbb{E}^2)$ was considered by Winternitz and Friš [43] in the context of a classification of quadratic differential operators which commute with the Laplace operator. We consider the problem of classification of the orthogonal coordinate webs generated by non-trivial elements of the vector space $\mathcal{K}_0^2(\mathbb{E}^2)$ in the framework of the above diagram. Note first that in this case a Killing tensor $\mathbf{K} \in \mathcal{K}_0^2(\mathbb{E}^2)$ with pointwise distinct eigenvalues has necessarily hypersurface-forming eigenvectors, which generate one of the four orthogonal coordinate webs, namely elliptic-hyperbolic, polar, parabolic and cartesian. The latter depends on the corresponding orbit of $\mathcal{K}_0^2(\mathbb{E}^2)/SE(2)$ that $\mathbf{K}$ belongs to. More specifically, we fix at $\mathbf{x} \in \mathbb{E}^2$ the following orthonormal frame $(\mathbf{x}; E_1, E_2)$, where E_1, E_2 are the eigenvectors of $\mathbf{K}$. We also assume that $\mathbf{x} \in \mathbb{E}^2$ is not a singular point of $\mathbf{K}$. In this view, $\mathbf{x} = \pi_1 : SE(2) \to SE(2)/SO(2)$. The vectors E_1, E_2 form a basis of $T_{\mathbf{x}}\mathbb{E}^2$. Let E^1, E^2 be the corresponding dual basis of $T^*_{\mathbf{x}}\mathbb{E}^2$. Then E^1, E^2 are horizontal in the fibration and in this frame the components of the metric $\mathbf{g}$ of $\mathbb{E}^2$ and $\mathbf{K}$ are given by

$$g_{ab} = \delta_{ab} E^a \odot E^b \quad \text{and} \quad K_{ab} = \lambda_a \delta_{ab} E^a \odot E^b, \quad a, b = 1, 2 \tag{3.1}$$

respectively, where $\odot$ is the symmetric tensor product, δ_{ab} is the Kronecker delta and λ_1, λ_2 are the eigenvalues of $\mathbf{K}$. Note that we have used the metric to orthonormalize the frame (E_1, E_2), while the formulas (3.1) represent the pull-back under $\mathbf{x} = \pi_1$ of the metric $\mathbf{g}$ and the Killing tensor $\mathbf{K}$ to the bundle of frames. Moreover, we also note that by fixing a frame (E_1, E_2) we have fixed the corresponding map $f : \mathcal{K}_0^2(\mathbb{E}^2)/SE(2) \to SE(2)$. Then the problem of classification of the orthogonal coordinate webs generated by the non-trivial elements $\mathcal{K}^2(\mathbb{E}^2)$ can be solved by making use of the Cartan theory of exterior differential forms [4, 38]. Thus, upon introduction of the connection coefficients Γ of the Levi-Civita connection ∇ via:

$$\nabla_{E_a} E_b = \Gamma_{ab}{}^c E_c, \quad \nabla_{E_c} E^b = \Gamma_{cd}{}^b E^d, \quad a, b, c, d = 1, 2, \tag{3.2}$$

we arrive at the Cartan structure equations

$$dE^a + \omega^a{}_b \wedge E^b = T^a = 0, \quad d\omega^a{}_b + \omega^a{}_c \wedge \omega^c{}_b = \Omega^a{}_b = 0, \tag{3.3}$$

where $\omega^a{}_b := \Gamma_{cb}{}^a E^c$ are the connection one-forms, $T^a = \frac{1}{2} T^a{}_{bc} E^b \wedge E^a$ is the vanishing (∇ is Levi-Civita) torsion two-form and $\Omega^a{}_b := \frac{1}{2} R^a{}_{bcd} E^c \wedge E^d$ is the vanishing (in view of flatness of $\mathbb{E}^2$) curvature two-form. The choice of the connection is not arbitrary. As is well-known from Riemannian geometry, given a connection ∇ on a manifold M one can parallel propagate frames. For any path τ between two points of M, parallel transport along τ defines a linear mapping $L(\tau)$ between the tangent spaces of two points. This linear map is an *isometry* if the connection ∇ is a Levi-Civita connection. Clearly, the linear map $L(\tau)$ induced by a Levi-Civita connection ∇ maps orthonormal frames to orthonormal frames. The conditions $E^a \wedge dE^a = 0$, $a = 1, 2$ are automatically satisfied, hence by (one of the corollaries of) the Frobenius theorem one can introduce curvilinear coordinates (u, v) and functions f_1 and f_2 such that $E^1 = f_1(u, v)du$ and $E^2 = f_2(u, v)dv$. Furthermore, it is possible to show that the functions f_1 and f_2 are such that $f_1^2 = f_2^2 = A(u) + B(v)$, where $A(u)$ and $B(v)$ are functions of one variable, from which it follows that the metric can be rewritten as $ds^2 = (A(u) + B(v))(du^2 + dv^2)$. Note that the explicit forms for $A(u)$ and $B(v)$ follow from the vanishing of the curvature two-form. Then the following (differential) *invariants* can be used to solve the classification problem (for a complete solution refer to [4, 38]): $\Delta_1(u, v) = -\Gamma_{12}{}^1$,

$\Delta_2(u,v) = -\Gamma_{22}{}^1$. In addition, we introduce in this paper the following differential invariant: $\Delta_3(u,v) := \frac{F_{,uu}}{F}$, where $F = \Delta_2/\Delta_1$. The information provided by $\Delta_3(u,v)$ simplifies the classification considerably. Ultimately, we arrive at the following classification for the orthogonal coordinate webs (orbits) generated by the non-trivial elements of the vector space $\mathcal{K}_0^2(\mathbb{E}^2)$.

$$\begin{array}{rl} \Delta_1 = 0, \Delta_2 = 0 & \text{cartesian} \\ \Delta_1\Delta_2 = 0 & \text{polar} \\ \Delta_1\Delta_2 \neq 0, \Delta_3 = 0 & \text{parabolic} \\ \Delta_1\Delta_2 \neq 0, \Delta_3 \neq 0 & \text{elliptic-hyperbolic} \end{array} \tag{3.4}$$

However in applications, particularly, those pertinent to the study of Hamiltonian systems of classical mechanics [2, 10, 17, 26, 29, 38], the Killing tensors in question arise as functions on the cotangent bundle of the configuration space that define principal parts of first integrals of motion. Commonly the components of such Killing tensors are given in terms of (cartesian) position coordinates. In this view it is convenient to employ a different version of the moving frames method [12, 13, 33] upon noticing that the Lie group $SE(2)$ acts transitively on the bundle of frames. We conclude therefore that the same procedure can be carried out "in the group". Indeed, let $\mathbf{x} = (x_1, x_2)$ be cartesian coordinates of $\mathbb{E}^2$. Solving the generalized Killing tensor equation (2.1) for $p = 2$, $n = 0$ with respect to these coordinates yields:

$$\begin{aligned} \mathbf{K} \; = \; & (\beta_1 + 2\beta_4 x_2 + \beta_6 x_2^2)\partial_1 \odot \partial_1 \\ & +(\beta_3 - \beta_4 x_1 - \beta_5 x_2 - \beta_6 x_1 x_2)\partial_1 \odot \partial_2 \\ & +(\beta_2 + 2\beta_5 x_1 + \beta_6 x_1^2)\partial_2 \odot \partial_2, \end{aligned} \tag{3.5}$$

where $\partial_1 = \frac{\partial}{\partial x_1}$, $\partial_2 = \frac{\partial}{\partial x_2}$ and $\beta_1, \ldots, \beta_6$ are arbitrary constants (of integration) that represent the dimension of the vector space $\mathcal{K}_0^2(\mathbb{E}^2)$. The three-dimensional Lie group $SE(2)$ of (orientation-preserving) isometries of $\mathbb{E}^2$ can be represented in matrix form (using the notations adapted in [18]): $SE(2) = \left\{ M \in GL(3,\mathbb{R}) \,|\, M = \begin{pmatrix} 1 & 0 \\ \mathbf{t} & R \end{pmatrix}, \mathbf{t} \in \mathbb{R}^2\, R \in SO(2) \right\}$. In view of the formula (3.5), the action $SE(2) \circlearrowright \mathbb{E}^2$ given by

$$\begin{aligned} \tilde{x} &= x\cos p_3 - y \sin p_3 + p_1, \\ \tilde{y} &= x \sin p_3 + y\cos p_3 + p_2 \end{aligned} \tag{3.6}$$

induces the corresponding action $SE(2) \circlearrowright \mathcal{K}_0^2(\mathbb{E}^2)$ represented by the following formulas [43, 26]:

$$\begin{aligned}
\tilde{\beta}_1 &= \beta_1\cos^2 p_3 - 2\beta_3\cos p_3\sin p_3 + \beta_2\sin^2 p_3 - 2p_2\beta_4\cos p_3\\
&\quad -2p_2\beta_5\sin p_3 + \beta_6 p_2^2,\\
\tilde{\beta}_2 &= \beta_1\sin^2 p_3 - 2\beta_3\cos p_3\sin p_3 + \beta_2\cos^2 p_3 - 2p_1\beta_5\cos p_3\\
&\quad +2p_1\beta_4\sin p_3 + \beta_6 p_1^2,\\
\tilde{\beta}_3 &= (\beta_1-\beta_2)\sin p_3\cos p_3 + \beta_3(\cos^2 p_3 - \sin^2 p_3)\\
&\quad +(p_1\beta_4 + p_2\beta_5)\cos p_3 + (p_1\beta_5 - p_2\beta_4)\sin p_3 - \beta_6 p_1 p_2,\\
\tilde{\beta}_4 &= \beta_4\cos p_3 + \beta_5\sin p_3 - \beta_6 p_2,\\
\tilde{\beta}_5 &= \beta_5\cos p_3 - \beta_4\sin p_3 - \beta_6 p_1,\\
\tilde{\beta}_6 &= \beta_6.
\end{aligned} \tag{3.7}$$

Using the method of moving frames "à la Fels and Olver" [12, 13, 33], one arrives at the following (algebraic) *invariants* (see [38] and the relevant references therein for more details):

$$\begin{aligned}
\Delta_1' &= \beta_6,\\
\Delta_2' &= \beta_6(\beta_1+\beta_2) - \beta_4^2 - \beta_5^2,\\
\Delta_3' &= (\beta_6(\beta_1-\beta_2) - \beta_4^2 + \beta_5^2)^2 + 4(\beta_6\beta_3 + \beta_4\beta_5)^2.
\end{aligned} \tag{3.8}$$

The fact that $\tilde{\Delta}_1$ is an invariant is obvious, the fundamental invariant $\tilde{\Delta}_3$ was derived for the first time in [43] (see also [26] where $\tilde{\Delta}_3$ was rederived by making use of Lie's method of infinitesimal generators). Combining the invariants (3.8), one can also make use of the invariant k^2 which is (half-) the distance between the singular points (foci) in the elliptic-hyperbolic case and is given by

$$k^2 = \frac{\sqrt{\Delta_3'}}{(\Delta_1')^2}. \tag{3.9}$$

We also note that the singular points are the foci of the confocal conics (ellipses and hyperbolas) that form the coordinate web in this case. In terms of the algebraic invariants (3.8) the classification (3.4) can be equivalently reformulated as follows [43, 26] (see also [36])

$$\begin{array}{ll}
\Delta_1' = 0, \Delta_2' = 0 & \text{cartesian}\\
\Delta_1' = 0, \Delta_3' \neq 0 & \text{parabolic}\\
\Delta_1' \neq 0, \Delta_3' = 0 & \text{polar}\\
\Delta_1' \neq 0, \Delta_3' \neq 0 & \text{elliptic-hyperbolic.}
\end{array} \tag{3.10}$$

4. Main theorem. The next natural step in this study is to consider the action of G on the product of the vector space and the underlying manifold: $\mathcal{K}_0^2(M)\times M$ or on n copies of the vector space: $\mathcal{K}_0^2(M)\times\mathcal{K}_0^2(M)\times\cdots\times\mathcal{K}_0^2(M)$. In the former case the action leads to the *covariants* of Killing tensors, while in the latter - the *joint invariants* of Killing tensors (see [36] for more details). This aspect of the theory also has much similarity with the study of covariants and joint invariants in classical invariant theory (see Boutin [3], Olver [33], Weyl [42] and the relevant references therein for more details). More specifically, in this work we extend the study of joint invariants introduced in [36]. To this end, we shall introduce an analogue of the concept of a *resultant* in classical invariant theory, as well as to formulate and prove an analogue of the Weyl theorem on joint invariants [33, 42] concerning the joint action of $SE(n)$ or $O(n)$ on n copies of $\mathbb{R}^n$: $\mathbb{R}^n \times \mathbb{R}^n \times \cdots \times \mathbb{R}^n$.

DEFINITION 4.1. *A three-dimensional orbit $\mathcal{O} \in \mathcal{K}^2(\mathbb{E}^2)/SE(2)$ is said to be* non-degenerate *iff along $\mathcal{O}$ the invariant $k^2 = \sqrt{\Delta_3'}/(\Delta_1')^2 \neq 0$, where Δ_1' and Δ_3' are given by (3.10). The action $SE(2) \circlearrowright \mathcal{K}^2(\mathbb{E}^2)$ for which $k^2 \neq 0$ is also said to be* non-degenerate.

Recall, that all other orbits of $\mathcal{K}^2(\mathbb{E}^2)/SE(2)$ are various degeneracies of the non-degenerate orbits defined above [26, 38]. The (eigenvectors of) Killing two-tensors corresponding to the non-degenerate orbits generate elliptic-hyperbolic coordinate webs [26, 38, 43]. In what follows we shall investigate geometric meaning of the joint invariants of non-degenerate orbits of the orbit space $(\mathcal{K}^2(\mathbb{E}^2) \times \mathcal{K}^2(\mathbb{E}^2) \times \cdots \times \mathcal{K}^2(\mathbb{E}^2))/SE(2)$. For simplicity we investigate first the joint invariants of the non-degenerate action $SE(2) \circlearrowright \mathcal{K}^2(\mathbb{E}^2) \times \mathcal{K}^2(\mathbb{E}^2)$. In this case the group $SE(2)$ acts on each copy of $\mathcal{K}^2(\mathbb{E}^2)$ with three-dimensional non-degenerate orbits. The group action is given by

$$
\begin{aligned}
\tilde{\alpha_1} &= \alpha_1 \cos^2 p_3 - 2\alpha_3 \cos p_3 \sin p_3 + \alpha_2 \sin^2 p_3 - 2p_2\alpha_4 \cos p_3 \\
&\quad -2p_2\alpha_5 \sin p_3 + \alpha_6 p_2^2, \\
\tilde{\alpha_2} &= \alpha_1 \sin^2 p_3 - 2\alpha_3 \cos p_3 \sin p_3 + \alpha_2 \cos^2 p_3 - 2p_1\alpha_5 \cos p_3 \\
&\quad +2p_1\alpha_4 \sin p_3 + \alpha_6 p_1^2, \\
\tilde{\alpha_3} &= (\alpha_1 - \alpha_2) \sin p_3 \cos p_3 + \alpha_3(\cos^2 p_3 - \sin^2 p_3) \\
&\quad +(p_1\alpha_4 + p_2\alpha_5) \cos p_3 + (p_1\alpha_5 - p_2\alpha_4) \sin p_3 - \alpha_6 p_1 p_2, \\
\tilde{\alpha_4} &= \alpha_4 \cos p_3 + \alpha_5 \sin p_3 - \alpha_6 p_2, \\
\tilde{\alpha_5} &= \alpha_5 \cos p_3 - \alpha_4 \sin p_3 - \alpha_6 p_1, \\
\tilde{\alpha_6} &= \alpha_6, \\
\tilde{\beta_1} &= \beta_1 \cos^2 p_3 - 2\beta_3 \cos p_3 \sin p_3 + \beta_2 \sin^2 p_3 - 2p_2\beta_4 \cos p_3 \\
&\quad -2p_2\beta_5 \sin p_3 + \beta_6 p_2^2, \\
\tilde{\beta_2} &= \beta_1 \sin^2 p_3 - 2\beta_3 \cos p_3 \sin p_3 + \beta_2 \cos^2 p_3 - 2p_1\beta_5 \cos p_3 \\
&\quad +2p_1\beta_4 \sin p_3 + \beta_6 p_1^2, \\
\tilde{\beta_3} &= (\beta_1 - \beta_2) \sin p_3 \cos p_3 + \beta_3(\cos^2 p_3 - \sin^2 p_3) \\
&\quad +(p_1\beta_4 + p_2\beta_5) \cos p_3 + (p_1\beta_5 - p_2\beta_4) \sin p_3 - \beta_6 p_1 p_2, \\
\tilde{\beta_4} &= \beta_4 \cos p_3 + \beta_5 \sin p_3 - \beta_6 p_2, \\
\tilde{\beta_5} &= \beta_5 \cos p_3 - \beta_4 \sin p_3 - \beta_6 p_1, \\
\tilde{\beta_6} &= \beta_6,
\end{aligned} \tag{4.1}
$$

where α_i, β_i, $i = 1, \ldots, 6$ are the parameters of the respective vector spaces and the conditions $k_1^2 = \frac{\sqrt{(\alpha_4^2-\alpha_5^2+\alpha_6(\alpha_2-\alpha_1))^2+4(\alpha_6\alpha_3+\alpha_4\alpha_5)^2}}{\alpha_6} \neq 0$, $k_2^2 = \frac{\sqrt{(\beta_4^2-\beta 5^2+\beta_6(\beta_2-\beta_1))^2+4(\beta_6\beta_3+\beta_4\beta_5)^2}}{\beta_6} \neq 0$ hold true. Six joint invariants of the action $SE(2) \circlearrowleft \mathcal{K}^2(\mathbb{E}^2) \times \mathcal{K}^2(\mathbb{E}^2)$ follow immediately from (3.8):

$$\begin{aligned}
\Delta_1' &= \alpha_6, \\
\Delta_2' &= \alpha_6(\alpha_1 + \alpha_2) - \alpha_4^2 - \alpha_5^2, \\
\Delta_3' &= (\alpha_6(\alpha_1 - \alpha_2) - \alpha_4^2 + \alpha_5^2)^2 + 4(\alpha_6\alpha_3 + \alpha_4\alpha_5)^2, \\
\Delta_4' &= \beta_6, \\
\Delta_5' &= \beta_6(\beta_1 + \beta_2) - \beta_4^2 - \beta_5^2, \\
\Delta_6' &= (\beta_6(\beta_1 - \beta_2) - \beta_4^2 + \beta_5^2)^2 + 4(\beta_6\beta_3 + \beta_4\beta_5)^2.
\end{aligned} \tag{4.2}$$

The joint invariants are functionally independent (they are obtained via the moving frames method), any analytic function of the invariants (4.2) is a joint invariant of the non-degenerate group action $SE(2) \circlearrowleft \mathcal{K}_0^2(\mathbb{E}^2) \times \mathcal{K}_0^2(\mathbb{E}^2)$ (see Olver [33] for more details). In view of the Fundamental Theorem on invariants of regular Lie group action [33], we need to produce in total 12 (the dimension of the product space $\mathcal{K}_0^2(\mathbb{E}^2) \times \mathcal{K}_0^2(\mathbb{E}^2)$) - 3 (the dimension of the orbits) = 9 fundamental joint invariants. To derive the remaining three joint invariants, we employ geometric reasoning. Recall that each element of a non-degenerate orbit of the action $SE(2) \circlearrowleft \mathcal{K}_0^2(\mathbb{E}^2)$ corresponds to a Killing tensor whose eigenvectors generate an elliptic-hyperbolic coordinate web. In turn, the coordinate web is characterized by the foci of the confocal conics (the two families of ellipses and hyperbolas). If $F_1, F_2 \in SE(2)/SO(2) = \mathbb{E}^2$ are the two foci of such a coordinate web, then their respective coordinates $(x_1, y_1)_{F_1}$ and $(x_2, y_2)_{F_2}$ are given in terms of the parameters β_i, $i = 1, \ldots, 6$ that determine the Killing tensor via the formula (3.5) by [26]:

$$\begin{aligned}
&(x_1, y_1)_{F_1} = \\
&\left(\frac{-\beta_5}{\beta_6} + \frac{1}{\beta_6} \left(\frac{\sqrt{\Delta_6'} - \sigma_1}{2} \right)^{1/2}, \frac{-\beta_4}{\beta_6} + \frac{1}{\beta_6} \left(\frac{\sqrt{\Delta_6'} + \sigma_1}{2} \right)^{1/2} \right), \\
&(x_2, y_2)_{F_2} = \\
&\left(\frac{-\beta_5}{\beta_6} - \frac{1}{\beta_6} \left(\frac{\sqrt{\Delta_6'} - \sigma_1}{2} \right)^{1/2}, \frac{-\beta_4}{\beta_6} - \frac{1}{\beta_6} \left(\frac{\sqrt{\Delta_6'} + \sigma_1}{2} \right)^{1/2} \right),
\end{aligned} \tag{4.3}$$

where $\sigma_1 = \beta_4^2 - \beta_5^2 + \beta_6(\beta_2 - \beta_1)$ and Δ_6' is given by (4.2). In the case of the non-degenerate action $SE(2) \circlearrowleft \mathcal{K}_0^2(\mathbb{E}^2) \times \mathcal{K}_0^2(\mathbb{E}^2)$, we have two more (generic) foci F_3 and F_4 corresponding to the second copy of $\mathcal{K}_0^2(\mathbb{E}^2)$. Their coordinates are given accordingly by

$$
\begin{aligned}
&(x_3, y_3)_{F_3} = \\
&\left(\frac{-\alpha_5}{\alpha_6} + \frac{1}{\alpha_6}\left(\frac{\sqrt{\Delta_3'} - \sigma_2}{2}\right)^{1/2}, \frac{-\alpha_4}{\alpha_6} + \frac{1}{\alpha_6}\left(\frac{\sqrt{\Delta_3'} + \sigma_2}{2}\right)^{1/2},\right. \\
&(x_4, y_4)_{F_4} = \\
&\left(\frac{-\alpha_5}{\alpha_6} - \frac{1}{\alpha_6}\left(\frac{\sqrt{\Delta_3'} - \sigma_2}{2}\right)^{1/2}, \frac{-\alpha_4}{\alpha_6} - \frac{1}{\alpha_6}\left(\frac{\sqrt{\Delta_3'} + \sigma_2}{2}\right)^{1/2},\right.
\end{aligned}
\tag{4.4}
$$

where $\sigma_2 = \alpha_4^2 - \alpha_5^2 + \alpha_6(\alpha_2 - \alpha_1)$ and Δ_3' is given by (4.2). Now we can treat the non-degenerate action $SE(2) \circlearrowleft \mathcal{K}_0^2(\mathbb{E}^2) \times \mathcal{K}_0^2(\mathbb{E}^2)$, as the free and regular action $SE(2) \circlearrowleft \mathbb{E}^2 \times \mathbb{E}^2 \times \mathbb{E}^2 \times \mathbb{E}^2$, where the foci F_1, F_2, F_3 and F_4 belong to the corresponding copies of $\mathbb{E}^2$. Recall [33, 42], that any joint invariant $\mathcal{J}(F_1, F_2, F_3, F_4)$ of the non-transitive action so defined can be written as a function of the interpoint distances $d(F_1, F_2)$, $d(F_2, F_3)$, $d(F_3, F_4)$, $d(F_4, F_1)$ (this result is also known as "the Weyl theorem on joint invariants" [33, 42]). Therefore we choose the following square distances as the remaining three fundamental joint invariants:

$$
\begin{aligned}
\Delta_7' &= d^2(F_2, F_3) = (x_2 - x_3)^2 + (y_2 - y_3)^2, \\
\Delta_8' &= d^2(F_1, F_3) = (x_1 - x_3)^2 + (y_1 - y_3)^2, \\
\Delta_9' &= d^2(F_2, F_4) = (x_2 - x_4)^2 + (y_2 - y_4)^2,
\end{aligned}
\tag{4.5}
$$

where $(x_i, y_i), i = 1, 2, 3, 4$ are specified by (4.3) and (4.4). Direct verification using MAPLE shows that $\mathrm{d}\Delta_1' \wedge \mathrm{d}\Delta_2' \wedge \ldots \wedge \mathrm{d}\Delta_9' \neq 0$, that is the nine joint invariants given by (4.2) and (4.5) are functionally independent at a generic point (this conclusion also follows from a geometric argument). Therefore we arrive at the following result.

THEOREM 4.1. *Every joint invariant of the non-degenerate action $SE(2) \circlearrowleft \mathcal{K}_0^2(\mathbb{E}^2) \times \mathcal{K}_0^2(\mathbb{E}^2)$ is a function of the nine fundamental joint invariants $\Delta_i'(\alpha_j, \beta_j)$, $i = 1, \ldots, 9$, $j = 1, \ldots, 6$ given by (4.2) and (4.5).*

Theorem 4.1 can naturally be extended to the general case of the non-degenerate group action of $SE(2)$ on $n > 2$ copies of the vector space $\mathcal{K}_0^2(\mathbb{E}^2)$ as well as, more generally, to the case of Killing two-tensors defined on pseudo-Riemannian spaces of constant curvature of higher dimensions. It must also be mentioned that other joint invariants having a geometric meaning are the various angles, as well as areas within the quadrilateral $F_1F_2F_3F_4$. For example, the angle ϕ given by

$$\cos\phi = \frac{\overrightarrow{F_1F_3}\cdot\overrightarrow{F_1F_2}}{d(F_1,F_3)d(F_1,F_2)}$$

is a joint invariant.

We recall now the notion of a *resultant* in classical invariant theory (see Olver [33] for more details). Thus, let

$$\begin{aligned} P(\mathbf{x}) &= \tilde{a}_m x^m + \tilde{a}_{m-1}x^{m-1}y + \cdots + \tilde{a}_0 y^m, \\ Q(\mathbf{x}) &= \tilde{b}_n x^n + \tilde{b}_{n-1}x^{n-1}y + \cdots + \tilde{b}_0 y^n \end{aligned} \tag{4.6}$$

be two homogeneous polynomials of degrees m and n respectively. Then a joint invariant of the polynomials given by (4.6) is a function $J(\tilde{\mathbf{a}}, \tilde{\mathbf{b}})$ of the coefficients $a_0, \ldots, a_m, b_0, \ldots, b_n$ preserved under the action of $GL(2,\mathbb{R}^2)$ (or its subgroups). Particularly important joint invariants in this study are those whose vanishing is equivalent to the fact that P and Q have common roots. Such joint invariants are said to be *resultants* of the system (4.6). This important concept can be transferred naturally to the study of Killing tensors.

DEFINITION 4.2. *Consider the non-degenerate action $SE(2) \circlearrowright \mathcal{K}_0^2(\mathbb{E}^2) \times \mathcal{K}_0^2(\mathbb{E}^2)$. Let $\mathbf{K}_1$, $\mathbf{K}_2 \in \mathcal{K}_0^2(\mathbb{E}^2)$ be two Killing tensors belonging to non-degenerate orbits. Then a* resultant *$\mathcal{R}[\mathbf{K}_1, \mathbf{K}_2]$ is a joint invariant of the action with the property that the vanishing $\mathcal{R}[\mathbf{K}_1, \mathbf{K}_2] = 0$ is equivalent to the fact that the orthogonal coordinate webs generated by $\mathbf{K}_1$ and $\mathbf{K}_2$ have a common focus.*

Thus, for example, the joint invariants given by (4.5) are resultants, while $\Delta_3' + \Delta_4'$, where Δ_3' and Δ_4' are given by (4.2), is not. Another resultant is

$$\Delta_{10}' = d^2(F_1, F_4) = (x_1 - x_4)^2 + (y_1 - y_4)^2. \tag{4.7}$$

In the next section we shall demonstrate how the joint invariants, in particular, the resultants manifest themselves in the study of superintegrable systems.

5. An application to the theory of superintegrable systems. In order to link the results presented in the preceding sections with the theory of superintegrable systems let us consider now a Hamiltonian system defined in $\mathbb{E}^2$ by a natural Hamiltonian

$$H(\mathbf{x}, \mathbf{p}) = \frac{1}{2}(p_1^2 + p_2^2) + V(\mathbf{x}), \tag{5.1}$$

where $\mathbf{x} = (x_1, x_2)$ (position coordinates) $\mathbf{p} = (p_1, p_2)$ (momenta coordinates). We assume that the Hamiltonian system (5.1) admits two *distinct* in a certain sense (see below) first integrals F_1, F_2, which are quadratic in the momenta

$$F_\ell(\mathbf{x}, \mathbf{p}) = K_\ell^{ij}(\mathbf{x})p_ip_j + U_\ell(\mathbf{x}), \quad \ell, i, j = 1, 2 \tag{5.2}$$

and functionally-independent with H. It is well-known that the vanishing of Poisson brackets $\{F_1, H\} = \{F_2, H\} = 0$ yields two sets of conditions, namely the Killing tensor equations

$$[\mathbf{K}_1, \mathbf{g}] = [\mathbf{K}_2, \mathbf{g}] = 0,$$

and the compatibility conditions (also known as the Bertrand-Darboux PDEs)

$$\mathrm{d}(\hat{\mathbf{K}}_1 \mathrm{d}V) = \mathrm{d}(\hat{\mathbf{K}}_2 \mathrm{d}V) = 0,$$

where the components of the Killing tensors $\mathbf{K}_1$, $\mathbf{K}_2$ are determined by the quadratic in the momenta terms of the first integrals F_1 and F_2 respectively given by (5.2), while the components of the $(1,1)$-tensors $\hat{\mathbf{K}}_1$, $\hat{\mathbf{K}}_2$ are as follows: $\hat{K}_{\ell}{}^j_i = \hat{K}_{\ell i m} g^{mj}$, $\ell = 1, 2$, where g^{ij} are the components of the metric tensor that determines the kinetic part of the Hamiltonian (5.1). These assumptions afford orthogonal separation of variables in the associated Hamilton-Jacobi equation of the Hamiltonian system defined by (5.1). More specifically, the orthogonal coordinates are determined by the eigenvectors (eigenvalues) of the Killing tensors $\mathbf{K}_1$ and $\mathbf{K}_2$. Thus, the orthogonal separable coordinate systems can be used to find exact solutions to the Hamiltonian system defined by (5.1) via solving the Hamilton-Jacobi equation by separation of variables. Moreover, in view of the above, the Hamiltonian system is superintegrable and multiseparable.

Furthermore, we assume that the Killing tensors $\mathbf{K}_1$ and $\mathbf{K}_2$ belong to two *distinct* non-degenerate orbits, namely their respective eigenvectors (eigenvalues) generate elliptic-hyperbolic coordinate webs. Let F_1, F_2 be the foci of the first elliptic-hyperbolic coordinate system generated by $\mathbf{K}_1$, while F_3, F_4 - the foci of the second one, generated by $\mathbf{K}_2$. Thus, all four foci are distinct. We can now treat the pair $\{\mathbf{K}_1, \mathbf{K}_2\}$ as an element of the product space $\mathcal{K}_0^2(\mathbb{E}^2) \times \mathcal{K}_0^2(\mathbb{E}^2)$. Moreover, it is an element of a non-degenerate orbit of the action $SE(2) \circlearrowleft \mathcal{K}_0^2(\mathbb{E}^2) \times \mathcal{K}_0^2(\mathbb{E}^2)$. Indeed, since $F_1 \neq F_2$ and $F_3 \neq F_4$, the corresponding invariants k_1^2 and k_2^2 given by (3.9) do not vanish. Moreover, $k_1^2 \neq k_2^2$. In addition, we also have three joint invariants given by (4.5). In the most general case all of the five joint invariants $k_1^2, k_2^2, \Delta_7', \Delta_8', \Delta_9'$ that completely characterize the pair $\{\mathbf{K}_1, \mathbf{K}_2\}$ are, in view of Theorem 4.1, functionally independent. At the same time we can treat the Killing tensors $\mathbf{K}_1$, $\mathbf{K}_2$, as well as any linear combination $c_1\mathbf{K}_1 + c_2\mathbf{K}_2$ of them as elements of the vector space $\mathcal{K}_0^2(\mathbb{E}^2)$. Consider now the following Killing tensor:

$$\mathbf{K}_g = \mathbf{K}_1 + \mathbf{K}_2 + \ell \mathbf{g}, \quad \ell \in \mathbb{R}, \tag{5.3}$$

where $\mathbf{g}$ is the Euclidean metric of the Hamiltonian (5.1). Clearly

$$\mathrm{d}(\hat{\mathbf{K}}_g \mathrm{d}V) = 0, \tag{5.4}$$

where V is the potential part of (5.1). Our next observation is that the Killing tensor $\mathbf{K}_g$ depends upon exactly *six* parameters, that is the five joint invariants and ℓ. This is only possible if $\mathbf{K}_g$ is the most general Killing tensor (3.5). But then the compatibility condition (5.4) yields that V is constant. Since $k_1^2 k_2^2 \neq 0$, we conclude therefore that one of the joint invariants $\Delta_7', \Delta_8', \Delta_9'$ must be a vanishing resultant in order for V to be non-constant. At the same time two of the joint invariants cannot be resultants simultaneously, since $k_1^2 \neq k_2^2$. Without loss of generality let us assume that $\Delta_8' = \mathcal{R}[\mathbf{K}_1, \mathbf{K}_2] = 0$. It follows that $\Delta_7' = k_1^2$ and the Killing tensor $\mathbf{K}_1 + \mathbf{K}_2$ in (5.3) depends upon three parameters, namely $k_1^2 = \Delta_7'$, k_1^2 and Δ_9'. We also note that since Δ_8' is a vanishing resultant, the elliptic-hyperbolic coordinate webs generated by $\mathbf{K}_1$ and $\mathbf{K}_2$ have a common focus. Now let us set in (3.5) the parameters $\beta_3 = 0$ and $\beta_1 = \beta_2$. The resulting Killing tensor is of the form

$$\mathbf{K}_g' = \mathbf{K}_1' + \beta_1 \mathbf{g}, \tag{5.5}$$

where as before $\mathbf{g}$ is the Euclidean metric of the Hamiltonian (5.1). Thus, the Killing tensor $\mathbf{K}_1'$ that appears in (5.5) is given by

$$\begin{aligned} \mathbf{K}_1' \;=\; & (2\beta_4 x_2 + \beta_6 x_2^2)\partial_1 \odot \partial_1 \\ & +(-\beta_4 x_1 - \beta_5 x_2 - \beta_6 x_1 x_2)\partial_1 \odot \partial_2 \\ & +(2\beta_5 x_1 + \beta_6 x_1^2)\partial_2 \odot \partial_2. \end{aligned} \tag{5.6}$$

We assume that the parameters $\beta_4, \beta_5, \beta_6$ are such that both $\beta_4^2 + \beta_5^2 \neq 0$ and $\beta_6 \neq 0$. Note that $\beta_4^2 + \beta_5^2$ and β_6 are invariants of the action $SE(2) \circlearrowright \mathcal{K}_0^2(\mathbb{E}^2)$ in this case. We also note that in this case the Killing tensor $\mathbf{K}_1'$ generates an elliptic-hyperbolic web characterized by two foci: F_1 with the coordinates $(0,0)$ and F_2 with the coordinates $\left(\frac{-2\beta_5}{\beta_6}, \frac{-2\beta_4}{\beta_6}\right)$. Consider now another Killing tensor of the same type, namely given by

$$\begin{aligned} \mathbf{K}_2' \;=\; & (2\beta_4' x_2 + \beta_6' x_2^2)\partial_1 \odot \partial_1 \\ & +(-\beta_4' x_1 - \beta_5' x_2 - \beta_6' x_1 x_2)\partial_1 \odot \partial_2 \\ & +(2\beta_5' x_1 + \beta_6' x_1^2)\partial_2 \odot \partial_2, \end{aligned} \tag{5.7}$$

under the additional assumptions that the three parameters $\beta_4', \beta_5', \beta_6'$ are such that $(\beta_4')^2 + (\beta_5')^2 \neq 0$ and $\beta_6' \neq 0$. Moreover, we assume in addition that $(\beta_4^2 + \beta_5^2)/\beta_6 \neq [(\beta_4')^2 + (\beta_5')^2]/\beta_6'$, so that the Killing tensors $\mathbf{K}_1'$, $\mathbf{K}_2''$ are truly distinct. Note that the coordinates of the foci F_3, F_4 of the elliptic-hyperbolic coordinate web generated by $\mathbf{K}_2'$ are $(0,0)$ and $\left(\frac{-2\beta_5'}{\beta_6'}, \frac{-2\beta_4'}{\beta_6'}\right)$ respectively. Thus, it follows that the elliptic-hyperbolic coordinate webs generated by $\mathbf{K}_1'$ and $\mathbf{K}_2'$ have one common focus at $(0,0)$, or, in other words, the pair $\{\mathbf{K}_1', \mathbf{K}_2'\} \in \mathcal{K}_0^2(\mathbb{E}^2) \times \mathcal{K}_0^2(\mathbb{E}^2)$ admits a vanishing resultant. Therefore $\{\mathbf{K}_1', \mathbf{K}_2'\}$ have exactly the same geometric properties as the pair of Killing tensors $\{\mathbf{K}_1, \mathbf{K}_2\}$ given by (5.2), namely the elliptic-hyperbolic coordinate webs share one focus and the system is characterized by three parameters. Hence, we can *identify* $\mathbf{K}_1 + \mathbf{K}_2$ with $\mathbf{K}_1' + \mathbf{K}_2'$. Moreover, we see that the sum $\mathbf{K}_1' + \mathbf{K}_2'$ is given by the same formula as either of the Killing tensors (5.6) or (5.7). In this view in order to determine the potential V in (5.1), we substitute the formula (5.6) into the compatibility condition (5.4) and solve the resulting PDE for V. More specifically, after the substitution we set successively in (5.4) $\beta_4 = \beta_5 = 0$, which defines a polar web with singular point at $(0,0)$, then $\beta_4 = \beta_6 = 0$, which defines a parabolic web with singular point at $(0,0)$ and x-axis as the focal axis, and finally $\beta_5 = \beta_6 = 0$, which defines a parabolic web with singular point at $(0,0)$ and focal axis the y-axis. This results in three PDEs whose unique solution can be easily found. Indeed, the PDEs enjoy the following forms:

$$2x_2V_{x_1} - 2x_1V_{x_2} - x_1x_2(V_{x_2x_2} - V_{x_1x_1}) + (x_2^2 - x_1^2)V_{x_1x_2} = 0, \tag{5.8}$$

$$3V_{x_2} + x_2(V_{x_2x_2} - V_{x_1x_1}) + 2x_1V_{x_1x_2} = 0, \tag{5.9}$$

$$3V_{x_1} + x_1(V_{x_2x_2} - V_{x_1x_1}) + 2x_2V_{x_1x_2} = 0 \tag{5.10}$$

The PDEs (5.8)-(5.10) yield

$$x_2V_{x_1} - x_1V_{x_2} = 0, \tag{5.11}$$

which implies that (in terms of polar coordinates)

$$V(r,\theta) = V(r\cos\theta, r\sin\theta) \tag{5.12}$$

is independent of θ. Now the general solution of (5.10) in polar coordinates has the form

$$V(r,\theta) = F(r) + \frac{G(\theta)}{r^2}, \tag{5.13}$$

where F and G are arbitrary functions. Thus by (5.11) we have

$$G(\theta) = \ell \tag{5.14}$$

where ℓ is an arbitrary constant. Substituting (5.13) into (5.11) (in polar coordinates) and taking (5.14) into account, we find that

$$F(r) = -\frac{\ell}{r^2} - \frac{m}{r} + n, \tag{5.15}$$

where ℓ as in (5.14) and m, n are arbitrary constants. Substituting (5.14) and (5.15) into (5.13) and transforming back to cartesian coordinates, we arrive at the potential of the *Kepler problem*:

$$V(x_1,x_2) = \frac{1}{\sqrt{x_1^2 + x_2^2}}, \tag{5.16}$$

where without loss of generality we set $m = -1$, $n = 0$. The inverse problem, namely when one starts with the Kepler potential (5.16) and then finds the most general Killing tensor compatible with (5.16) via (5.4), can be solved in the same manner, that is by solving the corresponding (Bertrand-Darboux) PDE. The calculations are straightforward and we present here the result only. Thus, the most general Killing tensor compatible with the Kepler potential (5.16) is precisely the three-parameter family of Killing tensors given by (5.6). We conclude therefore that we have proven the following theorem.

THEOREM 5.1. *Let the potential V of the general Hamiltonian (5.1) be compatible via (5.4) with any two non-degenerate Killing tensors $\mathbf{K}_1, \mathbf{K}_2 \in \mathcal{K}_0^2(\mathbb{E}^2)$. Then the following statements are equivalent.*

(1) The pair of Killing tensors $\{\mathbf{K}_1, \mathbf{K}_2\} \in \mathcal{K}_0^2(\mathbb{E}^2) \times \mathcal{K}_0^2(\mathbb{E}^2)$ admits one vanishing resultant $\mathcal{R}[\mathbf{K}_1, \mathbf{K}_2]$.

(2) The potential V given by (5.1) is the Kepler potential (5.16).

6. Conclusions. The results presented in this article can naturally be generalized by increasing the dimension of the underlying (pseudo-) Riemannian manifold, changing its curvature and the signature of its metric. Some of these cases will be investigated by the authors in the forthcoming papers on the subject.

REFERENCES

[1] C. ADLAM, *A Lie Group Theory Approach to the Problem of Classification of Superintegrable Potentials in the Euclidean Plane,* MSc thesis, Dalhousie University, 2005.

[2] S. BENENTI, *Intrinsic characterization of the variable separation in the Hamilton-Jacobi equation,* J. Math. Phys. (1997), **38**: 6578–6602.

[3] M. BOUTIN, *On orbit dimensions under a simultaneous Lie group action on n copies of a manifold,* J. Lie Theory (2002), **12**: 191–203.

[4] A. BRUCE, R. MCLENAGHAN, AND R. SMIRNOV, *A geometric approach to the problem of integrability of Hamiltonian systems by separation of variables,* J. Geom. Phys. (2001), **39**: 301–322.

[5] C. CHANU, L. DEGIOVANNI, AND R. MCLENAGHAN, *Geometrical classification of Killing tensors on bi-dimensional flat manifolds,* J. Math. Phys. (2006), **47**, 073506, 20 pp.

[6] R. DEELEY, J. HORWOOD, R. MCLENAGHAN, AND R. SMIRNOV, *Theory of algebraic invariants of vector spaces of Killing tensors: Methods for computing the fundamental invariants.* In Proceedings of the Conference on Symmetry in Nonlinear Mathematical Physics, 2004, pp. 1079–1086,

[7] R. DELONG, JR., *Killing Tensors and the Hamilton-Jacobi Equation,* PhD thesis, University of Minnesota, 1982.

[8] M. EASTWOOD, *Representations via overdetermined systems,* Contemp. Math., AMS (2005), **368**: 201–210.

[9] M. EASTWOOD, *Higher symmetries of the Laplacian,* Ann. of Math. (2005), **161**: 1645–1665.

[10] L. EISENHART, *Separable systems of Stäckel,* Ann. of Math. (1934), **35**: 284–305.

[11] L. FATIBENE AND M. FRANCAVIGLIA, *Natural and Gauge Natural Formalism for Classical Field Theories. A Geometric Perspective Including Spinors and Gauge Theories,* Kluwer Academic Publishers, Dordrecht, 2003.

[12] M. Fels and P. Olver, *Moving coframes. I. A practical algorithm*, Acta. Appl. Math. (1998), **51**: 161–213.

[13] M. Fels and P. Olver, *Moving coframes. II. Regularization and theoretical foundations*, Acta. Appl. Math. (1999), **55**: 127–208.

[14] I. Friš, V. Mandrosov, Ya. Smorodinsky, M. Uhlíř, and P. Winternitz, *On higher order symmetries in quantum mechanics*, Phys. Lett. (1965), **16**: 354–356.

[15] P. Griffiths, *On Cartan's method of Lie groups and moving frames as applied to uniqueness and existence questions in differential geometry*, Duke Math. J. (1974), **41**: 775–814.

[16] J. Horwood, R. McLenaghan, R. Smirnov, and D. The, *Fundamental covariants in the invariant theory of Killing tensors*. In Proceedings of the Conference Symmetry and Perturbation Theory - SPT2004 (Cala Gonone, May 30–June 6, 2004), World Scientific, 2005, pp. 124–131.

[17] J. Horwood, R. McLenaghan, and R. Smirnov, *Invariant classification of orthogonally separable Hamiltonians systems in Euclidean space*, Comm. Math. Phys. (2005), **259**: 679–709.

[18] T. Ivey and J. Landsberg, *Cartan for Beginners: Differential Geometry via Moving Frames and Exterior Differential Forms*, AMS, Providence, 2003.

[19] E. Kalnins, J. Kress, and W. Miller, Jr., *Second-order superintegrable systems in conformally flat spaces. I. Two-dimensional classical structure theory*, J. Math. Phys. (2005), **46**, 053509, 28 pp.

[20] E. Kalnins, J. Kress, and W. Miller, Jr., *Second order superintegrable systems in conformally flat spaces. II. The classical two-dimensional Stckel transform*, J. Math. Phys. (2005), **46**, 053510, 15 pp.

[21] E. Kalnins, J. Kress, and W. Miller, Jr., *Second order superintegrable systems in conformally flat spaces. III. Three-dimensional classical structure theory*, J. Math. Phys. (2005), **46**, 103507, 28 pp.

[22] E. Kalnins, J. Kress, and W. Miller, Jr., *Second order superintegrable systems in conformally flat spaces. IV. The classical 3D Stäckel transform and 3D classification theory*, J. Math. Phys. (2006), **47**, 043514, 28 pp.

[23] E. Kalnins, J. Kress, and W. Miller, Jr., *Second order superintegrable systems in conformally flat spaces. V: 2D and 3D quantum systems*, J. Math. Phys.(2006), **47**, 093501, 25 pp.

[24] J. MacArthur, *The Equivalence Problem in Differential Geometry*, MSc thesis, Dalhousie University, 2005.

[25] A. Makarov, Ya. Smorodinsky, Kh. Valiev, and P. Winternitz, *A systematic approach for nonrelativistic systems with dynamical symmetries*, Nuovo Cim. (1967), **52**: 1061–1084.

[26] R. McLenaghan, R. Smirnov, and D. The, *Group invariant classification of separable Hamiltonian systems in the Euclidean plane and the $O(4)$-symmetric Yang-Mills theories of Yatsun*, J. Math. Phys. (2002), **43**: 1422–1440.

[27] R. McLenaghan, R. Smirnov, and D. The, *Group invariants of Killing tensors in the Minkowski plane*. In the Proceedings of the Conference on Symmetry and Perturbation Theory - SPT2002 (Cala Gonone, May 19–26, 2002), World Scientific, 2002, pp. 153–161.

[28] R. McLenaghan, R. Smirnov and D. The, *Group invariant classification of orthogonal coordinate webs*, In the Proceedings of Recent Advances in Riemannian and Lorentzian Geometries (Baltimore, MD, 2003), 109–120, Contemp. Math. **337**, AMS, Providence, RI, 2003, pp. 109–120.

[29] R. McLenaghan, R. Smirnov, and D. The, *An extension of the classical theory of algebraic invariants to pseudo-Riemannian geometry and Hamiltonian mechanics*, J. Math. Phys. (2004), **45**: 1079–1120.

[30] R. McLenaghan, R. Milson and R. Smirnov, *Killing tensors as irreducible representations of the general linear group*, C. R. Math. Acad. Sci. Paris (2004), **339**: 621–624.

[31] R. McLenaghan, R. Smirnov, and D. The, *Towards a classification of cubic integrals of motion. Superintegrability in classical and quantum systems,* CRM Proc. Lecture Notes **37**, AMS Providence, RI, 2004, pp. 199–209.

[32] A. Nikitin and O. Prylypko, *Generalized Killing tensors and symmetry of Klein-Gordon equations*, www.arxiv.org/abs/math-ph/0506002, 1990.

[33] P. Olver, *Classical Invariant Theory*, Cambridge University Press, 1999.

[34] J. Schouten, *Über Differentalkomitanten zweier kontravarianter Grössen,* Proc. Kon. Ned. Akad. Amsterdam (1940), **43**: 449–452.

[35] R. Sharpe, *Differential Geometry. Cartan's Generalization of Klein's Erlangen Program,* Springer-Verlag, NY, 1996.

[36] R. Smirnov and J. Yue, *Covariants, joint invariants and the problem of equivalence in the invariant theory of Killing tenosors defined in pseudo-Riemannian spaces of constant curvature,* J. Math. Phys. (2004), **45**: 4141–4163.

[37] R. Smirnov and J. Yue, *A moving frames technique and the invariant theory of Killing tensors.* In Proceedings of the 9th International Conference on Differential Geometry and its Applications (Prague, August 30–September 3, 2004) (2005), pp. 549–558.

[38] R. Smirnov, *The classical Bertrand-Darboux problem,* `www.arxiv.org`: math-ph/0604038, to appear in Fund. Appl. Math. (in Russian), 2006.

[39] R. Smirnov and P. Winternitz, *A class of superintegrable potentials of Calogero type,* J. Math. Phys. . (2006), **47**, 093505, 8 pp.

[40] M. Takeuchi, *Killing tensor fields on spaces of constant curvature,* Tsukuba J. Math. (1983), **7**: 233–255.

[41] G. Thompson, *Killing tensors in spaces of constant curvature,* J. Math. Phys. (1986), **27**: 2693–2699.

[42] H. Weyl, *The Classical Groups. Their Invariants and Representations,* Princeton University Press, Princeton, 2nd ed., 1946.

[43] P. Winternitz and I. Friš, *Invariant expansions of relativistic amplitudes and subgroups of the proper Lorenz group*, Soviet J. Nuclear Phys. (1965), **1**: 636–643.

[44] J. Yue, *The 1856 lemma of Cayley revisited. I. Infinitesimal generators,* J. Math. Phys. (2005), **46**, 073511, 15 pp., 53C50 (53C20).

[45] J. Yue, *Development of the Invariant Theory of Killing Tensors Defined in Pseudo-Riemannian Spaces of Constant Curvature,* PhD thesis, Dalhousie University, 2005.

HAMILTONIAN CURVE FLOWS IN LIE GROUPS $G \subset U(N)$ AND VECTOR NLS, mKdV, SINE-GORDON SOLITON EQUATIONS

STEPHEN C. ANCO*

Abstract. A bi-Hamiltonian hierarchy of complex vector soliton equations is derived from geometric flows of non-stretching curves in the Lie groups $G = SO(N+1)$, $SU(N) \subset U(N)$, generalizing previous work on integrable curve flows in Riemannian symmetric spaces $G/SO(N)$. The derivation uses a parallel frame and connection along the curves, involving the Klein geometry of the group G. This is shown to yield the two known $U(N-1)$-invariant vector generalizations of both the nonlinear Schrödinger (NLS) equation and the complex modified Korteweg-de Vries (mKdV) equation, as well as $U(N-1)$-invariant vector generalizations of the sine-Gordon (SG) equation found in recent symmetry-integrability classifications of hyperbolic vector equations. The curve flows themselves are described in explicit form by chiral wave maps, chiral variants of Schrödinger maps, and mKdV analogs.

Key words. bi-Hamiltonian, soliton equation, recursion operator, Lie group, curve flow, wave map, Schrödinger map, mKdV map.

AMS(MOS) subject classifications. 37K05,37K10,37K25,35Q53,53C35.

1. Introduction. The theory of integrable partial differential equations has many deep links to the differential geometry of curves and surfaces. For instance the famous sine-Gordon (SG) and modified Korteveg-de Vries (mKdV) soliton equations along with their common hierarchy of symmetries, conservation laws, and associated recursion operators all can be encoded in geometric flows of non-stretching curves in Euclidean plane geometry [1, 2] by looking at the induced flow equation of the curvature invariant of such curves. A similar encoding is known to hold [3] in spherical geometry.

Recent work [4] has significantly generalized this geometric origin of fundamental soliton equations to encompass vector versions of mKdV and SG equations by considering non-stretching curve flows in Riemannian symmetric spaces of the form $M = G/SO(N)$ for $N \geq 2$. (Here G represents the isometry group of M, and $SO(N)$ acts as a gauge group for the frame bundle of M, such that $SO(N) \subset G$ is an invariant subgroup under an involutive automorphism of G.) Such spaces [5] are exhausted by the groups $G = SO(N+1), SU(N)$ and describe curved G-invariant geometries that are a natural generalization of Euclidean spaces. In particular, for $N = 2$, the local isomorphism $SO(3) \simeq SU(2)$ implies both of these spaces are isometric to the standard 2-sphere geometry, $S^2 \simeq G/SO(2)$.

As main results in [4], it was shown firstly that there is a geometric encoding of $O(N-1)$-invariant bi-Hamiltonian operators in the Cartan

*Department of Mathematics, Brock University, St. Catharines, ON L2S 3A1, CANADA (sanco@brocku.ca).

structure equations for torsion and curvature of a moving parallel frame and its associated frame connection 1-form for non-stretching curves in the spaces $G/SO(N)$ viewed as Klein geometries [6]. The group $O(N-1)$ here arises as the isotropy subgroup in the gauge group $SO(N)$ preserving the parallel property of the moving frame. Secondly, this bi-Hamiltonian structure generates a hierarchy of integrable flows of curves in which the frame components of the principal normal along the curve satisfy $O(N-1)$-invariant vector soliton equations related by a hereditary recursion operator. These normal components in a parallel moving frame have the geometrical meaning of curvature covariants of curves relative to the isotropy group $O(N-1)$. Thirdly, the two isometry groups $G = SO(N+1), SU(N)$ were shown to give different hierarchies whose $O(N-1)$-invariant vector evolution equations of lowest-order are precisely the two known vector versions of integrable mKdV equations found in the symmetry-integrability classifications presented in [7]. In addition these hierarchies were shown to also contain $O(N-1)$-invariant vector hyperbolic equations given by two different vector versions of integrable SG equations that are known from a recent generalization of symmetry-integrability classifications to the hyperbolic case [8]. Finally, the geometric curve flows corresponding to these vector SG and mKdV equations in both hierarchies were found to be described by wave maps and mKdV analogs of Schrodinger maps into the curved spaces $SO(N+1)/SO(N)$, $SU(N)/SO(N)$.

The present paper extends the same analysis to give a geometric origin for $U(N-1)$-invariant vector soliton equations and their bi-Hamiltonian integrability structure from considering flows of non-stretching curves in the Lie groups $G = SO(N+1), SU(N) \subset U(N)$. A main idea will be to view these Lie groups as Klein geometries carrying the structure of a Riemannian symmetric space [5] given by $G \simeq G \times G/\operatorname{diag}(G \times G) = M$ for $N \geq 2$ (with G thus representing both the isometry group of M as well as the gauge group for the frame bundle of M). Note for $N = 2$ both these spaces locally describe a 3-sphere geometry, $S^3 \simeq SO(3) \simeq SU(2)$.

In this setting the parallel moving frame formulation of non-stretching curves developed in [4] for the Riemannian symmetric spaces $SO(N+1)/SO(N)$, $SU(N)/SO(N)$ can be applied directly to the Lie groups $SO(N+1)$, $SU(N)$ themselves, where the isotropy subgroup $O(N-1) \subset SO(N)$ of such frames is replaced by $U(N-1) \subset SU(N)$ and $U(1) \times O(N-1) \subset SO(N+1)$ in the two respective cases. As a result, it will be shown that the frame structure equations geometrically encode $U(N-1)$-invariant bi-Hamiltonian operators that generate a hierarchy of integrable flows of curves in both spaces $G = SO(N+1), SU(N)$. Moreover, the frame components of the principal normal along the curves in the two hierarchies will be seen to satisfy $U(N-1)$-invariant vector soliton equations that exhaust the two known vector versions of integrable NLS equations and corresponding complex vector versions of integrable mKdV equations, as well as the two known complex vector versions of integrable

SG equations, found respectively in the symmetry-integrability classifications stated in [7] and [8]. Lastly, the geometric curve flows arising from these vector SG, NLS, and mKdV equations in both hierarchies will be shown to consist of chiral wave maps, chiral variants of Schrodinger maps and their mKdV analogs, into the curved spaces $SO(N+1)$, $SU(N)$.

Taken together, the results here and in [4] geometrically account for the existence and the bi-Hamiltonian integrability structure of all known vector generalizations of NLS, mKdV, SG soliton equations.

Related work [9–12] has obtained geometric derivations of the KdV equation and other scalar soliton equations along with their hereditary integrability structure from non-stretching curve flows in planar Klein geometries (which are group-theoretic generalizations of the Euclidean plane such that the Euclidean group is replaced by a different isometry group acting locally and effectively on $\mathbb{R}^2$).

Previous results on integrable vector NLS and mKdV equations geometrically associated to Lie groups in the Riemannian case appeared in [13–16, 3]. Earlier work on deriving vector SG equations from Riemannian symmetric spaces and Lie groups can be found in [17, 18, 3]. The basic idea of studying curve flows via parallel moving frames appears in [19–21].

2. Curve flows and parallel frames. Compact semisimple Lie groups G are well-known to have the natural structure of a Riemannian manifold whose metric tensor g arises in a left-invariant fashion [22] from the Cartan-Killing inner product $<\cdot,\cdot>_{\mathfrak{g}}$ on the Lie algebra $\mathfrak{g}$ of G. This structure can be formulated in an explicit way by the introduction of a left-invariant orthonormal frame e_a on G, satisfying the commutator property $[e_a, e_b] = c_{ab}{}^c e_c$ where $c_{ab}{}^c$ denotes the structure constants of $\mathfrak{g}$, and frame indices a, b, etc. run $1, \ldots, n$ where $n = \dim G$. The Riemannian metric tensor g on G is then given by

$$g(e_a, e_b) = -\frac{1}{2} c_{ac}{}^d c_{bd}{}^c = \delta_{ab} \tag{2.1}$$

while

$$R_c{}^d(e_a, e_b) = c_{ab}{}^f c_{cf}{}^d \tag{2.2}$$

yields the Riemannian curvature tensor of G expressed as a linear map $[{}^g\nabla, {}^g\nabla] = R(\cdot,\cdot)$. The frame vectors e_a also determine connection 1-forms

$$\omega^{ab} = c^{ab}{}_c e^c \tag{2.3}$$

in terms of coframe 1-forms e^a dual to e_a obeying the standard frame structure equations [22]

$${}^g\nabla e^a = \omega_b{}^a \otimes e^b, \tag{2.4}$$

$$[{}^g\nabla, {}^g\nabla] e^a = R_b{}^a(\cdot,\cdot) e^b, \tag{2.5}$$

with

$$R_b{}^a(\cdot,\cdot) = d\omega_b{}^a + \omega_b{}^c \wedge \omega_c{}^a \tag{2.6}$$

where d denotes the exterior total derivative on G and ${}^g\nabla$ denotes the Riemannian covariant derivative on G. Note that frame indices are raised and lowered using $\delta_{ab} = \mathrm{diag}(+1,\dots,+1)$.

Now let $\gamma(t,x)$ be a flow of a non-stretching curve in G. Write $Y = \gamma_t$ for the evolution vector of the curve and write $X = \gamma_x$ for the tangent vector along the curve normalized by $g(X,X) = 1$, which is the condition for γ to be non-stretching, so thus x represents arclength. Suppose e_a is a moving parallel frame [23] along the curve γ. Specifically, in the two-dimensional tangent space $T_\gamma M$ of the flow, e_a is assumed to be adapted to γ via

$$e^a := X \ (a=1), \quad (e^a)_\perp \ (a=2,\dots,n) \tag{2.7}$$

where $g(X,(e^a)_\perp) = 0$, such that the covariant derivative of each of the $n-1$ normal vectors $(e^a)_\perp$ in the frame is tangent to γ,

$${}^g\nabla_x (e^a)_\perp = -v^a X \tag{2.8}$$

holding for some functions v^a, while the covariant derivative of the tangent vector X in the frame is normal to γ,

$${}^g\nabla_x X = v^a (e_a)_\perp. \tag{2.9}$$

Equivalently, in the notation of [4], the components of the connection 1-forms of the parallel frame along γ are given by the skew matrix

$$\omega_{xa}{}^b := X \lrcorner \omega_a{}^b = e_{xa} v^b - e_x{}^b v_a = \begin{pmatrix} 0 & v^b \\ -v_a & \mathbf{0} \end{pmatrix} \tag{2.10}$$

where

$$e_x{}^a := g(X, e^a) = (1, \vec{0}) \tag{2.11}$$

is the row matrix of the frame in the tangent direction. (Throughout, upper/lower frame indices will represent row/column matrices.) This description gives a purely algebraic characterization [4] of a parallel frame: $e_x{}^a$ is a fixed unit vector in $\mathbb{R}^n \simeq T_x G$ preserved by a $SO(n-1)$ rotation subgroup of the local frame structure group $SO(n)$ with G viewed as being a n-dimensional Riemannian manifold, while $\omega_{xa}{}^b$ belongs to the orthogonal complement of the corresponding rotation subalgebra $\mathfrak{so}(n-1)$ in the Lie algebra $\mathfrak{so}(n)$ of $SO(n)$.

However, taking into account the left-invariance property (2.3), note $\omega_x{}^{ab} = c^{ab}{}_c e_x{}^c$ and consequently $c^{ab}{}_c = 2\delta_c{}^{[a} v^{b]}$ which implies degeneracy of the structure constants, namely $c^{ab}{}_c v^c = 0$ and $c_{[abc]} = 0$. But such conditions are impossible in a non-abelian semisimple Lie algebra [24], and hence there do not exist parallel frames that are left-invariant. This difficulty can be by-passed if moving parallel frames are introduced in a setting that uses the structure of G as a Klein geometry rather than a left-invariant Riemannian manifold, which will relax the precondition for parallel frames to be left-invariant.

The Klein geometry of a compact semisimple Lie group G is given by [6, 22] the Riemannian symmetric space $M = K/H \simeq G$ for $K = G \times G \supset H = \operatorname{diag} K \simeq G$ in which K is regarded as a principle G bundle over M. Note H is a Lie subgroup of K invariant under an involutive automorphism σ given by a permutation of the factors G in K. The Riemannian structure of M is isomorphic with that of G itself. In particular, under the canonical mapping of G into $K/H \simeq G$, the Riemannian curvature tensor and metric tensor on M pull back to the standard ones $R(\cdot,\cdot)$ and g on G, both of which are covariantly constant and G-invariant. The primary difference in regarding G as a Klein geometry is that its frame bundle [6] will naturally have G for the gauge group, which is a reduction of the $SO(n)$ frame bundle of G as a purely Riemannian manifold. [1]

In the same manner as for the Klein geometries considered in [4], the frame structure equations for non-stretching curve flows in the space $M = K/H \simeq G$ can be shown to directly encode a bi-Hamiltonian structure based on geometrical variables, utilizing a moving parallel frame combined with the property of the geometry of M that its frame curvature matrix is constant on M. In addition the resulting bi-Hamiltonian structure will be invariant under the isotropy subgroup of H that preserves the parallel property of the frame thereby leaving X invariant. Since in the present work we are seeking $U(N-1)$-invariant bi-Hamiltonian operators, the simplest situation will be to have $U(N-1) \subset SU(N) = H$. Hence we first will consider the Klein geometry $M = K/H \simeq SU(N)$ given by the Lie group $G = SU(N)$.

The frame bundle structure of this space $M = K/H \simeq SU(N)$ is tied to a zero-curvature connection 1-form ω_K called the Cartan connection [6] which is identified with the left-invariant $\mathfrak{k}$-valued Maurer-Cartan form on the Lie group $K = SU(N) \times SU(N)$. Here $\mathfrak{k} = \mathfrak{su}(N) \oplus \mathfrak{su}(N)$ is the Lie algebra of K and $\mathfrak{h} = \operatorname{diag}\mathfrak{k} \simeq \mathfrak{su}(N)$ is the Lie subalgebra in $\mathfrak{k}$ corresponding to the gauge group $H = \operatorname{diag}(SU(N) \times SU(N)) \simeq SU(N)$ in K. The involutive automorphism σ of K induces the decomposition $\mathfrak{k} = \mathfrak{p} \oplus \mathfrak{h}$ where $\mathfrak{h}$ and $\mathfrak{p}$ are respective eigenspaces $\sigma = \pm 1$ in $\mathfrak{k}$, where σ permutes the $\mathfrak{su}(N)$ factors of $\mathfrak{k}$. The subspace $\mathfrak{p} \simeq \mathfrak{su}(N)$ has the Lie bracket relations

$$[\mathfrak{p}, \mathfrak{p}] \subset \mathfrak{h} \simeq \mathfrak{su}(N), \quad [\mathfrak{h}, \mathfrak{p}] \simeq [\mathfrak{su}(N), \mathfrak{p}] \subset \mathfrak{p}. \tag{2.12}$$

[1] More details will be given elsewhere [25].

In particular there is a natural action of $\mathfrak{h} \simeq \mathfrak{su}(N)$ on $\mathfrak{p}$. To proceed, in the group K regarded as a principal $SU(N)$ bundle over M, fix any local section and pull-back ω_K to give a $\mathfrak{k}$-valued 1-form ${}^{\mathfrak{k}}\omega$ at x in M. The effect of changing the local section is to induce a $SU(N)$ gauge transformation on ${}^{\mathfrak{k}}\omega$. We now decompose ${}^{\mathfrak{k}}\omega$ with respect to σ. It can be shown that [6] the symmetric part

$$\omega := \frac{1}{2}{}^{\mathfrak{k}}\omega + \frac{1}{2}\sigma({}^{\mathfrak{k}}\omega) \tag{2.13}$$

defines a $\mathfrak{su}(N)$-valued connection 1-form for the group action of $SU(N)$ on the tangent space $T_xM \simeq \mathfrak{p}$, while the antisymmetric part

$$e := \frac{1}{2}{}^{\mathfrak{k}}\omega - \frac{1}{2}\sigma({}^{\mathfrak{k}}\omega) \tag{2.14}$$

defines a $\mathfrak{p}$-valued coframe for the Cartan-Killing inner product $< \cdot, \cdot >_{\mathfrak{p}}$ on $T_xK \simeq \mathfrak{k}$ restricted to $T_xM \simeq \mathfrak{p}$. This inner product [2] provides a Riemannian metric

$$g = < e \otimes e >_{\mathfrak{p}} \tag{2.15}$$

on $M \simeq SU(N)$, such that the squared norm of any vector $X \in T_xM$ is given by $|X|_g^2 = g(X,X) = < X \lrcorner e, X \lrcorner e >_{\mathfrak{p}}$. Note e and ω will be left-invariant with respect to the group action of $H \simeq SU(N)$ if and only if the local section of the $SU(N)$ bundle K used to define the 1-form ${}^{\mathfrak{k}}\omega$ is a left-invariant function. In particular, if $h : M \to H$ is an $SU(N)$ gauge transformation relating a left-invariant section to an arbitrary local section of K then $\tilde{e} = \mathrm{Ad}(h^{-1})e$ and $\tilde{\omega} = \mathrm{Ad}(h^{-1})\omega + h^{-1}dh$ will be a left-invariant coframe and connection on $M \simeq SU(N)$.

Moreover, associated to this structure provided by the Maurer-Cartan form, there is a $SU(N)$-invariant covariant derivative ∇ whose restriction to the tangent space $T_\gamma M$ for any curve flow $\gamma(t,x)$ in $M \simeq SU(N)$ is defined via

$$\nabla_x e = [e, \gamma_x \lrcorner \omega] \qquad \text{and} \qquad \nabla_t e = [e, \gamma_t \lrcorner \omega]. \tag{2.16}$$

These covariant derivatives obey the Cartan structure equations obtained from a decomposition of the zero-curvature equation of the Maurer-Cartan form

$$0 = d\omega_K + \frac{1}{2}[\omega_K, \omega_K]. \tag{2.17}$$

Namely ∇_x, ∇_t have zero torsion

$$0 = (\nabla_x \gamma_t - \nabla_t \gamma_x) \lrcorner e = D_x e_t - D_t e_x + [\omega_x, e_t] - [\omega_t, e_x] \tag{2.18}$$

[2]The sign convention that $< \cdot, \cdot >_{\mathfrak{p}}$ is positive-definite will be used for convenience.

and carry $SU(N)$-invariant curvature

$$R(\gamma_x,\gamma_t)e = [\nabla_x,\nabla_t]e = D_x\omega_t - D_t\omega_x + [\omega_x,\omega_t] \\ = -[e_x,e_t] \tag{2.19}$$

where

$$e_x := \gamma_x \lrcorner e, \qquad e_t := \gamma_t \lrcorner e, \qquad \omega_x := \gamma_x \lrcorner \omega, \qquad \omega_t := \gamma_t \lrcorner \omega. \tag{2.20}$$

REMARK 2.1. *The soldering relations (2.13) and (2.14) together with the canonical identifications $\mathfrak{p} \simeq \mathfrak{su}(N)$ and $\mathfrak{h} \simeq \mathfrak{su}(N)$ provide an isomorphism between the Klein geometry of $M \simeq SU(N)$ and the Riemannian geometry of the Lie group $G = SU(N)$. This isomorphism allows e and ω to be regarded hereafter as an $\mathfrak{su}(N)$-valued coframe and its associated $\mathfrak{su}(N)$-valued connection 1-form introduced on $G = SU(N)$ itself, without the property of left-invariance.*

Geometrically, it thus follows that e_x and ω_x represent the tangential part of the coframe and the connection 1-form along γ. For a non-stretching curve γ, where x is the arclength, note e_x has unit norm in the inner product, $< e_x, e_x >_{\mathfrak{p}} = 1$. This implies $\mathfrak{p} \simeq \mathfrak{su}(N)$ has a decomposition into tangential and normal subspaces $\mathfrak{p}_{\parallel}$ and $\mathfrak{p}_{\perp}$ with respect to e_x such that $< e_x, \mathfrak{p}_{\perp} >_{\mathfrak{p}} = 0$, with $\mathfrak{p} = \mathfrak{p}_{\perp} \oplus \mathfrak{p}_{\parallel} \simeq \mathfrak{su}(N)$ and $\mathfrak{p}_{\parallel} \simeq \mathbb{R}$.

The isotropy subgroup in $H \simeq SU(N)$ preserving e_x is clearly the unitary group $U(N-1) \subset SU(N)$ acting on $\mathfrak{p} \simeq \mathfrak{su}(N)$. This motivates an algebraic characterization of a parallel frame [4] defined by the property that ω_x belongs to the orthogonal complement of the $U(N-1)$ unitary rotation Lie subalgebra $\mathfrak{u}(N-1)$ contained in the Lie algebra $\mathfrak{su}(N)$ of $SU(N)$, in analogy to the Riemannian case. Its geometrical meaning, however, generalizes the Riemannian properties (2.8) and (2.9), as follows. Let e_a be a frame whose dual coframe is identified with the $\mathfrak{su}(N)$-valued coframe e in a fixed orthonormal basis for $\mathfrak{p} \simeq \mathfrak{su}(N)$. Decompose the coframe into parallel/perpendicular parts with respect to e_x in an algebraic sense as defined by the kernel/cokernel of Lie algebra multiplication $[e_x, \cdot\ ]_{\mathfrak{p}} = \mathrm{ad}(e_x)$. Thus we have $e = (e_C, e_{C^\perp})$ where the $\mathfrak{su}(N)$-valued covectors $e_C, e_{C^\perp}$ satisfy $[e_x, e_C]_{\mathfrak{p}} = 0$, and $e_{C^\perp}$ is orthogonal to e_C, so $[e_x, e_{C^\perp}]_{\mathfrak{p}} \neq 0$. Note there is a corresponding algebraic decomposition of the tangent space $T_xG \simeq \mathfrak{su}(N) \simeq \mathfrak{p}$ given by $\mathfrak{p} = \mathfrak{p}_C \oplus \mathfrak{p}_{C^\perp}$, with $\mathfrak{p}_{\parallel} \subseteq \mathfrak{p}_C$ and $\mathfrak{p}_{C^\perp} \subseteq \mathfrak{p}_{\perp}$, where $[\mathfrak{p}_{\parallel}, \mathfrak{p}_C] = 0$ and $< \mathfrak{p}_{C^\perp}, \mathfrak{p}_C >_{\mathfrak{p}} = 0$, so $[\mathfrak{p}_{\parallel}, \mathfrak{p}_{C^\perp}] \neq 0$ (namely, $\mathfrak{p}_C$ is the centralizer of e_x in $\mathfrak{p} \simeq \mathfrak{su}(N)$). This decomposition is preserved by $\mathrm{ad}(\omega_x)$ which acts as an infinitesimal unitary rotation, $\mathrm{ad}(\omega_x)\mathfrak{p}_C \subseteq \mathfrak{p}_{C^\perp}$, $\mathrm{ad}(\omega_x)\mathfrak{p}_{C^\perp} \subseteq \mathfrak{p}_C$. Hence, from Equation (2.16), the derivative ∇_x of a covector perpendicular (respectively parallel) to e_x lies parallel (respectively perpendicular) to e_x, namely $\nabla_x e_C$ belongs to $\mathfrak{p}_{C^\perp}$, $\nabla_x e_{C^\perp}$ belongs to $\mathfrak{p}_C$. In the Riemannian setting, these properties correspond to ${}^g\nabla_x(e^a)_C = v^a{}_b(e^b)_{C^\perp}$, ${}^g\nabla_x(e^a)_{C^\perp} = -v_b{}^a(e^b)_C$ for some functions $v^{ab} = -v^{ba}$, without the left-invariance property (2.3). Such a frame will be called $SU(N)$*-parallel* and defines a strict generalization of a Riemannian parallel frame whenever $\mathfrak{p}_C$ is larger than $\mathfrak{p}_{\parallel}$.

It should be noted that the existence of a $SU(N)$-parallel frame for curve flows in the Klein geometry $M = K/H \simeq SU(N)$ is guaranteed by the $SU(N)$ gauge freedom on e and ω inherited from the local section of $K = SU(N) \times SU(N)$ used to pull back the Maurer-Cartan form to M.

All these developments carry over to the Lie group $G = SO(N+1)$ viewed as a Klein geometry $M = K/H \simeq SO(N+1)$ for $K = SO(N+1) \times SO(N+1) \supset H = \operatorname{diag} K \simeq SO(N+1)$. The only change is that the isotropy subgroup of H leaving X fixed is given by $U(1) \times O(N-1)$ which is a proper subgroup of $U(N-1)$. Nevertheless the Cartan structure equations of a $SO(N+1)$-parallel frame for non-stretching curve flows in $M \simeq SO(N+1)$ will actually turn out to exhibit a larger invariance under unitary rotations $U(N-1)$.

REMARK 2.2. *We will set up parallel frames for curve flows in the Lie groups $G = SU(N), SO(N+1)$ using the same respective choice of unit vector e_x in $\mathfrak{g}/\mathfrak{so}(N) \subset \mathfrak{g} = \mathfrak{su}(N), \mathfrak{so}(N+1)$ as was made in [4] for curve flows in the corresponding symmetric spaces $G/SO(N)$.*

3. Bi-Hamiltonian operators and vector soliton equations for $SU(N)$. Let $\gamma(t,x)$ be a flow of a non-stretching curve in $G = SU(N)$. We consider a $SU(N)$-parallel coframe $e \in T^*_\gamma G \otimes \mathfrak{su}(N)$ and its associated connection 1-form $\omega \in T^*_\gamma G \otimes \mathfrak{su}(N)$ along γ [3] given by

$$\omega_x = \gamma_x \lrcorner \omega = \begin{pmatrix} 0 & \vec{v} \\ -\vec{v}^{\dagger} & \mathbf{0} \end{pmatrix} \in \mathfrak{p}_{C^{\perp}}, \qquad \vec{v} \in \mathbb{C}^{N-1}, \qquad \mathbf{0} \in \mathfrak{u}(N-1) \quad (3.1)$$

and

$$\begin{gathered} e_x = \gamma_x \lrcorner e = \kappa \tfrac{\mathrm{i}}{N} \begin{pmatrix} 1-N & \vec{0} \\ \vec{0}^T & \mathbf{1} \end{pmatrix} \in \mathfrak{p}_{\parallel}, \\ \vec{0} \in \mathbb{R}^{N-1}, \qquad \mathrm{i}\mathbf{1} \in \mathfrak{u}(N-1) \end{gathered} \quad (3.2)$$

up to a normalization factor κ fixed as follows. Note the form of e_x indicates that the coframe is adapted to γ provided e_x is scaled to have unit norm in the Cartan-Killing inner product,

$$< e_x, e_x >_{\mathfrak{p}} = -\frac{1}{2} \operatorname{tr} \left(\kappa^2 \begin{pmatrix} N^{-1} - 1 & 0 \\ 0 & N^{-1}\mathbf{1} \end{pmatrix}^2 \right) = \kappa^2 \frac{N-1}{2N} = 1 \quad (3.3)$$

by putting $\kappa^2 = 2N(N-1)^{-1}$. As a consequence, all $\mathfrak{su}(N)$ matrices will have a canonical decomposition into tangential and normal parts relative to e_x,

[3] Note ω is related to e by the Riemannian covariant derivative (2.16) on the surface swept out by the curve flow $\gamma(t,x)$.

$$\mathfrak{su}(N) = \begin{pmatrix} (N^{-1}-1)p_{\parallel}\mathrm{i} & \vec{p}_{\perp} \\ -\vec{p}_{\perp}{}^{\dagger} & \boldsymbol{p}_{\perp} - N^{-1}p_{\parallel}\mathrm{i}\mathbf{1} \end{pmatrix}$$
$$= \frac{\mathrm{i}}{N}\begin{pmatrix} (1-N)p_{\parallel} & \vec{0} \\ \vec{0}^T & p_{\parallel}\mathbf{1} \end{pmatrix} + \begin{pmatrix} 0 & \vec{p}_{\perp} \\ -\vec{p}_{\perp}{}^{\dagger} & \boldsymbol{p}_{\perp} \end{pmatrix} \simeq \mathfrak{p} \tag{3.4}$$

parameterized by the matrix $\boldsymbol{p}_{\perp} \in \mathfrak{su}(N-1)$, the vector $\vec{p}_{\perp} \in \mathbb{C}^{N-1}$, and the scalar $\mathrm{i}p_{\parallel} \in \mathbb{R}$, corresponding to $\mathfrak{p} = \mathfrak{p}_{\parallel} \oplus \mathfrak{p}_{\perp}$ where $\mathfrak{p}_{\parallel} \simeq \mathfrak{u}(1)$. In this decomposition the centralizer of e_x consists of matrices parameterized by $(p_{\parallel}, \boldsymbol{p}_{\perp})$ and hence $\mathfrak{p}_C \simeq \mathfrak{u}(N-1) \supset \mathfrak{p}_{\parallel} \simeq \mathfrak{u}(1)$ while its perp space $\mathfrak{p}_{C^{\perp}} \subset \mathfrak{p}_{\perp}$ is parameterized by $\vec{p}_{\perp}$. Note $\boldsymbol{p}_{\perp}$ is empty only if $N=2$, so consequently for $N>2$ the $SU(N)$-parallel frame (3.1) and (3.2) is a strict generalization of a Riemannian parallel frame.

In the flow direction we put

$$e_t = \gamma_t \lrcorner e = \kappa h_{\parallel}\frac{\mathrm{i}}{N}\begin{pmatrix} 1-N & \vec{0} \\ \vec{0}^T & \mathbf{1} \end{pmatrix} + \kappa\begin{pmatrix} 0 & \vec{h}_{\perp} \\ -\vec{h}_{\perp}{}^{\dagger} & \boldsymbol{h}_{\perp} \end{pmatrix} \in \mathfrak{p}_{\parallel} \oplus \mathfrak{p}_{\perp}$$
$$= \kappa\begin{pmatrix} (N^{-1}-1)h_{\parallel}\mathrm{i} & \vec{h}_{\perp} \\ -\vec{h}_{\perp}{}^{\dagger} & \boldsymbol{h}_{\perp} + N^{-1}h_{\parallel}\mathrm{i}\mathbf{1} \end{pmatrix} \tag{3.5}$$

and

$$\omega_t = \gamma_t \lrcorner \omega = \begin{pmatrix} -\mathrm{i}\theta & \vec{\varpi} \\ -\vec{\varpi}^{\dagger} & \boldsymbol{\Theta} \end{pmatrix} \in \mathfrak{p}_C \oplus \mathfrak{p}_{C^{\perp}}, \tag{3.6}$$

with

$$h_{\parallel} \in \mathbb{R}, \qquad \vec{h}_{\perp} \in \mathbb{C}^{N-1}, \qquad \boldsymbol{h}_{\perp} \in \mathfrak{su}(N-1),$$
$$\vec{\varpi} \in \mathbb{C}^{N-1}, \qquad \boldsymbol{\Theta} \in \mathfrak{u}(N-1), \qquad \theta = -\mathrm{i}\,\mathrm{tr}\boldsymbol{\Theta} \in \mathbb{R}.$$

The components $h_{\parallel}, (\vec{h}_{\perp}, \boldsymbol{h}_{\perp})$ correspond to decomposing $e_t = g(\gamma_t, \gamma_x)e_x + (\gamma_t)_{\perp} \lrcorner e_{\perp}$ into tangential and normal parts relative to e_x. We then have

$$[e_x, e_t] = -\kappa^2\mathrm{i}\begin{pmatrix} 0 & \vec{h}_{\perp} \\ \vec{h}_{\perp}{}^{\dagger} & \mathbf{0} \end{pmatrix} \in \mathfrak{p}_{C^{\perp}}, \tag{3.7}$$

$$[\omega_x, e_t] = \kappa\begin{pmatrix} \vec{h}_{\perp}\cdot\bar{\vec{v}} - \vec{v}\cdot\bar{\vec{h}}_{\perp} & \vec{v}\lrcorner\boldsymbol{h}_{\perp} + \mathrm{i}h_{\parallel}\vec{v} \\ -(\vec{v}\lrcorner\boldsymbol{h}_{\perp} + \mathrm{i}h_{\parallel}\vec{v})^{\dagger} & \bar{\vec{h}}_{\perp}\otimes\vec{v} - \bar{\vec{v}}\otimes\vec{h}_{\perp} \end{pmatrix} \in \mathfrak{p}_C \oplus \mathfrak{p}_{C^{\perp}}, \tag{3.8}$$

$$[\omega_t, e_x] = \kappa\mathrm{i}\begin{pmatrix} 0 & \vec{\varpi} \\ \vec{\varpi}^{\dagger} & \mathbf{0} \end{pmatrix} \in \mathfrak{p}_{C^{\perp}}. \tag{3.9}$$

Here $\otimes$ denotes the outer product of pairs of vectors ($1 \times N-1$ row matrices), producing $N-1 \times N-1$ matrices $\vec{A}\otimes\vec{B} = \vec{A}^T\vec{B}$, and $\lrcorner$ denotes multiplication of $N-1 \times N-1$ matrices on vectors ($1 \times N-1$ row matrices), $\vec{A}\lrcorner(\vec{B}\otimes\vec{C}) := (\vec{A}\cdot\vec{B})\vec{C}$ which is the transpose of the standard matrix product on column vectors, $(\vec{B}\otimes\vec{C})\vec{A} = (\vec{C}\cdot\vec{A})\vec{B}$.

Hence the curvature equation (2.19) reduces to

$$D_t\vec{v} - D_x\vec{\varpi} - \vec{v}\lrcorner\mathbf{\Theta} - \mathrm{i}\theta\vec{v} = -\kappa^2\mathrm{i}\vec{h}_\perp, \tag{3.10}$$

$$D_x\mathbf{\Theta} + \bar{\vec{\varpi}}\otimes\vec{v} - \bar{\vec{v}}\otimes\vec{\varpi} = 0, \tag{3.11}$$

$$\mathrm{i}D_x\theta + \vec{v}\cdot\bar{\vec{\varpi}} - \vec{\varpi}\cdot\bar{\vec{v}} = 0, \tag{3.12}$$

while the torsion equation (2.18) yields

$$(\frac{1}{N} - 1)\mathrm{i}D_x h_\parallel + \vec{h}_\perp\cdot\bar{\vec{v}} - \vec{v}\cdot\bar{\vec{h}}_\perp = 0, \tag{3.13}$$

$$D_x\boldsymbol{h}_\perp + \bar{\vec{h}}_\perp\otimes\vec{v} - \bar{\vec{v}}\otimes\vec{h}_\perp - \frac{1}{N-1}(\vec{v}\cdot\bar{\vec{h}}_\perp - \vec{h}_\perp\cdot\bar{\vec{v}})\mathbf{1} = 0, \tag{3.14}$$

$$\mathrm{i}\vec{\varpi} - D_x\vec{h}_\perp - \mathrm{i}h_\parallel\vec{v} - \vec{v}\lrcorner\boldsymbol{h}_\perp, = 0. \tag{3.15}$$

To proceed, we use Equations (3.11)–(3.14) to eliminate

$$\mathbf{\Theta} = D_x^{-1}(\bar{\vec{v}}\otimes\vec{\varpi} - \bar{\vec{\varpi}}\otimes\vec{v}), \tag{3.16}$$

$$\theta = \mathrm{i}D_x^{-1}(\bar{\vec{\varpi}}\cdot\vec{v} - \bar{\vec{v}}\cdot\vec{\varpi}), \tag{3.17}$$

$$\boldsymbol{h}_\perp + \frac{1}{N}h_\parallel\mathrm{i}\mathbf{1} = D_x^{-1}(\bar{\vec{v}}\otimes\vec{h}_\perp - \bar{\vec{h}}_\perp\otimes\vec{v}), \tag{3.18}$$

$$(1 - \frac{1}{N})h_\parallel = \mathrm{i}D_x^{-1}(\bar{\vec{h}}_\perp\cdot\vec{v} - \bar{\vec{v}}\cdot\vec{h}_\perp), \tag{3.19}$$

in terms of the variables $\vec{v}$, $\vec{h}_\perp$, $\vec{\varpi}$. Then Equation (3.10) gives a flow on $\vec{v}$,

$$\vec{v}_t = D_x\vec{\varpi} + D_x^{-1}(\vec{\varpi}\cdot\bar{\vec{v}} - \vec{v}\cdot\bar{\vec{\varpi}})\vec{v} + \vec{v}\lrcorner D_x^{-1}(\bar{\vec{v}}\otimes\vec{\varpi} - \bar{\vec{\varpi}}\otimes\vec{v}) - \kappa^2\mathrm{i}\vec{h}_\perp \tag{3.20}$$

with

$$\vec{\varpi} = -\mathrm{i}D_x\vec{h}_\perp + \mathrm{i}D_x^{-1}(\bar{\vec{h}}_\perp\cdot\vec{v} - \bar{\vec{v}}\cdot\vec{h}_\perp)\vec{v} + \mathrm{i}\vec{v}\lrcorner D_x^{-1}(\bar{\vec{h}}_\perp\otimes\vec{v} - \bar{\vec{v}}\otimes\vec{h}_\perp) \tag{3.21}$$

obtained from Equation (3.15). We now read off the obvious operators

$$\mathcal{H} = D_x - 2\mathrm{i}D_x^{-1}(\bar{\vec{v}}\cdot^\dagger\mathrm{i}\ )\vec{v} + \vec{v}\lrcorner D_x^{-1}(\bar{\vec{v}}\wedge^\dagger\ ), \qquad \mathcal{I} = \mathrm{i}, \tag{3.22}$$

and introduce the related operator

$$\mathcal{J} = \mathcal{I}^{-1}\circ\mathcal{H}\circ\mathcal{I} = D_x + 2D_x^{-1}(\bar{\vec{v}}\cdot^\dagger\ )\vec{v} + \vec{v}\lrcorner D_x^{-1}(\bar{\vec{v}}\odot^\dagger\ ), \tag{3.23}$$

where $\vec{A}\wedge^\dagger\vec{B} := \vec{A}\otimes\vec{B} - \bar{\vec{B}}\otimes\bar{\vec{A}}$ is a Hermitian version of the wedge product $\vec{A}\wedge\vec{B} = \vec{A}\otimes\vec{B} - \vec{B}\otimes\vec{A}$, and where $\vec{A}\odot^\dagger\vec{B} := \vec{A}\otimes\vec{B} + \bar{\vec{B}}\otimes\bar{\vec{A}}$ and $\vec{A}\cdot^\dagger\vec{B} := \frac{1}{2}\vec{A}\cdot\vec{B} + \frac{1}{2}\bar{\vec{B}}\cdot\bar{\vec{A}}$ are Hermitian versions of the symmetric product $\vec{A}\odot\vec{B} = \vec{A}\otimes\vec{B} + \vec{B}\otimes\vec{A}$ and dot product $\vec{A}\cdot\vec{B}$. Note the intertwining $\vec{A}\odot^\dagger\mathrm{i}\vec{B} = \mathrm{i}(\vec{A}\wedge^\dagger\vec{B})$ and vice versa $\vec{A}\wedge^\dagger\mathrm{i}\vec{B} = \mathrm{i}(\vec{A}\odot^\dagger\vec{B})$, which imply the identities $\bar{\vec{A}}\wedge^\dagger\vec{A} = \bar{\vec{A}}\odot^\dagger\mathrm{i}\vec{A} = 0$, $\bar{\vec{A}}\cdot^\dagger\mathrm{i}\vec{A} = 0$, and $\mathrm{tr}(\bar{\vec{A}}\odot^\dagger\vec{A}) = -\mathrm{i}\,\mathrm{tr}(\bar{\vec{A}}\wedge^\dagger\mathrm{i}\vec{A}) = 2\vec{A}\cdot\bar{\vec{A}} = 2|\vec{A}|^2$.

The Hamiltonian structure determined by these operators $\mathcal{H}, \mathcal{J}, \mathcal{I}$ and variables $\vec{v}, \vec{\varpi}, \vec{h}_\perp$ in the space $G = SU(N)$ is somewhat different in comparison to the space $G/SO(N)$. Use of the methods in [21] establishes the following main result.

THEOREM 3.1. *$\mathcal{H}, \mathcal{I}$ are a Hamiltonian pair of $U(N-1)$-invariant cosymplectic operators with respect to the Hamiltonian variable $\vec{v}$, while $\mathcal{J}, \mathcal{I}^{-1} = -\mathcal{I}$ are compatible symplectic operators. Consequently, the flow equation takes the Hamiltonian form*

$$\begin{aligned} \vec{v}_t = \mathcal{H}(\vec{\varpi}) - \chi\mathcal{I}(\vec{h}_\perp) = \mathcal{R}^2(\mathrm{i}\vec{h}_\perp) - \chi\mathrm{i}\vec{h}_\perp, \\ -\vec{\varpi} = \mathcal{J}(\mathrm{i}\vec{h}_\perp) = \mathrm{i}\mathcal{H}(\vec{h}_\perp) \end{aligned} \tag{3.24}$$

where $\mathcal{R} = \mathcal{H} \circ \mathcal{I} = \mathcal{I} \circ \mathcal{J}$ is a hereditary recursion operator.

Here $\chi = \kappa^2$ is a constant related to the Riemannian scalar curvature of the space $G = SU(N)$.

On the x-jet space of $\vec{v}(t,x)$, the variables $\mathrm{i}\vec{h}_\perp$ and $\vec{\varpi}$ have the respective meaning of a Hamiltonian vector field $\mathrm{i}\vec{h}_\perp \lrcorner \partial/\partial\vec{v}$ and covector field $\vec{\varpi} \lrcorner d\vec{v}$ (see [26, 27] and the appendix of [3]). Thus the recursion operator

$$\mathcal{R} = \mathrm{i}(D_x + 2D_x^{-1}(\bar{\vec{v}}\cdot^\dagger\ )\vec{v} + \vec{v}\lrcorner D_x^{-1}(\bar{\vec{v}}\odot^\dagger\ )) \tag{3.25}$$

generates a hierarchy of commuting Hamiltonian vector fields $\mathrm{i}\vec{h}_\perp^{(k)}$ starting from $\mathrm{i}\vec{h}_\perp^{(0)} = \mathrm{i}\vec{v}$ defined by the infinitesimal generator of phase rotations on $\vec{v}$, and followed by $\mathrm{i}\vec{h}_\perp^{(1)} = -\vec{v}_x$ which is the infinitesimal generator of x-translations on $\vec{v}$ (in terms of arclength x along the curve).

The adjoint operator

$$\mathcal{R}^* = \mathrm{i}D_x + 2D_x^{-1}(\bar{\vec{v}}\cdot^\dagger\mathrm{i}\ )\vec{v} + \mathrm{i}\vec{v}\lrcorner D_x^{-1}(\bar{\vec{v}}\wedge^\dagger\ )) \tag{3.26}$$

generates a related hierarchy of involutive Hamiltonian covector fields $\vec{\varpi}^{(k)} = \delta H^{(k)}/\delta\bar{\vec{v}}$ in terms of Hamiltonians $H = H^{(k)}(\vec{v}, \vec{v}_x, \vec{v}_{2x}, \ldots)$ starting from $\vec{\varpi}^{(0)} = -\vec{v}$, $H^{(0)} = -\vec{v}\cdot\bar{\vec{v}} = -|\vec{v}|^2$, and followed by $\vec{\varpi}^{(1)} = -\mathrm{i}\vec{v}_x$, $H^{(1)} = \frac{1}{2}\mathrm{i}(\vec{v}\cdot\bar{\vec{v}}_x - \vec{v}_x\cdot\bar{\vec{v}}) = \vec{v}\cdot^\dagger\mathrm{i}\vec{v}_x$. These hierarchies are related by

$$\mathrm{i}\vec{h}_\perp^{(k+1)} = \mathcal{H}(\vec{\varpi}^{(k)}), \qquad \vec{\varpi}^{(k+1)} = -\mathcal{J}(\mathrm{i}\vec{h}_\perp^{(k)}), \qquad \vec{h}_\perp^{(k)} = -\vec{\varpi}^{(k)}, \tag{3.27}$$

for $k = 0, 1, 2, \ldots$, so thus $\mathrm{i}\vec{\varpi}^{(k)}$ can be also interpreted as a Hamiltonian vector field and $\vec{h}_\perp^{(k)}$ as a corresponding covector field. Both hierarchies share the NLS scaling symmetry $x \to \lambda x$, $\vec{v} \to \lambda^{-1}\vec{v}$, under which we see $\vec{h}_\perp^{(k)}$ and $\vec{\varpi}^{(k)}$ have scaling weight $1+k$, while $H^{(k)}$ has scaling weight $2+k$.

COROLLARY 3.1. *Associated to the recursion operator $\mathcal{R}$ there is a corresponding hierarchy of commuting bi-Hamiltonian flows on $\vec{v}$ given by $U(N-1)$-invariant vector evolution equations*

$$\vec{v}_t = \mathrm{i}(\vec{h}_\perp^{(k+2)} - \chi\vec{h}_\perp^{(k)}) = \mathcal{H}(\delta H^{(k+1,\chi)}/\delta\bar{\vec{v}}) = -\mathcal{I}(\delta H^{(k+2,\chi)}/\delta\bar{\vec{v}}), \tag{3.28}$$

with Hamiltonians $H^{(k+2,\chi)} = H^{(k+2)} - \chi H^{(k)}$, $k = 0, 1, 2, \ldots$ *In the terminology of [4],* $\vec{h}_\perp^{(k)}$ *will be said to produce a* $+(k+1)$ *flow on* $\vec{v}$.

The $+1$ flow given by $\vec{h}_\perp = \vec{v}$ yields

$$\mathrm{i}\vec{v}_t = \vec{v}_{2x} + 2|\vec{v}|^2\vec{v} + \chi\vec{v} \tag{3.29}$$

which is a vector NLS equation up to a phase term that can be absorbed by a phase rotation $\vec{v} \to \exp(-\mathrm{i}\chi t)\vec{v}$. Higher-order versions of this equation are produced by the $+(1+2k)$ odd-flows, $k \geq 1$.

The $+2$ flow is given by $\vec{h}_\perp = \mathrm{i}\vec{v}_x$, yielding a complex vector mKdV equation

$$\vec{v}_t = \vec{v}_{3x} + 3|\vec{v}|^2\vec{v}_x + 3(\vec{v}_x \cdot \bar{\vec{v}})\vec{v} + \chi\vec{v}_x \tag{3.30}$$

up to a convective term that can be absorbed by a Galilean transformation $x \to x + \chi t$, $t \to t$. The $+(2+2k)$ even-flows, $k \geq 1$, give higher-order versions of this equation.

There is also a 0 flow $\vec{v}_t = \vec{v}_x$ arising from $\vec{h}_\perp = 0$, $h_\parallel = 1$, which falls outside the hierarchy generated by $\mathcal{R}$.

All these flows correspond to geometrical motions of the curve $\gamma(t, x)$, given by

$$\gamma_t = f(\gamma_x, \nabla_x\gamma_x, \nabla_x^2\gamma_x, \ldots) \tag{3.31}$$

subject to the non-stretching condition

$$|\gamma_x|_g = 1. \tag{3.32}$$

The equation of motion for γ is obtained from the identifications $\gamma_t \leftrightarrow e_t$, $\nabla_x\gamma_x \leftrightarrow \mathcal{D}_x e_x = [\omega_x, e_x]$, and so on, using $\nabla_x \leftrightarrow D_x + [\omega_x, \cdot] = \mathcal{D}_x$. These identifications correspond to $T_xG \leftrightarrow \mathfrak{su}(N) \simeq \mathfrak{p}$ as defined via the parallel coframe along γ in $G = SU(N)$. Note we have

$$[\omega_x, e_x] = \kappa\mathrm{i}\begin{pmatrix} 0 & \vec{v} \\ \vec{v}^\dagger & \mathbf{0} \end{pmatrix}, \tag{3.33}$$

$$[\omega_x, [\omega_x, e_x]] = 2\kappa\mathrm{i}\begin{pmatrix} |\vec{v}|^2 & \vec{0} \\ \vec{0} & -\bar{\vec{v}} \otimes \vec{v} \end{pmatrix},$$

and so on. In addition,

$$\mathrm{ad}([\omega_x, e_x])e_x = -\kappa^2\begin{pmatrix} 0 & \vec{v} \\ -\vec{v}^\dagger & \mathbf{0} \end{pmatrix} = -\mathrm{ad}(e_x)[\omega_x, e_x], \tag{3.34}$$

$$\begin{aligned} \mathrm{ad}([\omega_x, e_x])^2 e_x &= 2\kappa^3\mathrm{i}\begin{pmatrix} |\vec{v}|^2 & \vec{0} \\ \vec{0} & -\bar{\vec{v}} \otimes \vec{v} \end{pmatrix} \\ &= \chi[\omega_x, [\omega_x, e_x]], \end{aligned} \tag{3.35}$$

where ad$(\cdot)$ denotes the standard adjoint representation acting in the Lie algebra $\mathfrak{su}(N)$.

For the +1 flow,

$$\vec{h}_\perp = \vec{v}, \qquad h_\parallel = 0, \qquad \boldsymbol{h}_\perp = \boldsymbol{0}, \tag{3.36}$$

we have (from Equation (3.5))

$$\begin{aligned} e_t &= \kappa \begin{pmatrix} (N^{-1}-1)h_\parallel \mathrm{i} & \vec{h}_\perp \\ -\vec{h}_\perp{}^\dagger & \boldsymbol{h}_\perp + N^{-1}h_\parallel \mathrm{i}\boldsymbol{1} \end{pmatrix} = \kappa \begin{pmatrix} 0 & \vec{v} \\ -\vec{v}^\dagger & \boldsymbol{0} \end{pmatrix} \\ &= \chi^{-1/2}\mathrm{ad}(e_x)[\omega_x, e_x] \end{aligned} \tag{3.37}$$

which we identify as $J_\gamma \nabla_x \gamma_x$ where $J_\gamma \leftrightarrow \mathrm{ad}(e_x)$ is an algebraic operator in $T_x G \leftrightarrow \mathfrak{su}(N)$ obeying $J_\gamma^2 = -\mathrm{id}$. Hence the frame equation (3.37) describes a geometric nonlinear PDE for $\gamma(t,x)$

$$\gamma_t = \chi^{-1/2} J_\gamma \nabla_x \gamma_x, \qquad J_\gamma = \mathrm{ad}(\gamma_x) \tag{3.38}$$

analogous to the well-known Schrödinger map. This PDE (3.38) will be called a *chiral Schrödinger map* equation on the Lie group $G = SU(N)$. (A different derivation was given in [3] using left-invariant frames, which needed a Lie algebraic restriction on the curve γ. Here we see any such restrictions are unnecessary, due to the use of a parallel frame.)

Next, for the +2 flow

$$\begin{aligned} \vec{h}_\perp = \mathrm{i}\vec{v}_x, \qquad h_\parallel = N(N-1)^{-1}|\vec{v}|^2, \\ \boldsymbol{h}_\perp = \mathrm{i}(\bar{\vec{v}} \otimes \vec{v} + (1-N)^{-1}|\vec{v}|^2\boldsymbol{1}), \end{aligned} \tag{3.39}$$

we obtain (again via Equation (3.5))

$$\begin{aligned} e_t &= \kappa \begin{pmatrix} (N^{-1}-1)h_\parallel \mathrm{i} & \vec{h}_\perp \\ -\vec{h}_\perp{}^\dagger & \boldsymbol{h}_\perp + N^{-1}h_\parallel \mathrm{i}\boldsymbol{1} \end{pmatrix} = \kappa\mathrm{i} \begin{pmatrix} -|\vec{v}|^2 & \vec{v}_x \\ \vec{v}_x{}^\dagger & \bar{\vec{v}} \otimes \vec{v} \end{pmatrix} \\ &= D_x[\omega_x, e_x] - \frac{1}{2}[\omega_x, [\omega_x, e_x]]. \end{aligned} \tag{3.40}$$

Then writing these expressions in terms of $\mathcal{D}_x$ and $\mathrm{ad}([\omega_x, e_x])$, we get

$$e_t = \mathcal{D}_x[\omega_x, e_x] - \frac{3}{2}\chi^{-1}\mathrm{ad}([\omega_x, e_x])^2 e_x. \tag{3.41}$$

The first term is identified as $\mathcal{D}_x[\omega_x, e_x] \leftrightarrow \nabla_x(\nabla_x \gamma_x) = \nabla_x^2 \gamma_x$. For the second term we observe $\mathrm{ad}([\omega_x, e_x])^2 \leftrightarrow \chi^{-1}\mathrm{ad}(\nabla_x \gamma_x)^2$. Thus, $\gamma(t,x)$ satisfies a geometric nonlinear PDE

$$\gamma_t = \nabla_x^2 \gamma_x - \frac{3}{2}\chi^{-1}\mathrm{ad}(\nabla_x \gamma_x)^2 \gamma_x \tag{3.42}$$

called the non-stretching mKdV map equation on the Lie group $G = SU(N)$. We will refer to it as the *chiral mKdV map.* The same PDE was found to arise from curve flows in the corresponding symmetric space $G/SO(N)$.

All higher odd- and even- flows on $\vec{v}$ in the hierarchy respectively determine higher-order chiral Schrödinger map equations and chiral mKdV map equations for γ. The 0 flow on $\vec{v}$ directly corresponds to

$$\gamma_t = \gamma_x \tag{3.43}$$

which is just a convective (linear traveling wave) map equation.

In addition there is a -1 flow contained in the hierarchy, with the property that $\vec{h}_\perp$ is annihilated by the symplectic operator $\mathcal{J}$ and hence gets mapped into $\mathcal{R}(\vec{h}_\perp) = 0$ under the recursion operator. Geometrically this flow means simply $\mathcal{J}(\vec{h}_\perp) = \vec{\varpi} = 0$ which implies $\omega_t = 0$ from Equations (3.6), (3.16), (3.17), and hence $0 = [\omega_t, e_x] = \mathcal{D}_t e_x$ where $\mathcal{D}_t = D_t + [\omega_t, \cdot]$. The correspondence $\nabla_t \leftrightarrow \mathcal{D}_t$, $\gamma_x \leftrightarrow e_x$ immediately leads to the equation of motion

$$0 = \nabla_t \gamma_x \tag{3.44}$$

for the curve $\gamma(t, x)$. This nonlinear geometric PDE is recognized to be a non-stretching wave map equation on the Lie group $G = SU(N)$, which also was found to arise [4] in the same manner from curve flows in $G/SO(N)$.

The -1 flow equation produced on $\vec{v}$ is a nonlocal evolution equation

$$\vec{v}_t = -\chi \mathrm{i} \vec{h}_\perp, \qquad \chi = \kappa^2 \tag{3.45}$$

with $\vec{h}_\perp$ satisfying

$$0 = \mathrm{i}\vec{\varpi} = D_x \vec{h}_\perp + \mathrm{i} h \vec{v} + \vec{v} \lrcorner \boldsymbol{h}, \tag{3.46}$$

where it is convenient to introduce the variables

$$\boldsymbol{h} = \boldsymbol{h}_\perp + N^{-1} h_\parallel \mathrm{i}\mathbf{1}, \qquad h = N^{-1}(N-1) h_\parallel = -\mathrm{i}\,\mathrm{tr}\boldsymbol{h}, \tag{3.47}$$

which satisfy

$$D_x h = \mathrm{i}(\bar{\vec{h}}_\perp \cdot \vec{v} - \bar{\vec{v}} \cdot \vec{h}_\perp), \tag{3.48}$$

$$D_x \boldsymbol{h} = \bar{\vec{v}} \otimes \vec{h}_\perp - \bar{\vec{h}}_\perp \otimes \vec{v}. \tag{3.49}$$

Note these variables $\vec{h}_\perp$, h, $\boldsymbol{h}$ will be nonlocal functions of $\vec{v}$ (and its x derivatives) as determined by Equations (3.46) to (3.49). To proceed, as in the case $G/SO(N)$, we seek an inverse local expression for $\vec{v}$ arising from an algebraic reduction of the form

$$\boldsymbol{h} = \alpha \mathrm{i} \bar{\vec{h}}_\perp \otimes \vec{h}_\perp + \beta \mathrm{i}\mathbf{1} \tag{3.50}$$

for some expressions $\alpha(h), \beta(h) \in \mathbb{R}$. Similarly to the analysis for the case $G/SO(N)$, substitution of $\boldsymbol{h}$ into Equation (3.49) followed by the use of Equations (3.46) and (3.48) yields

$$\alpha = -(h+\beta)^{-1}, \qquad \beta = \text{const.} \tag{3.51}$$

Next, by taking the trace of $\boldsymbol{h}$ from Equation (3.50), we obtain

$$|\vec{h}_\perp|^2 = N\beta(h+\beta) - (h+\beta)^2 \tag{3.52}$$

which enables h to be expressed in terms of $\vec{h}_\perp$ and β. To determine β we use the conservation law

$$0 = D_x(|\vec{h}_\perp|^2 + \frac{1}{2}(h^2 + |\boldsymbol{h}|^2)), \tag{3.53}$$

admitted by Equations (3.46) to (3.49), corresponding to a wave map conservation law

$$0 = \nabla_x |\gamma_t|_g^2 \tag{3.54}$$

where

$$|\gamma_t|_g^2 = < e_t, e_t >_{\mathfrak{p}} = \kappa^2(|\vec{h}_\perp|^2 + \frac{1}{2}(h^2 + |\boldsymbol{h}|^2)) \tag{3.55}$$

and

$$|\boldsymbol{h}|^2 := -\mathrm{tr}(\boldsymbol{h}^2) = \alpha^2|\vec{h}_\perp|^4 + 2\alpha\beta|\vec{h}_\perp|^2 + \beta^2(N-1). \tag{3.56}$$

A conformal scaling of t can now be used to make $|\gamma_t|_g$ equal to a constant. To simplify subsequent expressions we put $|\gamma_t|_g = 2$, so then

$$(2/\kappa)^2 = |\vec{h}_\perp|^2 + \frac{1}{2}(|\boldsymbol{h}|^2 + h^2). \tag{3.57}$$

Substitution of Equations (3.50) to (3.52) into this expression yields

$$\beta^2 = (2/N)^2 \tag{3.58}$$

from which we obtain via Equation (3.52)

$$h = 2N^{-1} - 1 \pm \sqrt{1 - |\vec{h}_\perp|^2}, \qquad \alpha = |\vec{h}_\perp|^{-2}(1 \pm \sqrt{1 - |\vec{h}_\perp|^2}). \tag{3.59}$$

These variables then can be expressed in terms of $\vec{v}$ through the flow equation (3.45),

$$|\vec{h}_\perp|^2 = \chi^{-2}|\vec{v}_t|^2. \tag{3.60}$$

In addition, by substitution of Equations (3.50) and (3.51) into Equation (3.46) combined with the relation (3.45), we obtain

$$\vec{h}_\perp = \mathrm{i}D_x^{-1}(\alpha^{-1}\vec{v} - \chi^{-2}\alpha(\vec{v}\cdot\bar{\vec{v}}_t)\vec{v}_t). \tag{3.61}$$

Finally, the same equations also yield the inverse expression

$$\mathrm{i}\vec{v} = \alpha(\vec{h}_{\perp x} + \mathrm{i}\alpha(\bar{\vec{h}}_\perp\cdot\vec{v})\vec{h}_\perp), \quad \mathrm{i}\bar{\vec{h}}_\perp\cdot\vec{v} = \alpha(1-\alpha^2\bar{\vec{h}}_\perp\cdot\vec{h}_\perp)^{-1}\bar{\vec{h}}_\perp\cdot\vec{h}_{\perp x} \tag{3.62}$$

after a dot product is taken with $\bar{\vec{h}}_\perp$.

Hence, with the factor χ absorbed by a scaling of t, the -1 flow equation on $\vec{v}$ becomes the nonlocal evolution equation

$$\vec{v}_t = D_x^{-1}(A_\mp\vec{v} - A_\pm|\vec{v}_t|^{-2}(\vec{v}\cdot\bar{\vec{v}}_t)\vec{v}_t) \tag{3.63}$$

where

$$A_\pm := 1 \pm \sqrt{1-|\vec{v}_t|^2} = |\vec{v}_t|^2/A_\mp. \tag{3.64}$$

In hyperbolic form

$$\vec{v}_{tx} = A_\pm\vec{v} - A_\mp|\vec{v}_t|^{-2}(\vec{v}\cdot\bar{\vec{v}}_t)\vec{v}_t \tag{3.65}$$

gives a complex variant of a vector SG equation, found in [8]. Equivalently, through relations (3.62) and (3.59), $\vec{h}_\perp$ is found to obey a complex vector SG equation

$$\left(\alpha(\vec{h}_{\perp x} \mp \frac{1}{2}\alpha(1-|\vec{h}_\perp|^2)^{-1/2}(\bar{\vec{h}}_\perp\cdot\vec{h}_{\perp x})\vec{h}_\perp)\right)_t = \vec{h}_\perp. \tag{3.66}$$

It is known from the symmetry-integrability classification results in [8] that the hyperbolic vector equation (3.65) admits the vector NLS equation (3.29) as a higher symmetry. As a consequence of Corollary 3.1, the hierarchy of vector NLS/mKdV higher symmetries

$$\vec{v}_t^{(0)} = \mathrm{i}\vec{v}, \tag{3.67}$$
$$\vec{v}_t^{(1)} = \mathcal{R}(\mathrm{i}\vec{v}) = -\vec{v}_x, \tag{3.68}$$
$$\vec{v}_t^{(2)} = \mathcal{R}^2(\mathrm{i}\vec{v}) = -\mathrm{i}(\vec{v}_{2x} + 2|\vec{v}|^2\vec{v}), \tag{3.69}$$
$$\vec{v}_t^{(3)} = \mathcal{R}^2(-\vec{v}_x) = \vec{v}_{3x} + 3|\vec{v}|^2\vec{v}_x + 3(\vec{v}_x\cdot\bar{\vec{v}})\vec{v}, \tag{3.70}$$

and so on, generated by the recursion operator (3.25), all commute with the -1 flow

$$\vec{v}_t^{(-1)} = -\mathrm{i}\vec{h}_\perp \tag{3.71}$$

associated to the vector SG equation (3.66). Therefore all these symmetries are admitted by the hyperbolic vector equation (3.65). The corresponding hierarchy of NLS/mKdV Hamiltonians (modulo total derivatives)

$$\begin{aligned}
H^{(0)} &= |\vec{v}|^2,\\
H^{(1)} &= \mathrm{i}\vec{v}_x \cdot \bar{\vec{v}},\\
H^{(2)} &= -|\vec{v}_x|^2 + |\vec{v}|^4,\\
H^{(3)} &= \mathrm{i}\vec{v}_x \cdot (\bar{\vec{v}}_{2x} + 3|\vec{v}|^2\bar{\vec{v}}),
\end{aligned}$$

and so on, generated from the adjoint recursion operator, are all conserved densities for the -1 flow and thereby determine conservation laws admitted for the hyperbolic vector equation (3.65).

Viewed as flows on $\vec{v}$, the entire hierarchy of vector PDEs (3.67) to (3.70), etc., including the -1 flow (3.71), possesses the NLS scaling symmetry $x \to \lambda x$, $\vec{v} \to \lambda^{-1}\vec{v}$, with $t \to \lambda^k t$ for $k = -1, 0, 1, 2$, etc.. As well, the flows for $k \geq 0$ will be local polynomials in the variables $\vec{v}, \vec{v}_x, \vec{v}_{2x}, \ldots$ as established by general results in [28–30] concerning nonlocal operators.

THEOREM 3.2. *In the Lie group $SU(N)$ there is a hierarchy of bi-Hamiltonian flows of curves $\gamma(t,x)$ described by geometric map equations. The* 0 *flow is a convective (traveling wave) map* (3.43), *while the* +1 *flow is a non-stretching chiral Schrödinger map* (3.38) *and the* +2 *flow is a non-stretching chiral mKdV map* (3.42), *and the other odd- and even- flows are higher order analogs. The kernel of the recursion operator* (3.25) *in the hierarchy yields the* −1 *flow which is a non-stretching chiral wave map* (3.44). *Moreover the components of the principal normal vector along the* $+1, +2, -1$ *flows in a $SU(N)$-parallel frame respectively satisfy a vector NLS equation* (3.29), *a complex vector mKdV equation* (3.30) *and a complex vector hyperbolic equation* (3.65).

4. Bi-Hamiltonian operators and vector soliton equations for $SO(N+1)$. Let $\gamma(t,x)$ be a flow of a non-stretching curve in $G = SO(N+1)$. We introduce a $SO(N+1)$-parallel coframe $e \in T^*_\gamma G \otimes \mathfrak{so}(N+1)$ and its associated connection 1-form $\omega \in T^*_\gamma G \otimes \mathfrak{so}(N+1)$ along γ [4]

$$\omega_x = \gamma_x \rfloor \omega = \begin{pmatrix} 0 & 0 & \vec{v}_1 \\ 0 & 0 & \vec{v}_2 \\ -\vec{v}_1{}^T & -\vec{v}_2{}^T & \mathbf{0} \end{pmatrix} \in \mathfrak{p}_{C^\perp}, \qquad \vec{v}_1, \vec{v}_2 \in \mathbb{R}^{N-1} \tag{4.1}$$

and

$$e_x = \gamma_x \rfloor e = \begin{pmatrix} 0 & 1 & \vec{0} \\ -1 & 0 & \vec{0} \\ \vec{0}^T & \vec{0}^T & \mathbf{0} \end{pmatrix} \in \mathfrak{p}_{\parallel}, \qquad \vec{0} \in \mathbb{R}^{N-1}, \quad \mathbf{0} \in \mathfrak{so}(N-1) \tag{4.2}$$

[4] As before, ω is related to e by the Riemannian covariant derivative (2.16) on the surface swept out by the curve flow $\gamma(t,x)$.

normalized so that e_x has unit norm in the Cartan-Killing inner product, $< e_x, e_x >_{\mathfrak{p}} = -\frac{1}{2}\mathrm{tr}\left(\begin{pmatrix} 0 & 1 \\ -1 & 0 \end{pmatrix}^2\right) = 1$ indicating that the coframe is adapted to γ. Consequently, all $\mathfrak{so}(N+1)$ matrices will have a canonical decomposition into tangential and normal parts relative to e_x,

$$\mathfrak{so}(N+1) = \begin{pmatrix} 0 & p_{\|} & \vec{p}_{1\perp} \\ -p_{\|} & 0 & \vec{p}_{2\perp} \\ -\vec{p}_{1\perp}{}^T & -\vec{p}_{2\perp}{}^T & \boldsymbol{p}_{\perp} \end{pmatrix} \simeq \mathfrak{p} \tag{4.3}$$

parameterized by the matrix $\boldsymbol{p}_{\perp} \in \mathfrak{so}(N-1)$, the pair of vectors $\vec{p}_{1\perp}, \vec{p}_{2\perp} \in \mathbb{R}^{N-1}$, and the scalar $p_{\|} \in \mathbb{R}$, corresponding to $\mathfrak{p} = \mathfrak{p}_{\|} \oplus \mathfrak{p}_{\perp}$ where $\mathfrak{p}_{\|} \simeq \mathfrak{so}(2) \simeq \mathfrak{u}(1)$. The centralizer of e_x thus consists of matrices parameterized by $(p_{\|}, \boldsymbol{p}_{\perp})$ and hence $\mathfrak{p}_C \simeq \mathfrak{u}(1) \oplus \mathfrak{so}(N-1) \supset \mathfrak{p}_{\|} \simeq \mathfrak{u}(1)$ while its perp space $\mathfrak{p}_{C^\perp} \subset \mathfrak{p}_{\perp}$ is parameterized by $\vec{p}_{1\perp}, \vec{p}_{2\perp}$. As before, $\boldsymbol{p}_{\perp}$ is empty only if $N = 2$, so consequently for $N > 2$ the $SO(N+1)$-parallel frame (4.1) and (4.2) is a strict generalization of a Riemannian parallel frame.

In the flow direction we put

$$\begin{aligned} e_t = \gamma_t \lrcorner e &= h_{\|} \begin{pmatrix} 0 & 1 & \vec{0} \\ -1 & 0 & \vec{0} \\ \vec{0}^T & \vec{0}^T & \mathbf{0} \end{pmatrix} + \begin{pmatrix} 0 & 0 & \vec{h}_{1\perp} \\ 0 & 0 & \vec{h}_{2\perp} \\ -\vec{h}_{1\perp}{}^T & -\vec{h}_{2\perp}{}^T & \boldsymbol{h}_{\perp} \end{pmatrix} \in \mathfrak{p}_{\|} \oplus \mathfrak{p}_{\perp} \\ &= \begin{pmatrix} 0 & h_{\|} & \vec{h}_{1\perp} \\ -h_{\|} & 0 & \vec{h}_{2\perp} \\ -\vec{h}_{1\perp}{}^T & -\vec{h}_{2\perp}{}^T & \boldsymbol{h}_{\perp} \end{pmatrix} \end{aligned} \tag{4.4}$$

and

$$\omega_t = \gamma_t \lrcorner \omega = \begin{pmatrix} 0 & \theta & \vec{\varpi}_1 \\ -\theta & 0 & \vec{\varpi}_2 \\ -\vec{\varpi}_1{}^T & -\vec{\varpi}_2{}^T & \boldsymbol{\Theta} \end{pmatrix} \in \mathfrak{p}_C \oplus \mathfrak{p}_{C^\perp}, \tag{4.5}$$

with

$$\begin{gathered} h_{\|} \in \mathbb{R}, \qquad \vec{h}_{1\perp}, \vec{h}_{2\perp}, \in \mathbb{R}^{N-1}, \qquad \boldsymbol{h}_{\perp} \in \mathfrak{so}(N-1), \\ \vec{\varpi}_1, \vec{\varpi}_2 \in \mathbb{R}^{N-1}, \qquad \boldsymbol{\Theta} \in \mathfrak{so}(N-1), \qquad \theta \in \mathbb{R}. \end{gathered}$$

Here the components $h_{\|}, (\vec{h}_{1\perp}, \vec{h}_{2\perp}, \boldsymbol{h}_{\perp})$ correspond to decomposing $e_t = g(\gamma_t, \gamma_x)e_x + (\gamma_t)_{\perp} \lrcorner e_{\perp}$ into tangential and normal parts relative to e_x. We now have, in the same notation used before,

$$[e_x, e_t] = \begin{pmatrix} 0 & 0 & \vec{h}_{2\perp} \\ 0 & 0 & -\vec{h}_{1\perp} \\ -\vec{h}_{2\perp}{}^T & \vec{h}_{1\perp}{}^T & \mathbf{0} \end{pmatrix} \in \mathfrak{p}_{C^\perp}, \tag{4.6}$$

$$[\omega_x, e_t] =$$

$$\begin{pmatrix} 0 & \vec{v}_2 \cdot \vec{h}_{1\perp} - \vec{v}_1 \cdot \vec{h}_{2\perp} & \vec{v}_1 \lrcorner \boldsymbol{h}_\perp - h_\parallel \vec{v}_2 \\ \vec{v}_1 \cdot \vec{h}_{2\perp} - \vec{v}_2 \cdot \vec{h}_{1\perp} & 0 & \vec{v}_2 \lrcorner \boldsymbol{h}_\perp + h_\parallel \vec{v}_1 \\ -(\vec{v}_1 \lrcorner \boldsymbol{h}_\perp)^T + h_\parallel \vec{v}_2{}^T & -(\vec{v}_2 \lrcorner \boldsymbol{h}_\perp)^T - h_\parallel \vec{v}_1{}^T & \begin{matrix} \vec{h}_{1\perp} \otimes \vec{v}_1 + \vec{h}_{2\perp} \otimes \vec{v}_2 \\ -\vec{v}_1 \otimes \vec{h}_{1\perp} - \vec{v}_2 \otimes \vec{h}_{2\perp} \end{matrix} \end{pmatrix}$$

$$\in \mathfrak{p}_C \oplus \mathfrak{p}_{C^\perp}, \tag{4.7}$$

$$[\omega_t, e_x] = \begin{pmatrix} 0 & 0 & -\vec{\varpi}_2 \\ 0 & 0 & \vec{\varpi}_1 \\ \vec{\varpi}_2{}^T & -\vec{\varpi}_1{}^T & \mathbf{0} \end{pmatrix} \in \mathfrak{p}_{C^\perp}. \tag{4.8}$$

The resulting torsion and curvature equations can be simplified if we adopt a complex variable notation

$$\vec{v} := \vec{v}_1 + \mathrm{i}\vec{v}_2, \qquad \vec{\varpi} := \vec{\varpi}_1 + \mathrm{i}\vec{\varpi}_2, \qquad \vec{h}_\perp := \vec{h}_{1\perp} + \mathrm{i}\vec{h}_{2\perp}. \tag{4.9}$$

Hence the curvature equation (2.19) becomes

$$D_t\vec{v} - D_x\vec{\varpi} - \vec{v} \lrcorner \boldsymbol{\Theta} - \mathrm{i}\theta\vec{v} = -\mathrm{i}\vec{h}_\perp, \tag{4.10}$$

$$2D_x\boldsymbol{\Theta} + \bar{\vec{\varpi}} \otimes \vec{v} + \vec{\varpi} \otimes \bar{\vec{v}} - \bar{\vec{v}} \otimes \vec{\varpi} - \vec{v} \otimes \bar{\vec{\varpi}} = 0, \tag{4.11}$$

$$2\mathrm{i}D_x\theta + \vec{v} \cdot \bar{\vec{\varpi}} - \vec{\varpi} \cdot \bar{\vec{v}} = 0, \tag{4.12}$$

and the torsion equation (2.18) reduces to

$$2\mathrm{i}D_x h_\parallel - \vec{h}_\perp \cdot \bar{\vec{v}} + \vec{v} \cdot \bar{\vec{h}}_\perp = 0, \tag{4.13}$$

$$2D_x\boldsymbol{h}_\perp + \bar{\vec{h}}_\perp \otimes \vec{v} + \vec{h}_\perp \otimes \bar{\vec{v}} - \bar{\vec{v}} \otimes \vec{h}_\perp - \vec{v} \otimes \bar{\vec{h}}_\perp = 0, \tag{4.14}$$

$$\mathrm{i}\vec{\varpi} - D_x\vec{h}_\perp - \mathrm{i}h_\parallel\vec{v} - \vec{v} \lrcorner \boldsymbol{h}_\perp, = 0. \tag{4.15}$$

These equations are nearly the same as those for the space $G = SU(N)$, except that both $\boldsymbol{\Theta}$ and $\boldsymbol{h}_\perp$ are now real (skew matrices) instead of complex (antihermitian matrices). This similarity is a result of the homomorphism

$$\begin{pmatrix} 0 & p_\parallel & \vec{p}_{1\perp} \\ -p_\parallel & 0 & \vec{p}_{2\perp} \\ -\vec{p}_{1\perp}{}^T & -\vec{p}_{2\perp}{}^T & \boldsymbol{p}_\perp \end{pmatrix} \mapsto \begin{pmatrix} -\mathrm{i}p_\parallel & \vec{p}_{1\perp} + \mathrm{i}\vec{p}_{2\perp} \\ -\vec{p}_{1\perp} + \mathrm{i}\vec{p}_{2\perp}{}^\dagger & \boldsymbol{p}_\perp \end{pmatrix} \tag{4.16}$$

of $\mathfrak{so}(N+1)$ into $\mathfrak{u}(N)$, such that $[\mathfrak{so}(N+1), \mathfrak{so}(N+1)] \subset \mathfrak{su}(N) \subset \mathfrak{u}(N)$.

Proceeding as before, we use Equations (4.11)–(4.14) to eliminate

$$\Theta = \frac{1}{2}D_x^{-1}(\bar{\vec{v}} \otimes \vec{\varpi} + \vec{v} \otimes \bar{\vec{\varpi}} - \bar{\vec{\varpi}} \otimes \vec{v} - \vec{\varpi} \otimes \bar{\vec{v}}), \tag{4.17}$$

$$\theta = \frac{\mathrm{i}}{2}D_x^{-1}(\bar{\vec{\varpi}} \cdot \vec{v} - \bar{\vec{v}} \cdot \vec{\varpi}), \tag{4.18}$$

$$\boldsymbol{h}_\perp = \frac{1}{2}D_x^{-1}(\bar{\vec{v}} \otimes \vec{h}_\perp + \vec{v} \otimes \bar{\vec{h}}_\perp - \bar{\vec{h}}_\perp \otimes \vec{v} - \vec{h}_\perp \otimes \bar{\vec{v}}), \tag{4.19}$$

$$h_\| = \frac{\mathrm{i}}{2}D_x^{-1}(\bar{\vec{h}}_\perp \cdot \vec{v} - \bar{\vec{v}} \cdot \vec{h}_\perp), \tag{4.20}$$

in terms of the variables $\vec{v}$, $\vec{h}_\perp$, $\vec{\varpi}$. Then Equation (4.10) gives a flow on $\vec{v}$,

$$\vec{v}_t = D_x\vec{\varpi} + \frac{1}{2}D_x^{-1}(\vec{\varpi} \cdot \bar{\vec{v}} - \vec{v} \cdot \bar{\vec{\varpi}})\vec{v} \\ + \frac{1}{2}\vec{v}\lrcorner D_x^{-1}(\bar{\vec{v}} \otimes \vec{\varpi} + \vec{v} \otimes \bar{\vec{\varpi}} - \bar{\vec{\varpi}} \otimes \vec{v} - \vec{\varpi} \otimes \bar{\vec{v}}) - \mathrm{i}\vec{h}_\perp \tag{4.21}$$

with

$$\vec{\varpi} = -\mathrm{i}D_x\vec{h}_\perp + \frac{\mathrm{i}}{2}D_x^{-1}(\bar{\vec{h}}_\perp \cdot \vec{v} - \bar{\vec{v}} \cdot \vec{h}_\perp)\vec{v} \\ + \frac{\mathrm{i}}{2}\vec{v}\lrcorner D_x^{-1}(\bar{\vec{h}}_\perp \otimes \vec{v} + \vec{h}_\perp \bar{\otimes} \vec{v} - \bar{\vec{v}} \otimes \vec{h}_\perp - \vec{v} \otimes \bar{\vec{h}}_\perp) \tag{4.22}$$

obtained from Equation (4.15). We thus read off the operators

$$\mathcal{H} = D_x - \mathrm{i}D_x^{-1}(\bar{\vec{v}} \cdot^\dagger \mathrm{i}\ )\vec{v} + \frac{1}{2}\vec{v}\lrcorner D_x^{-1}((\bar{\vec{v}} \wedge^\dagger\ ) - (\ \wedge^\dagger \bar{\vec{v}})), \qquad \mathcal{I} = \mathrm{i}, \tag{4.23}$$

and define the related operator

$$\mathcal{J} = \mathcal{I}^{-1} \circ \mathcal{H} \circ \mathcal{I} = D_x + D_x^{-1}(\bar{\vec{v}} \cdot^\dagger\ )\vec{v} + \frac{1}{2}\vec{v}\lrcorner D_x^{-1}((\bar{\vec{v}} \odot^\dagger\ ) - (\ \odot^\dagger \bar{\vec{v}})) \tag{4.24}$$

using the Hermitian versions of the wedge product $\vec{A} \wedge^\dagger \vec{B} := \vec{A} \otimes \vec{B} - \bar{\vec{B}} \otimes \bar{\vec{A}}$, the symmetric product $\vec{A} \odot^\dagger \vec{B} := \vec{A} \otimes \vec{B} + \bar{\vec{B}} \otimes \bar{\vec{A}}$, and the dot product $\vec{A} \cdot^\dagger \vec{B} := \frac{1}{2}\vec{A} \cdot \vec{B} + \frac{1}{2}\bar{\vec{B}} \cdot \bar{\vec{A}}$, introduced before.

These operators $\mathcal{H}, \mathcal{J}, \mathcal{I}$ and variables $\vec{v}, \vec{\varpi}, \vec{h}_\perp$ determine a very similar Hamiltonian structure in the space $G = SO(N+1)$ compared to $G = SU(N)$

PROPOSITION 4.1. *Theorem 3.1 and Corollary 3.1 apply verbatim here (with the same method of proof) to the flow equation* (4.21) *and to the operators (4.23) and (4.24), apart from a change in the scalar curvature factor* $\chi = 1$ *connected with the Riemannian geometry of* $SO(N+1)$.

Thus,

$$\mathcal{R} = \mathrm{i}(D_x + D_x^{-1}(\bar{\vec{v}} \cdot^\dagger\ )\vec{v} + \frac{1}{2}\vec{v}\lrcorner D_x^{-1}((\bar{\vec{v}} \odot^\dagger\ ) - (\ \odot^\dagger \bar{\vec{v}}))) \tag{4.25}$$

yields a hereditary recursion operator generating a hierarchy of $U(N-1)$-invariant commuting bi-Hamiltonian flows on $\vec{v}$, corresponding to commuting Hamiltonian vector fields $\mathrm{i}\vec{h}_\perp^{(k)}$ and involutive covector fields $\vec{\varpi}^{(k)} = \delta H^{(k)}/\delta\bar{\vec{v}}$, $k = 0, 1, 2, \ldots$. The hierarchy starts from $\mathrm{i}\vec{h}_\perp^{(0)} = \mathrm{i}\vec{v}$, $\vec{\varpi}^{(0)} = -\vec{v}$, which generates phase rotations, and is followed by $\mathrm{i}\vec{h}_\perp^{(1)} = -\vec{v}_x$, $\vec{\varpi}^{(1)} = -\mathrm{i}\vec{v}_x$, which generates x-translations. All these flows have the same recursion relations (3.27) as in the space $G = SU(N)$, and they also share the same NLS scaling symmetry $x \to \lambda x$, $\vec{v} \to \lambda^{-1}\vec{v}$.

The $+1$ and $+2$ flows given by $\vec{h}_\perp = \vec{v}$ and $\vec{h}_\perp = \mathrm{i}\vec{v}_x$ respectively yield a vector NLS equation

$$\mathrm{i}\vec{v}_t = \vec{v}_{2x} + |\vec{v}|^2\vec{v} - \frac{1}{2}\vec{v}\cdot\vec{v}\bar{\vec{v}} + \chi\vec{v} \tag{4.26}$$

up to a phase term (which can be absorbed by a phase rotation on $\vec{v}$), and a complex vector mKdV equation

$$\vec{v}_t = \vec{v}_{3x} + \frac{3}{2}(|\vec{v}|^2\vec{v}_x + (\vec{v}_x\cdot\bar{\vec{v}})\vec{v} - (\vec{v}_x\cdot\vec{v})\bar{\vec{v}}) + \chi\vec{v}_x \tag{4.27}$$

up to a convective term (which can be absorbed by a Galilean transformation). Note these two equations differ compared to the ones arising in the space $G = SU(N)$. The higher odd- and even- flows yield higher-order versions of Equations (4.26) and (4.27).

This hierarchy of flows corresponds to geometrical motions of the curve $\gamma(t,x)$ obtained from Equation (4.4) in a similar fashion to the ones for $G = SU(N)$ via identifying $\gamma_t \leftrightarrow e_t$, $\gamma_x \leftrightarrow e_x$, $\nabla_x\gamma_x \leftrightarrow [\omega_x, e_x] = \mathcal{D}_x e_x$, and so on as before, where $\nabla_x \leftrightarrow \mathcal{D}_x = D_x + [\omega_x, e_x]$. Here we have

$$[\omega_x, e_x] = \begin{pmatrix} 0 & 0 & -\vec{v}_2 \\ 0 & 0 & \vec{v}_1 \\ \vec{v}_2{}^T & -\vec{v}_1{}^T & \mathbf{0} \end{pmatrix}, \tag{4.28}$$

$$\mathrm{ad}([\omega_x, e_x])e_x = -\mathrm{ad}(e_x)[\omega_x, e_x] = \begin{pmatrix} 0 & 0 & -\vec{v}_1 \\ 0 & 0 & -\vec{v}_2 \\ \vec{v}_1{}^T & \vec{v}_2{}^T & \mathbf{0} \end{pmatrix}, \tag{4.29}$$

$$\begin{aligned}[\omega_x, [\omega_x, e_x]] &= \begin{pmatrix} 0 & \vec{v}_1{}^2 + \vec{v}_2{}^2 & \vec{0} \\ -(\vec{v}_1{}^2 + \vec{v}_2{}^2) & 0 & \vec{0} \\ \vec{0}^T & \vec{0}^T & 2\vec{v}_2\otimes\vec{v}_1 - 2\vec{v}_1\otimes\vec{v}_2 \end{pmatrix} \\ &= \mathrm{ad}([\omega_x, e_x])^2 e_x,\end{aligned} \tag{4.30}$$

and so on, where $\mathrm{ad}(\cdot)$ denotes the standard adjoint representation acting in the Lie algebra $\mathfrak{so}(N+1)$.

The $+1$ flow

$$\vec{h}_\perp = \vec{v}, \qquad h_\parallel = 0, \qquad \boldsymbol{h}_\perp = \mathbf{0}, \tag{4.31}$$

gives the frame equation

$$e_t = \begin{pmatrix} 0 & 0 & \vec{v}_1 \\ 0 & 0 & \vec{v}_2 \\ -\vec{v}_1{}^T & -\vec{v}_2{}^T & \mathbf{0} \end{pmatrix} = \mathrm{ad}(e_x)[\omega_x\,, e_x], \tag{4.32}$$

so thus $\gamma(t,x)$ satisfies the chiral Schrödinger map equation (3.38) on the Lie group $G = SO(N+1)$. All higher odd-flows on $\vec{v}$ in the hierarchy determine higher-order chiral Schrödinger map equations.

Next, the +2 flow

$$\vec{h}_\perp = \mathrm{i}\vec{v}_x, \qquad h_\parallel = \frac{1}{2}|\vec{v}|^2, \qquad \boldsymbol{h}_\perp = \frac{\mathrm{i}}{2}(\bar{\vec{v}} \otimes \vec{v} - \vec{v} \otimes \bar{\vec{v}}) \tag{4.33}$$

yields the frame equation

$$\begin{aligned} e_t &= \begin{pmatrix} 0 & \frac{1}{2}(\vec{v}_1{}^2 + \vec{v}_2{}^2) & -(\vec{v}_2)_x \\ -\frac{1}{2}(\vec{v}_1{}^2 + \vec{v}_2{}^2) & 0 & (\vec{v}_1)_x \\ (\vec{v}_2)_x{}^T & -(\vec{v}_1)_x{}^T & \vec{v}_2 \otimes \vec{v}_1 - \vec{v}_1 \otimes \vec{v}_2 \end{pmatrix} \\ &= D_x[\omega_x\,, e_x] - \frac{1}{2}[\omega_x\,, [\omega_x\,, e_x]] \end{aligned} \tag{4.34}$$

which gives the same frame equation as in the space $G = SU(N)$,

$$e_t = \mathcal{D}_x[\omega_x\,, e_x] - \frac{3}{2}\chi^{-1}\mathrm{ad}([\omega_x\,, e_x])^2 e_x \tag{4.35}$$

up to the change in the scalar curvature factor, $\chi = 1$. Thus, $\gamma(t,x)$ satisfies the chiral mKdV map equation (3.42) on the Lie group $G = SO(N+1)$. All higher even- flows on $\vec{v}$ in the hierarchy yield higher-order chiral mKdV map equations for γ.

These same geometric nonlinear PDEs were found to arise [4] from curve flows in the corresponding symmetric space $G/SO(N) \simeq S^N$.

The hierarchy also contains a -1 flow in which $\vec{h}_\perp$ is annihilated by the symplectic operator $\mathcal{J}$ so it lies in the kernel $\mathcal{R}(\vec{h}_\perp) = 0$ of the recursion operator. This flow has the same geometrical meaning as in the space $G = SU(N)$, namely $\mathcal{J}(\vec{h}_\perp) = \vec{\varpi} = 0$ whence $\omega_t = 0$ which implies $0 = [\omega_t\,, e_x] = \mathcal{D}_t e_x$ where $\mathcal{D}_t = D_t + [\omega_t\,, \cdot]$. Thus, the correspondence $\nabla_t \leftrightarrow \mathcal{D}_t$, $\gamma_x \leftrightarrow e_x$ directly yields the chiral wave map equation (3.44) on the Lie group $G = SO(N+1)$. The resulting -1 flow equation on $\vec{v}$ is given by

$$\vec{v}_t = -\chi\mathrm{i}\vec{h}_\perp, \qquad \chi = 1 \tag{4.36}$$

where $\vec{h}_\perp$ satisfies the equation

$$0 = \mathrm{i}\vec{\varpi} = D_x\vec{h}_\perp + \mathrm{i}h_\parallel\vec{v} + \vec{v}\lrcorner\boldsymbol{h}_\perp \tag{4.37}$$

together with Equations (4.13) and (4.14). Similarly to the case $G = SU(N)$, these three equations determine $\vec{h}_\perp$, $h_\parallel$, $\boldsymbol{h}_\perp$ as nonlocal functions of $\vec{v}$ (and its x derivatives). Proceeding as before, we seek an inverse local expression for $\vec{v}$ obtained through an algebraic reduction

$$\boldsymbol{h}_\perp = \alpha\mathrm{i}(\bar{\vec{h}}_\perp \otimes \vec{h}_\perp - \vec{h}_\perp \otimes \bar{\vec{h}}_\perp) \tag{4.38}$$

for some expression $\alpha(h_\parallel) \in \mathbb{R}$. Substitution of $\boldsymbol{h}_\perp$ into Equation (4.14), followed by the use of Equations (4.13) and (4.37), gives

$$\alpha = -\frac{1}{2}h_\parallel{}^{-1}. \tag{4.39}$$

We next use the wave map conservation law (3.54) where now $|\gamma_t|_g^2 =< e_t, e_t >_\mathfrak{p}= |\vec{h}_\perp|^2 + h_\parallel{}^2 + \frac{1}{2}|\boldsymbol{h}|^2$, corresponding to the conservation law

$$0 = D_x\Big(|\vec{h}_\perp|^2 + h_\parallel{}^2 + \frac{1}{2}|\boldsymbol{h}_\perp|^2\Big) \tag{4.40}$$

admitted by Equations (4.37), (4.13), (4.14) with

$$|\boldsymbol{h}_\perp|^2 := -\mathrm{tr}(\boldsymbol{h}_\perp{}^2) = 2\alpha^2(|\vec{h}_\perp|^4 - \vec{h}_\perp{}^2\bar{\vec{h}}_\perp{}^2) \tag{4.41}$$

where $\vec{h}_\perp{}^2 := \vec{h}_\perp \cdot \vec{h}_\perp$, $\bar{\vec{h}}_\perp{}^2 := \bar{\vec{h}}_\perp \cdot \bar{\vec{h}}_\perp$. As before, a conformal scaling of t can be used to make $|\gamma_t|_g$ equal to a constant. By putting $|\gamma_t|_g = 1$ we obtain

$$1 = |\vec{h}_\perp|^4 + h_\parallel{}^2 + \frac{1}{2}|\boldsymbol{h}_\perp|^2 = \frac{1}{4}h_\parallel{}^{-2}(|\vec{h}_\perp|^4 - \vec{h}_\perp{}^2\bar{\vec{h}}_\perp{}^2) + h_\parallel{}^2 + |\vec{h}_\perp|^2 \tag{4.42}$$

from Equations (4.39) and (4.41). This yields a quadratic equation

$$0 = h_\parallel{}^4 + (|\vec{h}_\perp|^2 - 1)h_\parallel{}^2 + |\vec{h}_\perp|^4 - \vec{h}_\perp{}^2\bar{\vec{h}}_\perp{}^2 \tag{4.43}$$

determining

$$2h_\parallel{}^2 = 1 - |\vec{h}_\perp|^2 \pm \sqrt{1 - 2|\vec{h}_\perp|^2 + \vec{h}_\perp{}^2\bar{\vec{h}}_\perp{}^2}, \tag{4.44}$$

$$2\alpha^2 = (|\vec{h}_\perp|^4 - \vec{h}_\perp{}^2\bar{\vec{h}}_\perp{}^2)^{-1}\Big(1 - |\vec{h}_\perp|^2 \mp \sqrt{1 - 2|\vec{h}_\perp|^2 + \vec{h}_\perp{}^2\bar{\vec{h}}_\perp{}^2}\Big). \tag{4.45}$$

The flow equation (4.36) allows these variables to be expressed in terms of $\vec{v}$,

$$|\vec{h}_\perp|^2 = |\vec{v}_t|^2, \qquad \vec{h}_\perp{}^2 = \vec{v}_t{}^2, \qquad \bar{\vec{h}}_\perp{}^2 = \bar{\vec{v}}_t{}^2. \tag{4.46}$$

Similarly, Equation (4.37) combined with Equations (4.38) and (4.39) yields

$$\vec{h}_\perp = \mathrm{i}D_x^{-1}\Big(\frac{1}{2}\alpha^{-1}\vec{v} + \alpha((\vec{v}\cdot\vec{v}_t)\bar{\vec{v}}_t - (\vec{v}\cdot\bar{\vec{v}}_t)\vec{v}_t)\Big). \tag{4.47}$$

Thus the -1 flow equation on $\vec{v}$ becomes the nonlocal evolution equation

$$\vec{v}_t = \frac{1}{\sqrt{2}} D_x^{-1}\Big(B_{\mp}\vec{v} + |B|^{-1}B_{\pm}((\vec{v}\cdot\bar{\vec{v}}_t)\vec{v}_t - (\vec{v}\cdot\vec{v}_t)\bar{\vec{v}}_t)\Big) \tag{4.48}$$

where

$$B_{\pm}^2 := 1 - |\vec{v}_t|^2 \pm \sqrt{1 - 2|\vec{v}_t|^2 - \vec{v}_t{}^2\bar{\vec{v}}_t{}^2} = |B|^2/B_{\mp}^2, \tag{4.49}$$

$$|B|^2 := |\vec{v}_t|^4 - \vec{v}_t{}^2\bar{\vec{v}}_t{}^2 = B_+^2 B_-^2. \tag{4.50}$$

The hyperbolic form of this vector PDE is a complex variant of a vector SG equation (with a factor $\sqrt{2}$ absorbed into a scaling of t)

$$\vec{v}_{tx} = B_{\mp}\vec{v} + |B|^{-1}B_{\pm}((\vec{v}\cdot\vec{v}_t)\bar{\vec{v}}_t - (\vec{v}\cdot\bar{\vec{v}}_t)\vec{v}_t) \tag{4.51}$$

which was found in [8].

There is an equivalent hyperbolic equation on $\vec{h}_{\perp}$ given by an inverse for expression (4.47) as follows. Substitution of Equation (4.38) into Equation (4.37) first yields the relation

$$\mathrm{i}\vec{v} = 2\alpha(\vec{h}_{\perp x} + \mathrm{i}\alpha((\vec{v}\cdot\bar{\vec{h}}_{\perp})\vec{h}_{\perp} - (\vec{v}\cdot\vec{h}_{\perp})\bar{\vec{h}}_{\perp})). \tag{4.52}$$

Then by taking its dot product separately with $\vec{h}_{\perp}$ and $\bar{\vec{h}}_{\perp}$, we obtain the additional relations

$$\mathrm{i}(1 - |\vec{h}_{\perp}|^2 - 2h_{\parallel}{}^2)\vec{v}\cdot\vec{h}_{\perp} = \Big(\alpha|\vec{h}_{\perp}|^2 - \frac{1}{2}\alpha^{-1}\Big)\vec{h}_{\perp}\cdot\vec{h}_{\perp x} - \alpha\vec{h}_{\perp}{}^2\bar{\vec{h}}_{\perp}\cdot\vec{h}_{\perp x},$$

$$\mathrm{i}(1 - |\vec{h}_{\perp}|^2 - 2h_{\parallel}{}^2)\vec{v}\cdot\bar{\vec{h}}_{\perp} = \alpha\bar{\vec{h}}_{\perp}{}^2\vec{h}_{\perp}\cdot\vec{h}_{\perp x} - \Big(\alpha|\vec{h}_{\perp}|^2 + \frac{1}{2}\alpha^{-1}\Big)\bar{\vec{h}}_{\perp}\cdot\vec{h}_{\perp x},$$

which thus determines $\vec{v}\cdot\vec{h}_{\perp}$ and $\vec{v}\cdot\bar{\vec{h}}_{\perp}$ and hence $\vec{v}$ in terms of $\vec{h}_{\perp}$, $\bar{\vec{h}}_{\perp}$, and $\vec{h}_{\perp x}$. Finally, substitution of these expressions into the flow equation (4.36) yields the complex vector SG equation

$$\begin{aligned}\vec{h}_{\perp} = \Big(\alpha\Big(2\vec{h}_{\perp x} &\mp (1 - 2|\vec{h}_{\perp}|^2 + \vec{h}_{\perp}{}^2\bar{\vec{h}}_{\perp}{}^2)^{-1/2} \\ &\times\Big(((1 - 2\alpha^2|\vec{h}_{\perp}|^2)\vec{h}_{\perp}\cdot\vec{h}_{\perp x} + 2\alpha^2\vec{h}_{\perp}{}^2\bar{\vec{h}}_{\perp}\cdot\vec{h}_{\perp x})\bar{\vec{h}}_{\perp} \\ &-((1 + 2\alpha^2|\vec{h}_{\perp}|^2)\bar{\vec{h}}_{\perp}\cdot\vec{h}_{\perp x} + 2\alpha^2\bar{\vec{h}}_{\perp}{}^2\vec{h}_{\perp}\cdot\vec{h}_{\perp x})\vec{h}_{\perp}\Big)\Big)\Big)_t.\end{aligned} \tag{4.53}$$

Note that, as written, the hyperbolic PDEs (4.51) and (4.53) for $G = SO(N+1)$ are valid only when $|B| \neq 0$, which holds precisely in the vector case, $N > 2$. The scalar case $N = 2$ becomes a singular limit $|B| = 0$, as seen from the quadratic equation (4.43) whose solutions (4.44) degenerate to ${h_\parallel}^2 = \frac{1}{2}B_\pm^2 = 1 - |\vec{h}_\perp|^2, 0$ in the $+/-$ cases respectively. Thus $\alpha = -\frac{1}{2}{h_\parallel}^{-1}$ is well-defined only in the $+$ case, with

$$ {h_\parallel}^2 = \frac{1}{2}B_+^2 = 1 - |\vec{h}_\perp|^2, \qquad N = 2, \tag{4.54} $$

where we have the corresponding limit

$$ \alpha^2 = \lim_{|B|\to 0} \frac{1}{2}|B|^{-2}B_-^2 = \frac{1}{4}(1 - |\vec{h}_\perp|^2)^{-1}, \qquad N = 2, \tag{4.55} $$

and where we identify the 1-component complex vectors $\vec{h}_\perp, \vec{v}, \vec{\varpi} \in \mathbb{C}$ with complex scalars. (This settles the questions raised in [8] concerning the existence of a scalar limit for the hyperbolic vector equation (4.51).)

In this limit the hyperbolic PDEs (4.51) and (4.53) for $G = SO(3)$ reduce to the scalar case of the hyperbolic PDEs (3.65) and (3.66) for $G = SU(2)$, due to the local isomorphism of the Lie groups $SU(2) \simeq SO(3)$. The same happens for the evolutionary PDEs in the hierarchies for $G = SO(3)$ and $G = SU(2)$, namely the scalar case of the NLS equations (4.26) and (3.29) and the mKdV equations (4.27) and (3.30) each coincide (up to scalings of the variables).

The symmetry-integrability classification results in [8] show that the hyperbolic vector equation (4.51) admits the vector NLS equation (4.26) as a higher symmetry. We see that, from Corollary 3.1 applied to the recursion operator (4.25), there is a hierarchy of vector NLS/mKdV higher symmetries

$$ \vec{v}_t^{(0)} = \mathrm{i}\vec{v}, \tag{4.56} $$

$$ \vec{v}_t^{(1)} = \mathcal{R}(\mathrm{i}\vec{v}) = -\vec{v}_x, \tag{4.57} $$

$$ \vec{v}_t^{(2)} = \mathcal{R}^2(\mathrm{i}\vec{v}) = -\mathrm{i}(\vec{v}_{2x} + |\vec{v}|^2\vec{v} - \frac{1}{2}\vec{v}\cdot\vec{v}\bar{\vec{v}}), \tag{4.58} $$

$$ \vec{v}_t^{(3)} = \mathcal{R}^2(-\vec{v}_x) = \vec{v}_{3x} + \frac{3}{2}(|\vec{v}|^2\vec{v}_x + (\vec{v}_x\cdot\bar{\vec{v}})\vec{v} - (\vec{v}_x\cdot\vec{v})\bar{\vec{v}}), \tag{4.59} $$

and so on, generated by this operator $\mathcal{R}$, while the adjoint operator $\mathcal{R}^*$ generates a corresponding hierarchy of NLS/mKdV Hamiltonians (modulo total derivatives)

$$ \begin{aligned} H^{(0)} &= |\vec{v}|^2, \\ H^{(1)} &= \mathrm{i}\vec{v}_x\cdot\bar{\vec{v}}, \\ H^{(2)} &= -|\vec{v}_x|^2 + \frac{1}{2}|\vec{v}|^2 - \frac{1}{4}\vec{v}^2\bar{\vec{v}}^2, \\ H^{(3)} &= \mathrm{i}\vec{v}_x\cdot(\bar{\vec{v}}_{2x} + \frac{3}{2}|\vec{v}|^2\bar{\vec{v}}) - \mathrm{i}\frac{3}{8}(\vec{v}^2)_x\bar{\vec{v}}^2, \end{aligned} $$

and so on. All of Hamiltonians are conserved densities for the -1 flow

$$ \vec{v}_t^{(-1)} = -\mathrm{i}\vec{h}_\perp \tag{4.60} $$

associated to the vector SG equation (4.53), and hence they determine a hierarchy of conservation laws admitted for the hyperbolic vector equation (4.51). Likewise all of the symmetries comprise a hierarchy that commutes with the -1 flow and are therefore admitted symmetries of the hyperbolic vector equation (4.51).

All of the vector PDEs (4.56) to (4.59), etc., viewed as flows on $\vec{v}$, including the -1 flow (4.60), possess the NLS scaling symmetry $x \to \lambda x$, $\vec{v} \to \lambda^{-1}\vec{v}$, with $t \to \lambda^k t$ for $k = -1, 0, 1, 2, \ldots$, where these PDEs for $k \geq 0$ will be local polynomials in the variables $\vec{v}, \vec{v}_x, \vec{v}_{2x}, \ldots$ in the same manner as before.

THEOREM 4.1. *In the Lie group $SO(N+1)$ there is a hierarchy of bi-Hamiltonian flows of curves $\gamma(t,x)$ described by geometric map equations. The 0 flow is a convective (traveling wave) map* (3.43), *while the $+1$ flow is a non-stretching chiral Schrödinger map* (3.38) *and the $+2$ flow is a non-stretching chiral mKdV map* (3.42), *and the other odd- and even-flows are higher order analogs. The kernel of the recursion operator* (4.25) *in the hierarchy yields the -1 flow which is a non-stretching chiral wave map* (3.44). *Moreover the components of the principal normal vector along the $+1, +2, -1$ flows in a $SU(N)$-parallel frame respectively satisfy a vector NLS equation* (4.26), *a complex vector mKdV equation* (4.27) *and a complex vector hyperbolic equation* (4.51).

5. Concluding remarks. The Lie groups $SO(N+1)$ and $SU(N)$ each contain a hierarchy of integrable bi-Hamiltonian curve flows described by a chiral Schrödinger map equation (3.38) for the $+1$ flow, a chiral mKdV map equation (3.42) for the $+2$ flow, and a chiral wave map equation (3.44) for the -1 flow coming from the kernel of the recursion operator of each hierarchy. The principal normal components in a parallel frame along these flows in each Lie group satisfy $U(N-1)$-invariant soliton equations respectively given by a vector NLS equation, a complex vector mKdV equation, and a hyperbolic vector equation related to a complex vector SG equation.

These two Lie groups are singled out as exhausting the isometry groups G that arise for Riemannian symmetric spaces of the type $G/SO(N)$ as known from Cartan's classification [5]. Moreover, since $G = SO(N+1)$ is locally isomorphic to $G = SU(N)$ when (and only when) $N = 2$, the integrable hierarchies of curve flows in the spaces $SO(3) \simeq SU(2) \simeq S^3$ therefore coincide precisely in the scalar case, with the $+1, +2, -1$ flows reducing to $U(1)$-invariant scalar soliton equations consisting of the NLS equation $iv_t = v_{2x} + 2|v|^2 v$ and complex versions of mKdV and SG equations $v_t = v_{3x} + 6|v|^2 v_x$ and $v_{tx} = (2\sqrt{1-|v_t|^2})v$ (up to rescalings of v and t).

The present results thus account for the existence of the two unitarily-invariant vector generalizations of the NLS equation and the complex mKdV and SG equations that are known from symmetry-integrability classifications [7, 8]. Moreover, their bi-Hamiltonian integrability structure as summarized by the operators $\mathcal{R} = \mathcal{H} \circ \mathcal{I} = \mathcal{I} \circ \mathcal{J}$ is shown to be geometrically encoded in the frame structure equations for the corresponding curve flows in the two Lie groups $G = SO(N+1), SU(N) \subset U(N)$. This encoding utilizes a parallel moving frame formulation based on earlier work [4] studying integrable curve flows in the Riemannian symmetric spaces $G/SO(N)$. Indeed, the bi-Hamiltonian operator structure derived in [4] for curve flows in $G/SO(N)$ can be recovered from $\mathcal{H}$ and $\mathcal{J}$ if the connection variables $\vec{v}$ and $\vec{\varpi}$ are restricted to be real while the flow-direction variable $\vec{h}_\perp$ is restricted to be imaginary, in which case the -1 flow and all even-flows reduce to the hierarchy of flows in $G/SO(N)$. More particularly, the operator $\mathcal{R}^2 = -\mathcal{H} \circ \mathcal{J}$ acts as a vector NLS/mKdV recursion operator which is (up to a sign) a complex version of the vector mKdV recursion operator coming from $G/SO(N)$.

Finally, there is a broad generalization [25] of these results yielding hierarchies of group-invariant soliton equations associated to integrable curve flows described by geometric map equations in all semisimple Lie groups and Riemannian symmetric spaces.

Acknowledgments. S.C.A is supported by an N.S.E.R.C. grant.

REFERENCES

[1] R.E. Goldstein and D.M. Petrich, *The Korteweg-de Vries hierarchy as dynamics of closed curves in the plane*, Phys. Rev. Lett. 1991, **67**: 3203–3206.

[2] K. Nakayama, H. Segur, and M. Wadati, *Integrability and the motion of curves*, Phys. Rev. Lett. 1992, **69**: 2603–2606.

[3] S.C. Anco, *Bi-Hamiltonian operators, integrable flows of curves using moving frames, and geometric map equations*, J. Phys. A: Math. Gen. 2006, **39**: 2043–2072.

[4] S.C. Anco, *Hamiltonian flows of curves in symmetric spaces $G/SO(N)$ and vector soliton equations of mKdV and sine-Gordon type*, SIGMA 2006, **2**(044), 18 pages.

[5] S. Helgason, *Differential Geometry, Lie Groups, and Symmetric Spaces*, Amer. Math. Soc., Providence, 2001.

[6] R.W. Sharpe, *Differential Geometry*, Springer-Verlag, New York, 1997.

[7] V.V. Sokolov and T. Wolf, *Classification of integrable vector polynomial evolution equations*, J. Phys. A: Math. Gen. 2001, **34**: 11139–11148.

[8] S.C. Anco and T. Wolf, *Some symmetry classifications of hyperbolic vector evolution equations*, J. Nonlinear Math. Phys. 2005, **12**: 13–31; ibid. 607–608.

[9] K.-S. Chou and C. Qu, *Integrable equations arising from motions of plane curves*, Physica D 2002, **162**: 9–33.

[10] K.-S. Chou and C. Qu, *Integrable motion of space curves in affine geometry*, Chaos, Solitons and Fractals 2002, **14**: 29–44.

[11] K.-S. Chou and C. Qu, *Integrable equations arising from motions of plane curves. II*, J. Nonlinear Sci. 2003, **13**: 487–517.

[12] K.-S. Chou and C. Qu, *Motion of curves in similarity geometries and Burgers-mKdV hierarchies*, Chaos, Solitons and Fractals 2004, **19**: 47–53.

[13] A. Fordy and P. Kulish, *Nonlinear Schrodinger equations and simple Lie algebras*, Commun. Math. Phys. 1983, **89**: 427–443.

[14] A. Fordy, *Derivative nonlinear Schrodinger equations and Hermitian symmetric spaces*, J. Phys. A: Math. Gen. 1984, **17**: 1235–1245.

[15] C. Athorne and A. Fordy, *Generalised KdV and mKdV equations associated with symmetric spaces*, J. Phys. A: Math. Gen. 1987, **20**: 1377-1386.

[16] C. Athorne, *Local Hamiltonian structures of multicomponent KdV equations*, J. Phys. A: Math. Gen. 1988, **21**: 4549–4556.

[17] I. Bakas, Q.-H. Park, and H.-J. Shin, *Lagrangian formulation of symmetric space sine-Gordon models*, Phys. Lett. B 1996, **372**: 45–52.

[18] K. Pohlmeyer and K.-H. Rehren, *Reduction of the two-dimensional* $O(n)$ *nonlinear* σ*-model*, J. Math. Phys. 1979, **20**: 2628–2632.

[19] J. Langer and R. Perline, *Curve motion inducing modified Korteweg-de Vries systems*, Phys. Lett. A 1998, **239**: 36–40.

[20] G. Mari Beffa, J. Sanders, and J.-P. Wang, *Integrable systems in three-dimensional Riemannian geometry*, J. Nonlinear Sci. 2002, **12**: 143–167.

[21] J. Sanders and J.-P. Wang, *Integrable systems in* n *dimensional Riemannian geometry*, Moscow Mathematical Journal 2003, **3**: 1369–1393.

[22] S. Kobayashi and K. Nomizu, *Foundations of Differential Geometry* Volumes I and II, Wiley, 1969.

[23] R. Bishop, *There is more than one way to frame a curve*, Amer. Math. Monthly 1975, **82**: 246–251.

[24] A. Sagle and R. Walde, *Introduction to Lie Groups and Lie Algebras*, Academic Press, 1973.

[25] S.C. Anco, in preparation (2006).

[26] I. Dorfman, *Dirac Structures and Integrability of Nonlinear Evolution Equations*, Wiley, 1993.

[27] P.J. Olver, *Applications of Lie Groups to Differential Equations*, Springer, New York, 1986.

[28] J.-P. Wang, *Symmetries and Conservation Laws of Evolution Equations*, PhD Thesis, Vrije Universiteit, Amsterdam 1998.

[29] A. Sergyeyev, *Why nonlocal recursion operators produce local symmetries: new results and applications*, J. Phys. A: Math. Gen. 2005, **38**: 3397–3407.

[30] A. Sergyeyev, *The structure of cosymmetries and a simple proof of locality for hierarchies of symmetries of odd order evolution equations*, in Proceedings of Institute of Mathematics of NAS of Ukraine (conference on "Symmetry in Nonlinear Mathematical Physics") 2004, **50**(Part I): 238–245.

CONFORMAL KILLING SPINORS AND THE HOLONOMY PROBLEM IN LORENTZIAN GEOMETRY – A SURVEY OF NEW RESULTS –

HELGA BAUM*

Abstract. This paper is a survey of recent results about conformal Killing spinors in Lorentzian geometry based on a lecture given during the Summer Program *Symmetries and Overdetermined Systems of Partial Differential Equations* at IMA, Minnesota, 17.07.06 - 04.08.06. In particular, we will focus on a special class of geometries admitting conformal Killing spinors – on Brinkmann spaces with parallel spinors. We will discuss their holonomy groups and the global realizability as globally hyperbolic spaces with complete Cauchy surfaces.

Key words. Lorentzian manifold, spinor, conformal Killing spinor, twistor spinor, twistor equation, Brinkmann space, holonomy group, parallel spinor, globally hyperbolic manifold.

AMS(MOS) subject classifications. 53C29, 53C27, 53B50, 53C15.

1. Introduction. Conformal Killing spinors are solutions of a special system of overdetermined PDE's - the twistor equation on spinors - which is conformally invariant. Conformal Killing spinors give rise to conformal Killing vector fields and conformal Killing forms. In the last years much progress was made in classification of special Lorentzian structures which admit conformal Killing spinors. In 2004, F Leitner gave a local classification of all Lorentzian geometries with 'generic' conformal Killing spinors ([20]). Conformal Killing spinors appear on Lorentzian Einstein Sasaki manifolds and on Fefferman spaces. These both classes of geometries are well known and well studied, cf. for example [2, 3, 9, 6, 13]. Besides these geometries, Brinkmann spaces with parallel spinors appear, a class of Lorentzian manifolds with special holonomy. For that reason, I will focus in this survey on recent results on the classification of Lorentzian holonomy groups, in particular those which admit parallel spinors, and their realization as globally hyperbolic Lorentzian manifolds.

In Section 2 I will introduce conformal Killing spinors, describe some basic properties and the local classification result of F. Leitner. In Section 3 I give an outline on holonomy groups, in particular on the relation between parallel spinors and holonomy groups of metrics. Section 4 explains the classification results for holonomy groups of Riemannian and of Lorentzian manifolds, the latter recently obtained by Th. Leistner ([18]). In the final section I will describe own results on the realization of Lorentzian holonomy groups by globally hyperbolic manifolds, proved in [7]. This includes a con-

*Department of Mathematics, Humboldt University of Berlin, 10099 Berlin, Unter den Linden 6, GERMANY (baum@mathematik.hu-berlin.de).

struction principle for globally hyperbolic Brinkmann spaces with parallel spinors and complete Cauchy surfaces.

2. Conformal Killing spinors and overdetermined PDE's. In this section we give an introduction to conformally invariant differential equations on spinors. We define the notion of conformal Killing spinors, which are special solutions of a system of overdetermined PDE's on spinors and describe recent results on conformal Killing spinors on Lorentzian spin manifolds.

2.1. Conformally invariant operators on spinors. Let $(M^{p,q}, g)$ be a space- and time-oriented pseudo-Riemannian manifold of signature (p, q) and dimension $n = p + q \geq 3$. We denote by P the bundle of all space- and time-oriented g-orthonormal frames of (M, g). We suppose, that the second Stiefel-Whitney class of M vanishes: $w_2(M) = 0$. Then (M, g) admits a *spin structure*, that is a principal fibre bundle Q over M with structure group $Spin_0(p, q)$ together with a double covering map $f : Q \to P$ which commutes with the group actions and the projections of the principal fibre bundles. Then we have a special complex vector bundle on (M, g), the *spinor bundle*, which is the associated vector bundle

$$S := Q \times_{Spin_0(p,q)} \Delta_{p,q},$$

where $\Delta_{p,q}$ denotes the standard spinor modul. Details on these notions can be found in [17].

The spinor bundle S is equipped with a hermitian (in general indefinite) inner product $\langle \cdot, \cdot \rangle$, with the spinor derivative $\nabla^S : \Gamma(S) \to \Gamma(T^*M \otimes S)$ and the Clifford multiplication

$$\begin{array}{rccc} \mu : & TM \otimes S & \longrightarrow & S \\ & (X, \varphi) & \longmapsto & X \cdot \varphi , \end{array}$$

which satisfy the following rules:

- $(X \cdot Y + Y \cdot X) \cdot \varphi = -2g(X, Y)\, \varphi$
- $\langle X \cdot \varphi, \psi \rangle = (-1)^{p-1} \langle \varphi, X \cdot \psi \rangle$
- $\nabla^S_X (Y \cdot \varphi) = (\nabla^g_X Y) \cdot \varphi + Y \cdot \nabla^S_X \varphi$
- $X(\langle \varphi, \psi \rangle) = \langle \nabla^S_X \varphi, \psi \rangle + \langle \varphi, \nabla^S_X \psi \rangle.$

For $\sigma \otimes \varphi \in T^*M \otimes S$ we define the Clifford multiplication of σ and φ by

$$\mu(\sigma \otimes \varphi) := \sigma^\sharp \cdot \varphi,$$

where $\sigma^\sharp \in TM$ is given by $g(\sigma^\sharp, \cdot) = \sigma$. Then we can decompose the bundle $T^*M \otimes S$ of 1-forms with values in the spinor bundle into two subbundles

$$T^*M \otimes S \simeq Im\, \mu \oplus Ker\, \mu \simeq S \oplus Ker\, \mu\,.$$

This gives us two differential operators of first order by composing the spinor derivative ∇^S with the orthogonal projections onto each of these subbundles, the *Dirac operator* D

$$D : \Gamma(S) \xrightarrow{\nabla^S} \Gamma(T^*M \otimes S) \simeq \Gamma(S \oplus Ker\,\mu) \xrightarrow{pr_S} \Gamma(S)$$

and the *twistor operator* P

$$P : \Gamma(S) \xrightarrow{\nabla^S} \Gamma(T^*M \otimes S) \simeq \Gamma(S \oplus Ker\,\mu) \xrightarrow{Ker\,\mu} \Gamma(Ker\,\mu).$$

Locally, these operators are given by the following formulas

$$D\varphi = \sum_{i=1}^{n} \sigma^i \cdot \nabla^S_{s_i} \varphi$$

$$P\varphi = \sum_{i=1}^{n} \sigma^i \otimes (\nabla^S_{s_i} \varphi + \frac{1}{n} s_i \cdot D\varphi),$$

where $(s_1, \ldots, s_n)$ is a local g-orthonormal basis and $(\sigma^1, \ldots, \sigma^n)$ denotes the dual basis. Both operators, which depend on the metric g by definition, are conformally covariant. More exactly, if $\tilde{g} = e^{2\sigma} g$ is a conformal change of the metric, the Dirac and the twistor operator satisfy

$$D(\tilde{g}) = e^{-\frac{n+1}{2}\sigma} D(g) e^{\frac{n-1}{2}} \sigma$$

$$P(\tilde{g}) = e^{-\frac{\sigma}{2}} P(g) e^{-\frac{\sigma}{2}}.$$

2.2. Conformal Killing spinors and special Lorentzian geometries. The spinor fields $\varphi \in \Gamma(S)$ we are interested in are solutions of the conformally invariant *twistor equation*

$$P\varphi = 0$$

and are called *conformal Killing spinors.* Using the local formula for the twistor operator one obtains that the twistor equation $P\varphi = 0$ is equivalent to the following overdetermind system of PDE

$$\nabla^S_X \varphi + \frac{1}{n} X \cdot D\varphi = 0 \qquad \text{for all vector fields } X \text{ on } M. \tag{2.1}$$

Obviously, *parallel spinors*, spinor fields φ with $\nabla^S \varphi = 0$, are conformal Killing spinors.

As for conformal vector fields, conformal Killing tensors or conformal Killing forms, the dimension of the space of conformal Killing spinors is finite. If (M^n, g) is conformally flat and simply connected, this dimension is maximal and equals $2^{[\frac{n}{2}]+1}$. Moreover, any metric g with this maximal number of independent solutions of the conformal Killing spinor equation is conformally flat. In the case of the flat pseudo-Euklidean space $\mathbb{R}^{p,q}$, the spinor fields can be identified with smooth functions $C^\infty(\mathbb{R}^{p,q}, \Delta_{p,q})$. Using the conformal Killing spinor equation (2.1) it is easy to check that a function φ is a conformal Killing spinor if and only if it has the form

$$\varphi(x) = u + x \cdot v\,, \qquad x \in \mathbb{R}^{p,q}\,,$$

where u and v are fixed spinors in $\Delta_{p,q}$.

Now, we focus on the Lorentzian case and describe (non-conformally flat) Lorentzian geometries which admit conformal Killing spinors. Let us note, that any spinor field $\varphi \in \Gamma(S)$ defines a vector field V_φ on M, the so-called *Dirac current*, by

$$g(V_\varphi, X) = -\langle X \cdot \varphi, \varphi\rangle \qquad \text{for all vector fields } X.$$

In the Lorentzian case the Dirac current V_φ is timelike or lightlike in any point, and has the same zeros as the spinor field φ. If φ is a conformal Killing spinor, than the Dirac current V_φ is a conformal vector field. This explains the name 'conformal Killing spinor'. We call a conformal Killing spinor φ on the Lorentzian manifold (M, g) *generic* if φ has no zeros, V_φ does not change the causal type and the dual 1-form $V_\varphi^\flat$ has constant rank, where the rank of a 1-form σ is the number $rank\,\sigma = max\{k \mid \sigma \wedge (d\sigma)^k \neq 0\}$. Recently, Felipe Leitner obtained the following local classification result for Lorentzian manifolds with *generic* conformal Killing spinors (cf. [20]).

PROPOSITION 2.1 (F. Leitner, 2004). *Let (M, g) be a Lorentzian manifold with a generic conformal Killing spinor, then (M, g) is locally conformal equivalent to one of the following spaces.*

1. *A product of $(\mathbb{R}, -dt^2)$ with a Ricci-flat Riemannian manifold with parallel spinors.*
2. *A Lorentzian Einstein-Sasaki manifold.*
3. *A product of a Lorentzian Einstein-Sasaki manifold with a Riemannian manifold (N, h), where (N, h) is Riemannian Einstein-Sasaki, 3-Sasaki, nearly Kähler or a Riemannian sphere.*
4. *A Fefferman space.*
5. *A Brinkman space with parallel spinor.*

There are partial results on *non-generic* conformal Killing spinors. Conformal Killing spinors with zeros were studied by F. Leitner and Ch. Frances (cf. [21, 22, 14]). By a result of Ch. Frances, in a neighborhood of any zero of a conformal Killing spinor the Lorentzian manifold is conformally flat. In particular, an analytic Lorentzian manifold which admits a conformal Killing spinor with zeros is conformally flat. F. Leitner constructed a (non-analytic) non-conformally flat 5-dimensional Lorentzian manifold with a conformal Killing spinor with zero.

All Lorentzian geometries appearing in Proposition 2.1 in points 1.–4. admit global solutions of the conformal Killing spinor equation. They are well studied and quite well understood (cf. for example [2–6, 9]). The cases 1. and 5. in Proposition 2.1 are those, where the conformal Killing spinor φ is parallel. In this case the Dirac current V_φ is parallel as well. If it is timelike, the Lorentzian manifold splits into a product of a timelike line $(\mathbb{R}, -dt^2)$ and a Riemannian manifold with a parallel spinor. Such a Riemannian manifold is always Ricci-flat (case 1. of Proposition 2.1). If the parallel vector field V_φ is lightlike, we are in case 5. of Proposition 2.1. The aim of the following sections is to study this 5. type of geometries, the Brinkmann spaces with parallel spinors.

A *Brinkmann space* is a Lorentzian manifold with a parallel light-like vector field. Such a Lorentzian manifold has special holonomy. In order to describe the Brinkmann spaces with parallel spinors, one has to study holonomy groups for Lorentzian metrics. In the last three years much progress was made in the classification of Lorentzian holonomy groups. Meanwhile, the holonomy groups of simply connected Lorentzian manifolds are completely known as well as those holonomy groups, which admit parallel spinors. We will describe these results in section 4.2.

3. Holonomy groups of spin connections and metrics. In this section we want to explain shortly, how parallel spinors are related to holonomy groups of metrics.
Let E be a vector bundle over M with covariant derivative ∇ and let $x \in M$. For any piecewise smooth loop γ which is closed in x we denote by $\mathcal{P}^{\nabla}_{\gamma} : E_x \longrightarrow E_x$ the parallel transport along γ. The *holonomy group of* (E, ∇) *with respect to* x is the Lie group

$$Hol_x(E, \nabla) := \{\mathcal{P}^{\nabla}_{\gamma} \mid \gamma \text{ loop in } x\} \subset GL(E_x).$$

Let P be a G-principal fibre bundle over M with principal bundle connection ω and let $p \in P_x$. For a loop $\gamma : [0,1] \to M$, closed in x, we denote by γ^*_p the ω-horizontal lift with starting point p. The *holonomy group of* (P, ω) *with respect to* p is the Lie group

$$Hol_p(P, \omega) = \{g \in G \mid \exists \text{ loop in } x \text{ such that } \gamma^*_p(1) = p \cdot g\} \subset G.$$

Now, let $\rho : G \longrightarrow GL(V)$ be a representation of the Lie group G, and consider for E the associated vector bundle $E := P \times_G V$ and for ∇ the covariant derivative $\nabla := \nabla^{\omega}$ on E induced by ω. The point $p \in P_x$ gives an isomorphism between the vector spaces E_x and V and, using this identification, the holonomy groups are isomorphic:

$$Hol_x(E, \nabla^{\omega}) \simeq \rho(Hol_p(P, \omega)).$$

A key fact is the
HOLONOMY PRINCIPLE: There is a 1-1 correspondence between the space of parallel sections

$$\{\varphi \in \Gamma(E) \mid \nabla^{\omega}\varphi = 0\}$$

and the space of holonomy invariant vectors

$$\{v \in V \mid \rho(Hol_p(P,\omega))v = v\}.$$

If M is simply connected, the holonomy group is connected and therefore

$$\{v \in V \mid \rho(Hol_p(P,\omega))v = v\} = \{v \in V \mid \rho_*(\mathfrak{hol}_p(P,\omega))v = 0\}.$$

Let us look more closely to the spin situation. We denote by ω^{LC} the Levi-Civita connection on the frame bundle P and by $\tilde{\omega}^{LC}$ the lift of ω^{LC} to the principal fibre bundle Q, which is canonically given by the 2-fold covering $f : Q \longrightarrow P$. By $\lambda : Spin(p,q) \longrightarrow SO(p,q)$ we denote the 2-fold covering of the pseudo-orthogonal group $SO(p,q)$ by the Spin group $Spin(p,q)$ and by $\kappa : Spin(p,q) \longrightarrow GL(\Delta_{p,q})$ the spinor representation. We have

$$TM = P \times_{SO_0(p,q)} \mathbb{R}^{p,q} \qquad \text{and} \qquad S = Q \times_{Spin_0(p,q)} \Delta_{p,q}.$$

Let $p \in P_x$ be an orthonormal frame in $x \in M$ and $q \in Q_x$ a spin frame with $f(q) = p$. For our special situation we obtain

$$Hol_x(TM, \nabla^g) \simeq Hol_p(P, \omega^{LC}) \subset SO(p,q)$$
$$Hol_x(S, \nabla^S) \simeq \kappa(Hol_q(Q, \omega^{LC})) \subset \kappa(Spin(p,q)).$$

Moreover, it holds

$$\lambda(Hol_q(Q, \tilde{\omega}^{LC})) = Hol_x(TM, \nabla^g)$$
$$\mathfrak{hol}_q(Q, \tilde{\omega}^{LC}) = \lambda_*^{-1}\mathfrak{hol}_x(TM, \nabla^g).$$

This shows that one can decide the existence of parallel spinors on a spin manifold (M,g) if one knows the holonomy group

$$Hol(M,g) := Hol(TM, \nabla^g)$$

of (M,g).

PROPOSITION 3.1. *Let $(M^{p,q}, g)$ be a simply connected (pseudo-)Riemannian spin manifold. Then there is a 1-1 correspondence between the space of parallel spinors*

$$\{\varphi \in \Gamma(S) \mid \nabla^S\varphi = 0\}$$

and the space of holonomy invariant vectors in the spin modul

$$\{v \in \Delta_{p,q} \mid \kappa_*(\lambda_*^{-1}(\mathfrak{hol}(M,g)))\, v = 0\}.$$

This proposition shows the use of holonomy groups in studying special geometries with parallel spinors. In the next section we give a survey of new results on holonomy groups of Lorentzian manifolds. Since we will need it, we first recall the known results on Riemannian holonomy groups.

4. Holonomy groups of Riemannian and Lorentzian manifolds.

4.1. Riemannian manifolds. Let (M^n, g) be a complete and simply-connected *Riemannian* manifold. We want to describe the possible holonomy groups. First, by the DeRham Splitting Theorem, we can reduce this problem to irreducible manifolds, i.e. to manifolds with an irreducible holonomy representation.

PROPOSITION 4.1 (G. DeRham 1952). *Let (M^n, g) be a complete, simply-connected Riemannian manifold. Then (M^n, g) is isometric to a Riemannian product*

$$(M^n, g) \simeq \mathbb{R}^k \times (M_1, g_1) \times \cdots \times (M_k, g_k),$$

where (M_i, g_i) are irreducible.

The irreducible holonomy representations are collected in the 'Berger list'.

PROPOSITION 4.2 (M. Berger 1955). *Let (M^n, g) be an irreducible simply-connected non-locally symmetric Riemannian manifold. Then the holonomy group $Hol(M, g)$ is (up to conjugation) one of the following once*

$SO(n)$	*generic type*	0
$U(\frac{n}{2})$	*Kähler*	0
$SU(\frac{n}{2})$	*Ricci-flat, Kähler*	2
$Sp(\frac{n}{4})$	*Hyperkähler*	$\frac{n}{4} + 1$
$Sp(\frac{n}{4}) \cdot Sp(1)$	*quaternionic Kähler*	0
G_2	$n = 7$, *special parallel 3-form*	1
$Spin(7)$	$n = 8$, *special parallel 4-form*	1

The last column shows the number of linearly independent parallel spinors on (M^n, g).

For the symmetric case one has

PROPOSITION 4.3. *Let (M, g) be a simply-connected Riemannian symmetric space with $M = G/K$, where $G \subset Isom(M, g)$ is the transvection group of M and $K = G_x$ the stabilizer of a point $x \in M$. Then*

1. *$Hol_x(M, g) \simeq K$.*
2. *The holonomy representation $Hol_x(M, g) \longrightarrow SO(T_xM, g_x)$ is given by the isotropy representation of K.*
3. *If (M, g) is non-flat, there are no parallel spinors on (M, g).*

Since all Riemannian symmetric spaces are classified, the isotropy representation of K and hence the holonomy groups are known also in this case. The third statement of Proposition 4.3 is a consequence of the following two facts: Any Riemannian manifold with a parallel spinor is Ricci-flat and any Ricci-flat Riemannian homogeneous space is flat.

4.2. Lorentzian manifolds. Now, let (M^n, g) be a complete and simply-connected *Lorentzian* manifold. As in the Riemannian case we can reduce the problem to weakly irreducible manifolds by the Splitting Theorem of Wu.

PROPOSITION 4.4 (H. Wu 1967). *Let (M^n, g) be a complete, simply-connected Lorentzian manifold. Then (M^n, g) is isometric to a product*

$$(M^n, g) \simeq (N, h) \times (M_1, g_1) \times \cdots \times (M_k, g_k),$$

where (M_i, g_i) are flat or irreducible Riemannian manifolds and (N, h) is a Lorentzian manifold which is either

- *flat*
- *irreducible or*
- *weakly irreducible and non-irreducible, i.e. the holonomy representation $\rho : Hol(N, h) \longrightarrow SO(T_xM, g_x)$ has no non-degenerate invariant subspace, but a degenerate invariant one.*

The next Proposition shows, that there is no special irreducible Lorentzian holonomy group. In fact, $SO_0(1, n-1)$ is the only irreducible, connected subgroup of $O(1, n-1)$. There are different proofs of this statement, a geometric one by Di Scala and Olmos ([12]), a dynamical one by Boubel and Zeghib ([11]), and an algebraic one by Benoist and de la Harpe ([10]). For the special situation of Lorentzian holonomy groups it follows also from Berger's list of irreducible holonomy groups of pseudo-Riemannian manifolds.

PROPOSITION 4.5. *If the holonomy group $Hol(N, h)$ of a simply connected Lorentzian manifold (N, h) acts irreducible, then*

$$Hol(N, h) = SO_0(1, n-1) .$$

It remains to discuss the case of weakly-irreducibly but non-irreducibly acting holonomy groups. Let $Hol(N, h)$ act weakly-irreducible and non-irreducible. If W is a degenerate invariant subspace, then $W \cap W^\perp = \mathbb{R}v_0$ for a light-like vector v_0. Hence

$$Hol(N, h) \subset SO(1, n-1)_{\mathbb{R}v_0} \simeq (\mathbb{R}^* \times SO(n-2)) \ltimes \mathbb{R}^{n-2}.$$

Weakly irreducible subalgebras in $\mathfrak{so}(1, n-1)_{\mathbb{R}v_0}$ were described by L. Berard-Bergery and A. Ikemakhen in [8]. Another proof was given by A. Galaev in [15].

PROPOSITION 4.6 (Berard-Bergery/Ikemakhen'93, Galaev'04)). *Let $\mathfrak{h} \subset \mathfrak{so}(1, n-1)_{\mathbb{R}v_0} = (\mathbb{R} \oplus \mathfrak{so}(n-2)) \ltimes \mathbb{R}^{n-2}$ be a weakly-irreducible subalgebra and $\mathfrak{g} := proj_{\mathfrak{so}(n-2)}(\mathfrak{h}) = \mathfrak{z}(\mathfrak{g}) \oplus [\mathfrak{g}, \mathfrak{g}] \subset \mathfrak{so}(n-2)$.*
Then there are 4 cases.

1. $\mathfrak{h} = (\mathbb{R} \oplus \mathfrak{g}) \ltimes \mathbb{R}^{n-2}$.
2. $\mathfrak{h} = \mathfrak{g} \ltimes \mathbb{R}^{n-2}$.

3. $\mathfrak{h} = (graph(\varphi) \oplus [\mathfrak{g},\mathfrak{g}]) \ltimes \mathbb{R}^{n-2}$,

where $\varphi : \mathfrak{z}(\mathfrak{g}) \to \mathbb{R}$ *is linear and surjective.*

4. $\mathfrak{h} = ([\mathfrak{g},\mathfrak{g}] \oplus graph(\psi)) \ltimes \mathbb{R}^{r}$,

where $\mathbb{R}^{n-2} = \mathbb{R}^r \oplus \mathbb{R}^s\ ,\ 0 < r, s < n-2$,
$\mathfrak{g} \subset so(\mathbb{R}^r)$,
$\psi : \mathfrak{z}(\mathfrak{g}) \to \mathbb{R}^s$ *linear and surjective.*

If the Lorentzian manifold is a Brinkmann space, only type 2. and 4. can occur for the holonomy algebra, since this are the types with a lightlike parallel vector field.

Proposition 4.6 reduces the holonomy problem to the 'orthogonal part' $\mathfrak{g} = proj_{so(n-2)}\mathfrak{hol}(N,h)$ of the holonomy algebra resp. to its Lie group. For this part, Thomas Leistner proved in 2003 the following classification result, cf. [18].

PROPOSITION 4.7 (Th. Leister'03). *Let (N^n,h) be a simply-connected Lorentzian manifold with a weakly irreducibly and non-irreducibly acting holonomy group $Hol(N,h)$ and let*

$$G := proj_{SO(n-2)}Hol(N,h) \subset SO(n-2).$$

Then

1. G is a product of Riemannian holonomy groups.

2. (N,h) has parallel spinors if and only if

$$Hol(N,h) = G \ltimes \mathbb{R}^{n-2},$$

where G is trivial or a product of groups of the form $SU(k)$, $Sp(l)$, G_2 or $Spin(7)$.

Any group appearing in Propositions 4.6 and 4.7 is in fact a holonomy group of a Lorentian manifold. For the uncoupled types 1. and 2. in Proposition 4.6 such a metric can be easily given, cf. for example [18] and [19]. For the coupled types 3. and 4. of Proposition 4.6 this was proved by A. Galaev in 2005 (cf. [16]).

PROPOSITION 4.8 (Galaev'05). *For any of the groups H appearing in Propositions 4.6 and 4.7 there is an analytic metric h on $N = \mathbb{R}\times\mathbb{R}\times\mathbb{R}^{n-2}$ with holonomy group H. The metric h has the form*

$$h_{(t,s,x)} = 2dtds + f(t,s,x)ds^2 + 2\sum_{j=1}^{n_0} u^j(s,x)dx^j ds + \sum_{j=1}^{n-2}(dx^j)^2,$$

where $f(t,s,x)$ is a function of special form in the four cases, which can be expressed by the coupling functions φ and ψ and $u^j(s,x) = A^j_{\alpha ik}x^i x^k s^{\alpha-1}$, where the coefficients $A^{\cdot}_{\dots}$ come from a basis of the orthogonal part $\mathfrak{g} = proj_{so(n-2)}\mathfrak{hol}(N,h)$.

The explicit formulas for f and $A^{\cdot}_{\dots}$ and examples can be found in [16].

Due to these results the classification of the holonomy groups of simply-connected Lorentzian manifolds is finished. It remains to find global geometric models for given Lorentzian holonomy groups. We address to the question

Which Lorentzian holonomy groups can be realized by globally hyperbolic manifolds with complete Cauchy surfaces?

In the final section we answer this question for all Lorentzian holonomy groups which admit parallel spinors.

5. Globally hyperbolic Lorentzian manifolds with special holonomy and parallel spinors. Let (M, g) be a time-oriented Lorentzian manifold and $p \in M$. We denote by $J^{\pm}(p)$ the set of all points of M which lie on causal future (+) resp. past (-) directed curves starting in p.

A Lorentzian manifold (M, g) is called *globally hyperbolic* iff

- (M, g) is strongly causal (for example if there exists a continuous function f on M which is strictly increasing along any future directed causal curve) and
- $J^+(p) \cap J^-(q) \subset M$ is compact for all $p, q \in M$.

Globally hyperbolic manifolds are of special interest for various reasons. For example, normally hyperbolic operators on globally hyperbolic manifolds have a global and unique forward and backward fundamental solution. There exist Cauchy surfaces. Moreover, for any causal related pair of points $p, q \in M$, $p \leq q$, there exists a causal geodesic of maximal length from p to q. In this respect, globally hyperbolic Lorentzian manifolds behave similar to complete Riemannian ones.

The following Proposition gives a partial answer to our question. In [7] we proved that any Lorentzian holonomy group that allows parallel spinors, can be realized by globally hyperbolic Lorentzian manifolds.

PROPOSITION 5.1 (Baum/Müller05). *Any Lorentzian holonomy group of the form*

$$G \ltimes \mathbb{R}^{n-2} \subset SO(1, n-1),$$

where $G \subset SO(n-2)$ is trivial or a product of groups $SU(k)$, $Sp(l)$, G_2 or $Spin(7)$, can be realized by a globally hyperbolic Lorentzian manifold (M^n, g) with complete Cauchy surfaces.

The idea for the construction of such metrics was inspired by the paper [1] of Ch. Bär, P. Gauduchon and A. Moroianu, who studied the spin geometry of generalized pseudo-Riemannian cylinders. We adapted the method for our purpose. First, we consider a special kind of spinor fields.

Let (M, g_0) be a Riemannian spin manifold with a Codazzi tensor A, which is a symmetric (1,1)-tensor field satisfying

$$(\nabla^{g_0}_X A)(Y) = (\nabla^{g_0}_Y A)(X) \qquad \text{for all vector fields } X, Y.$$

If A is uniformally bounded, we denote by $\mu_+(A)$ the supremum of the positive eigenvalues of A or zero if all eigenvalues are non-positive, and by $\mu_-(A)$ the infimum of the negative eigenvalues of A or zero if all eigenvalues are non-negative.

A spinor field $\varphi \in \Gamma(S)$ is called *A-Codazzi spinor* if

$$\nabla^S_X \varphi = iA(X) \cdot \varphi \qquad \text{for all vector fields } X.$$

PROPOSITION 5.2 (Bär/Gauduchon/Moroianu'04, Baum/Müller'05). *Let (M, g_0) be a complete Riemannian spin manifold with a uniformally bounded Codazzi tensor A and a non-trivial A-Codazzi spinor. Then the Lorentzian cylinder*

$$C := I \times M \ , \ g_C := -dt^2 + (1 - 2tA)^* g_0,$$

with the interval $I = ((2\mu_-(A))^{-1}, (2\mu_+(A))^{-1})$, is globally hyperbolic with special holonomy and parallel spinors and has complete Cauchy surfaces.

In order to obtain such cylinders, we have to ensure the existence of Codazzi spinors. We studied this problem for the case of invertible Codazzi tensors A.

PROPOSITION 5.3 (Baum/Müller'05). *Let (M, g_0) be a complete Riemannian manifold with an A-Codazzi spinor and let all eigenvalues of the Codazzi tensor A be uniformally bounded away from zero. Then (M, g_0) is isometric to*

$$(\mathbb{R} \times F \ , \ (A^{-1})^*(ds^2 + e^{-4s} g_F)),$$

where (F, h) is a complete Riemannian manifold with parallel spinors, and A^{-1} is a Codazzi-tensor on the warped product $(\mathbb{R} \times F, ds^2 + e^{-4s} g_F)$.

Vice versa, let (F, h) be a complete Riemannian manifold with parallel spinors and a Codazzi tensor T which has eigenvalues uniformally bounded from below. Then there is a Codazzi tensor B on the warped product

$$M = \mathbb{R} \times F \ , \ g_{wp} = ds^2 + e^{-4s}\, g_F$$

with eigenvalues uniformally bounded away from zero, B^{-1} is a Codazzi tensor on $(M, g_0 := (B^{-1})^* g_{wp})$, the Riemannian manifold (M, g_0) is complete and has B^{-1}-Codazzi spinors.

A Codazzi tensor B on a warped product

$$M = \mathbb{R} \times F \ , \ g_{wp} = ds^2 + f(s)^2\, g_F$$

with properties mentioned in Proposition 5.3 can be constructed from a Codazzi tensor T on (F, g_F) in the following way. We set

$$B := \begin{pmatrix} b \cdot Id & 0 \\ 0 & E \end{pmatrix}$$

with respect to the decomposition $TM = \mathbb{R} \oplus TF$, where b is a function depending only on s and E is given by

$$E(s) = \frac{1}{f(s)}\Big(T + \int_0^s b(\sigma)\dot{f}(\sigma)d\sigma \cdot Id_F\Big).$$

This yields a construction principle for globally hyperbolic manifolds with special holonomy and parallel spinors.

PROPOSITION 5.4 (Baum/Müller'05). *Let (F, g_F) be a complete Riemannian manifold with parallel spinors and a Codazzi tensor T with eigenvalues bounded from below. Then there are Codazzi tensors B on the warped product $(\mathbb{R} \times F, ds^2 + e^{-4s}g_F)$ with eigenvalues uniformally bounded away from zero. Let*

$$C(F,B) := I \times \mathbb{R} \times F \ , \ g_C := -dt^2 + (B - 2t)^*(ds^2 + e^{-4s}g_F).$$

Then

- *(C, g_C) is a globally hyperbolic Brinkman space with complete Cauchy surfaces.*
- *If (F, h) has a flat factor, then $C(F, B)$ is decomposable.*
- *If (F, h) is (locally) a product of irreducible factors, then $C(F, B)$ is weakly irreducible and*

$$Hol^0_{(0,0,x)}(C, g_C) = (B^{-1} \circ Hol^0_x(F, g_F) \circ B) \ltimes \mathbb{R}^{dimF}.$$

Our construction of Riemannian manifolds with Codazzi spinors and of globally hyperbolic Brinkmann spaces (Proposition 5.4) is based on the existence of Codazzi tensors on Riemannian manifolds with parallel spinors. Let us discuss some examples for that.

Example 1. On the flat space $\mathbb{R}^k$ the endomorphism

$$T_h^{\mathbb{R}^k}(X) = \nabla_X^{\mathbb{R}^k}(grad(h)) = X(\partial_1 h, \ldots, \partial_k h))$$

is a Codazzi tensor for any function h on $\mathbb{R}^k$, and every Codazzi tensor is of this form. In this case the cylinder $C(F, B)$ is flat for any Codazzi tensor B on the warped product that is constructed out of T as described above.

Example 2. Let (F_1, g_{F_1}) be a complete simply connected irreducible Riemannian spin manifold with parallel spinors and (F, g_F) its Riemannian product with a flat $\mathbb{R}^k$. Then (F, g_F) is complete and has parallel spinors. Let B be a Codazzi tensor on the warped product $\mathbb{R} \times_{e^{-2s}} F$ constructed out of the Codazzi tensor $\lambda Id_{F_1} + T_h^{\mathbb{R}^k}$ of F, where $T_h^{\mathbb{R}^k}$ is taken from Example 1. Then the cylinder $C(F, B)$ is globally hyperbolic and decomposable with special holonomy

$$Hol(F_1, g_{F_1}) \ltimes \mathbb{R}^{dimF_1}.$$

Example 3. Let us consider the metric cone

$$(F^{n-1}, g_F) := (\mathbb{R}^+ \times N, dr^2 + r^2 g_N)$$

where (N, g_N) is simply connected and a Riemannian Einstein-Sasaki manifold, a nearly Kähler manifold, a 3-Sasakian manifold or a 7-dimensional manifold with vector product. Then (F, g_F) is irreducible and has parallel spinors (but fails to be complete). Furthermore, $T := \nabla^F \partial_r$ is a Codazzi tensor on (F, g_F), since ∂_r lies in the kernel of the curvature endomorphism. Then the cylinder $C(F, B)$, where the Codazzi tensor B is constructed out of T as described above, has special holonomy

$$Hol(C, g_C) \simeq G \ltimes \mathbb{R}^{n-1}$$

where

$$G = \begin{cases} SU((n-1)/2) & \text{if} \quad N \text{ is Einstein-Sasaki} \\ Sp((n-1)/4) & \text{if} \quad N \text{ is 3-Sasakian} \\ G_2 & \text{if} \quad N \text{ is nearly Kähler} \\ Spin(7) & \text{if} \quad N \text{ 7-dimensional with vector product.} \end{cases}$$

Example 4. Let $(F, g_F) = (F_1, g_{F_1}) \times \cdots \times \ldots (F_k, g_{F_k})$ be a Riemannian product of simply connected complete irreducible Riemannian manifolds with parallel spinors. Let T be the Codazzi tensor $T = \lambda_1 \mathbf{1}_{F_1} + \ldots \lambda_k \mathbf{1}_{F_k}$ and B constructed out of T as described above. Then $C(F, B)$ is globally hyperbolic, weakly irreducible and the holonomy group is isomorphic to

$$(Hol(F_1, g_{F_1}) \times \cdots \times Hol(F_k, g_{F_k})) \ltimes \mathbb{R}^{dimF}.$$

Example 5. Eguchi-Hansen space. Eguchi-Hansen spaces are complete, irreducible Riemannian 4-manifolds with holonomy $SU(2)$. Hence they have 2 linearly independent parallel spinors. Any Codazzi tensor on a Eguchi-Hansen space has the form $T = \lambda Id$ for a constant λ. (cf. [7]).

REFERENCES

[1] CH. BÄR, P. GAUDUCHON, AND A. MOROIANY, *Generalized cylinders in semi-Riemannian and spin geometry*, Math. Zeitschrift, **249** (2005), 545–580.

[2] H. BAUM, *Lorentzian twistor spinors and CR-geometry*, Diff. Geom. and its Appl., **11** (1999), 69–96.

[3] H. BAUM, *Conformal Killing spinors and special geometric structures in Lorentzian geometry – A survey*, Proc. of the Workshop on Special Geometric Structures in String Theory, Bonn, Sept. 2001, Proc. Archive of the EMS Electronic Library of Math.

[4] H. BAUM, *The Conformal Analog of Calabi Yau manifolds*, To appear in *Handbook of pseudo-Riemannian Geometry and Supersymmetry*, Publishing House of the EMS.

[5] H. Baum, T. Friedrich, R. Grunewald, and I. Kath, *Twistors and Killing Spinors on Riemannian Manifolds*, Volume **124** of *Teubner-Texte zur Mathematik*, Teubner-Verlag, Stuttgart/Leipzig, 1991.

[6] H. Baum and F. Leitner, *The twistor equation in Lorentzian spin geometry*, Math. Zeitschrift, **247** (2004), 795–812.

[7] H. Baum and O. Müller, *Codazzi spinors ans Globally hyperbolic Lorentzian manifolds with special holonomy*, Vienna, Preprint ESI **1757** (2005).

[8] L. Berard-Bergery and A. Ikemakhen, *On the holonomy of Lorentzain manifolds*, In: *Differential Geometry: Geometry in Mathematical Physics and Related Topics*, Volume **54** of *Proc. Symp. Pure Math.* , 27–40, AMS, 1993.

[9] Ch. Bohle, *Killing spinors on Lorentzian manifolds*, Journal Geometry and Physics, **45** (2003), 285–308.

[10] Y. Benoist and P. de la Harps, *Adherence de Zariski des groupes de Coxeter*, Compos. Math., **140** (2004), 1357–1366.

[11] C. Boubel and A. Zeghib, *Isometric actions of Lie subgroups of the Möbius group*, Nonlinearity, **17** (2004), 1677–1688.

[12] A. J. Di Scala and C. Olmos, *The geometry of homogeneous submanifolds of hyperbolic space*, Math. Zeitschrift, **237** (2001), 199–209.

[13] S. Dragomir and G. Tomassini, *Differential Geometry and Analyis on CR manifolds*, Progress in Mathematics, Vol. **246**, Birkhäuser 2006.

[14] Ch. Frances, *Causal conformal vector fields, and singularities of twistor spinors*, preprint 2006.

[15] A. Galaev, *Isometry groups of Lobachewskian spaces, similarity transformation groups of Euklidean spaces and Lorentzian holonomy groups*, arXiv:mathDG/0404426, 2004.

[16] A. Galaev, *Metrics that realize all types of Lorentzian holonomy algebras*, arXiv:mathDG/0502575, 2005.

[17] H.B. Lawson and M.-L. Michelsohn, *Spin Geometry*, Princeton University Press, 1989.

[18] Th. Leistner, *Holonomy and parallel spinors in Lorentzian geometry*, PhD-Thesis, Humboldt University of Berlin, 2003. arXiv:mathDG/0305139 and arXiv:mathDG/0309274.

[19] Th. Leistner, *Screen bundles of Lorentzian manifolds and some generalizations of pp-waves*, J. Geom. Phys., **56** (2006), 2117–2134.

[20] F. Leitner, *Conformal Killing forms with normalization condition*, Rend.Circ. Mat. Palermo, suppl. Ser II, **75** (2005), 279–292.

[21] F. Leitner, *A note on twistor spinors with zeros in Lorentzain geometry*, arXiv:mathDG/0406298.

[22] F. Leitner, *Twistor spinors with zeros on Lorentzian 5-spaces*, arXiv: mathDG/0602622.

PROJECTIVE-TYPE DIFFERENTIAL INVARIANTS FOR CURVES AND THEIR ASSOCIATED PDES OF KDV TYPE

GLORIA MARÍ BEFFA*

Abstract. In this paper we present an overview of the direct relation between differential invariants of projective type for curves in flat semisimple homegenous spaces and PDEs of KdV type. We describe the progress in the proof of a conjectured Theorem stating that for any such space there are geometric evolutions of curves that induce completely integrable evolutions on their invariants of projective-type. The Theorem also states that these evolutions decouple always into either complexly coupled KdV equations (conformal type), decoupled KdV equations (Lagrangian type) or Adler-Gelfand-Dikii generalized KdV types (projective type). The paper also describes the fundamental role of group-based moving frames in this study.

Key words. Completely integrable systems, moving frames, projective differential invariants, curves in semisimple homogeneous spaces, differential invariants of curves, infinite dimensional Poisson geometry.

AMS(MOS) subject classifications. 37K25, 37K30, 37K10.

1. Moving frames and differential invariants of projective-type for curves in G/H. The classical concept of moving frame was developed by Élie Cartan ([2], [3]). A classical moving frame along a curve in a manifold M is a curve in the frame bundle of the manifold over the curve, invariant under the action of the transformation group under consideration. This method is a very powerful tool but its explicit application relied on intuitive choices that were not clear on a general setting. Some ideas in Cartan's work and later work of Griffiths ([7]), Green ([8]) and others laid the foundation for the concept of a group-based moving frame, that is, an equivariant map between the jet space of curves in the manifold and the group of transformations. Recent work by Fels and Olver ([4, 5]) finally gave the precise definition of the group based moving frame and extended its application beyond its original geometric picture to an astonishingly large group of applications. In this section we will describe Fels and Olver moving frame together with its relation to the classical moving frame. We will also introduce some definitions useful to our study of geometric Poisson brackets and the relation to completely integrable systems of KdV type. From now on we will assume $M = G/H$ with G acting on M via left multiplication. We will also assume that curves in M are *parametrized* and, therefore, the group G does not act on the parameter.

DEFINITION 1.1. Let $J^k(\mathbb{R}, M)$ the k-jet space of curves, that is, the set of equivalence classes of curves in M up to k^{th} order of contact. If we denote by $u(x)$ a curve in M and by u_r the r derivative of u with respect

*Department of Mathematics, University of Wisconsin, Madison, WI 53506 (maribeff@math.wisc.edu).

to the parameter x, $u_r = \frac{d^r u}{dx^r}$, the jet space has local coordinates that can be represented by $u^{(k)} = (x, u, u_1, u_2, \dots, u_k)$. The group G acts naturally on parametrized curves, therefore it acts naturally on the jet space via the formula

$$g \cdot u^{(k)} = (x, g \cdot u, (g \cdot u)_1, (g \cdot u)_2, \dots)$$

where by $(g \cdot u)_k$ we mean the formula obtained when one differentiates $g \cdot u$ and then writes the result in terms of g, u, u_1, etc. This is usually called the *prolonged* action of G on $J^k(\mathbb{R}, M)$.

DEFINITION 1.2. A function

$$I : J^k(\mathbb{R}, M) \to \mathbb{R}$$

is called a kth order *differential invariant* if it is invariant with respect to the prolonged action of G.

DEFINITION 1.3. A map

$$\rho : J^k(\mathbb{R}, M) \to G$$

is called a left (resp. right) *moving frame* if it is equivariant with respect to the prolonged action of G on $J^k(\mathbb{R}, M)$ and the left (resp. right) action of G on itself.

The direct relation between classical moving frames and group-based moving frames is contained in the following theorem.

THEOREM 1.1. *([13]) Let $\Phi_g : G/H \to G/H$ be defined by multiplication by g. That is $\Phi_g([x]) = [gx]$. Let ρ be a group-based left moving frame with $\rho \cdot u = o$ where $o = [H] \in G/H$. Identify $d\Phi_\rho(o)$ with an element of $GL(n)$, where n is the dimension of M.*

Then, the matrix $d\Phi_\rho(o)$ contains in its columns a classical moving frame.

Intuitively, this Theorem illustrates how classical moving frames are described only by the action of the group on first order frames, while group-based moving frames are described by the complete action. Next we will introduce the equivalent to the classical Serre–Frenet equations. This concept if fundamental in our Poisson geometry study. Recall that the traditional road to finding a generating set of differential invariants for a curve is to differentiate the classical moving frame along the curve. The differential invariants are found as coefficients of these equations. Unfortunately this approach only works if the action of the group is affine. For example, in the case of $M = O(n+1,1)/H$, the *Möbius sphere* or local model for flat conformal manifolds, classical Serre–Frenet equations fail to produce two differential invariants for curves ([6]), while in the case $M = \mathrm{PSL}(n)/H \cong \mathbb{RP}^n$ the Serre–Frenet equations fail to produce *any* invariant at all. The missing invariants are those generated by the action of the group on higher order frames.

DEFINITION 1.4. Consider Kdx to be the horizontal component of the pullback of the left-invariant Maurer-Cartan form of the group G via a group-based left moving frame ρ. That is

$$K = \rho^{-1}\rho_x \in \mathfrak{g}$$

(K is the coefficient matrix of the equation satisfied by ρ). We call K the *Serre–Frenet equations* for the moving frame ρ.

The crucial property that makes the group-based picture useful to our study is the fact that, if the moving frame is properly chosen, a complete set of generating differential invariants can always be found among the coefficients of the Serre–Frenet equations. By a moving frame of minimal order we mean one for which $\rho \cdot u^{(k)}$ has as many constant coordinates as is allowed by the rank of the prolonged action on $J^k(\mathbb{R}, M)$ (for example, if the rank is full, $\rho \cdot u^{(k)}$ should be constant). For more information on normalization constants, please see [4, 5].

THEOREM 1.2. *(A direct consequence from [4, 5]) Let ρ be a (left or right) moving frame of minimal order for a curve u. Then, the coefficients of the (left or right) Serre–Frenet equations for ρ contain a basis for the set of differential invariants of the curve. That is, any other differential invariant for the curve is a function of the entries of K and their derivatives with respect to x.*

From now on we will assume that G/H is flat and G is semisimple. Then ([20]) the Lie algebra $\mathfrak{g}$ has a gradation of length 1, that is

$$\mathfrak{g} = \mathfrak{g}_{-1} \oplus \mathfrak{g}_0 \oplus \mathfrak{g}_1 \tag{1.1}$$

with $\mathfrak{h} = \mathfrak{g}_0 \oplus \mathfrak{g}_1$, where $\mathfrak{h}$ is the Lie algebra of H. Therefore, locally G splits as $G_{-1} \cdot G_0 \cdot G_1$ with H given by $G_0 \cdot G_1$. The subgroup G_0 is called *the isotropic subgroup* of G and it is the component of G that acts linearly on G/H (for more information see [20]).

Choose a minimal order moving frame ρ with $\rho \cdot u = o$ and let $K = \rho^{-1}\rho_x$ be its Serre–Frenet equations. We split K according to the gradation (1.1), that is

$$K = K_1 + K_0 + K_{-1}.$$

It is known ([12]) that, if the moving frame is minimal, the term K_{-1} will always be constant.

DEFINITION 1.5. We call *differential invariants of projective type* those invariants that appear in the component K_1.

These invariants are those that are failed to be produced by the classical Serre–Frenet equations because they are generated by the action of the group on higher frames.

2. Geometric Poisson brackets. In this section we will introduce the definition of geometric Poisson bracket, that is, Hamiltonian structures on the space of differential invariants of curves. For a complete description and proofs see [12, 13].

Consider the group of Loops $\mathcal{L}G = C^\infty(S^1, G)$ and its Lie algebra $\mathcal{L}\mathfrak{g} = C^\infty(S^1, \mathfrak{g})$. One can define two natural Poisson brackets on $\mathcal{L}\mathfrak{g}^*$ (see [22] for more information), namely, if $\mathcal{H}, \mathcal{F} : \mathcal{L}\mathfrak{g}^* \to \mathbb{R}$ are two functionals defined on $\mathcal{L}\mathfrak{g}^*$ and if $L \in \mathcal{L}\mathfrak{g}^*$, we define

$$\{\mathcal{H}, \mathcal{F}\}_1(L) = \int_{S^1} \left\langle -\left(\frac{\delta \mathcal{H}}{\delta L}(L)\right)_x + ad^*(L)\left(\frac{\delta \mathcal{H}}{\delta L}(L)\right), \frac{\delta \mathcal{F}}{\delta L}(L)\right\rangle dx \quad (2.1)$$

where $\langle , \rangle$ is the natural coupling between $\mathfrak{g}$ and $\mathfrak{g}^*$, and where $\frac{\delta \mathcal{H}}{\delta L}(L)$ is the variational derivative of $\mathcal{H}$ at L identified, as usual, with an element of $\mathcal{L}\mathfrak{g}$.

One also has a family of second brackets, namely

$$\{\mathcal{H}, \mathcal{F}\}_2(L) = \int_{S^1} \left\langle ad^*(L_0)\left(\frac{\delta \mathcal{H}}{\delta L}(L)\right), \frac{\delta \mathcal{F}}{\delta L}(L)\right\rangle dx \quad (2.2)$$

where $L_0 \in \mathfrak{g}^*$ is any constant element. Since $\mathfrak{g}$ is semisimple we can identify $\mathfrak{g}$ with its dual $\mathfrak{g}^*$ and we will do so from now on.

The basis for the definition of a geometric Poisson bracket is contained in the following Theorem.

THEOREM 2.1. *([12]) Let $M = G/H$ be a flat semisimple manifold, with $\mathfrak{g}$ having a 1-gradation as in (1.1). Assume that for every curve in M we choose a minimal moving frame ρ with $\rho \cdot u = o$ and $\rho \cdot u_1 = \Lambda \in \mathfrak{g}_{-1}$ (we are identifying the tangent with $\mathfrak{g}_{-1}$). Then, locally around a given curve u, the space $\mathcal{K}$ of elements $K \in \mathcal{L}\mathfrak{g}$ defined by the Serre–Frenet equations for ρ can be described as a quotient $U/\mathcal{N}$, where U is an open set of $\Lambda \oplus \mathcal{L}\mathfrak{g}_0 \oplus \mathcal{L}\mathfrak{g}_1$ and where $\mathcal{N} = \mathcal{L}\mathcal{N}_0 \cdot \mathcal{L}G_1 \subset \mathcal{L}G_0 \cdot \mathcal{L}G_1$ acts on U via the Kac-Moody action*

$$a(n)(L) = n^{-1} n_x + n^{-1} L n. \quad (2.3)$$

The subgroup $\mathcal{N}_0$ is the isotropy subgroup of Λ in G_0.

Once we see that $\mathcal{K}$ (a space we can call the *space of differential invariants*) can be written as a quotient, the idea of Poisson reduction is unavoidable. Indeed, the *symplectic leaves* (or orbits where Hamiltonian flows lie) of bracket (2.1) are given by the orbits of $\mathcal{L}\mathfrak{g}^*$ under the action of $\mathcal{L}G$ defined by (2.3). Thus, one gets the following Theorem.

THEOREM 2.2. *([12]) The Poisson bracket (2.1) can be reduced to $\mathcal{K}$ so that there exists a well-defined Poisson bracket $\{,\}$ defined on a generating set of independent differential invariants. We call this bracket a geometric Poisson bracket.*

It is not true in general that (2.2) is also reducible to $\mathcal{K}$. In fact, one finds that for $M = \mathbb{RP}^n$ and $G = \mathrm{PSL}(n+1)$ the second bracket (2.2) is also reducible to $\mathcal{K}$ when $L_0 = \Gamma^* \in \mathfrak{g}^*$, while if M is the so-called Lagrangian Grassmannian, $G = \mathrm{Sp}(4)$, the second bracket is never reducible to $\mathcal{K}$ ([15]). A similar construction exists for affine geometries $M = G \ltimes \mathbb{R}^n/G$, G semisimple. Again, the first bracket is always reducible while the reduction of the second depends on the manifold. On the other hand, in all known cases, once we know that the second bracket is reducible, a biHamiltonian PDE (that is, a completely integrable system) exists for both reduced brackets. For example, in the case of $M = \mathbb{RP}^n$ both brackets are reducible for $L_0 = \Gamma^*$ and the resulting brackets are the first and second Hamiltonian structures for generalized KdV equations. In the Euclidean space both structures can be reduced and the results are equivalent to the first and second Hamiltonian structures for the nonlinear Shrödinger equation. And so on. But there is more to it, one can also find a geometric evolution of curves of the form

$$u_t = F(u, u_1, u_2, \dots)$$

invariant under the group G, and such that the induced evolution on the generating system of differential invariants determined by $\mathcal{K}$ is the completely integrable system that was found using the geometric brackets. The precise relation between both equations is given by the following Theorem.

THEOREM 2.3. *([12]) Assume ρ is as in the statement of Theorem 2.1. Assume $u(t,x)$ is a solution of the evolution*

$$u_t = d\Phi_\rho(o)\mathbf{r} = r_1 V_1 + \cdots + r_n V_n \tag{2.4}$$

where $\mathbf{r} = (r_i)$ is a vector of differential invariants (recall that $d\Phi_\rho(o) = (V_1 \dots V_n)$ is a classical moving frame). Then, the evolution induced on the generating system of differential invariants defined by $\mathcal{K}$, call them $\mathbf{k}$, is Hamiltonian with associated Hamiltonian functional $h : \mathcal{K} \to \mathbb{R}$ whenever there exists an extension of h, $\mathcal{H} : U \to \mathbb{R}$, constant of the leaves of $\mathcal{N}$, such that $\left[\frac{\delta \mathcal{H}}{\delta L}(L)\right]_{-1} = \mathbf{r}$ (the subindex -1 indicates the component in K_{-1}).

Given a Hamiltonian system $\mathbf{k_t} = \xi_h(\mathbf{k})$ we will call its associated geometric evolution (2.4) *its geometric realization in M*. Thus, it is known that the Vortex filament evolution is a Euclidean geometric realization of the nonlinear Shrödinger equation, as it was shown by Hasimoto in [9]. The Schwarzian KdV equation is a projective geometric realization of the KdV equation, and so on.

A completely integrable PDE could, in principle, have more than one geometric realization. Indeed, modified KdV systems have geometric realizations in several manifolds (see [11], [18], [23], [1]) but all these manifolds share a common Riemannian structure that has a fundamental role in the generation of the Hamiltonian structures and of their associated geometric realizations. Furthermore, the known results indicate that the reduction of the second bracket (2.2) points at the existence of a completely integrable PDE induced by geometric realizations, but the geometric character of the manifold determines what type of PDE can have a geometric realization in the manifold. Indeed, the Sowada-Koterra equation has a geometric realization in the equi-affine case $M = SL(2) \ltimes \mathbb{R}^2/SL(2)$ ([19], [13]) while the KdV equation has no such realization, it has a projective one instead. But both equi-affine and projective *bi-Hamiltonian geometric brackets are the same.* The Hamiltonian structures are known to be the same for both Sowada–Koterra and KdV, but the affine character of the equi-affine manifold allows for the existence of a geometric realization of Sowada–Koterra, and not for KdV, while the projective character of $\mathbb{RP}^1$ allows for the existence of a geometric realization for KdV but not for Sowada–Koterra.

3. Completely integrable systems of KdV type associated to differential invariants of projective type. Finally, in our last section we would like to drive home the last point of our previous section by describing how general projective-type differential invariants have associated integrable PDEs, in the sense of our previous section, always of KdV type. They also have geometric realizations in G/H, assuming we restrict our initial curves to curves whose non-projective differential invariants vanish. The following Theorem was conjecture in [14].

THEOREM 3.1. *The geometric Poisson bracket associated to curves on G/H can be further restricted to the submanifold of projective differential invariants, that is, the submanifold of $\mathcal{K}$ for which all non-projective differential invariants vanish. The resulting Poisson structures and the behavior of the invariants of projective type under properly chosen geometric flows are either* projective, conformal or Lagrangian like, *always of KdV-type.*

In what remains of this paper we will describe the three types of behaviour, projective, conformal and Lagrangian. We will also describe the progress made in proving this Theorem.

3.1. The Lagrangian case. Let $G = \mathrm{Sp}(2n)$. If $g \in G$ then, locally

$$g = \begin{pmatrix} I & u \\ 0 & I \end{pmatrix} \begin{pmatrix} \Theta & 0 \\ 0 & \Theta^{-T} \end{pmatrix} \begin{pmatrix} I & 0 \\ S & I \end{pmatrix}$$

with u and S symmetric and $\Theta \in \mathrm{GL}(n)$. This factorization corresponds to the splitting given by the gradation (1.1). The subgroup H is locally defined by matrices of the form

$$\begin{pmatrix} \Theta & 0 \\ 0 & \Theta^{-T} \end{pmatrix} \begin{pmatrix} I & 0 \\ S & I \end{pmatrix}.$$

The manifold G/H is usually called the n dimensional *Lagrangian Grassmanian* and it is identified with the manifold of Lagrangian planes in $\mathbb{R}^{2n}$. The following Theorem describes the manifold $\mathcal{K}$ for curves of Lagrangian planes in $\mathbb{R}^{2n}$ under the action of $\mathrm{Sp}(2n)$.

THEOREM 3.2. *([15]) There exists a minimal order moving frame ρ along a curve of Lagrangian planes such that its Serre–Frenet equations are given by*

$$K = \rho^{-1}\rho_x = \begin{pmatrix} K_0 & I \\ K_1 & K_0 \end{pmatrix}$$

where K_0 is skew-symmetric and contains all differential invariants of order 4, and where $K_1 = -\frac{1}{2}\mathcal{S}_d$ with $\mathcal{S}_d$ diagonal and containing along the diagonal the eigenvalues of the Lagrangian Schwarzian derivative *(Ovsienko [21])*

$$\mathcal{S}(u) = u_1^{-1/2}\left(u_3 - \frac{3}{2}u_2 u_1^{-1} u_2\right) u_1^{-1/2}.$$

The entries of K_0 and K_1 are functionally independent differential invariants for curves of Lagrangian planes in $\mathbb{R}^{2n}$ under the action of Sp$(2n)$ *generating all other differential invariants. The n differential invariants that appear in K_1 are the invariants of projective type.*

Now we describe some of the geometric flows that preserve the value of K_0. Therefore, geometric flows as below will affect only invariants of projective type.

THEOREM 3.3. *([15]) Assume $u : J \subset \mathbb{R}^2 \to$* Sp$(2n)/H$ *is a flow solution of*

$$u_t = \Theta^T u_1^{1/2}\, \mathbf{h}\, u_1^{1/2}\Theta \tag{3.1}$$

where $\Theta(x,t) \in O(n)$ diagonalizes $\mathcal{S}(u)$ and where $\mathbf{h}$ *is a symmetric matrix of differential invariants. Assume* $\mathbf{h}$ *is diagonal. Then the flow preserves K_0.*

Finally, the integrable PDEs associated to geometric flows of Lagrangian planes are given in our next Theorem.

THEOREM 3.4. *([15]) Let $\mathcal{K}_1$ be the submanifold of $\mathcal{K}$ given by $K_0 = 0$. Then, the geometric bracket restricts to $\mathcal{K}_1$ to induce a decoupled system of n second Hamiltonian structures for KdV. Bracket (2.2) also reduces to $\mathcal{K}_1$ (even though it does not in general reduce to $\mathcal{K}$ for any value of L_0). The reduction for a special value of $L_0 \in g_1$ is a decoupled system of n first KdV Hamitonian structures.*

Furthermore, assume $u(t,x)$ is a flow solution of (3.1) and assume $u(0,x)$ is an initial condition for which $K_0 = 0$. Then the differential invariants $\mathcal{S}_d$ of the flow satisfies the equation

$$(\mathcal{S}_d)_t = \left(\mathbf{D}^3 + \mathcal{S}_d\mathbf{D} + (\mathcal{S}_d)_x\right)\mathbf{h}$$

where $\mathbf{D}$ *is the diagonal matrix with $\frac{d}{dx}$ down its diagonal.*

From this Theorem, it is clear that if $\mathbf{h} = \mathcal{S}_d$, then $\mathcal{S}_d$ is the solution of a *decoupled system of n KdV equations*

$$(\mathcal{S}_d)_t = (\mathcal{S}_d)_{xxx} + 3\mathcal{S}_d(\mathcal{S}_d)_x.$$

This would also be the situation for curves in another different case, that of the manifold $M = O(2n, 2n)/H$. In this particular case it was shown in [14] that the differential invariants of projective type are represented in $\mathcal{K}$ by skew-symmetric matrices that diagonalise in off-diagonal blocks. The geometric Poisson brackets behave in exactly the same way as those in the Lagrangian Grassmannian manifold. For more information see [14].

3.2. Conformal case. Consider $G = O(n+1,1)$. If $g \in G$ then, locally,

$$g = \begin{pmatrix} 1 & u^T & \frac{1}{2}u^T u \\ 0 & I & u \\ 0 & 0 & 1 \end{pmatrix} \begin{pmatrix} \alpha & 0 & 0 \\ 0 & A & 0 \\ 0 & 0 & \alpha^{-1} \end{pmatrix} \begin{pmatrix} 1 & 0 & 0 \\ v & I & 0 \\ \frac{1}{2}v^T v & v^T & 1 \end{pmatrix}$$

with $A \in O(n)$, $u, v \in \mathbb{R}^n$ and $\alpha \in \mathbb{R}$. Again, the above splitting corresponds to the gradation (1.1) and H is given by matrices with factorization

$$\begin{pmatrix} \alpha & 0 & 0 \\ 0 & A & 0 \\ 0 & 0 & \alpha^{-1} \end{pmatrix} \begin{pmatrix} 1 & 0 & 0 \\ v & I & 0 \\ \frac{1}{2}v^T v & v^T & 1 \end{pmatrix}.$$

The manifold $O(n+1,1)/H$ is *the Möbius sphere*, the local model for flat conformal manifolds. Differential invariants for conformal curves were first found in [6] and its group-based representation was described in [16]. The following Theorem describes the Serre-Frenet equations associated to minimal moving frames.

THEOREM 3.5. *([15]) There exists a minimal order (left) moving frame ρ along a curve in the Möbius sphere with $\rho \cdot u = o, \rho \cdot u_1 = e_1$ and such that its Serret-Frenet equations are given by*

$$K = \begin{pmatrix} 0 & e_1^T & 0 \\ k_1 e_1 + k_2 e_2 & K_0 & e_1 \\ 0 & k_1 e_1^T + k_2 e_2^T & 0 \end{pmatrix}$$

where k_1 and k_2 are third order conformal invariants of projective type *and where K_0 are fourth order and higher. The vectors e_i represent the standard basis in $\mathbb{R}^n$.*

In this case there are only two generating differential invariants of projective type, namely k_1 and k_2. Again, the behavior of the Poisson brackets (2.1) and (2.2) with respect to this submanifold is spotless.

THEOREM 3.6. *([15]) Let $\mathcal{K}_1$ be the submanifold of $\mathcal{K}$ given by $K_0 = 0$. Then, the geometric bracket restricts to $\mathcal{K}_1$ to induce the second Hamiltonian structure for a complexly coupled KdV system. Bracket (2.2) also reduced to $\mathcal{K}_1$ to produce the first Hamiltonian structure for this system.*

And, again, the geometric realization for complexly coupled KdV is found.

THEOREM 3.7. *([14]) Assume $u : J \subset \mathbb{R}^2 \to O(n+1,1)/H$ is a solution of*

$$u_t = h_1 T + h_2 N$$

where T and N are conformal tangent and normal (see [16]). Then the flow preserves K_0.

If $K_0 \to 0$, the evolution of k_1 and k_2 becomes

$$\begin{pmatrix} k_1 \\ k_2 \end{pmatrix}_t = \begin{pmatrix} -\frac{1}{2}D^3 + k_1 D + Dk_1 & k_2 D + Dk_2 \\ k_2 D + Dk_2 & \frac{1}{2}D^3 - k_1 D - Dk_1 \end{pmatrix} \begin{pmatrix} h_1 \\ h_2 \end{pmatrix}.$$

In particular, if $h_1 = k_1$ and $h_2 = k_2$, then k_1 and k_2 are solutions of a *complexly coupled system of KdV equations.*

As it was shown in [14], a very similar situation holds true for the case $G = O(p+q, q)$. In this case one also finds only two differential invariants of projective type and the same reductions and similar geometric realizations can be found.

Finally, our last case is the projective one.

3.3. Projective case. Let $G = \mathrm{PSL}(n+1)$. If $g \in G$ then, locally

$$g = \begin{pmatrix} I & u \\ 0 & 1 \end{pmatrix} \begin{pmatrix} A & 0 \\ 0 & (\det A)^{-1} \end{pmatrix} \begin{pmatrix} I & 0 \\ v^T & 1 \end{pmatrix}$$

with $u, v \in \mathbb{R}^n$ and $A \in \mathrm{GL}(n)$.

The manifold $\mathrm{PSL}(n+1)/H$ can be identified with the n *projective space* $\mathbb{RP}^n$. We will simply state the results in this case.

THEOREM 3.8. *(reformulation of Wilczynski [24]) There exists a minimal order moving frame ρ along a curve in $\mathbb{RP}^n$ such that its Serret-Frenet equations are defined by matrices of the form*

$$K = \begin{pmatrix} 0 & 1 & 0 & \dots & 0 \\ 0 & 0 & 1 & \dots & 0 \\ \vdots & \vdots & \ddots & \ddots & \vdots \\ 0 & 0 & \dots & 0 & 1 \\ k_1 & k_2 & \dots & k_n & 0 \end{pmatrix}$$

where k_i, $i = 1, \dots, n$ are in general a combination of the Wilczynski projective invariants.

THEOREM 3.9. *([12]) The n-projective geometric Poisson bracket is given by the Adler–Gel'fand–Dikii bracket or second Hamiltonian structure for KdV. The bracket (2.2) reduces for $L_0 = e_n^*$ and its reduction is the first Hamiltonian structure for generalised KdV equations.*

THEOREM 3.10. *([17]) Assume $u : J \subset \mathbb{R}^2 \to PSL(n)/H$ is a solution of*

$$u_t = h_1 T_1 + h_2 T_2 + \dots + h_n T_n$$

where T_i form a projective classical moving frame.

Then, $\mathbf{k} = (k_i)$ satisfy an equation of the form

$$\mathbf{k}_t = P\mathbf{h}$$

where $\mathbf{k} = (k_1, \ldots, k_n)^T$, $\mathbf{h} = (h_1, \ldots, h_n)^T$ *and where* P *is the Poisson tensor defining the* Adler–Gel'fand–Dikii Hamiltonian structure. *In particular, if* $\mathbf{h} = \mathbf{k}$ *then* $\mathbf{k}$ *satisfies a* generalized KdV system *of equations.*

These are the three cases that seem to describe the pattern for other cases. In [10] Kobayashi showed how the action of the group G on a homogeneous space G/H with a gradation as in (1.1) decouples on actions by its simple terms. The classification of all the simple cases was found in [10]. They correspond to all cases mentioned in this paper plus $G = O(2n+1, 2n+1), \mathrm{PSL}(p+q)$ and two with exceptional Lie algebras. These four cases are still open, although the conjecture is that $O(2n+1, 2n+1)$ might behave as a Lagrangian type while $\mathrm{PSL}(p+q)$ is probably projective type. The principal obstacle in most of these studies is finding minimal moving frames and the classification of differential invariants of projective type in a simple enough way that the resulting integrable equations can be identified. The moving frames and differential invariants are unknown for the remaining cases and need to be found as a previous step to the study of their geometric brackets.

REFERENCES

[1] Stephen Anco, *Hamiltonian flows of curves in G/SO(N) and vector soliton equations of mKdV and sine-Gordon type*, SIGMA, Vol. **2** (2006).

[2] E. Cartan, *La Méthode du Repère Mobile, la Théorie des Groupes Continus, et les Espaces Généralisés*, Exposés de Géométrie n. 5, Hermann, Paris, 1935.

[3] E. Cartan, *La Théorie des Groupes Finis et Continus et la Géométrie Différentielle Traitées par la Méthode du Repère Mobile*, Cahiers Scientifiques, Vol. **18**, Gauthier-Villars, Paris, 1937.

[4] M. Fels and P.J. Olver, *Moving coframes. I. A practical algorithm*, Acta Appl. Math. (1997), pp. 99–136.

[5] M. Fels and P.J. Olver, *Moving coframes. II. Regularization and theoretical foundations*, Acta Appl. Math. (1999), pp. 127–208.

[6] A. Fialkov, *The Conformal Theory of Curves*, Transactions of the AMS, **51** (1942), pp. 435–568.

[7] P.A. Griffiths, *On Cartan's method of Lie groups and moving frames as applied to uniqueness and existence questions in differential geometry*, Duke Math. J., **41** (1974), 775–814.

[8] M.L. Green, *The moving frame, differential invariants and rigidity theorems for curves in homogeneous spaces*, Duke Math. J., **45** (1978), 735–779.

[9] R. Hasimoto, *A soliton on a vortex filament*, J. Fluid Mechanics, **51** (1972), 477–485.

[10] S. Kobayashi and T. Nagano, *On filtered Lie Algebras and Geometric Structures I*, Journal of Mathematics and Mechanics, **13**(5) (1964), 875–907.

[11] J. Langer and R. Perline, *Poisson geometry of the filament equation*, J. Nonlinear Sci., **1**(1), (1991), 71–93.

[12] G. Marí Beffa, *Hamiltonian Structures on the space of differential invariants of curves in flat semisimple homogenous manifolds*, submitted.

[13] G. Marí Beffa, *Poisson geometry of differential invariants of curves in some nonsemisimple homogenous spaces*, Proc. Amer. Math. Soc., **134** (2006), 779–791.

[14] G. Marí Beffa, *Projective-type differential invariants and geometric curve evolutions of KdV-type in flat homogeneous manifolds*, submitted.

[15] G. Marí Beffa, *On completely integrable geometric evolutions of curves of Lagrangian planes*, accepted for publication in the Proceedings of the Royal academy of Edinburg, 2006.

[16] G. Marí Beffa, *Poisson brackets associated to the conformal geometry of curves*, Trans. Amer. Math. Soc., **357** (2005), 2799–2827.

[17] G. Marí Beffa, *The theory of differential invariants and KdV Hamiltonian evolutions*, Bull. Soc. Math. France, **127**(3), (1999), 363–391.

[18] G. Marí Beffa, J. Sanders, and J.P. Wang, *Integrable Systems in Three-Dimensional Riemannian Geometry*, J. Nonlinear Sc. (2002), pp. 143–167.

[19] P. Olver , *Invariant variational problems and integrable curve flows*, presentation, Cocoyoc, Mexico (2005).

[20] T. Ochiai, *Geometry associated with semisimple flat homogeneous spaces*, Transactions of the AMS, **152** (1970), pp. 159–193.

[21] V. Ovsienko, *Lagrange Schwarzian derivative and symplectic Sturm theory*, Ann. de la Fac. des Sciences de Toulouse, **6**(2), (1993), pp. 73–96.

[22] A. Pressley and G. Segal, *Loop groups*, Graduate Texts in Mathematics, Springer, 1997.

[23] J. Sanders and J.P. Wang, *Integrable Systems in n-dimensional Riemannian Geometry*, Moscow Mathematical Journal, **3**, 2004.

[24] E.J. Wilczynski, *Projective differential geometry of curves and ruled surfaces*, B.G. Teubner, Leipzig, 1906.

ALGEBRAIC CONSTRUCTION OF THE QUADRATIC FIRST INTEGRALS FOR A SPECIAL CLASS OF ORTHOGONAL SEPARABLE SYSTEMS

SERGIO BENENTI*

Abstract. With the notion of L-pencil, based on the notion of L-tensor, we construct a new class of Stäckel systems such that the quadratic first integrals associated with the orthogonal separation are computed by a coordinate-independent algebraic process.

Key words. Completely integrable systems, separation of variables, algebraic computation of first integrals.

AMS(MOS) subject classifications. 37K10, 37K25, 70H20.

1. Introduction.

1.1. Stäckel systems. A **Stäckel system** (S-system) on a Riemannian manifold (Q_n, g) is an orthogonal coordinate system (q^i) which allows the integration by (additive) separation of variables of the Hamilton-Jacobi equation of the geodesic flow. More precisely, a S-system is an equivalence class of such coordinates, being equivalent two coordinate systems related by a rescaling (i.e., by a coordinate transformation with a diagonal Jacobian matrix).

A celebrated theorem of Eisenhart, revised in [3, 4], shows that the existence of a S-system is equivalent to the (local) existence of a n-dimensional linear space of Killing two-tensors with common normal eigenvectors; we call such a space a **Killing-Stäckel space** (KS-space). This is equivalent to the existence of a complete system of quadratic first integrals in involution of the geodesic flow. Any separable coordinate system (q^i) associated with a KS-space is such that the differentials dq^i are common eigenforms.

It is also known that a KS-space is completely determined by one of its elements, called **characteristic Killing tensor** (although not unique) with normal eigenvectors and simple eigenvalues. However, if a characteristic tensor is given, the full KS-space can be determined by integrating a system of PDE's.

1.2. L-systems. There exists a special class of S-systems for which the whole KS-space can be generated by a single symmetric two-tensor **L** through a pure algebraic process. Such a tensor **L** is a *torsionless conformal Killing tensor with simple eigenvalues* and it has been called **L-tensor** [2, 3]. This process is based on the following theorem (see [1] for a proof and further details):

*Department of Mathematics, University of Turin. Research supported by G.N.F.M. (Gruppo Nazionale di Fisica Matematica) of I.N.d.A.M. (Istituto Nazionale di Alta Matematica, Roma).

THEOREM 1.1. *Let* $\mathbf{L}$ *be a symmetric 2-tensor. The tensors*

$$(\mathbf{K}_a) = (\mathbf{K}_0, \mathbf{K}_1, \ldots, \mathbf{K}_{n-1})$$

defined by the sequence

$$\begin{aligned} \mathbf{K}_0 &= \mathbf{I}, \\ \mathbf{K}_a &= \frac{1}{a}\,\mathrm{tr}\,(\mathbf{K}_{a-1}\mathbf{L})\,\mathbf{I} - \mathbf{K}_{a-1}\,\mathbf{L}, \quad 1 < a < n. \end{aligned} \tag{1.1}$$

form a basis of a KS-space if and only if $\mathbf{L}$ *is a L-tensor.*

Since we are on a Riemannian manifold, any symmetric 2-tensor (covariant or contravariant) can be interpreted as a $(1,1)$-tensor i.e., as a linear endomorphism on the space of vector fields or on the space of one-forms. In the recursive formula (1.1) all tensors must be interpreted in this sense. In particular, we observe that the identity operator $\mathbf{I}$ is the $(1,1)$-tensor associated with the metric tensor.

We call **L-sequence** (or **L-chain**) a sequence of the kind (1.1) and **L-system** a S-system having this property.

The geodesic flow of an asymmetric ellipsoid (Jacobi) as well as of any asymmetric hyperquadrics of a Euclidean space, are examples of L-systems. Within this framework, we can also deal with cofactor and bi-cofactor systems (see [5] and the bibliography therein).

1.3. L-pencils. Although L-systems form a very special class of S-systems, they can be used for defining other classes. This idea has been recently developed by Błaszak [6]. In the present note, I point out the existence of a further class of S-systems for which the KS-spaces can be constructed by an L-sequences. We call them **LP-systems**, since they are based on the notion of **L-pencil**:

DEFINITION 1.1. *A* **L-pencil** *is a linear combination*

$$\mathbf{L}_m = \mathbf{M} + m\,\mathbf{N} \tag{1.2}$$

which is a L-tensor for all values of $m \in \mathbb{R}$.

If we compute the L-sequence (1.1) of a L-pencil (1.2), then each $\mathbf{K}_a(m)$ is a polynomial at most of degree a in the parameter m. It is clear that $\mathbf{K}_a(0)$ form a L-system and that all the coefficients of these polynomials are Killing tensors. On the other hand, since the Killing tensors generated by a L-sequence commute as linear operators and are in involution as first integrals, we get

THEOREM 1.2. *Let* $\mathbf{H}_a$ *denote the coefficient of maximal degree in the parameter* m *of* $\mathbf{K}_a(m)$. *Then:* (i) *all tensors* $\mathbf{H}_a$ *are Killing tensors;* (ii) *they commute as linear operators,* $\mathbf{H}_a\mathbf{H}_b = \mathbf{H}_b\mathbf{H}_a$*;* (iii) *they commute in the Lie-Schouten brackets,* $[\mathbf{H}_a, \mathbf{H}_b] = 0$.

Item (iii) means that the quadratic functions on T^*Q associated with these tensors, $P_1 = H_1^{ij} p_i p_j$ and $P_2 = H_2^{ij} p_i p_j$, are in involution i.e., $\{P_1, P_2\} = 0$.

DEFINITION 1.2. *We call* **effective** *a L-pencil for which the tensors* $\mathbf{H}_a$ *are linearly independent.*

In this case they form a KS-space. Starting from this theorem and this definition we can get the following two main results.

THEOREM 1.3. *A L-pencil* $\mathbf{L}_m = \mathbf{M} + m\mathbf{N}$ *is effective if and only if* $\mathbf{M}$ *is a L-tensor and* $\mathbf{N}$ *has the form* $\mathbf{N} = \mathbf{X} \otimes \mathbf{X}^\flat$, *where* $\mathbf{X}$ *is a conformal Killing vector field whose associated one-form* $\mathbf{X}^\flat$ *is closed,* $d\mathbf{X}^\flat = 0$.

THEOREM 1.4. *Let* $\mathbf{L}_m = \mathbf{M} + m\mathbf{X} \otimes \mathbf{X}^\flat$ *be an effective L-pencil. Then:* (i) *the CKV* $\mathbf{X}$ *is a translation or a dilatation;* (ii) *all tensors* $\mathbf{K}_a$ *of the L-sequence are linear in* m *i.e., of the form*

$$\mathbf{K}_a = \mathbf{K}_{a0} + m\,\mathbf{H}_a;$$

(iii) $\mathbf{X}$ *is an eigenvector with zero eigenvalues of all* $\mathbf{H}_a$, $\mathbf{H}_a\mathbf{X} = 0$; (iv) *the restrictions of the tensors* $\mathbf{H}_a$ *to any leaf of the foliation orthogonal to* $\mathbf{X}$ *form a L-system on that leaf.*

We call **LP-system** a Stäckel system generated by a L-pencil.

REMARK 1.1. A fundamental example of LP-system, that inspired the definition of L-pencil and guided this research, is the asymmetric spherical-conical web in $\mathbb{R}^n$, where $\mathbf{M}$ is a symmetric constant matrix (note that in this case $\mathbf{M}$ is a Killing tensor) with distinct eigenvalues, and $\mathbf{X} = \mathbf{r}$ is the radius vector, whose components at a point $(x_1, \ldots, x_n)$ are just $(x_1, \ldots, x_n)$ [2, 3]. ◇

REMARK 1.2. As shown by Theorem 1.4, LP-systems are of two types, according to the type of the vector $\mathbf{X}$: **dilatational** or **translational.** ◇

REMARK 1.3. These two theorems have an intrinsic character: any L-sequence can be computed by using any suitable coordinate systems. However, they have a local meaning, not only because the global existence of special objects, like conformal Killing vectors (in particular, dilatations, etc.), is known to be impossible on certain kinds of Riemannian manifolds, but also because of the structure itself of a KS-space. For instance, the independence of the tensors $\mathbf{H}_a$ may occur in an open subset of Q, with the exception of a closed **singular set.** ◇

2. Differential conditions. The torsion $\mathbf{H}(\mathbf{A})$ of a $(1,1)$-tensor $\mathbf{A}$ is defined by

$$H^k_{ij}(\mathbf{A}) \doteq 2\big(A^h_{[i}\partial_{|h|}A^k_{j]} - A^k_m\partial_{[i}A^m_{j]}\big). \tag{2.1}$$

For the torsion the following **additive rule** holds

$$\mathbf{H}(\mathbf{A}+\mathbf{B}) = \mathbf{H}(\mathbf{A}) + \mathbf{H}(\mathbf{B}) + 2\,\mathbf{T}(\mathbf{A},\mathbf{B}). \tag{2.2}$$

where[1]

$$T^k_{ij}(\mathbf{A},\mathbf{B}) \doteq A^h_{[i}\partial_{|h|}B^k_{j]} - A^k_m\partial_{[i}B^m_{j]} + B^h_{[i}\partial_{|h|}A^k_{j]} - B^k_m\partial_{[i}A^m_{j]}. \tag{2.3}$$

[1]The tensor $\mathbf{T}(\mathbf{A},\mathbf{B})$ hab been introduced by Okubo, [12], formula 3.9.

In the definition (2.1), as well as in (2.3), the partial derivatives $\partial_i = \partial/\partial q^i$ can be replaced by the covariant derivatives ∇_i associated with *any symmetric linear connection* (in particular, the Levi-Civita connection):

$$H^k_{ij}(\mathbf{A}) \doteq 2\big(A^h_{[i}\nabla_{|h|}A^k_{j]} - A^k_m\nabla_{[i}A^m_{j]}\big),$$
$$T^k_{ij}(\mathbf{A},\mathbf{B}) = A^h_{[i}\nabla_{|h|}B^k_{j]} - A^k_m\nabla_{[i}B^m_{j]} + B^h_{[i}\nabla_{|h|}A^k_{j]} - B^k_m\nabla_{[i}A^m_{j]}.$$

By applying the additive rule (2.2) to equation $\mathbf{H}(\mathbf{M} + m\,\mathbf{N}) = 0$ we get

THEOREM 2.1. *The tensor* $\mathbf{L}_m = \mathbf{M} + m\,\mathbf{N}$ *is a L-pencil if and only if* (i) $\mathbf{M}$ *is a L-tensor,* (ii) $\mathbf{N}$ *is a torsionless CKT and* (iii)

$$\mathbf{T}(\mathbf{M},\mathbf{N}) = 0. \tag{2.4}$$

3. Algebraic conditions. Let us compute the first elements of the L-chain (1.1) for a L-pencil $\mathbf{L} = \mathbf{M} + m\,\mathbf{N}$. The first step of the L-chain (1.1) gives

$$\begin{aligned}\mathbf{K}_1 &= \operatorname{tr}\mathbf{L}_m\,\mathbf{I} - \mathbf{L}_m = (\operatorname{tr}\mathbf{M} + m\operatorname{tr}\mathbf{N})\,\mathbf{I} - \mathbf{M} - m\,\mathbf{N}\\ &= (\mu + m\,\nu)\,\mathbf{I} - \mathbf{M} - m\,\mathbf{N},\end{aligned}$$

where

$$\mu \doteq \operatorname{tr}\mathbf{M}, \qquad \nu \doteq \operatorname{tr}\mathbf{N}.$$

Hence,

$$\mathbf{K}_1 = \mathbf{K}_{10} + m\,\mathbf{K}_{11} \qquad \begin{cases}\mathbf{K}_{10} \doteq \mu\,\mathbf{I} - \mathbf{M},\\ \mathbf{K}_{11} \doteq \nu\,\mathbf{I} - \mathbf{N},\end{cases} \tag{3.1}$$

and

$$\mathbf{H}_1 = \mathbf{K}_{11} = \nu\,\mathbf{I} - \mathbf{N} \tag{3.2}$$

The second step gives

$$\mathbf{K}_2 = \mathbf{K}_{20} + m\,\mathbf{K}_{21} + m^2\,\mathbf{K}_{22}, \tag{3.3}$$

with

$$\begin{aligned}\mathbf{K}_{20} &\doteq \frac{1}{2}\,(\mu^2 - \operatorname{tr}\mathbf{M}^2)\,\mathbf{I} - \mu\,\mathbf{M} + \mathbf{M}^2,\\ \mathbf{K}_{21} &\doteq (\mu\nu - \operatorname{tr}\mathbf{M}\mathbf{N})\,\mathbf{I} - \nu\mathbf{M} - \mu\,\mathbf{N} + \mathbf{M}\mathbf{N} + \mathbf{N}\mathbf{M},\\ \mathbf{K}_{22} &\doteq \frac{1}{2}\,(\nu^2 - \operatorname{tr}\mathbf{N}^2)\,\mathbf{I} - \nu\,\mathbf{N} + \mathbf{N}^2.\end{aligned}$$

REMARK 3.1. These equations shows that we have to deal with two types of L-systems,

$$\textbf{Type-1}: \quad \mathbf{K}_{22} \doteq \frac{1}{2}\,(\nu^2 - \operatorname{tr}\mathbf{N}^2)\,\mathbf{I} - \nu\,\mathbf{N} + \mathbf{N}^2 = 0, \tag{3.4}$$

$$\textbf{Type-2}: \quad \mathbf{K}_{22} \doteq \frac{1}{2}\,(\nu^2 - \operatorname{tr}\mathbf{N}^2)\,\mathbf{I} - \nu\,\mathbf{N} + \mathbf{N}^2 \neq 0. \tag{3.5}$$

These are algebraic conditions involving the tensor $\mathbf{N}$ only. For a type-1 L-pencil we have

$$\mathbf{H}_2 = \mathbf{K}_{21} = (\mu\nu - \operatorname{tr}\mathbf{MN})\,\mathbf{I} - \nu\mathbf{M} - \mu\,\mathbf{N} + \mathbf{MN} + \mathbf{NM}, \tag{3.6}$$

and for a type-2 L-pencil,

$$\begin{aligned} \mathbf{H}_2 \quad &= \mathbf{K}_{22} = \frac{1}{2}\,(\nu^2 - \operatorname{tr}\mathbf{N}^2)\,\mathbf{I} - \nu\,\mathbf{N} + \mathbf{N}^2 \\ &= \frac{1}{2}\operatorname{tr}(\mathbf{H}_1\,\mathbf{N})\,\mathbf{I} - \mathbf{H}_1\,\mathbf{N}. \end{aligned}$$

This last expression compared with (3.2) shows that $\mathbf{H}_1$ and $\mathbf{H}_2$ are the first two elements of the L-chain generated by the torsionless tensor $\mathbf{N}$. Going back to Eq. (3.3) we observe that the coefficient of m^3 of

$$\mathbf{K}_3 = \frac{1}{3}\operatorname{tr}(\mathbf{K}_2\mathbf{L}_m)\,\mathbf{I} - \mathbf{K}_2\mathbf{L}_m$$

is the tensor

$$\mathbf{K}_{33} = \frac{1}{3}\operatorname{tr}(\mathbf{K}_{22}\,\mathbf{N})\,\mathbf{I} - \mathbf{K}_{22}\,\mathbf{N} = \frac{1}{3}\operatorname{tr}(\mathbf{H}_2\,\mathbf{N})\,\mathbf{I} - \mathbf{H}_2\,\mathbf{N}.$$

This is sufficient to show that *a type-2 L-pencil* $\mathbf{L}_m = \mathbf{M} + m\,\mathbf{N}$ *generates the L-chain of* $\mathbf{N}$. This means that if $\mathbf{N}$ has distinct eigenvalues (i.e., it is a L-tensor) the L-pencil $\mathbf{L}_m$ generates nothing new. In the case of non-distinct eigenvalues the $\mathbf{N}$-chain generates a space of dimension $< n$ of Killing tensors (see Appendix B for an example). Hence, in both cases, *the type-2 L-pencils have no interest*, and hereafter we consider type-1 L-pencils only. ◇

4. The eigenvalues of N. Eq. (3.4) written in the form

$$\mathbf{N}^2 - \nu\,\mathbf{N} = \frac{1}{2}\,(\operatorname{tr}\mathbf{N}^2 - \nu^2)\,\mathbf{I}, \tag{4.1}$$

shows that

$$\operatorname{tr}\mathbf{N}^2 - \nu^2 = \frac{n}{2}\,(\operatorname{tr}\mathbf{N}^2 - \nu^2),$$

i.e., for $n > 2$, $\operatorname{tr}\mathbf{N}^2 = \nu^2 = (\operatorname{tr}\mathbf{N})^2$. Thus, from (4.1), $\mathbf{N}^2 = \nu\,\mathbf{N}$. Let ν_i be the eigenvalues of $\mathbf{N}$. Then the diagonalization of this equation[2] yields equation $\nu_i^2 = \nu\,\nu_i$, which is equivalent to

$$\nu_i \textstyle\sum_{k\neq i} \nu_k = 0, \quad \forall i.$$

Let us consider the simplest case $n = 3$. Then we have

$$\nu_1(\nu_2 + \nu_3) = 0, \quad \nu_2(\nu_3 + \nu_1) = 0, \quad \nu_3(\nu_1 + \nu_2) = 0.$$

Assume $\nu_2 \neq 0$ and $\nu_3 \neq 0$. Then,

$$\nu_1(\nu_2 + \nu_3) = 0, \quad \nu_3 = -\nu_1, \quad \nu_2 = -\nu_1.$$

If we replace the last two equations into the first one, we get $\nu_1 = 0$. Hence, $\nu_2 = \nu_3 = 0$: absurd.

For $n = 4$,

$$\begin{aligned}
\nu_1(\nu_2 + \nu_3 + \nu_4) &= 0,\\
\nu_2(\nu_3 + \nu_4 + \nu_1) &= 0,\\
\nu_3(\nu_4 + \nu_1 + \nu_2) &= 0,\\
\nu_4(\nu_1 + \nu_2 + \nu_3) &= 0.
\end{aligned}$$

Assume $\nu_3, \nu_4 \neq 0$:

$$\begin{aligned}
&\nu_1(\nu_2 + \nu_3 + \nu_4) = 0,\\
&\nu_2(\nu_3 + \nu_4 + \nu_1) = 0,\\
&\nu_4 = -\nu_1 - \nu_2,\\
&\nu_3 = -\nu_1 - \nu_2.
\end{aligned}$$

Replace the last two equations into the first two:

$$\begin{aligned}
&\nu_1(\nu_2 + 2\nu_1) = 0,\\
&\nu_2(2\nu_2 + \nu_1) = 0,\\
&\nu_4 = -\nu_1 - \nu_2,\\
&\nu_3 = -\nu_1 - \nu_2.
\end{aligned}$$

Take the difference of the first two equations:

$$\nu_1(\nu_2 + 2\nu_1) - (\nu_2(2\nu_2 + \nu_1)) = 2(\nu_1^2 + \nu_2^2) = 0.$$

[2] We are in a pure algebraic context. If N_i^j are the components of $\mathbf{N}$ with respect to any basis, the tensor $N_{ik} = g_{jk}\,N_i^j$ is symmetric. Hence, N_{ij} and g_{ij} can be simultaneously diagonalized in a canonical basis.

This implies $\nu_1 = \nu_2 = 0$, and consequently $\nu_3 = \nu_4 = 0$: absurd. The above calculations can be extended to any dimension n. The result is

PROPOSITION 4.1. *In a L-pencil the tensor* $\mathbf{N}$ *cannot have two eigenvalues different from zero.*

Since the case $\mathbf{N} = 0$ is excluded, we have proved

PROPOSITION 4.2. *In a L-pencil the tensor* $\mathbf{N}$ *has only one eigenvalue different from zero.*

It follows that if $\mathbf{U}$ is a unit eigenvector corresponding to the non-zero eigenvalue ν_1, then $\mathbf{N} = \nu_1 \mathbf{U} \otimes \mathbf{U}^\flat$, i.e., $N_i^j = \nu_1 \, U_i U^j$. It is not restrictive to assume hereafter

$$\nu_1 > 0.$$

This corresponds to replace $\mathbf{N}$ by $-\mathbf{N}$, or m by $-m$, in the L-pencil. Hence, if we introduce the vector field

$$\mathbf{X} \doteq \sqrt{\nu_1}\, \mathbf{U},$$

then it is proved that

THEOREM 4.1. *In a L-pencil* $\mathbf{L}_m = \mathbf{M} + m\mathbf{N}$, *the tensor* $\mathbf{N}$ *has the form*

$$\mathbf{N} = \mathbf{X} \otimes \mathbf{X}^\flat, \quad N_i^j = X_i X^j,$$

and

$$\nu \doteq \operatorname{tr} \mathbf{N} = \nu_1 = \mathbf{X} \cdot \mathbf{X}.$$

5. The differential properties of the vector X. The aim of this section is to prove

THEOREM 5.1. *The vector field* $\mathbf{X}$ *is a dilatation or a translation.*

Let us recall that a **conformal Killing vector** (CKV) is a vector field $\mathbf{X}$ on a Riemannian manifold characterized by equation

$$\{P(\mathbf{X}), P(\mathbf{G})\} = \psi \, P(\mathbf{G}), \tag{5.1}$$

where $P(\mathbf{X}) \doteq X^i p_i$, $P(\mathbf{G}) = g^{ij}\, p_i p_j$, and f a scalar field on the manifold Q. If $\psi = 0$, $\mathbf{X}$ is a **Killing vector**. If $\psi =$ constant $\neq 0$, $\mathbf{X}$ is a **dilatation**. If $\psi = 0$ and $\mathbf{X} \cdot \mathbf{X} =$ constant, $\mathbf{X}$ is a **translation**.[3]

To prove Theorem 5.1 we need preliminary statements.

THEOREM 5.2. *The vector field* $\mathbf{X}$ *is a conformal Killing vector.*

Proof. A conformal Killing tensor (CKT) $\mathbf{N}$ is characterized by equation

$$\{P(\mathbf{N}), P(\mathbf{G})\} = \; - 2\, P(\mathbf{C}) P(\mathbf{G})$$

[3] An *infinitesimal translation* is a Killing vector with constant length or, equivalently, a vector field whose integral curves are geodesics [9], §72.

where $\mathbf{C}$ is a vector field and $P(\mathbf{C}) = C^i p_i$. Being $\mathbf{N} = \mathbf{X} \otimes \mathbf{X}^\flat$ a CKT, and $P(\mathbf{N}) = (X^i p_i)^2 = P^2(\mathbf{X})$, equation

$$\{P^2(\mathbf{X}), P(\mathbf{G})\} = -2\,P(\mathbf{C})P(\mathbf{G}) \tag{5.2}$$

holds. However,

$$\{P^2(\mathbf{X}), P(\mathbf{G})\} = 2P(\mathbf{X})\,\{P(\mathbf{X}), P(\mathbf{G})\}$$

so that Eq. (5.2) becomes

$$P(\mathbf{X})\,\{P(\mathbf{X}), P(\mathbf{G})\} = -\,P(\mathbf{C})P(\mathbf{G}).$$

This polynomial equation shows that $P(\mathbf{X}) = -\,\psi\,P(\mathbf{C})$, where ψ is a function on Q, and consequently that Eq. (5.1) holds. □

THEOREM 5.3. *The one-form $\mathbf{X}^\flat = (X_i)$ is closed, $\partial_i X_j = \partial_j X_i$.*

Proof. This property is due to the torsionless condition $\mathbf{H}(\mathbf{N}) = \mathbf{H}(\mathbf{X} \otimes \mathbf{X}^\flat) = 0$. Indeed, by the definition of torsion (2.1), the tensor

$$S_{ij}^k(\mathbf{N}) \doteq N_i^h \nabla_h N_j^k - N_m^k \nabla_i N_j^m$$

must be symmetric in the lower indices. For $N_i^j = X_i X^j$ we have

$$\begin{aligned} S_{ij}^k &= X^h X_i \nabla_h (X_j X^k) - X_m X^k \nabla_i (X_j X^m) \\ &= X_i X^k X^h\, \nabla_h X_j + X_i X_j X^h\, \nabla_h X^k - X_m X^m X^k\, \nabla_i X_j \\ &\quad - X^k X_j X_m\, \nabla_i X^m \\ &= X^k\,[X_i X^h\, \nabla_h X_j - \nu_1\, \nabla_i X_j - X_j X^m\, \nabla_i X_m] + \ldots \end{aligned}$$

where $\ldots$ denote terms symmetric in the indices i, j. Now we recall that a CKV is also characterized by equation

$$\nabla_i X_j + \nabla_j X_i = \psi\, g_{ij}. \tag{5.3}$$

Thus,

$$\begin{aligned} S_{ij}^k(\mathbf{N}) &= X^k\,[X_i X^h\, \nabla_h X_j - \nu_1\, \nabla_i X_j + X_j X^m\, (\nabla_m X_i - \psi\, g_{im}] + \ldots \\ &= -\,\nu_1\, X^k\, \nabla_i X_j. \end{aligned}$$

This shows that the torsionless condition is equivalent to $\nabla_i X_j - \nabla_j X_i = \partial_i X_j - \partial_j X_i = 0$. □

By summing this last equation to Eq. (5.3) we get ($\kappa \doteq \psi/2$)

$$\nabla_i X_j = \kappa\, g_{ij}. \tag{5.4}$$

Moreover, There exist local functions ρ such that

$$\mathbf{X} = \nabla \rho, \qquad X_i = \partial_i \rho. \tag{5.5}$$

As a consequence, we have proved that

LEMMA 5.1. *The vector field* **X** *is* **normal**.

Recall that a vector field **X** is called **normal** if it is orthogonal to a local web of submanifolds of codimension 1.[4] In this case the leaves of the web (which we call **X-web**) are defined by equation $\rho =$ constant. As a consequence of this fact, there exist local coordinates $(q^i) = (q^1, q^a)$ such that $q^1 = \rho$ and $g^{1a} = 0$. We call them **X-coordinates**. From Eq. (5.5) it follows that in these coordinates

$$X_1 = 1, \quad X_a = 0. \tag{5.6}$$

Hence,

$$\begin{aligned} X^1 = g^{1i} X_i = g^{11}, \quad X^a = 0, \qquad g_{1a} = 0, \\ \nu_1 = X^i X_i = X^1, \qquad \mathbf{X} = \nu_1 \, \partial_1. \end{aligned} \tag{5.7}$$

Now we prove Theorem 5.1.

Proof. In any coordinate system Eq. (5.1) reads

$$\{X^i p_i, g^{hk} p_h p_k\} = \psi \, g^{hk} p_h p_k. \tag{5.8}$$

In **X**-coordinates,

$$\{X^1 p_1, g^{11} p_1^2 + g^{ab} p_a p_b\} = \psi \, (g^{11} p_1^2 + g^{ab} p_a p_b),$$

$$\Longrightarrow \quad \begin{aligned} & X^1 (\partial_1 g^{11} p_1^2 + \partial_1 g^{ab} p_a p_b) - \partial_i X^1 p_1 \partial^i (g^{11} p_1^2 + g^{ab} p_a p_b) \\ & = \psi \, (g^{11} p_1^2 + g^{ab} p_a p_b), \end{aligned}$$

$$\Longrightarrow \quad \begin{aligned} & X^1 (\partial_1 g^{11} p_1^2 + \partial_1 g^{ab} p_a p_b) - 2 \, p_1^2 \partial_1 X^1 g^{11} - 2 \, p_1 \, \partial_a X^1 g^{ab} p_b \\ & = \psi \, (g^{11} p_1^2 + g^{ab} p_a p_b). \end{aligned}$$

This polynomial equation in the momenta is equivalent to equations

$$\begin{aligned} & X^1 \partial_1 g^{11} - 2 \, g^{11} \partial_1 X^1 = \psi \, g^{11}, \\ & \partial_a X^1 = 0, \\ & X^1 \partial_1 g^{ab} = \psi \, g^{ab}. \end{aligned} \tag{5.9}$$

Recall that $X^1 = g^{11} = \nu_1$. The first equation shows that

$$\psi = \, - \, \partial_1 \nu_1, \tag{5.10}$$

and the second one that ν_1 depends on the coordinate $q^1 = \rho$ only.

$$\nu_1 = \nu_1(q^1).$$

---

[4]It it also called **surface-forming** of **orthogonally integrable**.

Eq. (5.10) shows that $\mathbf{X}$ is a KV i.e., $\psi = 0$ if and only if $\nu_1 =$ constant. Since $\nu_1 = \mathbf{X} \cdot \mathbf{X}$, if $\mathbf{X}$ is a KV, then it is a translation. In the case of a non-constant ν_1, it is convenient to replace the coordinate q^1 by a new coordinate r such that $\mathbf{X} = r\,\partial_r$ and $\partial_r \cdot \partial_r = 1$. These two conditions imply $\mathbf{X} \cdot \mathbf{X} = \nu_1 = r^2$. Hence, the coordinate transformation can be defined by

$$r = \sqrt{\nu_1(q^1)}.$$

Since

$$\mathbf{X} = r\,\partial_r = \nu_1 \partial_1, \tag{5.11}$$

we get equation

$$r\,\partial_r \nu_1 = \nu_1 \partial_1 \nu_1. \tag{5.12}$$

It follows that $\psi = -\partial_1 \nu_1 = -r\,\partial_r \log \nu_1 = -r\,\partial_r \log r^2 = -2$. This proves that $\mathbf{X}$ is a dilatation. □

REMARK 5.1. Note that the third equation (5.9) is a further property of the vector field $\mathbf{X}$ not considered in Theorem 5.1. Due to (5.10) it can be written

$$\partial_1 g^{ab} = -\partial_1 \log \nu_1 \; g^{ab}.$$

It represents a law of evolution of the metric components g^{ab} along the flow of $\mathbf{X} = \nu_1 \partial_1$. It can be written

$$\partial_1 g^{ab} = \phi_1 \; g^{ab} \tag{5.13}$$

where $\phi_1 \doteq -\partial_1 \log \nu_1$ is a function of q^1 only. Eq. (5.13) is equivalent to

$$\partial_1 g_{ab} = \partial_1 \log \nu_1 \; g_{ab}, \tag{5.14}$$

and, for $g^{ab} \neq 0$, to

$$\partial_1 \log g^{ab} = \phi_1(q^1). \tag{5.15}$$

Note that the right hand side does not depend on the indices a, b. ◇

REMARK 5.2. If $\nu_1 =$ constant, $\mathbf{X}$ is a translation, Eq. (5.13) reduces to $\partial_1 g^{ab} = 0$. If $\mathbf{X}$ is a dilatation, then from Eq. (5.12) we get

$$\partial_1 \log \nu_1 = \frac{r}{\nu_1}\,\partial_r \log \nu_1 = \frac{r}{r^2}\,\partial_r \log r^2 = \frac{2}{r^2}.$$

Moreover, due to Eq. (5.11), $\nu_1 \partial_1 r = r$, so that

$$\partial_1 g^{ab} = \partial_1 r\,\partial_r g^{ab} = \frac{r}{\nu_1}\,\partial_r g^{ab} = \frac{1}{r}\,\partial_r g^{ab},$$

and Eq. (5.13) becomes

$$\partial_r g^{ab} = -\frac{2}{r}\,g^{ab}.$$

◇

6. CKV in orthogonal coordinates. In orthogonal coordinates the characteristic equation (5.1) of a CKV reads

$$(X^i\,\partial_i g^{hh} - \psi\, g^{hh})\, p_h^2 - 2\,\partial_k X^h\, g^{kk}\, p_h p_k = 0. \tag{6.1}$$

The diagonal part of this equation gives

$$X^i\,\partial_i g^{hh} - \psi\, g^{hh} - 2\,\partial_h X^h\, g^{hh} = 0, \quad X^i\,\partial_i \log g^{hh} - \psi - 2\,\partial_h X^h = 0.$$

Hence,

$$\psi = \textstyle\sum_i X^i\,\partial_i \log g^{hh} - 2\,\partial_h X^h. \tag{6.2}$$

Note that this last equation holds for any choice of the index h, which is not summed. The non-diagonal part of Eq. (6.1) gives

$$g^{ii}\partial_i X^j + g^{jj}\partial_j X^i = 0, \quad j \neq i. \tag{6.3}$$

PROPOSITION 6.1. *If* $\partial_i X_j = \partial_j X_i$, *then Eq.* (6.3) *is equivalent to*

$$2\,\partial_i X_j + X_j\partial_i \log g^{jj} + X_i\partial_j \log g^{ii} = 0. \tag{6.4}$$

Proof. Let us translate Eq. (6.3) in covariant components of $\mathbf{X}$,

$$g_{jj}\partial_i(g^{jj}X_j) + g_{ii}\partial_j(g^{ii}X_i) = 0.$$

It follows that

$$X_j\partial_i \log g^{jj} + \partial_j X_i + X_i\partial_j \log g^{ii} + \partial_i X_j = 0.$$

Since $\partial_i X_j = \partial_j X_i$, we get Eq. (6.4). □

PROPOSITION 6.2. *Eq.* (6.2)*is equivalent to*

$$\begin{aligned}\psi &= \textstyle\sum_i X^i\,\partial_i \log g^{hh} - 2\,X^h\,\partial_h \log g^{hh} - 2\,g^{hh}\,\partial_h X_h \\ &= \textstyle\sum_{i\neq h} X^i\,\partial_i \log g^{hh} - X^h\,\partial_h \log g^{hh} - 2\,g^{hh}\,\partial_h X_h.\end{aligned} \tag{6.5}$$

Proof. Let us use to Eq. (6.2),

$$\begin{aligned}\psi &= \textstyle\sum_i X^i\,\partial_i \log g^{hh} - 2\,\partial_h X^h \\ &= \textstyle\sum_i X^i\,\partial_i \log g^{hh} - 2\,\partial_h(g^{hh}\,X_h) \\ &= \textstyle\sum_i X^i\,\partial_i \log g^{hh} - 2\,X_h\,\partial_h g^{hh} - 2\,g^{hh}\,\partial_h X_h \\ &= \text{etc.}\end{aligned}$$

□

7. The condition $\mathbf{T}(\mathbf{M},\mathbf{N})=0$. In this section we prove the remarkable (and rather surprising) fact: *the differential condition* $\mathbf{T}(\mathbf{M},\mathbf{N})=0$ *is identically satisfied.* For this we shall use another special kind of coordinates associated with a L-pencil. Since $\mathbf{M}$ is a L-tensor, after the results of [5], there exist local coordinate systems, which we call **M-coordinates** in the following, such that

$$\begin{aligned} g^{ij} &= M^{ij}=0, \quad i\neq j,\\ M^{ii} &= \mu^i\, g^{ii},\\ M_i^j &= \mu^i\,\delta_i^j=\mu^j\,\delta_i^j, \end{aligned} \tag{7.1}$$

$$\begin{aligned} \partial_i\mu^j &= 0, \quad i\neq j,\\ \partial_i\mu^i &= (\mu^j-\mu^i)\partial_i\log g^{jj}, \end{aligned} \tag{7.2}$$

where the eigenvalues μ^i of $\mathbf{M}$ are all distinct. Note that the M-coordinates are orthogonal and separable.

Due to the second equation (7.2), in M-coordinates Eq. (6.4) reads

$$2\,\partial_i X_j = -X_j\,\partial_i\log g^{jj}-X_i\,\partial_j\log g^{ii}=-X_j\frac{\partial_i\mu^i}{\mu^j-\mu^i}-X_i\frac{\partial_j\mu^j}{\mu^i-\mu^j}.$$

Hence,

$$2\,\partial_i X_j=\frac{X_j\,\partial_i\mu^i-X_i\,\partial_j\mu^j}{\mu^i-\mu^j}. \tag{7.3}$$

Now we prove the following general statement.

THEOREM 7.1. *If* $\mathbf{M}$ *is a L-tensor and* $\mathbf{X}$ *is a CKV such that* $\mathbf{X}^\flat$ *is closed, then* $\mathbf{T}(\mathbf{M},\mathbf{X}\otimes\mathbf{X}^\flat)=0$.

LEMMA 7.1. *In* $\mathbf{M}$*-coordinates condition* $\mathbf{T}(\mathbf{M},\mathbf{N})=0$ *is equivalent to equations*

$$(\mu^i-\mu^k)\,\partial_i N_j^k=(\mu^j-\mu^k)\,\partial_j N_i^k, \qquad i,j,k\neq, \tag{7.4}$$

$$(\mu^j-\mu^i)\,\partial_j N_i^i+N_j^i\partial_i\mu^i=0, \qquad i\neq j. \tag{7.5}$$

Proof. Due to the definition (2.3) this condition is equivalent to equation (of course with $i\neq j$)

$$\begin{aligned} &M_i^h\partial_h N_j^k-M_m^k\partial_i N_j^m+N_i^h\partial_h M_j^k-N_m^k\partial_i M_j^m\\ &=M_j^h\partial_h N_i^k-M_m^k\partial_j N_i^m+N_j^h\partial_h M_i^k-N_m^k\partial_j M_i^m, \end{aligned} \tag{7.6}$$

sum over h and m, in any coordinate systems. In M-coordinates, due to the third line of (7.1), this equation becomes

$$\begin{aligned} &M_i^i\partial_i N_j^k-M_k^k\partial_i N_j^k+N_i^h\partial_h M_j^k-N_j^k\partial_i M_j^j\\ &=M_j^j\partial_j N_i^k-M_k^k\partial_j N_i^k+N_j^h\partial_h M_i^k-N_i^k\partial_j M_i^i, \end{aligned} \qquad i\neq j,$$

and consequently

$$\begin{aligned} &\mu^i\partial_i N_j^k - \mu^k\partial_i N_j^k + N_i^h\partial_h M_j^k - N_j^k\partial_i\mu^j \\ &= \mu^j\partial_j N_i^k - \mu^k\partial_j N_i^k + N_j^h\partial_h M_i^k - N_i^k\partial_j\mu^i, \end{aligned} \qquad i \neq j.$$

We get a further simplification by the first line of (7.2):

$$(\mu^i - \mu^k)\,\partial_i N_j^k + N_i^h\partial_h M_j^k = (\mu^j - \mu^k)\,\partial_j N_i^k + N_j^h\partial_h M_i^k,$$

for $i \neq j$. For $i, j, k \neq$, taking into account (7.1) we get Eq. (7.4). For $k = i \neq j$,

$$0 = (\mu^j - \mu^i)\,\partial_j N_i^i + N_j^h\partial_h M_i^i,$$

and also Eq. (7.5) is proved. □

Let us apply this result to the case $N_i^j = X_i X^j$.

LEMMA 7.2. *The compatibility condition* $\mathbf{T}(\mathbf{M}, \mathbf{X}\otimes\mathbf{X}^\flat) = 0$ *is equivalent to the cyclic equation*

$$\begin{aligned} &X_k\,(X_j\,\partial_i\mu^i - X_i\,\partial_j\mu^j) + X_i\,(X_k\,\partial_j\mu^j - X_j\,\partial_k\mu^k) \\ &+ X_j\,(X_i\,\partial_k\mu^k - X_k\,\partial_i\mu^i) = 0, \qquad i, j, k \neq . \end{aligned} \tag{7.7}$$

Proof. For $N_i^j = X_i X^j$, Eqs. (7.4) and (7.5) read

$$(\mu^i - \mu^k)\,\partial_i(X_j X^k) - (\mu^j - \mu^k)\,\partial_j(X_i X^k) = 0, \tag{7.8}$$

$$(\mu^j - \mu^i)\,\partial_j(X_i X^i) + X^i X_j\,\partial_i\mu^i = 0, \tag{7.9}$$

respectively, for all distinct indices i, j, k. (i) Since

$$\begin{aligned} \partial_j(X_i X^i) &= X_i\,\partial_j X^i + X^i\partial_j X_i = X_i\,\partial_j(g^{ii}X_i) + X^i\partial_j X_i \\ &= (X_i)^2\,\partial_j g^{ii} + X_i g^{ii}\partial_j X_i + X^i\partial_j X_i \\ &= X^i\,(X_i\,\partial_j \log g^{ii} + 2\,\partial_j X_i), \end{aligned}$$

Eq. (7.9) becomes

$$(\mu^j - \mu^i)\,(X_i\,\partial_j \log g^{ii} + 2\,\partial_j X_i) + X_j\,\partial_i\mu^i = 0.$$

This implies

$$(\mu^j - \mu^i)\,(X_i\,\frac{\partial_j\mu^j}{\mu^i - \mu^j} + 2\,\partial_j X_i) + X_j\,\partial_i\mu^i = 0,$$

i.e.,

$$2\,(\mu^j - \mu^i)\,\partial_j X_i - X_i\partial_j\mu^j + X_j\,\partial_i\mu^i = 0.$$

This equation is identically satisfied, due to (7.3). (ii) Eq. (7.8) can be written

$$(\mu^i - \mu^k)\,[X^k\,\partial_i X_j + X_j\,\partial_i X^k] - (\mu^j - \mu^k)\,[X^k\,\partial_j X_i + X_i\,\partial_j X^k] = 0.$$

Since $\partial_i X_j = \partial_j X_i$, we get

$$(\mu^i - \mu^j)\,X^k\,\partial_i X_j + (\mu^k - \mu^j)\,X_i\,\partial_j X^k + (\mu^i - \mu^k) X_j\,\partial_i X^k = 0.$$

However,

$$\begin{aligned} \partial_i X^k &= \partial_i (g^{kk} X_k) = g^{kk}\,\partial_i X_k + X_k\,\partial_i g^{kk} \\ &= g^{kk}\,\partial_i X_k + X^k\,\partial_i \log g^{kk} \\ &= g^{kk}\,\partial_i X_k + X^k\,\frac{\partial_i \mu^i}{\mu^k - \mu^i}, \end{aligned}$$

and we get equation

$$\begin{aligned} &(\mu^i - \mu^j)\,X^k\,\partial_i X_j + (\mu^k - \mu^j) X_i \left(g^{kk}\,\partial_j X_k + X^k\,\frac{\partial_j \mu^j}{\mu^k - \mu^j}\right) \\ &+ (\mu^i - \mu^k) X_j \left(g^{kk}\,\partial_i X_k + X^k\,\frac{\partial_i \mu^i}{\mu^k - \mu^i}\right) = 0, \end{aligned}$$

which can be put in the form

$$\begin{aligned} &g^{kk}\left((\mu^i - \mu^j)\,X_k\,\partial_i X_j + (\mu^k - \mu^j) X_i\,\partial_j X_k + (\mu^i - \mu^k) X_j\,\partial_i X_k\right) \\ &+ X^k\,(X_i\,\partial_j \mu^j - X_j\,\partial_i \mu^i) = 0, \end{aligned}$$

Due to (7.3),

$$2\,\partial_i X_j = \frac{X_j\,\partial_i \mu^i - X_i\,\partial_j \mu^j}{\mu^i - \mu^j},$$

we get

$$\begin{aligned} &g^{kk}[(\mu^i - \mu^j)\,X_k\,\partial_i X_j + (\mu^k - \mu^j) X_i\,\partial_j X_k + (\mu^i - \mu^k) X_j\,\partial_i X_k] \\ &+ 2\,X^k\,(\mu^j - \mu^i)\,\partial_i X_j = 0, \end{aligned}$$

and finally

$$(\mu^j - \mu^i)\,X_k\,\partial_i X_j + (\mu^k - \mu^j) X_i\,\partial_j X_k + (\mu^i - \mu^k) X_j\,\partial_i X_k = 0.$$

This is an equivalent form of the cyclic equation (7.7). □

Proof of Theorem 7.1.

Proof. If we apply Eq. (7.3), $2\,(\mu^i - \mu^j)\,\partial_i X_j = X_j\,\partial_i \mu^i - X_i\,\partial_j \mu^j$, to the cyclic equation (7.7), we see that it is identically satisfied. □

Let us examine equation $\mathbf{T}(\mathbf{M}, \mathbf{X} \otimes \mathbf{X}^\flat) = 0$ in $\mathbf{X}$-coordinates.

THEOREM 7.2. *In* **X**-*coordinates equation* $\mathbf{T}(\mathbf{M}, \mathbf{X}\otimes\mathbf{X}^\flat) = 0$ *is equivalent to the following equations,*

$$\partial_i M_j^1 = \partial_j M_i^1, \qquad i, j \neq 1, \tag{7.10}$$

$$\partial_1 M_j^k = 0, \qquad j, k \neq 1, \tag{7.11}$$

$$M_j^1 \partial_1 \nu_1 = \nu_1 \partial_j M_1^1, \qquad j \neq 1. \tag{7.12}$$

Proof. In any coordinate system, equation $\mathbf{T}(\mathbf{M}, \mathbf{X}\otimes\mathbf{X}^\flat) = 0$ is equivalent to Eq. (7.6), with $N_i^j = X_i X^j$:

$$\begin{aligned}
&M_i^h \partial_h (X_j X^k) - M_m^k \partial_i (X_j X^m) \\
&+ X^h X_i \partial_h M_j^k - X_m X^k \partial_i M_j^m \\
&- M_j^h \partial_h (X_i X^k) + M_m^k \partial_j (X_i X^m) \\
&- X^h X_j \partial_h M_i^k + X_m X^k \partial_j M_i^m = 0.
\end{aligned} \tag{7.13}$$

In **X**-coordinates $X^1 = \nu_1$ and $X_1 = 1$ are the only non-vanishing components of **X**, and they depend on the coordinate q^1 only. Thus, Eq. (7.13) becomes

$$\begin{aligned}
&M_i^h \partial_h (X_j X^k) - M_1^k \partial_i (X_j X^1) \\
&+ X^1 X_i \partial_1 M_j^k - X_1 X^k \partial_i M_j^1 \\
&- M_j^h \partial_h (X_i X^k) + M_1^k \partial_j (X_i X^1) \\
&- X^1 X_j \partial_1 M_i^k + X_1 X^k \partial_j M_i^1 = 0.
\end{aligned} \tag{7.14}$$

For $i, j \neq 1$, Eq. (7.14) reduces to

$$- X_1 X^k \partial_i M_j^1 + X_1 X^k \partial_j M_i^1 = 0.$$

This proves Eq. (7.10). For $i = 1$ Eq. (7.14) reduces to

$$\begin{aligned}
&M_1^h \partial_h (X_j X^k) - M_1^k \partial_1 (X_j X^1) + X^1 X_1 \partial_1 M_j^k - X_1 X^k \partial_1 M_j^1 \\
&- M_j^h \partial_h (X_1 X^k) + M_1^k \partial_j (X_1 X^1) - X^1 X_j \partial_1 M_1^k + X_1 X^k \partial_j M_1^1 = 0.
\end{aligned}$$

For $j = 1$ this equation is of course identically satisfied. For $j \neq 1$ it reads

$$\begin{aligned}
&X^1 X_1 \partial_1 M_j^k - X_1 X^k \partial_1 M_j^1 - M_j^h \partial_h (X_1 X^k) + M_1^k \partial_j (X_1 X^1) \\
&+ X_1 X^k \partial_j M_1^1 = 0,
\end{aligned}$$

Now we recall that $\nu_1(q^1) = X_1 X^1 > 0$:

$$\nu_1 \partial_1 M_j^k - X_1 X^k \partial_1 M_j^1 - M_j^h \partial_h (X_1 X^k) + X_1 X^k \partial_j M_1^1 = 0.$$

For $k \neq 1$ and $k = 1$ we find Eqs. (7.11) and (7.12), respectively. □

8. Further algebraic properties. Let us return to the explicit calculation of the first elements of a L-pencil. Of course, $\mathbf{K}_1$ is of first degree, Eq. (3.1),

$$\mathbf{K}_1 = \mu\,\mathbf{I} - \mathbf{M} + m\,(\nu\,\mathbf{I} - \mathbf{N}),$$

as well as $\mathbf{K}_2$, as shown in Section 3,

$$\mathbf{K}_2 = \mathbf{K}_{20} + m\,\mathbf{K}_{21},$$

with

$$\begin{aligned}\mathbf{K}_{20} &\doteq \frac{1}{2}\,(\mu^2 - \operatorname{tr}\mathbf{M}^2)\,\mathbf{I} - \mu\,\mathbf{M} + \mathbf{M}^2,\\ \mathbf{K}_{21} &\doteq (\mu\nu - \operatorname{tr}\mathbf{MN})\,\mathbf{I} - \nu\mathbf{M} - \mu\,\mathbf{N} + \mathbf{MN} + \mathbf{NM}.\end{aligned}$$

Let us introduce the vector field

$$\mathbf{Y} \doteq \mathbf{M}\mathbf{X}.$$

Then Eqs. (3.6) become

$$\begin{aligned}\mathbf{H}_1 &= \nu_1\,\mathbf{I} - \mathbf{X}\otimes\mathbf{X}^\flat,\\ \mathbf{H}_2 &= (\mu\nu_1 - \mathbf{Y}\cdot\mathbf{X})\,\mathbf{I} - \nu_1\mathbf{M} - \mu\,\mathbf{X}\otimes\mathbf{X}^\flat + \mathbf{MN} + \mathbf{NM}.\end{aligned}$$

In any coordinate system,

$$\begin{aligned}(\mathbf{MN})_i^j &= M_k^j N_i^k = M_k^j X^k X_i = Y^j X_i,\\ (\mathbf{NM})_i^j &= N_k^j M_i^k = X^j X_k M_i^k = X^j Y_i.\end{aligned}$$

Hence,

$$\mathbf{MN} = \mathbf{Y}\otimes\mathbf{X}^\flat, \qquad \mathbf{NM} = \mathbf{X}\otimes\mathbf{Y}^\flat, \qquad \operatorname{tr}(\mathbf{MN}) = \mathbf{X}\cdot\mathbf{Y}.$$

This shows that

PROPOSITION 8.1. *In a L-pencil the two tensors* $\mathbf{H}_1$ *and* $\mathbf{H}_2$ *have the following expressions:*

$$\begin{aligned}\mathbf{H}_1 &= \nu_1\,\mathbf{I} - \mathbf{X}\otimes\mathbf{X}^\flat,\\ \mathbf{H}_2 &= (\mu\nu_1 - \mathbf{Y}\cdot\mathbf{X})\,\mathbf{I} - \nu_1\mathbf{M} - \mu\,\mathbf{X}\otimes\mathbf{X}^\flat + \mathbf{Y}\otimes\mathbf{X}^\flat + \mathbf{X}\otimes\mathbf{Y}^\flat.\end{aligned} \tag{8.1}$$

As a consequence,

PROPOSITION 8.2. *The vector field* $\mathbf{X}$ *is an eigenvector of both* $\mathbf{H}_1$ *and* $\mathbf{H}_2$, *with zero eigenvalue.*

Proof. Since $\mathbf{X}\cdot\mathbf{X} = \nu_1$, we have $\mathbf{H}_1\mathbf{X} = \nu_1\,\mathbf{X} - \mathbf{X}\nu_1 = 0$, and $\mathbf{H}_2\mathbf{X} = (\mu\nu_1 - \mathbf{Y}\cdot\mathbf{X})\,\mathbf{X} - \nu_1\mathbf{Y} - \mu\,\nu_1\mathbf{X} + \nu_1\,\mathbf{Y} + \mathbf{Y}\cdot\mathbf{X}\,\mathbf{X} = 0.$ □

REMARK 8.1. In any coordinate system Eqs. (8.1) read

$$
\begin{aligned}
(\mathbf{H}_1)_i^j &= \nu_1\,\delta_i^j - X_iX^j, \\
(\mathbf{H}_2)_i^j &= (\mu\nu_1 - \eta)\,\delta_i^j - \nu_1\,M_i^j - \mu\,X_iX^j + X_iY^j + Y_iX^j,
\end{aligned}
\tag{8.2}
$$

where $\eta \doteq \mathbf{Y}\cdot\mathbf{X} = X^iY_i$. Recall that in $\mathbf{X}$-coordinates Eqs. (5.6) and (5.7) hold. so that

$$
\begin{aligned}
&Y^1 = g^{11}Y_1 = M_i^1X^i = M_1^1\,X^1 = \nu_1\,M_1^1, \\
&\eta = X_iY^i = X_1Y^1 = \nu_1\,M_1^1 = Y^1, \\
&Y^a = M_i^a\,X^i = \nu_1\,M_1^a, \\
&Y_a = M_{ai}\,X^i = \nu_1\,M_{a1} = g^{11}\,M_{a1} = M_a^1.
\end{aligned}
$$

Hence, from Eqs. (8.2) it follows that $(a, b = 2, \dots, n)$

$$
\begin{aligned}
(\mathbf{H}_1)_1^1 &= \nu_1 - X_1X^1 = 0, \\
(\mathbf{H}_2)_1^1 &= (\mu\nu_1 - \eta) - \nu_1\,M_1^1 - \mu\,X_1X^1 + X_1Y^1 + Y_1X^1, \\
&= (\mu\nu_1 - Y^1) - \nu_1\,M_1^1 - \mu\,\nu_1 + Y^1 + Y_1g^{11} = 0.
\end{aligned}
$$

$$
\begin{aligned}
(\mathbf{H}_1)_1^a &= 0, \\
(\mathbf{H}_1)_a^1 &= 0, \\
(\mathbf{H}_1)_a^b &= \nu_1\,\delta_a^b.
\end{aligned}
$$

$$
\begin{aligned}
(\mathbf{H}_2)_1^a &= -\,\nu_1\,M_1^a + \nu_1\,M_1^a = 0, \\
(\mathbf{H}_2)_a^1 &= -\,\nu_1\,M_a^1 + \nu_1\,Y_a = 0, \\
(\mathbf{H}_2)_a^a &= \nu_1\,(\mu - M_1^1 - M_a^a), \\
(\mathbf{H}_2)_a^b &= -\,\nu_1\,M_a^b, \quad a \neq b.
\end{aligned}
$$

In matrix form,

$$
\mathbf{H}_1 = \nu_1 \begin{bmatrix} 0 & 0 \\ 0 & \bar{\mathbf{I}} \end{bmatrix}, \qquad \mathbf{H}_2 = \nu_1 \begin{bmatrix} 0 & 0 \\ 0 & \bar{\mathbf{H}}_1 \end{bmatrix},
\tag{8.3}
$$

where

$$
\begin{aligned}
\bar{\mathbf{I}} &= \mathbf{I}_{n-1}, & \bar{\mathbf{H}}_1 &\doteq \bar{\mu}\,\bar{\mathbf{I}} - \bar{\mathbf{M}}, \\
\bar{\mathbf{M}} &\doteq [M_a^b], & \bar{\mu} &\doteq \operatorname{tr}\bar{\mathbf{M}} = \mu - M_1^1.
\end{aligned}
\tag{8.4}
$$

$\diamondsuit$

The matrix forms (8.3) of $\mathbf{H}_1$ and $\mathbf{H}_2$ show once more the property expressed by Proposition 8.2: $\mathbf{H}_1\mathbf{X} = \mathbf{H}_2\mathbf{X} = 0$. For proving this we observe that

PROPOSITION 8.3. *If* $\mathbf{H}_a\mathbf{X} = 0$, *then all* $\mathbf{K}_a$ *are of first degree in* m,

$$\mathbf{K}_a = \mathbf{K}_{a0} + m\,\mathbf{K}_a.$$

Proof. Recall the recursive formula (1.1) of a L-chain,

$$\mathbf{K}_{a+1} = \frac{1}{a+1}\,\mathrm{tr}\,(\mathbf{K}_a\mathbf{L}_m)\,\mathbf{I} - \mathbf{K}_a\,\mathbf{L}_m.$$

Assume that $\mathbf{K}_a$ is first-degree, $\mathbf{K}_a = \mathbf{K}_{a0} + m\,\mathbf{H}_a$. Let us denote by a prime $'$ the derivative w.r.to m. For instance, $\mathbf{L}'_m = \mathbf{N}$. Note that $\mathbf{K}'_a = \mathbf{H}_a$ is of zero-degree. Then,

$$\begin{aligned}(\mathbf{K}_a\mathbf{L}_m)' &= \mathbf{K}'_a\mathbf{L}_m + \mathbf{K}_a\mathbf{N} = \mathbf{H}_a(\mathbf{M} + m\,\mathbf{N}) + (\mathbf{K}_{a0} + m\mathbf{H}_a)\mathbf{N}\\ &= \mathbf{H}_a\mathbf{M} + \mathbf{K}_{a0}\mathbf{N} + 2m\,\mathbf{H}_a\mathbf{N} = \mathbf{H}_a\mathbf{M} + \mathbf{K}_{a0}\mathbf{N}.\\ (\mathbf{K}_a\mathbf{L}_m)'' &= 2\,\mathbf{H}_a\mathbf{N} = 2\,\mathbf{H}_a\mathbf{X}\cdot\mathbf{X} = 0.\end{aligned}$$

It follows that $\mathbf{K}_{a+1}$ is first-degree. □

9. The induced L-systems. What we are going to do now is suggested by the remarkable formulas (8.4). Indeed, the $(n-1)\times(n-1)$ matrix

$$\bar{\mathbf{H}}_1 = \bar{\mu}\,\bar{\mathbf{I}} - \bar{\mathbf{M}}, \qquad \bar{\mu} = \mathrm{tr}\,\bar{\mathbf{M}},$$

has the same form of the first step of a L-sequence. To interpret this analogy in a right way, let us consider any leaf of the $\mathbf{X}$-web i.e., any $n-1$-dimensional submanifold S orthogonal to $\mathbf{X}$. Such a submanifold is defined by an equation $q^1 =$ constant (in $\mathbf{X}$-coordinates). Then the bar-tensors like $\bar{\mathbf{I}}$, $\bar{\mathbf{M}}$, $\bar{\mathbf{H}}_1$, etc., can be interpreted as $(1,1)$-tensors on any S.

THEOREM 9.1. *On each leaf* S *the tensor* $\bar{\mathbf{M}} = [M^b_a]$ *is torsionless.*

Proof. The tosionless condition of $\mathbf{M}$ is equivalent to the symmetry of

$$S^k_{ij}(\mathbf{M}) \doteq M^h_i\partial_h M^k_j - M^k_m\partial_i M^m_j.$$

A special case is

$$\begin{aligned}0 &= S^c_{ab}(\mathbf{M}) = M^h_a\partial_h M^c_b - M^c_m\partial_a M^m_b\\ &= M^d_a\partial_d M^c_b - M^c_d\partial_a M^d_b + M^1_a\partial_1 M^c_b - M^c_1\partial_a M^1_b\\ &= S^c_{ab}(\bar{\mathbf{M}}) + M^1_a\partial_1 M^c_b - M^c_1\partial_a M^1_b.\end{aligned}$$

Due to Eq. (7.11), $\partial_1 M^c_b = 0$. Due to Eq. (7.10), $\partial_a M^1_b = \partial_b M^1_a$. This shows that $S^c_{ab}(\bar{\mathbf{M}})$ is symmetric. □

THEOREM 9.2. *On each leaf* S *the tensor* $\bar{\mathbf{M}}$ *is a trace-type CKT.*

Proof. Since $M^{ab} = g^{bi}M^a_i = g^{bc}M^a_c$, the contravariant components of $\mathbf{M}'$ are just the contravariant components M^{ab} (with $a, b > 1$) of $\mathbf{M}$. The tensor $\mathbf{M}$ is a trace-type CKT i.e., the polynomial equation

$$\{M^{ij}\,p_ip_j, g^{hk}\,p_hp_k\} = -2\,C^ip_i\,g^{hk}\,p_hp_k$$

holds with

$$C^i \doteq g^{ij}\, \partial_i \mu.$$

In **X**-coordinates, we get

$$\begin{aligned}&\{M^{11}\, p_1^2 + 2M^{1a}\, p_1 p_a + M^{ab}\, p_a p_b, g^{11}\, p_1^2 + g^{ab}\, p_a p_b\}\\ &= -2\,(C^1 p_1 + C^a p_a)(g^{11}\, p_1^2 + g^{ab}\, p_a p_b).\end{aligned} \tag{9.1}$$

The left hand side is

$$\begin{aligned} L &\doteq \{M^{11}\, p_1^2 + 2M^{1a}\, p_1 p_a + M^{ab}\, p_a p_b, g^{11}\, p_1^2 + g^{ab}\, p_a p_b\}\\ &= 2\,(M^{11}\, p_1 + M^{1a}\, p_a)(\partial_1 g^{11}\, p_1^2 + \partial_1 g^{bc}\, p_b p_c)\\ &+ 2\,(M^{1a}\, p_1 + M^{ab}\, p_b)(\partial_a g^{11}\, p_1^2 + \partial_a g^{bc}\, p_b p_c)\\ &- 2\,(\partial_1 M^{11}\, p_1^2 + 2\partial_1 M^{1a}\, p_1 p_a + \partial_1 M^{ab}\, p_a p_b)\; g^{11}\, p_1\\ &- 2\,(\partial_a M^{11}\, p_1^2 + 2\partial_a M^{1b}\, p_1 p_b + \partial_a M^{bc}\, p_b p_c)\; g^{ab}\, p_b.\end{aligned}$$

This is a homogeneous third-degree polynomial, which can be written in the following form:

$$\begin{aligned}\frac{1}{2}\,L &= p_1^3\,[M^{11}\partial_1 g^{11} + M^{1a}\,\partial_a g^{11} - g^{11}\,\partial_1 M^{11}]\\ &+ p_1^2 p_a\,[M^{1a}\,\partial_1 g^{11} + M^{ab}\,\partial_b g^{11} - g^{ab}\partial_b M^{11} - 2\,g^{11}\,\partial_1 M^{1a}]\\ &+ p_1 p_a p_b\,[M^{11}\,\partial_1 g^{ab} + M^{1c}\,\partial_c g^{ab} - g^{11}\,\partial_1 M^{ab} - 2\,g^{ac}\partial_c M^{1b}]\\ &+ p_a p_b p_c\,[M^{1a}\,\partial_1 g^{bc} + M^{da}\,\partial_d g^{bc} - g^{da}\,\partial_d M^{bc}].\end{aligned}$$

A further evolution of this expression is due to equations $g^{11} = \nu_1$ and $\partial_a g^{11} = \partial_a \nu_1 = 0$,

$$\begin{aligned}\frac{1}{2}\,L &= p_1^3\,[M^{11}\partial_1 \nu_1 - \nu_1\,\partial_1 M^{11}]\\ &+ p_1^2 p_a\,[M^{1a}\,\partial_1 \nu_1 - g^{ab}\partial_b M^{11} - 2\,\nu_1\,\partial_1 M^{1a}]\\ &+ p_1 p_a p_b\,[M^{11}\,\partial_1 g^{ab} + M^{1c}\,\partial_c g^{ab} - \nu_1\,\partial_1 M^{ab} - 2\,g^{ac}\partial_c M^{1b}]\\ &+ p_a p_b p_c\,[M^{1a}\,\partial_1 g^{bc} + M^{da}\,\partial_d g^{bc} - g^{da}\,\partial_d M^{bc}].\end{aligned}$$

For the right hand side R we have

$$\begin{aligned}-\frac{1}{2}\,R &= (C^1 p_1 + C^a p_a)\,(g^{11}\, p_1^2 + g^{ab}\, p_a p_b)\\ &= p_1^3\, C^1\, g^{11} + p_1^2 p_a\, C^a\, g^{11} + p_1 p_a p_b\, C^1\, g^{ab} + p_a p_b p_c\, C^a g^{bc}\\ &= p_1^3\, C^1\, \nu_1 + p_1^2 p_a\, C^a\, \nu_1 + p_1 p_a p_b\, C^1\, g^{ab}\\ &+ p_a p_b p_c\, C^a g^{bc}.\end{aligned}$$

Hence, Eq. (9.1) is equivalent to the following equations:

$$\begin{aligned}
&\text{(I)} \quad M^{11}\partial_1\nu_1 - \nu_1\,\partial_1 M^{11} + C^1\nu_1 = 0, \\
&\text{(II)} \quad M^{1a}\,\partial_1\nu_1 - g^{ab}\partial_b M^{11} - 2\,\nu_1\,\partial_1 M^{1a} + C^a\nu_1 = 0, \\
&\text{(III)} \quad p_a p_b\,[M^{11}\,\partial_1 g^{ab} + M^{1c}\,\partial_c g^{ab} \\
&\qquad - \nu_1\,\partial_1 M^{ab} - 2\,g^{ac}\partial_c M^{1b} - C^1\,g^{ab}] = 0, \\
&\text{(IV)} \quad p_a p_b p_c\,[M^{1a}\,\partial_1 g^{bc} + M^{da}\,\partial_d g^{bc} - g^{da}\,\partial_d M^{bc} \\
&\qquad + C^a\,g^{bc}] = 0.
\end{aligned} \tag{9.2}$$

Let us develop these equations taking into account that in $\mathbf{X}$-coordinates

$$\begin{aligned}
M^{11} &= g^{1i}\,M_i^1 = g^{11}\,M_1^1 = \nu_1\,M_1^1, \\
M^{1a} &= g^{ai}M_i^1 = g^{ab}\,M_b^1, \\
M^{1a} &= g^{1i}M_i^a = g^{11}\,M_1^a = \nu_1\,M_1^a, \\
C^1 &= g^{11}\,C_1 = \nu_1\,\partial_1\mu = \nu_1\,\partial_1(\bar{\mu} + M_1^1), \\
C^a &= g^{ab}\,\partial_b\mu = g^{ab}\,\partial_b(\bar{\mu} + M_1^1).
\end{aligned}$$

In fact, for proving the theorem it is sufficient to develop Eq. (IV), which reads

$$p_a p_b p_c\,[M^{1a}\,\partial_1 g^{bc} + M^{da}\,\partial_d g^{bc} - g^{da}\,\partial_d M^{bc} + g^{ad}\,\partial_d(\bar{\mu} + M_1^1)\,g^{bc}] = 0.$$

It implies

$$\begin{aligned}
&p_a p_b p_c\,[M^{da}\,\partial_d g^{bc} - g^{da}\,\partial_d M^{bc} + g^{ad}\,\partial_d\bar{\mu}\,g^{bc}] \\
&+ p_a p_b p_c\,[M^{1a}\,\partial_1 g^{bc} + g^{ad}\,\partial_d M_1^1\,g^{bc}] = 0,
\end{aligned}$$

i.e.,

$$\begin{aligned}
&\frac{1}{2}\,\{P(\bar{\mathbf{M}}), P(\bar{\mathbf{G}}\}_S + P(\nabla\bar{\mu})P(\bar{\mathbf{G}}) \\
&+ p_a p_b p_c\,[M^{1a}\,\partial_1 g^{bc} + g^{ad}\,\partial_d M_1^1\,g^{bc}] = 0.
\end{aligned}$$

Hence, $\bar{\mathbf{M}}$ is a trace-type CKT on S if and only if

$$p_a p_b p_c\,[M^{1a}\,\partial_1 g^{bc} + g^{ad}\,\partial_d M_1^1\,g^{bc}] = 0,$$

i.e.,

$$p_a p_b p_c\,g^{ad}\,[M_d^1\,\partial_1 g^{bc} + \partial_d M_1^1\,g^{bc}] = 0.$$

Due to (5.13) it follows that

$$p_a p_b p_c\,g^{ad}\,g^{bc}\,[M_d^1\,\phi_1 + \partial_d M_1^1] = 0,$$

where $\phi_1(q^1) \doteq -\partial_1 \log \nu_1$. But Eq. (7.12) shows that $M_b^1 \partial_1 \nu_1 - \nu_1 \partial_b M_1^1 = 0$ i.e., $M_b^1 \phi_1 + \partial_b M_1^1 = 0$. □

REMARK 9.1. Equations (I), (II), (III) in (9.2) have not been used. They provide further necessary conditions on $\mathbf{M}$, whose analysis is here omitted. ◊

According to the two theorems above, $\bar{\mathbf{M}}$ is a torsionless CKT on each submanifold S orthogonal to $\mathbf{X}$. Its eigenvalues are distinct (see Remark 9.3 below) with the exception of a closed subset (may be empty) of S. Thus, $\bar{\mathbf{M}}$ is a L-tensor on any S and the recursive formula

$$\bar{\mathbf{H}}_a = \frac{1}{a} \operatorname{tr}(\bar{\mathbf{H}}_{a-1}\bar{\mathbf{M}})\,\bar{\mathbf{I}} - \bar{\mathbf{H}}_{a-1}$$

for $a = 1, 2, \ldots, n-2$ and $\bar{\mathbf{H}}_0 = \bar{\mathbf{I}}$, form a L-system on S (whose dimension is $n-1$).

Now we return to the whole space Q_n and consider the $n-1$ tensors

$$\mathbf{H}_a = \nu_1 \begin{bmatrix} 0 & 0 \\ 0 & \bar{\mathbf{H}}_{a-1} \end{bmatrix} \tag{9.3}$$

with $a = 1, \ldots, n-1$. Note that for $a = 1$ and $a = 2$ we get the matrices (8.3). The $n-1$ (1,1)-tensors $\mathbf{H}_a$ are linearly independent, commute and have common eigenvectors tangent to S, since the same properties hold for $\bar{\mathbf{H}}_a$. But they have also $\mathbf{X}$ as common eigenvector, with zero eigenvalue.

In order to get a basis for a KS-space we have to add to them the identity

$$\mathbf{H}_0 = \mathbf{I} = \begin{bmatrix} 1 & 0 \\ 0 & \bar{\mathbf{I}} \end{bmatrix}$$

and to prove that

THEOREM 9.3. *The tensors $\mathbf{H}_a$ are Killing tensors.*

Proof. The $\mathbf{X}$-components of $\mathbf{H}_*$ are $H^{11} = 0$, $H^{1a} = 0$, $H^{bc} = \nu_1 \, g^{bd} \, \bar{H}_d^c$. Hence, $P(\mathbf{H}_*) = \nu_1 \, \bar{H}^{bc} \, p_b p_c$. We have to show that

$$\{P(\mathbf{H}_*), g^{ij} p_i p_j\} = 0.$$

In $\mathbf{X}$-coordinates $g^{11} = \nu_1$ is a function of q^1 only, so that

$$\begin{aligned}
&\{P(\mathbf{H}_*)\,,\, g^{ij} p_i p_j\} = \{\nu_1 \, \bar{H}^{bc} \, p_b p_c \,,\, g^{11} p_1^2 + g^{ab} \, p_a p_b\} \\
&= 2\,\nu_1 \, \bar{H}^{bc} \, p_c \, \partial_b g^{ad} \, p_a p_d - 2 p_1 \nu_1 \, \partial_1 (\nu_1 \, \bar{H}^{bc} \, p_b p_c) \\
&- 2 g^{ab} p_b \, \partial_a (\nu_1 \, \bar{H}^{bc} \, p_b p_c) \\
&= 2\,\nu_1 \, \bar{H}^{bc} \, p_c \, \partial_b g^{ad} \, p_a p_d - 2\,\nu_1 \, g^{ab} p_b \, \partial_a (\bar{H}^{bc} \, p_b p_c) \\
&- 2 p_1 \nu_1 \, \partial_1 (\nu_1 \, \bar{H}^{bc} \, p_b p_c) \\
&= \nu_1 \, \{P(\bar{\mathbf{H}}_*), P(\bar{\mathbf{G}})\} - 2 p_1 \nu_1 \, \partial_1 (\nu_1 \, \bar{H}^{bc} \, p_b p_c) \\
&= \; - 2 p_1 \nu_1 \, \partial_1 (\nu_1 \, \bar{H}^{bc} \, p_b p_c) = \ldots
\end{aligned}$$

Since $\bar{H}^{bc} = g^{ba}\bar{H}_a^c$, we have $\partial_1 \bar{H}^{bc} = \partial_1 g^{ba}\bar{H}_a^c + g^{ba}\,\partial_1 \bar{H}_a^c$. However, $\partial_1 \bar{H}_a^c = 0$, since $\bar{H}_a^c$ is constructed by an algebraic process from M_a^b and $\partial_1 M_a^b = 0$, Eq. (7.11). Hence, due to Eqs. (5.13) and (5.14),

$$\partial_1 \bar{H}^{bc} = \partial_1 g^{ba}\bar{H}_a^c = \phi_1\, g^{ba}\bar{H}_a^c = \phi_1\, \bar{H}^{bc}.$$

It follows that

$$\partial_1(\nu_1 \bar{H}^{bc}) = \partial_1 \nu_1\, \bar{H}^{bc} + \nu_1 \phi_1\, \bar{H}^{bc} = 0.$$

Thus, $\ldots = 0$. □

REMARK 9.2. This theorem proves that $\mathbf{H}_0 = \mathbf{I}$ and the $n-1$ Killing tensors $\mathbf{H}_a$ defined in (9.3) form a basis of a KS-space on Q. Since $\mathbf{H}_a\mathbf{X} = 0$ $(a > 0)$, due to Proposition 8.3, the tensors $\mathbf{K}_a$ are of first degree in m. ◇

REMARK 9.3. Let us consider in $\mathbb{R}^n$ a diagonal matrix

$$\mathbf{M} = \mathrm{diag}\,[\mu_1, \mu_2, \ldots, \mu_n]$$

with all distinct $\mu_i \neq 0$, and a vector $\mathbf{X} = [X^1, X^2, \ldots, X^n]$. The matrix

$$\bar{\mathbf{M}} \doteq \mathbf{M} - \alpha\,\mathbf{M}(\mathbf{X}) \otimes \mathbf{X}^\flat,$$

with $\alpha^{-1} \doteq \langle \mathbf{X}, \mathbf{X}^\flat \rangle = X^i X_i$, satisfies equation $\bar{\mathbf{M}}(\mathbf{X}) = 0$. Its components are (we consider for simplicity $n = 4$).

For a vector $\mathbf{X}$ with only a non-zero component, say $X^1 \neq 0$, we get the diagonal form

$$\bar{\mathbf{M}} = \mathrm{diag}\,[0, \mu_2, \ldots, \mu_n]$$

which shows that the eigenvalues are distinct. Then, in an open cone around $\mathbf{X}$ this property is preserved. We have n open cones generated in this way. On the other hand it is known (after Sylvester) that the condition that $\bar{\mathbf{M}}$ has non-simple eigenvalues is expressed by an algebraic equation of order $2n+2$ in the variables X^i. This equation defines a surface (or the union of surfaces) in the space (X^i), or the empty set. ◇

10. Conclusion. The necessity of the conditions listed in Theorems 1.3 and 1.4 are proved:

- $\mathbf{M}$ is a L-tensor: Theorem 2.1.
- $\mathbf{N} = \mathbf{X} \otimes \mathbf{X}^\flat$: Theorem 4.1.
- $\mathbf{X}$ is a conformal vector field: Theorem 5.2.
- $\mathbf{X}$ is a translation or a dilatation: Theorem 5.1.
- All tensors $\mathbf{K}_a$ of the L-sequence are linear in m: Remark 9.2.
- $\mathbf{H}_a\mathbf{X} = 0$: Eq. (9.3).
- The restrictions of $\mathbf{H}_a$ to any leaf of the foliation orthogonal to $\mathbf{X}$ form a L-system on that leaf: Section 9.

It remains to prove the sufficiency in Theorem 1.3: *If* $\mathbf{M}$ *is a L-tensor and* $\mathbf{X}$ *is a CKV such that* $d\mathbf{X}^\flat = 0$, *then* $\mathbf{L}_m = \mathbf{M} + m\,\mathbf{X} \otimes \mathbf{X}^\flat$ *is an effective L-pencil.*

Proof. Since $\mathbf{X}$ is a CKV, $\mathbf{N} = \mathbf{X} \otimes \mathbf{X}^\flat$ is a CKT. Since $d\mathbf{X}^\flat = 0$, $\mathbf{N}$ is torsionless: see the proof of Theorem 5.3. Then apply Theorems 7.1 and 2.1. □

11. Final comments. About the existence of L-pencils, we recall the following properties (to be applied to the L-tensor $\mathbf{M}$ of a L-pencil).

- *If a Stäckel web has a foliation orthogonal to a proper CKV, then it is not a L-web* (Theorem 9.4 in [5]).
- *If a Stäckel web has* $m < n$ *foliations orthogonal to translations, then it is not a L-web* (Theorem 9.6 in [5]).

As a consequence,

- *All translational Stäckel webs in a Euclidean* n*-space, different from the Cartesian web (for which* $m = n$*) are not L-webs* (Remark 9.2 in [5]).

Moreover, let us recall that ([9], p. 249):

- *If a Killing vector is normal then the lines of curvature of the orthogonal submanifolds are indeterminate.*
- *The orbits of a translation form a flat submanifold.*
- *If a translation is normal, then the orthogonal submanifolds are totally geodesic.*

In [9] – formula (69.5) – it is proved that for a CKV $\mathbf{X}$ equation

$$\Delta_2 \psi = \frac{2}{n-1} \left(X_m \nabla_i R^{mi} + \nabla_i X_m \, R^{mi} \right)$$

holds. Thus, for a Killing vector ($\psi = 0$) and for a dilatation ($\psi =$ constant),

$$X_m \nabla_i R^{mi} + \nabla_i X_m \, R^{mi} = 0.$$

For a manifold of constant curvature $K_0 \neq 0$ this equation reduces to

$$\nabla_i \nabla_j \psi + K_0 \, g_{ij} \, \psi = 0.$$

For a Killing vector this is identically satisfied, but for a dilatation it gives $K_0 = 0$: absurd. Thus,

THEOREM 11.1. *A manifold with non-zero constant curvature does not admit dilatations.*

But it can also be shown that a manifold with non-zero constant curvature does not admit translations. So, for instance, the sphere $\mathbb{S}_n$ and the pseudosphere $\mathbb{H}_n$, does not have L-pencils.

All statements listed above represent strong obstructions for the existence of translations and dilatations, hence, for the existence of L-pencils. This list is certainly incomplete, since further results should be present in the ancient–may be also recent–literature.

As said in Remark 1.1, the basic example of L-pencil is the spherical-conical (asymmetric) web in $\mathbf{r}^n$, and, according to the above remarks, it is the only L-pencil existing in $\mathbb{R}^n$. In spite to this rather restrictive result, futher arguments of research arise:

- To find examples of L-pencils form Riemannian manifolds with non-constant curvature.
- To extend the notion of L-pencil to pseudo-Euclidean spaces. Indeed, we know the general form of a L-tensor in these spaces (Appendix B in [5], see also [8]).
- To introduce and study the notion of **multipencil**. A 2-pencil is for instance

$$\mathbf{L}_{m_1,m_2} = \mathbf{M} + m_1\,\mathbf{X}_1 \otimes \mathbf{X}_1^\flat + m_2\,\mathbf{X}_2 \otimes \mathbf{X}_2^\flat, \quad m_1 \neq m_2.$$

  Since the spherical-conical webs play an important role in the diagrammatic classification of Stäckel systems due to Kalnins and Miller [11, 10], the notion of multipencil should be useful for this classification.
- To extend the notions of L-tensor and of L-pencil by dropping out the requirement of 'simple eigenvalues', and including the cases of S-webs invariant w.r.to Killing tensors.

APPENDIX

A. Induced L-systems. It is known that a S-system induces a S-system on each leaf of its web [3, 7].

In Section 9 we have seen that a LP-system induces a L-system on each leaf of its web. A similar property holds for L-systems:

THEOREM A.1. *A L-system induces a L-system on each leaf of its web.*

Proof. Let (u^i) be the eigenvalues of a L-tensor $\mathbf{L}$. Let us order the indices in such a way that

$$u^1 < u^2 < \ldots < u^n,$$

and set

$$\Delta_i \doteq \prod_{k=1}^{i-1}(u^i - u^k) \prod_{k=i+1}^{n} (u^k - u^i).$$

Then $\Delta_i > 0$ for each index i, and the metric tensor components in $\mathbf{L}$-coordinates can be written

$$g^{ii} = \frac{\phi_i(q^i)}{\Delta_i}$$

with $\phi_i(q^i) > 0$. As a consequence, up to a rescaling of the coordinates, we get

$$g^{ii} = \frac{1}{\Delta_i}, \qquad g_{ii} = \Delta_i.$$

These components are all positive. For the special case $n = 3$:

$$\begin{aligned} \Delta_1 &= (u^2 - u^1)(u^3 - u^1), \\ \Delta_2 &= (u^2 - u^1)(u^3 - u^2), \\ \Delta_3 &= (u^3 - u^1)(u^3 - u^2). \end{aligned} \tag{A.1}$$

For $n = 4$:

$$\begin{aligned} \Delta_1 &= (u^2 - u^1)(u^3 - u^1)(u^4 - u^1), \\ \Delta_2 &= (u^2 - u^1)(u^3 - u^2)(u^4 - u^2), \\ \Delta_3 &= (u^3 - u^1)(u^3 - u^2)(u^4 - u^3), \\ \Delta_4 &= (u^4 - u^1)(u^4 - u^2)(u^4 - u^3). \end{aligned} \tag{A.2}$$

Let us consider for simplicity the case $n = 4$ and the leaf S defined by $u^4 = 1$. The components g_{11}, g_{22}, g_{33} restricted to S become

$$\begin{aligned} \Delta_1 &= (u^2 - u^1)(u^3 - u^1)(1 - u^1), \\ \Delta_2 &= (u^2 - u^1)(u^3 - u^2)(1 - u^2), \\ \Delta_3 &= (u^3 - u^1)(u^3 - u^2)(1 - u^3). \end{aligned}$$

They can be written in the form

$$\begin{aligned} \Delta_1 &= (u^2 - u^1)(u^3 - u^1)\phi_1(q^1), \\ \Delta_2 &= (u^2 - u^1)(u^3 - u^2)\phi_2(q^2), \\ \Delta_3 &= (u^3 - u^1)(u^3 - u^2)\phi_3(q^3), \end{aligned}$$

with $\phi_i(q^i) > 0$. Then the coordinates can be normalized in order to get

$$\begin{aligned} \Delta_1 &= (u^2 - u^1)(u^3 - u^1), \\ \Delta_2 &= (u^2 - u^1)(u^3 - u^2), \\ \Delta_3 &= (u^3 - u^1)(u^3 - u^2). \end{aligned}$$

This is a form of the kind (A.1). □

REMARK A.1. For $n = 4$ we get

$$\begin{aligned} \Delta_1 &= (u^2 - u^1)(u^3 - u^1)(u^4 - u^1) \\ &= (u^2 - u^1)(u^3 u^4 - u^3 u^1 - u^1 u^4 + (u^1)^2) \\ &= u^2 u^3 u^4 - u^2 u^3 u^1 - u^2 u^1 u^4 + u^2 (u^1)^2 \\ &\quad - u^1 u^3 u^4 + u^3 (u^1)^2 + (u^1)^2 u^4 - (u^1)^3) \\ &= \sigma_3^1 - u^1\, \sigma_2^1 + (u^1)^2\, \sigma_1^1 - (u^1)^3. \end{aligned}$$

Thus, in this special case,

$$g_{11} = \sigma_3^1 - u^1\,\sigma_2^1 + (u^1)^2\,\sigma_1^1 - (u^1)^3.$$

We get the general formula

$$g_{ii} = \sigma_3^i - u^i\,\sigma_2^i + (u^i)^2\,\sigma_1^i - (u^i)^3.$$

Similar formulas can be obtained for any dimension n. ◊

B. An example of type-2 L-pencil. In $\mathbb{R}^3$ the parabolic web is determined by the L-tensor [2]

$$\mathbf{L}_m = \mathbf{M} + m\,\mathbf{u}\odot\mathbf{r}, \qquad \mathbf{u}^2 = 1,$$

where $\mathbf{u}$ is a constant unit vector. Let us look at this tensor as a L-pencil with

$$\mathbf{N} = \mathbf{u}\odot\mathbf{r} = \frac{1}{2}(\mathbf{u}\otimes\mathbf{r} + \mathbf{r}\otimes\mathbf{u}).$$

Since $\operatorname{tr}\mathbf{L}_m = \mu + m\,\mathbf{u}\cdot\mathbf{r}$, $\nu = \mathbf{u}\cdot\mathbf{r}$, we get

$$\mathbf{K}_{22} = \frac{1}{2}\,(\nu^2 - \operatorname{tr}\mathbf{N}^2)\,\mathbf{I} - \nu\,\mathbf{N} + \mathbf{N}^2.$$

Moreover,

$$\begin{aligned}
\mathbf{N}^2 &= \frac{1}{4}(\mathbf{u}\otimes\mathbf{r} + \mathbf{r}\otimes\mathbf{u})(\mathbf{u}\otimes\mathbf{r} + \mathbf{r}\otimes\mathbf{u}) \\
&= \frac{1}{4}\,(\mathbf{r}\cdot\mathbf{u}\,\mathbf{u}\otimes\mathbf{r} + r^2\,\mathbf{u}\otimes\mathbf{u} + \mathbf{r}\otimes\mathbf{r} + \mathbf{u}\cdot\mathbf{r}\,\mathbf{r}\otimes\mathbf{u}) \\
&= \frac{1}{4}\,(2\,\mathbf{r}\cdot\mathbf{u}\,\mathbf{u}\odot\mathbf{r} + r^2\,\mathbf{u}\otimes\mathbf{u} + \mathbf{r}\otimes\mathbf{r}),
\end{aligned}$$

and

$$\operatorname{tr}\mathbf{N}^2 = \frac{1}{4}\,[2(\mathbf{r}\cdot\mathbf{u})^2 + 2r^2] = \frac{1}{2}\,[(\mathbf{r}\cdot\mathbf{u})^2 + r^2],$$

so that

$$\frac{1}{2}\,(\nu^2 - \operatorname{tr}\mathbf{N}^2) = \frac{1}{2}\left[(\mathbf{u}\cdot\mathbf{r})^2 - \frac{1}{2}\,((\mathbf{r}\cdot\mathbf{u})^2 + r^2)\right] = \frac{1}{2}\left[\frac{1}{2}\,((\mathbf{r}\cdot\mathbf{u})^2 + r^2)\right],$$

and

$$\begin{aligned}
\mathbf{N}^2 - \nu\,\mathbf{N} &= \frac{1}{4}\,(2\,\mathbf{r}\cdot\mathbf{u}\,\mathbf{u}\odot\mathbf{r} + r^2\,\mathbf{u}\otimes\mathbf{u} + \mathbf{r}\otimes\mathbf{r}) - (\mathbf{u}\cdot\mathbf{r})(\mathbf{u}\odot\mathbf{r}) \\
&= \frac{1}{4}\,(r^2\,\mathbf{u}\otimes\mathbf{u} - 2\,\mathbf{r}\cdot\mathbf{u}\,\mathbf{u}\odot\mathbf{r} + \mathbf{r}\otimes\mathbf{r}).
\end{aligned}$$

We observe that

$$\begin{aligned}&(\mathbf{r}\otimes\mathbf{u}-\mathbf{u}\otimes\mathbf{r})(\mathbf{r}\otimes\mathbf{u}-\mathbf{u}\otimes\mathbf{r})\\ &=\mathbf{u}\cdot\mathbf{r}\,\mathbf{r}\otimes\mathbf{u}-\mathbf{r}\otimes\mathbf{r}-r^2\,\mathbf{u}\otimes\mathbf{u}+\mathbf{r}\cdot\mathbf{u}\,\mathbf{u}\otimes\mathbf{r}\\ &=2\mathbf{u}\cdot\mathbf{r}\,\mathbf{r}\odot\mathbf{u}-\mathbf{r}\otimes\mathbf{r}-r^2\,\mathbf{u}\otimes\mathbf{u},\end{aligned}$$

and we get the final expression

$$\mathbf{N}^2-\nu\,\mathbf{N}=\ -\frac{1}{4}\,(\mathbf{r}\otimes\mathbf{u}-\mathbf{u}\otimes\mathbf{r})^2,$$

which allows the computation of $\mathbf{K}_{22}$:

$$\begin{aligned}\mathbf{K}_{22}\ &=\frac{1}{2}\,(\nu^2-\operatorname{tr}\mathbf{N}^2)\,\mathbf{I}-\nu\,\mathbf{N}+\mathbf{N}^2\\ &=\frac{1}{2}\left[\frac{1}{2}\,((\mathbf{r}\cdot\mathbf{u})^2+r^2)\right]\mathbf{I}-\frac{1}{4}\,(\mathbf{r}\otimes\mathbf{u}-\mathbf{u}\otimes\mathbf{r})^2\neq 0.\end{aligned}$$

This shows that $\mathbf{L}_m$ is a L-pencil of type 2.

Let us check the validity of Remark 3.1. The first element of the L-sequence generated by $\mathbf{L}_m$ is

$$\mathbf{K}_1=(\mu+m\,\mathbf{u}\cdot\mathbf{r})\,\mathbf{I}-\mathbf{M}-m\,\mathbf{u}\odot\mathbf{r}.$$

Thus, $\mathbf{H}_1=\mathbf{u}\cdot\mathbf{r}\,\mathbf{I}-\mathbf{u}\odot\mathbf{r}$. For computing $\mathbf{K}_2$ we need to compute $\mathbf{K}_1\mathbf{L}_m$,

$$\mathbf{K}_1\mathbf{L}_m=[(\mu+m\,\mathbf{u}\cdot\mathbf{r})\,\mathbf{I}-\mathbf{M}-m\,\mathbf{u}\odot\mathbf{r}]\,[\mathbf{M}+m\,\mathbf{u}\odot\mathbf{r}].$$

After a straightforward calculation we get

$$\begin{aligned}\mathbf{K}_1\mathbf{L}_m\ &=\mu\,\mathbf{M}-\mathbf{M}^2+m\,(\mathbf{u}\cdot\mathbf{r}\,\mathbf{M}-\mathbf{u}\odot\mathbf{M}\mathbf{r}-\mathbf{r}\odot\mathbf{M}\mathbf{u}+\mu\,\mathbf{u}\odot\mathbf{r}\\ &\quad-\frac{1}{4}\,r^2\,\mathbf{u}\otimes\mathbf{u}-\frac{1}{4}\,\mathbf{r}\otimes\mathbf{r})+\frac{1}{2}\,m^2\,\mathbf{u}\cdot\mathbf{r}\,\mathbf{u}\odot\mathbf{r}.\end{aligned}$$

The coefficient of m^2 is $\dfrac{1}{2}\,\mathbf{u}\cdot\mathbf{r}\,\mathbf{u}\odot\mathbf{r}$. Since

$$\mathbf{K}_2=\frac{1}{2}\operatorname{tr}(\mathbf{K}_1\mathbf{L}_m)\,\mathbf{I}-\mathbf{K}_1\mathbf{L}_m,$$

we find

$$\mathbf{H}_2=\frac{1}{4}\,(\mathbf{u}\cdot\mathbf{r})^2\,\mathbf{I}-\frac{1}{2}\,\mathbf{u}\cdot\mathbf{r}\,\mathbf{u}\odot\mathbf{r}=\frac{1}{4}\,(\mathbf{u}\cdot\mathbf{r})\,(\mathbf{u}\cdot\mathbf{r}\,\mathbf{I}-2\,\mathbf{u}\odot\mathbf{r}).$$

For $n=3$, the tensors

$$\mathbf{H}_0=\mathbf{I},\quad \mathbf{H}_1=\mathbf{u}\cdot\mathbf{r}\,\mathbf{I}-\mathbf{u}\odot\mathbf{r},\quad \mathbf{H}_2=\frac{1}{4}\,(\mathbf{u}\cdot\mathbf{r})\,(\mathbf{u}\cdot\mathbf{r}\,\mathbf{I}-2\,\mathbf{u}\odot\mathbf{r})$$

are linearly dependent: $\mathbf{H}_2=\dfrac{1}{4}\,(\mathbf{u}\cdot\mathbf{r})\,(2\,\mathbf{H}_1-\mathbf{u}\cdot\mathbf{r}\,\mathbf{H}_0)$.

REFERENCES

[1] Benenti S., *Separability in Riemannian manifolds*, article submitted (2004) to Royal Society for a special issue on 'The State of the Art of the Separation of Variables', available on the personal page of the site www.dm.unito.it.

[2] ———, *Inertia tensors and Stäckel systems in the Euclidean spaces*, Rend. Sem. Mat. Univ. Politec. Torino **50** (1992), no. 4, 315–341, Differential geometry (Turin, 1992).

[3] ———, *Orthogonal separable dynamical systems*, Differential geometry and its applications (Opava, 1992), Math. Publ., Vol. **1**, Silesian Univ., Opava, 1993, pp. 163–184.

[4] ———, *Intrinsic characterization of the variable separation in the Hamilton-Jacobi equation*, J. Math. Phys. **38** (1997), no. 12, 6578–6602.

[5] ———, *Special symmetric two-tensors, equivalent dynamical systems, cofactor and bi-cofactor systems*, Acta Appl. Math. **87** (2005), no. 1–3, 33–91.

[6] Błaszak M., *Separable systems with quadratic in momenta first integrals*, J. Phys. A: Math. Gen. **38** (2005), 1667–1685.

[7] Boyer C.P., Kalnins E.G., and Miller W., *Stäckel-equivalent integrable Hamiltonian systems*, SIAM J. Math. Anal. **17** (1986), no. 4, 778–797.

[8] Crampin M., Sarlet W., and Thompson G., *Bi-differential calculi, bi-Hamiltonian systems and conformal Killing tensors*, J. Phys. A: Math. Gen. **33** (2000), 8755–8770.

[9] Eisenhart R.P., *Riemannian Geometry*, Princeton University Press, Fifth printing, 1964.

[10] Kalnins E.G., *Separation of variables for Riemannian spaces of constant curvature*, Pitman Monographs and Surveys in Pure and Applied Mathematics, Vol. **28**, Longman Scientific & Technical, Harlow, 1986.

[11] Kalnins E.G. and Miller W., *Separation of variables on n-dimensional Riemannian manifolds. I. The n-sphere $\mathbb{S}_n$ and Euclidean n-space $\mathbb{R}_n$*, J. Math. Phys. **27** (1986), no. 7, 1721–1736.

[12] Okubo S., *Integrable dynamical systems with hierarchy. I. Formulation*, J. Math. Phys. **30** (1989), 834–843.

GEOMETRY OF NON-REGULAR SEPARATION

CLAUDIA CHANU*

Abstract. A geometrical setting for the notion of non-regular additive separation for a PDE, introduced by Kalnins and Miller, is given. This general picture contains as special cases both fixed-energy separation and constrained separation of Helmholtz and Schödinger equations (not necessarily orthogonal). The geometrical approach to non-regular separation allows to explain why it is possible to find some coordinates in Euclidean 3-space where the R-separation of Helmholtz equation occurs, but it depends on a lower number of parameters than in the regular case, and it is apparently not related to the classical Stäckel form of the metric.

Key words. Separation of Variables, R-separation, Helmholtz and Schrödinger equations.

AMS(MOS) subject classifications. Primary 70H06, 35J05.

1. Introduction.

The notion of separation of variables (SoV) for a PDE can be interpreted in (at least) two ways:

- Separation of variables is the reduction of a PDE to a set of separated ODEs involving a single unknown function (Jacobi).
- Separation of variables is the search of (families of) solutions of a special form e.g., additive separation of variables, multiplicative separation of variables, etc.

For a Hamilton-Jacobi equation (i.e., a first order PDE without explicit dependence on the unknown function u), the first approach is equivalent to the second (additive SoV), when the separated solution is *complete*, that is when the number of the constants (c_a) parameterizing the solution $u = u(q^i, c_a)$ is equal to the number of independent variables (q^i) and the completeness condition

$$\det\left(\frac{\partial^2 u}{\partial c_a \partial q^i}\right) \neq 0$$

is satisfied. Conditions for additive separation for a Hamiltonian in a given coordinate system were found by Levi-Civita in 1904 [7]. For higher order PDEs it is not immediate to give a suitable definition of what we mean by complete separable solution and which is the test for having separation. In the 80's Kalnins and Miller extended the Levi-Civita conditions for additive separation of higher order PDEs [5] and the there called *regular separation* replaces the idea of complete solution. In the same paper they briefly consider another case of separation of variables that they call *non-regular*

*Department of Mathematics, University of Torino, via Carlo Alberto 10, 10123 Torino, ITALY. The work of the author was supported by Progetto Lagrange of Fondazione CRT and by National Research Project "Geometry of dynamical system".

separation: it occurs when the test for regular separation is not identically satisfied, but still there exists a separable solution of the PDE.

Following their approach, we reformulate their conditions in a geometrical way, as it has been done in [2] for the regular separation (there called *free separation*). The geometrical picture allows to give an effective definition of the non-regular separation, that we call *constrained separation*. A criterion for testing constrained separation is provided and some examples and applications are given, showing the interest of the approach. Indeed, this more general setting includes and unifies important special cases of separation of variables, like the fixed-energy separation [1, 3, 5] and the reduced (non-orthogonal) separation of the Schrödinger equation [2]. Two examples of non-regular R-separation for the Helmholtz equation on $\mathbb{R}^3$, recently described in [9], fit in this approach. Even if at first sight they appear unrelated to the standard R-separation involving Stäckel matrices [8, 6, 5], we show that the separability conditions are in fact an extension of the well-known Stäckel conditions.

2. Non-regular separation. Let us consider the separability of the l-th order PDE

$$\mathcal{H}(q^i, u, u_i, \dots, u_i^{(l)}) = h \qquad (h \in \mathbb{R}) \tag{2.1}$$

in the coordinates (q^i) on the n-dimensional manifold Q. Here and in the following we denote by $u = u(q^i)$ the unknown function and set

$$u_i = \frac{\partial u}{\partial q^i}, \qquad u_i^{(2)} = u_{ii} = \frac{\partial^2 u}{(\partial q^i)^2}, \quad \dots \quad u_i^{(l)} = \frac{\partial^l u}{(\partial q^i)^l}.$$

For the sake of simplicity, we assume that the maximal order of derivatives involved in H is the same for each index $i = 1, \dots, n$. Let Z be the $(nl+1)$-dimensional space of the dependent variable and its separated derivatives: coordinates on Z are given by $(u, u_i, u_i^{(2)}, \dots, u_i^{(l)})$. We consider the trivial bundle over Q, $M = Q \times Z$. In [2] *free separation* is defined as the existence of an additively separated solution u of (2.1), depending on $nl + 1$ parameters (c_A), satisfying the completeness condition

$$\operatorname{rank} \left[\frac{\partial u}{\partial c_A} \middle| \frac{\partial u_i}{\partial c_A} \middle| \dots \middle| \frac{\partial u_i^{(l)}}{\partial c_A} \right] = nl + 1.$$

Free separation occurs if and only if the n vector fields of the form

$$D_i = \frac{\partial}{\partial q^i} + u_i \frac{\partial}{\partial u} + u_i^{(2)} \frac{\partial}{\partial u_i} + \dots + u_i^{(l)} \frac{\partial}{\partial u_i^{(l-1)}} + R_i \frac{\partial}{\partial u_i^{(l)}}, \tag{2.2}$$

where $R_i(q^j, u, u_j, \dots, u_j^{(l)})$ are functions on M, are commuting symmetries of (2.1) i.e., conditions

$$D_i \mathcal{H} = 0, \tag{2.3}$$

$$[D_i, D_j] = 0 \tag{2.4}$$

are satisfied (cf. [2]). Eqs (2.3) determine the functions R_i, while (2.4) are equivalent to

$$D_i R_j = 0 \qquad (i \neq j). \tag{2.5}$$

By expanding conditions (2.5), we get exactly the conditions given in [5] for the *regular separation.* Thus, free separation corresponds to regular separation.

REMARK 2.1. The geometrical interpretation of the free separation is summarized in the following items (cf. [2]):

- $\Delta = \text{span}(D_i)$ is an integrable distribution of rank n on M.
- The foliation of n-dimensional integral manifolds of Δ is described by a *complete separated solution* of $\mathcal{H} = h$:

$$\begin{cases} u = S \\ u_i = S_i \\ u_{ii} = S_{ii} \\ \dots \end{cases} \qquad \text{with } \partial_i S = S_i, \ \partial^2_{ij} S = \delta_{ij} S_{ii}, \ \dots$$

- Free (regular) separated solutions depend on $nl + 1$ parameters.
- Completeness means that for any point P_0 with coordinates (q^i_0) there is a separated solution of $\mathcal{H} = h$ for each choice of the value of u and its nl derivatives $u_i, u_i^{(2)}, \dots, u_i^{(l)}$ at P_0.

By considering the case in which condition (2.5) is not identically satisfied, Kalnins and Miller introduce in [5] the so called *non-regular separation.* The authors say that also in this case separable solutions still may exist, but they depend on less than $nl + 1$ parameters. Such a kind of separation is called non-regular. However, in [5] it is not specified under which conditions separable solutions do exist and how to determine the number of the involved parameters, even if some examples are given, where explicit case by case non-regular solutions of the PDE are shown. One of these examples is examined in detail in Example 1.

In the following we give a geometric interpretation of the situation that naturally leads to an effective definition of non-regular separation.

Let S be a submanifold of $M = Q \times Z$ locally described by the r equations

$$f_a = 0 \qquad (a = 1, \dots, r).$$

If the vectors D_i commute on S, that is

$$D_i R_j|_S = 0 \qquad (i \neq j, \ i, j = 1, \dots, n),$$

and the vector fields D_i are tangent to S, that is

$$D_i f_a|_S = 0 \quad (i = 1, \dots, n, \ a = 1, \dots r),$$

then we can restrict the distribution Δ generated by the D_i to an involutive distribution on S. Thus, on S we have a complete separated solution of $\mathcal{H} = h$ depending on $nl + 1 - r$ parameters (c_α), which can be considered as a *constrained separated* solution on M. Hence, it is natural to introduce the following definition

DEFINITION 2.1. The PDE $\mathcal{H} = h$ admits a *constrained additive separation* on a submanifold S of M defined by the r equations $f_a = 0$, if

1. $u = \sum_i S^{(i)}(q^i, c_\alpha)$ is a solution of $\mathcal{H} = h$;
2. u depends on $nl+1-r$ parameters (c_α) satisfying the completeness conditions

$$\operatorname{rank}\left[\frac{\partial u}{\partial c_\alpha}\middle|\frac{\partial u_i}{\partial c_\alpha}\middle|\dots\middle|\frac{\partial u_i^{(l)}}{\partial c_A}\right] = nl + 1 - r;$$

3. u and its derivatives satisfy $f_\alpha(q^j, u, u_i ...) = 0$ for all (admissible) values of the parameters (c_α).

From the above discussion we get the following criterion for the constrained separation

PROPOSITION 2.1. *In a given coordinate system* (q^i) *equation* (2.1) *admits a constrained separation on the submanifold* S *defined by equations* $f_a = 0$ *if and only if the vector fields* D_i (2.2) *are symmetries of* (2.1), *tangent to* S *and commute on* S, *that is*

$$D_i\mathcal{H} = 0, \quad D_iR_j|_S = 0, \quad D_if_a|_S = 0 \quad (i \neq j = 1, \dots, n, \ a = 1, \dots, r).$$

REMARK 2.2. If D_iR_j are everywhere different from zero, then equation (2.1) has no additive separable solutions. A possible choice for S is the set of points satisfying equations $D_iR_j = 0$, but in many cases this set is not a well-defined manifold or the vectors D_i are not tangent to it.

REMARK 2.3. In analogy with Remark 2.1, the geometric interpretation of the constrained separation on a submanifold S is sketched in the following items.

- $\Delta_S = \operatorname{span}(D_i)$ is an integrable n-dimensional distribution on S whose integral manifolds are described by a constrained complete separated solution of $\mathcal{H} = h$.
- constrained separated solutions depend on $nl + 1 - r$ parameters.
- completeness means that for any $P_0 = (q_0^i)$ there is a separated solution of $\mathcal{H} = h$ for each choice of the value of u and its nl separated derivatives at P_0 satisfying $f_\alpha = 0$ (i.e., the initial condition belongs to S).

We apply our approach to one of the examples of [5]:

EXAMPLE 1. Let us consider the 2-nd order PDE in two variables $(x_i) = (x_1, x_2)$

$$H = x_2u_{11} + x_1u_{22} + u_1 + u_2 = h.$$

In this case we have

$$D_i = \frac{\partial}{\partial x_i} + u_i \frac{\partial}{\partial u} + u_{ii} \frac{\partial}{\partial u_i} + R_i \frac{\partial}{\partial u_{ii}} \qquad (i = 1, 2).$$

Being

$$D_i H = u_{ii} + u_{jj} + R_i x_j \qquad (i = 1, 2,\ j \neq i),$$

we get

$$R_i = -\frac{u_{ii} + u_{jj}}{x_j} \qquad (i = 1, 2,\ j \neq i).$$

By expanding $D_i R_j$ we get

$$D_i R_j = \frac{(u_{ii} + u_{jj})(x_i + x_j)}{x_i^2 x_j} \qquad (i = 1, 2,\ j \neq i)$$

which do not vanish everywhere. For $u_{11} + u_{22} = 0$, Kalnins and Miller show a non-regular separable solution involving 4 parameters (h, c_1, c_2, c_3)

$$u = c_1 x_1^2 + c_2 x_1 - c_1 x_2^2 + (h - c_2) x_2 + c_3.$$

However, we have that $D_i R_j$ both vanish for $u_{11} + u_{22} = 0$, and $x_1 + x_2 = 0$, but this alternative constrain (on initial positions) is not mentioned in [5]. By using the definition of constrained separation, it is easy to point out the difference between these two situations: let us call S and $\tilde{S}$ the submanifolds defined by $u_{11} + u_{22} = 0$ and $x_1 + x_2 = 0$, respectively. Since we have

$$D_i(u_{11} + u_{22})|_S = \frac{u_{11} + u_{22}}{x_i}\Big|_S = 0 \qquad (i = 1, 2),$$

and

$$D_i(x_1 + x_2)|_{\tilde{S}} = 1 \qquad (i = 1, 2),$$

the generator D_i are tangent to S but not to $\tilde{S}$. Thus, on S we have a complete separated solution depending on $\dim S = 5 - 1$ parameters, which is the one explicitly shown in [5], while constrained separation does not occur on $\tilde{S}$.

3. Applications and examples. The definition of constrained separation is effective and useful since it recovers many known "exceptional cases" of separation. In the following subsections we examine in details some of them.

3.1. Fixed-energy separation. Fixed-energy separation is the separation of the single PDE $\mathcal{H} = E$ [5, 3] (e.g. the Laplace equation $\Delta\psi = 0$). The condition for this kind of separation is that the functions $D_i R_j$ are of the form

$$D_i R_j = (\mathcal{H} - E)\mu_{ij} \qquad (i \neq j = 1, \dots, n), \tag{3.1}$$

for some functions μ_{ij} on M. Fixed energy separation corresponds to constrained separation on the submanifold S defined by $\mathcal{H} = E$. Indeed, Eq. (3.1) means that the D_i commute on the level set $\mathcal{H} = E$. The additional condition that D_i are tangent to S is automatically satisfied since the symmetries D_i are tangent, by construction, to each submanifold $\mathcal{H} = h$, $h \in \mathbb{R}$.

3.2. Reduced separation of the Schrödinger equation. Also the *reduced separation* for the Schrödinger equation introduced in [2] can be considered as a case of constrained separation. The multiplicatively separated solutions of the Schrödinger equation

$$\Delta\psi - (E - V)\psi = 0 \tag{3.2}$$

correspond via the change of variable $u = \ln\psi$ to the additively separated solutions of

$$\mathcal{H} = g^{ii}u_{ii} + g^{ij}u_i u_j - \Gamma^i u_i - V = -E \tag{3.3}$$

where $\Gamma^i = g^{ij}\Gamma_j = g^{hk}\Gamma^i_{hk}$ are the *contracted Christoffel symbols.* Since the equation is independent of u, we can disregard u as variable on M: this corresponds to ignore the inessential constant that can be added to any solution u, which is a multiplicative constant factor in ψ. Moreover, we assume that $g^{ii} \neq 0$ for all i, that is we do not consider null coordinates. Under these assumptions, the vector fields D_i are

$$D_i = \frac{\partial}{\partial q^i} + u_{ii}\frac{\partial}{\partial u_i} + R_i\frac{\partial}{\partial u_{ii}},$$

with

$$R_i = -\frac{1}{g^{ii}}\left(\partial_i\mathcal{H} + (2g^{ik}u_k - \Gamma^i)u_{ii}\right). \tag{3.4}$$

If equations $D_i R_j = 0$ for all $i \neq j$ are identically satisfied, we have the *free* or *regular* separation. The reduced separation is defined [2] as a separation where $r \leq n$ of the separated factors ψ_α are of the form

$$\psi_\alpha = \exp(k_\alpha q^\alpha), \quad k_\alpha \in \mathbb{R}.$$

In this situation the coordinates split into two kinds: $(q^i) = (q^a, q^\alpha)$ $(a = 1, \dots, m,\ \alpha = m+1, \dots, n)$.

In [2] it is proved that

PROPOSITION 3.1. *Reduced separation of the Schrödinger equation occurs, up to equivalence classes of separable coordinates, if and only if*

(*i*) *the* q^α *are ignorable* ($\partial_\alpha g^{ij} = 0$, $\partial_\alpha V = 0$);
(*ii*) *the* q^a *are pairwise orthogonal* ($g^{ab} = 0$ *for all* $a \neq b = 1, \dots m$);
(*iii*) *the metric is in standard form*

$$g^{ij} = \begin{pmatrix} \delta^a_b g^{ab} & 0 \\ 0 & g^{\alpha\beta} \end{pmatrix};$$

(*iv*) *the corresponding Hamilton-Jacobi equation is separable;*
(*v*) *the contracted Christoffel symbols satisfy* $\partial_a \Gamma_b = 0$ *for all* $a \neq b$, *or equivalently the Ricci tensor is in standard form (reduced Robertson condition).*

We show that

PROPOSITION 3.2. *The reduced separation of the Schrödinger equation* (3.2) *is the constrained separation of* (3.3) *on the manifold* S *defined by the* r *equations* $u_{\alpha\alpha} = 0$.

Proof. Let S be the submanifold of M defined by the r equations $u_{\alpha\alpha} = 0$. Clearly, the reduced separated solutions satisfy $u_{\alpha\alpha} = 0$ for all α. We want to prove that, by assuming the conditions for the occurrence of constrained separation on S, that is

$$D_i u_{\alpha\alpha}|_S = 0 \qquad (i = 1, \dots, n,\ \alpha = m+1, \dots, n), \tag{3.5}$$

$$D_i R_j|_S = 0 \qquad (i \neq j = 1, \dots, n), \tag{3.6}$$

we deduce exactly the five conditions of Proposition 3.1. We have that $D_i u_{\alpha\alpha} = 0$ identically, for $i \neq \alpha$, and $D_\alpha u_{\alpha\alpha} = R_\alpha$. Thus, for $i = \alpha$, Eq. (3.5) becomes $R_\alpha|_S = 0$ that is, by (3.4),

$$\frac{1}{g^{\alpha\alpha}} \partial_\alpha \mathcal{H}|_S = 0.$$

By expanding the partial derivative and inserting the conditions $u_{\beta\beta} = 0$ for $\beta = m+1, \dots, n$, we get

$$\partial_\alpha g^{aa} u_{aa} + \partial_\alpha g^{ij} u_i u_j - \partial_\alpha \Gamma^i u_i - \partial_\alpha V = 0,$$

which has to be satisfied for all values of u_{aa} and u_i. Hence, we find that

$$\partial_\alpha g^{ij} = 0, \qquad \partial_\alpha V = 0$$

i.e., that the coordinates $u_{\alpha\alpha}$ are ignorable and condition (*i*) of Proposition 3.1 is proved. Moreover, the above calculations imply that R_α is identically null and

$$D_\alpha = \partial_\alpha + u_{\alpha\alpha} \frac{\partial}{\partial u_\alpha}.$$

Thus, we have $D_\alpha R_\beta = 0$ and

$$D_\alpha R_a|_S = \left(u_{\alpha\alpha} \frac{\partial R_a}{\partial u_\alpha} \right) \Big|_S = 0.$$

Hence, we have only to impose that the remaining conditions $D_a R_b|_S = 0$ are satisfied, where

$$R_b = \frac{1}{g^{bb}}[-2\, g^{bh} u_{bb} u_h - \partial_b g^{hh} u_{hh} - \partial_b g^{hk} u_h u_k + \partial_b \Gamma^h u_h + \Gamma^b u_{bb} + \partial_b V].$$

Since R_b (and all its partial derivatives) contains terms at most linear in the variables (u_{aa}) and being

$$D_a R_b = \partial_a R_b + u_{aa} \frac{\partial R_b}{\partial u_a} - \frac{\partial_b g^{aa}}{g^{bb}} R_a,$$

$D_a R_b$ is a quadratic polynomial in the u_{hh}, whose quadratic part is

$$-2\, \frac{g^{ab}}{g^{bb}} u_{aa} u_{bb}, \qquad a, b \text{ n.s.} \quad a \neq b,$$

which is independent of $u_{\alpha\alpha}$. Hence, the condition $D_a R_b|_S = 0$ implies $g^{ab} = 0$ for $a \neq b$, and (ii) is proved. In order to expand $D_a R_b$ in a simple way we set

$$\hat{\Gamma}^a = \Gamma^a - g^{\alpha\alpha} u_\alpha, \qquad \hat{V} = V + \Gamma^\alpha u_\alpha - g^{\alpha\beta} u_\alpha u_\beta.$$

Then, the functions $\mathcal{H}$ and R_b become

$$\mathcal{H} = g^{aa}(u_{aa} + u_a^2) + g^{\alpha\alpha} u_{\alpha\alpha} - \hat{\Gamma}^a u_a - \hat{V}, \qquad R_b = -\frac{1}{g^{bb}}(\partial_b \mathcal{H} - u_{bb} \hat{\Gamma}^b)$$

and

$$D_a R_b = \frac{1}{g^{bb}}\big(- \partial^2_{ab} \mathcal{H} + \partial_a \ln g^{bb} \partial_b \mathcal{H} + \partial_b \ln g^{aa} \partial_a \mathcal{H}$$
$$+ u_{aa}(\partial_b \ln g^{aa} \hat{\Gamma}^a - \partial_b \hat{\Gamma}^a) + u_{bb}(\partial_a \ln g^{bb} \hat{\Gamma}^b - \partial_a \hat{\Gamma}^b)\big),$$

which is a polynomial in the variables (u_a, u_{aa}) when it is restricted to S. Thus, we expand the partial derivatives and we impose that all coefficients of (u_a, u_{aa}) vanish. By introducing the Stäckel operators S_{ab} associated with the diagonal submatrix g^{ab} and acting on a function f over Q as

$$S_{ab}(f) \doteq \partial^2_{ab} f - \partial_a \ln g^{bb} \partial_b f - \partial_b \ln g^{aa} \partial_a f \qquad (a \neq b = 1, \ldots, m),$$

we get that $D_a R_b|_S = 0$ if and only if the following conditions hold for all $a \neq b = 1, \ldots, m$:

$$\begin{cases} S_{ab}(g^{cc}) = 0 & (c = 1, \ldots, m), \\ S_{ab}(\hat{\Gamma}^c) = 0 & (c = 1, \ldots, m), \\ S_{ab}(\hat{V}) = 0, & \\ \partial_a g^{bb} \hat{\Gamma}^b = g^{bb} \partial_a. & \end{cases} \tag{3.7}$$

From these equations, repeating the same proof as in [2], we get (iii), (iv), (v). □

3.3. Constrained orthogonal R-separation of Helmholtz equation. In [9], two orthogonal coordinate systems in $\mathbb{R}^3$ in which non-regular R-separation of the Helmholtz equation occurs are shown. These coordinates do not allow regular R-separation and the metric coefficients g^{ii} are not of Stäckel type. We show that these examples perfectly fit in the general frame of constrained separation and we give the general differential conditions that the components of the metric tensor g^{ii} must satisfy, in order to have this kind of constrained separation.

We recall R-separation of the Schrödinger (3.2) or Helmholtz equation ($V = 0$)

$$\Delta\psi - (E - V)\psi = 0 \tag{3.8}$$

is the ansatz of assuming that there exists a family of solutions depending on a set of parameters (c_A) of the form

$$\psi(q^j, c_A) = R(q^j)\prod_i \phi_i(q^i, c_A),$$

where ϕ_i are separated factors, and R is a not vanishing function on Q, independent from the parameters (c_A). It generalizes the ordinary multiplicative separation where $R = 1$. By the change of unknown

$$u = \ln\psi - \ln R = \ln(\prod_i \phi_i(q^i)),$$

and under the assumption that the separated coordinates are orthogonal, the associated PDE that has to be solved via additive SoV is (see [5, 3])

$$H = g^{ii}(u_{ii} + u_i^2) - g^{ii}(\Gamma_i - 2\partial_i \ln R)u_i - V + R^{-1}\Delta R = -E,$$

which can be written in a way similar to (3.3) as

$$H = g^{ii}(u_{ii} + u_i^2) - g^{ii}\hat{\Gamma}_i u_i - U = -E, \tag{3.9}$$

where

$$\hat{\Gamma}_i = \Gamma_i - 2\partial_i \ln R, \qquad U = V - R^{-1}\Delta R,$$

and $E \in \mathbb{R}$ is the energy constant. We remark (see [3] for further details) that the function R is considered as a part of the equation, playing the same role as the metric tensor and the potential V. We study the additive separation of (3.9), by the method exposed in the previous section. By using the Stäckel operator (3.7) associated with the diagonal metric (g^{ii}), the functions D_iR_j becomes

$$\begin{aligned} D_iR_j = \big(&- S_{ij}(g^{hh})(u_{hh} - u_h^2) + S_{ij}(g^{hh}\hat{\Gamma}_h)u_h \\ &+ S_{ij}(V - R^{-1}\Delta R) + u_{ii}(\partial_j \ln g^{ii}\hat{\Gamma}^i - \partial_j\hat{\Gamma}^i) \\ &+ u_{jj}(\partial_i \ln g^{jj}\hat{\Gamma}^j - \partial_i\hat{\Gamma}^j)\big)/g^{jj} \qquad (i \neq j \text{ n.s.}). \end{aligned} \tag{3.10}$$

The following result is well known (see [6, 3]).

PROPOSITION 3.3. *Free (regular) R-separation (in particular for all values of E) for the Schrödinger equation occurs if and only if*

(*i*) *the corresponding geodesic Hamilton-Jacobi equation is separable, i.e.,* g^{hh} *are in Stäckel form:*

$$S_{ij}(g^{hh}) = 0 \qquad (i, j, h = 1, \dots, n,\ i \neq j) \tag{3.11}$$

(*ii*) *R is (up to separated factors) a solution of*

$$\partial_i \ln R = \tfrac{1}{2}\Gamma_i, \tag{3.12}$$

(*iii*) *U is a* separated factor, *that is there are functions of a single variable* $f_h(q^h)$ *such that*

$$U = V - \frac{\Delta R}{R} = g^{hh} f_h(q^h). \tag{3.13}$$

In the examples given in [9] they restrict themselves to the Helmholtz equation with $V = 0$ and, even if the first condition does not hold, nevertheless the metric coefficients and the function R satisfy the second and the third. This fact suggests to examine what happens if only conditions (*ii*) and (*iii*) hold.

REMARK 3.1. When (3.11) holds there exists a Stäckel matrix $\mathbf{S}$ such that the metric coefficients (g^{ii}) form a row of $\mathbf{S}^{-1}$. Then, the differential system (3.12) can be written in terms of $\mathbf{S}$ as (see [5, 8])

$$\partial_i \ln R = \tfrac{1}{2}\partial_i(\textstyle\prod_h g^{hh} / \det \mathbf{S}),$$

and its integrability condition is automatically satisfied. Indeed, by expanding $\partial_i\Gamma_j = \partial_j\Gamma_i$ we get

$$\frac{S_{ij}(g^{jj})}{g^{jj}} = \frac{S_{ij}(g^{ii})}{g^{ii}}, \tag{3.14}$$

which is a consequence of (3.11). Moreover, the condition (3.13) is equivalent to $S_{ij}(U) = 0$ for any $i \neq j$, and U is called *Stäckel factor.* Since we are interested to the case when (3.11) does not hold, we have used the above formulation which does not depend on the Stäckel matrix.

PROPOSITION 3.4. *If* (q^i) *are orthogonal coordinates on Q such that*

$$\partial^2_{ij} \log\left(g^{ii}/g^{jj}\right) = 0, \tag{3.15}$$

$$\exists f_1(q^1), \dots, f_n(q^n) \mid g^{hh}(2\partial_h\Gamma_h - \Gamma_h^2) = g^{hh} f_h, \tag{3.16}$$

$$\partial_\alpha g^{cc} = 0 \qquad (\alpha = 1, \dots, r), \tag{3.17}$$

$$S_{ab}(g^{cc}) = 0 \qquad (a, b, c = r+1, \dots, n,\ a \neq b), \tag{3.18}$$

then on the submanifold S defined by the $r = n - m$ *equations*

$$u_{\alpha\alpha} - u_\alpha^2 + \tfrac{1}{4} f_\alpha = 0 \qquad (\alpha = 1, \dots r), \tag{3.19}$$

constrained R-separation of the Helmholtz equation occurs, depending on $m+n$ free parameters. The function R is a solution of the system (3.12).

Proof. The integrability condition $\partial_i\Gamma_j = \partial_j\Gamma_i$ of system (3.12) is equivalent to (3.15). If R is a solution of (3.12) we have for all h $\hat{\Gamma}_h = 0$. Then, the coefficients of u_{ii} and u_{jj} in D_iR_j vanish and $S_{ij}(g^{hh}\hat{\Gamma}_h)u_h = 0$. Moreover, for R satisfying (3.12), the term $\Delta R/R$ becomes

$$\frac{\Delta R}{R} = \frac{g^{hh}}{4}(2\partial_h\Gamma_h - \Gamma_h^2).$$

Hence, for this choice of R, (3.10) becomes

$$D_iR_j = \frac{1}{g^{jj}}\Big(- S_{ij}(g^{hh})(u_{hh} - u_h^2) + S_{ij}\left(V - \frac{g^{hh}}{4}(2\partial_h\Gamma_h - \Gamma_h^2)\right)\Big).$$

For the Helmholtz equation we have $V = 0$, and by assuming (3.16) we get

$$H = g^{hh}(u_{hh} + u_h^2 + f_h(q^h)/4) = -E,$$
$$D_iR_j = \frac{1}{g^{jj}}\left[-S_{ij}(g^{hh})(u_{hh} - u_h^2 + \tfrac{1}{4}f_h)\right].$$

Let us call S the submanifold of $M = Q \times Z$ defined by the r equations (3.19). By applying the criterion of Proposition 2.1, we see that the constrained separation on S occurs if and only if (3.17) and (3.18) hold. Indeed, the generators D_i are tangent to S if and only if (3.17) are satisfied: we have

$$D_i(u_{\alpha\alpha} - u_\alpha^2 + \tfrac{1}{4}f_\alpha) = 0 \qquad (i \neq \alpha),$$
$$D_\alpha(u_{\alpha\alpha} - u_\alpha^2 + \tfrac{1}{4}f_\alpha) = \tfrac{1}{g^{\alpha\alpha}}\partial_\alpha g^{hh}(u_{hh} - u_h^2 + \tfrac{1}{4}f_h),$$

which is identically zero on S if and only if $\partial_\alpha g^{cc} = 0$. Moreover, by (3.17) we have $S_{\alpha\beta}(g^{cc}) = 0$ for all $\alpha, \beta = 1, \ldots, r$. Thus, by (3.18), we have $S_{ij}(g^{cc}) = 0$ for all $i \neq j = 1, \ldots, n$ and this implies that $D_iR_j|_S = 0$. □

REMARK 3.2. Condition (3.15) is equivalent to (3.14) which is weaker than the request of the Stäckel form of the metric (g^{ii}). However, conditions (3.18) show that Stäckel matrices are strictly related with this kind of constrained separation, but they are $m \times m$ instead of $n \times n$. Indeed, conditions (3.17), (3.18) mean that (g^{aa}) are independent of (q^α) and they form the row of the inverse of a $m \times m$ Stäckel matrix in the (q^a). We remark that (3.17) is weaker than the request of the presence of ignorable coordinates: the components $g^{\beta\beta}$ in general depend on the (q^α).

REMARK 3.3. In the case of the Schrödinger equation it is always possible to choose the potential V in such a way that the modified potential has the form

$$U = V - \frac{g^{hh}}{4}(2\partial_h\Gamma_h - \Gamma_h^2) = g^{hh}f_h(q^h),$$

for some functions (f_i). However, if we have $V = 0$ (Helmholtz equation) or if V is given and it is a separated factor ($V = g^{hh}\varphi_h(q^h)$), then condition (3.16) is essential for this kind of constrained separation.

As illustrative examples of coordinates satisfying the above Proposition we consider the coordinates on $\mathbb{R}^3$ considered in [9] and previously introduced in a paper of Friedlander [4]. For both those cases, we have $m = 1$, $r = 2$, and the two types of coordinates are $(q^a) = (q^3)$, $(q^\alpha) = (q^1, q^2)$. In the first coordinate system the metric is

$$\begin{aligned} g^{11} &= \frac{(a\cosh q^2 - c\cos q^1)^2}{(a^2 - c^2)(a\cosh q^2 - q^3)^2}, \\ g^{22} &= \frac{(a\cosh q^2 - c\cos q^1)^2}{(a^2 - c^2)(c\cos q^1 - q^3)^2}, \qquad (0 < c < a \in \mathbb{R}), \\ g^{33} &= 1. \end{aligned}$$

The metric components satisfy condition (3.15) and $(g^{aa}) = (g^{33})$ which is a 1×1 Stäckel matrix in the coordinate q^3. moreover, we have

$$R = [(q^3 - c\cos q^1)(a\cosh q^2 - q^3)]^{-1/2},$$

$$\frac{\Delta R}{R} = \frac{1}{4}g^{11} - \frac{1}{4}g^{22}.$$

Thus, the constraints that the separated solution has to satisfy (i.e., the equations of the submanifold S) are

$$u_{11} + u_1^2 + 1/4 = 0, \qquad u_{22} + u_2^2 - 1/4 = 0.$$

For the second coordinate system, the components of the metric are

$$\begin{aligned} g^{11} &= \frac{[(q^1)^2 + (q^2)^2 + 16a^2]^2}{[8a^2 + (q^2)^2 + 8aq^3]^2}, \\ g^{22} &= \frac{[(q^1)^2 + (q^2)^2 + 16a^2]^2}{[8a^2 + (q^1)^2 - 8aq^3]^2}, \qquad (0 < a \in \mathbb{R}) \\ g^{33} &= 1. \end{aligned}$$

The metric components satisfy condition (3.15) and $(g^{aa}) = (g^{33})$ which is a 1×1 Stäckel matrix in q^3; moreover, we have

$$R = [(8a^2 + (q^1)^2 - 8aq^3)(8a^2 + (q^2)^2 + 8aq^3)]^{-1/2}, \qquad \frac{\Delta R}{R} = 0.$$

Hence, the separated solution has to satisfy

$$u_{11} + u_1^2 = 0, \qquad u_{22} + u_2^2 = 0.$$

By looking at the form of the metric tensor in the two cases, we see many analogies. Indeed, by applying Proposition 3.4 for $r = 2$ to a three-dimensional flat metric we get

PROPOSITION 3.5. *If an orthogonal coordinate system* (q^i) *on* $\mathbb{R}^3$ *allows constrained R-separation of Helmholtz equation with* $r = 2$, $(q^\alpha) = (q^1, q^2)$, $(q^a) = (q^3)$, *then the components of the metric tensor are (up to rescaling)*

$$\begin{aligned} g^{11} &= \frac{(F_1 H_2 - H_1 F_2)^2}{(F_2 q^3 + H_2)^2}, \\ g^{22} &= \frac{(F_1 H_2 - H_1 F_2)^2}{(F_1 q^3 + H_1)^2}, \\ g^{33} &= 1, \end{aligned} \tag{3.20}$$

with $F_i = F_i(q^i)$, $H_i = H_i(q^i)$, *and the function R has the form*

$$R = [(F_1 q^3 + H_1)(F_2 q^3 + H_2)]^{-1/2}.$$

REMARK 3.4. The form (3.20) of the metric tensor is a necessary but not sufficient condition. For instance, the functions F_i and H_i must satisfy a non linear second order differential condition in order to get that the component R_{1212} of the Riemann tensor of the metric is null, while the other components are automatically zero for a metric of the kind (3.20).

REFERENCES

[1] S. BENENTI, C. CHANU, AND G. RASTELLI, *Variable separation theory for the null Hamilton–Jacobi equation*, J. Math. Phys. (2005), **46**, 042901/29.

[2] S. BENENTI, C. CHANU, AND G. RASTELLI, *Remarks on the connection between the additive separation of the Hamilton-Jacobi equation and the multiplicative separation of the Schrödinger equations. I. The completeness and Robertson conditions*, J. Math. Phys. (2002), **43**: 5183–5222.

[3] C. CHANU AND G. RASTELLI, *Fixed Energy R-separation for Schrödinger equation* International Journal on Geometric Methods in Modern Physics (2006), **3**(3): 489–508.

[4] F.G. FRIEDLANDER, *Simple progressive solutions of the wave equation*, Proc. Cambridge Philos. Soc. (1946), **43**: 360–373.

[5] E.G. KALNINS AND W. MILLER JR., *Intrinsic characterization of variable separation for the partial differential equations of Mechanics*, in Proceedings of IUTAM-ISIMM Symposium on Modern Developments in Analytical Mechanics, Torino 1982, Atti Accad. Sci. Torino (1983), **117**(2): 511–533.

[6] E.G. KALNINS AND W. MILLER JR., *The theory of orthogonal R-separation for Helmholtz equation*, Adv. in Math. (1984), **51**: 91–106.

[7] T. LEVI CIVITA, *Sulla integrazione della equazione di HamiltonJacobi per separazione di variabili*, Math. Ann. (1904), **59**: 3383–3397.

[8] P. MOON AND D.E. SPENCER, *Separability conditions for the Laplace and Helmholtz equations*, J. Franklin Inst. (1952), **253**: 585–600.

[9] R. PRUS AND A. SYM, *Non-regular and non-Stäckel R-separation for 3-dimensional Helmholtz equation and cyclidic solitons of wave equation*, Physics Letters A (2005), **336**: 459–462.

HIGHER SYMMETRIES OF THE SQUARE OF THE LAPLACIAN*

MICHAEL EASTWOOD† AND THOMAS LEISTNER‡

(In memory of Thomas Branson)

Abstract. The symmetry operators for the Laplacian in flat space were recently described and here we consider the same question for the square of the Laplacian. Again, there is a close connection with conformal geometry. There are three main steps in our construction. The first is to show that the symbol of a symmetry is constrained by an overdetermined partial differential equation. The second is to show existence of symmetries with specified symbol (using a simple version of the AdS/CFT correspondence). The third is to compute the composition of two first order symmetry operators and hence determine the structure of the symmetry algebra. There are some interesting differences as compared to the corresponding results for the Laplacian.

Key words. Symmetry algebra, Laplacian, conformal geometry.

AMS(MOS) subject classifications. Primary 58J70; Secondary 16S32, 53A30, 70S10.

1. Introduction. The second order symmetry operators for the Laplacian on $\mathbb{R}^3$ were determined by Boyer, Kalnins, and Miller [2]. The higher order symmetries for the Laplacian on $\mathbb{R}^n$ were found in [7] and the structure of the resulting algebra was also described. Here were prove the corresponding results for the square of the Laplacian. The other aspect of [2], namely the relation between second order symmetries and separation of variables, is unclear for the square of the Laplacian.

We are grateful to Ernie Kalnins who suggested the square of the Laplacian as a candidate for having interesting symmetries. We would also like to acknowledge pertinent comments from Petr Somberg, Vladimír Souček, and Misha Vasiliev.

2. Definitions and statements of results. We shall always work on n-dimensional Euclidean space $\mathbb{R}^n$ for $n \geq 3$ and adopt the usual convention of writing vectors and tensors adorned with indices, which we shall raise and lower with the standard (flat) metric g_{ab}. Let us also write $\nabla_a = \partial/\partial x^a$ for differentiation in coördinates. Then $\nabla^a = g^{ab}\nabla_b$ and the Laplacian is given

*This work was undertaken in preparation for and during the 2006 Summer Program at the Institute for Mathematics and its Applications at the University of Minnesota. The authors would like to thank the IMA for hospitality during this time. The authors are supported by the Australian Research Council.

†School of Mathematical Sciences, University of Adelaide, SA 5005, Australia (meastwoo@member.ams.org).

‡School of Mathematical Sciences, University of Adelaide, SA 5005, Australia (tleistne@maths.adelaide.edu.au); Current address: Department Mathematik, Universität Hamburg, Bundesstraße 55, D-20146 Hamburg, Germany (leistner@math.uni-hamburg.de).

by $\Delta = \nabla^a \nabla_a$. All functions and tensors in this article will be smooth. All differential operators will be linear with smooth coefficients.

DEFINITION 2.1. *A differential operator $\mathcal{D}$ is a symmetry of Δ^2 if and only if there is another differential operator δ such that $\Delta^2 \mathcal{D} = \delta \Delta^2$.*
Obviously, any operator of the form $\mathcal{P}\Delta^2$ is a symmetry of Δ^2 because we can take $\delta = \Delta^2 \mathcal{P}$. Therefore one introduces the following equivalence relation.

DEFINITION 2.2. *Two symmetries $\mathcal{D}_1$ and $\mathcal{D}_2$ of Δ^2 are equivalent, $\mathcal{D}_1 \sim \mathcal{D}_2$, if and only if $\mathcal{D}_1 - \mathcal{D}_2 = \mathcal{P}\Delta^2$ for some differential operator $\mathcal{P}$.*
Of course, this equivalence relation only effects symmetries of order $s \geq 4$. The composition of two symmetries is again a symmetry. Also, composition is compatible with the equivalence relation, i.e. if $\mathcal{D}_1 \sim \mathcal{D}_2$ and $\mathcal{D}_3 \sim \mathcal{D}_4$, then $\mathcal{D}_1 \mathcal{D}_3 \sim \mathcal{D}_2 \mathcal{D}_4$. This allows us to define an algebra:

DEFINITION 2.3. *The algebra $\mathcal{B}_n$ consists of all symmetries of Δ^2 on $\mathbb{R}^n$ considered modulo equivalence and with algebra operation induced by composition.*
In the following we shall study this algebra and describe its structure. To this end we need the notion of conformal Killing tensors and their generalisations as studied in [4] and in [11, 12]. We shall write $\phi^{(ab\cdots c)}$ for the symmetric part of a tensor $\phi^{ab\cdots c}$.

DEFINITION 2.4. *A conformal Killing tensor $V^{bcd\cdots e}$ is a symmetric trace-free tensor such that*

$$\text{the trace-free part of } \nabla^{(a} V^{bcd\cdots e)} = 0, \tag{2.1}$$

equivalently that

$$\nabla^{(a} V^{bcd\cdots e)} = g^{(ab} \phi^{cd\cdots e)}$$

for some tensor $\phi^{cd\cdots e}$. A conformal Killing tensor with one index is called a conformal Killing vector. A conformal Killing tensor with no indices is simply a constant.

DEFINITION 2.5. *A generalised conformal Killing tensor $W^{d\cdots e}$ of order 3 is a symmetric trace-free tensor such that*

$$\text{the trace-free part of } \nabla^{(a} \nabla^b \nabla^c W^{d\cdots e)} = 0,$$

equivalently that

$$\nabla^{(a} \nabla^b \nabla^c W^{d\cdots e)} = g^{(ab} \phi^{cd\cdots e)}$$

for some tensor $\phi^{cd\cdots e}$.
Though it is clear how to define a generalised conformal Killing tensor of any order, we shall only need order 3. This should be taken as read for the rest of this article.

Our main theorems on the existence and uniqueness of symmetries are as follows.

THEOREM 2.1. *Any zeroth order symmetry of Δ^2 is of the form*

$$f \longmapsto Vf \quad \text{for } V \text{ constant.}$$

Any first order symmetry of Δ^2 is of the form

$$V^b\nabla_b + \text{lower order terms,}$$

where V^b is a conformal Killing vector. Any higher symmetry, say of degree s, of Δ^2 is canonically equivalent to one of the form

$$V^{bcd\ldots e}\nabla_b\nabla_c\nabla_d\cdots\nabla_e + W^{d\cdots e}\Delta\nabla_d\cdots\nabla_e + \text{lower order terms,}$$

where $V^{bcd\cdots e}$ is a conformal Killing tensor of valency s and $W^{d\cdots e}$ is a generalised conformal Killing tensor of valency $s-2$.

THEOREM 2.2. *Suppose that $V^{bcd\cdots e}$ is a conformal Killing tensor on $\mathbb{R}^n$. Then there is a canonically defined differential operator*

$$\mathcal{D}_V = V^{bcd\ldots e}\nabla_b\nabla_c\nabla_d\cdots\nabla_e + \text{lower order terms}$$

that is a symmetry of Δ^2. Suppose that $W^{d\cdots e}$ is a generalised conformal Killing tensor. Then there is a canonically defined differential operator

$$\mathcal{D}_W = W^{d\ldots e}\Delta\nabla_d\cdots\nabla_e + \text{lower order terms}$$

that is a symmetry of Δ^2.

The proof of this theorem will be given in Section 4 by using the ambient metric construction. Here we only want to give the first and second order symmetries. As a special case of (3.5), any first order symmetry is given by

$$\mathcal{D}f = V^a\nabla_a f + \tfrac{n-4}{2n}(\nabla_a V^a)f + cf,$$

for a conformal Killing vector V^a and arbitrary constant c. The canonical ones $\mathcal{D}_V$ are those with $c = 0$. The canonical second order symmetries are

$$\mathcal{D}_V f = V^{ab}\nabla_a\nabla_b f + \tfrac{n-2}{n+2}(\nabla_a V^{ab})\nabla_b f + \tfrac{(n-2)(n-4)}{4(n+1)(n+2)}(\nabla_a\nabla_b V^{ab})f, \quad (2.2)$$

for V^{ab} a conformal Killing tensor and

$$\mathcal{D}_W f = W\Delta f - (\nabla^a W)\nabla_a f - \tfrac{n-4}{2(n+2)}(\Delta W)f, \quad (2.3)$$

for W a generalised conformal Killing scalar, i.e. $\nabla^a\nabla^b\nabla^c W = g^{(ab}\phi^{c)}$. Of course, there is no freedom in equivalence until we consider fourth order operators. Hence, we can use Theorems 2.1 and 2.2 to write down all second order symmetries as follows. Suppose that $\mathcal{D}$ is a second order symmetry operator. According to Theorem 2.1, it has the form

$$\mathcal{D} = V^{ab}\nabla_a\nabla_b + W\Delta + \text{lower order terms}$$

where V^{ab} is a conformal Killing tensor and W is a generalised conformal Killing scalar. According to Theorem 2.2, however, there are canonically defined symmetry operators of the same form, which we can subtract to obtain a first order symmetry. Iterating this procedure we conclude that

$$\mathcal{D} = \mathcal{D}_{V_2} + \mathcal{D}_W + \mathcal{D}_{V_1} + \mathcal{D}_{V_0},$$

where $\mathcal{D}_{V_s}$ are the canonically defined differential operators associated to conformal Killing tensors of valency s and $\mathcal{D}_W$ is the operator associated to a generalised conformal Killing scalar. As above, we have explicit formulae for these operators and, of course, $\mathcal{D}_{V_0} f = V_0 f$ for constant V_0. We shall soon see that the space of (generalised) conformal Killing tensors is finite-dimensional. (In particular, the space of second order symmetries of Δ^2 has dimension $(n+1)(n+2)(n^2+5n+12)/12$.)

More generally, let $\mathcal{K}_{n,s}$ denote the vector space of conformal Killing tensors on $\mathbb{R}^n$ with s indices and, for $s \geq 2$, let $\mathcal{L}_{n,s}$ denote the vector space of generalised conformal Killing tensors on $\mathbb{R}^n$ with $s-2$ indices. Reasoning as we just did for second order symmetries, but now taking into account the equivalence necessitated by Theorem 2.1 in general, we conclude that any symmetry of Δ^2 may be canonically thrown into an equivalent one of the form

$$\mathcal{D}_{V_s} + \mathcal{D}_{W_s} + \cdots + \mathcal{D}_{V_2} + \mathcal{D}_{W_2} + \mathcal{D}_{V_1} + \mathcal{D}_{V_0}, \quad \text{for } V_s \in \mathcal{K}_{n,s},\ W_s \in \mathcal{L}_{n,s}.$$

Another way of stating this is:

COROLLARY 2.1. *There is the following canonical isomorphism of vector spaces.*

$$\mathcal{B}_n \simeq \mathcal{K}_{n,0} \oplus \mathcal{K}_{n,1} \oplus \bigoplus_{s=2}^{\infty} (\mathcal{K}_{n,s} \oplus \mathcal{L}_{n,s}). \tag{2.4}$$

In order the present the algebra structure on $\mathcal{B}_n$ we need more detail on the spaces $\mathcal{K}_{n,s}$ and $\mathcal{L}_{n,s}$. It is well-known and given explicitly in (4.3), that the space of conformal Killing vectors on $\mathbb{R}^n$ is isomorphic as a Lie algebra to $\mathfrak{so}(n+1,1)$. The spaces $\mathcal{K}_{n,s}$ and $\mathcal{L}_{n,s}$ are irreducible finite-dimensional representations of $\mathfrak{so}(n+1,1)$. Specifically,

$$\mathcal{K}_{n,s} \simeq \overbrace{\begin{array}{|c|c|c|c|c|c|c|}\hline \; & \; & \; & \cdots & \; & \; & \; \\ \hline \; & \; & \; & \cdots & \; & \; & \; \\ \hline \end{array}}^{s}{}_{\circ} \tag{2.5}$$

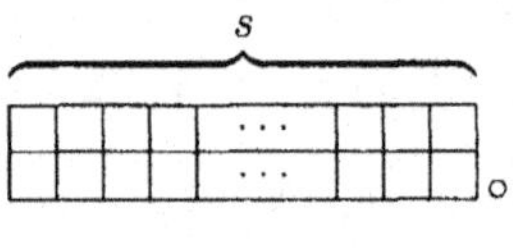

and

$$\mathcal{L}_{n,s} \simeq \overbrace{\begin{array}{|c|c|c|c|c|c|c|}\hline \; & \; & \; & \cdots & \; & \; & \; \\ \hline \; & \; & \; & \cdots & \; & \multicolumn{2}{c}{} \\ \cline{1-5} \end{array}}^{s}{}^{\circ} \tag{2.6}$$

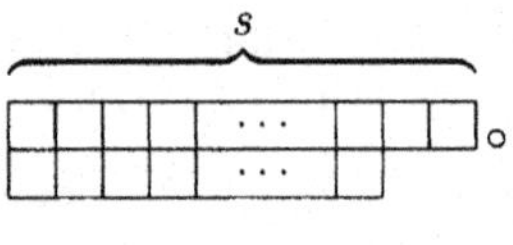

as Young tableau, where $\circ$ denotes the trace-free part. These isomorphisms may be derived from results concerning induced modules in representation theory, namely Lepowsky's generalisation [10] of the Bernstein-Gelfand-Gelfand resolution. A proof in the language of partial differential operators appears in [3].

From now on, let us write $\mathfrak{g}$ for the Lie algebra $\mathfrak{so}(n+1,1)$. Then

$$\mathfrak{g}\otimes\mathfrak{g} = \ydiagram{1,1} \otimes \ydiagram{1,1} = \ydiagram{2,2}_{\circ} \oplus \ydiagram{2}_{\circ} \oplus \mathbb{R} \oplus \ydiagram{2,1,1}_{\circ} \oplus \ydiagram{1,1} \oplus \ydiagram{1,1,1,1} \tag{2.7}$$

and, following [7], we shall write $X \circledcirc Y$ for the projection

$$\mathfrak{g}\otimes\mathfrak{g} \ni V\otimes W \longmapsto V\circledcirc W \in \ydiagram{2,2}_{\circ}.$$

It is shown in [7] that the symmetry algebra $\mathcal{A}_n$ for the Laplacian is isomorphic to the tensor algebra $\bigotimes\mathfrak{g}$ modulo the two-sided ideal generated by

$$V\otimes W - V\circledcirc W - \tfrac{1}{2}[V,W] + \tfrac{n-2}{4n(n+1)}\langle V,W\rangle \quad \text{for } V,W\in\mathfrak{g}, \tag{2.8}$$

where $\langle\ ,\ \rangle$ is the Killing form (normalised in the usual way, not as in [7]). To state the corresponding result for $\mathcal{B}_n$ we also need a notation for the projection onto $\ydiagram{2}_{\circ}$ and we shall write this as $V\otimes W \mapsto V\bullet W$ meaning as an idempotent homomorphism of $\mathfrak{g}\otimes\mathfrak{g}$ into itself. We also need to observe that

$$\ydiagram{4}_{\circ} \hookrightarrow \mathfrak{g}\odot\mathfrak{g}\odot\mathfrak{g}\odot\mathfrak{g} \subset \mathfrak{g}\otimes\mathfrak{g}\otimes\mathfrak{g}\otimes\mathfrak{g}$$

meaning that there is a unique irreducible summand of the symmetric tensor product $\bigodot^4\mathfrak{g}$ of the indicated type. With these conventions in place, we have:

THEOREM 2.3. *The algebra $\mathcal{B}_n$ is isomorphic to the tensor algebra $\bigotimes\mathfrak{g}$ modulo the 2-sided ideal generated by*

$$V\otimes W - V\circledcirc W - V\bullet W - \tfrac{1}{2}[V,W] + \tfrac{(n-4)(n+4)}{4n(n+1)(n+2)}\langle V,W\rangle \quad \textit{for } V,W\in\mathfrak{g} \tag{2.9}$$

and the image of $\ydiagram{4}_{\circ}$ *in* $\bigotimes^4\mathfrak{g}$.

As noted in [7] for $\mathcal{A}_n$, we can quotient firstly by $V\wedge W - \frac{1}{2}[V,W]$ to realise $\mathcal{B}_n$ as a quotient of $\mathfrak{U}(\mathfrak{g})$, the universal enveloping algebra of $\mathfrak{g}$. Compared to $\mathcal{A}_n$, the appearance of additional generators at 4th order is new.

In fact, there is a more precise statement from which Theorem 2.3 easily follows. It appears as Theorem 5.1 in Section 5.

3. The proof of Theorem 2.1.

LEMMA 3.1. *Suppose $V^{bcd\cdots ef}$ is a conformal Killing tensor with s indices. If we define $\phi^{cd\cdots ef}$ according to*

$$\nabla^{(a}V^{bcd\cdots ef)} = g^{(ab}\phi^{cd\cdots ef)}, \tag{3.1}$$

then

$$\Delta V^{bcd\cdots ef} = (s-1)g^{(bc}\nabla_a\phi^{d\cdots ef)a} - (n+2s-4)\nabla^{(b}\phi^{cd\cdots ef)} \tag{3.2}$$

and

$$\text{the trace-free part of } \nabla^{(a}\nabla^{b}\phi^{cd\cdots ef)} = 0.$$

Proof. Taking the trace of (3.1) gives

$$\tfrac{2}{s+1}\nabla_b V^{bcd\cdots ef} = \tfrac{2n}{s(s+1)}\phi^{cd\cdots ef} + \tfrac{4(s-1)}{s(s+1)}\phi^{cd\cdots ef}$$

hence

$$\phi^{cd\cdots ef} = \tfrac{s}{n+2s-2}\nabla_b V^{bcd\cdots ef}. \tag{3.3}$$

If we apply ∇_a to (3.1) we obtain

$$\tfrac{1}{s+1}\Delta V^{bcd\cdots ef} + \tfrac{s}{s+1}\nabla^{(b}\nabla_a V^{cd\cdots ef)a} = \tfrac{2}{s+1}\nabla^{(b}\phi^{cd\cdots ef)} + \tfrac{s-1}{s+1}g^{(bc}\nabla_a\phi^{d\cdots ef)a}.$$

In combination with (3.3), this completes the proof of (3.2). Since $n \geq 3$ and $s \geq 1$, the coefficient $n+2s-4$ is always non-zero and the final conclusion now follows by differentiating (3.2). □

Now, we are in a position to prove Theorem 2.1. Let us write

$$\mathcal{D} = T^{abcde\cdots f}\nabla_a\nabla_b\nabla_c\nabla_d\nabla_e\cdots\nabla_f + \text{lower order terms},$$

where $T^{abcde\cdots f}$ is a non-zero symmetric tensor, namely the symbol of $\mathcal{D}$. This tensor splits uniquely as

$$T^{abcde\cdots f} = V^{abcde\cdots f} + g^{(ab}W^{cde\cdots f)} + g^{(ab}g^{cd}X^{e\cdots f)}$$

where $V^{abcde\cdots f}$ and $W^{cde\cdots f}$ are symmetric trace-free and $X^{e\cdots f}$ is symmetric. By subtracting

$$\Delta^2 X^{e\cdots f}\nabla_e\cdots\nabla_f = X^{e\cdots f}\Delta^2\nabla_e\cdots\nabla_f + \text{lower order terms}$$

from $\mathcal{D}$ we have found a canonically equivalent symmetry of the form

$$\begin{aligned}\mathcal{D} = {} & V^{abcde\cdots f}\nabla_a\nabla_b\nabla_c\nabla_d\nabla_e\cdots\nabla_f + W^{cde\cdots f}\Delta\nabla_c\nabla_d\nabla_e\cdots\nabla_f \\ & + \text{lower order terms}\end{aligned}$$

and we claim that $V^{abcde\cdots f}$ must be a conformal Killing tensor and $W^{cde\cdots f}$ must be a generalised conformal Killing tensor. To see this we simply compute $\Delta^2\mathcal{D}$ and for this task it is convenient to use the formula

$$\begin{aligned}\Delta^2(fg) &= f\Delta^2 g + 4(\nabla^a f)\Delta\nabla_a g + 2(\Delta f)\Delta g + 4(\nabla^a\nabla^b f)\nabla_a\nabla_b g\\ &\quad + 4(\Delta\nabla^a f)\nabla_a g + (\Delta^2 f)g\end{aligned}$$

and its evident extension to tensor expressions. If we write

$$\begin{aligned}\mathcal{D} &= V^{hijkl\cdots m}\nabla_h\nabla_i\nabla_j\nabla_k\nabla_l\cdots\nabla_m + W^{jkl\cdots m}\Delta\nabla_j\nabla_k\nabla_l\cdots\nabla_m\\ &\quad + Y^{ijkl\cdots m}\nabla_i\nabla_j\nabla_k\nabla_l\cdots\nabla_m + Z^{jkl\cdots m}\nabla_j\nabla_k\nabla_l\cdots\nabla_m\\ &\quad + \text{lower order terms}\end{aligned}$$

where $Y^{ijkl\cdots m}$ and $Z^{jkl\cdots m}$ are symmetric, then

$$\begin{aligned}\Delta^2\mathcal{D} &= V^{hij\cdots m}\Delta^2\nabla_h\nabla_i\nabla_j\cdots\nabla_m + W^{j\cdots m}\Delta^3\nabla_j\cdots\nabla_m\\ &\quad + 4(\nabla^{(a}V^{hij\cdots m)})\Delta\nabla_a\nabla_h\nabla_i\nabla_j\cdots\nabla_m\\ &\quad + 4(\nabla^{(a}W^{j\cdots m)})\Delta^2\nabla_a\nabla_j\cdots\nabla_m\\ &\quad + Y^{ij\cdots m}\Delta^2\nabla_i\nabla_j\cdots\nabla_m\\ &\quad + \text{lower order terms.}\end{aligned}$$

Moving the Laplacian to the right hand side of each of these terms gives

$$\Delta^2\mathcal{D} = \mathcal{P}\Delta^2 + 4(\nabla^{(a}V^{hij\cdots m)})\nabla_a\nabla_h\nabla_i\nabla_j\cdots\nabla_m\Delta + \text{lower order terms}$$

for some differential operator $\mathcal{P}$ and for this to be of the form $\delta\Delta^2$ for some differential operator δ forces (2.1) to hold, as required.

To find a constraint on $W^{jik\cdots m}$ we should consider lower order terms:

$$\begin{aligned}\Delta^2\mathcal{D} &= \mathcal{Q}\Delta^2 + 4(\nabla^{(a}V^{hijkl\cdots m)})\nabla_a\nabla_h\nabla_i\nabla_j\nabla_k\nabla_l\cdots\nabla_m\Delta\\ &\quad + 2(\Delta V^{hijkl\cdots m})\nabla_h\nabla_i\nabla_j\nabla_k\nabla_l\cdots\nabla_m\Delta\\ &\quad + 4(\nabla^{(a}\nabla^b V^{hijkl\cdots m)})\nabla_a\nabla_b\nabla_h\nabla_i\nabla_j\nabla_k\nabla_l\cdots\nabla_m\\ &\quad + 4(\nabla^{(a}\nabla^b W^{jkl\cdots m)})\nabla_a\nabla_b\nabla_j\nabla_k\nabla_l\cdots\nabla_m\Delta\\ &\quad + 4(\nabla^{(a}Y^{ijkl\cdots m)})\nabla_a\nabla_i\nabla_j\nabla_k\nabla_l\cdots\nabla_m\Delta\\ &\quad + \text{lower order terms.}\end{aligned}$$

If we write $V^{hijkl\cdots m}$ according to (3.1) and substitute from (3.2), we obtain

$$\begin{aligned}\Delta^2\mathcal{D} &= \mathcal{R}\Delta^2 - 2(n+2s-4)(\nabla^{(h}\phi^{ijkl\cdots m)})\nabla_h\nabla_i\nabla_j\nabla_k\nabla_l\cdots\nabla_m\Delta\\ &\quad + 4(\nabla^{(h}\phi^{ijkl\cdots m)})\nabla_h\nabla_i\nabla_j\nabla_k\nabla_l\cdots\nabla_m\Delta\\ &\quad + 4(\nabla^{(h}\nabla^i W^{jkl\cdots m)})\nabla_h\nabla_i\nabla_j\nabla_k\nabla_l\cdots\nabla_m\Delta\\ &\quad + 4(\nabla^{(h}Y^{ijkl\cdots m)})\nabla_h\nabla_i\nabla_j\nabla_k\nabla_l\cdots\nabla_m\Delta\\ &\quad + \text{lower order terms}\end{aligned}$$

from which we deduce that

$$2(\nabla^{(h}\nabla^i W^{jkl\cdots m)}) + 2(\nabla^{(h}Y^{ijkl\cdots m)}) = (n+2s-6)(\nabla^{(h}\phi^{ijkl\cdots m)}). \quad (3.4)$$

Passing to the next order, we find

$$\begin{aligned}\Delta^2\mathcal{D} &= \mathcal{R}\Delta^2 + 4(\Delta\phi^{ajkl\cdots m)})\nabla_a\nabla_j\nabla_k\nabla_l\cdots\nabla_m\Delta \\ &\quad + 4(\Delta\nabla^{(a}W^{jkl\cdots m)})\nabla_a\nabla_j\nabla_k\nabla_l\cdots\nabla_m\Delta \\ &\quad + 2(\Delta Y^{ijkl\cdots m})\nabla_i\nabla_j\nabla_k\nabla_l\cdots\nabla_m\Delta \\ &\quad + 4(\nabla^{(a}\nabla^h Y^{ijkl\cdots m)})\nabla_a\nabla_h\nabla_i\nabla_j\nabla_k\nabla_l\cdots\nabla_m \\ &\quad + 4(\nabla^{(a}Z^{jkl\cdots m)})\nabla_a\nabla_j\nabla_k\nabla_l\cdots\nabla_m\Delta \\ &\quad + \text{lower order terms}\end{aligned}$$

from which we deduce that

$$\text{the trace-free part of } \nabla^{(a}\nabla^h Y^{ijkl\cdots m)} = 0.$$

Together with Lemma 3.1, it follows from (3.4) that

$$\text{the trace-free part of } \nabla^{(a}\nabla^h\nabla^i W^{jkl\cdots m)} = 0,$$

as required. This completes the proof of Theorem 2.1. □

In principle, computations such as these are all that is needed to find $\mathcal{D}_V$ and $\mathcal{D}_W$ as in Theorem 2.2. For example, we may arrange that (3.4) holds by taking

$$Y^{ijkl\cdots m} = \tfrac{n+2s-6}{2}\phi^{ijkl\cdots m}$$

and from (3.3) we see that $\mathcal{D}_V$ must take the form

$$\mathcal{D}_V f = V^{ab\cdots c}\nabla_a\nabla_b\cdots\nabla_c f + \tfrac{s(n+2s-6)}{2(n+2s-2)}(\nabla_a V^{ab\cdots c})\nabla_b\cdots\nabla_c f + \cdots. \tag{3.5}$$

This direct approach, however, is difficult. Fortunately, there is a much easier way of constructing the operators $\mathcal{D}_V$ and $\mathcal{D}_W$ and this is done in the next section.

4. The ambient metric and the proof of Theorem 2.2. The constructions in this section closely follow those of [7] and so we shall be brief. Let us consider the Lorentzian quadratic form

$$\tilde{g}_{AB}x^A x^B = 2x^0x^\infty + g_{ab}x^a x^b = (x^0, x^a, x^\infty)\begin{pmatrix} 0 & 0 & 1 \\ 0 & g_{ab} & 0 \\ 1 & 0 & 0\end{pmatrix}\begin{pmatrix} x^0 \\ x^b \\ x^\infty\end{pmatrix}$$

on $\mathbb{R}^{n+2}$. If we embed $\mathbb{R}^n \hookrightarrow \mathbb{RP}_{n+1}$ according to

$$x^a \mapsto \begin{bmatrix} 1 \\ x^a \\ -x^a x_a/2 \end{bmatrix},$$

then the action of $\mathrm{SO}(n+1,1)$ on $\mathbb{R}^{n+2}$ preserves $\mathcal{N}$ the null cone of $\tilde{g}_{AB}$ and the corresponding infinitesimal action of $\mathfrak{g} = \mathfrak{so}(n+1,1)$ on the space of null directions gives rise to conformal Killing vectors on $\mathbb{R}^n$. Explicitly, if $\mathfrak{g}$ is realised as skew tensors V^{BQ} on $\mathbb{R}^{n+2}$ in the usual way, then one may check that

$$V^{BQ} \mapsto V^b = \Phi_B V^{BQ} \Psi^b{}_Q, \tag{4.1}$$

where

$$\Phi_B = (-x^b x_b/2, x_b, 1), \quad \Psi^b{}_Q = (-x^b, \delta^b{}_q, 0), \quad \delta^b{}_q = \text{Kronecker delta.} \tag{4.2}$$

In other words,

$$V^{BQ} = \begin{pmatrix} V^{00} & V^{0q} & V^{0\infty} \\ V^{b0} & V^{bq} & V^{b\infty} \\ V^{\infty 0} & V^{\infty q} & V^{\infty\infty} \end{pmatrix} = \begin{pmatrix} 0 & r^q & \lambda \\ -r^q & m^{bq} & s^b \\ -\lambda & -s^q & 0 \end{pmatrix}$$

corresponds to the conformal Killing vector

$$V^b = -s^b - m^b{}_q x^q + \lambda x^b + r_q x^q x^b - (1/2) x_q x^q r^b. \tag{4.3}$$

As in [7], the formula (4.1) generalises

$$\begin{aligned} V^{BQ\cdots CRDSET} &\mapsto V^{b\cdots cde} \\ &= \Phi_B \cdots \Phi_C \Phi_D \Phi_E V^{BQ\cdots CRDSET} \Psi^b{}_Q \cdots \Psi^c{}_R \Psi^d{}_S \Psi^e{}_T. \end{aligned} \tag{4.4}$$

to provide an explicit realisation of the isomorphism (2.5). Here, $V^{BQ\cdots CRDSET}$ is skew in each pair of indices $BQ, \ldots, CR, DS, ET$, is totally trace-free, and is such that skewing over any three indices gives zero. Similarly, we may take

$$W^{BQ\cdots CRDE} \in \begin{array}{|c|c|c|c|c|c|c|}\hline & & & \cdots & & & \\ \hline & & & \cdots & \\ \cline{1-5} \end{array}\circ \tag{4.5}$$

as totally trace-free, skew in each pair $BQ, \ldots, CR$, symmetric in DE, and such that skewing over any three indices gives zero. Then (2.6) is realised by

$$W^{BQ\cdots CRDE} \mapsto W^{b\cdots c} = \Phi_B \cdots \Phi_C \Phi_D \Phi_E W^{BQ\cdots CRDE} \Psi^b{}_Q \cdots \Psi^c{}_R. \tag{4.6}$$

Following Fefferman and Graham [9], we shall use the term 'ambient' to refer to objects defined on $\mathbb{R}^{n+2}$. For example, there is the ambient wave operator

$$\tilde{\Delta} = \tilde{g}^{AB} \frac{\partial^2}{\partial x^A \partial x^B}$$

where $\tilde{g}^{AB}$ is the inverse of $\tilde{g}_{AB}$. Let $r = \tilde{g}_{AB} x^A x^B$ so that $\mathcal{N} = \{r = 0\}$. Suppose that g is an ambient function homogeneous of degree $w - 2$. A simple calculation gives

$$\tilde{\Delta}(rg) = r\tilde{\Delta} g + 2(n + 2w - 2)g. \tag{4.7}$$

In particular, if $w = 1 - n/2$, then $\tilde{\Delta}(rg) = r\tilde{\Delta}g$. Therefore, if f is homogeneous of degree $1 - n/2$, then $\tilde{\Delta}f|_{\mathcal{N}}$ depends only on $f|_{\mathcal{N}}$ (since rg provides the freedom in extending such a function off $\mathcal{N}$). This defines a differential operator on $\mathbb{R}^n$ and, as detailed in [8], one may easily verify that it is the Laplacian. This construction is due to Dirac [5] and the main point is that it is manifestly invariant under the action of $\mathfrak{so}(n+1,1)$. We say that Δ is conformally invariant acting on conformal densities of weight $1 - n/2$ on $\mathbb{R}^n$. This ambient construction of the Laplacian is a simple example of the 'AdS/CFT correspondence' in physics. A principal feature of this correspondence is that calculations are simplified by doing them ambiently or 'in the bulk'. This feature pervades all that follows.

Invariance may also be viewed as follows. Recall that $\mathfrak{g} = \mathfrak{so}(n+1,1)$ is realised as skew tensors V^{BQ}. Each gives rise to an ambient differential operator

$$\mathfrak{D}_V = V^{BQ} x_B \frac{\partial}{\partial x^Q} \quad \text{where } x_B = x^A \tilde{g}_{AB}. \tag{4.8}$$

It is easily verified that, for g and f of any homogeneity,

$$\mathfrak{D}_V(rg) = r\mathfrak{D}_V g \quad \text{and} \quad \tilde{\Delta}\mathfrak{D}_V f = \mathfrak{D}_V \tilde{\Delta} f. \tag{4.9}$$

The first of these implies that $\mathfrak{D}_V$ induces differential operators on $\mathbb{R}^n$ for densities of any conformal weight: simply extend the corresponding homogeneous function on $\mathcal{N}$ into $\mathbb{R}^{n+2}$, apply $\mathfrak{D}_V$, and restrict back to $\mathcal{N}$. In particular, let us denote by $\mathcal{D}_V$ and δ_V the differential operators so induced on densities of weight $1 - n/2$ and $-1 - n/2$, respectively. Bearing in mind the ambient construction of the Laplacian, it follows immediately from the second equation of (4.9) that $\Delta\mathcal{D}_V = \delta_V\Delta$. In other words, the infinitesimal conformal invariance of Δ gives rise to the symmetries $\mathcal{D}_V$ as differential operators.

The formula (4.8) generalises to provide further symmetries. It is shown in [7] that the ambient differential operator

$$\mathfrak{D}_V = V^{BQ\cdots CRDSET} x_B \cdots x_C x_D x_E \frac{\partial^s}{\partial x^Q \cdots \partial x^R \partial x^S \partial x^T} \tag{4.10}$$

provides a symmetry of the Laplacian for all

$$V^{BQ\cdots CRDSET} \in \begin{array}{|c|c|c|c|c|c|c|}\hline \; & \; & \; & \cdots & \; & \; & \; \\ \hline \; & \; & \; & \cdots & \; & \; & \; \\ \hline \end{array}_{\circ}.$$

The proof of Theorem 2.2 for conformal Killing tensors is essentially contained in the following:

PROPOSITION 4.1. *The ambient operator* (4.10) *induces a symmetry of* Δ^2. *The symbol of this operator is given by* (4.4).

Proof. Firstly, we need to know the ambient description of Δ^2. Iterating (4.7) gives

$$\tilde{\Delta}^2(rg) = r\tilde{\Delta}^2 g + 4(n + 2w - 4)\tilde{\Delta}g.$$

In particular, if $w = 2 - n/2$, then $\tilde{\Delta}^2(rg) = r\tilde{\Delta}^2 g$. Therefore, if f is homogeneous of degree $2 - n/2$, then $\tilde{\Delta}^2 f|_{\mathcal{N}}$ depends only on $f|_{\mathcal{N}}$ and it is shown in [8] that the resulting differential operator on $\mathbb{R}^n$ is Δ^2. It is it easily verified that (4.9) holds more generally for $\mathfrak{D}_V$ of the form (4.10). It follows that

$$\mathfrak{D}_V(rg) = r\mathfrak{D}_V g \quad \text{and} \quad \tilde{\Delta}^2\mathfrak{D}_V f = \mathfrak{D}_V\tilde{\Delta}^2 f \tag{4.11}$$

for f and g of any homogeneity. Arguing as for the Laplacian shows that the operator $\mathcal{D}_V$ on $\mathbb{R}^n$ obtained from $\mathfrak{D}_V$ acting on functions homogeneous of degree $w = 2 - n/2$ is a symmetry of Δ^2. There are some details to be verified to make sure that the symbol of $\mathcal{D}_V$ is given by (4.4). However, similar verifications are done in [7] and we leave them to the interested reader. □

The ambient construction of symmetries from tensors $W^{BQ\cdots CRDE}$ as in (4.5) is less obvious. The following proposition completes the proof of Theorem 2.2.

PROPOSITION 4.2. *For any tensor $W^{BQ\cdots CRDE}$ satisfying the symmetries of* (4.5) *the ambient differential operator*

$$W^{BQ\cdots CRDE}x_B\cdots x_C\Big(x_Dx_E\tilde{\Delta} - 2x_D\frac{\partial}{\partial x^E}\Big)\frac{\partial^{s-2}}{\partial x^Q\cdots\partial x^R}$$

induces a symmetry $\mathcal{D}_W$ of Δ^2 of the form

$$\mathcal{D}_W = W^{b\ldots c}\Delta\nabla_b\cdots\nabla_c + \text{lower order terms}$$

where $W^{b\cdots c}$ is given by (4.6).

It is not too hard to prove this Proposition by direct calculation along the lines of Proposition 4.1. There is a difference, however, in that the analogue of the first equation of (4.11) holds only for g of homogeneity $-n/2$. Moreover, for an analogue of the second equation, one needs to use the ambient operator

$$W^{BQ\cdots CRDE}x_B\cdots x_C\Big(x_Dx_E\tilde{\Delta} + 6x_D\frac{\partial}{\partial x^E}\Big)\frac{\partial^{s-2}}{\partial x^Q\cdots\partial x^R}$$

on the right hand side and, even so, it is valid only for homogeneity $2-n/2$. Of course, these homogeneities are exactly what we need for Δ^2 but there is a more satisfactory ambient construction giving rise to exactly the same symmetries, which we shall defer to the following section. The operators in this more satisfactory construction enjoy the proper generalisation of (4.11), namely (5.4).

5. The proof of Theorem 2.3. We shall prove Theorem 2.3 by a new method, improving on [7]. As a side effect, we shall obtain a straightforward proof of Proposition 4.2.

In the previous section, we found in (4.8) a linear mapping

$$\mathfrak{so}(n+1,1)=\mathfrak{g}\ni V\longmapsto \mathfrak{D}_V, \tag{5.1}$$

where $\mathfrak{D}_V$ is an ambient differential operator acting on functions homogeneous of degree w for any w. By dint of (4.9), we obtain an induced linear mapping

$$\mathfrak{g}\ni V\longmapsto \mathcal{D}_V,$$

where $\mathcal{D}_V$ is a differential operator acting on conformal densities of weight w. In fact, it is shown in [7] that

$$\mathcal{D}_V f = V^a\nabla_a f - \tfrac{w}{n}(\nabla_a V^a)f \tag{5.2}$$

where V^a is the conformal Killing vector associated to $V\in\mathfrak{g}$ according to (4.1).

The mapping (5.1) immediately extends to the whole tensor algebra $\bigotimes\mathfrak{g}$ by

$$\mathfrak{g}\otimes\mathfrak{g}\otimes\cdots\otimes\mathfrak{g}\ni U\otimes\cdots\otimes V=X\mapsto\mathfrak{D}_X\equiv\mathfrak{D}_U\cdots\mathfrak{D}_V \tag{5.3}$$

and extended by linearity. It follows from (4.9) that

$$\mathfrak{D}_X(rg)=r\mathfrak{D}_X g\quad\text{and}\quad\tilde{\Delta}\mathfrak{D}_X=\mathfrak{D}_X\tilde{\Delta} \tag{5.4}$$

and hence that there is an induced series of operators $\mathcal{D}_X$ acting on densities of weight w on $\mathbb{R}^n$ and providing symmetries of Δ^k when $w=k-n/2$. Of course, for simple tensors X these operators are obtained by composing the basic operators $\mathcal{D}_V$ for $V\in\mathfrak{g}$. However, usefully to compute even the basic composition $\mathcal{D}_U\mathcal{D}_V$ from (5.2) for $U,V\in\mathfrak{g}$ is difficult. Though this is done in [7], the simpler approach adopted there is to compute $\mathfrak{D}_U\mathfrak{D}_V$ instead. The object is to see how this composition breaks up under (2.7) but, for this purpose, the following argument is even more straightforward.

We compute

$$\begin{aligned}\mathfrak{D}_U\mathfrak{D}_V &= U^{BQ}x_B\frac{\partial}{\partial x^Q}V^{CR}x_C\frac{\partial}{\partial x^R}\\ &= U^{BQ}V^{CR}x_Bx_C\frac{\partial^2}{\partial x^Q\partial x^R}+U^{BQ}V_Q{}^Rx_B\frac{\partial}{\partial x^R},\end{aligned}$$

which extends by linearity to give

$$\mathfrak{D}_X = X^{BQCR}x_Bx_C\frac{\partial^2}{\partial x^Q\partial x^R}+X^{BQ}{}_Q{}^Rx_B\frac{\partial}{\partial x^R},\quad\text{for } X^{BQCR}\in\mathfrak{g}\otimes\mathfrak{g}. \tag{5.5}$$

We can simply apply this formula to tensors X from each of the summands on the right hand side of (2.7). All X^{BQCR} in $\mathfrak{g}\otimes\mathfrak{g}$ are skew in BQ and CR but the various summands of (2.7) are characterised as follows.

$$\begin{smallmatrix}\square\square\\\square\square\end{smallmatrix}_\circ \longleftrightarrow \begin{cases} X^{BQCR}+X^{BCRQ}+X^{BRQC}=0 \\ X^{BQCR} \text{ is totally trace-free.}\end{cases}$$

Therefore, the second term in (5.5) vanishes and $\mathfrak{D}_X$ is given by (4.10), as expected.

Next we have

$$\square\square_\circ \leftrightarrow X^{BQCR}=W^{BC}\tilde{g}^{QR}-W^{QC}\tilde{g}^{BR}-W^{BR}\tilde{g}^{QC}+W^{QR}\tilde{g}^{BC} \quad (5.6)$$

where W^{BC} is symmetric trace-free. Therefore,

$$\begin{aligned}\mathfrak{D}_X &= W^{BC}x_Bx_C\tilde{\Delta}-2W^{QC}x_Cx^R\frac{\partial^2}{\partial x^R\partial x^Q}\\ &\quad -nW^{BR}x_B\frac{\partial}{\partial x^R}+W^{QR}r\frac{\partial^2}{\partial x^Q\partial x^R}\end{aligned}$$

and, when acting on functions homogeneous of degree w,

$$\mathfrak{D}_X=W^{BC}\Big(x_Bx_C\tilde{\Delta}-(n+2w-2)x_B\frac{\partial}{\partial x^C}+r\frac{\partial^2}{\partial x^B\partial x^C}\Big).$$

There are two immediate consequences of this formula. Firstly, when $w=1-n/2$, the appropriate homogeneity for the Laplacian, we obtain

$$\mathfrak{D}_X=W^{BC}\Big(x_Bx_C\tilde{\Delta}+r\frac{\partial^2}{\partial x^B\partial x^C}\Big)$$

and the induced operator $\mathcal{D}_X$ on $\mathbb{R}^n$ is clearly of the form $\mathcal{P}\Delta$. Therefore, this summand in the decomposition (2.7) is contained in the annihilator ideal. This is confirmed by (2.8). On the other hand, when $w=2-n/2$, we obtain

$$\mathfrak{D}_X=W^{BC}\Big(x_Bx_C\tilde{\Delta}-2x_B\frac{\partial}{\partial x^C}+r\frac{\partial^2}{\partial x^B\partial x^C}\Big)$$

and the induced operator on $\mathbb{R}^n$ coincides with the statement of Proposition 4.2 in this case. It is also easy to compute the symbol of the induced operator on $\mathbb{R}^n$ as follows.

LEMMA 5.1.

$$\tilde{g}^{QR}\Psi^b{}_Q\Psi^c{}_R=g^{bc}\qquad \Phi_B\tilde{g}^{BQ}\Psi^b{}_Q=0\qquad \Phi_B\Phi_C\tilde{g}^{BC}=0.$$

Proof. These are simple computations from (4.2). □

From this lemma, if X^{BQCR} is of the form given in (5.6), then

$$\Phi_B \Phi_C X^{BQCR} \Psi^b{}_Q \Psi^c{}_R = \Phi_B \Phi_C W^{BC} g^{bc} = W g^{bc}$$

where W is given by (4.6). It follows that the induced operator on $\mathbb{R}^n$ is of the form

$$W g^{ab} \nabla_a \nabla_b + \text{lower order terms} = W\Delta + \text{lower order terms}.$$

Therefore, we have proved Proposition 4.2 for second order operators. Let us return to analysing the effect of the various summands of $\mathfrak{g} \otimes \mathfrak{g}$ in (5.5).

Next we have

$$\mathbb{R} \leftrightarrow X^{BQCR} = V \tfrac{1}{n(n+1)(n+2)} \left(\tilde{g}^{QC} \tilde{g}^{BR} - \tilde{g}^{BC} \tilde{g}^{QR} \right) \quad \text{for constant } V.$$

The normalisation is arranged so that the Killing form $\langle\ ,\ \rangle : \mathfrak{g} \otimes \mathfrak{g} \to \mathbb{R}$ gives

$$\begin{aligned} X^{BQCR} \longmapsto -nX^{BQ}{}_{BQ} &= -nV \tfrac{1}{n(n+1)(n+2)} \left(\delta^Q{}_B \delta^B{}_Q - \delta^B{}_B \delta^Q{}_Q \right) \\ &= -nV \tfrac{1}{n(n+1)(n+2)} \left((n+2) - (n+2)^2 \right) = V. \end{aligned}$$

We compute

$$\mathfrak{D}_X = V \tfrac{1}{n(n+1)(n+2)} \left(x^Q x^R \frac{\partial^2}{\partial x^R \partial x^Q} + r\tilde{\Delta} + (n+1) x^R \frac{\partial}{\partial x^R} \right).$$

Therefore, when acting on functions homogeneous of degree w,

$$\mathfrak{D}_X = V \tfrac{1}{n(n+1)(n+2)} \left(r\tilde{\Delta} + w(n+w) \right).$$

Hence, the corresponding action of $\mathcal{D}_X$ on $\mathbb{R}^n$ is

$$\mathcal{D}_X f = \tfrac{w(n+w)}{n(n+1)(n+2)} V f$$

for conformal densities of weight w. In particular, if $w = 1 - n/2$ then

$$\mathcal{D}_X f = -\tfrac{(n-2)}{4n(n+1)} V f.$$

This is exactly as predicted in (2.8). If $w = 2 - n/2$, however, then

$$\mathcal{D}_X f = -\tfrac{(n-4)(n+4)}{4n(n+1)(n+2)} V f,$$

in agreement with (2.9).

Next we have

$$\begin{array}{|c|c|}\hline & \\ \hline & \multicolumn{1}{c}{} \\ \cline{1-1} & \multicolumn{1}{c}{} \\ \cline{1-1} \end{array}_{\circ} \longleftrightarrow \left\{ \begin{array}{l} X^{BQCR} + X^{CRBQ} = 0 \\ X^{BQCR} \text{ is totally trace-free.} \end{array} \right.$$

In this case both terms in (5.5) evidently vanish.

Next we have

$$\begin{array}{|c|}\hline \\ \hline \\ \hline\end{array} \leftrightarrow X^{BQCR} = \tfrac{1}{2n}\left(V^{BR}\tilde{g}^{QC} - V^{QR}\tilde{g}^{BC} - V^{BC}\tilde{g}^{QR} + V^{QC}\tilde{g}^{BR}\right)$$

where V^{BR} is skew. The normalisation is arranged so that the Lie bracket $\mathfrak{g} \otimes \mathfrak{g} \to \mathfrak{g}$ gives

$$X^{BQCR} \longmapsto X^{B}{}_{Q}{}^{QR} - X^{R}{}_{Q}{}^{QB} = \tfrac{1}{2}V^{BR} - \tfrac{1}{2}V^{RB} = V^{BR}.$$

Since $X^{BQCR} = -X^{CRBQ}$, the first term in (5.5) vanishes. Therefore,

$$\mathfrak{D}_X = X^{BQ}{}_{Q}{}^{R}x_B\frac{\partial}{\partial x^R} = \tfrac{1}{2}V^{BR}x_B\frac{\partial}{\partial x^R} = \tfrac{1}{2}\mathfrak{D}_V.$$

This accounts for the term $\frac{1}{2}\mathfrak{D}_{[V,W]}$ in both (2.8) and (2.9).

The final summand corresponds to totally skew tensors X^{BQCR} and for these it is clear that $\mathfrak{D}_X$ given by (5.5) vanishes. It accounts for the presence of this summand in the ideals defined by (2.8) or (2.9).

In summary, by considering the effect in (5.5) of tensors from the various summands in the decomposition (2.7) of $\mathfrak{g} \otimes \mathfrak{g}$, we have verified (2.8) and (2.9). It is also worthwhile recording what we have shown for a general conformal weight w.

THEOREM 5.1. *Suppose that X^a and Y^a are conformal Killing vector fields on $\mathbb{R}^n$ corresponding to X^{AP} and Y^{AP}, respectively, in $\mathfrak{g} = \mathfrak{so}(n+1,1)$. Then*

$$\mathcal{D}_X\mathcal{D}_Y f = \mathcal{D}_{X\circledcirc Y} f + \mathcal{D}_{X\bullet Y} f + \tfrac{1}{2}\mathcal{D}_{[X,Y]} f + \tfrac{w(n+w)}{n(n+1)(n+2)}\langle X, Y\rangle f$$

on densities of weight w. Here,

* $\mathcal{D}_X f = X^a\nabla_a f - \frac{w}{n}(\nabla_a X^a)f$.
* $(X \circledcirc Y)^{ab} = \frac{1}{2}X^aY^b + \frac{1}{2}X^bY^a - \frac{1}{n}X^cY_cg^{ab}$ *is a conformal Killing tensor and for a general conformal Killing tensor V^{ab},*

$$\mathcal{D}_V f = V^{ab}\nabla_a\nabla_b f - \tfrac{2(w-1)}{n+2}(\nabla_a V^{ab})\nabla_b f + \tfrac{w(w-1)}{(n+2)(n+1)}(\nabla_a\nabla_b V^{ab})f.$$

* $X \bullet Y = \frac{1}{n}X^aY_a = W$ *satisfies* $\nabla_a\nabla_b\nabla_c W = g_{(ab}\phi_{c)}$ *and, for such a field in general,*

$$\mathcal{D}_W f = W\Delta f - \tfrac{n+2w-2}{2}(\nabla^a W)\nabla_a f + \tfrac{w(n+2w-2)}{2(n+2)}(\Delta W)f.$$

* $[X,Y]^a = X^b\nabla_b Y^a - Y^b\nabla_b X^a$ *is a conformal Killing field.*
* $\langle X, Y\rangle = n(\nabla_b X^a)(\nabla_a Y^b) - \frac{n-2}{n}(\nabla_a X^a)(\nabla_b Y^b) - 2X^a\nabla_a\nabla_b Y^b - 2Y^a\nabla_a\nabla_b X^b$ *is constant.*

Within $(\mathfrak{g}\otimes\mathfrak{g})\oplus\mathfrak{g}\oplus\mathbb{R}$, *however, these operations are defined as in Section 2.*

Proof. Apart from $W = X \bullet Y$ the various formulae have just been established or are taken from [7] (with a minor rearrangement for $\langle X, Y\rangle$ on $\mathbb{R}^n$). To complete the proof, therefore, it remains to establish the formula on $\mathbb{R}^n$ for $X \bullet Y$ and the formula for $\mathcal{D}_W$ in general. One possibility is to compute the composition $\mathcal{D}_X\mathcal{D}_Y$ in full and collect terms. Though this certainly works, there is a short cut as follows. Certainly,

$$\begin{aligned}\mathcal{D}_X\mathcal{D}_Y &= X^aY^b\nabla_a\nabla_b + \text{lower order terms}\\ &= \left(\tfrac{1}{2}X^aY^b + \tfrac{1}{2}X^bY^a - \tfrac{1}{n}X^cY_cg^{ab}\right)\nabla_a\nabla_b + \tfrac{1}{n}X^cY_c\Delta\\ &\quad + \text{lower order terms}\\ &= (X \circledcirc Y)^{ab}\nabla_a\nabla_b + \tfrac{1}{n}X^cY_c\Delta + \text{lower order terms.}\end{aligned}$$

It follows that $X \bullet Y = \frac{1}{n}X^aY_b$, as advertised. Rather than find the lower order terms by direct computation, we claim that they are forced by invariance under the conformal action of $\mathfrak{so}(n+1,1)$. This is essentially the argument used in [7, §5] to find explicit formulae for $\mathcal{D}_V$ in case of an arbitrary conformal Killing tensor $V^{bc\cdots d}$ acting on densities of any weight. In our case, the argument is as follows. We are looking for a differential operator of the form

$$\mathcal{D}_Wf = W\Delta f + \alpha(\nabla^aW)\nabla_af + \beta(\Delta W)f \tag{5.7}$$

and it remains to determine the constants α and β in order that such an operator be conformally invariant under flat-to-flat rescalings of the standard metric on $\mathbb{R}^n$. We shall follow the conventions of [1] concerning conformal geometry. If $\hat{g}_{ab} = \Omega^2g_{ab}$ is also a flat metric, then

$$\nabla^a\Upsilon_a = -\tfrac{n-2}{2}\Upsilon^a\Upsilon_a \quad \text{for } \Upsilon_a = (\nabla_a\Omega)/\Omega.$$

Now W has conformal weight 2 and we are supposing that f has conformal weight w. It follows that

$$\begin{aligned}\hat{\nabla}_af &= \nabla_af + w\Upsilon_af\\ \hat{\nabla}^aW &= \nabla^aW + 2\Upsilon^aW\\ \hat{\Delta}f &= \Delta f + (n+2w-2)\big(\Upsilon^a\nabla_af + \tfrac{w}{2}\Upsilon^a\Upsilon_af\big)\\ \hat{\Delta}W &= \Delta W + (n+2)\big(\Upsilon^a\nabla_aW + \Upsilon^a\Upsilon_aW\big)\end{aligned}$$

whence (5.7) satisfies $\hat{\mathcal{D}}_W = \mathcal{D}_W$ if and only if

$$\alpha = -\tfrac{n+2w-2}{2} \quad \text{and} \quad \beta = \tfrac{w(n+2w-2)}{2(n+2)},$$

as required. □

Notice that (2.2) and (2.3) are special cases of Theorem 5.1. Also notice that $\mathcal{D}_W = W\Delta$ when $w = 1/n - 2$, which explains the absence of $X \bullet Y$ in the generators (2.8) of the annihilator ideal in this case.

We shall now complete the proof of Proposition 4.2 (and, hence, of Theorem 2.2).

LEMMA 5.2. *Suppose that*

$$X^{BQ\cdots CRDSET} \in \begin{array}{|c|}\hline \\ \hline \\ \hline\end{array} \odot \cdots \odot \begin{array}{|c|}\hline \\ \hline \\ \hline\end{array} \odot \begin{array}{|c|}\hline \\ \hline \\ \hline\end{array} \odot \begin{array}{|c|}\hline \\ \hline \\ \hline\end{array} = \bigodot^s \mathfrak{g} \subset \bigotimes^s \mathfrak{g}.$$

In other words, $X^{BQ\cdots CRDSET}$ *is skew in each pair of indices* $BQ, \ldots, CR, DS, ET$, *has* $2s$ *indices in total, and is invariant under permutations of the paired indices. Then the operator defined by* (5.3) *is*

$$\begin{aligned}
\mathfrak{D}_X &= X^{BQ\cdots CRDSET} x_B \cdots x_C x_D x_E \frac{\partial^s}{\partial x^Q \cdots \partial x^R \partial x^S \partial x^T} \\
&\quad + \tfrac{s(s-1)}{2} X^{BQ\cdots CRDS}{}_S{}^T x_B \cdots x_C x_D \frac{\partial^{s-1}}{\partial x^Q \cdots \partial x^R \partial x^T} \\
&\quad + \text{lower order terms.}
\end{aligned}$$

Proof. The derivation of (5.5) from (4.8) is easily extended by induction. □

Suppose that $W^{BQ\cdots CRDE}$ satisfies the symmetries of (4.5) as in the statement of Proposition 4.2. Generalising (5.6), let $X^{BQ\cdots CRDSET}$ be obtained by forming

$$W^{BQ\cdots CRDE}\tilde{g}^{ST} - W^{BQ\cdots CRSE}\tilde{g}^{DT} - W^{BQ\cdots CRSE}\tilde{g}^{DT} + W^{BQ\cdots CRST}\tilde{g}^{DE}$$

and then symmetrising over the paired indices $BQ, \ldots, CR, DS, ET$. From Lemma 5.2, a short calculation gives

$$\begin{aligned}
\mathfrak{D}_X &= W^{BQ\cdots CRDE} x_B \cdots x_C \Big(x_D x_E \tilde{\Delta} - 2 x_D x^S \frac{\partial^2}{\partial x^S \partial x^E} \\
&\quad + r \frac{\partial^2}{\partial x^D \partial x^E} \Big) \frac{\partial^{s-2}}{\partial x^Q \cdots \partial x^R} \\
&\quad - (n + 2s - 4) W^{BQ\cdots CRDE} x^B \cdots x^C x^D \frac{\partial^{s-1}}{\partial x^E \partial x^Q \cdots \partial x^R},
\end{aligned}$$

where $W^{BQ\cdots CRDE}$ being trace-free ensures that there are no lower order terms. Therefore, when acting on functions homogeneous of degree w, we find

$$\begin{aligned}
\mathfrak{D}_X &= W^{BQ\cdots CRDE} x_B \cdots x_C \Big(x_D x_E \tilde{\Delta} - (n + 2w - 2) x_D \frac{\partial}{\partial x^E} \\
&\quad + r \frac{\partial^2}{\partial x^D \partial x^E} \Big) \frac{\partial^{s-2}}{\partial x^Q \cdots \partial x^R}
\end{aligned}$$

and, in particular, if $w = 2 - n/2$ then we have completed the proof of Proposition 4.2.

It remains to finish the proof of Theorem 2.3. As in [7], this is done by considering the corresponding graded algebra $\mathrm{gr}(\mathcal{B}_n)$. From Corollary 2.1 we know the structure of this algebra—as a vector space it is (2.4) and its algebra structure arises from its being a quotient of the tensor algebra $\bigotimes \mathfrak{g}$. Theorem 5.1 with $w = 2-n/2$ implies that the elements (2.9) are contained in the ideal defining $\mathcal{B}_n$, namely the kernel of the mapping $\bigotimes \mathfrak{g} \to \mathcal{B}_n$. From these elements alone, the corresponding graded ideal contains

$$V \otimes W - V \circledcirc W - V \bullet W \quad \text{for } V, W \in \mathfrak{g}.$$

Let us consider the ideal generated by these elements alone, i.e. generated by $\mathcal{I}_2$ where we have grouped the decomposition (2.7) according to

$$\begin{smallmatrix}\square\\\square\end{smallmatrix} \otimes \begin{smallmatrix}\square\\\square\end{smallmatrix} = \begin{smallmatrix}\square\square\\\square\square\end{smallmatrix}_{\circ} \oplus \square\square_{\circ} \oplus \mathcal{I}_2.$$

In particular, $\mathcal{I}_2$ contains $\mathfrak{g} \wedge \mathfrak{g}$ and so $\mathrm{gr}(\mathcal{B}_n)$ is a quotient of the symmetric tensor algebra $\bigodot^s \mathfrak{g}$. We have just seen how the differential operators $\mathcal{D}_W$ in Proposition 4.2 and hence Theorem 2.2 arise—the representation (4.5) is realised as a specific submodule of $\bigodot^s \mathfrak{g}$ and, indeed, this is the unique submodule of this type. Hence, as a vector space, for $s \geq 2$ we may write $\mathcal{K}_{n,s} \oplus \mathcal{L}_{n,s} \subset \bigodot^s \mathfrak{g}$ and the corresponding symmetry operators are given by the ambient construction in a uniform fashion (as developed earlier in this section). More specifically,

$$\begin{smallmatrix}\square\square\square\square\cdots\square\square\square\\\square\square\square\square\cdots\square\square\square\end{smallmatrix} \subset \textstyle\bigodot^s \mathfrak{g}$$

consists of those $X^{BQ\cdots CRDSET}$ such that skewing over any three indices gives zero and then

$$\mathcal{K}_{n,s} \oplus \mathcal{L}_{n,s} = \left\{X \in \begin{smallmatrix}\square\square\square\square\cdots\square\square\square\\\square\square\square\square\cdots\square\square\square\end{smallmatrix} \text{ s.t. } \mathrm{trace}(\mathrm{trace}(X)) = 0\right\}.$$

For convenience, let us write $\mathcal{K}_{n,s} \oplus \mathcal{L}_{n,s} \equiv \mathcal{M}_{n,s}$. From this viewpoint it is easy to see that the two-sided ideal generated by $\mathcal{I}_2$ in $\bigotimes \mathfrak{g}$ is not big enough to have (2.4) as its quotient and the problem is when $s = 4$. Arguing as in [7], or more specifically as in [6, Theorem 3], we would like to show that

$$\mathcal{M}_{n,s} = \left(\mathcal{M}_{n,s-1} \otimes \begin{smallmatrix}\square\\\square\end{smallmatrix}\right) \cap \left(\begin{smallmatrix}\square\\\square\end{smallmatrix} \otimes \mathcal{M}_{n,s-1}\right) \quad \text{for } s \geq 3 \tag{5.8}$$

but this is not true when $s = 4$ and the problem is with traces. More specifically, it is shown in [6, Theorem 2] and, in any case, is easily verified as in [7] that

$$\begin{array}{|c|c|c|}\hline \; & \; & \; \\ \hline \; & \; & \; \\ \hline \end{array} = \left(\begin{array}{|c|c|c|}\hline \; & \; & \; \\ \hline \; & \; & \; \\ \hline \end{array} \otimes \begin{array}{|c|}\hline \; \\ \hline \; \\ \hline \end{array}\right) \cap \left(\begin{array}{|c|}\hline \; \\ \hline \; \\ \hline \end{array} \otimes \begin{array}{|c|c|c|}\hline \; & \; & \; \\ \hline \; & \; & \; \\ \hline \end{array}\right). \tag{5.9}$$

Therefore, we are asking whether a tensor $X^{BQCRDSET}$ enjoying the symmetries of the left hand side of (5.9) and such that

$$X^{BQ}{}_{BQ}{}^{DSET} = 0 \quad \text{and} \quad X^{BQCRDS}{}_{DS} = 0 \tag{5.10}$$

has the property that all its second traces vanish. This is not the case. Indeed, a counterexample may be constructed from any trace-free symmetric tensor Z^{BCDE}. Let $\tilde{g}^{QRST} \equiv \tilde{g}^{(QR}\tilde{g}^{ST)}$ and then

$$X^{BQCRDSET} = \operatorname{skew}(Z^{BCDE}\tilde{g}^{QRST}) \tag{5.11}$$

where 'skew' means to take the skew part in the index pairs BQ, CR, DS, ET (thus generalising (5.6) to $\begin{array}{|c|c|c|c|}\hline \; & \; & \; & \; \\ \hline \end{array}_{\circ}$). It is readily verified that (5.10) are satisfied but that

$$X^{BQC}{}_{Q}{}^{DSE}{}_{S} = \text{a non-zero multiple of } Z^{BCDE}.$$

The proof of Theorem 2.3 now reduces to the following two facts. The first is that the tensor $X \in \bigodot^4 \mathfrak{g}$ constructed in (5.11) induces a non-zero multiple of the differential operator

$$W^{BCDE} x_B x_C x_D x_E \Delta^2$$

on $\mathbb{R}^n$, which we decreed to be equivalent to zero in Definition 2.2. On the one hand this shows that $\begin{array}{|c|c|c|c|}\hline \; & \; & \; & \; \\ \hline \end{array}_{\circ}$ should be included in the annihilator ideal for $\mathcal{B}_n$ as stated in Theorem 2.3. On the other hand, the second easy fact is that (5.8) is true for $s \neq 4$ and this implies that no further additions to the ideal are necessary. The first fact is an elementary calculation. Both will be left to the reader.

REFERENCES

[1] R.J. Baston and M.G. Eastwood, *Invariant operators*, Twistors in Mathematics and Physics, Lond. Math. Soc. Lecture Notes vol. 156, Cambridge University Press, 1990, pp. 129–163.

[2] C.P. Boyer, E.G. Kalnins, and W. Miller, Jr., *Symmetry and separation of variables for the Helmholtz and Laplace equations*, Nagoya Math. Jour. (1976), **60**: 35–80.

[3] T.P. Branson, A. Čap, M.G. Eastwood, and A.R. Gover, *Prolongations of geometric overdetermined systems*, Int. Jour. Math. (2006), **17**: 641–664.

[4] R.P. Delong, Jr., *Killing tensors and the Hamilton-Jacobi equation*, PhD thesis, University of Minnesota, 1982.

[5] P.A.M. Dirac, *The electron wave equation in de-Sitter space*, Ann. Math. (1935) **36**: 657–669.
[6] M.G. Eastwood, *The Cartan product*, Bull. Belg. Math. Soc. (2004), **11**: 641–651.
[7] M.G. Eastwood, *Higher symmetries of the Laplacian*, Ann. Math. (2005), **161**: 1645–1665.
[8] M.G. Eastwood and C.R. Graham, *Invariants of conformal densities*, Duke Math. Jour. (1991), **63**: 633–671.
[9] C. Fefferman and C.R. Graham, *Conformal invariants*, Élie Cartan et les Mathématiques d'Aujourdui, Astérisque, 1985, pp. 95–116.
[10] J. Lepowsky, *A generalization of the Bernstein-Gelfand-Gelfand resolution*, Jour. Alg. (1977), **49**: 496–511.
[11] A.G. Nikitin, *Generalized Killing tensors of arbitrary rank and order*, Ukrain. Mat. Zh. (1991), **43**: 786–795.
[12] A.G. Nikitin and A.I. Prilipko, *Generalized Killing tensors and the symmetry of the Klein-Gordon-Fock equation*, Akad. Nauk Ukrain. SSR Inst. Mat., preprint, 1990.

METRIC CONNECTIONS IN PROJECTIVE DIFFERENTIAL GEOMETRY*

MICHAEL EASTWOOD† AND VLADIMIR MATVEEV‡

(In memory of Thomas Branson)

Abstract. We search for Riemannian metrics whose Levi-Civita connection belongs to a given projective class. Following Sinjukov and Mikeš, we show that such metrics correspond precisely to suitably positive solutions of a certain projectively invariant finite-type linear system of partial differential equations. Prolonging this system, we may reformulate these equations as defining covariant constant sections of a certain vector bundle with connection. This vector bundle and its connection are derived from the Cartan connection of the underlying projective structure.

Key words. Projective differential geometry, metric connection, tractor.

AMS(MOS) subject classifications. Primary 53A20; Secondary 58J70.

1. Introduction. We shall always work on a smooth oriented manifold M of dimension n. Suppose that ∇ is a torsion-free connection on the tangent bundle of M. We may ask whether there is a Riemannian metric on M whose geodesics coincide with the geodesics of ∇ as unparameterised curves. We shall show that there is a linear system of partial differential equations that precisely controls this question.

To state our results, we shall need some terminology, notation, and preliminary observations. Two torsion-free connections ∇ and $\hat{\nabla}$ are said to be projectively equivalent if they have the same geodesics as unparameterised curves. A projective structure on M is a projective equivalence class of connections. In these terms, we are given a projective structure on M and we ask whether it may be represented by a metric connection. Questions such as this have been addressed by many authors. Starting with a metric connection, Sinjukov [9] considered the existence of other metrics with the same geodesics. He found a system of equations that controls this question and Mikeš [7] observed that essentially the same system pertains when starting with an arbitrary projective structure.

We shall use Penrose's abstract index notation [8] in which indices act as markers to specify the type of a tensor. Thus, ω_a denotes a 1-form whilst X^a denotes a vector field. Repeated indices denote the canonical

*This work was undertaken during the 2006 Summer Program at the Institute for Mathematics and its Applications at the University of Minnesota. The authors would like to thank the IMA for hospitality during this time. The first author is supported by the Australian Research Council.

†Department of Mathematics, University of Adelaide, SA 5005, Australia (meastwoo@member.ams.org).

‡Mathematisches Institut, Fakultät für Mathematik und Informatik, Friedrich-Schiller-Universität Jena, 07737 Jena, Germany (matveev@minet.uni-jena.de).

pairing between vectors and co-vectors. Thus, we shall write $X^a\omega_a$ instead of $X \lrcorner\, \omega$. The tautological 1-form with values in the tangent bundle is denoted by the Kronecker delta $\delta_a{}^b$.

As is well-known [5], the geometric formulation of projective equivalence may be re-expressed as

$$\hat{\nabla}_a X^b = \nabla_a X^b + \Upsilon_a X^b + \delta_a{}^b \Upsilon_c X^c \tag{1.1}$$

for an arbitrary 1-form Υ_a. We shall also adopt the curvature conventions of [5]. In particular, it is convenient to write

$$(\nabla_b \nabla_a - \nabla_a \nabla_b) X^b = R_{ab} X^b,$$

where R_{ab} is the usual Ricci tensor, as

$$(\nabla_b \nabla_a - \nabla_a \nabla_b) X^b = (n-1)\mathrm{P}_{ab} X^b - \beta_{ab} X^b \quad \text{where } \beta_{ab} = \mathrm{P}_{ba} - \mathrm{P}_{ab}.$$

If a different connection is chosen in the projective class according to (1.1), then

$$\hat{\beta}_{ab} = \beta_{ab} + \nabla_a \Upsilon_b - \nabla_b \Upsilon_a.$$

Therefore, as a 2-form β_{ab} changes by an exact form. On the other hand, the Bianchi identity implies that β_{ab} is closed. Thus, there is a well-defined de Rham cohomology class $[\beta] \in H^2(M, \mathbb{R})$ associated to any projective structure.

PROPOSITION 1.1. *The class $[\beta] \in H^2(M, \mathbb{R})$ is an obstruction to the existence of a metric connection in the given projective class.*

Proof. The Ricci tensor is symmetric for a metric connection. □

In searching for a metric connection in a given projective class, we may as well suppose that the obstruction $[\beta]$ vanishes. For the remainder of this article we suppose that this is the case and we shall consider only representative connections with symmetric Ricci tensor. In other words, all connections from now on enjoy

$$(\nabla_b \nabla_a - \nabla_a \nabla_b) X^b = (n-1)\mathrm{P}_{ab} X^b \quad \text{where } \mathrm{P}_{ab} = \mathrm{P}_{ba}. \tag{1.2}$$

A convenient alternative characterisation of such connections as follows.

PROPOSITION 1.2. *A torsion-free affine connection has symmetric Ricci tensor if and only if it induces the flat connection on the bundle of n-forms.*

Proof. If $\epsilon^{pqr\cdots s}$ has n indices and is totally skew then

$$(\nabla_a \nabla_b - \nabla_b \nabla_a)\epsilon^{pqr\cdots s} = \kappa_{ab}\epsilon^{pqr\cdots s}$$

for some 2-form κ_{ab}. But, by the Bianchi symmetry,

$$(\nabla_a \nabla_b - \nabla_b \nabla_a)\epsilon^{abr\cdots s} = -2R_{ab}\epsilon^{abr\cdots s},$$

which vanishes if and only if R_{ab} is symmetric. □

Having restricted our attention to affine connections that are flat on the bundle of n-forms, we may as well further restrict to connections ∇_a for which there is a volume form $\epsilon_{bc\cdots d}$ (necessarily unique up to scale) with $\nabla_a\epsilon_{bc\cdots d} = 0$. We shall refer to such connections as special. The freedom in special connections within a given projective class is given by (1.1) where $\Upsilon_a = \nabla_a f$ for an arbitrary smooth function f. Following [5], the full curvature of a special connection may be conveniently decomposed:

$$(\nabla_a\nabla_b - \nabla_b\nabla_a)X^c = W_{ab}{}^c{}_d X^d + \delta_a{}^c \mathrm{P}_{bd}X^d - \delta_b{}^c \mathrm{P}_{ad}X^d, \qquad (1.3)$$

where $W_{ab}{}^c{}_d$ is totally trace-free and P_{ab} is symmetric. The tensor $W_{ab}{}^c{}_d$ is known as the Weyl curvature and is projectively invariant.

2. A linear system of equations. In this section we present, as Proposition 2.1, an alternative characterisation of the Levi-Civita connection. The advantage of this characterisation is that it leads, almost immediately, to a system of linear equations that controls the metric connections within a given projective class. The precise results are Theorems 2.1 and 2.2.

PROPOSITION 2.1. *Suppose g^{ab} is a metric on M with volume form $\epsilon_{bc\cdots d}$. Then a torsion-free connection ∇_a is the metric connection for g^{ab} if and only if*

- $\nabla_a g^{bc} = \delta_a{}^b\mu^c + \delta_a{}^c\mu^b$ *for some vector field μ^a*
- $\nabla_a\epsilon_{bc\cdots d} = 0$.

Proof. Write D_a for the metric connection of g^{ab}. Then

$$\nabla_a\omega_b = D_a\omega_b - \Gamma_{ab}{}^c\omega_c \qquad (2.1)$$

for some tensor $\Gamma_{ab}{}^c = \Gamma_{ba}{}^c$. We compute

$$\epsilon^{bc\cdots d}\nabla_a\epsilon_{bc\cdots d} = -n\epsilon^{bc\cdots d}\Gamma_{ab}{}^e\epsilon_{ec\cdots d} = -n!\,\Gamma_{ab}{}^b$$

and so $\Gamma_{ab}{}^b = 0$. Similarly,

$$\nabla_a g^{bc} = \Gamma_{ad}{}^b g^{dc} + \Gamma_{ad}{}^c g^{bd}$$

and so

$$\Gamma_{ad}{}^b g^{dc} + \Gamma_{ad}{}^c g^{bd} = \delta_a{}^b\mu^c + \delta_a{}^c\mu^b. \qquad (2.2)$$

Let g_{ab} denote the inverse of g^{ab} and contract (2.2) with g_{bc} to conclude that

$$2\Gamma_{ab}{}^b = 2\mu_a \quad \text{where } \mu_a = g_{ab}\mu^b$$

and hence that $\mu^a = 0$. If we let $\Gamma_{abc} = \Gamma_{ab}{}^d g_{cd}$, then (2.2) now reads

$$\Gamma_{acb} + \Gamma_{abc} = 0.$$

Together with $\Gamma_{abc} = \Gamma_{bac}$, this implies that $\Gamma_{abc} = 0$. From (2.1) we see that $\nabla_a = D_a$, which is what we wanted to show. □

THEOREM 2.1. *Suppose ∇_a is a special torsion-free connection and there is a metric tensor σ^{ab} such that*

$$\nabla_a \sigma^{bc} = \delta_a{}^b \mu^c + \delta_a{}^c \mu^b \quad \textit{for some vector field } \mu^a. \tag{2.3}$$

Then ∇_a is projectively equivalent to a metric connection.

Proof. Consider the projectively equivalent connection

$$\hat{\nabla}_a X^b = \nabla_a X^b + \Upsilon_a X^b + \delta_a{}^b \Upsilon_c X^c \quad \text{where } \Upsilon_a = \nabla_a f$$

for some function f. If we let $\hat{\sigma}^{ab} \equiv e^{-2f} \sigma^{ab}$, then

$$\begin{aligned} \hat{\nabla}_a \hat{\sigma}^{bc} &= e^{-2f} \left(-2\Upsilon_a \sigma^{bc} + \nabla_a \sigma^{bc} + 2\Upsilon_a \sigma^{bc} + \delta_a{}^b \Upsilon_d \sigma^{dc} + \delta_a{}^c \Upsilon_d \sigma^{bd} \right) \\ &= e^{-2f} \left(\delta_a{}^b \mu^c + \delta_a{}^c \mu^b + \delta_a{}^b \Upsilon_d \sigma^{dc} + \delta_a{}^c \Upsilon_d \sigma^{bd} \right) \end{aligned}$$

and so

$$\hat{\nabla}_a \hat{\sigma}^{bc} = \delta_a{}^b \hat{\mu}^c + \delta_a{}^c \hat{\mu}^b \quad \text{where } \hat{\mu}^a = e^{-2f} \left(\mu^a + \Upsilon_b \sigma^{ab} \right). \tag{2.4}$$

Similarly, if we choose a volume form $\epsilon_{bc\cdots d}$ killed by ∇_a and let $\hat{\epsilon}_{bc\cdots d} \equiv e^{(n+1)f} \epsilon_{bc\cdots d}$, then

$$\hat{\nabla}_a \hat{\epsilon}_{bc\cdots d} = e^{(n+1)f} \left(\nabla_a \epsilon_{bc\cdots d} + \Upsilon_{[a} \epsilon_{bc\cdots d]} \right) = e^{(n+1)f} \nabla_a \epsilon_{bc\cdots d} = 0. \tag{2.5}$$

Define

$$\det(\sigma) \equiv \epsilon_{a\cdots b} \epsilon_{c\cdots d} \sigma^{ac} \cdots \sigma^{bd}$$

and compute

$$\begin{aligned} \widehat{\det}(\hat{\sigma}) &= \hat{\epsilon}_{a\cdots b} \hat{\epsilon}_{c\cdots d} \hat{\sigma}^{ac} \cdots \hat{\sigma}^{bd} \\ &= e^{2(n+1)f} e^{-2nf} \epsilon_{a\cdots b} \epsilon_{c\cdots d} \sigma^{ac} \cdots \sigma^{bd} = e^{2f} \det(\sigma). \end{aligned}$$

Therefore, if we take

$$f = -\tfrac{1}{2} \log \det(\sigma),$$

then we have arranged that $\widehat{\det}(\hat{\sigma}) = 1$. This is precisely the condition that $\hat{\epsilon}_{bc\cdots d}$ be the volume form for the metric $\hat{\sigma}^{ab}$. With (2.4) and (2.5) we are now in a position to use Proposition 2.1 to conclude that $\hat{\nabla}_a$ is the metric connection for $\hat{\sigma}^{ab}$. We have shown that our original connection ∇_a is projectively equivalent to the Levi-Civita connection for the metric $g^{ab} \equiv \det(\sigma)\, \sigma^{ab}$. □

Evidently, the equations (2.3) precisely control the metric connections within a given special projective class. Precisely, if g_{ab} is a Riemannian metric with associated Levi-Civita connection ∇_a, then

$$\hat{\nabla}_a \hat{g}^{bc} = \delta_a{}^b \hat{\mu}^c + \delta_a{}^c \hat{\mu}^b,$$

where $\hat{\nabla}_a$ is projectively equivalent to ∇_a according to (1.1) with $\Upsilon_a = \nabla_a f$ and where $\hat{g}^{bc} = e^{-2f} g^{bc}$. In other words, we have shown (cf. [7, 9]):–

THEOREM 2.2. *There is a one-to-one correspondence between solutions of* (2.3) *for positive definite* σ^{bc} *and metric connections that are projectively equivalent to* ∇_a.

3. Prolongation. Let us consider the system of equations (2.3) in more detail. It is a linear system for any symmetric contravariant 2-tensor σ^{bc}. Specifically, we may write (2.3) as

$$\text{the trace-free part of } (\nabla_a \sigma^{bc}) = 0 \tag{3.1}$$

or, more explicitly, as

$$\nabla_a \sigma^{bc} - \tfrac{1}{n+1} \delta_a{}^b \nabla_d \sigma^{cd} - \tfrac{1}{n+1} \delta_a{}^c \nabla_d \sigma^{bd} = 0.$$

According to [2], this equation is of finite-type and may be prolonged to a closed system as follows. According to (1.2) and (1.3) we have

$$(\nabla_a \nabla_b - \nabla_b \nabla_a) \sigma^{bc} = W_{ab}{}^c{}_d \sigma^{bd} + \delta_a{}^c \mathrm{P}_{bd} \sigma^{bd} - n \mathrm{P}_{ad} \sigma^{cd}.$$

On the other hand, from (2.3) we have

$$(\nabla_a \nabla_b - \nabla_b \nabla_a) \sigma^{bc} = (n+1) \nabla_a \mu^c - \nabla_b (\delta_a{}^b \mu^c + \delta_a{}^c \mu^b) = n \nabla_a \mu^c - \delta_a{}^c \nabla_b \mu^b.$$

We conclude that

$$n \nabla_a \mu^c = \delta_a{}^c \left(\nabla_b \mu^b + \mathrm{P}_{bd} \sigma^{bd} \right) - n \mathrm{P}_{ad} \sigma^{cd} + W_{ab}{}^c{}_d \sigma^{bd}$$

or, equivalently, that

$$\nabla_a \mu^c = \delta_a{}^c \rho - \mathrm{P}_{ad} \sigma^{cd} + \tfrac{1}{n} W_{ab}{}^c{}_d \sigma^{bd}, \tag{3.2}$$

for some function ρ. To complete the prolongation, we use (1.2) to write

$$(\nabla_c \nabla_a - \nabla_a \nabla_c) \mu^c = (n-1) \mathrm{P}_{ac} \mu^c$$

whereas from (3.2) we also have

$$\begin{aligned}(\nabla_c \nabla_a - \nabla_a \nabla_c) \mu^c = \nabla_c &\left(\delta_a{}^c \rho - \mathrm{P}_{ad} \sigma^{cd} + \tfrac{1}{n} W_{ab}{}^c{}_d \sigma^{bd} \right) \\ &- \nabla_a \left(n\rho - \mathrm{P}_{cd} \sigma^{cd} \right).\end{aligned}$$

Therefore,

$$\begin{aligned}(n-1) \mathrm{P}_{ac} \mu^c = \nabla_c &\left(-\mathrm{P}_{ad} \sigma^{cd} + \tfrac{1}{n} W_{ab}{}^c{}_d \sigma^{bd} \right) \\ &- (n-1) \nabla_a \rho + \nabla_a (\mathrm{P}_{cd} \sigma^{cd}).\end{aligned} \tag{3.3}$$

The terms involving Weyl curvature

$$\nabla_c(W_{ab}{}^c{}_d\sigma^{bd}) = (\nabla_c W_{ab}{}^c{}_d)\sigma^{bd} + W_{ab}{}^c{}_d\nabla_c\sigma^{bd}$$

may be dealt with by (2.3) and a Bianchi identity

$$\nabla_c W_{ab}{}^c{}_d = (n-2)(\nabla_a \mathrm{P}_{bd} - \nabla_b \mathrm{P}_{ad}).$$

We see that

$$\nabla_c(W_{ab}{}^c{}_d\sigma^{bd}) = (n-2)(\nabla_a \mathrm{P}_{bd} - \nabla_b \mathrm{P}_{ad})\sigma^{bd}$$

and (3.3) becomes

$$\begin{aligned}(n-1)\mathrm{P}_{ac}\mu^c = {} & \tfrac{n-2}{n}(\nabla_a \mathrm{P}_{bd} - \nabla_b \mathrm{P}_{ad})\sigma^{bd} - \nabla_c(\mathrm{P}_{ad}\sigma^{cd}) \\ & -(n-1)\nabla_a\rho + \nabla_a(\mathrm{P}_{cd}\sigma^{cd})\end{aligned}$$

or, equivalently,

$$\begin{aligned}\mathrm{P}_{ac}\mu^c = {} & \tfrac{2}{n}(\nabla_a \mathrm{P}_{bd} - \nabla_b \mathrm{P}_{ad})\sigma^{bd} - \tfrac{1}{n-1}\mathrm{P}_{ad}\nabla_c\sigma^{cd} \\ & -\nabla_a\rho + \tfrac{1}{n-1}\mathrm{P}_{cd}\nabla_a\sigma^{cd}.\end{aligned} \tag{3.4}$$

Again, we substitute from (2.3) to rewrite

$$\mathrm{P}_{cd}\nabla_a\sigma^{cd} - \mathrm{P}_{ad}\nabla_c\sigma^{cd} = \mathrm{P}_{cd}(\delta_a{}^c\mu^d + \delta_a{}^d\mu^c) - (n+1)\mathrm{P}_{ad}\mu^d = -(n-1)\mathrm{P}_{ad}\mu^d$$

and (3.4) becomes

$$\mathrm{P}_{ac}\mu^c = \tfrac{2}{n}(\nabla_a \mathrm{P}_{bd} - \nabla_b \mathrm{P}_{ad})\sigma^{bd} - \mathrm{P}_{ad}\mu^d - \nabla_a\rho,$$

which we may rearrange as

$$\nabla_a\rho = -2\mathrm{P}_{ab}\mu^b + \tfrac{2}{n}(\nabla_a \mathrm{P}_{bd} - \nabla_b \mathrm{P}_{ad})\sigma^{bd}.$$

Together with (2.3) and (3.2), we have a closed system, essentially as in [7, 9]:

$$\begin{aligned}\nabla_a\sigma^{bc} &= \delta_a{}^b\mu^c + \delta_a{}^c\mu^b \\ \nabla_a\mu^b &= \delta_a{}^b\rho - \mathrm{P}_{ac}\sigma^{bc} + \tfrac{1}{n}W_{ac}{}^b{}_d\sigma^{cd} \\ \nabla_a\rho &= -2\mathrm{P}_{ab}\mu^b + \tfrac{4}{n}Y_{abc}\sigma^{bc}\end{aligned} \tag{3.5}$$

where $Y_{abc} = \frac{1}{2}(\nabla_a \mathrm{P}_{bc} - \nabla_b \mathrm{P}_{ac})$, the Cotton-York tensor. The three tensors σ^{bc}, μ^b, and ρ may be regarded together as a section of the vector bundle

$$\mathcal{T} = \bigodot^2 TM \oplus TM \oplus \mathbb{R}$$

where $\bigodot$ denotes symmetric tensor product and $\mathbb{R}$ denotes the trivial bundle. We have proved:

THEOREM 3.1. *If we endow $\mathcal{T}$ with the connection*

$$\begin{pmatrix} \sigma^{bc} \\ \mu^b \\ \rho \end{pmatrix} \longmapsto \begin{pmatrix} \nabla_a\sigma^{bc} - \delta_a{}^b\mu^c - \delta_a{}^c\mu^b \\ \nabla_a\mu^b - \delta_a{}^b\rho + \mathrm{P}_{ac}\sigma^{bc} - \frac{1}{n}W_{ac}{}^b{}_d\sigma^{cd} \\ \nabla_a\rho + 2\mathrm{P}_{ab}\mu^b - \frac{4}{n}Y_{abc}\sigma^{bc} \end{pmatrix} \tag{3.6}$$

then there is a one-to-one correspondence between covariant constant sections of $\mathcal{T}$ and solutions σ^{bc} of (2.3).

4. Projective invariance. The equation (2.3) is projectively invariant in the following sense. Following [5], let $\mathcal{E}^{(ab)}(w)$ denote the bundle of symmetric contravariant 2-tensors of projective weight w. Thus, in the presence of a volume form $\epsilon_{bc\cdots d}$, a section $\sigma^{ab} \in \Gamma(M, \mathcal{E}^{(ab)}(w))$ is an ordinary symmetric contravariant 2-tensor but if we change volume form

$$\epsilon_{bc\cdots d} \mapsto \hat{\epsilon}_{bc\cdots d} = e^{(n+1)f}\epsilon_{bc\cdots d} \quad \text{for any smooth function } f,$$

then we are obliged to rescale σ^{ab} according to $\hat{\sigma}^{ab} = e^{wf}\sigma^{ab}$. Equivalently, we are saying that $\mathcal{E}^{(ab)}(w) = \bigodot^2 TM \otimes (\Lambda^n)^{-w/(n+1)}$, where Λ^n is the line-bundle of n-forms on M. The projectively weighted irreducible tensor bundles are fundamental objects on a manifold with projective structure.

PROPOSITION 4.1. *The differential operator*

$$\mathcal{E}^{(ab)}(-2) \longrightarrow \text{the trace-free part of } \mathcal{E}_a{}^{(bc)}(-2) \tag{4.1}$$

defined by (3.1) *is projectively invariant.*

Proof. This is already implicit in the proof of Theorem 2.1. Explicitly, however, we just compute from (1.1):

$$\hat{\nabla}_a\hat{\sigma}^{bc} = \nabla_a\hat{\sigma}^{bc} + 2\Upsilon_a\hat{\sigma}^{bc} + \delta_a{}^b\Upsilon_d\hat{\sigma}^{dc} + \delta_a{}^c\Upsilon_d\hat{\sigma}^{bd},$$

where $\Upsilon_a = \nabla_a f$ whilst

$$\nabla_a\hat{\sigma}^{bc} = \nabla_a(e^{-2f}\sigma^{bc}) = e^{-2f}\left(\nabla_a\sigma^{bc} - 2\Upsilon_a\sigma^{bc}\right) = \widehat{\nabla_a\sigma^{bc}} - 2\Upsilon_a\hat{\sigma}^{bc}.$$

It follows that

$$\hat{\nabla}_a\hat{\sigma}^{bc} = \widehat{\nabla_a\sigma^{bc}} + \text{trace terms},$$

which is what we wanted to show. □

In hindsight, it is not too difficult to believe that (3.1) should control the metric connections within a given projective class. There are very few projectively invariant operators. In fact, there are precisely two finite-type first order invariant linear operators on symmetric 2-tensors. One of them is (4.1) and the other is

$$\mathcal{E}_{(ab)}(4) \to \mathcal{E}_{(abc)}(4) \quad \text{given by} \quad \sigma_{ab} \mapsto \nabla_{(a}\sigma_{bc)}. \tag{4.2}$$

In two dimensions, (4.2) and (4.1) coincide. In higher dimensions, however, being in the kernel of (4.2) for positive definite σ_{ab} corresponds to having a metric g_{ab} and a totally trace-free tensor Γ_{abc} with

$$\Gamma_{abc} = \Gamma_{bac} \quad \text{and} \quad \Gamma_{abc} + \Gamma_{bca} + \Gamma_{cab} = 0$$

such that the connection

$$\omega_b \longmapsto D_a\omega_b - \Gamma_{ab}{}^c\omega_c$$

belongs to the projective class of ∇_a, where D_a is the Levi-Civita connection of g_{ab}. The available tensors Γ_{abc} for a given metric have dimension $n(n+2)(n-2)/3$.

5. Relationship to the Cartan connection. On a manifold with projective structure, it is shown in [5] how to associate vector bundles with connection to any irreducible representation of $\mathrm{SL}(n+1,\mathbb{R})$. These are the tractor bundles following their construction by Thomas [10]. Equivalently, they are induced by the Cartan connection [4] of the projective structure. The relevant tractor bundle in our case is induced by $\bigodot^2\mathbb{R}^{n+1}$ where $\mathbb{R}^{n+1}$ is the defining representation of $\mathrm{SL}(n+1,\mathbb{R})$. It has a composition series

$$\mathcal{E}^{(BC)} = \mathcal{E}^{(bc)}(-2) + \mathcal{E}^b(-2) + \mathcal{E}(-2)$$

and in the presence of a connection is simply the direct sum of these bundles. Under projective change of connection according to (1.1), however, we decree that

$$\widehat{\begin{pmatrix} \sigma^{bc} \\ \mu^b \\ \rho \end{pmatrix}} = \begin{pmatrix} \sigma^{bc} \\ \mu^b + \Upsilon_c\sigma^{bc} \\ \rho + 2\Upsilon_b\mu^b + \Upsilon_b\Upsilon_c\sigma^{bc} \end{pmatrix}. \tag{5.1}$$

Following [5], the tractor connection on $\mathcal{E}^{(AB)}$ is given by

$$\nabla_a \begin{pmatrix} \sigma^{bc} \\ \mu^b \\ \rho \end{pmatrix} = \begin{pmatrix} \nabla_a\sigma^{bc} - \delta_a{}^b\mu^c - \delta_a{}^c\mu^b \\ \nabla_a\mu^b - \delta_a{}^b\rho + \mathrm{P}_{ac}\sigma^{bc} \\ \nabla_a\rho + 2\mathrm{P}_{ab}\mu^b \end{pmatrix}.$$

Therefore, we have proved:

THEOREM 5.1. *The solutions of* (2.3) *are in one-to-one correspondence with solutions of the following system:*

$$\nabla_a \begin{pmatrix} \sigma^{bc} \\ \mu^b \\ \rho \end{pmatrix} - \frac{1}{n}\begin{pmatrix} 0 \\ W_{ac}{}^b{}_d\sigma^{cd} \\ 4Y_{abc}\sigma^{bc} \end{pmatrix} = 0. \tag{5.2}$$

COROLLARY 5.1. *There is a one-to-one correspondence between solutions of* (5.2) *for positive definite* σ^{bc} *and metric connections that are projectively equivalent to* ∇_a.
Notice that the extra terms in (5.2) are projectively invariant as they should be. Specifically, it is observed in [5] that

$$\hat{Y}_{abc} = Y_{abc} + \tfrac{1}{2} W_{ab}{}^{d}{}_{c} \Upsilon_d$$

and so

$$4\hat{Y}_{abc}\sigma^{bc} = 4Y_{abc}\sigma^{bc} + 2\Upsilon_b W_{ac}{}^{b}{}_{d}\sigma^{cd}$$

in accordance with (5.1).

It is clear from Theorem 3.1 that, generically, (2.3) has no solutions. Indeed, this is one reason why the prolonged from is so helpful. More generally, we should compute the curvature of the connection (3.6) and the form (5.2) is useful for this task. A model computation along these lines is given in [5]. In our case, the tractor curvature is given by

$$(\nabla_a\nabla_b - \nabla_b\nabla_a)\begin{pmatrix} \sigma^{cd} \\ \mu^c \\ \rho \end{pmatrix} = \begin{pmatrix} W_{ab}{}^{c}{}_{e}\sigma^{de} + W_{ab}{}^{d}{}_{e}\sigma^{ce} \\ W_{ab}{}^{c}{}_{d}\mu^d + 2Y_{abd}\sigma^{cd} \\ 4Y_{abc}\mu^c \end{pmatrix}$$

and we obtain:

PROPOSITION 5.1. *The curvature of the connection* (3.6) *is given by*

$$\begin{pmatrix} \sigma^{cd} \\ \mu^c \\ \rho \end{pmatrix} \mapsto \begin{pmatrix} W_{ab}{}^{c}{}_{e}\sigma^{de} + W_{ab}{}^{d}{}_{e}\sigma^{ce} \\ W_{ab}{}^{c}{}_{d}\mu^d + 2Y_{abd}\sigma^{cd} \\ 4Y_{abc}\mu^c \end{pmatrix} + \frac{1}{n}\begin{pmatrix} \delta_a{}^c U_b{}^d + \delta_a{}^d U_b{}^c - \delta_b{}^c U_a{}^d - \delta_b{}^d U_a{}^c \\ * \\ * \end{pmatrix},$$

where $U_b{}^d = W_{be}{}^{d}{}_{f}\sigma^{ef}$ *and* $*$ *denotes expressions that we shall not need.*

COROLLARY 5.2. *The curvature of the connection* (3.6) *vanishes if and only if the projective structure is flat.*

Proof. Let us suppose that $n \geq 3$. The uppermost entry of the curvature is given by

$$\sigma^{cd} \longmapsto \text{the trace-free part of } (W_{ab}{}^{c}{}_{e}\sigma^{de} + W_{ab}{}^{d}{}_{e}\sigma^{ce})$$

and it is a matter of elementary representation theory to show that if this expression is zero for a fixed $W_{ab}{}^{c}{}_{d}$ and for all σ^{cd}, then $W_{ab}{}^{c}{}_{d} = 0$. Specifically, the symmetries of $W_{ab}{}^{c}{}_{d}$, namely

$$W_{ab}{}^{c}{}_{d} + W_{ba}{}^{c}{}_{d} = 0, \qquad W_{ab}{}^{c}{}_{d} + W_{bd}{}^{c}{}_{a} + W_{da}{}^{c}{}_{b} = 0, \qquad W_{ab}{}^{a}{}_{d} = 0, \quad (5.3)$$

constitute an irreducible representation of $\mathrm{SL}(n,\mathbb{R})$. Hence, the submodule

$$\{W_{ab}{}^c{}_d \text{ s.t. the trace-free part of } (W_{ab}{}^c{}_e\sigma^{de} + W_{ab}{}^d{}_e\sigma^{ce}) = 0,\ \forall\, \sigma^{cd}\}$$

must be zero since it is not the whole space. We have shown that if the curvature of the connection (3.6) vanishes, then $W_{ab}{}^c{}_d = 0$. For $n \geq 3$ this is exactly the condition that the projective structure be flat. For $n = 2$, the Weyl curvature $W_{ab}{}^c{}_d$ vanishes automatically since the symmetries (5.3) are too severe a constraint. Instead, a similar calculation shows that $Y_{abc} = 0$ and this is the condition that the projective structure be flat. □

Following Mikeš [7], the dimension of the space of solutions of (2.3) is called the degree of mobility of the projective structure. Theorem 3.1 implies that the degree of mobility is bounded by $(n+1)(n+2)/2$ and Corollary 5.2 implies that this bound is achieved only for the flat projective structure. Of course, the flat projective structure may as well be represented by the flat connection $\nabla_a = \partial/\partial x^a$ on $\mathbb{R}^n$, which is the Levi-Civita connection for the standard Euclidean metric. In this case, we may use (3.5) find the general solution of (2.3):

$$\sigma^{ab} = s^{ab} + x^a m^b + x^b m^a + x^a x^b r. \tag{5.4}$$

This form is positive definite near the origin if and only if s^{ab} is positive definite. We conclude that the general projectively flat metric near the origin in $\mathbb{R}^n$ is

$$g^{ab} = \det(\sigma)\,\sigma^{ab},$$

where σ^{ab} is as in (5.4) for some positive definite quadratic form s^{ab}. In fact, these metrics are constant curvature. Rather than prove this by calculation, there is an alternative as follows. As already observed, the Weyl curvature $W_{ab}{}^c{}_d$ corresponds to an irreducible representation of $\mathrm{SL}(n,\mathbb{R})$ characterised by (5.3). In the presence of a metric g_{ab}, however, we should decompose $W_{ab}{}^c{}_d$ further under $\mathrm{SO}(n)$.

PROPOSITION 5.2. *In the presence of a metric* g_{ab}

$$\begin{aligned} W_{ab}{}^c{}_d = C_{ab}{}^c{}_d &+ \tfrac{1}{(n-1)(n-2)}\,(\delta_a{}^c\Phi_{bd} - \delta_b{}^c\Phi_{ad}) \\ &+ \tfrac{1}{n-2}\,(\Phi_a{}^c g_{bd} - \Phi_b{}^c g_{ad}) \end{aligned} \tag{5.5}$$

where $C_{ab}{}^c{}_d$ *is the Weyl part of the Riemann curvature tensor and* Φ_{ab} *is the trace-free part of the Ricci tensor.*

Proof. According to (1.3),

$$R_{ab}{}^c{}_d = W_{ab}{}^c{}_d + \delta_a{}^c \mathrm{P}_{bd} - \delta_b{}^c \mathrm{P}_{ad} \tag{5.6}$$

but the Riemann curvature decomposes according to

$$R_{abcd} = C_{abcd} + g_{ac}Q_{bd} - g_{bc}Q_{ad} + Q_{ac}g_{bd} - Q_{bc}g_{ad}, \tag{5.7}$$

where Q_{ab} is the Schouten tensor

$$Q_{ab} = \frac{1}{n-2}\Phi_{ab} + \frac{1}{2n(n-1)}Rg_{ab}.$$

Comparing (5.6) and (5.7) leads, after a short computation, to (5.5). □

COROLLARY 5.3. *A projectively flat metric is constant curvature.*

Proof. If $n \geq 3$ and the projective Weyl tensor vanishes then the only remaining part of the Riemann curvature tensor is the scalar curvature. As usual, a separate proof based on Y_{abc} is needed for the case $n = 2$. □

This corollary is usually stated as follows. If a local diffeomorphism between two Riemannian manifolds preserves geodesics and one of them is constant curvature, then so is the other. This is a classical result due to Beltrami [1].

6. Concluding remarks. Results such as Theorem 2.2 and Theorem 5.1 are quite common in projective, conformal, and other parabolic geometries. It is shown in [5], for example, that the Killing equation in Riemannian geometry is projectively invariant and its solutions are in one-to-one correspondence with covariant constant sections of the tractor bundle $\mathcal{E}_{[AB]}$ equipped with a connection that is derived from (but not quite equal to) the tractor connection. The situation is completely parallel for conformal Killing vectors in conformal geometry and, more generally, for the infinitesimal automorphisms of parabolic geometries [3]. It is well-known that having an Einstein metric in a given conformal class is equivalent to having a suitably positive covariant constant section of the standard tractor bundle $\mathcal{E}^A$ equipped with its usual tractor connection. Gover and Nurowski [6] use this observation systematically to find obstructions to the existence of an Einstein metric within a given conformal class. We anticipate a similar use for Theorem 5.1 in establishing obstructions to the existence of a metric connection within a given projective class.

REFERENCES

[1] E. BELTRAMI, *Rizoluzione del problema: riportare i punti di una superficie sopra un piano in modo che le linee geodetiche vengano rappresentate da linee rette*, Ann. Mat. Pura Appl. (1865), **7**: 185–204.

[2] T.P. BRANSON, A. ČAP, M.G. EASTWOOD, AND A.R. GOVER, *Prolongations of geometric overdetermined systems*, Int. Jour. Math. (2006), **17**: 641–664.

[3] A. ČAP, *Infinitesimal automorphisms and deformations of parabolic geometries*, preprint ESI 1684 (2005), Erwin Schrödinger Institute, available at http://www.esi.ac.at.

[4] E. CARTAN, *Sur les variétés à connexion projective*, Bull. Soc. Math. France (1924), **52**: 205–241.

[5] M.G. EASTWOOD, *Notes on projective differential geometry*, this volume.

[6] A.R. GOVER AND P. NUROWSKI, *Obstructions to conformally Einstein metrics in n dimensions*, Jour. Geom. Phys. (2006), **56**: 450–484.

[7] J. MIKEŠ, *Geodesic mappings of affine-connected and Riemannian spaces*, Jour. Math. Sci. (1996), **78**: 311–333.

[8] R. Penrose and W. Rindler, *Spinors and Space-time, Vol. 1*, Cambridge University Press 1984.
[9] N.S. Sinjukov, *Geodesic mappings of Riemannian spaces* (Russian), "Nauka," Moscow 1979.
[10] T.Y. Thomas, *Announcement of a projective theory of affinely connected manifolds*, Proc. Nat. Acad. Sci. (1925), **11**: 588–589.

EXTERIOR DIFFERENTIAL SYSTEMS WITH SYMMETRY

MARK E. FELS*

Abstract. The symmetry group of an exterior differential system is used to simplify finding integral manifolds in the case of Pfaffian systems. This leads to the definition of a moment map for Pfaffian systems.

Key words. Pfaffian systems, symmetry reduction, group invariant solutions.

AMS(MOS) subject classifications. Primary 34A26; Secondary 58A15.

1. Introduction. Let $\Delta = 0$ be a system of differential equations, and let $\mathcal{S}$ be its solution space. A symmetry group of the differential equations $\Delta = 0$ is group G which acts on the space of solutions $\mathcal{S}$. Typically a symmetry group of a differential equation is known, even though the solution space is not. A basic idea, which originated with Sophus Lie, is that it might be easier to determine $\mathcal{S}/G$ than it is to determine $\mathcal{S}$. This idea (simplifying finding solutions to differential equations using a group) is one of the principle motivating problems in studying exterior differential systems with symmetry.

To describe the quotient $\mathcal{S}/G$ we first partition $\mathcal{S}$ into orbits having inequivalent stabilizers,

$$\mathcal{S} = \mathcal{S}^G \cup \mathcal{S}^1 \cup \ldots \cup \mathcal{S}^{free}. \tag{1.1}$$

The set $\mathcal{S}^G$ corresponds to fixed points of the action of G. The solutions in $\mathcal{S}^G$ are the *G-invariant solutions*, which are also sometimes called equivariant solutions. The set $\mathcal{S}^{free}$ correspond to points in $\mathcal{S}$ on which G acts freely. Solutions in $\mathcal{S}^{free}$ have no symmetry. The intermediate terms in (1.1) are solutions which are fixed (invariant) respect to some subgroup of G. The quotient can then be partitioned,

$$\mathcal{S}/G = (\mathcal{S}^G/G) \cup (\mathcal{S}^1/G) \cup \ldots \cup (\mathcal{S}^{free}/G). \tag{1.2}$$

where $\mathcal{S}^G/G = \mathcal{S}^G$.

EXAMPLE 1.1. *Laplace's equation on* $\mathbb{R}^2 - (0,0)$,

$$u_{xx} + u_{yy} = 0$$

admits $SO(2)$ *acting on the punctured plane in the usual way*

$$\begin{pmatrix} x \\ y \end{pmatrix} \to \begin{pmatrix} \cos\theta & -\sin\theta \\ \sin\theta & \cos\theta \end{pmatrix} \begin{pmatrix} x \\ y \end{pmatrix},$$

*Department of Mathematics and Statistics, Utah State University, Logan, Utah 84322 (mark.fels@usu.edu).

as a symmetry group. If $u = f(x,y)$ is a solution to Laplace's equation, then it is easy to check that

$$\hat{f}(x,y) = f(x\cos\theta + y\sin\theta, -x\sin\theta + y\cos\theta)$$

is also a solution, and so $SO(2)$ acts on $\mathcal{S}$. The set $\mathcal{S}^G$ are the rotationally invariant solutions to Laplace's equation. These satisfy $f(x,y) = \hat{f}(x,y)$, and so f must have the form

$$f = f(\sqrt{x^2+y^2}).$$

This leads to the fundamental solution. The set $\mathcal{S}^{free}$ are the solutions to Laplace's equation without any rotational symmetry, which would be most solutions.

Lie's original way of thinking was to find a family of *quotient differential equations* $\bar{\Delta} = 0$ whose solutions would be in 1-1 correspondence with the different terms in the quotient $\mathcal{S}/G$ in Equation (1.2). A serious problem with this idea is that it is unclear whether the quotient of a differential equation is another differential equation. For example, what would the quotient of Laplace's equation by $SO(2)$ in Example 1.1 be? The difficultly here lies in the usual coordinate description of differential equations. However exterior differential systems (EDS) are a coordinate invariant way to represent differential equations and in this context the idea of a quotient differential equation can be easily realized.

A second important problem which arises in implementing a quotient is the following. Suppose that $\bar{\Delta}$ was a set of differential equations whose solutions represented one of the quotients in Equation (1.2). We then need to take a solution $\bar{s}$ to this quotient equation $\bar{\Delta}$ and construct a solution s to the original equation $\Delta = 0$. This is sometimes called the *reconstruction problem.*

In this article I will focus on how to find the quotient $\bar{\Delta}$, and what is the "reconstruction problem". I would like to list two other important problems which should be kept in mind in the theory of quotients.

1) The *inverse problem.* Given a differential equation $\delta = 0$, is there "simple" differential equation $\Delta = 0$ such that $\delta = \bar{\Delta}$?

2) How are the geometric properties of $\Delta = 0$ related to those of $\bar{\Delta} = 0$? For example suppose $\Delta = 0$ are the Euler-Lagrange equations for some Lagrangian λ. Are $\bar{\Delta}$ the Euler-Lagrange equations for some quotient Lagrangian $\bar{\lambda}$? For $\mathcal{S}^G$ the question is known as the principle of symmetric criticality [4, 1]. For $\mathcal{S}^{free}$ this is sometimes known as Lagrangian Reduction [9, 10]. Symmetric criticality and Lagrangian reduction lie at the *opposite* ends of the symmetry spectrum!

2. Symmetries and quotients of EDS.

DEFINITION 2.1. *An exterior differential system (EDS) is a differential ideal $\mathcal{I} \subset \Omega^*(M)$.*

Differential equations $\Delta = 0$ give rise to exterior differential systems (EDS). Examples are given below.

DEFINITION 2.2. *An integral manifold of $\mathcal{I}$ is an immersion $s : N \to M$ such that $s^*\mathcal{I} = 0$.*

Solutions to $\Delta = 0$ are integral manifolds to a corresponding EDS $\mathcal{I}$.

DEFINITION 2.3. *A symmetry of an exterior differential system is a diffeomorphism $\phi : M \to M$ such that $\phi^*\mathcal{I} = \mathcal{I}$. A symmetry group of $\mathcal{I}$ will be a Lie group G acting smoothly on M where each diffeomorphism $g : M \to M$ is a symmetry of $\mathcal{I}$.*

Symmetries of differential equations determine symmetries of their corresponding EDS $\mathcal{I}$. Symmetries of EDS behave like symmetries of differential equations. If ϕ is a symmetry of the EDS $\mathcal{I}$ and $s : N \to M$ is an integral manifold of $\mathcal{I}$, then

$$(\phi \circ s)^*\mathcal{I} = s^*\phi^*\mathcal{I} = s^*\mathcal{I} = 0.$$

Therefore symmetries map integral manifolds to integral manifolds. Given a symmetry group G of an EDS $\mathcal{I}$, the *symmetry group of an integral manifold* is

$$G_s = \{\ g \in G \mid gs(N) = s(N)\ \}.$$

If $G_s = G$, then N is a G-invariant integral manifold. The G-invariant solutions to a differential equation correspond to G-invariant integral manifolds.

We use the following notation. If $\{\theta^i\} \in \Omega^*(M)$ is a set of differential forms, then the differential ideal they generate is

$$\mathcal{I} = <\theta^i, d\theta^i>,$$

where $<\,,\,>$ means the algebraic ideal in $\Omega^*(M)$. If $\theta^i \in \Omega^1(M)$ (one-forms) then $\mathcal{I}$ is a *Pfaffian System*. The EDS which will be of main interest through this article are constant rank Pfaffian systems.

DEFINITION 2.4. *A (constant) rank r Pfaffian system $\mathcal{I}$ is an exterior differential system generated by the sections of a rank r subbundle $I \subset T^*M$.*

We will usually refer to the bundle I as the Pfaffian system, and we will also denote by

$$I = \{\theta^i\}$$

the subbundle of T^*M which is the point-wise span of the differential one-forms $\theta^i \in \Omega^1(M)$.

The annihilator $I^\perp \subset TM$ of a rank r Pfaffian system $I \subset T^*(M)$ is the rank $n - r$ subbundle defined point-wise by

$$I_p^\perp = \{V \in T_pM \mid \theta(V) = 0, \quad \forall\, \theta \in I_p,\ p \in M\ \}.$$

An integral manifold of a constant rank Pfaffian system I is then an immersion $s : N \to M$ such that

$$s_* T_x N \in I^{\perp}_{s(x)}.$$

Symmetries of a constant rank Pfaffian system I are diffeomorphisms $\phi : M \to M$ which preserve the subbundles I and $I^{\perp}$.

REMARK 2.1. Infinitesimal methods will be used when discussing symmetry [13].

EXAMPLE 1.1 (Continued). *Laplace's equation on* $\mathbb{R}^2 - (0,0)$, *gives rise to a rank three Pfaffian system* $I = \{\theta_u, \theta_{u_x}, \theta_{u_y}\}$ *on a seven manifold* $M_7 = (x, y, u, u_x, u_y, u_{xy}, u_{yy})$, *where*

$$\begin{aligned} \theta_u &= du - u_x dx - u_y dy, \\ \theta_{u_x} &= du_x + u_{yy} dy - u_{xy} dy, \\ \theta_{u_y} &= du_y - u_{xy} dx - u_{yy} dy. \end{aligned} \tag{2.1}$$

Solutions $u = f(x,y)$ *to Laplace's equation define integral manifolds* $s : \mathbb{R}^2 - (0,0) \to M_7$,

$$s(x,y) = \Big(x, y, u = f,\ u_x = f_x,\ u_y = f_y,\ u_{xy} = f_{xy},\ u_{yy} = f_{yy}\Big).$$

The infinitesimal generator of the prolonged $SO(2)$ *action on* M_7 *is*

$$X = x\partial_y - y\partial_x + u_x \partial_{u_y} - u_y \partial_{u_x} - 2u_{yy}\partial_{u_{xy}} + 2u_{xy}\partial_{u_{yy}}, \tag{2.2}$$

and is an infinitesimal symmetry of I *(or* $\mathcal{I}$*).*

2.1. The quotient. Let G be a Lie group acting smoothly on M, with infinitesimal generators Γ. Let $\mathbf{\Gamma} \subset TM$, be the corresponding point-wise span of elements of Γ. In the following discussion we will assume that the action of G on M if sufficiently regular so that $\mathbf{q} : M \to M/G = \bar{M}$ is a smooth submersion. With this hypothesis on the action, $\mathbf{\Gamma} \subset TM$ is a rank q subbundle where q is the dimension of the orbits. Furthermore

$$\mathbf{\Gamma} = \ker \mathbf{q}_*$$

which is also the vertical bundle for $\mathbf{q}$.

DEFINITION 2.5. *The quotient of an EDS* $\mathcal{I} \subset \Omega^*(M)$, *is the EDS* $\bar{\mathcal{I}} \subset \Omega^*(\bar{M})$ *defined by*

$$\bar{\mathcal{I}} = \{\ \bar{\theta} \in \Omega^*(\bar{M}) \mid \mathbf{q}^* \bar{\theta} \in \mathcal{I}\ \}.$$

If $\mathcal{I}$ is a constant rank Pfaffian system (with bundle $I \subset T^*M$), the *quotient of the bundle* I is the subset $\bar{I} \subset T^*\bar{M}$ given point-wise by

$$\bar{I}_{\bar{x}} = \{\ \bar{\theta} \in T^*_{\bar{x}}\bar{M} \mid \mathbf{q}^*\bar{\theta}_{\bar{x}} \in I_x \text{ where } \mathbf{q}(x) = \bar{x}\ \}.$$

This can be computed in terms of $I^\perp \subset TM$ by

$$\bar{I} = \left(\mathbf{q}_*(I^\perp)\right)^\perp .$$

Necessary and sufficient conditions that the subsets $\bar{I} \subset T^*\bar{M}$, or $\mathbf{q}_*(I^\perp) \subset T\bar{M}$ are subbundles are well known [8].

LEMMA 2.1. *The subsets $\bar{I} \subset T^*\bar{M}$, and $\mathbf{q}_*(I^\perp) \subset T\bar{M}$ are subbundles if and only if there exists a non-negative integer k such that*

$$\dim\left(I_x^\perp \cap (\ker \mathbf{q}_*)_x\right) = \dim\left(I_x^\perp \cap \mathbf{\Gamma}_x\right) = k\,, \quad \forall\, x \in M. \tag{2.3}$$

Then rank $\mathbf{q}_*(I^\perp) = n - r - k$ *and* rank $\bar{I} = r + k - q$, *where* rank $\mathbf{\Gamma} = q$.

It is handy to write out the intersection $I_x^\perp \cap \mathbf{\Gamma}_x$ in Equation (2.3) in a basis. Let $x \in M$, $\{X_a\}_{1\le a\le q}$ be a basis for $\mathbf{\Gamma}_x$ and let $\{\theta^i\}_{1\le i\le r}$ be a basis for I_x. Then

$$I_x^\perp \cap \mathbf{\Gamma}_x = \ker \theta^i(X_a), \tag{2.4}$$

where the kernel is computed on the index a.

The first issue that occurs in EDS reduction is that even when condition (2.3) is satisfied, bundle reduction and Pfaffian system reduction do not necessarily commute.

EXAMPLE 2.2. *Consider the three dimensional manifold $M = (x, u, u_x)$, with rank one Pfaffian system*

$$I = \{\ du - u_x dx\ \}, \ \text{and} \quad \mathcal{I} = < du - u_x dx,\ du_x \wedge dx >.$$

The group $G = \mathbb{R}$ acting on M by $c\cdot(x, u, u_x) = (x, u+c, u_x)$ is a symmetry of I. The projection map is $\mathbf{q}(x, u, u_x) = (x, u_x)$. The bundle quotient is

$$\mathbf{q}_*(I^\perp) = T\mathbb{R}^2, \quad \text{therefore} \quad \bar{I} = 0.$$

While the EDS quotient is,

$$\bar{\mathcal{I}} = < du_x \wedge dx >.$$

If a Pfaffian systems $\mathcal{I}$ is completely integrable then EDS reduction and bundle reduction commute [5]. For Pfaffian systems which are not completely integrable, sufficient conditions are given in [2].

The intersection condition in Equation (2.3) will be the focus for the remainder of the article. We begin with a simple application, see [5].

THEOREM 2.1. *Suppose $\mathcal{I}$ is a constant rank Pfaffian system with symmetry group G. If there exists $X_{x_0} \in I_{x_0}^\perp \cap \mathbf{\Gamma}_{x_0}$ with $X_{x_0} \neq 0$, then $e^{tX}x_0$, $X \in \mathbf{g}$ is a one-dimensional integral manifold.*

The integral manifold $e^{tX}x_0$ in this theorem is the integral curve of the infinitesimal generator X through the point x_0. Equivalently, it is the orbit of the one-parameter subgroup e^{tX} through the point x_0.

We now turn to a few examples before continuing with the development of the theory.

3. Examples.

EXAMPLE 3.3. *The Chazy equation*

$$y_{xxx} = 2yy_{xx} - 3y_x^2$$

gives rise to the completely integrable rank three Pfaffian system $I = \{\theta^1, \theta^2, \theta^3\}$ *on a four dimensional manifold* $M_4 = (x, y, y_x, y_{xx})$, *where*

$$\theta^1 = dy - y_x dx, \quad \theta^2 = dy_x - y_{xx} dx, \quad \theta^3 = dy_{xx} - (2yy_{xx} - 3y_x^2)dx \quad (3.1)$$

Solutions to the Chazy equation are integral manifolds of I. *The Pfaffian system (3.1) is invariant with respect to the (infinitesimal) action of* $SL(2)$ *on* M_4 *given by* $\Gamma = \{X_1, X_2, X_3\}$ *where*

$$X_1 = \partial_x,$$
$$X_2 = 2x\partial_x - 2y\partial_y - 4y_x\partial_{y_x} - 6y_{xx}\partial_{y_{xx}},$$
$$X_3 = -x^2\partial_x + 2(xy+3)\partial_y + 2(2xy_x + y)\partial_{y_x} + 6(y_x + xy_{xx})\partial_{y_{xx}}.$$

By Equation (2.4), there exists $X_p \in I_p^\perp \cap \mathbf{\Gamma}_p$ *with* $X_p \neq 0$ *if and only if the determinant* $\det(\theta^i(X_j)) = 0$. *This occurs at the points* $p \in M_4$ *satisfying,*

$$y_{xx} = yy_x - \frac{1}{9}y^3 \pm \frac{1}{9}(y^2 - 6y_x)^{\frac{3}{2}}. \quad (3.2)$$

For initial conditions $(x^0, y^0, y_x^0, y_{xx}^0)$ *satisfying this constraint, the (unique) solution obtained from Theorem 2.10 is the one-dimensional orbit,*

$$x = x^0 + 2t\frac{(\delta^0 + 3y_x^0 + y^0\sqrt{\delta^0})}{ty_x^0(y^0 + \sqrt{\delta^0}) - 1},$$
$$y = y^0 + 2ty_x^0\left((3t(y_x^0)^2 - y^0 + ty_x^0\delta^0)\sqrt{\delta^0} + ty_x^0 y^0\delta^0 - 3y_x^0 - \delta^0\right)$$

where $\delta^0 = (y^0)^2 - 6y_x^0$. *This is a 2 parameter family of invariant solutions to the Chazy equation, which are easily written as a graph.*

EXAMPLE 3.4. *The standard Pfaffian system for the ordinary differential equation*

$$u_{xxxxx} = \frac{5u_{xxx}(9u_{xxxx}u_{xx} - 8u_{xxx}^2)}{9u_{xx}^2},$$

is invariant with respect to the five dimensional special affine group

$$G = SA(2) = \{\ (A, b) \mid A \in SL(2, \mathbb{R}),\ b \in \mathbb{R}^2\ \},$$

with affine action on (x, u) *and then prolonged. Every solution is the orbit of a one-parameter subgroup, from which the general solution,*

$$u = c_0 + c_1 x \pm \sqrt{c_3 x^2 + c_x x + c_2}(4c_3 c_2 - c_x^2)$$

where c_0, c_1, c_2, c_3, c_x are constants, can be found. See [5] for the details.

EXAMPLE 3.5. *Consider the pseudo-Riemannian metric on $\mathbb{R}^4$,*

$$\eta = e^{-\frac{4}{3}x_4}(dx_1dx_3 - dx_2dx_2) + e^{\frac{2}{3}x_4}dx_3dx_3 + c\, dx_4dx_4$$

where if $c < 0$ the metric is Lorentzian, and if $c > 0$ the metric has split signature. The geodesic equation define a completely integrable rank eight Pfaffian system I on a nine-dimensional manifold $M_9 = \{(t, x_i, \dot{x}_i),\ 1 \leq i \leq 4\}$ given by

$$\begin{aligned} I = \Big\{ & dx_i - \dot{x}_i dt,\quad d\dot{x}_1 - \frac{2}{3}\dot{x}_4(2\dot{x}_1 - 3\dot{x}_3 e^{2x_4})dt,\quad d\dot{x}_2 - \frac{4}{3}\dot{x}_2\dot{x}_4 dt, \\ & d\dot{x}_3 - \frac{4}{3}\dot{x}_3\dot{x}_4 dt,\quad d\dot{x}_4 + \frac{1}{3c}\left(e^{-\frac{4}{3}x_4}(4\dot{x}_1\dot{x}_3 - 2\dot{x}_2^2) - e^{\frac{2}{3}x_4}\dot{x}_3^2\right)dt \Big\}. \end{aligned} \tag{3.3}$$

The geodesics are integral manifolds. The Pfaffian system (3.3) is invariant with respect to time translations and the induced action of the isometry group which has Killing vector-fields

$$\begin{aligned} & X_1 = \partial_{x_1},\ X_2 = \partial_{x_2},\ X_3 = \partial_{x_3},\ X_4 = x_2\partial_{x_1} + x_3\partial_{x_2}, \\ & X_5 = 5x_1\partial_{x_1} + 2x_2\partial_{x_2} - x_3\partial_{x_3} + 3\partial_{x_4}. \end{aligned}$$

We work at the point $p \in M_9$ given by

$$t = 0,\ \mathbf{x} = (0,0,0,0),\ \dot{\mathbf{x}} = (\frac{k}{4}, 0, k, 0), \tag{3.4}$$

where $k \neq 0$. The vector-field $X \in \Gamma$,

$$X = \partial_t + \frac{k}{4}X_1 + kX_3 = \partial_t + \frac{k}{4}\partial_{x_1} + k\partial_{x_3}$$

satisfies $\theta(X) = 0$, $\forall\ \theta \in I_p$ where p is the point (3.4). The integral curve of X in M through the point (3.4) is

$$x_1 = \frac{k}{4}t,\ x_2 = 0,\ x_3 = kt,\ x_4 = 0, \tag{3.5}$$

which by Theorem 2.10 is an integral manifold. The curve (3.5) is a geodesic which is the orbit of a one parameter subgroup corresponding to X.

REMARK 3.1. Geodesics which are orbits of the isometry group are called homogeneous geodesics. Homogeneous geodesics always exist for homogeneous Riemannian manifolds [11], but it is an open question whether every homogeneous pseudo-Riemannian manifold admits a homogeneous geodesic.

EXAMPLE 3.6. *Every geodesic on a Riemannian symmetric space is homogeneous.*

EXAMPLE 3.7. *The contact system on $J^2(\mathbb{R},\mathbb{R}^2)$ is a rank four Pfaffian system $I = \{\theta_x, \theta_y, \theta_{\dot{x}}, \theta_{\dot{y}}\}$ on a seven dimensional manifold $M_7 = (t, x, y, \dot{x}, \dot{y}, \ddot{x}, \ddot{y})$ where*

$$\theta_x = dx - \dot{x}dt,\ \theta_y = dy - \dot{y}dt,\ \theta_{\dot{x}} = d\dot{x} - \ddot{x}dt,\ \theta_{\dot{y}} = d\dot{y} - \ddot{y}dt. \tag{3.6}$$

Any prolonged graph $(x(t), y(t))$ is an integral manifold. The action of the oriented Euclidean group $E(2)^+$,

$$\begin{aligned} \begin{pmatrix} \hat{x} \\ \hat{y} \end{pmatrix} &= \begin{pmatrix} \cos\theta & -\sin\theta \\ \sin\theta & \cos\theta \end{pmatrix} \begin{pmatrix} x \\ y \end{pmatrix} + \begin{pmatrix} a \\ b \end{pmatrix}, \\ \begin{pmatrix} \hat{\dot{x}} \\ \hat{\dot{y}} \end{pmatrix} &= \begin{pmatrix} \cos\theta & -\sin\theta \\ \sin\theta & \cos\theta \end{pmatrix} \begin{pmatrix} \dot{x} \\ \dot{y} \end{pmatrix}, \\ \begin{pmatrix} \hat{\ddot{x}} \\ \hat{\ddot{y}} \end{pmatrix} &= \begin{pmatrix} \cos\theta & -\sin\theta \\ \sin\theta & \cos\theta \end{pmatrix} \begin{pmatrix} \ddot{x} \\ \ddot{y} \end{pmatrix} \end{aligned} \tag{3.7}$$

is a symmetry group of I. The only possible integral curves which are orbits satisfy $\dot{x} = \dot{y} = 0$, and are single points. (Re-parameterization of t is not being allowed as part of the symmetries.)

4. Quotients.

4.1. Invariant integral manifolds. In order to compute the quotient of a constant rank Pfaffian system I, we need to partition M into G-invariant subsets on which we can control the behavior of the quotient. One key is the intersection condition

$$I_x^\perp \cap \mathbf{\Gamma}_x$$

from Equation (2.3) of Lemma 2.1, or (2.4). Let $K \subset M$ be the G-invariant subset

$$K = \{\ x \in M \mid \mathbf{\Gamma}_x \subset I_x^\perp\ \}. \tag{4.1}$$

This is the set of points $x \in M$ where every form $\theta \in I_x$ vanishes on every $X_x \in \mathbf{\Gamma}_x$. It is also the set of points where the matrix in (2.4) has the smallest possible rank (zero).

Assume $\iota : K \to M$ is an embedded submanifold and that $T_xK \cap I_x^\perp$ has constant dimension independent of x. Then $\mathcal{I}_K = \iota^*\mathcal{I}$ is a constant rank Pfaffian system. Further assume that the action of G on K is sufficiently regular so that $\mathbf{q} : K \to K/G$ is a smooth submersion. At points $x \in K$, $\mathbf{\Gamma}_x \subset (I_{K,x})^\perp$, and so by Lemma 2.1 the quotient $\bar{I}_K$ has the same rank as I_K. It is also easy to show that bundle reduction and Pfaffian system reduction commute in this case, and therefore $\bar{\mathcal{I}}_K$ is constant rank Pfaffian system.

The Pfaffian system $\bar{I}_K$ determines the "G-invariant solutions".

THEOREM 4.1. *Let $\bar{N} \subset M$ be an embedded integral manifold of $\bar{I}_K$. Then $N = \mathbf{q}^{-1}(\bar{N}) \subset M$ is a G-invariant integral manifold of I.*

Proof. The manifold N is clearly G-invariant, so we only need to show it is an integral manifold. Let $x \in N$, $X \in T_xN$, and $\bar{x} = \mathbf{q}(x)$, $\bar{X} = \mathbf{q}_*X$. Note $\bar{x} \in \bar{N}$ and $\bar{X} \in T_{\bar{x}}\bar{N}$.

Choose an open set $\bar{U} \subset \bar{K}$ containing $\bar{x}$, and a cross-section $\sigma : \bar{M} \to M$ with $\sigma(\bar{x}) = x$. Then

$$X = \sigma_*\bar{X} + V$$

for some $V \in \mathbf{\Gamma}_x$ (the vertical bundle at x). Evaluating on $\theta \in \mathcal{I}$,

$$\begin{aligned}\theta(X) &= \theta(\sigma_*\bar{X}_{\bar{x}} + V)\\ &= \sigma^*\theta(\bar{X}) + \theta(V)\\ &= 0.\end{aligned}$$

The first term vanishes because $\sigma^*\theta \in \bar{\mathcal{I}}_K$, and $\bar{N}$ is an integral manifold. The second term vanishes because we are at point of K (4.1). □

A few important remarks about this theorem are appropriate.

REMARK 4.1. The reconstruction problem is algebraic. The inverse image process in the theorem provides the reconstruction.

REMARK 4.2. The converse of this theorem is also true. Every invariant integral manifold factors though the set K, and projects to an integral manifold of $\bar{I}_K$.

REMARK 4.3. Integral manifolds of $\mathcal{I}_K$ can always be enlarged (locally) to be invariant.

REMARK 4.4. The set K is the subset of M on which $\mathbf{\Gamma}$ are Cauchy-characteristics for $\mathcal{I}_K$. Perhaps this makes Theorem 4.1 not so surprising.

EXAMPLE 1.1 (Continued). *For Laplace's equation with Pfaffian system (2.1), we determine the set K. With the forms in (2.1) and X in (2.2) we get*

$$\theta^\alpha(X) = (yu_x - xu_y, -u_y - xu_{xy} - uu_{yy}, u_x - xu_{yy} + yu_{xy})\,.$$

The set K is then given by $\theta^\alpha(X) = 0$. Solving these equations we find K is four dimensional and choosing coordinates $K = (x, y, u, a)$, the inclusion $\iota : K \to M_7$ is

$$\left(x{=}x, y{=}y, u{=}u,\ u_x{=}\ ax,\ u_y{=}\ ay,\ u_{xy}{=}-\frac{2xya}{x^2{+}y^2}, u_{yy}{=}\ \frac{(x^2{-}y^2)a}{x^2{+}y^2}\right).$$

The one-forms in $\mathcal{I}$ pullback by ι to give,

$$\begin{aligned}\iota^*\theta_u &= du - axdx - aydy,\\ \iota^*\theta_{u_x} &= xda + \frac{2x^2a}{x^2+y^2}dx + \frac{2xya}{x^2+y^2}dy,\\ \iota^*\theta_{u_y} &= yda + \frac{2xya}{x^2+y^2}dx + \frac{2y^2a}{x^2+y^2}dy.\end{aligned}$$

Therefore I_K *is rank two. The quotient* $K/G = \{ (r,u,a) \,|\, r > 0 \}$ *is three dimensional and the rank two quotient Pfaffian system is,*

$$\bar{I}_K = \{ \ du - ardr, \quad rda + 2adr \ \}.$$

The submanifolds

$$s(r) = (\, r,\, c_1 + c_2 \log r,\, c_2 r^{-2} \,) \tag{4.2}$$

are integral manifolds of $\bar{I}_K$. *By Theorem 4.1, the inverse image of (4.2) leads to the invariant solutions*

$$\left(x,\, y,\, u = c_1 + \frac{c_2}{2} \log(x^2 + y^2),\, a = c_2(x^2 + y^2)^{-1} \right).$$

4.2. The transverse set. A second G-invariant subset of M is the transverse set

$$M^t = \{ \ x \in M \mid I_x^\perp \cap \mathbf{\Gamma}_x = 0 \ \}. \tag{4.3}$$

This is the set of points in M where the rank of the matrix in (2.4) is as large as possible.

Suppose M^t is an embedded submanifold and that G acts regularly on M^t. Let

$$\mathcal{I}_{M^t} = \mathcal{I}|_{M^t}$$

be the restriction of $\mathcal{I}$ to M^t. The transversality condition $I_x^\perp \cap \mathbf{\Gamma}_x = 0$ implies that $\mathcal{I}_{M^t}$ is a constant rank Pfaffian system with the same rank as $\mathcal{I}$. Unlike the case for the invariant integral manifolds in section 4.1, the quotient $\bar{\mathcal{I}}_{M^t}$ is not necessarily a Pfaffian system.

The set M^t was studied in detail in [2]. We recall a few things from that reference. First, the integral manifolds of $\mathcal{I}_{M^t}$ have essentially no continuous symmetry. Second, if $s : N \to M$, an integral manifold of $\mathcal{I}_{M^t}$, then $\mathbf{q} \circ s : N \to \bar{M}^t$ is an integral manifold of $\bar{\mathcal{I}}_{M^t}$. (The immersion property still holds).

A generalization of Proposition 6.1 from [2], solves the reconstruction problem.

THEOREM 4.2. *Let* $\bar{N} \to \bar{M}^t$ *be an embedded integral manifold of* $\bar{\mathcal{I}}_{M^t}$. *Then* $\mathcal{I}|_{\mathbf{q}^{-1}(\bar{N})}$ *is completely integrable, and the leaves are integrable manifolds of* $\mathcal{I}$.

An immediate consequence of this theorem is that the integral manifolds of $\mathcal{I}$ are surjective (locally) by $\mathbf{q}$ onto the integral manifolds of $\bar{\mathcal{I}}_{M^t}$.

There is a particularly nice geometric way to think about the reconstruction problem when the action of G on M^t is free. Starting with an integral manifold $s : \bar{N} \to \bar{M}^t$ of $\bar{\mathcal{I}}_{M^t}$, let $\hat{s} : \bar{N} \to M$ be any cover of $\bar{N}$. Any other cover $s : \bar{N} \to M$ of $\bar{s}$ is of the form

$$s(t) = \mu(A(t), \hat{s}(t))$$

where $A : \bar{N} \to G$ is unique. If we require $s(t)$ to be an integral manifold of $\mathcal{I}$, then $A(t)$ satisfies a generalized equation of Lie type. Equations of Lie type are differential equations on Lie groups which have many applications and interesting properties. For example, the equations are integrable by quadratures for (simply connected) solvable Lie groups.

EXAMPLE 3.7 (Continued). *The infinitesimal generators of the action of the oriented Euclidean group $E(2)^+$ in (3.7) on the 7 dimensional manifold $J^2(\mathbb{R}, \mathbb{R}^2)$ are*

$$\Gamma = \text{ span } \{ \ \partial_x, \ \partial_y, \ x\partial_y - y\partial_x + \dot{x}\partial_{\dot{y}} - \dot{y}\partial_{\dot{x}} + \ddot{x}\partial_{\ddot{y}} - \ddot{y}\partial_{\ddot{x}} \ \}.$$

Using the forms in Equation (3.6) for I, the transverse subset M^t in Equation (4.3) for the $E^+(2)$ action is

$$M^t = \{p \in J^2(\mathbb{R}, \mathbb{R}^2) \mid (\dot{x}, \dot{y}) \neq (0,0)\}.$$

The group $E(2)^+$ acts freely on M^t and the quotient is $\bar{M}^t = M^t/G$ is 4 dimensional, $\bar{M}^t = \{(t, v, k_1, k_2), v \neq 0\}$. The quotient EDS is

$$\bar{\mathcal{I}}_{M^t} = < dv - k_2 dt, dk_2 \wedge dt, k_1 dk_1 \wedge dt >,$$

which is not a Pfaffian system. A typical integral manifold for $\bar{\mathcal{I}}_{M^t}$ is

$$\bar{s}(t) = (t, v = v(t), k_2 = \frac{dv}{dt}, k_1 = k(t)), \ \ v(t) \neq 0.$$

An integral manifold in M^t which projects to $\bar{s}$ is of the form

$$s(t) = \mu(A(t), \sigma \circ \bar{s}(t))$$

where $\sigma(t, v, k_1, k_2) = (t, 0, 0, 0, v, k_1, k_2)$, and $A : \mathbb{R} \to E(2)^+$ satisfies

$$\frac{da}{dt} = -v(t) \sin\theta(t) = 0, \quad \frac{db}{dt} = v(t)\cos\theta(t), \quad \frac{d\theta}{dt} = -\frac{k_1(t)}{v(t)}.$$

This is an equation of Lie type for the curve $\alpha : \mathbb{R} \to \mathbf{g}$,

$$\alpha(t) = \left(0, -v(t), \frac{k_1(t)}{v(t)}\right).$$

REMARK 4.5. The example above can be generalized to $J^k(\mathbb{R}, G/H)$ with the following interpretation. Compute the quotient of the contact structure $\mathcal{I}$ on $J^k(\mathbb{R}, G/H)/G$ on the transverse set for k sufficiently large. Let $\bar{s}$ be an integral manifold to the quotient system. By finding an integral manifold to $\mathcal{I}$ projecting to $\bar{s}$, we have solved the prescribed "curvature" problem for curves in a homogeneous space. The curve $\bar{s}$ is the prescribed curvature. The reconstruction is done (in general) by solving an equation of Lie type on the group G.

EXAMPLE 1.1 (Continued). *Laplace's equation with the basis for I in Equation (2.1), and infinitesimal generator (2.2), the matrix $\theta^\alpha(X)$ is*

$$(yu_x - xu_y, -u_y - xu_{xy} - uu_{yy}, u_x - xu_{yy} + yu_{xy}) .$$

On the $SO(2)$-invariant set $M_0^t = \{yu_x - xu_y \neq 0\}$, which is a subset of M^t, the matrix $\theta^\alpha(X)$ is full rank. The quotient is six dimensional $M_0^t/SO(2) = \{(r,v,p,q,s,t) \mid r > 0\}$ with projection map

$$\begin{aligned} r &= \frac{1}{2}\log(x^2+y^2),\ v = u,\ p = xu_x + yu_y,\ q = xu_y - yu_x, \\ s &= \frac{xu_y - yu_x + 2yxu_{yy} + (x^2-y^2)u_{xy}}{xu_y - yu_x}, \\ t &= \frac{u_{yy}(x^2-y^2) - 2xyu_{xy} - xu_x - yu_y}{xu_y - yu_x}. \end{aligned}$$

The quotient $\bar{I}_{M_0^t}$ is the rank two Pfaffian system

$$\bar{I}_{M_0^t} = \{\ dp - sdv + (tq+ps)dr,\ dq - tdu + (tp - sq)dr\ \}.$$

The submanifold

$$(\ r = r,\ v = v,\ p = \frac{v}{r},\ q = r,\ s = r^{-1},\ t = 0\ )$$

is an integral manifold of $\bar{I}_{M_0^t}$.

The reconstruction problem leads to the completely integrable system of partial differential equations for $\theta(r,v)$,

$$r^2\partial_r\theta = -v, \quad r\partial_v\theta = 1.$$

The solution is $\theta = vr^{-1} + c_0$. This leads to the integral manifolds (as a graph)

$$u = \frac{1}{2}\log(x^2+y^2)(\arctan\frac{y}{x} - c),$$

which are not $SO(2)$ invariant.

REMARK 4.6. It is a good exercise to compute the equation of Lie type on $SL(2)$ for the Chazy equation on the set of transverse initial conditions. The transverse initial conditions will be the complement to those in Equation (3.2).

EXAMPLE 4.8. *In this last example, we demonstrate an inverse problem. Consider the "Cartan-Hilbert equation"*[1] *in the form of the rank three Pfaffian system $I = \{\theta^1, \theta^2, \theta^3\}$ on the five dimensional manifold $M_5 = (z_1, z_2, z_3, t_1, t_2)$, where*

$$\theta^1 = \frac{1}{2}dz_1 - t_2dt_1, \quad \theta^2 = dz_2 - \frac{1}{2}t_1^2dt_2, \quad \theta^3 = dz_3 + \frac{1}{2}t_2^2dt_1. \tag{4.4}$$

[1] The usual form of Cartan Hilbert equation is $z' = (y'')^2$.

The derived flag for this Pfaffian system is (3, 2, 0), and the Lie algebra of the symmetry group is the split form of G_2. The integral manifolds are easily determined with $t = t_1$ to be,

$$z_1 = f(t), \ t_2 = \frac{1}{2}\frac{df}{dt}, \ z_2 = F_1(t), \ z_3 = F_2(t)$$

where F_1, F_2 satisfy,

$$\frac{dF_1}{dt} = \frac{1}{4}t^2\frac{d^2f}{dt^2}, \quad \frac{dF_2}{dt} = -\frac{1}{8}\left(\frac{df}{dt}\right)^2. \tag{4.5}$$

The group $G = \mathbb{R}^3$ acting on M_5,

$$(a, b, c) \cdot (z_1, z_2, z_3, t_1, t_2) = (z_1 + a, z_2 + b, z_3 + c, t_1, t_2) \tag{4.6}$$

is a symmetry group of I_1.

Now let $I = I_1 \oplus I_2$ be the direct sum two copies of the "Cartan-Hilbert" Pfaffian system (4.4), on the ten-dimensional manifold $M_5 \times M_5$ where,

$$I_2 = \left\{\frac{1}{2}dw_1 - s_2 ds_1, \ ds_2 - \frac{1}{2}s_1^2 ds_2, \ ds_3 + \frac{1}{2}s_2^2 ds_1\right\}$$

is on the second five-dimensional manifold $M_5 = (w_1, w_2, w_3, s_1, s_2)$. Let $G = \mathbb{R}^3$ from (4.6) act diagonally on $M_5 \times M_5$ by

$$\mathbf{a}(\mathbf{z}, \mathbf{t}) \times (\mathbf{w}, \mathbf{s}) = (\mathbf{z} + \mathbf{a}, \mathbf{t}) \times (\mathbf{w} + \mathbf{a}, \mathbf{s})$$

where $\mathbf{a} \in \mathbb{R}^3$. The quotient $(M_5 \times M_5)/G$ is a seven dimensional manifold $M_7 = (\bar{z}_1, \bar{z}_2, \bar{z}_3, t_1, t_2, s_1, s_2)$, and the quotient map $\mathbf{q} : M_5 \times M_5 \to M_7$ is

$$\mathbf{q} : (\mathbf{z}, \mathbf{t}) \times (\mathbf{w}, \mathbf{s}) = (\bar{\mathbf{z}} = \mathbf{z} - \mathbf{w}, \mathbf{t}, \mathbf{s}),$$

where $\bar{z}_i = z_i - w_i, 1 \le i \le 3$. The quotient Pfaffian system $(I_1 \oplus I_2)/G = \bar{I}$ is easy to compute and is the rank three Pfaffian system,

$$\bar{I} = \left\{\frac{1}{2}d\bar{z}_1 - t_2 dt_1 + s_2 ds_1, \ d\bar{z}_2 - \frac{1}{2}t_1^2 dt_2 + \frac{1}{2}s_1^2 ds_2, d\bar{z}_3 + \frac{1}{2}t_2^2 dt_1 - \frac{1}{2}s_2^2 ds_1\right\}.$$

By making a change of coordinates

$$\bar{z}_1 = \frac{4(u_y - yu_{yy} - xu_{xy})}{u_{yy}}, \quad \bar{z}_2 = \frac{6u_{yy}^3 u_x - x - 3xu_{xy}^2 u_{yy}^2}{u_{yy}^3},$$

$$\bar{z}_3 = \frac{x^2(1 + 3u_{yy}^2 u_{xy}^2)}{3u_{yy}^3} + 2(xu_x - u) + \frac{u_y(u_y - 2xu_{xy})}{u_{yy}},$$

$$t_1 = \frac{u_{xy}u_{yy} - 1}{u_{yy}}, \quad t_2 = \frac{xu_{xy}u_{yy} - u_y u_{yy} - x}{u_{yy}},$$

$$s_1 = \frac{u_{xy}u_{yy} + 1}{u_{yy}}, \quad s_2 = \frac{xu_{xy}u_{yy} - u_y u_{yy} + x}{u_{yy}},$$

we get

$$\bar{I} = \left\{ du - u_x dx - u_y dy,\ du_x + \frac{1}{3u_{yy}^3} dx - u_{xy} dy,\ du_y - u_{xy} dx - u_{yy} dy \right\}. \quad (4.7)$$

This is the standard Pfaffian system for the non-Monge-Ampere partial differential equation,

$$3u_{xx}u_{yy}^3 + 1 = 0. \quad (4.8)$$

By taking integral manifolds to I_1 and I_2, the map $\mathbf{q}$ produces the general solution to this non-Monge-Ampere equation by a non-linear superposition of solutions to the Cartan-Hilbert system. The solution is given implicitly by

$$x = \frac{(g' - \dot{f})}{2(t - s)}, \quad y = \frac{1}{8}(\dot{f} + g')(t - s) + \frac{1}{4}(g - f)$$

$$u = \frac{1}{2}(G_2 - F_2) - \frac{(tg' - s\dot{f})^2}{24(t - s)} - \frac{(sg' + t\dot{f})((2s - t)g' + (2t - s)\dot{f})}{48(t - s)} + \frac{ts\dot{f}g'}{12(t - s)} - \frac{(F_1 - G_1)(g' - \dot{f})}{4(t - s)}$$

where

$$(f(t_1), F_1(t_1), F_2(t_1)), \quad \text{and} \quad (g(s_1), G_1(s_1), G_2(s_1))$$

are integral manifolds of the corresponding system (4.5).

REMARK 4.7. The Pfaffian system in Equation (4.7) for the non Monge-Ampere partial differential equation in the plane (4.8) is the quotient of two fairly simple Pfaffian systems. The quotient allows us to find the general solution to the partial differential equation (4.8). The Pfaffian system (4.7) is called *Darboux Integrable.* It can be shown that a Darboux integrable EDS can be given explicitly by a non-linear superposition (or G-quotient) of two "simple" EDS. Darboux integrability also occurs for systems of equations, such as the harmonic map and the Toda Molecule equation. See [3] for more details.

5. The moment map. The procedure in Sections 4.1 and 4.2 for finding integral manifolds involves two steps. The first step is to restrict the Pfaffian system to a G-invariant set (K or M^t). The second step is to compute the quotient Pfaffian system on the invariant set.

This two step process is very similar to symplectic [13] or contact reduction [12] where a moment map is used. In the reduction process we have outlined, there is a moment map which is a direct generalization of the moment map in contact geometry (a contact manifold is a particular rank one Pfaffian system).

We begin with a local description of this map. Suppose I is constant rank r Pfaffian and $\{\theta^i\}_{1\leq i\leq r}$ is a basis of local sections. The moment $\mu : M \to \mathbb{R}^r \otimes \mathbf{g}^*$ given by

$$\theta^i(X)\,, \quad X \in \mathbf{g}.$$

In this equation we have identified $\mathbf{g}$ with the infinitesimal generators Γ. The set K in Section 4.1 is then the zero-set of the moment map, while M^t is the full rank set for the moment map. The reduction process we have described is then similar to that for symplectic or contact reduction - restriction then quotient.

A global description of μ can be given [12]. Let $L = TM/I^\perp$ be the quotient vector-bundle, and let $\Sigma : TM \to TM/I^\perp$ be the vector-bundle projection map. The moment map μ is a section of $Hom(\mathbf{g}, L) = \mathbf{g}^* \otimes L$ given by

$$\mu_x(X) = \Sigma(X_x) \quad X \in \mathbf{g}.$$

The argument in [12] for contact manifolds proves the following theorem.

THEOREM 5.1. *The moment map $\mu : M \to Hom(\mathbf{g}, L)$, is equivariant.*

Acknowledgement. The author would like to thank Ian Anderson for numerous helpful suggestions. The Maple package *Vessiot* available at *www.math.usu.edu/˜fg_mp* and developed by Ian Anderson was used in the examples.

REFERENCES

[1] I.M. ANDERSON AND M.E. FELS, *Symmetry reduction of variational bicomplexes and the principle of symmetric criticality*, Amer. J. Math. (1997), **3**: 609–670.
[2] I.M. ANDERSON AND M.E. FELS, *Exterior Differential Systems with Symmetry*, Acta. Appl. Math. (2005), **87**(1): 3–31.
[3] I.M. ANDERSON, M.E. FELS, AND P.J. VASSILIOU, *Darboux Integrability*, In preparation.
[4] I.M. ANDERSON, M.E. FELS, AND C.G. TORRE, *Group invariant solutions without transversality and the principle of symmetric criticality*, CRM Proceedings and Lecture Notes, 2001, **29**: 95–108.
[5] M.E. FELS, *Integrating Scalar Ordinary Differential Equations with Symmetry Revisited.*, Found. Comput. Math., to appear.
[6] M.E. FELS, *A moment map for Pfaffian systems*, In preparation.
[7] R.L. BRYANT, S.S. CHERN, R.B. GARDNER, H.L. GOLDSCHMIDT, AND P.A. GRIFFITHS, *Exterior Differential Systems*, Spinger-Verlag, 1991.
[8] D. HUSEMOLLER, *Fibre Bundles*, GTM 20, Springer-Verlag, 1990.
[9] M. CASTRILLON LOPEZ, P.L. GARCIA, AND T.S. RATIU, *Euler-Poincaré reduction on principal bundles*, Lett. Math. Phys. (2001), **58**(2): 167–180.
[10] I.A. KOGAN AND P.J. OLVER, *Invariant Euler-Lagrange equations and the invariant variational bicomplex*, Acta. Appl. Math. (2003), **76**(2): 137–193.

[11] O. Kowalski and J. Szenthe, *On the existence of homogeneous geodesics in homogeneous Riemannian manifolds*, Geometriae Dedicata (2000), **1**(2): 209–214.
[12] F. Loose, *Reduction in contact geometry*, J. Lie Theory (2001), **11**: 9–22.
[13] P.J. Olver, *Applications of Lie Groups to Differential Equations*, Springer-Verlag, 1998.

PARTIALLY INVARIANT SOLUTIONS TO IDEAL MAGNETOHYDRODYNAMICS

SERGEY V. GOLOVIN*

Abstract. We present two partially invariant solutions to ideal magnetohydrodynamics equations. The solutions describe generalizations of the classical one-dimensional fluid motions with planar and spherical waves. In both cases an overdetermined system of determining equations is reduced to involutive system of equations with two independent variables and a finite relation. Clear geometrical interpretation of the finite relation allows describing picture of the flow in the whole 3D space.

Key words. Symmetry, partially invariant solution, overdetermined system, magnetohydrodynamics, trajectory, magnetic field line.

AMS(MOS) subject classifications. 76W05, 76M60, 35C05, 35N10.

1. Introduction. Euclidean group in a three-dimensional space is a common part of the admissible group for majority of mathematical models of fluid mechanics. Its Lie algebra L_6 of infinitesimal generators has the following representation:

$$\begin{aligned} &X_1 = \partial_x, \quad X_2 = \partial_y, \quad X_3 = \partial_z, \\ &X_4 = z\partial_y - y\partial_z + w\partial_v - v\partial_w, \\ &X_5 = x\partial_z - z\partial_x + u\partial_w - w\partial_u, \\ &X_6 = y\partial_x - x\partial_y + v\partial_u - u\partial_v. \end{aligned} \tag{1.1}$$

Here $\mathbf{u} = (u, v, w)$ is some vector field (velocity, magnetic, etc.) and $\mathbf{x} = (x, y, z)$ is a position vector of a particle in $\mathbb{R}^3$. The optimal system of subalgebras ΘL_6 is given in the table below [11].

#	Basis	Nor	Type of invariant solution
6.1	$X_1, \ldots, X_6$	=	
4.1	X_1, X_2, X_3, X_4	=	
3.1	X_4, X_5, X_6	=	1D, spherical waves
3.2	X_1, X_2, X_3	6.1	simple waves
3.3	$X_2, X_3, \alpha X_1 + X_4$	4.1	
2.1	X_1, X_4	=	1D, cylindrical waves
2.2	X_2, X_3	4.1	1D, planar waves
1.1	X_1	4.1	2D
1.2	$\alpha X_1 + X_4$	4.1	helical

*Department of Mathematics & Statictics, Queen's University, Canada K7L 3N6; Lavrentyev Institute of Hydrodynamics, Novosibirsk, Russia 630090.

Each row in the table contains a number of a subalgebra, its basis and normalizer in L_6 and also the type of the group-invariant solution, which is generated by the subalgebra. Symbol '=' in the 'Nor' column denotes self-normalized subalgebras; α is an arbitrary constant. The invariant solutions, generated by the one- and two-dimensional subalgebras are well-known. Subalgebra $L_{3.1}$ corresponds to the group of rotations $SO(3)$. It generates the well-known one-dimensional solution with spherical waves as a singular invariant solution [1]. Subalgebra $L_{3.2}$ generates only a trivial constant invariant solution, however, it is responsible for another classical solution, namely, for the so-called simple waves in fluid. These are waves of compression or rarefaction, which spread with the characteristic speed in an ideal gas or plasma. We are going to present partially invariant solutions, generated by subalgebras $L_{3.3}|_{\alpha=0} = \{X_2, X_3, X_4\}$ and $L_{3.1} = \{X_4, X_5, X_6\}$. The subalgebras correspond to the isometry groups of a two-dimensional plane and sphere. Neither subalgebra satisfies the necessary condition of existence of an invariant solution [1, 2]. We refer the reader to [1] for the theory of partial invariance and partially invariant solutions.

According to the general theory, the representation of a partially invariant solution is obtained by calculation of the finite invariants of the Lie groups and establishing functional dependencies between the invariants. This leads to the following representations of the partially invariant vector field:

Flow with planar waves

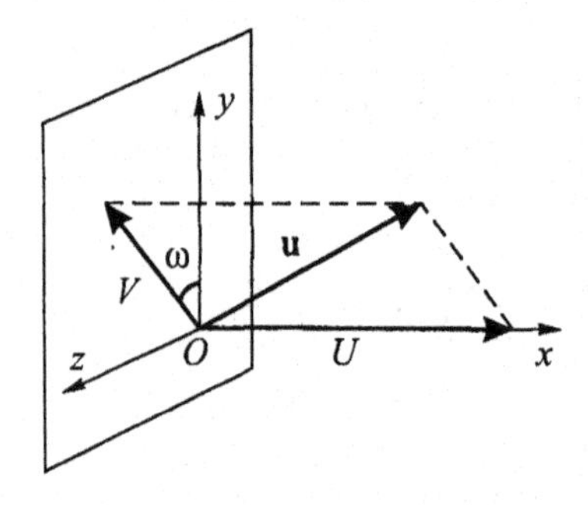

$$U = U(x), \quad V = V(x),$$
$$\omega = \omega(x, y, z).$$

Group of invariance:

$$\{X_2, X_3, X_4\}.$$

Flow with spherical waves

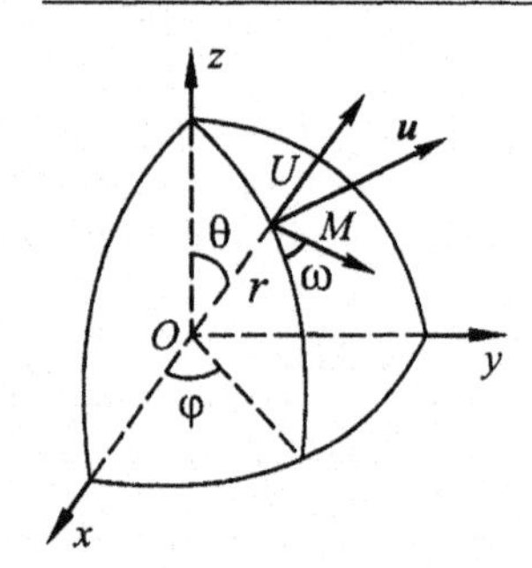

$$U = U(r), \quad M = M(r),$$
$$\omega = \omega(r, \theta, \varphi).$$

Group of invariance:

$$\{X_4, X_5, X_6\}.$$

In both cases there are two invariant components of the vector field and one non-invariant function denoted by ω. We will refer to planes $x = \text{const}$ and spheres $r =$ const as to the level surfaces of the corresponding solutions. The common properties of the representations are

- Absolute values of normal and tangential projections of vector field $\mathbf{u}$ to the level surface are invariant functions.
- Rotation angle of the vector field $\mathbf{u}$ about the normal to the level surface is a non-invariant function.

These representations generalize formulas for the velocity vector field in the well-known one-dimensional fluid motions with planar and spherical waves, where the whole vector field $\mathbf{u}$ depends only on the invariant variable x or r. We apply the representations above to the construction of exact solutions to ideal magnetohydrodynamics equations

$$\begin{aligned} &D\rho + \rho \operatorname{div} \mathbf{u} = 0, \\ &D\mathbf{u} + \rho^{-1}\nabla p + \rho^{-1}\mathbf{H} \times \operatorname{rot} \mathbf{H} = 0, \\ &Dp + A(p,\rho) \operatorname{div} \mathbf{u} = 0, \\ &D\mathbf{H} + \mathbf{H} \operatorname{div} \mathbf{u} - (\mathbf{H} \cdot \nabla)\mathbf{u} = 0, \\ &\operatorname{div} \mathbf{H} = 0, \quad D = \partial_t + \mathbf{u} \cdot \nabla. \end{aligned} \tag{1.2}$$

Here $\mathbf{u}$ is the velocity vector, p is the pressure, ρ is the density, and $\mathbf{H}$ is the magnetic field vector. All functions depend on time t and coordinates $\mathbf{x} = (x, y, z)$. The function $A(p,\rho) = \rho c^2$ ($c = \sqrt{\partial p/\partial \rho}$ is the thermodynamical speed of sound) is determined by the gas state equation $p = f(\rho, S)$, with entropy S. The construction of the solution with planar level surfaces will be given in detail, while calculations for the spherical solution will be shown briefly. We refer reader to the papers [4]–[10] for detailed investigation of the latter solution for gas– and magnetohydrodynamics equations.

2. Solutions with planar waves. In magnetohydrodynamics there are two vector fields at each particle, namely, its velocity $\mathbf{u}$ and its magnetic field vector $\mathbf{H}$. We will assume, that both vector fields satisfy the condition of partial invariance with respect to a group of planar isometries. This gives the following representation of solutions:

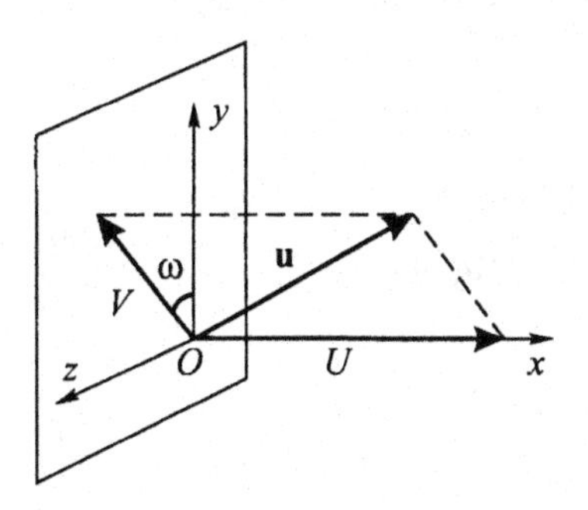

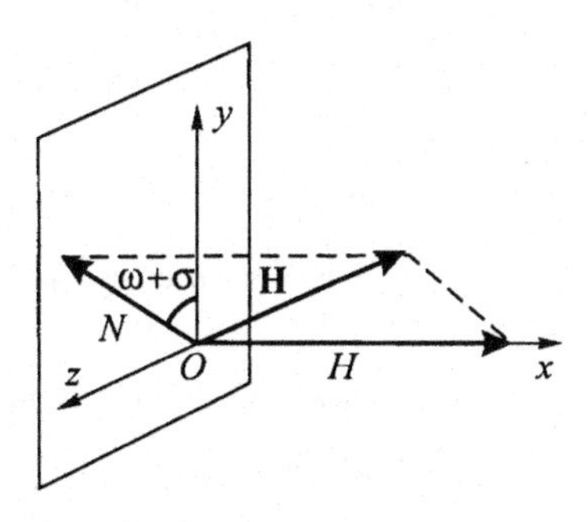

$$\begin{aligned} &u=U(t,x), && H^1=H(t,x),\\ &v=V(t,x)\cos\omega(t,x,y,z), && H^2=N(t,x)\cos\big(\omega(t,x,y,z)+\sigma(t,x)\big),\\ &w=V(t,x)\sin\omega(t,x,y,z), && H^3=N(t,x)\sin\big(\omega(t,x,y,z)+\sigma(t,x)\big),\\ &p=p(t,x),\quad \rho=\rho(t,x), && S=S(t,x). \end{aligned} \tag{2.1}$$

Here the function ω is non-invariant as long as it depends on all independent variables. All the other functions are invariant, depending only on invariant variables t and x. Substitution of the representation (2.1) into (1.2) gives the following. From the first (continuity) equation of the system (1.2) we have

$$\rho_t + U\rho_x + \rho(U_x - V\sin\omega\,\omega_y + V\cos\omega\,\omega_z) = 0. \tag{2.2}$$

Observe that the function h defined by

$$h(t,x) = -V^{-1}(\rho_t + U\rho_x + \rho U_x)$$

is invariant, i.e. it depends only on t and x. The rest of equation (2.2) gives a restriction for ω:

$$(\sin\omega)\,\omega_y - (\cos\omega)\,\omega_z + h = 0. \tag{2.3}$$

Besides, we have the following three equations, which relate only invariant functions:

$$\begin{aligned} \widetilde{D}U + \rho^{-1}p_x + \rho^{-1}NN_x &= 0,\\ \widetilde{D}H + hHV &= 0,\\ \widetilde{D}p + A(p,\rho)(U_x + hV) &= 0. \end{aligned}$$

Here and further $\widetilde{D} = \partial_t + U\partial_x$. The remaining equations of system (1.2) produce a highly-overdetermined system for the non-invariant function ω:

$$\begin{aligned} &\rho V\omega_t + \big(\rho UV - HN\cos\sigma\big)\,\omega_x + \big(\rho V^2\cos\omega - N^2\cos\sigma\cos(\omega+\sigma)\big)\,\omega_y\\ &+\big(\rho V^2\sin\omega - N^2\cos\sigma\sin(\omega+\sigma)\big)\,\omega_z - H(N_x\sin\sigma + N\cos\sigma\,\sigma_x) = 0. \end{aligned} \tag{2.4}$$

$$\begin{aligned} &HN\sin\sigma\,\omega_x + N^2\sin\sigma\cos(\omega+\sigma)\,\omega_y + N^2\sin\sigma\sin(\omega+\sigma)\,\omega_z\\ &+\rho\,\widetilde{D}V + HN\sin\sigma\,\sigma_x - HN_x\cos\sigma = 0. \end{aligned} \tag{2.5}$$

$$\begin{aligned} &N\omega_t + (NU - HV\cos\sigma)\,\omega_x + VN\sin\sigma\sin(\omega+\sigma)\,\omega_y\\ &-VN\sin\sigma\cos(\omega+\sigma)\,\omega_z + N\widetilde{D}\sigma + HV_x\sin\sigma = 0. \end{aligned} \tag{2.6}$$

$$\begin{aligned} &HV\sin\sigma\,\omega_x + NV\cos\sigma\sin(\omega+\sigma)\,\omega_y\\ &-NV\cos\sigma\cos(\omega+\sigma)\,\omega_z - \widetilde{D}N + HV_x\cos\sigma - NU_x = 0. \end{aligned} \tag{2.7}$$

$$N\big(\sin(\omega+\sigma)\omega_y - \cos(\omega+\sigma)\,\omega_z\big) - H_x = 0. \tag{2.8}$$

The system (2.3)–(2.8) for the function ω needs to be investigated for compatibility. For that we will look for solutions with functional arbitrariness in determination of function ω. This condition is referred to as "irreducibility condition" [1], which means that if it is satisfied then the partially invariant solution does not coincide with any invariant one.

PROPOSITION 2.1. *A partially invariant solution (2.1) is irreducible if and only if $\sigma \equiv 0$ or $\sigma \equiv \pi/2$. This means, that in an irreducible solution the velocity vector* $\mathbf{u}$ *at each particle is coplanar to the magnetic field vector* $\mathbf{H}$ *at the particle and the directional vector of Ox axis.*

In this case, equations for ω take the following form

$$\begin{aligned}
\sin\omega\,\omega_y - \cos\omega\,\omega_z + h &= 0,\\
\omega_t + U\omega_x + V\cos\omega\,\omega_y + V\sin\omega\,\omega_z &= 0,\\
H\omega_x + N\cos\omega\,\omega_y + N\sin\omega\,\omega_z &= 0.
\end{aligned} \tag{2.9}$$

Simplification of the system (2.4)–(2.8), verification of the compatibility conditions of equations (2.9), and integration of equations (2.9) lead to the following statements.

PROPOSITION 2.2. *In the main case $h \neq 0$ the invariant functions are determined by the following system of equations*

$$\begin{array}{ll}
\widetilde{D}\,\rho + \rho(U_x + hV) = 0, & \widetilde{D}\,U + \rho^{-1}p_x + \rho^{-1}NN_x = 0,\\
\widetilde{D}\,V - \rho^{-1}H_0hN_x = 0, & \widetilde{D}\,p + A(p,\rho)(U_x + hV) = 0,\\
\widetilde{D}\,N + NU_x - H_0hV_x + hNV = 0, & H = H_0h,\\
\widetilde{D}\,h + Vh^2 = 0, & H_0h_x + hN = 0.
\end{array} \tag{2.10}$$

The function ω satisfies the implicit equation

$$F(y - h^{-1}\cos\omega,\ z - h^{-1}\sin\omega) = 0 \tag{2.11}$$

with arbitrary smooth function F.

PROPOSITION 2.3. *In the case $h \equiv 0$ the invariant functions are determined by the system*

$$\begin{array}{ll}
\widetilde{D}\,\rho + \rho U_x = 0, & \widetilde{D}\,U + \rho^{-1}p_x + \rho^{-1}NN_x = 0,\\
\widetilde{D}\,V - \rho^{-1}H_0N_x = 0, & \widetilde{D}\,p + A(p,\rho)\,U_x = 0,\\
\widetilde{D}\,N + NU_x - H_0V_x = 0, & H = H_0.
\end{array} \tag{2.12}$$

The function ω satisfies an implicit relation

$$\jmath = f(\omega) - \varphi(t,x), \quad \jmath = y\cos\omega + z\sin\omega. \tag{2.13}$$

where the function φ *is such that*

$$\varphi_t + U\varphi_x = V, \quad H_0\,\varphi_x = N. \tag{2.14}$$

At this point the mathematical description of the solution is complete. Propositions 2 and 3 allows one to find the solution at any point by solving the reduced systems of equations (2.10) or (2.12) with suitable initial data and subsequently finding the non-invariant function from the finite relations (2.11) or (2.13). However, it is possible to obtain additional general information about the solution by analyzing formulas for the non-invariant function ω and representation (2.1) of the solution. From equations (2.9) it follows that ω conserves along particle trajectories and along magnetic field lines. This implies that these curves are flat in 3D-space. The following properties of the solution hold.

- Trajectories and magnetic field curves located in planes, which are orthogonal to the plane Oyz and turned on angle ω about Ox axis.
- All particles, which belong at some moment of time $t = t_0$ to a plane $x = x_0$, circumscribe the same trajectories in planes of each particle motion. Magnetic field lines passing through a plane $x = x_0$ are also the same planar curves.
- Angle of rotation about Ox axis of the plane containing the trajectory and the magnetic line of each particle is given by function ω, which satisfies equation (2.11) or (2.13).

Let us introduce Cartesian frame of reference $O'xl$ in the plane of particle's motion.

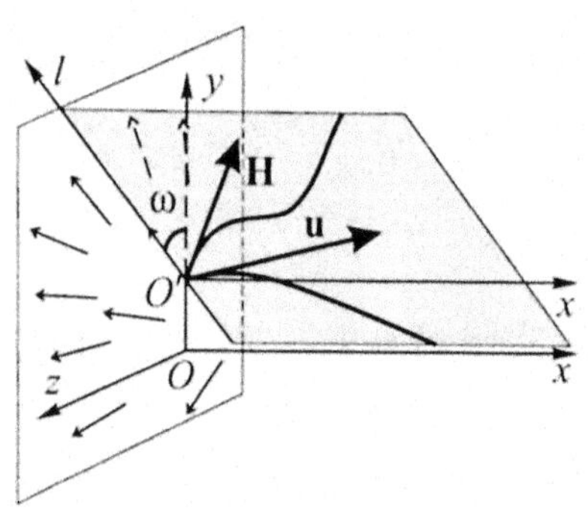

In order to find equations for the particle trajectory it is required to solve the following Cauchy problem

$$\frac{dx}{dt} = U(t,x), \quad x(t_0) = x_0.$$

The resulting dependence $x = x(t, x_0)$ allows one to determine $l = l(t)$ as

$$l(t) = \int_{t_0}^{t} V(t, x(t, x_0))dt.$$

In the original frame $Oxyz$ the particle's trajectory is given by

$$x = x(t, x_0), \quad y = y_0 + l(t)\cos\omega_0, \quad z = z_0 + l(t)\sin\omega_0. \tag{2.15}$$

The magnetic field line, which passes at $t = t_0$ through (x_0, y_0, z_0) is

$$y = y_0 + \cos\omega_0 \int_{x_0}^{x} \frac{N(t_0,s)}{H(t_0,s)}ds, \; z = z_0 + \sin\omega_0 \int_{x_0}^{x} \frac{N(t_0,s)}{H(t_0,s)}ds. \tag{2.16}$$

From the above calculations it follows that in order to determine the flow in the whole space it is required to set up an admissible vector field of directions in some plane $x = x_0$ (i.e. to determine the function ω consistent with equations (2.11) or (2.13)) and calculate the trajectory and magnetic field line for arbitrary particle in this plane. The whole picture of the flow is obtained by attaching the trajectory and the magnetic line pattern to each point on the plane $x = x_0$ in accordance with the vector field of directions.

Let us now turn to interpretation of implicit equation (2.11) for function ω. Here we observe only the case described by proposition 2, i.e. $h \neq 0$.

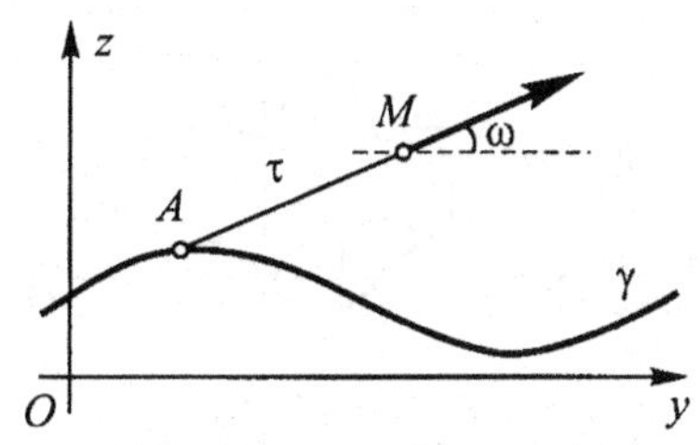

Let the function F in (2.11) be fixed. This define a curve $\gamma : F(y, z) = 0$ on plane Oyz. In order to find the angle ω at an arbitrary point $M = (y, z)$ one should draw a line segment AM of length $\tau = 1/h$ such that $A \in \gamma$. The direction of AM defines the angle ω as it is shown in the figure.

The main properties of the resulting vector field are the following

- The vector field is determined only inside the strip of width 2τ with curve $\gamma : \; F(y, z) = 0$ as a centerline.

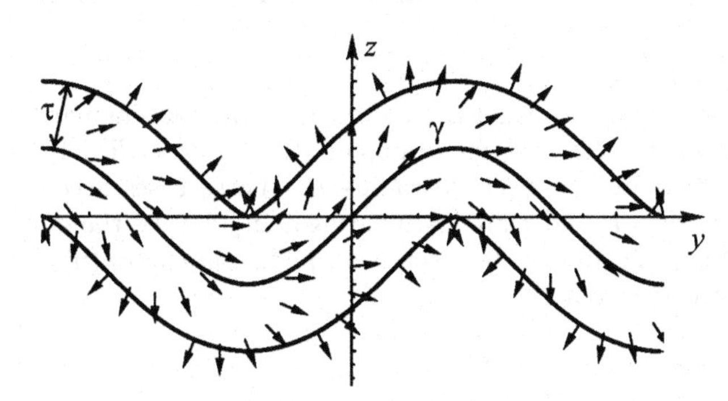

- The vector field is directed orthogonally to the boundaries of the strip.
- If $\tau > \min\limits_{\mathbf{x} \in \gamma} R(\mathbf{x})$, where $R(\mathbf{x})$ is a curvature radius of γ at point $\mathbf{x}$, then the boundary of the stripe has singularities of a "swallow tail" type. In this case the vector field can not be defined continuously over all boundaries of the swallow tail.

The latter statement is illustrated by the following set of figures. Here the boundary of the strip of the solution's determinacy has a swallow tail singularity. We determine the angle ω according to the given algorithm. As long as the point lies outside the swallow tail, the vector field has two possible directions (figures a and b). In figure c the point reaches the left boundary of the swallow tail — there appears a new branch of the solution.

Inside the swallow tail (figure *d*) the new branch splits into two branches, so we have four possible directions of the vector field at each point inside the swallow tail. At the right boundary of the swallow tail (figure *e*) two "old" branches of the solution stick together and disappear when the point leaves the swallow tail (figure *f*). One can chase, that the branches of solution we get on the right-hand side of the swallow tail are different from the ones we had on the left-hand side. Therefore, the solution should have a strong discontinuity over some curve inside the swallow tail. This type of singularity physically means a collision of particles, which move from different points along crossing trajectories.

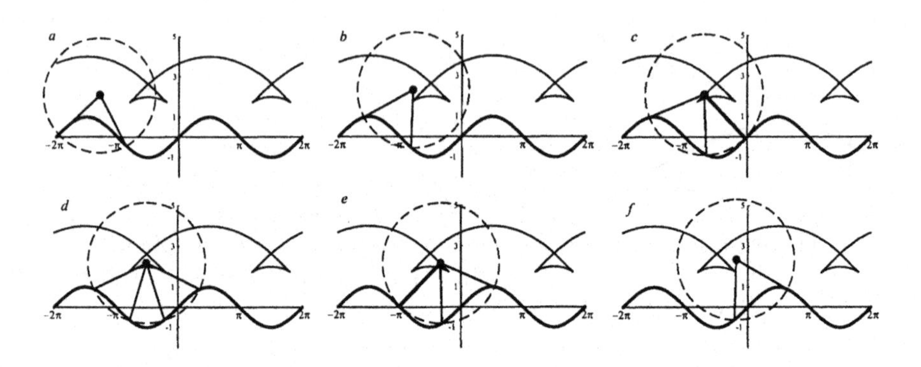

It is possible to construct a vector field without singularities for any value of τ. Let us choose, for example, $\gamma : y^2 + z^2 = R^2$. In the figures below γ is the middle circle of radius R. Three cases according to the relation between R and τ are distinguished. In all cases the domain of the vector fields is the annulus between two equidistant curves (inner and outer circles in the diagrams).

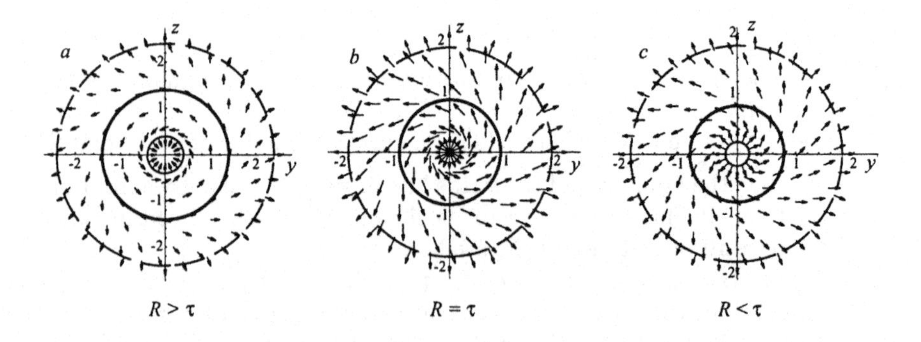

In order to show the corresponding 3D-motion of plasma we select a particular stationary solution of the invariant subsystem (2.10):

$$
\begin{aligned}
U &= \frac{H_0^2}{\cosh x}, \quad V = H_0^2 \tanh x, \quad \tau = \cosh x, \\
H &= \frac{H_0}{\cosh x}, \quad N = H_0 \tanh x, \quad \rho = H_0^{-2}, \quad S = S_0.
\end{aligned}
\tag{2.17}
$$

One can check that (2.17) represents a special case of the more general S. Chandrasekhar solution [12]. The streamlines and magnetic field lines here coincide and are given by the formula

$$
l(x) = \cosh x - 1. \tag{2.18}
$$

With the vector field in Oyz plane from previous example one can build solution, which describe a flow of ideal plasma in axisymmetric canal. The axial section of the canal is given in figures below. The uniform flow in

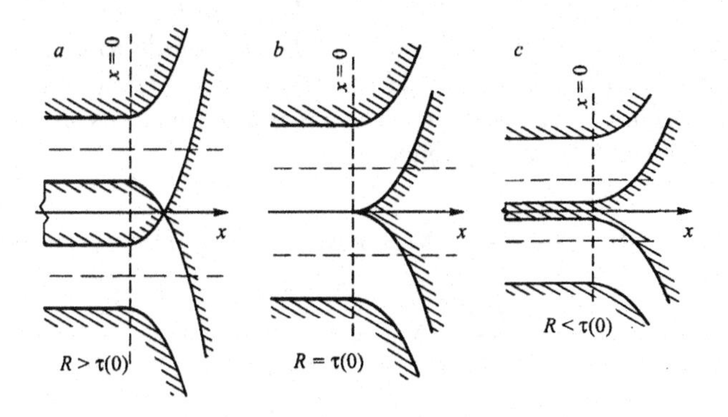

cylindrical canal for $x < 0$ switches at section $x = 0$ to the flow in the curvilinear canal for $x > 0$ described by the solution (2.17). The boundary of the canal is a rigid wall. Depending on the relation between $\tau(0)$ and R different pictures of motion are possible. Cases a, b and c correspond to the cases in the previous figure. In the diagrams a and c the canal has an inner cylindrical core.

3. Solution with spherical waves. In the similar manner we observe a partially invariant solution to equations (1.2) with respect to an admissible group of rotations $O(3)$. It has the following representation in the spherical frame of reference:

$$
\begin{aligned}
&U = U(t,r), \quad M = M(t,r), \quad H = H(t,r), \quad N = N(t,r), \\
&\Omega = \omega(t,r,\theta,\varphi), \quad \Sigma = \sigma(t,r) + \omega(t,r,\theta,\varphi), \\
&p = p(t,r), \quad \rho = \rho(t,r).
\end{aligned}
\tag{3.1}
$$

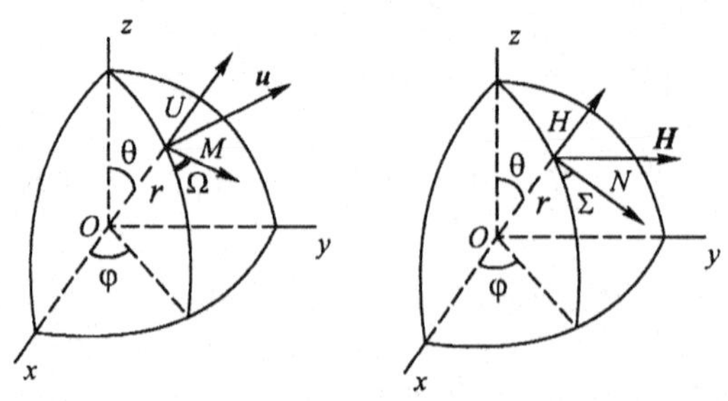

Substitution of the representation (3.1) provides a set of 3 equations for the invariant functions and an overdetermined system of 7 equations for the non-invariant function ω. The latter is investigated for involutivity. Again, we demand the solution to be irreducible, i.e. the overdetermined system is required to have functional arbitraryness in its general solution. The analysis is similar to the one presented in the previous section and can be found in [8]–[10]. The result is given by the following proposition.

PROPOSITION 3.1. *An irreducible solution to the ideal magnetohydrodynamics equations (1.2) of the form (3.1) exists only when $\sigma \equiv 0$. Then the velocity and the magnetic field vectors at any particle are coplanar to the radius-vector of the particle. The invariant functions are determined from the invariant system of equations*

$$
\begin{aligned}
&M_1 = \frac{M}{r}, \quad H = \frac{H_0}{r^2 \cos\tau}, \quad N_1 = r\,N, && \widetilde{D} = \partial_t + U\partial_r,\\
&\widetilde{D}\, M_1 + \frac{2}{r}\, U M_1 - \frac{H_0}{r^4 \rho \cos\tau}\, N_{1r} = 0, &&\\
&\widetilde{D}\, N_1 + N_1 U_r - \frac{H_0}{\cos\tau}\, M_{1r} - M_1 N_1 \tan\tau = 0, &&\\
&\widetilde{D}\, p + A(p,\rho)\left(U_r + \frac{2}{r}\, U - M_1 \tan\tau\right) = 0, &&\\
&\widetilde{D}\, U + \frac{1}{\rho}\, p_r + \frac{N_1 N_{1r}}{r^2 \rho} - r M_1^2 = 0, && H_0 \tau_r = N_1 \cos\tau,\\
&\widetilde{D}\, \rho + \rho\left(U_r + \frac{2}{r}\, U - M_1 \tan\tau\right) = 0, && \widetilde{D}\, \tau = M_1.
\end{aligned}
\tag{3.2}
$$

Here H_0 is an arbitrary constant, and $\tau \in (\pi/2, \pi/2)$ is some function of t and r. The non-invariant function ω is defined by the following implicit equation

$$\eta = f(\zeta). \tag{3.3}$$

with arbitrary smooth function f, and

$$\eta = \sin\theta\cos\omega\cos\tau - \cos\theta\sin\tau,$$
$$\zeta = \varphi + \arctan\frac{\sin\omega\cos\tau}{\cos\theta\cos\omega\cos\tau + \sin\theta\sin\tau}. \tag{3.4}$$

One can show that the particle trajectories and magnetic field lines are flat curves. All particles, which start from the same sphere $r = \text{const}$ circumscribe the same curves in 3D-space. The location and orientation of trajectories depend on the particle's initial position and its initial velocity vector.

The trajectories and magnetic field lines are given by solutions of the following Cauchy problem.

$$\frac{dr}{dt} = U(t, r), \; r(t_0) = r_*.$$

The particle trajectory in the polar coordinate system is determined by

$$r = r(t), \; \psi = \tau(t, r(t)) - \tau(t_0, r_*).$$

The magnetic force line at the time $t = t_0$ is

$$\psi = \tau(t_0, r) - \tau(t_0, r_*).$$

Again, as in the previous case, construction of the three-dimensional picture of motion involves two steps. For the first step one calculates some solution of the invariant system (3.2) and obtains the pattern of a trajectory and a magnetic field line using the formulas above. At the second step the obtained pattern is attached to each point on the sphere $r = r_0$ in accordance with the vector field of directions, determined by the implicit equation (3.3).

Let us give a geometrical interpretation of the finite relation (3.3). Here it is not straightforward, so we start from an auxiliary construction. For a given point M in the unit sphere, defined by its spherical coordinates (θ, φ) we introduce Cartesian frame of reference $O\xi_1\xi_2\xi_3$ according to the figure below. Here we pretend to know angle ω at M.

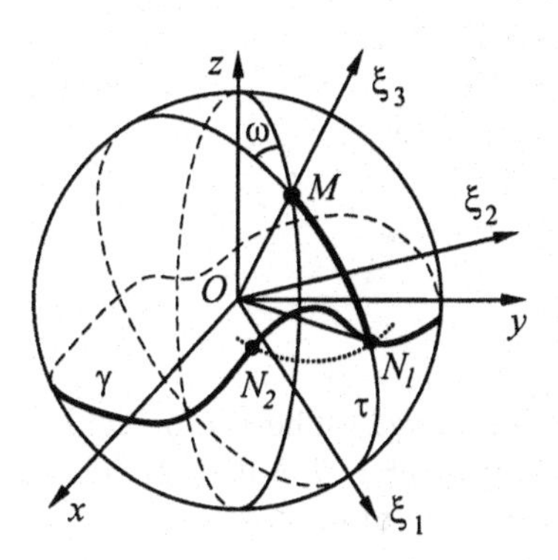

We put $\angle MON_1 = \pi/2 - \tau$. One can show that coordinates of point N_1 in $Oxyz$ frame of reference are given by $(x, y, z) = (\sigma\cos\zeta, \sigma\sin\zeta, \eta)$ with η and ζ from (3.4), and $\sigma = \sqrt{1-\eta^2}$. This observation allows one to give the following algorithm for finding ω at arbitrary point on a sphere for given dependence (3.3). Let a curve γ on the sphere $|\mathbf{x}| = R$ be determined by the relation $z = f(\varphi)$, where (r, φ, z) are cylindrical coordinates, and f is the same as in (3.3).

Given a point $M \in S^2$ we draw a circle S^1 on the sphere S^2 with the center at M and the geodesic radius $\pi/2 - \tau$. Note, that the angle τ is the same for each points of the sphere. We denote by N_i, $i = 1, \ldots, k$ points of intersection of the curve γ with the circle S^1.

PROPOSITION 3.2. *Angles between the meridian, which pass through M, and the geodesic curves connecting M with each N_i define all possible values of ω, which satisfy the relation (3.3).*

As in the planar case, the tangential vector field on a sphere is defined only inside of a stripe of width $\pi - 2\tau$ with the midline γ. For small or negative τ there could be singularities of a dovetail type on the boundaries of the stripe.

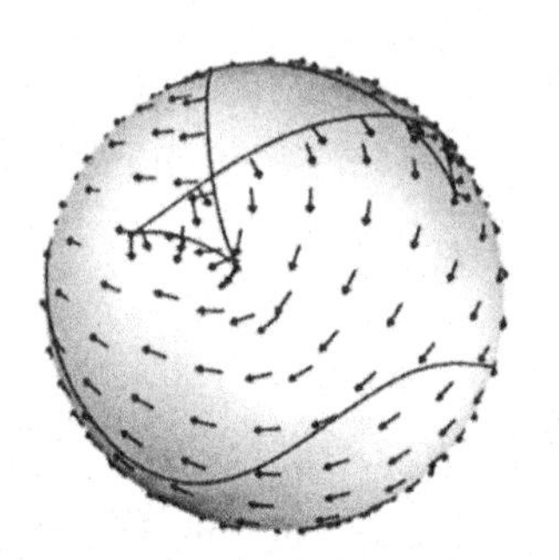

In this figure we give an example of a vector field, generated by the presented algorithm. The curve $\gamma : \theta = \pi/2 + 1/4 \sin 3\varphi$ and its northern equidistant $\gamma^+(\delta)$ are shown. One can see the multiple-valuedness of the vector field inside the dovetail. The dovetail appears only when value of $\pi/2 - \tau$ is large enough. The following theorem states how large τ can be.

PROPOSITION 3.3. *Let γ be a smooth curve on the sphere $|\mathbf{x}| = R$. The equidistants $\gamma^\pm(\delta)$ are smooth curves if and only if the following inequality holds*

$$\tan \delta < \min_{\mathbf{x} \in \gamma} R/k_g(\mathbf{x}). \tag{3.5}$$

Here k_g is a geodesic curvature of the curve γ.

The next figure gives examples of vector fields without singularities.

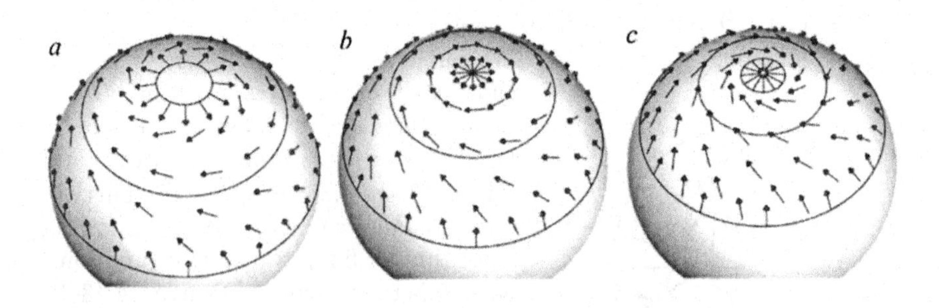

The middle circle γ is the parallel $\theta = \theta_* < \pi/2$. The vector field is determined in the strip between the equidistants $\gamma^\pm : \theta = \theta_* \mp \delta$. Three possibilities are shown: (a) $\theta_* > \delta$; (b) $\theta_* = \delta$; and (c) $\theta_* < \delta$.

We choose a particular stationary solution of invariant subsystem (3.2) in the form

$$M_1 = \frac{H_0^2 \tau'}{r^2 \cos\tau}, \quad N_1 = \frac{H_0 \tau'}{\cos\tau}, \quad U = \frac{H_0^2}{r^2 \cos\tau}, \quad \rho = \frac{1}{H_0^2}. \tag{3.6}$$

Here the function $\tau(r)$ satisfies the following equation

$$\tau'^2 = r^2 \cos^2\tau - r^{-2}. \tag{3.7}$$

For this solution the streamlines and magnetic field lines coincide and are shown in the figure below. Two different curves correspond to different initial data of equation (3.7). The shaded region is a spherical source of the plasma.

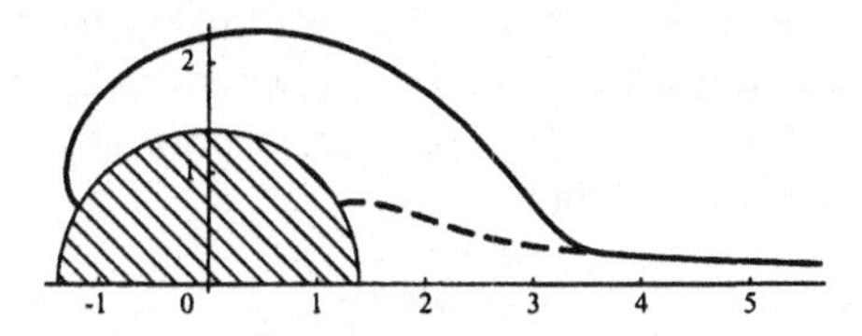

Now one can construct a 3D-picture of the flow. The following figure depicts typical magnetic field lines of the stationary flow. Here the tangential vector field on a sphere and magnetic field lines are taken from the previous figures. Domains of the latter flows are axially symmetric. Below

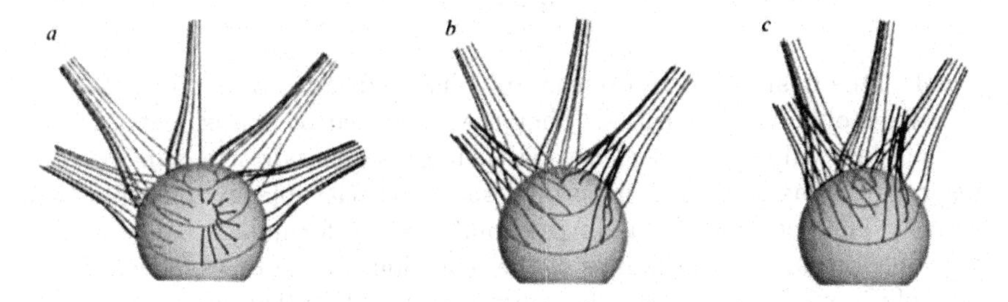

we provide figures of axial sections of the domains of the flows. The shaded region is the spherical source of the plasma. The flow is determined only inside the area bounded by the limiting magnetic field lines (solid lines in the diagrams).

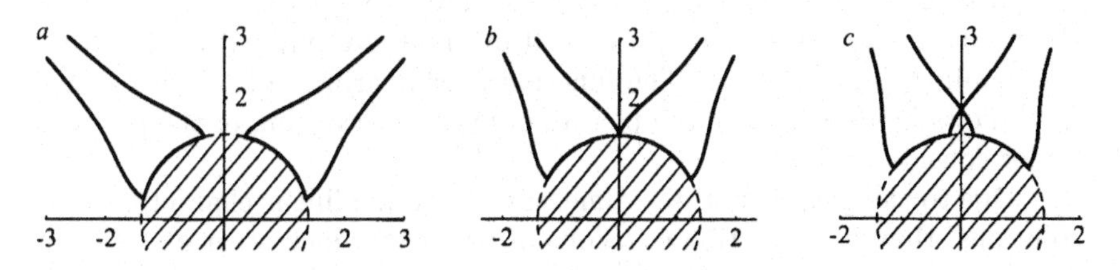

Another examples of magnetic field lines of the flow from the spherical source are shown below. The curve γ as the equator $\theta = \pi/2$. The magnetic field line is (a) the dashed curve; (b) the solid curve depicted in upper figure. The sphere is a source of the plasma. The solutions are determined outside

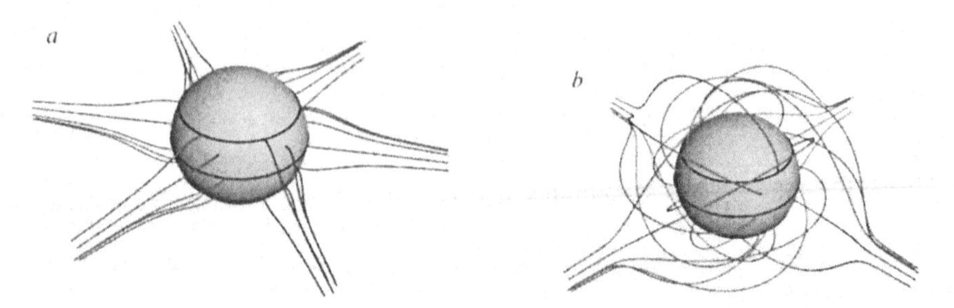

the shaded region in the area between the limiting magnetic curves. The solid curves are boundaries of the flow depicted in figure (a), the dashed ones bound the flow in figure (b). The solid part of the boundary of the shaded region is a source of the plasma.

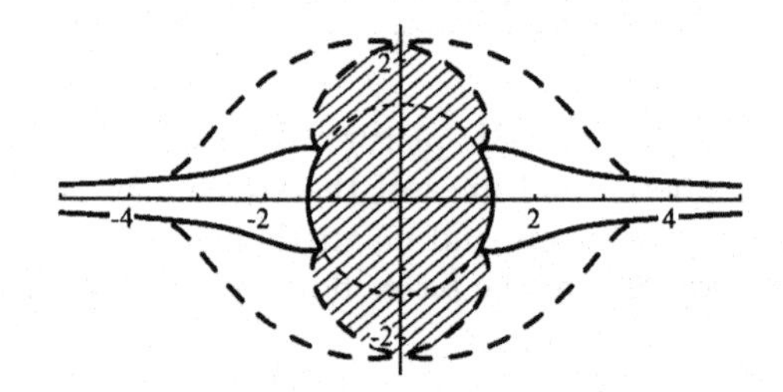

4. Discussion and conclusion. The solutions observed in this article have obvious similarities. They are generated by the isometry group of the plane and sphere and can be treated as generalizations of the corresponding classical one-dimensional solutions with planar and spherical waves. The representations of the solutions are formulated in a similar way. This produces similarities of the behaviour of the flows, governed by the solutions. In both cases the construction of the 3D picture of motion involves two steps. The first step is integration with suitable initial data of the reduced system of equations with two independent variables and construction of the pattern of a trajectory and a magnetic field line. The second step is a composition of the whole picture of motion by attaching the pattern to each point of the corresponding level surface according to the vector field of directions prescribed by the finite implicit equation for the non-invariant function. The functional arbitrariness of the finite relation allows modification of the picture of motion according to the problem under consideration.

Due to the genuinely three-dimensional and nonlinear character of the solutions, they can describe collisions of particles which start at different points and move along intersecting trajectories. Criteria for the presence or absence of the singularities is given in both cases, in terms of geometrical properties of the arbitrary functions involved in the finite relations.

The resulting solution may be used as a test for numerical modeling of complicated three-dimensional flows of ideal fluids. It also may be applied for theoretical investigations of genuinely three-dimensional singularities of the ideal fluid and plasma motions.

Acknowledgments. Author would like to thank O.I. Bogoyavlenskij for partial support of the travel to the 2006 IMA Summer Program "Symmetries and Overdetermined Systems of Partial Differential Equations". Research was partially supported by RFBR grant 05-01-00080 and by the Leading Scientific Schools Programme grant Sc.Sch.-5245.2006.1.

REFERENCES

[1] L.V. OVSYANNIKOV, *Group analysis of differential equations,* New York: Academic Press, 1982.
[2] P.J. OLVER, *Applications of Lie groups to differential equations,* New York: Springer-Verlag, 1986.
[3] L.V. OVSYANNIKOV, *Singular vortex,* J. Appl. Mech. Tech. Phys. (1995), **36**(3): 360–366.
[4] A.P. CHUPAKHIN, *Invariant submodels of the singular vortex,* J. Appl. Mech. Tech. Phys. (2003), **67**(3): 351–364.
[5] A.A. CHEREVKO AND A.P. CHUPAKHIN, *Homogeneous singular vortex,* J. Appl. Mech. Tech. Phys. (2004), **45**(2): 209–221.
[6] A.P. CHUPAKHIN, *Singular Vortex in Hydro- and Gas Dynamics.* In *Analytical Approaches to Multidimensional Balance Laws.,* Nova Sci. Publ., 2005.
[7] A.S. PAVLENKO, *Projective submodel of the Ovsyannikov vortex,* J. Appl. Mech. Tech. Phys. (2005), **46**(4): 459–470.
[8] S.V. GOLOVIN, *Singular vortex in magnetohydrodynamics,* J. Phys. A: Math. Gen. (2005), **38**(20): 4501–4516.
[9] S.V. GOLOVIN, *Invariant solutions of the singular vortex in magnetohydrodynamics,* J. Phys. A: Math. Gen. (2005), **38**(37): 8169–8184.
[10] S.V. GOLOVIN, *Generalization of the one-dimensional ideal plasma flow with spherical waves,* J. Phys. A: Math. Gen. (2006), **39**: 7579–7595.
[11] L.V. OVSYANNIKOV *On optimal systems of subalgebras,* Russ. Acad. Sci., Dokl., Math. (1994), **48**(3): 645–649.
[12] S. CHANDRASEKHAR *On the stability of the simplest solution of the equations of hydromagnetics,* Proc. Nat. Acad. Sci. U.S.A. (1956), **42**: 273–276.

INVARIANT PROLONGATION AND DETOUR COMPLEXES*

A. ROD GOVER†

(Dedicated to the memory of Thomas P. Branson (1953 - 2006))

Abstract. In these expository notes we draw together and develop the ideas behind some recent progress in two directions: the treatment of finite type partial differential operators by prolongation, and a class of differential complexes known as detour complexes. This elaborates on a lecture given at the IMA Summer Programme "Symmetries and overdetermined systems of partial differential equations".

1. Introduction. Differential complexes capture integrability conditions for linear partial differential operators. When they involve sequences of natural operators on geometric structures, such complexes typically encode deep geometric information about the structure. In these expository notes we survey some recent results concerning the construction of such complexes. A particular focus is the construction via prolonged differential systems and the links to the treatment of finite type differential operators. The broad issues arise in many contexts but we take our motivation here from conformal and Riemannian geometry.

Elliptic differential operators with good conformal behaviour play a special role in geometric analysis on Riemannian manifolds. The "Yamabe Problem" of finding, via conformal rescaling, constant scalar curvature metrics on compact manifolds is a case in point. This exploits heavily the conformal Laplacian, since it controls the conformal variation of the scalar curvature (see [39] for an overview and references). The higher order conformal Laplacian operators of Paneitz, Riegert, Graham et al. [34] (the *GJMS operators*) have been brought to bear on related problems by Branson, Chang, Yang and others [5, 19, 36]. However for many important tensor or spinor fields there is no natural conformally invariant elliptic operator (taking values in an irreducible bundle) available. This is true even in the conformally flat case, and indeed this claim follows easily from the classification of conformal differential operators on the sphere [3]. From this it is clear that for many bundles on the sphere the analogue, or replacement, for an elliptic operator is an elliptic complex of conformally invariant

*The author would like to thank the Institute for Mathematics & Its Applications, University of Minnesota, and also the organisers of the 2006 Summer Programme there "Symmetries and overdetermined systems of partial differential equations". Special thanks are due to Tom Branson, a friend, a colleague and a mentor. It is my understanding that his enthusiasm and work contributed significantly to the occurrence and organisation of the meeting. It is part of the tragedy of his departure that he did not get to enjoy the realisation of this event.

†Department of Mathematics, University of Auckland, Private Bag 92019, Auckland, New Zealand (gover@math.auckland.ac.nz).

differential operators. Ignoring the issue of ellipticity, the requirement that a sequence of operators be both conformally invariant and form a complex is already severe, and so the construction of such complexes is a delicate matter.

On the conformal sphere there is a class of conformal complexes known as Bernstein-Gelfand-Gelfand (BGG) complexes. Equivalents of these were first constructed in the representation theory of (generalised) Verma modules (see [40] and references therein). From there we see, more generally, versions of these complexes for a large class of homogeneous structures in the category of so-called parabolic geometries, that is manifolds of the form G/P where G is a semisimple Lie group and P a parabolic subgroup. Although these early constructions were motivated by geometric questions, the entry of the unifying picture of these complexes into the mainstream of Differential Geometry was pioneered by Eastwood, Baston and collaborators [22, 2]. In particular they developed techniques for constructing sequences of differential operators on general curved backgrounds that generalise the BGG sequences. Through work over the years the nature of many of the operators involved has become quite well developed [26, 14], and there has been spectacular recent progress in general approaches to constructing these complexes [18, 13]. However, as we shall discuss in Section 3.1, these sequences are in general not complexes on curved manifolds.

In the BGG sequences each operator may be viewed as an integrability condition for the preceding operator. The operators encode in part the geometry and it seems likely that in general, at least for conformal structures, one obtains a complex only in the conformally flat setting. In work with Tom Branson the ambient metric of Fefferman-Graham [23] and then variational ideas were used to give two constructions [11, 10] of what we have termed "detour complexes". These are related to BGG complexes, but involve fewer operators, weaker integrability conditions and yield complexes in curved settings. The central results and ideas (with some extensions) are reviewed and developed in Sections 3.1, 3.2 and 3.3. The complexes obtained there have interesting geometric applications and interpretations, but the constructions involved do not suggest how a full theory of such complexes could be developed. We should note that considerably earlier, following related constructions of Calabi and Gasqui-Goldschmidt, Gasqui constructed [24] a three step elliptic sequence which forms a complex if and only if the manifold is Einstein. This was studied further in [4]. In our current language this is a detour complex, exactly along lines we discuss, and may be viewed as a Riemannian analogue of the conformal complex of [10] as in Theorem 3.2.

Beginning at Section 3.4 we start the development of a model for a rather general construction principle for a class of detour complexes on pseudo-Riemannian manifolds. This is based on recent work with Petr Somberg and Vladimir Souček. We treat two examples, one in some detail, and the complexes we obtain are elliptic in the case that the background structure has Riemannian signature. In dimension 4 the complexes are conformally invariant. There are two main tools involved in the constructions. The first is general sequence which yields a detour complex for every Yang-Mills connection. The second ingredient is the use of prolonged differential systems.

The use of prolonged systems is related to the following problem: Given a suitable (i.e. finite type, as explained below) linear partial differential operator D, can one construct an equivalent first order prolonged system that is actually a vector bundle connection? Here there are close connections with the recent work [8] with Branson, Čap and Eastwood (for this, and further discussion of BGG complexes, see also the proceedings of A. Čap [15]). In fact we want more. The connection should be invariant in the sense that it should share the symmetry and invariance properties of D.

2. Invariant tractor connections for finite type PDE. First we introduce some notation to simplify later presentations. We shall write Λ^1 for the cotangent bundle, and Λ^p for its pth exterior power. Thus the trivial bundle (with fibre $\mathbb{R}$) may be denoted Λ^0 although, for reasons that will later become obvious, we will also write $\mathcal{E}$ for this. For simplicity the kth symmetric powers of Λ^1 will be denoted S^k (with the metric trace-free part of this denoted S_0^k); albeit that this introduces the redundancy that $S^1 = \Lambda^1$. In all cases we will use the same notation for a bundle as for its section spaces. All structures will be assumed smooth and we restrict to differential operators that take smooth sections to smooth sections.

The prolongations of a kth-order semilinear differential operator $D : E \to F$, between vector bundles, are constructed from its leading symbol $\sigma(D) : S^k \otimes E \to F$. At a point of M, denoting by K^0 the kernel of $\sigma(D)$, the spaces $V_i = (S^i \otimes E) \cap (S^{i-k} \otimes K^0)$, $i \geq k$, capture spaces of new variables to be introduced and the system closes up and is said to be of *finite type* if $V_i = 0$ for sufficiently large i. For example, on a Riemmanian manifold (M^n, g) a vector field k is an infinitesimal isometry, a so-called Killing vector field, if Lie differentiation along its flow preserves the metric g, that is $\mathcal{L}_k g = 0$. Rewriting this in terms of the Levi-Civita connection ∇ (i.e. the unique torsion free connection on TM preserving the metric) the Killing equation is seen to be a system of $n(n+1)/2$ equations in n unkowns, viz. $\nabla_a k_b + \nabla_b k_a = 0$, where we have used an obvious abstract index notation and also we have used the metric to identify k with the 1-form $g(k, \ )$. In this example K^0 is evidently Λ^2, the space of 2-forms, and we may rewrite the system as $\nabla_a k_b = \mu_{ab}$ where $\mu \in \Lambda^2$ (that is $\mu_{ab} = -\mu_{ba}$). Since $(S^2 \otimes \Lambda^1) \cap (\Lambda^1 \otimes \Lambda^2) = 0$, when we differentiate and compute consequences the system closes up algebraically after just one step. We obtain a prolonged system which is the equation of parallel transport for a connection:

$$\nabla_a^D \begin{pmatrix} k_b \\ \mu_{bc} \end{pmatrix} := \begin{pmatrix} \nabla_a k_b - \mu_{ab} \\ \nabla_a \mu_{bc} - R_{bc}{}^d{}_a k_d \end{pmatrix} = 0, \qquad (2.1)$$

where R is the curvature of ∇. We will regard this connection as being *equivalent* to the original equation, since solutions of the original equation are in 1-1 correspondence with sections of $\mathbb{T} := \Lambda^1 \oplus \Lambda^2$ that are parallel for ∇^D. It follows that the original equation has a solution space of dimension at most rank($\mathbb{T}$) $= n(n+1)/2$. The curvature of ∇^D evidently obstructs solutions and, in particular, the maximal number of solutions is achieved only if the connection ∇^D is flat.

Although this example is very simple it already brings us to several of the essential issues. The original equation $\mathcal{L}_k g = 0$ arose in a Riemannian setting. By the explicit formula given it is clear that that the connection ∇^D is naturally invariant for Riemannian structures. However comparing with section 3 in [1], or [16], one identifies this connection as an obvious curvature modification of the normal tractor connection on projective structures. The Levi-Civita connection ∇ in (2.1) may be viewed as an affine connection, so this makes sense. If ∇ is any torsion-free connection on Λ^1 then note that the equation $\nabla_a k_b + \nabla_b k_a = 0$ is invariant under the transformations

$$k_b \mapsto \widehat{k}_b = e^{2\omega} k_b \quad \text{and} \quad \nabla \mapsto \widehat{\nabla}$$

where, as an operator on 1-forms,

$$\widehat{\nabla}_a u_b = \nabla_a u_b - \Upsilon_a u_b - \Upsilon_b u_a \quad \text{with} \quad \Upsilon = d\omega \ .$$

These show that the equation is in fact *projectively invariant*, it is well defined on manifolds having only an equivalence class of torsion-free connections, where the equivalence relation is given by the transformations indicated, i.e. on projective manifolds. This leads us to question whether the tractor connection ∇^D in (2.1) shares this property. It does. It is readily verified directly that it is also projectively invariant; it differs from the invariant normal projective connection in [1, 16] by an action of the (projectively invariant) curvature of the normal connection. In this case it is straightforward to see that this was inevitable.

In general prolonged systems are more complicated. In [8] Kostant's algebraic Hodge theory [38] led to an explicit and uniform treatment of prolongations for a large class of overdetermined PDE. One of the simplest classes of examples is the set of equations controlling conformal Killing forms and via explicit calculations in terms of the Levi-Civita connection, prolonged systems for these were earlier calculated in [43]. However in neither of these works was the issue of invariance discussed. The equation for a conformal Killing p-form κ can be given as follows: for any tangent vector field u we have

$$\nabla_u \kappa = \varepsilon(u)\tau + \iota(u)\rho$$

where, on the right-hand side τ is a $(p-1)$-form, ρ is a $(p+1)$-form, and $\varepsilon(u)$ and $\iota(u)$ indicate, respectively, the exterior multiplication and (its formal adjoint) the interior multiplication of $g(u, \ )$. An important property of the conformal Killing equation is that it is conformally invariant (where we require the p-form κ to have conformal weight $p+1$). This can be phrased in similar terms to the projective invariance of the equation described above, but alternatively it simply means the equation descends to a well defined equation on manifolds where we do not have a metric, but rather only an equivalence class of conformally related metrics (this is a conformal structure). So it is natural to ask whether there is an equivalent prolonged system that may be realised as a conformally invariant connection. Using the framework of tractor calculus, this is answered in the affirmative in [32], where also the invariant connection is related to the normal conformal tractor connection.

The leading symbol determines whether or not an equation is of finite type; an operator $D : E \to F$ is of finite type if and only if its (complex) characteristic variety is empty [44]. One may ask whether any finite type overdetermined linear differential operator is equivalent, in the sense of prolongations, to a connection on vector bundle where the connection is "as invariant" as the original operator, that is it shares the same symmetries. This is obviously an important question in its right. For example if for some PDE one finds a connection on prolonged system which depends only on operator D and no other choices, then it is invariant in this sense and its curvature is really a geometric invariant of the equation D. In general this is probably more than one can hope for. It is probably only reasonable to hope to find a canonical connection after some well defined choices controlled by representation theory on finite dimensional vector spaces that are associated with the original equation. It turns out such questions are also important for the construction of certain natural differential complexes.

3. Detour complexes. We we will specialise our discussions here to the setting of oriented (pseudo-)Riemannian structures and conformal geometries. The restriction to oriented structures is just to simplify statements and is not otherwise required. It should also be pointed out that similar ideas apply in many other settings including, for example, CR geometry.

Recall that a *conformal manifold* of signature (p,q) on M is a smooth ray subbundle $\mathcal{Q} \subset S^2T^*M$ whose fibre over x consists of conformally related signature-(p,q) metrics at the point x. Sections of $\mathcal{Q}$ are metrics g on M. So we may equivalently view the conformal structure as the equivalence class $[g]$ of these conformally related metrics. The principal bundle $\pi : \mathcal{Q} \to M$ has structure group $\mathbb{R}_+$, and so each representation $\mathbb{R}_+ \ni x \mapsto x^{-w/2} \in \mathrm{End}(\mathbb{R})$ induces a natural line bundle on $(M,[g])$ that we term the conformal density bundle and denote $\mathcal{E}[w]$. As usual we use the same notation for its section space.

We write $\boldsymbol{g}$ for the *conformal metric*, that is the tautological section of $S^2[2] := S^2 \otimes \mathcal{E}[2]$ determined by the conformal structure. This will be used to identify T with $\Lambda^1[2]$. For many calculations we will use abstract indices in an obvious way. Given a choice of metric g from the conformal class, we write ∇ for the corresponding Levi-Civita connection. With these conventions the Laplacian Δ is given by $\Delta = \boldsymbol{g}^{ab}\nabla_a\nabla_b = \nabla^b\nabla_b$. Note $\mathcal{E}[w]$ is trivialised by a choice of metric g from the conformal class, and we write ∇ for the connection corresponding to this trivialisation. It follows immediately that (the coupled) ∇_a preserves the conformal metric.

Since the Levi-Civita connection is torsion-free, its curvature $R_{ab}{}^c{}_d$ (the Riemannian curvature) is given by $[\nabla_a, \nabla_b]v^c = R_{ab}{}^c{}_d v^d$ ($[\cdot,\cdot]$ indicates the commutator bracket). This can be decomposed into the totally trace-free Weyl curvature C_{abcd} and a remaining part described by the symmetric *Schouten tensor* P_{ab}, according to $R_{abcd} = C_{abcd} + 2\boldsymbol{g}_{c[a}P_{b]d} + 2\boldsymbol{g}_{d[b}P_{a]c}$, where $[\cdots]$ indicates antisymmetrisation over the enclosed indices. The Schouten tensor is a trace modification of the Ricci tensor Ric_{ab} and vice versa: $\mathrm{Ric}_{ab} = (n-2)P_{ab} + J\boldsymbol{g}_{ab}$, where we write J for the trace $P_a{}^a$ of P. The Weyl curvature is conformally invariant.

3.1. Detour sequences. In conformal geometry the de Rham complex is a prototype for a class of sequences of bundles and conformally invariant differential operators, each of the form

$$\mathcal{B}^0 \to \mathcal{B}^1 \to \cdots \to \mathcal{B}^n$$

where the vector bundles $\mathcal{B}^i$ are irreducible tensor-spinor bundles. For example, on the sphere with its standard conformal structure we have the following: these are complexes and there is one such complex for each irreducible module $\mathbb{V}$ for the group $G = SO(n+1,1)$ of conformal motions; the space of solutions of the first (finite type) conformal operator $\mathcal{B}^0 \to \mathcal{B}^1$ is isomorphic to $\mathbb{V}$; and the complex gives a resolution of this space viewed as a sheaf. The geometry of the manifold is partly encoded in the coefficients of the differential operators in these sequences, and in the conformally flat setting of the sphere the first PDE operator $D_0 : \mathcal{B}^0 \to \mathcal{B}^1$ is fully integrable. From representation theory (see [3] and references therein) one can deduce that there is a prolonged system and connection ∇^{D_0}, equivalent to D_0 in the sense discussed in Section 2. In this setting the space of solutions $\mathbb{V}$ is maximal in that $\dim \mathbb{V}$ achieves the dimension bound for the maximal size of the solution space. Since the sequence is a differential complex, the next operator in the sequence D_1 gives an integrability condition for D_0 and so on. That the complex gives a resolution means that these integrability conditions are in a sense maximally severe.

It turns out that there are "curved analogues" of these sequences, these are the conformal cases of the (generalised) Bernstein-Gelfand-Gelfand (BGG) sequences, a class of sequences of differential operators that exist on any parabolic geometry [18, 21]. Unfortunately in the general curved setting these sequences are no longer complexes, which limits their applications.

As well as the operators $D_i : \mathcal{B}^i \to \mathcal{B}^{i+1}$ of the BGG sequence, in even dimensions there are conformally invariant "long operators" $\mathcal{B}^k \to \mathcal{B}^{n-k}$ for $k = 1, \cdots, n/2 - 1$ [3]. Thus there are sequences of the form

$$\mathcal{B}^0 \xrightarrow{D_0} \mathcal{B}^1 \xrightarrow{D_1} \cdots \xrightarrow{D_{k-1}} \mathcal{B}^k \xrightarrow{L_k} \mathcal{B}^{n-k} \xrightarrow{D_{n-k}} \cdots \xrightarrow{D_{n-1}} \mathcal{B}^n . \tag{3.1}$$

and, following [11] (see also [7]), we term these detour sequences since, in comparison to the BGG sequence, the long operator here bypasses the middle of the BGG sequence and the operator L_k takes us (or "detours") directly from $\mathcal{B}^k$ to $\mathcal{B}^{n-k}$. Once again using the classification it follows immediately that these detour sequences are in fact complexes in the case that the structure is conformally flat. The operator L_k is again an integrability condition for D_{k-1} but since L_k has higher order than the operator D_k one does not expect that the detour complex is a resolution. In fact, by for example considering Taylor series expansions for solutions, it is straightforward to show that there is local cohomology at the bundle $\mathcal{B}^k$. We do not need the details here, the main point is that L_k is weaker, as an integrability condition, than D_k.

3.2. Curved detour complexes. An interesting direction is to try to find curved analogues of these complexes. That is detour sequences that are actually complexes on conformally curved structures. Remarkably this works in general in the de Rham case, that is the case where the bundles $\mathcal{B}^i$ are exterior powers of the cotangent bundle, Λ^i. We write Λ_k to denote the bundle of conformal density weighted k-forms $\Lambda^k[2k-n]$; sections of this pair conformally with Λ^k in integrals. Then the formal adjoint of the exterior derivative d acts conformally $\delta : \Lambda_i \to \Lambda_{i-1}$.

THEOREM 3.1. *[11] In even dimensions there are conformally invariant differential operators*

$$L_k : \Lambda^k \to \Lambda_k \qquad L_k = (\delta d)^{n/2-k} + \text{ lower order terms.}$$

so that

$$\Lambda^0 \xrightarrow{d} \cdots \xrightarrow{d} \Lambda^{k-1} \xrightarrow{d} \Lambda^k \xrightarrow{L_k} \Lambda_k \xrightarrow{\delta} \Lambda_{k-1} \xrightarrow{\delta} \cdots \xrightarrow{\delta} \Lambda_0$$

is a conformal complex. In Riemannian signature the complex is elliptic. The operator $L_{n/2-1}$ is the usual Maxwell operator δd, while L_0, included here (cf. above) as an extreme case, is the critical (i.e. dimension order) conformal Laplacian of GJMS [34]. For $k \neq 0$ the complex may be viewed as a differential form analogue of the critical GJMS operator. For $0 \leq k \leq n/2 - 2$ the L_k have the form $\delta Q_{k+1} d$ where the Q_ℓ are operators on closed forms that generalise the Q-curvature; for example they give conformal pairings that descend to pairings on de Rham cohomology [9].

3.3. Variational constructions. On curved backgrounds, obtaining complexes for almost any other case of (3.1) seems, at first, to be hopeless. The point is this. One of the main routes to constructing BGG sequences is via some variant of the curved translation principle of Eastwood and others [22, 2, 18, 13]. In the first step of this, one replaces the de Rham complex with the sequence obtained by twisting this sequence with an appropriate bundle and connection. In fact the connections used are usually so called *normal conformal tractor connections* as in [17]. These are connections on prolonged systems, along the lines of the connection constructed explicitly in Section 2, but these are natural to the conformal structure and are normalised in way that means that they are unique up to isomorphism. Appropriate differential splitting operators are then used to extract from the twisted de Rham sequence the sought BGG sequence. The full details are not needed here, the main point for our current discussion is as follows. If the structure is conformally flat then the tractor connection is flat and the twisted de Rham sequence sequence is still a complex. It follows from this that the BGG sequence is a complex. However in general the curvature of the tractor connection will obstruct the forming of a complex. This is evident even at the initial stages of the sequence: if we write $\Lambda^i(\mathcal{V})$ for the twistings of i-forms by some tractor bundle then, writing ∇ for the tractor connection, the composition

$$\Lambda^0(\mathcal{V}) \xrightarrow{d^\nabla} \Lambda^1(\mathcal{V}) \xrightarrow{d^\nabla} \Lambda^2(\mathcal{V}) \tag{3.2}$$

is simply the curvature of ∇ acting on $\mathcal{V}$.

Ignoring this difficulty for the moment, an interesting case from conformally flat structures is the conformal deformation complex of [25]. Some notation first. We write $\Lambda^{k,\ell}$ for the space of trace-free covariant $(k+\ell)$-tensors $t_{a_1\cdots a_k b_1 \cdots b_\ell}$ which are skew on the indices $a_1 \cdots a_k$ and also on the set $b_1 \cdots b_\ell$. Skewing over more than k indices annihilates t, as does symmetrising over any 3 indices. Thus for example $\Lambda^{1,1} = S^2_0$, but in the case of this bundle we will postpone changing to this notation. In dimensions $n \geq 5$ the initial part of this BGG complex is,

$$T \xrightarrow{\mathsf{K}_0} S^2_0[2] \xrightarrow{\mathsf{C}} \Lambda^{2,2}[2] \xrightarrow{\mathsf{Bi}} \Lambda^{3,2}[2] \to \cdots \tag{3.3}$$

where T is the tangent bundle. Here C is the linearisation, at a conformally flat structure, of the Weyl curvature as an operator on conformal structure; Bi is a conformal integrability condition arising from the Bianchi identity; the operator K_0 is the conformal Killing operator, viz the operator which takes infinitesimal diffeomorphisms to their action on conformal structure. Since infinitesimal conformal variations take values in $S^2_0[2]$, it follows from these interpretations that the cohomology of the complex, at this bundle, may be interpreted as the formal tangent space to the moduli space of conformally flat structures.

From this picture, and also from the discussion around (3.2), we do not expect a curved generalisation of the complex (3.3). One way to potentially avoid the issues brought up with (3.2) is to consider detour sequences of the form (3.1) with $k = 1$; these will be termed *short detour sequences.* (In [31] there is some discussion and applications of detour complexes for this BGG in the conformally flat setting.) It turns out that this idea is fruitful. One construction is based around the so-called obstruction tensor $\mathcal{B}_{ab}$ of Fefferman and Graham [23] which generalises to even dimensions the Bach tensor of Bach's gravity theory. For our current purposes we need only to know some key facts about this, and we shall introduce these as required. It is a trace-free conformal conformal 2-tensor with leading term $\Delta^{n/2-2}C$ (where, recall, C denotes the Weyl curvature). Let us write K_0^* for the formal adjoint of K_0 and B for the linearisation, at the metric g, of the operator which takes metrics g to $\mathcal{B}^g$. By taking the Lie derivative of $\mathcal{B}_{ab}$, and using the fact [35] that $\mathcal{B}_{ab}$ is the total metric variation of an action (viz. $\int Q$ where Q is Branson's Q curvature [5]) we obtain the following (where to simplify notation we have omitted the conformal weights).

THEOREM 3.2. *[10] On even dimensional pseudo-Riemannian manifolds with the Fefferman-Graham obstruction tensor vanishing everywhere, the sequence of operators*

$$T \xrightarrow{K_0} S_0^2 \xrightarrow{B} S_0^2 \xrightarrow{K_0^*} T \tag{3.4}$$

is a formally self-adjoint complex of conformally invariant operators. In Riemannian signature the complex is elliptic.

As with the deformation complex (3.3) there is an immediate interpretation of the cohomology at S_0^2. By construction it is the formal tangent space to the moduli space of obstruction-flat structures. There are determinant quantities, with interesting conformal behaviour, for detour complexes [12, 7]; this detour torsion should be especially interesting for (3.4).

Leaving aside the potential applications it is already interesting that there is a differential complex along these lines in such a general setting. Although the obstruction tensor is rather mysterious, and at this point has not been fully explored, it is known that it vanishes on conformally Einstein manifolds [23, 31, 35], certain products of Einstein manifolds [29], and on half-flat structures in dimension 4.

3.4. The Yang-Mills detour complex. The construction in Theorem 3.2 and subsequent observations suggest the possibility of a rich theory of (short) detour complexes. On a large class of manifolds, the formally self-adjoint operator B is evidently sufficiently "weak" as an integrability condition for K_0 that we obtain a complex. The construction may obviously be generalised by using different Lagrangian densities (cf. Q) in the first step. On the other hand, if we use directly the ideas from the *proof* of Theorem 3.2, then it seems that we may only obtain detour complexes with either the Killing operator on vector fields, $k \mapsto \mathcal{L}_k g$ (with g the metric), or its conformal analogue K_0, as the first operator $\mathcal{B}^0 \to \mathcal{B}^1$. This motivates a rather different approach.

The simplest example of a conformal detour complex is the Maxwell detour complex in dimension 4

$$\Lambda^0 \xrightarrow{d} \Lambda^1 \xrightarrow{\delta d} \Lambda_1 \xrightarrow{\delta} \Lambda_0 \,. \tag{3.5}$$

In other dimensions this is also a complex, but the middle operator is no longer conformally invariant. Let us temporarily relax the condition of conformal invariance and consider this complex on a pseudo-Riemannian n-manifold (M, g) of signature (p, q), with $n \geq 2$.

For any vector vector bundle V, with connection $\tilde{\nabla}$, we might consider twisting the Maxwell detour. Of course the result would not be a complex. However by a minor variation on this theme we make an interesting observation. Recall that we write $d^{\tilde{\nabla}}$ for the connection-coupled exterior derivative operator $d^{\tilde{\nabla}} : \Lambda^k(V) \to \Lambda^{k+1}(V)$. Of course we could equally consider the coupled exterior derivative operator $d^{\tilde{\nabla}} : \Lambda^k(V^*) \to \Lambda^{k+1}(V^*)$ and for the formal adjoint of this we write $\delta^{\tilde{\nabla}} : \Lambda_{k+1}(V) \to \Lambda_k(V)$.

Denote by F the curvature of $\tilde{\nabla}$ and write $F\cdot$ for the action of the curvature on the twisted 1-forms, $F\cdot : \Lambda^1(V) \to \Lambda_1(V)$ given by

$$(F \cdot \phi)_a := F_a{}^b \phi_b,$$

where we have indicated the abstract form indices explicitly, whereas the standard $\mathrm{End}(V)$ action of the curvature on the V-valued 1-form is implicit. Using this we construct a differential operator

$$M^{\tilde{\nabla}} : \Lambda^1(V) \to \Lambda_1(V)$$

by

$$M^{\tilde{\nabla}} \phi = \delta^{\tilde{\nabla}} d^{\tilde{\nabla}} \phi - F \cdot \phi \,.$$

By a direct calculation, the composition $M^{\tilde{\nabla}} d^{\tilde{\nabla}}$ amounts to an exterior algebraic action by the "Yang-Mills current" $\delta^{\tilde{\nabla}} F$, thus we have the first statement of the following result. The other claims are also easily verified.

THEOREM 3.3. *[33] The sequence of operators,*

$$\Lambda^0(V) \xrightarrow{d^{\tilde{\nabla}}} \Lambda^1(V) \xrightarrow{M^{\tilde{\nabla}}} \Lambda_1(V) \xrightarrow{\delta^{\tilde{\nabla}}} \Lambda_0(V) \tag{3.6}$$

is a complex if and only if the curvature F of the connection $\tilde{\nabla}$ satisfies the (pure) Yang-Mills equation

$$\delta^{\tilde{\nabla}} F = 0.$$

In addition:

(i) If $\tilde{\nabla}$ is an orthogonal or unitary connection then the sequence is formally self-adjoint.
(ii) In Riemannian signature the sequence is elliptic.
(iii) In dimension 4 the complex is conformally invariant.

This obviously yields a huge class of complexes. For example, taking V to be any tensor (or spin) bundle on a (spin) manifold with harmonic curvature (i.e. a pseudo-Riemannian structure where the Riemannian curvature satisfies the Yang-Mills equations) yields a complex. Einstein metrics are harmonic so this is large class of structures. To be specific if we take, for example, V to be the second exterior power of the tangent bundle T^2, and use ∇ to denote the Levi-Civita connection then we obtain the complex

$$T^2 \xrightarrow{d^\nabla} \Lambda^1 \otimes T^2 \xrightarrow{M^\nabla} \Lambda^1 \otimes T^2 \xrightarrow{\delta^\nabla} T^2,$$

where M^∇ is given by

$$S_b{}^{cd} \mapsto -2\nabla^a \nabla_{[a} S_{b]}{}^{cd} - R_{ba}{}^c{}_e S^{aed} - R_{ba}{}^d{}_e S^{ace} .$$

This example reminds us that part (iii) of the theorem means that the complex (3.6) is conformally invariant in dimension 4, provided the connection $\tilde{\nabla}$ is conformally invariant. For applications in conformal geometry we need suitable conformal $(V, \tilde{\nabla})$.

3.5. Short detour complexes and tractor connections. We may use Theorem 3.3 to construct more differential complexes. Consider the following general situation. Suppose that there are vector bundles (or rather section spaces thereof) $\mathcal{B}^0$, $\mathcal{B}^1$, $\mathcal{B}_1$ and $\mathcal{B}_0$ and differential operators L_0, L_1, L^1, L^0, D and $\overline{D}$ which act as indicated in the following diagram:

$$\begin{array}{ccccccc} \Lambda^0(V) & \xrightarrow{d^{\tilde{\nabla}}} & \Lambda^1(V) & \xrightarrow{\quad M^{\tilde{\nabla}} \quad} & \Lambda_1(V) & \xrightarrow{\delta^{\tilde{\nabla}}} & \Lambda_0(V) \\ {\scriptstyle L_0}\uparrow & & {\scriptstyle L_1}\uparrow & & \downarrow{\scriptstyle L^1} & & \downarrow{\scriptstyle L^0} \\ \mathcal{B}^0 & \xrightarrow{D} & \mathcal{B}^1 & \xrightarrow{\quad M^{\mathcal{B}} \quad} & \mathcal{B}_1 & \xrightarrow{\overline{D}} & \mathcal{B}_0 \end{array} \qquad [D]$$

The top sequence is (3.6) for a connection $\tilde{\nabla}$ with curvature F and the operator $M^{\mathcal{B}} : \mathcal{B}^1 \to \mathcal{B}_1$ is defined to be the composition $L^1 M^{\tilde{\nabla}} L_1$. Suppose that the squares at each end commute, in the sense that as operators $\mathcal{B}^0 \to \Lambda^1(V)$ we have $d^{\tilde{\nabla}} L_0 = L_1 D$ and as operators $\Lambda_1(V) \to \mathcal{B}_0$ we have $L^0 \delta^{\tilde{\nabla}} = \overline{D} L^1$. Then on $\mathcal{B}^0$ we have

$$M^{\mathcal{B}} D = L^1 M^{\tilde{\nabla}} L_1 D = L^1 M^{\tilde{\nabla}} d^{\tilde{\nabla}} L_0 = L^1 \epsilon(\delta^{\tilde{\nabla}} F) L_0,$$

and similarly $\overline{D} M^{\mathcal{B}} = -L^0 \iota(\delta^{\tilde{\nabla}} F) L_0$. Here $\epsilon(\delta^{\tilde{\nabla}} F)$ and $\iota(\delta^{\tilde{\nabla}} F)$ indicate exterior and interior actions of the Yang-Mills current $\delta^{\tilde{\nabla}} F$. Thus if $\tilde{\nabla}$ is

Yang-Mills then the lower sequence, viz.

$$\mathcal{B}^0 \xrightarrow{D} \mathcal{B}^1 \xrightarrow{M^{\mathcal{B}}} \mathcal{B}_1 \xrightarrow{\overline{D}} \mathcal{B}_0 \ , \tag{3.7}$$

is a complex.

Remark. Note that if the connection $\tilde{\nabla}$ preserves a Hermitian or metric structure on V then we need only the single commuting square $d^{\tilde{\nabla}} L_0 = L_1 D$ on $\mathcal{B}^0$ to obtain such a complex; by taking formal adjoints we obtain a second commuting square $(L^0 \delta^{\tilde{\nabla}} = \overline{D} L^1) : \mathcal{B}_1 \to \mathcal{B}_0$ where $\mathcal{B}_0$ and $\mathcal{B}_1$ are appropriate density twistings of the bundles $\mathcal{B}^0$ and $\mathcal{B}^1$ respectively.

We are now in a position to link the various constructions and indicate the general prospects. Suppose that in the setting of Riemannian or conformal geometry, we have some finite type PDE operator $D : \mathcal{B}^0 \to \mathcal{B}^1$ that we wish to study. A question we raised in Section 2 is whether there is a corresponding connection ∇^D that is equivalent in the sense that solutions of D are in 1-1 correspondence with parallel sections of ∇^D. Consider the situation of the first square in the diagram. If $D\phi = 0$ then obviously $L_0 \phi$ is parallel for $\tilde{\nabla}$. Now suppose that the L_i are in fact differential splitting operators, that is for $i = 0, 1$ these may be inverted by differential operators L_i^{-1} in the sense that $L_i^{-1} \circ L_i = \mathrm{id}_{\mathcal{B}^i}$. Then $D = L_1^{-1}(L_1 D) = L_1^{-1}(\tilde{\nabla} L_0)$ and so solutions for D are mapped by L_0 injectively to sections of V that are parallel for $\tilde{\nabla}$. Finally if also L_0^{-1} is well defined on general sections of V and $L_0 \circ L_0^{-1}$ acts as the identity on the space of parallel sections of V, then L_0^{-1} maps parallel sections for $\tilde{\nabla}$ to D-solutions. All these conditions are somewhat more than is strictly necessary according to ideas of section 2. Nevertheless, as we indicate below, these are satisfied for some examples, and it seems likely that these are the simplest cases of a large class.

3.6. Examples. Here we treat mainly a single example (with a hint of a second example), but it illustrates well the general idea. We return to the setting of conformal n-manifolds. Modulo the trace part, the conformal transformation of the Schouten tensor is controlled by the equation

$$D\sigma = 0 \quad \text{where} \quad D\sigma := \mathrm{TF}(\nabla_a \nabla_b \sigma + P_{ab}\sigma), \tag{3.8}$$

which is written in terms of some metric g from the conformal class. In particular a metric $\sigma^{-2} g$ is Einstein if and only if the scale $\sigma \in \mathcal{E}[1]$ is non-vanishing and satisfies (3.8).

The conformal standard tractor bundle and connection arises from a prolongation of this overdetermined equation as follows, cf. [1, 16]. Recalling the notation developed in Section 3.3 we have $\Lambda^{1,1}$ as an alternative notation for S_0^2. Let us also write $\Lambda_{1,1} := \Lambda^{1,1} \otimes \mathcal{E}[4-n]$. As usual we use the same notation for the spaces of sections of these. The *standard tractor bundle* $\mathcal{T}$ may be defined as the quotient of $J^2\mathcal{E}[1]$ by the image of $\Lambda^{1,1}[1]$ in $J^2\mathcal{E}[1]$ through the jet exact sequence at 2-jets. Note that there is a tautological operator $\mathbb{D} : \mathcal{E}[1] \to \Lambda^0(\mathcal{T})$ which is simply the composition of the universal 2-jet operator $j^2 : \mathcal{E}[1] \to J^2\mathcal{E}[1]$ followed by the canonical projection $J^2\mathcal{E}[1] \to \Lambda^0(\mathcal{T})$. By construction both $\mathcal{T}$ and $\mathbb{D}$ are invariant, they depend only on the conformal structure and no other choices. Via a choice of metric g, and the Levi-Civita connection it determines, we obtain a differential operator

$$\mathcal{E}[1] \to \mathcal{E}[1] \oplus \Lambda^1[1] \oplus \mathcal{E}[-1] \quad \text{by} \quad \sigma \mapsto (\sigma, \nabla_a \sigma, -\frac{1}{n}(\Delta + J)\sigma). \tag{3.9}$$

Since this is second order it factors through a linear map $J^2\mathcal{E}[1] \to \mathcal{E}[1] \oplus \Lambda^1[1] \oplus \mathcal{E}[-1]$. Considering Taylor series one sees that the kernel of this is a copy of $\Lambda^{1,1}[1]$ and so the map determines an isomorphism

$$\mathcal{T} \stackrel{g}{\cong} \mathcal{E}[1] \oplus \Lambda^1[1] \oplus \mathcal{E}[-1] \, . \tag{3.10}$$

In terms of this the formula at the right extreme of the display (3.9) then tautologically gives an explicit formula for $\mathbb{D}$. This is a differential splitting operator since through the jet projections there is a conformally invariant surjection $X : \mathcal{E}(\mathcal{T}) \to \mathcal{E}[1]$ which inverts $\mathbb{D}$. In terms of the splitting (3.10) this is simply $(\sigma, \mu, \rho) \mapsto \sigma$.

Observe that if we change to a conformally related metric $\widehat{g} = e^{2\omega} g$ then the Levi-Civita connection has a corresponding transformation, and so we obtain a different isomorphism. For $t \in \mathcal{T}$ then via (3.10) we have $[t]_g = (\sigma, \mu, \rho)$. In terms of the analogous isomorphism for $\widehat{g}$ we have

$$[t]_{\widehat{g}} = (\widehat{\sigma}, \widehat{\mu}_a, \widehat{\rho}) = (\sigma, \mu_a + \sigma\Upsilon_a, \rho - \boldsymbol{g}^{bc}\Upsilon_b\mu_c - \tfrac{1}{2}\sigma\boldsymbol{g}^{bc}\Upsilon_b\Upsilon_c). \tag{3.11}$$

where $\Upsilon = d\omega$.

Now let us define a connection on $\mathcal{E}[1] \oplus \Lambda^1[1] \oplus \mathcal{E}[-1]$ by the formula

$$\nabla_a \begin{pmatrix} \sigma \\ \mu_b \\ \rho \end{pmatrix} := \begin{pmatrix} \nabla_a \sigma - \mu_a \\ \nabla_a \mu_b + g_{ab}\rho + P_{ab}\sigma \\ \nabla_a \rho - P_{ab}\mu^b \end{pmatrix} \tag{3.12}$$

where, on the right-hand-side ∇ is the Levi-Civita connection for g. Obviously this determines a connection on $\mathcal{T}$ via the isomorphism (3.10). What is more surprising is that if we repeat this using a different metric $\widehat{g}$, this induces the *same* connection on $\mathcal{T}$. Equivalently the connection in the display transforms according to (3.11). This canonical connection on $\mathcal{T}$ depends only on the conformal structure and is known as the *(standard) tractor connection.*

There are several ways we may understand the connection (3.12) and its invariance. On the one hand this may be viewed as a special case of a *normal* tractor connection for parabolic geometries. This construction is treated from that point of view in [16]. Such connections determine the normal Cartan connections on the corresponding adapted frame bundles for the given tractor bundle $\mathbb{T}$, [17]. On the other hand one may start directly with the operator D in (3.8). It is readily verified that, for $n \geq 3$, this is of finite type and so we may attempt to construct a prolonged system and connection ∇^D for the operator along the lines of the treatment of Killing operator in Section 2. This is done in detail in [1] and one obtains exactly (3.12). So the connection (3.12) is the sought ∇^D. Some key points are as follows: since the operator in (3.8) acts $D : \mathcal{E}[1] \to \Lambda^{1,1}[1]$ and is second order, it factors through a linear bundle map $D^{(0)} : J^2\mathcal{E}[1] \to \Lambda^{1,1}[1]$; from the formula for D it is clear that this is a splitting of the exact sequence

$$0 \to \Lambda^{1,1}[1] \to J^2\mathcal{E}[1] \to \mathcal{T} \to 0$$

that defines $\mathcal{T}$. It follows that the prolonged system will include $\mathcal{T}$. In this case the key to the closing up of the prolonged system is that any completely symmetric covariant 3 tensor that is also pure trace on some pair of indices is necessarily zero. From this it follows that the full prolonged system may be expressed in terms of $\mathcal{T}$.

There is also a differential splitting operator

$$E : \Lambda^{1,1}[1] \to \Lambda^1(\mathcal{T}) \qquad \psi_{ab} \mapsto (0, \psi_{ab}, -(n-1)^{-1}\nabla^b\psi_{ab})$$

(cf. [20]). An easy calculation verifies this is also conformally invariant, and from the formula it is manifest that we have an operator E^{-1} (in fact bundle map) that inverts E from the left. Crucially, we have the following.

PROPOSITION 3.1. *[33] As differential operators on $\mathcal{E}[1]$, we have*

$$\nabla^D \mathbb{D} = ED\ .$$

For $\sigma \in \mathcal{E}[1]$, $\mathbb{D}\sigma$ is parallel if and only if $D\sigma = 0$.

Proof. The second statement is immediate from the first. A straightforward calculation verifies that either composition applied to $\sigma \in \mathcal{E}[1]$ yields

$$\begin{pmatrix} 0 \\ \mathrm{TF}(\nabla_a\nabla_b\sigma + P_{ab}\sigma) \\ -\frac{1}{n}\nabla_a(\Delta\sigma + J\sigma) - P_a{}^c\nabla_c\sigma \end{pmatrix}$$

□

For our present purposes the main point of Proposition 3.1 is that it gives the first step in constructing a special case of a commutative detour diagram [D]. First note that we have exactly the situation discussed in the last part of Section 3.5: here $\mathbb{D}$ and E play the roles of, respectively, L_0 and L_1. They are differential splitting operators and have inverses exactly as discussed there. It is an easy exercise to verify that if I is a parallel section of $\mathcal{T}$ then $I = \mathbb{D}\sigma$ for some section σ of $\mathcal{E}[1]$, as observed in [30] (and so X as an inverse to $\mathbb{D}$ maps parallel tractors to solutions of D). A useful consequence of this last result is that a conformal manifold with a parallel tractor is *almost Einstein* in the sense that it has a section of $\mathcal{E}[1]$

that gives an an Einstein scale on an open dense subset (see [27] for further details).

Next we observe that there is a conformally invariant *tractor metric* h on $\mathcal{T}$ given (as a quadratic form) by $(\sigma,\ \mu,\ \rho) \mapsto \boldsymbol{g}^{-1}(\mu,\mu) + 2\sigma\rho$. This has signature $(p+1, q+1)$ (corresponding to $\boldsymbol{g}$ of signature (p,q)) and is preserved by the tractor connection. In view of this, and using our general observations in Section 3.5, the formal adjoints of the operators above give the other required commutative square of operators. That is with

$$\begin{array}{ll} D^* : \Lambda_{1,1}[-1] \to \Lambda_0[-1] & \phi_{ab} \mapsto \nabla^a\nabla^b\phi_{ab} + P^{ab}\phi_{ab} \\ E^* : \Lambda_1(\mathcal{T}) \to \Lambda_{1,1}[-1] & (\alpha_a,\ \nu_{ab},\ \tau_a) \mapsto \nu_{(ab)_0} + \frac{1}{n-1}\nabla_{(a}\alpha_{b)_0} \\ \mathbb{D}^* : \Lambda_0(\mathcal{T}) \to \Lambda_0[-1] & (\sigma,\ \mu_b,\ \rho) \mapsto \rho - \nabla^a\mu_a - \frac{1}{n}(\Delta\sigma + J\sigma) \\ \delta^{\nabla^D} : \Lambda_1(\mathcal{T}) \to \Lambda_0(\mathcal{T}) & \Phi_{aB} \mapsto -\nabla^a\Phi_{aB} \end{array}$$

we have $\mathbb{D}^*\delta^{\nabla^D} = D^*E^*$ on $\Lambda_1(\mathcal{T})$.

We now want to consider what it means for the tractor connection to satisfy the Yang-Mills equations. Using the explicit presentation (3.12) of the tractor connection at a metric, it is straightforward to calculate its curvature. This is

$$\Omega_{ab}{}^C{}_D = \begin{pmatrix} 0 & 0 & 0 \\ A^c{}_{ab} & C_{ab}{}^c{}_d & 0 \\ 0 & -A_{dab} & 0 \end{pmatrix}$$

where A is the *Cotton tensor*, $A_{abc} := 2\nabla_{[b}P_{c]a}$. Taking the required divergence we obtain $-\delta^{\nabla^D}\Omega$ (see e.g. [30] for further details),

$$\nabla^D_a\Omega^a{}_b{}^C{}_E = \begin{pmatrix} 0 & 0 & 0 \\ B^c{}_b & (n-4)A_b{}^c{}_e & 0 \\ 0 & -B_{eb} & 0 \end{pmatrix} \tag{3.13}$$

where, on the left-hand side, ∇^D is really the Levi-Civita connection coupled with the tractor connection on $\mathrm{End}(\mathcal{T})$ induced from ∇^D. Here B_{ab} is the *Bach tensor* $B_{ab} := \nabla^c A_{acb} + P^{dc}C_{dacb}$.

Let us say that a *pseudo-Riemannian* manifold is *semi-harmonic* if its tractor curvature is Yang-Mills, that is $\nabla^a\Omega_{ab}{}^C{}_D = 0$. Note that in dimensions $n \neq 4$ this is not a conformally invariant condition and a semi-harmonic space is a Cotton space that is also Bach-flat. From our observations above, the semi-harmonic condition is conformally invariant in dimension 4 and according to the last display we have the following result (which in one form or another has been known for many years e.g. [41, 37]).

PROPOSITION 3.2. *In dimension 4 the conformal tractor connection is a Yang-Mills connection if and only if the structure is Bach-flat.*

Writing $M^{\mathcal{T}}$ for the composition $E^*M^{\nabla}E$ from the construction in section 3.5 above, we have the following (except for the claim of ellipticity, which is straightforward).

THEOREM 3.4. *[33] The sequence*

$$\Lambda^0[1] \xrightarrow{P} \Lambda^{1,1}[1] \xrightarrow{M^{\mathcal{T}}} \Lambda_{1,1}[-1] \xrightarrow{P^*} \Lambda_0[-1] \tag{3.14}$$

has the following properties.
(i) It is a formally self-adjoint sequence of differential operators and, for $\sigma \in \mathcal{E}[1]$

$$(M^{\mathcal{T}} D\sigma)_{ab} = -TFS\big(B_{ab}\sigma - (n-4)A_{abc}\nabla^c\sigma\big), \tag{3.15}$$

where $TFS(\cdots)$ *indicates the trace-free symmetric part of the tensor concerned. In particular it is a complex on semi-harmonic manifolds.*
(ii) In the case of Riemannian signature the complex is elliptic.
(iii) In dimension 4, (3.14) is a sequence of conformally invariant operators and it is a complex if and only if the conformal structure is Bach-flat.

COROLLARY 3.1. *Einstein 4-manifolds are Bach-flat.*

Proof. If a non-vanishing density σ is an Einstein scale then, calculating in that scale, we have $M^{\mathcal{T}} D\sigma = -B\sigma$, where B is the Bach tensor. On the other hand if σ is an Einstein scale then $D\sigma = 0$ (see (3.8)). □

In fact, the result in the Corollary, and more general results, have been known by other means for some time (see e.g. [31, 35] and references therein). Nevertheless it seems the detour complex gives an interesting route to this.

We conclude here with just the statements of another example from [33]. This is also obtained from the diagram [D] and an appropriate tractor connection ∇ for an operator D, in this case the operator D is the twistor operator. Thus we assume here that we have a conformal spin structure. Following [6] we write $\mathbb{S}$ for the basic spinor bundle and $\overline{\mathbb{S}} = \mathbb{S}[-n]$ (i.e. the bundle that pairs globally in an invariant way with $\mathbb{S}$ on conformal n-manifolds). The weight conventions here give $\mathbb{S}$ a "neutral weight". In terms of, for example, the Penrose weight conventions $\mathbb{S} = E^{\lambda}[-\frac{1}{2}] = E_{\lambda}[\frac{1}{2}]$, where E^{λ} denotes the basic contravariant spinor bundle in [42].

We write Tw for the so-called twistor bundle, that is the subbundle of $\Lambda^1 \otimes \mathbb{S}[1/2]$ consisting of form spinors u_a such that $\gamma^a u_a = 0$, where γ_a is the usual Clifford symbol. We use $\mathbb{S}$ and Tw also for the section spaces of these bundles. The *twistor operator* is the conformally invariant Stein-Weiss gradient

$$\mathbf{T} : \mathbb{S}[1/2] \to \mathrm{Tw}$$

given explicitly by

$$\psi \mapsto \nabla_a\psi + \frac{1}{n}\gamma_a\gamma^b\nabla_b\psi\ .$$

This completes to a differential complex as follows.

THEOREM 3.5. *[33] On semi-harmonic pseudo-Riemannian n-manifolds $n \geq 4$ we have a differential complex*

$$\mathbb{S}[1/2] \xrightarrow{\mathbf{T}} \mathrm{Tw} \xrightarrow{\mathbf{N}} \overline{\mathrm{Tw}} \xrightarrow{\mathbf{T}^*} \overline{\mathbb{S}}[-1/2], \tag{3.16}$$

where $\mathbf{T}$ *is the usual twistor operator,* $\mathbf{T}^*$ *its formal adjoint, and* $\mathbf{N}$ *is third order. The sequence is formally self-adjoint and in the case of Riemannian signature the complex is elliptic.*

In dimension 4 the sequence (3.16) is conformally invariant and it is a complex if and only if the conformal structure is Bach-flat.

The operator $\mathbf{N}$ is a third order analogue of a Rarita-Schwinger operator. Of course on a fixed pseudo-Riemannian manifold we may ignore the conformal weights.

3.7. Outlook. BGG complexes do not generally have curved analogues. The weaker integrability conditions involved with detour complexes suggest the possibility of large classes of such objects.

In the terminology of physics, the formally self-adjoint short detour complexes $\mathcal{B}^0 \to \mathcal{B}^1 \to \mathcal{B}_1 \to \mathcal{B}_0$ give "classically consistent systems". The constraint equations, $\mathcal{B}_1 \to \mathcal{B}_0$ which give an integrability condition on the middle operator, are suitably dual to, the gauge transformations of $\mathcal{B}^1$ given by $\mathcal{B}^0 \to \mathcal{B}^1$. Thus if we take $\mathcal{B}^1 \to \mathcal{B}_1$ as the field equations then the first cohomology of the complex gives "observable" field quantities. It seems that there is some scope for quantising this picture [28].

As we see in Corollary 3.1 the first operator $\mathcal{B}^0 \to \mathcal{B}^1$ can itself have an important geometric or physical interpretation and the complexes provide a tool for studying these. The first cohomology of the complex gives a global invariant.

Generalising the examples given above requires two main steps. The first is to develop the theory of prolonged differential systems in a way which leads to prolonged systems that share the invariance properties of the original equation (this is currently the subject of joint work with M.G. Eastwood). The second part is to understand what operators may be used to replace M^{∇^D}, so that, for example, we may have conformally invariant examples in higher dimensions.

REFERENCES

[1] T.N. BAILEY, M.G. EASTWOOD, AND A.R. GOVER, *Thomas's structure bundle for conformal, projective and related structures*, Rocky Mountain J. Math. (1994), **24**: 1191–1217.

[2] R. BASTON, *Almost Hermitian symmetric manifolds I, II*, Duke Math. Jour. (1991), **63**: 81–111, 113–138.

[3] B.D. BOE AND D.H. COLLINGWOOD, *A comparison theory for the structure of induced representations, II.* Math. Z. (1985), **190**: 1–11.

[4] N. BOKAN, P. GILKEY, AND R. ŽIVALJEVIĆ, *An inhomogeneous elliptic complex*, J. Anal. Math. (1993), **61**: 367–393.

[5] T. BRANSON, *Sharp inequalities, the functional determinant, and the complementary series.* Trans. Amer. Math. Soc. (1995), **347**: 3671–3742.
[6] T. BRANSON, *Conformal Structure and spin geometry* in: Dirac operators: yesterday and today, Proceedings of the Summer School and Workshop held in Beirut, August 27–September 7, 2001. Edited by J.-P. Bourguignon, T. Branson, A. Chamseddine, O. Hijazi, and R.J. Stanton. International Press, Somerville, MA, 2005. viii+310 pp.
[7] T. BRANSON, *Q-curvature and spectral invariants*, Rend. Circ. Mat. Palermo (2) Suppl. (2005), **75**: 11–55.
[8] T. BRANSON, A. ČAP, M.G. EASTWOOD, AND A.R. GOVER, *Prolongations of Geometric Overdetermined Systems*, Int. J. Math. (2006), **17**: 641–664.
[9] T. BRANSON AND A.R. GOVER, *Pontrjagin forms and invariant objects related to the Q-curvature*, Commun. Contemp. Math., to appear. Preprint math.DG/0511311, http://www.arxiv.org
[10] T. BRANSON AND A.R. GOVER, *The conformal deformation detour complex for the obstruction tensor*, Proc. Amer. Math. Soc. to appear. Preprint Math.DG/0605192, http://www.arxiv.org
[11] T. BRANSON AND A.R. GOVER, *Conformally invariant operators, differential forms, cohomology and a generalisation of Q-curvature*, Comm. Partial Differential Equations (2005), **30**: 1611–1669.
[12] T. BRANSON AND A.R. GOVER, *Detour torsion.* In progress.
[13] D.M.J. CALDERBANK AND T. DIEMER, *Differential invariants and curved Bernstein-Gelfand-Gelfand sequences*, J. Reine Angew. Math. (2001), **537**: 67–103.
[14] D.M.J. CALDERBANK, T. DIEMER, AND V. SOUČEK, *Ricci-corrected derivatives and invariant differential operators*, Differential Geom. Appl. (2005), **23**: 149–175.
[15] A. ČAP, *Overdetermined systems, conformal geometry, and the BGG complex.* Preprint math.DG/0610225, http://www.arxiv.org.
[16] A. ČAP AND A.R. GOVER, *Tractor bundles for irreducible parabolic geometries.* Global analysis and harmonic analysis (Marseille-Luminy, 1999), pp. 129–154, Sémin. Congr., **4**, Soc. Math. France, Paris, 2000. Preprint Publications/SeminairesCongres, http://smf.emath.fr.
[17] A. ČAP AND A.R. GOVER, *Tractor calculi for parabolic geometries*, Trans. Amer. Math. Soc. (2002), **354**: 1511-1548.
[18] A. ČAP, J. SLOVÁK, AND V. SOUČEK, *Bernstein-Gelfand-Gelfand sequences*, Annals of Math., (2001), **154**: 97–113.
[19] S.-Y. A. CHANG, Non-linear elliptic equations in conformal geometry, Zurich Lectures in Advanced Mathematics, European Mathematical Society, Zürich, 2004, viii+92pp.
[20] M.G. EASTWOOD, *Notes on conformal differential geometry.* The Proceedings of the 15th Winter School "Geometry and Physics" (Srni, 1995). Rend. Circ. Mat. Palermo (2) Suppl. (1996), **43**: 57–76.
[21] M. EASTWOOD, *Variations on the de Rham complex*, Notices Amer. Math. Soc. (1999), **46**: 1368–1376.
[22] M.G. EASTWOOD AND J.W. RICE, *Conformally invariant differential operators on Minkowski space and their curved analogues*, Comm. Math. Phys. (1987), **109**: 207–228; Erratum: Comm. Math. Phys. (1992), **144**: 213.
[23] C. FEFFERMAN AND C.R. GRAHAM, *Conformal invariants*, in *Elie Cartan et les mathématiques d'aujourd'hui*, Astérisque, hors série (SMF, Paris, 1985), 95–116.
[24] J. GASQUI, *Sur la résolubilité locale des équations d'Einstein*, [On the local solvability of Einstein's equations], Compositio Math. (1982), **47**: 43–69.
[25] J. GASQUI AND H. GOLDSCHMIDT, *Déformations infinitésimales des structures conformes plates (French) [Infinitesimal deformations of flat conformal structures].* Progress in Mathematics, 52. Birkhäuser Boston, Inc., Boston, MA, 1984.

[26] A.R. Gover, *Conformally invariant operators of standard type*, Quart. J. Math. (1989), **40**: 197–207.
[27] A.R. Gover, *Almost conformally Einstein manifolds and obstructions*, in Differential Geometry and its Applications, proceedings of the 9th International Conference on Differential Geometry and its Applications, Prague 2004, Charles University, Prague, 2005, pp. 243–255. Electronic: math.DG/0412393, http://www.arxiv.org.
[28] A.R. Gover, K. Hallowell, and A. Waldron, *Higher Spin Gravitational Couplings and the Yang–Mills Detour Complex*, preprint hep-th/0606160, http://www.arxiv.org. Physical Review D, to appear.
[29] A.R. Gover and F. Leitner, *A sub-product construction of Poincaré-Einstein metrics*, preprint math.DG/0608044, http://www.arxiv.org.
[30] A.R. Gover and P. Nurowski, *Obstructions to conformally Einstein metrics in n dimensions*, J. Geom. Phys. (2006), **56**: 450–484.
[31] A.R. Gover and L.J. Peterson, *The ambient obstruction tensor and the conformal deformation complex*, Pacific J. Math. (2006), **226**: 309–351. Preprint math.DG/0408229, http://arXiv.org.
[32] A.R. Gover and J. Silhan, *The conformal Killing equation on forms – prolongations and applications*. Preprint math.DG/0601751, http://arxiv.org. Differential Geom. Appl., to appear.
[33] A.R. Gover, Petr Somberg, and Vladimir Souček, *Yang-Mills detour complexes and conformal geometry*. Preprint math.DG/0606401, http://arXiv.org.
[34] C.R. Graham, R. Jenne, L. Mason, and G. Sparling, *Conformally invariant powers of the Laplacian, I: existence*, J. London Math. Soc., (1992), **46**: 557–565.
[35] C.R. Graham and K. Hirachi, *The ambient obstruction tensor and Q-curvature*, in: AdS/CFT correspondence: Einstein metrics and their conformal boundaries, 59–71, IRMA Lect. Math. Theor. Phys., 8, Eur. Math. Soc., Zürich, 2005.
[36] C.R. Graham and M. Zworski, *Scattering matrix in conformal geometry*, Invent. Math. (2003), **152**: 89–118.
[37] M. Korzynski and J. Lewandowski, *The Normal Conformal Cartan Connection and the Bach Tensor*, Class. Quant. Grav. (2003), **20**: 3745–3764.
[38] B. Kostant, *Lie algebra cohomology and the generalized Borel-Weil theorem*, Ann. Math. (1961), **74**: 329–387.
[39] J.M. Lee and T.H. Parker, *The Yamabe problem*, Bull. Amer. Math. Soc. (1987), **17**: 37–91.
[40] J. Lepowsky, *A generalization of the Bernstein-Gelfand-Gelfand resolution*, J. Alg. (1977), **49**: 496–511.
[41] S. Merkulov, *A conformally invariant theory of gravitation and electromagnetism*, Class. Quantum Gravity (1984), **1**: 349–354.
[42] R. Penrose and W. Rindler, Wolfgang, Spinors and space-time, Vol 1 and Vol. 2. Spinor and twistor methods in space-time geometry, Cambridge Monographs on Mathematical Physics Cambridge University Press, Cambridge, 1987, 1988, x+458pp, x+501 pp.
[43] U. Semmelmann, *Conformal Killing forms on Riemannian manifolds*, Math. Z. (2003), **245**: 503–527.
[44] D.C. Spencer, Overdetermined systems of linear partial differential equations, *Bull. Amer. Math. Soc.* (1969), **75**: 179–239.

INHOMOGENEOUS AMBIENT METRICS*

C. ROBIN GRAHAM[†] AND KENGO HIRACHI[‡]

1. Introduction. The ambient metric, introduced in [FG1], has proven to be an important object in conformal geometry. To a manifold M of dimension n with a conformal class of metrics $[g]$ of signature (p,q) it associates a diffeomorphism class of formal expansions of metrics $\widetilde{g}$ of signature $(p+1, q+1)$ on a space $\widetilde{\mathcal{G}}$ of dimension $n+2$. This generalizes the realization of the conformal sphere S^n as the space of null lines for a quadratic form of signature $(n+1, 1)$, with associated Minkowski metric $\widetilde{g}$ on $\mathbb{R}^{n+2}$. The ambient space $\widetilde{\mathcal{G}}$ carries a family of dilations with respect to which $\widetilde{g}$ is homogeneous of degree 2. The other conditions determining $\widetilde{g}$ are that it be Ricci-flat and satisfy an initial condition specified by the conformal class $[g]$.

The ambient metric behaves differently in even and odd dimensions, reflecting an underlying distinction in the structure of the space of jets of metrics modulo conformal rescaling. When n is odd, the equations determining $\widetilde{g}$ have a smooth infinite-order formal power series solution, uniquely determined modulo diffeomorphism. But when n is even and ≥ 4, there is a obstruction at order $n/2$ to the existence of a smooth formal solution, which is realized as a conformally invariant natural tensor called the ambient obstruction tensor. It is possible to continue the expansion of $\widetilde{g}$ to higher orders by including log terms ([K]), but this destroys the differentiability of the solution so is problematic for applications requiring higher order differentiation.

In this article, we describe a modification to the form of the ambient metric in even dimensions which enables us to obtain invariantly defined, smooth infinite order "ambient metrics". We introduce what we call inhomogeneous ambient metrics, which are formally Ricci-flat and have an asymptotic expansion involving the logarithm of a defining function for the initial surface which is homogeneous of degree 2. Such metrics are themselves no longer homogeneous, and of course are not smooth. However, we are able to define the smooth part of such a metric in an invariant way, and the smooth part is homogeneous and of course smooth and can be used in applications just as the ambient metric itself is used in odd dimensions. A significant difference, though, is that an inhomogeneous ambient metric is no longer uniquely determined up to diffeomorphism by the conformal class

*This research was partially supported by NSF grant # DMS 0505701 and Grants-in-Aid for Scientific Research, JSPS.

[†]Department of Mathematics, University of Washington, Box 354350, Seattle, WA 98195-4350 (robin@math.washington.edu).

[‡]Graduate School of Mathematical Sciences, University of Tokyo, 3-8-1 Komaba, Meguro, Tokyo 153-8914, Japan (hirachi@ms.u-tokyo.ac.jp).

$[g]$ on M; there is a family of diffeomorphism classes of inhomogeneous ambient metrics, and therefore of their smooth parts. Upon choosing a metric g in the conformal class, this family can be parametrized by the choice of an arbitrary trace-free symmetric 2-tensor field on M which we call the ambiguity tensor relative to the representative g.

There are two main motivations for the introduction of terms involving the logarithm of a defining function homogeneous of degree 2 in the expansion of an ambient metric. The simplest is the observation that in flat space, the null cone of the Minkowski metric has an invariant defining function homogeneous of degree 2, namely the quadratic form Q itself, but no invariant defining function homogeneous of degree 0. In this flat space setting, the theory of invariants of conformal densities initiated in [EG] can be extended to all orders by the introduction of $\log |Q|$ terms in the expansion of the ambient harmonic extension of a density. We will report on this work on invariants of densities elsewhere.

Another motivation is the construction of the potential of the inhomogeneous CR ambient metric in [H], which also involves the log of a defining function homogeneous of degree 2 and an invariantly defined smooth part. The existence of this construction in the CR case was compelling evidence that there should be a conformal analogue. Nonetheless, despite having the CR construction as a guide, we did not find the conformal construction to be straightforward. The relation between the constructions is still not completely clear; further discussion of this issue is contained in §3.

We also indicate how inhomogeneous ambient metrics can be used to complete the description of scalar invariants of conformal structures in even dimensions. One can form scalar invariants as Weyl invariants, defined to be linear combinations of complete contractions of covariant derivatives of the curvature tensor of the smooth part of an inhomogeneous ambient metric, and also its volume form and a modified volume form in the case of odd invariants. Such invariants generally depend on the choice of ambiguity. Nonetheless, in dimensions $n \equiv 2 \mod 4$, all scalar conformal invariants arise as Weyl invariants which are independent of the ambiguity. If $n \equiv 0 \mod 4$, there are exceptional odd invariants which are not of this form. In this case, following [BG], we provide a construction of a finite set of basic exceptional invariants, which, together with ambiguity-independent Weyl invariants, span all scalar conformal invariants. The study of scalar invariants was initiated in [F2] in the CR case. Fundamental algebraic results were derived in [BEGr], resulting in the full description of scalar invariants in odd dimensional conformal geometry and below the order of the obstruction in CR and even dimensional conformal geometry. The completion of the description in the CR case was given in [H]. A different approach to the study of conformal invariants is developed in [Go] using tractor calculus. In [A1], [A2], Alexakis has recently derived a theory handling invariants of a density coupled with a conformal structure, below the order of the obstruction of the ambient metric in even dimensions, and below the order of the obstruction of harmonic extension of the density.

In §2, we review the construction of smooth homogeneous ambient metrics. Inhomogeneous ambient metrics are introduced in §3 and the main theorem asserting the existence and uniqueness of inhomogeneous ambient metrics with prescribed ambiguity tensor is formulated. We describe in some detail the transformation law for the ambiguity tensor under conformal change. We state a necessary and sufficient condition for the asymptotic expansion of an inhomogeneous ambient metric to have a simpler form, and show by explicit identification of the obstruction tensor that Fefferman metrics of nondegenerate integrable CR manifolds always satisfy this condition. We also briefly discuss the Poincaré metrics associated to inhomogeneous ambient metrics. In §4, we describe the results concerning scalar invariants and give some examples of invariants. With the exception of the identification of the obstruction tensor of a Fefferman metric, this paper does not contain proofs. Detailed proofs of the results described here will appear elsewhere.

2. Smooth homogeneous ambient metrics. In this section we review the usual ambient metric construction in odd dimensions and up to the obstruction in even dimensions. Let $[g]$ be a conformal class of metrics of signature (p,q) on a (paracompact) manifold M of dimension $n \geq 3$. Here g is a smooth metric of signature (p,q) on M and $[g]$ consists of all metrics of the form $\Omega^2 g$ with $0 < \Omega \in C^\infty(M)$. The metric bundle $\mathcal{G} \subset \bigodot^2 T^*M$ of $[g]$ is the set of all pairs $(x,\underline{g})$, where $x \in M$ and $\underline{g} \in \bigodot^2 T^*_xM$ is of the form $\underline{g} = t^2 g(x)$ for some $t > 0$. The fiber variable t on $\mathcal{G}$ so defined is associated to the choice of metric g and provides an identification $\mathcal{G} \cong \mathbb{R}_+ \times M$. Sections of $\mathcal{G}$ are precisely metrics in the conformal class. There is a tautological symmetric 2-tensor g_0 on $\mathcal{G}$ defined by $g_0(X,Y) = \underline{g}(\pi_*X, \pi_*Y)$, where $\pi : \mathcal{G} \to M$ is the projection and X, Y are tangent vectors to $\mathcal{G}$ at $(x,\underline{g}) \in \mathcal{G}$. The family of dilations $\delta_s : \mathcal{G} \to \mathcal{G}$ defined by $\delta_s(x,\underline{g}) = (x, s^2\underline{g})$ defines an $\mathbb{R}_+$ action on $\mathcal{G}$, and one has $\delta^* g_0 = s^2 g_0$. We denote by $T = \frac{d}{ds}\delta_s|_{s=1}$ the vector field on $\mathcal{G}$ which is the infinitesimal generator of the dilations δ_s.

The ambient space is $\widetilde{\mathcal{G}} = \mathcal{G} \times \mathbb{R}$; the coordinate in the $\mathbb{R}$ factor is typically written ρ. The dilations δ_s extend to $\widetilde{\mathcal{G}}$ acting on the $\mathcal{G}$ factor and we denote also by T the infinitesimal generator of the δ_s on $\widetilde{\mathcal{G}}$. We embed $\mathcal{G}$ into $\widetilde{\mathcal{G}}$ by $\iota : z \to (z,0)$ for $z \in \mathcal{G}$, and we identify $\mathcal{G}$ with its image under ι. As described above, a choice of representative metric g induces an identification $\mathcal{G} \cong \mathbb{R}_+ \times M$, so also an identification $\widetilde{\mathcal{G}} \cong \mathbb{R}_+ \times M \times \mathbb{R}$. We use capital Latin indices to label objects on $\widetilde{\mathcal{G}}$, '0' indices to label the $\mathbb{R}_+$ factor, lower case Latin indices to label the M factor, and '∞' indices for the $\mathbb{R}$ factor.

In the even-dimensional case, different components of the ambient metric are determined to different orders. If S_{IJ} is a smooth symmetric 2-tensor field on an open neighborhood of $\mathcal{G}$ in $\widetilde{\mathcal{G}}$, then for $m \geq 0$ we write $S_{IJ} = O^+_{IJ}(\rho^m)$ if:

(i) $S_{IJ} = O(\rho^m)$; and

(ii) For each point $z \in \mathcal{G}$, the symmetric 2-tensor $(\iota^*(\rho^{-m}S))(z)$ is of the form $\pi^* s$ for some symmetric 2-tensor s at $x = \pi(z) \in M$ satisfying $\operatorname{tr}_{g_x} s = 0$. The symmetric 2-tensor s is allowed to depend on z, not just on x.

In terms of components relative to a choice of representative metric g, $S_{IJ} = O^+_{IJ}(\rho^m)$ if and only if all components satisfy $S_{IJ} = O(\rho^m)$ and if in addition one has that S_{00}, S_{0i} and $g^{ij}S_{ij}$ are $O(\rho^{m+1})$. The condition $S_{IJ} = O^+_{IJ}(\rho^m)$ is easily seen to be preserved by diffeomorphisms on a neighborhood of $\mathcal{G}$ which restrict to the identity on $\mathcal{G}$.

We consider metrics $\widetilde{g}$ of signature $(p+1, q+1)$ defined on a neighborhood of $\mathcal{G}$ in $\widetilde{\mathcal{G}}$. We will assume that the neighborhood is homogeneous, i.e., it is invariant under the dilations δ_s for $s > 0$. A smooth diffeomorphism defined on such a neighborhood will be said to be homogeneous if it commutes with the δ_s. The metric $\widetilde{g}$ (or more generally a symmetric 2-tensor field) will be said to be homogeneous (of degree 2) if it satisfies $\delta_s^* \widetilde{g} = s^2 \widetilde{g}$.

The main existence and uniqueness result for smooth ambient metrics is the following.

THEOREM 2.1. *Let $[g]$ be a conformal class on a manifold M. If n is odd, then there is a smooth metric $\widetilde{g}$ on a homogeneous neighborhood of $\mathcal{G}$ in $\widetilde{\mathcal{G}}$, uniquely determined up to a homogeneous diffeomorphism of a neighborhood of $\mathcal{G}$ in $\widetilde{\mathcal{G}}$ which restricts to the identity on $\mathcal{G}$, and up to a homogeneous term vanishing to infinite order along $\mathcal{G}$, by the requirements:*

(1) $\delta_s^* \widetilde{g} = s^2 \widetilde{g}$

(2) $\iota^* \widetilde{g} = g_0$

(3) $\operatorname{Ric}(\widetilde{g}) = 0$ *to infinite order along $\mathcal{G}$.*

If $n \geq 4$ is even, the same statement holds except that (3) is replaced by $\operatorname{Ric}(\widetilde{g}) = O^+_{IJ}(\rho^{n/2-1})$, *and $\widetilde{g}$ is uniquely determined up to a homogeneous diffeomorphism and up to a smooth homogeneous term which is $O^+_{IJ}(\rho^{n/2})$.*

A metric $\widetilde{g}$ satisfying the conditions of Theorem 2.1 is called a smooth ambient metric for $(M, [g])$.

The diffeomorphism indeterminacy in $\widetilde{g}$ can be fixed by the choice of a metric g in the conformal class. As described above, the choice of such a metric g determines an identification $\widetilde{\mathcal{G}} \cong \mathbb{R}_+ \times M \times \mathbb{R}$.

DEFINITION 2.1. *A smooth metric $\widetilde{g}$ on $\widetilde{\mathcal{G}}$ satisfying (1) and (2) above is said to be in normal form relative to g if in the identification $\widetilde{\mathcal{G}} = \mathbb{R}_+ \times M \times \mathbb{R}$ induced by g,*

(a) $\widetilde{g} = 2t\,dt\,d\rho + g_0$ *at $\rho = 0$, and*

(b) The lines $\rho \to (t, x, \rho)$ are geodesics for $\widetilde{g}$ for each choice of $(t, x) \in \mathbb{R}_+ \times M$.

As in the construction of Gaussian coordinates relative to a hypersurface, it follows by straightening out geodesics that if $\widetilde{g}$ is any smooth metric satisfying (1) and (2) and g is a representative metric in the conformal class, then there is a unique homogeneous diffeomorphism Φ defined in a homogeneous neighborhood of $\mathcal{G}$ and which restricts to the identity on $\mathcal{G}$, such that $\Phi^*\widetilde{g}$ is in normal form relative to g. It follows that in Theorem 2.1, the metric $\widetilde{g}$ can be chosen to be in normal form relative to g, and it is then uniquely determined up to $O(\rho^\infty)$ for n odd, and up to $O^+_{IJ}(\rho^{n/2})$ for n even.

Theorem 2.1 is proved by a formal power series analysis of the equation $\operatorname{Ric}(\widetilde{g}) = 0$ for metrics $\widetilde{g}$ in normal form relative to some metric g in the conformal class. In carrying out this analysis, one finds that the solution has an additional property.

DEFINITION 2.2. *A metric $\widetilde{g}$ on a homogeneous neighborhood $\mathcal{U}$ of $\mathcal{G}$ in $\widetilde{\mathcal{G}}$ is said to be straight if for each $p \in \mathcal{U}$, the dilation orbit $s \to \delta_s p$ is a geodesic for $\widetilde{g}$.*

PROPOSITION 2.1. *In Theorem 2.1, the metric $\widetilde{g}$ can be chosen to be straight.*

Since the condition that $\widetilde{g}$ be straight is invariant under smooth homogeneous diffeomorphisms, it follows that any smooth ambient metric $\widetilde{g}$ agrees with a straight metric to infinite order when n is odd and mod $O^+_{IJ}(\rho^{n/2})$ when n is even. That is, any smooth ambient metric is straight to the order that it is determined.

When n is even, in general there is an obstruction to the existence of a smooth ambient metric solving $\operatorname{Ric}(\widetilde{g}) = O(\rho^{n/2})$. Let $\widetilde{g}$ be a smooth ambient metric. We denote by $r_\#$ the function $r_\# = \|T\|^2 = \widetilde{g}(T,T)$. Since $g_0(T,T) = 0$, it follows that $r_\# = 0$ on $\mathcal{G}$, and it turns out (as a consequence of the Einstein condition or the straightness condition) that $dr_\# \neq 0$ on $\mathcal{G}$. Thus $r_\#$ is a defining function for $\mathcal{G} \subset \widetilde{\mathcal{G}}$ invariantly associated to $\widetilde{g}$ which is homogeneous of degree 2: $\delta_s^* r_\# = s^2 r_\#$. Now since $\operatorname{Ric}(\widetilde{g}) = O^+_{IJ}(\rho^{n/2-1})$, the quantity $(r_\#^{1-n/2} \operatorname{Ric} \widetilde{g})|_{T\mathcal{G}}$ defines a tensor field on $\mathcal{G}$, homogeneous of degree $2-n$, which annihilates T. It therefore defines a symmetric 2-tensor-density on M of weight $2-n$, which is trace-free. If g is a metric in the conformal class, evaluating this tensor-density at the image of g viewed as a section of $\mathcal{G}$ defines a 2-tensor on M which we denote by $(r_\#^{1-n/2} \operatorname{Ric} \widetilde{g})|_g$. The obstruction tensor of the metric g is defined to be

$$\mathcal{O} = c_n (r_\#^{1-n/2} \operatorname{Ric} \widetilde{g})|_g, \qquad c_n = (-1)^{n/2-1} \frac{2^{n-2}(n/2-1)!^2}{n-2}. \tag{2.1}$$

We then have:

PROPOSITION 2.2. *Let $n \geq 4$ be even. The obstruction tensor $\mathcal{O}_{ij}$ of g is independent of the choice of smooth ambient metric $\widetilde{g}$ and has the following properties:*

1. *$\mathcal{O}$ is a natural tensor invariant of the metric g; i.e. in local coordinates the components of $\mathcal{O}$ are given by universal polynomials*

in the components of g, g^{-1} and the curvature tensor of g and its covariant derivatives. The expression for $\mathcal{O}_{ij}$ takes the form

$$\begin{aligned} \mathcal{O}_{ij} &= \Delta^{n/2-2}\left(P_{ij},_k{}^k - P_k{}^k,_{ij}\right) + lots \\ &= (3-n)^{-1}\Delta^{n/2-2}W_{kijl},^{kl} + lots, \end{aligned} \tag{2.2}$$

where $\Delta = \nabla^i\nabla_i$,

$$P_{ij} = \frac{1}{n-2}\left(R_{ij} - \frac{R}{2(n-1)}g_{ij}\right),$$

W_{ijkl} *is the Weyl tensor, and lots denotes quadratic and higher terms in curvature involving fewer derivatives.*

2. *One has*

$$\mathcal{O}_i{}^i = 0 \qquad\qquad \mathcal{O}_{ij},^j = 0.$$

3. $\mathcal{O}_{ij}$ *is conformally invariant of weight* $2-n$; *i.e. if* $0 < \Omega \in C^\infty(M)$ *and* $\widehat{g}_{ij} = \Omega^2 g_{ij}$, *then* $\widehat{\mathcal{O}}_{ij} = \Omega^{2-n}\mathcal{O}_{ij}$.
4. *If* g_{ij} *is conformal to an Einstein metric, then* $\mathcal{O}_{ij} = 0$.
5. *If* $n = 4$, *then* $\mathcal{O}_{ij} = B_{ij}$ *is the classical Bach tensor.*

Here the Bach tensor is defined in any dimension $n \geq 3$ by

$$B_{ij} = C_{ijk},^k - P^{kl}W_{kijl},$$

where C_{ijk} is the Cotton tensor

$$C_{ijk} = P_{ij},_k - P_{ik},_j\,.$$

Clearly, if the obstruction tensor is nonzero, then it is impossible to find a smooth ambient metric $\widetilde{g}$ solving $\mathrm{Ric}(\widetilde{g}) = O(\rho^{n/2})$.

3. Inhomogeneous ambient metrics. Restrict now to the case n even. In order to find ambient metrics beyond order $n/2$, we broaden the class of allowable metrics $\widetilde{g}$. The $\widetilde{g}$ that we consider are neither homogeneous nor smooth. However, $\widetilde{g}$ will always be required to satisfy the initial condition

$$\iota^*\widetilde{g} = g_0. \tag{3.1}$$

Let r be a smooth defining function for $\mathcal{G} \subset \widetilde{\mathcal{G}}$ homogeneous of degree 0 (one could for example choose $r = \rho$) and let $r_\#$ be a smooth defining function homogeneous of degree 2 (for example, $r_\# = 2\rho t^2$). Denote by $\mathcal{M}$ the space of formal asymptotic expansions along $\mathcal{G}$ of metrics on $\widetilde{\mathcal{G}}$ of signature $(p+1, q+1)$ of the form

$$\widetilde{g} \sim \widetilde{g}^{(0)} + \sum_{N\geq 1}\widetilde{g}^{(N)}r(r^{n/2-1}\log|r_\#|)^N \tag{3.2}$$

where each $\widetilde{g}^{(N)}$, $N \geq 0$, is a smooth symmetric 2-tensor field on $\widetilde{\mathcal{G}}$ satisfying $\delta_s^* \widetilde{g}^{(N)} = s^2 \widetilde{g}^{(N)}$, such that the initial condition (3.1) holds. Observe that $\widetilde{g} \in \mathcal{M}$ is the expansion of a smooth metric on $\widetilde{\mathcal{G}}$ if and only if $\widetilde{g}^{(N)} = 0$ to infinite order for $N \geq 1$. We will say in this case that $\widetilde{g}$ is smooth. Note that if $\widetilde{g}$ is smooth, then $\widetilde{g}$ agrees to infinite order with a homogeneous metric; we will also say that $\widetilde{g}$ is homogeneous. Denote by $\mathcal{A}$ the space of formal asymptotic expansions of scalar functions f on $\widetilde{\mathcal{G}}$ of the form

$$f \sim f^{(0)} + \sum_{N \geq 1} f^{(N)} r (r^{n/2-1} \log |r_\#|)^N,$$

where each $f^{(N)}$ is a smooth function on $\widetilde{\mathcal{G}}$ homogeneous of degree 0. Observe that these spaces of asymptotic expansions are independent of the choice of defining functions r, $r_\#$. If Φ is a formal smooth, homogeneous diffeomorphism of $\widetilde{\mathcal{G}}$ satisfying $\Phi|_{\mathcal{G}} = Id$, then pullback by Φ preserves $\mathcal{M}$ and $\mathcal{A}$. As already initiated above, in the following we will refer to such asymptotic expansions for metrics, functions, and formal smooth diffeomorphisms as if they were actually defined in a neighborhood of $\mathcal{G}$, and when we say that such an object satisfies a certain condition, we will mean that the condition holds formally to infinite order along $\mathcal{G}$. For example, we will say that $\widetilde{g}$ is straight if the ordinary differential equations which state that the orbits $s \to \delta_s p$ are geodesics hold formally to infinite order along $\mathcal{G}$.

For inhomogeneous ambient metrics, we impose the straightness condition at the outset.

DEFINITION 3.1. *An inhomogeneous ambient metric for* $(M, [g])$ *is a straight metric* $\widetilde{g} \in \mathcal{M}$ *satisfying* $\operatorname{Ric}(\widetilde{g}) = 0$.

The straightness condition is crucial in the inhomogeneous case because of the following proposition.

PROPOSITION 3.1. *Let* $\widetilde{g} \in \mathcal{M}$ *be straight. Then* $T \lrcorner \widetilde{g}$ *is smooth and* $\widetilde{g}(T, T)$ *is a smooth defining function for* $\mathcal{G}$ *homogeneous of degree 2.*

Observe that for $\widetilde{g} \in \mathcal{M}$, $T \lrcorner \widetilde{g}$ and $\widetilde{g}(T, T)$ will in general have asymptotic expansions involving $\log |r_\#|$, so will be neither smooth nor homogeneous. Proposition 3.1 asserts that the requirement that $\widetilde{g}$ be straight has as a consequence that no log terms occur in the expansions of these quantities. The proof of Proposition 3.1 is a straightforward analysis of the geodesic equations for the dilation orbits.

If $\widetilde{g} \in \mathcal{M}$ is straight, then $\widetilde{g}(T, T)$ is a canonically determined smooth defining function for $\mathcal{G}$ homogeneous of degree 2. We may therefore take $r_\# = \widetilde{g}(T, T)$ in (3.2). Having fixed $r_\#$, the smooth symmetric 2-tensor fields $\widetilde{g}^{(0)}$ and $r^{(n/2-1)N+1} \widetilde{g}^{(N)}$ for $N \geq 1$ are then uniquely determined by $\widetilde{g}$ independently of any choices.

DEFINITION 3.2. *Let* $\widetilde{g} \in \mathcal{M}$ *be straight. The smooth part of* $\widetilde{g}$ *is the metric* $\widetilde{g}^{(0)}$ *appearing in the expansion* (3.2) *when* $r_\#$ *is taken to be* $r_\# = \widetilde{g}(T, T)$.

It is clear that the smooth part of $\widetilde{g}$ is a smooth metric in $\mathcal{M}$. If Φ is a smooth homogeneous diffeomorphism satisfying $\Phi|_{\mathcal{G}} = Id$ and $\widetilde{g} \in \mathcal{M}$ is straight, then $(\Phi^*\widetilde{g})^{(0)} = \Phi^*(\widetilde{g}^{(0)})$.

Next we formulate the notion of normal form for straight metrics in $\mathcal{M}$.

DEFINITION 3.3. *Let g be a metric in the conformal class and let $\widetilde{g} \in \mathcal{M}$ be straight. Then $\widetilde{g}$ is said to be in normal form relative to g if its smooth part is in normal form relative to g.*

It is clear from the existence and uniqueness of a diffeomorphism putting a smooth metric into normal form that if $\widetilde{g} \in \mathcal{M}$ is straight, then there is a unique smooth homogeneous diffeomorphism Φ such that $\Phi|_{\mathcal{G}} = Id$ and such that $\Phi^*\widetilde{g}$ is in normal form relative to g.

The following proposition makes explicit the normal form condition. Its proof is an analysis of the geodesic equations for the straightness and normal form conditions using the initial condition (a) of Definition 2.1.

PROPOSITION 3.2. *Let g be a representative for the conformal structure. A straight metric $\widetilde{g} \in \mathcal{M}$ is in normal form relative to g if and only if in the identification $\mathcal{G} = \mathbb{R}_+ \times M \times \mathbb{R}$ induced by g, $\widetilde{g}$ takes the form:*

$$\widetilde{g}_{IJ} = \begin{pmatrix} 2\rho & 0 & t \\ 0 & t^2 g_{ij} & t^2 g_{i\infty} \\ t & t^2 g_{j\infty} & t^2 g_{\infty\infty} \end{pmatrix} \qquad \textit{with } g_{ij},\, g_{i\infty},\, g_{\infty\infty} \in \mathcal{A}, \tag{3.3}$$

where $g_{ij}|_{\rho=0}$ is the given metric g on M, and where the expansions for the components $g_{j\infty}$ and $g_{\infty\infty}$ have zero smooth part when expanded using $r_\# = 2\rho t^2$. That is, for $I = i, \infty$ we have

$$g_{I\infty} \sim \sum_{N \geq 1} g_{I\infty}^{(N)} \rho(\rho^{n/2-1} \log |2\rho t^2|)^N \tag{3.4}$$

where the $g_{I\infty}^{(N)}$ are smooth and homogeneous of degree 0.

It is a consequence of Proposition 3.2 that if $\widetilde{g} \in \mathcal{M}$ is straight, then its smooth part is also straight.

The next theorem is our main theorem concerning the existence and uniqueness of inhomogeneous ambient metrics. As described in §2, in odd dimensions, given a representative metric g in the conformal class, there is to infinite order a unique smooth ambient metric in normal form relative to g. For n even, inhomogeneous ambient metrics in normal form are no longer uniquely determined by a representative for the conformal class. One has exactly the freedom of a smooth trace-free symmetric 2-tensor field on M, which we call the ambiguity tensor.

THEOREM 3.1. *Let $(M^n, [g])$ be a manifold with a conformal structure, $n \geq 4$ even. Up to pull-back by a smooth homogeneous diffeomorphism which restricts to the identity on $\mathcal{G}$, the inhomogeneous ambient metrics for $(M, [g])$ are parametrized by the choice of an arbitrary trace-free symmetric 2-tensor field A_{ij} on M. Precisely, for each representative metric g and* *choice of ambiguity tensor* $A_{ij} \in \Gamma(\bigodot_0^2 T^*M)$, *there is a unique inhomogeneous ambient metric* $\widetilde{g}$ *in normal form relative to* g *such that*

$$\mathrm{tf}\left((\partial_\rho)^{n/2} g_{ij}^{(0)}|_{\rho=0}\right) = A_{ij}. \tag{3.5}$$

Here we have written $\widetilde{g}$ *in the form* (3.3), $g_{ij}^{(0)}$ *denotes the smooth part of* g_{ij}, *and* tf *the trace-free part.*

There is a natural pseudo-Riemannian invariant 1-form D_i *so that the metric* $\widetilde{g}$ *determined by initial metric* g *and ambiguity* A_{ij} *is smooth if and only if* $\mathcal{O}_{ij} = 0$ *and* $A_{ij},^j = D_i$, *where* $\mathcal{O}_{ij}$ *denotes the obstruction tensor of* g.

The 1-form D_i is defined as follows. A straight smooth ambient metric in normal form takes the form

$$\widetilde{g}_{IJ} = \begin{pmatrix} 2\rho & 0 & t \\ 0 & t^2 g_{ij} & 0 \\ t & 0 & 0 \end{pmatrix} \tag{3.6}$$

where $g_{ij} = g_{ij}(x,\rho)$ is a smooth 1-parameter family of metrics on M. The derivatives $\partial_\rho^m g_{ij}$ for $m \le n/2 - 1$ and $g^{ij}\partial_\rho^{n/2} g_{ij}$ are determined at $\rho = 0$ and are natural invariants of the initial metric $g_{ij}(x,0)$. Fix the indeterminacy of $g_{ij}(x,\rho)$ to higher orders by fixing $g_{ij}(x,\rho)$ to be the Taylor polynomial of degree $n/2$ whose $(n/2)^{nd}$ derivative is pure trace:

$$g_{ij}(x,\rho) = \sum_{m=0}^{n/2-1} \frac{1}{m!}\partial_\rho^m g_{ij}\rho^m + \frac{1}{n(n/2)!}(g^{kl}\partial_\rho^{n/2} g_{kl}) g_{ij}\rho^{n/2}, \tag{3.7}$$

where on the right hand side, g_{ij} and its derivatives are evaluated at $(x,0)$. The Ricci curvature component $\widetilde{R}_{i\infty}$ of $\widetilde{g}$ vanishes to order $n/2-1$, and D_i is given by:

$$D_i = -2(\partial_\rho^{n/2-1}\widetilde{R}_{i\infty})|_{\rho=0,\,t=1}.$$

When $n = 4$, this reduces to:

$$D_i = 4P^{jk}P_{ij,k} - 3P^{jk}P_{jk,i} + 2P_i{}^j P^k{}_{k,j}.$$

If the metric $\widetilde{g}$ in Theorem 3.1 is written in the form (3.3), (3.4) and g_{ij} is also expanded as in (3.4) (with expansion including a smooth term $g_{ij}^{(0)}$), then $g_{ij}^{(1)}|_{\rho=0} = c_1\mathcal{O}_{ij}$ and $g_{i\infty}^{(1)}|_{\rho=0} = c_2(A_{ij},^j - D_i)$ for nonzero constants c_1, c_2.

Theorem 3.1 is proved by an inductive perturbation analysis of the equation of vanishing Ricci curvature for metrics of the form given by Proposition 3.2. In the proof, one sees that the smooth part $\widetilde{g}^{(0)}$ of any inhomogeneous ambient metric is a smooth ambient metric for $(M,[g])$ in the sense of §2.

The ambiguity tensor has a well-defined transformation law under conformal change. Let $\widetilde{g}$ be the inhomogeneous ambient metric in normal form determined by initial metric g and ambiguity tensor A_{ij}. Suppose we choose another metric $\widehat{g} = e^{2\Upsilon} g$ in the conformal class. There is a uniquely determined smooth homogeneous diffeomorphism Φ with $\Phi|_{\mathcal{G}} = Id$ so that $\Phi^* \widetilde{g}$ is in normal form relative to $\widehat{g}$. Now $\Phi^* \widetilde{g}$ determines an ambiguity tensor which we denote by $\widehat{A}_{ij}$, defined by the version of the equation (3.5) for the Taylor coefficient of the smooth part of $\Phi^* \widetilde{g}$ in the identification $\widetilde{\mathcal{G}} = \mathbb{R}_+ \times M \times \mathbb{R}$ induced by $\widehat{g}$. We describe next the expression for $\widehat{A}_{ij}$ in terms of A_{ij} and Υ. This transformation law is best understood in terms of another trace-free symmetric 2-tensor on M which is a modification of the ambiguity tensor.

According to Proposition 3.2, the smooth part $\widetilde{g}^{(0)}_{IJ}$ of an inhomogeneous ambient metric $\widetilde{g}_{IJ}$ in normal form relative to g takes the form

$$\widetilde{g}^{(0)}_{IJ} = \begin{pmatrix} 2\rho & 0 & t \\ 0 & t^2 g^{(0)}_{ij} & 0 \\ t & 0 & 0 \end{pmatrix},$$

where $g^{(0)}_{ij}(x,\rho)$ is a smooth 1-parameter family of metrics on M. Since $\widetilde{g}^{(0)}$ is itself a smooth ambient metric, the derivatives $\partial^m_\rho g^{(0)}_{ij}|_{\rho=0}$ for $m < n/2$ and the trace $g^{kl}\partial^{n/2}_\rho g^{(0)}_{kl}|_{\rho=0}$ are the same natural tensors $\partial^m_\rho g_{ij}|_{\rho=0}$ and $g^{kl}\partial^{n/2}_\rho g_{kl}|_{\rho=0}$ which occur in the expansion (3.7). Consider the value at $\rho = 0$, $t = 1$ of the component $\widetilde{R}^{(0)}_{\infty ij\infty,\underbrace{\infty\ldots\infty}_{n/2-2}}$ of the iterated covariant derivative of the curvature tensor of $\widetilde{g}^{(0)}_{IJ}$. This component depends on derivatives of orders $\leq n/2$ of the components of $\widetilde{g}^{(0)}_{IJ}$. An inspection of the formula for this covariant derivative component yields the following proposition.

PROPOSITION 3.3.

$$\widetilde{R}^{(0)}_{\infty ij\infty,\underbrace{\infty\ldots\infty}_{n/2-2}}|_{\rho=0,\,t=1} = \tfrac{1}{2}(A_{ij} + K_{ij}),$$

where K_{ij} is a natural trace-free symmetric tensor which can be expressed algebraically in terms of the tensors $\partial^m_\rho g_{ij}|_{\rho=0}$, $m < n/2$.

The tensors $\partial^m_\rho g_{ij}|_{\rho=0}$, $m < n/2$, are determined by the inductive solution of the equation $\mathrm{Ric}(\widetilde{g}) = 0$ for a smooth ambient metric in normal form, and K_{ij} is expressed in terms of these via the formula for covariant differentiation. To the extent that K_{ij}, and therefore also its conformal transformation law, may be regarded as known, the conformal transformation law of A_{ij} is determined by that of $\widetilde{R}^{(0)}_{\infty ij\infty,\underbrace{\infty\ldots\infty}_{n/2-2}}|_{\rho=0,\,t=1}$. Henceforth we shall write $\mathbf{A}_{ij} = A_{ij} + K_{ij}$ for this modified ambiguity tensor. For $n = 4$, one has $K_{ij} = -2\,\mathrm{tf}\left(P_i{}^k P_{jk}\right)$ so that

$$\mathbf{A}_{ij} = A_{ij} - 2\,\mathrm{tf}\left(P_i{}^k P_{jk}\right). \tag{3.8}$$

The conformal transformation law of $\mathbf{A}_{ij}$ can be expressed succinctly. Define the strength of lists of indices in $\mathbb{R}^{n+2}$ as follows. Set $\|0\| = 0$, $\|i\| = 1$ for $1 \le i \le n$, and $\|\infty\| = 2$. For a list, write $\|I \ldots J\| = \|I\| + \cdots + \|J\|$. An inductive analysis of the formula for covariant differentiation shows that if $r \ge 0$ and $\|ABCDF_1 \ldots F_r\| \le n+1$, then the curvature component $\widetilde{R}^{(0)}_{ABCD,F_1\cdots F_r}|_{\rho=0,\,t=1}$ defines a natural tensor on M as the indices between 1 and n vary and those which are 0 or ∞ remain fixed. Set

$$p^A{}_I = \begin{pmatrix} 1 & \Upsilon_i & -\frac{1}{2}\Upsilon_k\Upsilon^k \\ 0 & \delta^a{}_i & -\Upsilon^a \\ 0 & 0 & 1 \end{pmatrix}.$$

Proposition 3.4. *Set $r = n/2 - 2$. Under the conformal change $\widehat{g} = e^{2\Upsilon} g$, the modified ambiguity tensor transforms by:*

$$e^{(n-2)\Upsilon}\widehat{\mathbf{A}}_{ij} = \mathbf{A}_{ij} + 2{\sum}' \widetilde{R}^{(0)}_{ABCD,F_1\cdots F_r}|_{\rho=0,\,t=1} p^A{}_\infty p^B{}_i p^C{}_j p^D{}_\infty p^{F_1}{}_\infty \cdots p^{F_r}{}_\infty,$$

where $\sum'$ indicates the sum over all indices except $ABCDF_1\cdots F_r = \infty ij\infty\infty\cdots\infty$.

The upper-triangular form of $p^A{}_I$ implies that any component $\widetilde{R}^{(0)}_{ABCD,F_1\cdots F_r}|_{\rho=0,\,t=1}$ which occurs with nonzero coefficient in $\sum'$ necessarily satisfies $\|ABCDF_1\cdots F_r\| \le n+1$. So all these components are natural tensors, which can be calculated algorithmically using the expansion of $g_{ij}(x,\rho)$ through order $< n/2$. Thus Proposition 3.4 expresses the conformal transformation law of $\mathbf{A}_{ij}$ in terms of "known" natural tensors and Υ and its first derivatives. For example, for $n = 4$ this becomes

$$e^{2\Upsilon}\widehat{\mathbf{A}}_{ij} = \mathbf{A}_{ij} - 2\Upsilon^l(C_{ijl} + C_{jil}) + 2\Upsilon^k\Upsilon^l W_{kijl}.$$

Fix an even integer $n \ge 4$. In dimension $d > n$, there is a trace-free symmetric natural tensor satisfying the transformation law of Proposition 3.4; namely $2\widetilde{R}^{(0)}_{\infty ij\infty,\underbrace{\infty\ldots\infty}_{n/2-2}}|_{\rho=0,\,t=1}$. For instance,

$$\widetilde{R}^{(0)}_{\infty ij\infty}|_{\rho=0,\,t=1} = -\frac{1}{d-4}B_{ij}$$

if $d \ne 4$. Considered formally in the dimension d, the component $\widetilde{R}^{(0)}_{\infty ij\infty,\underbrace{\infty\ldots\infty}_{n/2-2}}|_{\rho=0,\,t=1}$ has a simple pole at $d = n$ whose residue is a multiple of $\mathcal{O}_{ij}$. The formal continuation to $d = n$ of the transformation law for this component is the statement of conformal invariance of the obstruction tensor. The modified ambiguity tensor $\mathbf{A}_{ij}$ provides a substitute for the natural tensor $\widetilde{R}^{(0)}_{\infty ij\infty,\underbrace{\infty\ldots\infty}_{n/2-2}}|_{\rho=0,\,t=1}$, which does not occur in dimension n because of the conformal invariance of the obstruction tensor.

The transformation law of Proposition 3.4 can be reinterpreted in terms of tractors. Recall (see, for example, [BEGo]) that a cotractor field of weight w on a conformal manifold $(M,[g])$ can be expressed upon choosing a representative metric g as $v_I = (v_0, v_i, v_\infty)$, where v_0, v_∞ are functions on M and v_i is a 1-form, such that under the conformal change $\widehat{g} = e^{2\Upsilon}g$, v_I transforms by

$$\widehat{v}_I = e^{(w+1-2\delta_I^\infty)\Upsilon} v_A p^A{}_I. \tag{3.9}$$

For $r < n/2-2$, the components of the tensor $\widetilde{\nabla}^r \widetilde{R}^{(0)}|_{\rho=0,\,t=1}$ define natural tensors on M and satisfy the correct transformation laws under conformal change to define a (natural) cotractor field of rank $4+r$ and weight $-2-r$. (A detailed discussion of the relation between ambient curvature and tractors can be found in [CG].) For $r = n/2-2$, the same is true of all components except for $\widetilde{R}^{(0)}_{\infty ij\infty,\underbrace{\infty\ldots\infty}_{n/2-2}}|_{\rho=0,\,t=1}$ (and components obtained from this one via the symmetries of the curvature tensor). The transformation law of Proposition 3.4 is precisely that required by (3.9) so that the prescription $\widetilde{R}^{(0)}_{\infty ij\infty,\underbrace{\infty\ldots\infty}_{n/2-2}}|_{\rho=0,\,t=1} = \frac{1}{2}\mathbf{A}_{ij}$ completes $\widetilde{\nabla}^{n/2-2}\widetilde{R}^{(0)}|_{\rho=0,\,t=1}$ to a cotractor field of rank $n/2+2$ and weight $-n/2$. Thus, choosing an ambiguity tensor A_{ij} is entirely equivalent to completeing $\widetilde{\nabla}^{n/2-2}\widetilde{R}^{(0)}|_{\rho=0,\,t=1}$ to a cotractor field.

When n is even, one can construct homogeneous ambient metrics with asymptotic expansions involving $\log|r|$ ([K]). Such nonsmooth homogeneous ambient metrics all have asymptotic expansions of the form $\sum_{N\geq 0} \bar{g}^{(N)}(r^{n/2}\log|r|)^N$, where the $\bar{g}^N$ are smooth and homogeneous of degree 2 ([FG2]). This suggests that the inhomogeneous ambient metrics considered here might actually have expansions of the form

$$\sum_{N\geq 0} \bar{g}^{(N)}(r^{n/2}\log|r_\#|)^N, \tag{3.10}$$

again with smoooth homogeneous coefficients $\bar{g}^{(N)}$. But this is not the case: for inhomogeneous ambient metrics, the coefficient $\widetilde{g}^{(2)}$ does not in general vanish at $\rho = 0$. In fact, we have

PROPOSITION 3.5. *Let $\widetilde{g}_{IJ}$ be an inhomogeneous ambient metric in normal form. When $\widetilde{g}_{IJ}$ is written in the form* (3.3) *and its components are expanded as in* (3.4) *(with a corresponding expansion for g_{ij} including a smooth term $g_{ij}^{(0)}$), then $g_{ij}^{(2)}|_{\rho=0} = 0$, $g_{i\infty}^{(2)}|_{\rho=0} = 0$, and $g_{\infty\infty}^{(2)}|_{\rho=0} =$*

$c\mathcal{O}_{ij}\mathcal{O}^{ij}$, *for a nonzero constant* c. *Thus if* $\widetilde{g}$ *has the form* (3.10), *then* $\mathcal{O}^{ij}\mathcal{O}_{ij}=0$. *Conversely, if* $\mathcal{O}^{ij}\mathcal{O}_{ij}=0$, *then* $\widetilde{g}$ *has the form* (3.10) *for any choice of ambiguity.*

In Proposition 3.5, one would expect that there is a condition analogous to $\mathcal{O}^{ij}\mathcal{O}_{ij}=0$ arising from the coefficient of each of $(\log|2\rho t^2|)^N$ for $N\geq 2$. The fact that all of these are satisfied once the condition holds for $N=2$ is surprising.

Proposition 3.5 raises the question of whether there are interesting classes of conformal manifolds for which $\mathcal{O}_{ij}$ is nonvanishing but $\mathcal{O}^{ij}\mathcal{O}_{ij}=0$. Of course, this cannot happen in definite signature. But the Fefferman conformal structure of any nondegenerate integrable CR manifold satisfies $\mathcal{O}^{ij}\mathcal{O}_{ij}=0$, and typically they satisfy also that $\mathcal{O}_{ij}$ is nonvanishing. These observations follow from an identification of the obstruction tensor of a Fefferman metric.

The analogue of the obstruction tensor in CR geometry is the scalar CR invariant, denoted here by L, which obstructs the existence of smooth solutions of Fefferman's complex Monge-Ampère equation. It is defined as follows. Let $\mathcal{M}\subset\mathbb{C}^n$ be a hypersurface with smooth defining function u whose Levi form $-u_{i\bar{j}}|_{T^{1,0}\mathcal{M}}$ has signature (p,q), $p+q=n-1$. Fefferman showed in [F1] that there is such a u uniquely determined mod $O(u^{n+2})$ such that $J(u)=1+O(u^{n+1})$, where

$$J(u)=(-1)^{p+1}\det\begin{pmatrix}u & u_{\bar{j}}\\ u_i & u_{i\bar{j}}\end{pmatrix}_{1\leq i,j\leq n}.$$

The invariant L is defined to be a constant multiple of $(J(u)-1)/u^{n+1}|_{\mathcal{M}}$, and is independent of the choice of u satisfying $J(u)=1+O(u^{n+1})$. The fundamental properties of L are derived in [L], [Gr1], [Gr2].

PROPOSITION 3.6. *Let* θ *be a pseudohermitian form for an integrable nondegenerate CR manifold* $\mathcal{M}$ *and let* g *be the associated representative of the Fefferman conformal structure on the circle bundle* $\mathcal{C}$. *Then the obstruction tensor* $\mathcal{O}$ *of* g *is a nonzero constant multiple of the pullback to* $\mathcal{C}$ *of* $L\theta^2$.

Proof. It suffices to assume that $\mathcal{M}\subset\mathbb{C}^n$ is embedded. The circle bundle $\mathcal{C}=S^1\times\mathcal{M}$ has dimension $2n$. Let u be a smooth defining function for $\mathcal{M}$ satisfying $J(u)=1+O(u^{n+1})$ as above. According to Fefferman's original definition, the representative g associated to $\theta=\frac{i}{2}(\partial u-\bar{\partial}u)$ is the pullback to

$$S^1\times\mathcal{M}=\{(z^0,z):|z^0|=1,\ z\in\mathcal{M}\}\subset\mathbb{C}^*\times\mathbb{C}^n$$

of the Kähler metric $\widetilde{g}$ on a neighborhood of $\mathbb{C}^*\times\mathcal{M}$ in $\mathbb{C}^*\times\mathbb{C}^n$ given by

$$\widetilde{g}=\partial^2_{I\bar{J}}(-|z^0|^2u)dz^Id\bar{z}^J. \tag{3.11}$$

The Ricci curvature of $\widetilde{g}$ is

$$\partial^2_{I\bar{J}}(\log J(u))dz^Id\bar{z}^J=cLu^{n-1}\partial u\bar{\partial}u+O(u^n).$$

Since the pullback to $\mathcal{G} = \mathbb{C}^* \times \mathcal{M}$ of $\partial u \bar{\partial} u$ is a multiple of θ^2, which is trace-free with respect to g, it is evident that this Ricci curvature is $O^+_{IJ}(\rho^{n-1})$. Now $\widetilde{g}$ is clearly homogeneous, and it satisfies the initial condition (3.1) by the definition of the Fefferman metric g. Therefore $\widetilde{g}$ is a smooth ambient metric for the Fefferman conformal structure in the sense of our earlier definition. Now $T = 2\,\mathrm{Re}\, z^0 \partial_{z^0}$, and an easy calculation gives $\widetilde{g}(T,T) = -|z^0|^2 u$. The definition (2.1) of the obstruction tensor thus reduces to a constant multiple of $L\theta^2$. □

Since θ is null with respect to g, Proposition 3.6 implies that the obstruction tensor of a Fefferman metric satisfies $\mathcal{O}_i{}^k \mathcal{O}_{kj} = 0$. In particular $\mathcal{O}^{ij}\mathcal{O}_{ij} = 0$, so Fefferman metrics satisfy the condition in Proposition 3.5.

Upon applying Theorem 3.1 to the Fefferman conformal structure of a CR manifold, one obtains a family of inhomogeneous ambient metrics with ambiguity an arbitrary trace-free symmetric tensor on the circle bundle. In [H], a family of "inhomogeneous CR ambient metrics" associated to a nondegenerate hypersurface in $\mathbb{C}^n$ was constructed with ambiguity a scalar function on the hypersurface. The relation between these constructions is not clear. The metrics constructed in [H] are not straight in general. It seems reasonable to guess that there are inhomogeneous, nonsmooth diffeomorphisms with expansions involving $\log |r_\#|$ relating the constructions (for an appropriately restricted ambiguity in the conformal construction). We intend to investigate this issue in the future.

A homogeneous ambient metric for a conformal manifold $(M, [g])$ gives rise to an asymptotically hyperbolic "Poincaré" metric by a procedure described in [FG1]. Namely, if $\widetilde{g}$ is a homogeneous ambient metric, then the pullback g_+ of $\widetilde{g}$ to the hypersurface $\{\widetilde{g}(T,T) = -1\}$ is a metric of signature of $(p+1, q)$ which in suitable coordinates is asymptotically hyperbolic with conformal infinity $(M, [g])$. The homogeneity of $\widetilde{g}$ implies that the condition $\mathrm{Ric}(\widetilde{g}) = 0$ is equivalent to $\mathrm{Ric}(g_+) = -ng_+$. The same construction can also be carried out for an inhomogeneous ambient metric. Proposition 3.1 shows that $\widetilde{g}(T,T)$ is still a smooth homogeneous defining function. When we take $r_\# = \widetilde{g}(T,T)$, all log terms in the asymptotic expansion (3.2) vanish on the hypersurface $\widetilde{g}(T,T) = -1$, so the Poincaré metric g_+ which one obtains agrees with that defined by the smooth part $\widetilde{g}^{(0)}$. In particular, g_+ has a smooth conformal compactification with no log terms. The g_+ which arise as $\widetilde{g}$ varies over the family of inhomogeneous ambient metrics associated to $(M, [g])$ form a family of smoothly compactifiable Poincaré metrics invariantly associated to $(M, [g])$ up to choice of ambiguity and up to a smooth diffeomorphism restricting to the identity on M. However, $\widetilde{g}^{(0)}$ is not usually Ricci flat, so g_+ will not in general be Einstein. We do not know a direct characterization of either the smooth parts $\widetilde{g}^{(0)}$ or the resulting family of Poincaré metrics g_+, other than to say that they arise from inhomogeneous ambient metrics by taking smooth parts.

4. Scalar invariants. An original motivation for the ambient metric construction in [FG1] was to construct scalar conformal invariants. In even dimensions, the construction of invariants in [FG1] terminates at finite order owing to the failure of the existence of infinite order smooth ambient metrics. Inhomogeneous ambient metrics can be used to extend the construction of invariants to all orders. The new invariants generally depend both on the initial metric and the ambiguity.

By a scalar invariant $I(g)$ of metrics we mean a polynomial in the variables $(\partial^\alpha g_{ij})_{|\alpha|\geq 0}$ and $|\det g_{ij}|^{-1/2}$, which is coordinate-free in the sense that its value is independent of orientation-preserving changes of the coordinates used to express and differentiate g. Such a scalar invariant is said to be *even* if it is also unchanged under orientation-reversing changes of coordinates, and *odd* if it changes sign under orientation-reversing coordinate changes. It is said to be conformally invariant of weight w if $I(\Omega^2 g) = \Omega^w I(g)$ for smooth positive functions Ω. By a scalar conformal invariant we will mean a scalar invariant of metrics which is conformally invariant of weight w for some w.

First recall the characterization of scalar conformal invariants in odd dimensions. Denote by $\widetilde{\nabla}^r \widetilde{R}$ the r-th iterated covariant derivative of the curvature tensor of a smooth ambient metric, by $\widetilde{\mu} \in \bigwedge^{n+2} T^*\widetilde{\mathcal{G}}$ the volume form of $\widetilde{g}$, and set $\widetilde{\mu}_0 = T \lrcorner \widetilde{\mu}$. Consider complete contractions of these tensors:

$$\begin{gathered}
\operatorname{contr}(\widetilde{\nabla}^{r_1}\widetilde{R} \otimes \cdots \otimes \widetilde{\nabla}^{r_L}\widetilde{R}) \\
\operatorname{contr}(\widetilde{\mu} \otimes \widetilde{\nabla}^{r_1}\widetilde{R} \otimes \cdots \otimes \widetilde{\nabla}^{r_L}\widetilde{R}) \\
\operatorname{contr}(\widetilde{\mu}_0 \otimes \widetilde{\nabla}^{r_1}\widetilde{R} \otimes \cdots \otimes \widetilde{\nabla}^{r_L}\widetilde{R}),
\end{gathered} \tag{4.1}$$

where the contractions are taken with respect to $\widetilde{g}$. Each such complete contraction defines a homogeneous function on $\widetilde{\mathcal{G}}$ whose restriction to $\mathcal{G}$ is independent of the diffeomorphism indeterminacy in the smooth ambient metric. If g is a metric in the conformal class, evaluating this homogeneous function at the image of g viewed as a section of $\mathcal{G}$ defines a function on M which depends polynomially on the derivatives of g. Every such function is a scalar conformal invariant. Contractions of the first type in (4.1) define even invariants and contractions of the second and third types define odd invariants. A linear combination of such complete contractions, all having the same weight, is called a Weyl invariant. The characterization of conformal invariants for n odd states that every scalar conformal invariant is a Weyl invariant. This follows from the invariant theory of [BEGr] together with the jet isomorphism theorem of [FG2], as outlined in [BEGr]. Full details will appear in [FG2].

When n is even, we carry out the same construction, but now replacing the $\widetilde{\nabla}^r \widetilde{R}$ in the complete contractions (4.1) by the corresponding covariant derivatives of curvature of the smooth part of an inhomogeneous ambient metric. Of course, the resulting functions on M generally depend on the choice of ambiguity tensor as well as the metric g; they can be regarded as conformal invariants of the pair (g, A). We still refer to linear combinations of such complete contractions as Weyl invariants. Some such Weyl invariants may actually be independent of the ambiguity, in which case they define scalar conformal invariants. We call such Weyl invariants ambiguity-independent Weyl invariants. We do not have a systematic general way of constructing ambiguity-independent Weyl invariants, but we do have examples and conditions for Weyl invariants to be ambiguity-independent in interesting cases. Also, we can extend the jet isomorphism theorem of [FG2] and the invariant theory of [BEGr] to prove that all scalar conformal invariants are of this form if $n \equiv 2 \mod 4$:

THEOREM 4.1. *If $n \equiv 2 \mod 4$, then every scalar conformal invariant is an ambiguity-independent Weyl invariant.*

If n is a multiple of 4, there are exceptional odd invariants which do not arise as ambiguity-independent Weyl invariants. The existence of exceptional invariants for a linearized problem was observed in [BEGr] and in this context they were studied systematically in [BG]. We follow the approach of [BG] to complete the description of scalar conformal invariants when $n \equiv 0 \mod 4$ as follows.

Choose a representative metric g and corresponding identification $\widetilde{\mathcal{G}} = \mathbb{R}_+ \times M \times \mathbb{R}$. Define an n-form η on $\widetilde{\mathcal{G}}$ by $\eta = t^{-2}\partial_\rho \lrcorner \mu_0$. Consider a partial contraction

$$L = \operatorname{pcontr}(\eta \otimes \underbrace{\widetilde{R} \otimes \cdots \otimes \widetilde{R}}_{n/2}),$$

where all the indices of η are contracted. While η depends on the choice of g, it can be shown that $L|_{\mathcal{G}}$ is independent of this choice and also of the choice of ambiguity tensor (for $n > 4$ this is obvious since $\widetilde{R}|_{\mathcal{G}}$ is already independent of the ambiguity). A scalar invariant can be obtained from L by applying an ambient realization of the tractor D operator (see again [CG]). This is a differential operator D defined initially as a map $D : \Gamma(\otimes^p T^*\widetilde{\mathcal{G}})(w) \to \Gamma(\otimes^{p+1} T^*\widetilde{\mathcal{G}})(w)$, where $\Gamma(\otimes^p T^*\widetilde{\mathcal{G}})(w)$ denotes the space of covariant p-tensor fields homogeneous of degree w in the sense of the previous sections ($\delta_s^* f = s^w f$), by:

$$Df = \widetilde{\nabla} f - \frac{1}{2(n + 2(w - p - 1))} dr_\# \otimes \widetilde{\Delta} f.$$

Here $\widetilde{\nabla}$ and $\widetilde{\Delta} = \widetilde{\nabla}^I \widetilde{\nabla}_I$ are defined with respect to $\widetilde{g}^{(0)}$ and it is assumed that $w \neq p + 1 - n/2$. One checks that D acts tangentially to $\mathcal{G}$ in the sense that Df vanishes on $\mathcal{G}$ if f does, so D induces a map

$$D : \Gamma(\otimes^p T^*\widetilde{\mathcal{G}}|_{\mathcal{G}})(w) \to \Gamma(\otimes^{p+1} T^*\widetilde{\mathcal{G}}|_{\mathcal{G}})(w).$$

The homogeneities are such that the expression

$$D^I D^J \cdots D^K L_{(IJ\cdots K)}$$

is defined and the above discussion implies that it restricts to $\mathcal{G}$ to give a homogeneous function independent of the choice of ambiguity tensor. (Here $(IJ \cdots K)$ denotes symmetrization over the enclosed indices.) As in the case of Weyl invariants, when evaluated at the image of g as a section of $\mathcal{G}$, one obtains a scalar conformal invariant. Such conformal invariants are odd and are called *basic exceptional invariants.* It is easy to see that there are only finitely many basic exceptional invariants in any dimension $n \equiv 0 \mod 4$ (the construction works just as well when $n \equiv 2 \mod 4$, but all basic exceptional invariants vanish in this case). For $n = 4$ there are only two nonzero basic exceptional invariants:

$$\eta^{IJKL}\widetilde{R}_{IJ}{}^{AB}\widetilde{R}_{KLAB}$$

of weight -4, and

$$D^A D^B(\eta^{IJKL}\widetilde{R}_{IJAC}\widetilde{R}_{KLB}{}^C)$$

of weight -6. The first is a multiple of $|W^+|^2 - |W^-|^2$, where $W^\pm$ denote the $\pm$ self-dual parts of the Weyl tensor, and the second is due to Bailey, Eastwood, and Gover (see p. 1207 of [BEGo]).

By extending the jet isomorphism theorem and invariant theory of [FG2], [BEGr] and [BG], we can prove:

THEOREM 4.2. *If $n \equiv 0 \mod 4$, then every even scalar conformal invariant is an ambiguity-independent Weyl invariant, and every odd scalar conformal invariant is a linear combination of an ambiguity-independent Weyl invariant and basic exceptional invariants.*

The above constructions use only the smooth part $\widetilde{g}^{(0)}$ of an inhomogeneous ambient metric. It is also possible to construct invariants using the tensors $r^{(n/2-1)N+1}\widetilde{g}^{(N)}$ for $N \geq 1$ in (3.2) (with $r_\# = \widetilde{g}(T,T)$). However, the above theorems show that this is not necessary: the invariants already can be constructed using just the smooth part.

We conclude with some examples. The smooth part $\widetilde{g}^{(0)}$ of an inhomogeneous ambient metric is not Ricci flat; the leading part of its Ricci tensor can be identified with the obstruction tensor. So Weyl invariants involving the Ricci tensor of $\widetilde{g}^{(0)}$ give rise to conformal invariants involving the obstruction tensor. Define an ambient version of the obstruction tensor by

$$\widetilde{\mathcal{O}}_{IJ} = \tfrac{1}{n-2}\widetilde{\Delta}^{n/2-1}\widetilde{R}_{IJ},$$

where $\widetilde{\Delta}$ and $\widetilde{R}_{IJ}$ are with respect to $\widetilde{g}^{(0)}$. One checks that $T \lrcorner \widetilde{\mathcal{O}} = 0$ on $\mathcal{G}$ and that $\widetilde{\mathcal{O}}|_{T\mathcal{G}}$ is the tensor on $\mathcal{G}$ homogeneous of degree $2-n$ defined by $\mathcal{O}$. Otherwise put, in the identification $\widetilde{\mathcal{G}} = \mathbb{R}_+ \times M \times \mathbb{R}$ induced by a representative g, one has $\widetilde{\mathcal{O}}_{0I} = 0$ and $\widetilde{\mathcal{O}}_{ij} = t^{2-n}\mathcal{O}_{ij}$ at $\rho = 0$. From this it is easy to see that the contraction $\widetilde{\mathcal{O}}_{IJ}\widetilde{\mathcal{O}}^{IJ}$ determines the conformal invariant $\mathcal{O}_{ij}\mathcal{O}^{ij}$, and similarly that $\widetilde{R}^{IJKA}\widetilde{R}_{IJK}{}^B\widetilde{\mathcal{O}}_{AB}$ determines the conformal invariant $W^{ijka}W_{ijk}{}^{b}\mathcal{O}_{ab}$. A more interesting example is to consider $\widetilde{R}^{IJKL}\widetilde{\mathcal{O}}_{IK,JL}$. One finds that for even $n \geq 6$, the restriction of this quantity to $\mathcal{G}$ is ambiguity-independent and determines the conformal invariant

$$W^{ijkl}\mathcal{O}_{ik,jl} - (n-1)W^{ijkl}P_{ik}\mathcal{O}_{jl} + 2nC^{jkl}\mathcal{O}_{jk,l} + \tfrac{n(n-1)}{n-4}B^{jk}\mathcal{O}_{jk}.$$

For $n = 4$, the Weyl invariant determined by $\widetilde{R}^{IJKL}\widetilde{\mathcal{O}}_{IK,JL}$ depends on the ambiguity tensor. It is given by the same formula, except that in the last term $\frac{1}{n-4}B^{jk}$ is replaced by $-\frac{1}{2}\mathbf{A}^{jk}$ with $\mathbf{A}$ given by (3.8).

REFERENCES

[A1] S. Alexakis, *On conformally invariant differential operators in odd dimensions*, Proc. Natl. Acad. Sci. USA **100** (2003), 4409–4410.

[A2] S. Alexakis, *On conformally invariant differential operators*, math.DG/0608771.

[BEGo] T.N. Bailey, M.G. Eastwood, and A.R. Gover, *Thomas's structure bundle for conformal, projective and related structures*, Rocky Mountain J. Math. **24** (1994), 1191–1217.

[BEGr] T.N. Bailey, M.G. Eastwood, and C.R. Graham, *Invariant theory for conformal and CR geometry*, Ann. Math. **139** (1994), 491–552.

[BG] T.N. Bailey and A.R. Gover, *Exceptional invariants in the parabolic invariant theory of conformal geometry*, Proc. A.M.S. **123** (1995), 2535–2543.

[CG] A. Čap and A.R. Gover, *Standard tractors and the conformal ambient metric construction*, Ann. Global Anal. Geom. **24** (2003), 231–259.

[EG] M.G. Eastwood and C.R. Graham, *Invariants of conformal densities*, Duke Math. J. **63** (1991), 633–671.

[F1] C. Fefferman, *Monge-Ampère equations, the Bergman kernel, and geometry of pseudoconvex domains*, Ann. Math. **103** (1976), 395–416; Correction: Ann. Math. **104** (1976), 393–394.

[F2] C. Fefferman, *Parabolic invariant theory in complex analysis*, Adv. Math. **31** (1979), 131–262.

[FG1] C. Fefferman and C.R. Graham, *Conformal invariants*, in *The mathematical heritage of Élie Cartan (Lyon, 1984)*. Astérisque, 1985, Numero Hors Serie, pp. 95–116.

[FG2] C. Fefferman and C.R. Graham, *The ambient metric*, in preparation.

[Go] A.R. Gover, *Invariant theory and calculus for conformal geometries*, Adv. Math. **163** (2001), 206–257.

[Gr1] C.R. Graham, *Scalar boundary invariants and the Bergman kernel*, Complex Analysis II, Proceedings, Univ. of Maryland 1985-86, Springer Lecture Notes **1276**, 108–135.

[Gr2] C.R. Graham, *Higher asymptotics of the complex Monge-Ampère equation*, Comp. Math. **64** (1987), 133–155.

[H] K. Hirachi, *Construction of boundary invariants and the logarithmic singularity of the Bergman kernel*, Ann. Math. **151** (2000), 151-191.

[K] S. Kichenassamy, *On a conjecture of Fefferman and Graham*, Adv. Math. **184** (2004), 268–288.

[L] J. Lee, *Higher asymptotics of the complex Monge-Ampère equation and geometry of CR-manifolds*, MIT Ph.D. thesis, 1982.

PFAFFIAN SYSTEMS OF FROBENIUS TYPE AND SOLVABILITY OF GENERIC OVERDETERMINED PDE SYSTEMS*

CHONG-KYU HAN†

Abstract. Solving an overdetermined PDE system of generic type is equivalent to finding an integral manifold of the associated Pfaffian system of Frobenius type. Then the general solutions are found by finding all the compatibility conditions, which is equivalent to the prolongation of the corresponding Pfaffian system to an involutive system. We present an algorithm for these: Given an overdetermined PDE systems without side conditions we find all the compatibility conditions by constructing a Pfaffian system of Frobenius type and by setting the associated torsions to be zero. All our discussion is local and our proofs need generic hypotheses.

AMS(MOS) subject classifications. 2000 Mathematics Subject Classification: 35N10, 58H05, 32V99.

1. Prolongation of Pfaffian systems. In this section, we briefly review basic notions of the prolongation and the involutivity. We refer the readers to [BCGGG], [GJ] and [Kura]. Let M be a smooth (C^∞) manifold of dimension n and let $\theta^1, \cdots, \theta^s, \omega^1, \cdots, \omega^p$, $s+p \le n$, be a set of linearly independent smooth 1-forms on M . We are concerned with the problem of finding a smooth submanifold $N \subset M$ of dimension p which satisfies

$$\begin{aligned} &\theta^\alpha|_N = 0, \quad \alpha = 1, \cdots, s \quad \text{(Pfaffian system)} \\ &\Omega|_N \neq 0, \quad \text{where } \Omega = \omega^1 \wedge \ldots \wedge \omega^p \quad \text{(independence condition).} \end{aligned} \tag{1.1}$$

Such a submanifold N is called an integral manifold of dimension p satisfying the independence condition, or simply a 'solution' of (1.1). To find a solution of (1.1) we consider subbundles $I \subset J \subset T^*M$. Here $I = \langle \theta^1, \cdots, \theta^s \rangle$ and $J = \langle \theta^1, \cdots, \theta^s, \omega^1, \cdots, \omega^p \rangle$, where $\langle \cdots \rangle$ denotes the linear span of what are inside. Let $\mathcal{D}$ be the $(n-s)$-dimensional plane field annihilated by $\theta^1, \cdots, \theta^s$. For $k = 1, \cdots, p$, an integral manifold of (1.1) of dimension k is a submanifold of M of dimension k whose tangent spaces belong to $\mathcal{D}$. An integral manifold N of dimension p such that $\Omega|_N \neq 0$ is a solution of (1.1). If $N$is an integral manifold of (1.1) then $\theta^\alpha|_N = 0$, and therefore, $d\theta^\alpha|_N = 0$, for each $\alpha = 1, \cdots, s$. A k-dimensional integral element is a k-dimensional subspace (x, E) of T_xM, for some $x \in M$, on which $\theta^\alpha = 0$ and $d\theta^\alpha = 0$, for all $\alpha = 1, \cdots, s$. By $V(I, J)$ we denote the set of all p-dimensional integral elements (x, E) satisfying $\Omega|_E \neq 0$.

*Presented on August 2, at IMA Summer Program: Symmetry and overdetermined systems of partial differential equations, July 17 August 4, 2006. The author was partially supported by KRF 2005-070-C00007 of Korea Research Foundation.

†Department of Mathematics, Seoul National University, Seoul 151-742, Korea (ckhan@math.snu.ac.kr).

Basic idea of the theory is that we can find a solution by constructing k-dimensional integral manifold N^k with N^{k-1} as initial data, inductively for $k = 1, \cdots, p$, so that we have a nested sequence of integral manifolds

$$N^0 \subset N^1 \subset \cdots \subset N^p.$$

Let

$$\{x\} = E^0 \subset E^1 \subset \cdots \subset E^p = E$$

be the corresponding flag of integral elements. The notion of involutivity is the existence of such a flag for each element of $V(I, J)$ so that the Cauchy problem is well-posed in each step and the solutions to the $(k+1)^{st}$ Cauchy problem remain solutions to the family of kth Cauchy problem with data smoothly changing in $(k+1)^{st}$ direction . If the system is analytic (C^ω) one can construct such a nested sequence of integral manifolds by using the Cauchy-Kowalewski theorem. This is the idea of the Cartan-Kähler theorem which asserts that an involutive analytic Pfaffian system has analytic solutions. If (I, J) is not involutive we construct an involutive system which is equivalent to the original system by repeating the process of the following two steps:

Step 1. *Reduce (1.1) to a submanifold $M' \subset M$ so that $V'(I, J) \to M'$ is surjective:*

Let M_1 be the image of $V(I, J) \to M$. If $M = M_1$ then we do nothing. If $M_1 \neq M$ then we note that any integral manifold of (I, J) must lie in M_1, and so we set

$$V_1(I, J) = \{(x, E) \in V(I, J) : E \subset T_x M_1\}.$$

Now consider the projection

$$V_1(I, J) \to M_1$$

with image M_2. If $M_2 = M_1$ we stop; otherwise we continue as before. Eventually we arrive either at the empty set, in which case (I, J) has no integral manifolds, or else at M' with $V'(I, J) \to M'$ being surjective and with all $(x, E) \in V'(I, J)$ satisfying $E \subset T_x M'$.

Step 2. *Assuming $V(I, J) \to M$ is surjective we do prolongation.*

To recall the definitions, let $G_p(M)$ be the Grassmann bundle of p-planes in TM. Let $\pi^1, \cdots, \pi^r$ be a set of 1-forms so that

$$\theta^1, \cdots, \theta^s, \omega^1, \cdots, \omega^p, \pi^1, \cdots, \pi^r$$

form a basis of T^*M. Let $(x, E) \in V(I, J)$. Since $\Omega|_E \neq 0$, on a neighborhood of $(x, E) \in G_p(M)$ we have $\theta^\alpha = m^\alpha_\rho \omega^\rho$, $\pi^\epsilon = \ell^\epsilon_\rho \omega^\rho$, (summation convention for $\rho = 1, \cdots, p$) and $\Omega \neq 0$. Thus $\{m^\alpha_\rho, \ell^\epsilon_\rho\}$ are local fibre coordinates in $G_p(M)$. The canonical system on $G_p(M)$ is the set of the tautological 1-forms

$$\begin{aligned} &\theta^\alpha - m^\alpha_\rho \omega^\rho, \qquad && \alpha = 1, \cdots, s \\ &\pi^\epsilon - \ell^\epsilon_\rho \omega^\rho, \qquad && \epsilon = 1, \cdots, r, \end{aligned} \tag{1.2}$$

where the summation convention is used for $\rho = 1, \cdots, p$. The first prolongation $(I^{(1)}, J^{(1)})$ is the restriction to $M^{(1)} := V(I, J) \subset G_p(M)$ of the canonical system. Since $m^\alpha_\rho = 0$ on $V(I, J)$ the problem of finding a solution of (1.1) is equivalent to finding a submanifold $N^{(1)} \subset M^{(1)}$ of dimension p satisfying

$$\begin{aligned} &\theta^\alpha|_{N^{(1)}} = 0, \qquad (\pi^\epsilon - \ell^\epsilon_\rho \omega^\rho)|_{N^{(1)}} = 0 \\ &\Omega|_{N^{(1)}} \neq 0. \end{aligned} \tag{1.3}$$

Integral manifolds of (I, J) and those of $(I^{(1)}, J^{(1)})$ are in a one-to-one correspondence. The k-th prolongation $(I^{(k)}, J^{(k)})$ on $M^{(k)} = V(I^{(k-1)}, J^{(k-1)})$ is defined inductively to be the first prolongation of $(I^{(k-1)}, J^{(k-1)})$ on $M^{(k-1)}$. We have a version of the Cartan-Kuranishi theorem [Kura]:

THEOREM 1.1. *Let $(I^{(k)}, J^{(k)})$, $k = 1, 2, \cdots$, be the sequence of prolongations of (1.1). Suppose that, for each k, $M^{(k)}$ is a smooth submanifold of $G_p(M^{(k-1)})$ and that the projection $M^{(k)} \to M^{(k-1)}$ is a surjective submersion. Then there is k_0 such that prolongations $(I^{(k)}, J^{(k)})$ are involutive for $k \geq k_0$.*

2. Pfaffian systems of Frobenius type. (1.1) is said to be quasi-linear if $dI \subset J$. To say this more precisely, let $\Lambda = \bigoplus_{j=0}^n \Lambda^j(M)$ be the exterior algebra of differential forms of M. A subalgebra $\mathcal{I} \subset \Lambda$ is called an algebraic ideal if $\mathcal{I} \wedge \Lambda \subset \mathcal{I}$. Consider the algebraic ideals $\mathcal{I}$ and $\mathcal{J}$ generated by $\{\theta^1, \cdots, \theta^s\}$ and $\{\theta^1, \cdots, \theta^s, \omega^1, \cdots, \omega^p\}$, respectively. (1.1) is quasi-linear if $d\mathcal{I} \subset \mathcal{J}$, namely,

$$d\theta^\alpha = \sum_{\beta=1}^{s} \phi^\alpha_\beta \wedge \theta^\beta + \sum_{\rho=1}^{p} \psi^\alpha_\rho \wedge \omega^\rho,$$

for some 1-forms $\phi^\alpha_\beta, \psi^\alpha_\rho$, for each $\alpha = 1, \cdots, s$. Existence of solutions has been studied mainly for the quasi-linear systems. (1.1) is said to be of Frobenius type if $s + p = n$, that is, if $\{\theta^1, \cdots, \theta^s, \omega^1, \cdots, \omega^p\}$ is a coframe of M. It is easy to see that Frobenius types are quasi-linear. In this section we focus our interest to the systems of Frobenius type. In this case no further equations are obtained by prolongation and the existence of general integral manifolds is determined only by Step 1. The notion of involutivity is very subtle as was indicated in E. Mansfield's lecture [M]. However, for $V(I, J)$ of Frobenius type the following are equivalent (see [GJ] Chapter 3):

i) $V(I,J) \to M$ is surjective.

ii) (I,J) is integrable in the sense of the Frobenius theorem.

iii) (I,J) is invoultive.

It is easy to prove the following

LEMMA 2.1. *Let M be a smooth manifold of dimension n. Let $\theta := (\theta^1, \cdots, \theta^s)$ be a set of independent 1-forms on M and $\mathcal{D} :=< \theta >^{\perp}$ be the $(n-s)$-dimensional plane field annihilated by θ. Suppose that N is a submanifold of M of dimension $n-r$, for some $r \leq s$, defined by $T_1 = \cdots = T_r = 0$, where T_i are smooth real-valued functions of M such that $dT_1 \wedge \cdots \wedge dT_r \neq 0$. Then the following are equivalent:*

(i) $\mathcal{D}$ is tangent to N.

(ii) $dT_j \equiv 0, \mod \theta$, on N, for each $j = 1, \ldots, r$.

In (ii) mod θ means that modulo the algebraic ideal $\mathcal{I}$. Thus (ii) is equivalent to saying that for each $j = 1, \cdots, r$ we have $dT_j \wedge \theta^1 \wedge \cdots \wedge \theta^s = 0$, on N. Our basic observation is the following algorithm for Step 1: For each $\alpha = 1, \cdots, s$, set

$$d\theta^\alpha = T^\alpha_{ij}\omega^i \wedge \omega^j, \quad \mod \theta, \quad \text{(summation convention for} i,j = 1, \cdots, p)$$

where T^α_{ij} are skew symmetric in (ij). Let $\mathcal{T}_1$ be the set of functions $\{T^\alpha_{ij}\}$. If $\mathcal{T}_1$ are identically zero then $V(I,J) \to M$ is surjective, which is the Frobenius integrability condition for θ, and by Frobenius theorem we have $(n-p)$-parameter family of integral manifolds. Otherwise, let M_1 be the common zero set of $\mathcal{T}_1$ and set

$$dT^\alpha_{ij} = T^\alpha_{ij,k}\omega^k, \ \mod \theta.$$

Let $\mathcal{T}_2$ be the set of functions $T^\alpha_{ij,k}$. If $\mathcal{T}_2$ are identically zero on M_1 then $V_1(I,J) \to M_1$ is surjective, and by Frobenius theorem there exist $(\dim M_1 - p)$-parameter family of solutions. If $\mathcal{T}_2$ are not identically zero, let M_2 be the submanifold of M_1 defined by $\mathcal{T}_2 = 0$ and continue as before. Eventually we arrive either at an empty set, in which case there is no integral manifolds, or at an integrable Pfaffian system on a submanifold $M' \subset M$, in which case there exist $(\dim M' - p)$-parameter family of integral manifolds.

3. Complete prolongation of overdetermined PDE systems. Let $u = (u^1, \cdots, u^q)$ be a system of real-valued functions of independent variables $x = (x^1, \cdots, x^p)$. Consider a system of partial differential equations of order m

$$\Delta_\lambda(x, u^{(m)}) = 0, \qquad \lambda = 1, \cdots \ell, \tag{3.1}$$

where $u^{(m)}$ denotes all the partial derivatives of u of order up to m. We assume that (3.1) is over-determined, that is, $\ell > q$.

As we differentialte (3.1) μ times we have partial derivatives of u up to order $m + \mu$. Since

$$\frac{\text{number of equations}}{\text{number of partial derivatives}} \to \frac{\ell}{q}, \qquad \text{as } \mu \to \infty,$$

generically it is possible to solve for all the partial derivatives of u of a sufficiently high order, say k, as functions of derivatives of lower order, by the implicit function theorem, namely,

$$u_K^\alpha = H_K^\alpha(x, u^{(k-1)}), \tag{3.2}$$

for all multi-index K with $|K| = k$, and for all $\alpha = 1, \cdots, q$. (3.2) is called a complete system of order k and we say (3.1) admits prolongation to a complete system of order k. (3.2) can be regarded as a Pfaffian system of Frobenius type on submanifold M of the $(k-1)$th jet space $J^{(k-1)}(X, U)$, where X is the set of independent variables x and U is the space of dependent variables u. K. Yamaguchi and T. Yatsui studied in [YY1] and [YY2] the equivalence problem by point transformations and formulated the geometry of $J^{(k-1)}$ with the distribution of p planes $\mathcal{D}$ given by (3.2). Solving (3.1) is finding an integral manifold of $\mathcal{D}$ on the submanifold M that is defined by (3.1) and its prolongation. We do this by Step 1 as in §2. For examples, we consider a PDE systems of first order for one unknown function $u(x, y)$

$$\begin{cases} u_x = A(x, y, u) \\ u_y = B(x, y, u). \end{cases} \tag{3.3}$$

This is a complete system of first order. In this case M is the whole 0-th order jet space $\mathbb{R}^3 = \{(x, y, u)\}$ we consider the Pfaffian system

$$\theta := du - Adx - Bdy = 0 \tag{3.4}$$

with the independence condition

$$dx \wedge dy \neq 0. \tag{3.5}$$

Then $d\theta = Tdx \wedge dy, \quad \text{mod } \theta, \quad$ where

$$T = A_y + A_u B - B_x - B_u A. \tag{3.6}$$

If $T = 0$ on $\mathbb{R}^3$, then for any initial condition $u(x_0, y_0) = u_0$ there exists a unique solution satisfying the initial condition and thus there exists a 1-parameter family of solutions.

Example 3.1.

$$\begin{cases} u_x = a(x, y) + u \\ u_y = b(x, y). \end{cases} \tag{3.7}$$

By (3.6) $T = a_y + b - b_x$. If the functions a and b satisfy $a_y + b - b_x = 0$, then there exists 1-parameter family of solutions. If T does not vanish identically then $T = 0$ gives a relation between x and y, which violates the condition $dx \wedge dy \neq 0$, hence no function $u(x, y)$ can be a solution of (3.7).

EXAMPLE 3.2.

$$\begin{cases} u_x = a(x, y) + u^2 \\ u_y = b(x, y), \qquad b \neq 0. \end{cases} \tag{3.8}$$

In this case $\theta = du - (a + u^2)dx - bdy$ and $T = a_y + 2ub - b_x$. T cannot be identically zero for all (x, y, u). Thus $T = 0$ gives

$$u = \frac{1}{2b}(-a_y + b_x). \tag{3.9}$$

Thus, if there is a solution then (3.9) is the solution and (3.9) is indeed a solution if it satisfies (3.8), namely,

$$\begin{aligned} &\left\{\frac{1}{2b}(-a_y + b_x)\right\}_x = a + \left\{\frac{1}{2b}(-a_y + b_x)\right\}^2 \\ &\left\{\frac{1}{2b}(-a_y + b_x)\right\}_y = b. \end{aligned} \tag{3.10}$$

However, we derive (3.10) as follows: dT modulo θ on the submanifold $\{T = 0\}$ is

$$\begin{aligned} &\left\{a_{xy} + \frac{b_x}{b}(-a_y + b_x) - b_{xx} + 2b\left[a + \left(\frac{-a_y + b_x}{2b}\right)^2\right]\right\}dx \\ &\quad + \left\{a_{yy} + \frac{b_y}{b}(-a_y + b_x) - b_{xy} + 2b^2\right\}dy. \end{aligned}$$

By setting the coefficient of dx and the coefficient of dy to be zero, we obtain (3.10).

Now we consider systems of second order

$$\begin{cases} u_x + u_y = b(x, y) \\ u_{yy} = a(x, y, u, u_x, u_y). \end{cases} \tag{3.11}$$

Differentiate the first equation of (3.11) with respect to x and y, respectively, and solving for all the second order derivatives of u , we obtain

$$\begin{cases} u_{xx} = b_x - b_y + a \\ u_{xy} = b_y - a \\ u_{yy} = a. \end{cases} \tag{3.12}$$

Thus (3.11) admits prolongation to a complete system of second order. Let $J^1(\mathbb{R}^2, \mathbb{R}^1) = \{(x, y, u, p, q)\} = \mathbb{R}^5$ be the space of the first jets, where the first jet-graph of a function $u(x,y)$ is given by $p = u_x$ and $q = u_y$. Let M be a real submanifold of dimension 4 defined by the first equation of (3.11), that is, $p + q = b(x,y)$. We consider 1-forms

$$\begin{cases} \theta^0 = du - pdx - (b-p)dy \\ \theta^1 = dp - (b_x - b_y + a)dx - (b_y - a)dy \\ \theta^2 = dq - (b_y - a)dx - ady. \end{cases}$$

Observe that on M we have $\theta^2 = -\theta^1$ and that $\{\theta^0, \theta^1\}$ defines a 2-dimensional plane field $\mathcal{D}$ on M, whose integral manifolds are the first jet graph of solutions. To check the Frobenius integrability conditions we see that on M

$$\begin{cases} d\theta^0 = 0, \quad \mod \quad \{\theta^0, \theta^1\} \\ d\theta^1 = Tdx \wedge dy, \quad \mod \quad \{\theta^0, \theta^1\}, \end{cases}$$

where

$$T = -b_{yy} + a_y + a_u b + a_x + a_p b_x + a_q b_y. \tag{3.13}$$

If $T = 0$ holds identically on M, then there exists 2-parameter family of solutions. Otherwise, we restrict $\{\theta^0, \theta^1\}$ to the submanifold M_1 of M given by $T = 0$.

EXAMPLE 3.3 (Linear Case).

$$\begin{cases} u_x + u_y = b(x,y) \\ u_{yy} = \alpha(x,y) + c_1 u + c_2 u_x + c_3 u_y. \end{cases}$$

Then

$$T(b^{(2)}, \alpha^{(1)}, c_1, c_2, c_3) = -b_{yy} + \alpha_x + \alpha_y + c_1 b + c_2 b_x + c_3 b_y,$$

which depends only on (x,y). If the functions α, b and the constants c_1, c_2, c_3 satisfy $T = 0$ then there are 2-parameter family of solutions. If T is not identically zero, $T = 0$ gives a relation between x and y, which violates $dx \wedge dy \neq 0$, therefore, no function $u(x,y)$ can be a solution.

EXAMPLE 3.4 (NonLinear Case).

$$\begin{cases} u_x + u_y = b(x), \quad (b' \neq 0) \\ u_{yy} = \alpha(x,y) + u_x^2. \end{cases}$$

In this example we assume b depends only on x, for simplicity. Then on $M := \{p + q = b(x)\}$ we have independent 1-forms

$$\begin{cases} \theta^0 = du - pdx - (b-p)dy \\ \theta^1 = dp - (b' + \alpha + p^2)dx + (\alpha + p^2)dy. \end{cases}$$

Then on M

$$d\theta^1 \equiv Tdx \wedge dy, \quad \mod \quad \{\theta^0, \theta^1\}$$

where $T = \alpha_y + \alpha_x + 2pb'$. Hence $T = 0$ implies $p = -\frac{1}{2b'}(\alpha_x + \alpha_y)$, which defines a 3-dimensional submanifold M_1 of M. If $dT \equiv 0, \mathrm{mod}\{\theta^0, \theta^1\}$ on M_1 then there is a 1-parameter family of solutions. In fact, if $u(x, y)$ is a solution, so is $u +$ constant. On M_1

$$dT = (\alpha_x + \alpha_y)_x dx + (\alpha_x + \alpha_y)_y dy + 2b'dp + 2pb'' dx,$$

substituting $p = -\frac{\alpha_x + \alpha_y}{2b'}$ and $dp = (b' + \alpha + p^2)dx + (\alpha + p^2)dy$ and setting each of the coefficients to dx and to dy to be zero we obtain

$$\begin{cases} (\alpha_x + \alpha_y)_x + 2(b')^2 + 2b'\alpha + \dfrac{(\alpha_x + \alpha_y)^2}{2b'} - \dfrac{b''}{b'}(\alpha_x + \alpha_y) = 0 \\ (\alpha_x + \alpha_y)_y + 2b'\alpha + \dfrac{(\alpha_x + \alpha_y)^2}{2b'} = 0. \end{cases} \tag{3.14}$$

If the functions α and b satisfy (3.14) there is a 1-parameter family of solutions. Otherwise, no solutions exist.

4. Applications. Finding integral manifolds of a Pfaffian system of Frobenius type is essentially an ODE problem. Therefore, if (3.1) admits prolongation to a complete system of order k then by the fundamental theorem of ODE we have
i) Existence and finiteness of solutions: A solution is determined by its $(k-1)$th jet at a point. General solutions are $(\dim M' - p)$-parameter family of solutions. If M' is empty there are no solutions.
ii) Regularity of solutions: If each Δ of (3.1) is C^∞ in its argument then the solutions are C^∞. If Δ is C^ω then the solutions are C^ω.

As for the regularity results, D. Burns [B] first observed that the analyticity of CR mappings can be shown by constructing a complete system of finite order. The author proved in [H1] the analyticity for CR mappings between real hypersurfaces of the same dimension and in [H3] for CR embeddings into a CR manifold of non-degenerate Levi form. As for the finiteness results, the rigidity of isometric embeddings and conformal embeddings has been studied in [CH2] and the finiteness of multi-contact structures (cf. [CDKR]) in [HOS]. As for the existence problems, [HOS] discusses the existence of infinitesimal symmetries of multi-contact structures given by 2 vector fields in $\mathbb{R}^3$ and we study the existence of a complex hypersurface in a real submanifold in $\mathbb{C}^n$ in [T].

REFERENCES

[BCGGG] R. Bryant, S.S. Chern, R. Gardner, H. Goldschmidt, and P. Griffiths, *Exterior differential systems*, Springer-Verlag, New York, 1986.
[Burns] D. Burns, *CR Geometry*, U. of Michigan Lecture note, 1980.
[CDKR] M. Cowling, F. De Mari, A. Koranyi, and H.M. Reimann, *Contact and conformal maps on Iwasawa N groups*, Rend Mat. Acc. Lincei, 2002, **s9**, **13**: 219–232.

[CH1] C.K. Cho and C.K. Han, *Compatibility equations for isometric embeddings of Riemannian manifolds*, Rocky Mt. J. Math., 1993, **23**: 1231–1252.

[CH2] C.K. Cho and C.K. Han, *Finiteness of infinitesimal deformations of isometric embeddings and conformal embeddings*, Rocky Mt. J. Math., 2005, **33**, to appear.

[CjsH] J.S. Cho and C.K. Han, *Complete prolongation and the Frobenius integrability for overdetermined systems of partial differential equations*, J. Korean Math. Soc., 2001, **38**, to appear.

[GJ] P. Griffiths and G. Jensen, *Differential systems and isometric embeddings*, Ann. of Math. Studies, No. 114, Princeton U. Press, Princeton, NJ, 1987.

[H1] C.K. Han, *Analyticity of CR equivalence between real hypersurfaces in $\mathbb{C}^n$ with degenerate Levi form*, Invent. Math., 1983, **73**: 51–69.

[H2] ——— *A method of prolongation of tangential Cauchy-Riemann equations*, Adv. Stud. Pure Math., 1997, **25**: 158–166.

[H3] ——— *Complete differential system for the mappings of CR manifolds of nondegenerate Levi forms*, Math. Ann., 1997, **309**: 229–238.

[H4] ——— *Solvability of overdetermined pde systems that admit a complete prolongation and some local problems in CR geometry*, J. Korean Math. Soc., 2003, **40**: 695–708.

[HOS] C.K. Han, Jongwon Oh, and G. Schmalz, *Symmetry algebra for integral curves of $2n$ vector fields on $(2n+1)$-manifold*, preprint.

[Kura] M. Kuranishi, *On E. Cartan's prolongation theorem of exterior differential systems*, Amer. J. Math., 1957, **79**: 1–47.

[M] E. Mansfield, *A simple criterion for involutivity*, Talk presented during this IMA Summer Program.

[T] G. Tomassini, *Complex hypersurfaces in CR manifolds*, Personal communication.

[YY1] K. Yamaguchi and T. Yatsui, *Geometry of higher order differential equations of finite type associated with symmetric spaces*, Adv. Stud. Pure Math., 2002, **37**: 397–458.

[YY2] ——— *Parabolic geometries associated with differential equations of finite type*, Preprint.

EXACT AND QUASI - EXACT SOLVABILITY OF SECOND ORDER SUPERINTEGRABLE QUANTUM SYSTEMS

ERNIE G. KALNINS*, WILLARD MILLER, JR.†, AND GEORGE S. POGOSYAN‡

Abstract. Quasi-exactly solvable (QES) problems in quantum mechanics are eigenvalue problems for the Schrödinger operator where it is posssible to compute exactly a certain finite number of eigenvalues and eigenfunctions, even though exact algebraic expressions for the full set of eigenvalues do not exist. In the past, mathematical physicists have used hidden symmetry methods to attack these problems. We give a brief review of these methods and then show the increased insight into the structure of such problems provided by superintegrability theory and separation of variables.

1. Introduction. It is well known that N-dimensional nonrelativistic quantum systems described by the Hamiltonian

$$\mathcal{H} = -\frac{1}{2}\sum_{j=1}^{N} \frac{\partial^2}{\partial x_i^2} + V(x_1, x_2, ..., x_N) \tag{1.1}$$

are integrable if there exist N linearly independent and global differential operators $\mathcal{I}_\ell$, $\ell = 0, 1, ..N-1$ and $\mathcal{I}_0 = \mathcal{H}$, commuting with the Hamiltonian $\mathcal{H}$ and with each other

$$[\mathcal{I}_\ell, \mathcal{H}] = 0, \qquad [\mathcal{I}_\ell, \mathcal{I}_j] = 0, \qquad \ell, j = 1, 2, ...N-1. \tag{1.2}$$

An integrable system is called **superintegrable** (this term was introduced first time by S.Rauch-Wojciechowski in [1]) or **maximally integrable** if it is integrable and, also, possesses additional integrals $\mathcal{L}_k$, commuting with the Hamiltonian

$$[\mathcal{L}_k, \mathcal{H}] = 0, \qquad k = 1, 2, ...N-1 \tag{1.3}$$

but not necessarily with each other.

Two examples of this kind have been well-known a long time, namely the Kepler-Coulomb problem and the isotropic harmonic oscillator. Another famous example is the many-body Calogero-Moser model. The existence of additional integrals of motion for these systems (second order superintegrability) leads to many interesting properties not shared by standard integrable systems [2–6]

1. In quantum mechanics, there is the phenomenon of *accidental degeneracy* when the energy eigenvalues are multiply degenerate.

*Department of Mathematics, University of Waikato, Hamilton, New Zealand.

†School of Mathematics, University of Minnesota, Minneapolis, MN, 55455, U.S.A.

‡Yerevan State University, Yerevan, Armenia, and Departamento de Matematicas, Universidad de Guadalajara, Guadalajara, Mexico.

2. This property is intimately related to the existence of a dynamical symmetry group (or algebra), a so-called *hidden symmetry group (algebra)*.
3. In classical mechanics the additional integrals of motion have the consequence that in the case of superintegrable systems in two dimensions and maximally superintegrable systems in three dimensions all finite trajectories are found to be periodic.
4. One of the most important properties of (second order) superintegrable systems is *multiseparability*, the separation of variables for the Hamilton-Jacobi and Schrödinger equations in more than one orthogonal coordinate system.

1.1. Quasi-Exactly Solvable Systems. The crucial example that stimulated the investigation of quasi-exactly solvable systems is the quantum system with the anisotropic potential

$$V(x) = \frac{1}{2}\omega^2 x^6 + 2\beta\omega^2 x^4 + (2\beta^2\omega^2 - 2\delta\omega - \lambda)x^2 + 2\frac{(\delta - \frac{1}{4})(\delta - \frac{3}{4})}{x^2}, \quad (1.4)$$

where ω, β, $\delta > 1/2$ and λ are the constants.

As noticed by many authors [7, 8], this system admits polynomial solutions (which obviously do not form a basis) only for special values of the constant $\lambda = \omega(2n+1)$, $(n = 0, 1, 2...)$

$$\Psi(x) \approx x^{2\delta - \frac{1}{2}} \, \mathrm{e}^{-\frac{\omega}{4}x^4 - \beta\omega x^2} \, P_n(x^2). \quad (1.5)$$

where

$$P_n(x^2) = \sum_{s=0}^{n} A_s x^2.$$

is a polynomial of degree $2n$.

There are two different approaches to the investigation of quasi-exactly solvable systems. In the algebraic approach formulated by Turbiner [9] quasi-exactly solvability is explained in terms of a "hidden symmetry algebra" $sl(2, R)$. More precisely this means following: the one-dimensional Hamiltonian $\mathcal{H} = \partial_{x_i}^2 + V(x)$ after suitable changes of variable $z = \xi(x)$ and "gauge transformation" $H = e^{-\alpha(z)}\mathcal{H}e^{\alpha(z)}$ can be written in the form

$$H = \sum_{a,b=0,\pm} C_{ab} J_a J_b + \sum_{a=0,\pm} C_a J_a \quad (1.6)$$

where the first-order differential operators $\{J_\pm, J_0\}$ satisfy the commutation relations for the Lie algebra $sl(2, R)$.

The second approach, known as analytic, was formulated by Uschveridze [10, 11] and represents a one-dimensional reduction of the Niven-Stielties method for solving multiparameter spectral problems such as the generalized Lamé equation (or ellipsoidal equation). The solution in this method is determined by the zeros of polynomials $P_n(x^2)$. Indeed the wave function (1.5) can be rewritten in the form

$$\Psi(x) \approx x^{2\delta-\frac{1}{2}} e^{-\frac{\omega}{4}x^4-\beta\omega x^2} \Pi_{i=0}^{n}(x^2-\xi_i), \tag{1.7}$$

where the numbers $(\xi_1, \xi_2, ...\xi_n)$ satisfy a system of n algebraic equations.

According to the oscillation theorem, the number of zeros in the physical interval $\xi_i \in [0, \infty)$ enumerates the ground state and first n - excitations, described in terms of all zeros (complete solutions of the systems of algebraic equations and including the non physical section $\xi_i \in (-\infty, 0]$) as

$$E = 4\delta \left[\beta\omega + \sum_{i=1}^{n} \frac{1}{\xi_i}\right]. \tag{1.8}$$

A few natural questions occur in these two approaches:

(i) why does the constant $\lambda = \omega(2n+1)$ in potential (1.4) have this special form,

(ii) what is the physical meaning of the negative zeros ξ_i, and finally,

(iii) why in the correct formula for the energy spectrum (1.8) do n zeros of the polynomial $P_n(x^2)$ appear?

Let us now consider the quantum motion in the plane for a charged particle with two fixed Coulomb centers with coordinates $(\pm D/2, 0)$ (the so-called plane two center problem)

$$V(x,y) = -\frac{\alpha_1}{\sqrt{y^2+(x+D/2)^2}} - \frac{\alpha_2}{\sqrt{y^2+(x-D/2)^2}}. \tag{1.9}$$

This system is not superintegrable and separation of variables is possible only in two-dimensional elliptic coordinates

$$x = \frac{D}{2}\cosh\nu\cos\mu, \qquad y = \frac{D}{2}\sinh\nu\sin\mu.$$

Upon the substitution $\psi(\nu,\mu;D^2) = X(\nu;D^2)Y(\mu;D^2)$ with separation constant $A(D)$, the Schrödinger equation separates into a **system** of two ordinary differential equations

$$\frac{d^2X}{d\nu^2} + \left[\frac{D^2E}{2}\cosh^2\nu + D(\alpha_1+\alpha_2)\cosh\nu + A(D)\right]X = 0,$$

$$\frac{d^2Y}{d\mu^2} - \left[\frac{D^2E}{2}\cos^2\mu + D(\alpha_1-\alpha_2)\cos\mu + A(D)\right]Y = 0.$$

These equations belong to the class of **non-exactly solvable** problems. Polynomial solutions do not exist even for the case of discrete spectrum $E < 0$, and each of the wave functions $X(\nu;D^2)$ and $Y(\mu;D^2)$ is expressed as an infinite series with a three-term recurrence relation.

Let us now put $\alpha_2 = 0$. Then the two center potential transforms to the ordinary two-dimensional (2D) hydrogen atom problem, which is well-known as a superintegrable system with dynamical symmetry group $SO(3)$, and admits separation of variables in three systems of coordinates: polar, parabolic and elliptic [12]. In this case we can see that two separation equations in elliptic coordinates [13], namely

$$\frac{d^2X}{d\nu^2} + \left[\frac{D^2E}{2}\cosh^2\nu + D\alpha_1\cosh\nu + A(D)\right]X = 0, \tag{1.10}$$

$$\frac{d^2Y}{d\mu^2} - \left[\frac{D^2E}{2}\cos^2\mu + D\alpha_1\cos\mu + A(D)\right]Y = 0 \tag{1.11}$$

transform into each other by the change $\mu \leftrightarrow i\nu$. Thus separation of variables in elliptic coordinates for the 2D hydrogen atom gives **two functionally identical** one-dimensional Schrödinger type equations with two parameters: coupling constant E and energy $A(D)$ (correspondingly energy and separation constant for 2D hydrogen atom), but one defined on the real and the other on the imaginary axis.

In other words, instead of yielding a system of differential equations (1.10)–(1.11), the problem reduces to solving only one of Equations (1.10) or (1.11), for instance

$$\frac{d^2Z}{d\xi^2} - \left[\frac{D^2E}{2}\cos^2\xi + D\alpha_1\cos\xi + A(D)\right]Z = 0$$

for which the "domain of definition" of variable ξ is the complex plane (see Fig. 1).

The requirement of finiteness for the wave functions in the complex plane permits **only** polynomial solutions. As result we obtain *simultaneous* quantization of the energy

$$E_n = -\frac{\alpha_1^2}{2(n+1/2)^2}, \qquad n = 0, 1, 2, ... \tag{1.12}$$

and the elliptic separation constant $A_s(D)$ where $s = 0, 1, 2, ...n$ (as a solution of an nth-degree algebraic equation). The polynomial solutions are defined with the help of a finite series with a three-term recurrence relation for the coefficients. They cannot be considered as exactly-solvable and can be investigated only numerically.

We note that each of Equations (1.10) or (1.11) has the form of a one dimensional Schrödinger equation with parameter E and eigenvalue $A(D)$, and could separately be considered in the regions $\mu \in [0, 2\pi]$ or $\nu \in [0, \infty)$, correspondingly. Then for arbitrary values of constant E the solutions of Eqs. (1.10) or (1.11) expressed via infinite series and only on the "energy surface" of the 2D hydrogen atom (1.12) split into polynomial

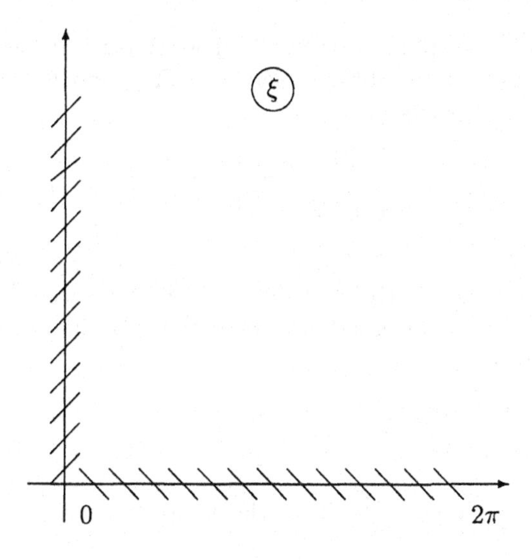

FIG. 1.

and nonpolynomial sectors (each of these sectors is non complete) and for fixed number n, only *some* of the eigenvalue $A_s(D)$, $(s = 0, 1, 2 \dots n)$ can be calculated from an nth-degree algebraic equation.

To understand the intimate connection between superintegrable and quasi-exactly solvable systems we consider one of the two dimensional superintegrable systems, namely the anisotropic oscillator.

2. Anisotropic oscillator. The potential

$$V(x, y) = \frac{1}{2}\omega^2(4x^2 + y^2) + k_1 x + \frac{k_2^2 - \frac{1}{4}}{2y^2}, \qquad k_2 > 0, \tag{2.1}$$

is known as the *singular anisotropic oscillator.* The Schrödinger equation separates in two coordinate systems: *Cartesian and parabolic.*

1. **Cartesian bases.** Separation of variables in Cartesian coordinates leads to two *independent* one-dimensional Schrödinger equations

$$\frac{d^2\psi_1}{dx^2} + (2\lambda_1 - 4\omega^2 x^2 - 2k_1 x)\psi_1 = 0, \tag{2.2}$$

$$\frac{d^2\psi_2}{dy^2} + \left(2\lambda_2 - \omega^2 y^2 - \frac{k_2^2 - \frac{1}{4}}{y^2}\right)\psi_2 = 0. \tag{2.3}$$

where

$$\Psi(x, y; k_1, \pm k_2) = \psi_1(x; k_1)\psi_2(y; \pm k_2) \tag{2.4}$$

and λ_1, λ_2 are *two Cartesian separation constants* with $\lambda_1 + \lambda_2 = E$.

Equation (2.3) represents the well-known linear singular oscillator system. It is an exactly-solvable problem and the complete set of orthonormalized eigenfunctions is

$$\psi_{n_2}(y;\pm k_2) = \sqrt{\frac{2\omega^{(1\pm k_2)}n_2!}{\Gamma(n_2 \pm k_2 + 1)}}\, y^{\frac{1}{2}\pm k_2}\, e^{-\frac{1}{2}\omega y^2}\, L_{n_2}^{\pm k_2}(\omega y^2) \qquad (2.5)$$

where $\lambda_2 = \omega(2n_2 + 1 \pm k_2)$. The positive sign at the k_2 occurs alone when $k_2 > \frac{1}{2}$ and both the positive and the negative signs occurs if $0 < k_2 < \frac{1}{2}$.

The second equation easily transforms to the ordinary one-dimensional oscillator problem:

$$\psi_{n_1}(x;k_1) = \left(\frac{2\omega}{\pi}\right)^{1/4} \frac{e^{-\omega z^2}}{\sqrt{2^{n_1}n_1!}} H_{n_1}(\sqrt{2\omega}z), \qquad z = x + \frac{k_1}{4\omega^2}, \qquad (2.6)$$

where $\lambda_1 = \omega(2n_1+1) - \frac{k_1^2}{8\omega^2}$. Thus the complete energy spectrum is

$$E = \lambda_1 + \lambda_2 = \omega[2n + 2 \pm k_2] - \frac{k_1^2}{8\omega^2}, \qquad n = n_1 + n_2 = 0, 1, 2, \ldots \qquad (2.7)$$

and the degree of degeneracy for fixed principal quantum number n is $(n+1)$.

2. **Parabolic bases.** In parabolic coordinates ξ and η

$$x = \frac{1}{2}(\xi^2 - \eta^2)\ , \qquad y = \xi\eta\ , \qquad \xi \in \mathbf{R},\ \eta > 0 \qquad (2.8)$$

upon the substitution

$$\Psi(\xi,\eta) = X(\xi)Y(\eta)$$

and the introduction of the parabolic separation constant λ, the Schrödinger equation splits into two ordinary differential equations

$$\frac{d^2X}{d\xi^2} + \left(2E\xi^2 - \omega^2\xi^6 - k_1\xi^4 - \frac{k_2^2 - \frac{1}{4}}{\xi^2}\right)X = -\lambda X, \qquad (2.9)$$

$$\frac{d^2Y}{d\eta^2} + \left(2E\eta^2 - \omega^2\eta^6 + k_1\eta^4 - \frac{k_2^2 - \frac{1}{4}}{\eta^2}\right)Y = +\lambda Y. \qquad (2.10)$$

Equations (2.9) and (2.10) are transformed into one another by the change $\xi \longleftrightarrow i\eta$. The complete wave function is

$$\Psi(\xi,\eta;E,\lambda) = C(E,\lambda)Z(\xi;E,\lambda)Z(i\eta;E,\lambda) \qquad (2.11)$$

where $C(E,\lambda)$ is the normalization constant and the wave function $Z(\mu;E,\lambda)$ is a solution of the equation

$$\left[-\frac{d^2}{d\mu^2} + \left(\omega^2\mu^6 + k_1\mu^4 - 2E\mu^2 + \frac{k_2^2 - \frac{1}{4}}{\mu^2}\right)\right] Z(\mu;E,\lambda) = \lambda Z(\mu;E,\lambda). \quad (2.12)$$

Thus, at $\mu \in (-\infty, \infty)$ we have Eq. (2.9) and at $\mu \in [0, i\infty)$ - Eq. (2.10) and in the complex μ domain the "physical" region is just the two lines Im $\mu = 0$ and Re $\mu = 0$, Im $\mu > 0$.

Our task is to find the solutions of Eq. (2.12) that are regular and decreasing as $\mu \to \pm\infty$ and $\mu \to i\infty$.

Requiring a solution in the form

$$Z(\mu; E, \lambda) = \exp\left(-\frac{\omega}{4}\mu^4 - \frac{k_1}{4\omega}\mu^2\right) \mu^{\frac{1}{2} \pm k_2} \psi(\mu; E, \lambda), \tag{2.13}$$

and passing to a new variable $z = \mu^2$

$$z\frac{d^2\psi}{dz^2} + \left[(1 \pm k_2) - \omega z\left(z + \frac{k_1}{2\omega^2}\right)\right]\frac{d\psi}{dz} + \left[\frac{1}{2}\tilde{E}z + \frac{1}{4}\tilde{\lambda}\right]\psi = 0 \tag{2.14}$$

where

$$\tilde{E} = E + \frac{k_1^2}{8\omega^2} - \omega(2 \pm k_2), \qquad \tilde{\lambda} = \lambda - \frac{k_1}{\omega}(1 \pm k_2), \tag{2.15}$$

we express the wave function $\psi(z)$ as

$$\psi(z; E, \lambda) = \sum_{s=0}^{\infty} A_s(E, \lambda) z^s. \tag{2.16}$$

The substitution (2.16) in Eq. (2.14) leads to the following three-term recurrence relation for the expansion coefficients $A_s \equiv A_s(E, \lambda)$,

$$\begin{aligned}(s+1)(s+1 \pm k_2)A_{s+1} &+ \frac{1}{4}\left[\lambda - \frac{k_1}{\omega}(2s + 1 \pm k_2)\right]A_s \\ &+ \frac{1}{2}\left[E + \frac{k_1}{8\omega^2} - \omega(2s \pm k_2)\right]A_{s-1} = 0,\end{aligned} \tag{2.17}$$

with the initial conditions $A_{-1} = 0$ and $A_0 = 1$.

Let us now examine the asymptotics for a function $\psi(z)$ on the real axis $z \in (-\infty, \infty)$. At large s, from (2.17) we have

$$[s^2 + s(2 \pm k_2)]\frac{A_{s+1}}{A_s}\frac{A_s}{A_{s-1}} - \frac{k_1}{2\omega}s\frac{A_s}{A_{s-1}} - \omega s = 0. \tag{2.18}$$

As $s^{-1} \ll 1$ the following expansions hold

$$\frac{A_{s+1}}{A_s} \sim c_0 + \frac{c_1}{\sqrt{s}} + \emptyset\left(\frac{1}{s}\right). \tag{2.19}$$

Using (2.19) in (2.18) we obtain

$$c_0 = 0, \qquad c_1 = \pm\sqrt{\omega}, \qquad A_s \sim \frac{(\pm\sqrt{\omega})^s}{\sqrt{s!}} \tag{2.20}$$

and therefore the function $\psi(z)$ may have two asymptotic regimes

$$\psi(z) \sim \sum \frac{(\pm\sqrt{\omega}z)^s}{\sqrt{s!}}. \tag{2.21}$$

Then we have for $z > 0$ [the case of Eq. (2.9)]

$$\sum \frac{(\sqrt{\omega}z)^s}{\sqrt{s!}} \sqrt{\sum \frac{(\omega z^2)^s}{s!}} = \exp\left(\frac{\omega}{2}z^2\right) \tag{2.22}$$

and the same for $z < 0$ [the case of Eq. (2.10)]

$$\sum \frac{(-\sqrt{\omega}z)^s}{\sqrt{s!}} = \sum \frac{(\sqrt{\omega}|z|)^s}{\sqrt{s!}} \sqrt{\sum \frac{(\omega z^2)^s}{s!}} = \exp\left(\frac{\omega}{2}z^2\right). \tag{2.23}$$

Thus the function $Z(\mu)$ cannot converge **simultaneously** at large μ for real and imaginary μ and therefore the series should be truncated. Note that a more detailed mathematical analysis of the three-term recurrence relations (2.17) has been carried out in our recent article [14].

The condition for the series to be truncated results in the energy spectrum, giving the already known formula. The coefficients $A_s \equiv A_s^{nq}(k_1, \pm k_2)$ satisfy the following relation

$$(s+1)(s+1\pm k_2)A_{s+1} + \beta_s A_s + \omega(n+1-s)A_{s-1} = 0, \tag{2.24}$$

$$\beta_s = \frac{\lambda}{4} - \frac{k_1}{4\omega}(2s+1\pm k_2).$$

The three-term recurrence relation (2.24) represents a homogeneous system of $n+1$ - algebraic equations for $n+1$ - coefficients $\{A_0, A_1, A_2, ...A_n\}$. The requirement for the existence of a non-trivial solution leads to a vanishing of the determinant

$$D_n(\lambda) = \begin{vmatrix} \beta_0 & 1\pm k_2 & & \cdot & & & \\ \omega n & \beta_1 & 2(2\pm k_2) & \cdot & & & \\ \cdot & \cdot & \cdot & \cdot & & \cdot & \cdot \\ & & & \cdot & 2\omega & \beta_{n-1} & n(n\pm k_2) \\ & & & \cdot & & \omega & \beta_n \end{vmatrix} = 0.$$

The roots of the corresponding algebraic equation give us the $(n+1)$ eigenvalues of the parabolic separation constant $\lambda_{nq}(k_1, \pm k_2)$ and can be enumerated with the help of the integer q, where $0 \le q \le n$. The degeneracy for the n - energy state, as in the Cartesian case, equals $(n+1)$.

The physical admissible solutions have the form

$$Mk_{nq}(z; k_1, \pm k_2) \equiv \psi(z, E, \lambda) = \sum_{s=0}^{n} A_s^{nq}(k_1, \pm k_2)\, z^s,$$

$$Z_{nq}(\mu; k_1, \pm k_2) = exp\left(-\frac{\omega}{4}\mu^4 - \frac{k_1}{4\omega}\mu^2\right) \mu^{\frac{1}{2}\pm k_2} Mk_{nq}(\mu^2; k_1, \pm k_2). \quad (2.25)$$

For $\mu = \xi$ the function $Z_{nq}(\mu; k_1, \pm k_2)$ gives the solution of Equation (2.9), and for $\mu = i\eta$ the solution for Equation (2.10). Thus the parabolic wave function (2.11) can be written in following way

$$\Psi_{nq}(\xi, \eta; k_1, \pm k_2) = C_{nq}(k_1, \pm k_2)\, Z_{nq}(\xi; k_1, \pm k_2)\, Z_{nq}(i\eta; k_1, \pm k_2).$$

2.1. Niven-Stilties approach. Let us express the solution of the Schrödinger equation for the anisotropic oscillator in the following form

$$\Psi(x, y) = e^{-\omega(x+\frac{k_1}{4\omega^2})^2 - \frac{1}{2}\omega y^2}\, y^{\frac{1}{2}\pm k_2}\, \Phi(x, y). \quad (2.26)$$

The function $\Phi(x, y)$ is polynomial in variables (x, y^2) in Cartesian coordinates and (ξ^2, η^2) in parabolic ones. It satisfies the equation

$$\mathcal{R}\Phi(x, y) = -2E\Phi(x, y), \quad (2.27)$$

where the operator $\mathcal{R}$ is

$$\mathcal{R} = \frac{\partial^2}{\partial x^2} + \frac{\partial^2}{\partial y^2} + \left[\frac{(1 \pm 2k_2)}{y} - 2\omega y\right]\frac{\partial}{\partial y}$$
$$-4\omega\left[x + \frac{k_1}{4\omega^2}\right]\frac{\partial}{\partial x} - \omega(2 \pm k_2) + \frac{k_1^2}{8\omega^2}.$$

Taking into account that

$$Mk_{nq}(z; k_1, \pm k_2) = \sum_{s=0}^{n} A_s^{nq}(k_1, \pm k_2)\, z^s \cong \Pi_{\ell=1}^{n}(z - \alpha_\ell), \quad (2.28)$$

where α_ℓ, $(\ell = 1, 2, ...n)$ are zeros of polynomials $Mk_{nq}(z)$ on the real axis $-\infty < z < \infty$, and that in parabolic coordinates

$$\frac{y^2}{\alpha} + 2x - \alpha = \frac{(\xi^2 - \alpha)(\eta^2 + \alpha)}{\alpha}, \quad (2.29)$$

we can choose a solution of Eq. (2.27) in the form

$$\begin{aligned}\Phi(x, y) &= Mk_{nq_1}(\xi^2; k_1, \pm k_2)\, Mk_{nq_2}(-\eta^2; k_1, \pm k_2) \\ &\cong \Pi_{\ell=1}^{n}\left(\frac{y^2}{\alpha_\ell} + 2x - \alpha_\ell\right).\end{aligned} \quad (2.30)$$

Then from (2.27) follows that zeros α_ℓ must satisfy the system of n - algebraic equations

$$\sum_{m\neq\ell}^{n} \frac{2}{\alpha_\ell - \alpha_m} + \frac{(1 \pm k_2)}{\alpha_\ell} - \omega\alpha_\ell = \frac{k_1}{2\omega}, \qquad \ell = 1, 2, ...n, \quad (2.31)$$

and for the energy spectrum we again have formula (2.7).

The system of algebraic equations (2.31) contains the n - set of solutions (zeros) $(\alpha_1^{(q)}, \alpha_2^{(q)}, ...\alpha_n^{(q)})$, $q = 1, 2, ...n$ and all zeros are real. The positive zeros $\alpha_\ell > 0$ define the nodes of wave functions for Equation (2.9), whereas negative zeros $\alpha_\ell < 0$ define the nodes of wave functions for Equation (2.10).

The eigenvalues for the parabolic separation constant can be calculated in the same way via the operator equation

$$\Lambda\Phi(x, y) = \lambda\Phi(x, y).$$

A more elegant way is to use the differential equation (2.14) directly. Rewriting first Eq. (2.14) in the form

$$\begin{aligned}\Bigg\{4z\frac{d^2}{dz^2} + 4\left[(1 \pm k_2) - \omega z\left(z + \frac{k_1}{2\omega^2}\right)\right]\frac{d}{dz} \\ + \left[4n\omega z - \frac{k_1}{\omega}(1 \pm k_2)\right]\Bigg\} Mk_{nq}(z; k_1, \pm k_2) = \lambda Mk_{nq}(z; k_1, \pm k_2)\end{aligned} \tag{2.32}$$

and expressing the wave function $Mk_{nq}(z; k_1, \pm k_2)$ in the form of (2.28), we obtain

$$\lambda_{nq}(k_1, \pm k_2) = 4(1 \pm k_2)\left[\frac{k_1}{4\omega} + \sum_{\ell=1}^{n} \frac{1}{\alpha_\ell^{(q)}}\right]. \tag{2.33}$$

2.2. Connection with quasi-exactly solvable systems. Substitution of the formula for the energy spectrum in Eq. (2.12) yields

$$\begin{aligned}\Bigg[-\frac{d^2}{d\mu^2} + \Bigg(\omega^2\mu^6 + k_1\mu^4 + \left[\frac{k_1^2}{4\omega^2} - \omega(4n + 4 \pm 2k_2)\right]\mu^2 \\ + \frac{k_2^2 - \frac{1}{4}}{\mu^2}\Bigg)\Bigg] Z_n(\mu) = \lambda Z_n(\mu),\end{aligned}$$

which on the real axis completely coincides for $k_1 = 4\beta\omega^2$ and $1 \pm k_2 = 2\delta$, with the one-dimensional spectral problem (1.4), and is called a quasi-exactly solvable system. Now it is easy to understand the origin of the occurence of quasi-exactly solvable systems. The requirement of convergence just in real space in the vicinity of the singular points $\mu = \pm\infty$ implies that there are polynomial solutions of the form (2.25) We also can shed light on the mystery of zeros of polynomial $P_n(x^2)$. Indeed, the substitution of the wave function

$$\Psi(x) \approx x^{2\delta - \frac{1}{2}}\, e^{-\frac{\omega}{4}x^4 - \beta\omega x^2}\, P_n(x^2).$$

into the one-dimensional Schrödinger equation with potential (1.4) leads to the differential equation for polynomial $P_n(x^2)$ in the same form as

Equation (2.32) (in variable $x^2 = z$), but with the difference that the physical region of Eq. (2.32) is whole real axis $z \in (-\infty, \infty)$, and therefore all zeros (for positive and negative x^2) of $P_n(x^2)$ correspond to the zeros of two-dimensional eigenfunction of singular anisotropic oscillator in parabolic coordinates.

The situation is repeated in the case of the potential ($k_1, k_2 > 0$)

$$V(x,y) = \frac{1}{2}\omega^2(x^2+y^2) + \frac{1}{2}\left(\frac{k_1^2 - \frac{1}{4}}{x^2} + \frac{k_2^2 - \frac{1}{4}}{y^2}\right).$$

The corresponding Schrödinger equation separates in three different orthogonal coordinate systems: **Cartesian, polar and elliptical coordinates.**

Separation of variables in two-dimensional elliptic coordinates leads to a Schrödinger type equation

$$\begin{aligned} \frac{d^2 Z(\zeta)}{d\zeta^2} + \Bigg[\frac{D^4\omega^2}{64}\cos^4\zeta - \frac{D^2 E}{4}\cos^2\zeta \\ - \frac{k_1^2 - \frac{1}{4}}{\cos^2\zeta} - \frac{k_2^2 - \frac{1}{4}}{\sin^2\zeta} - \lambda(D^2)\Bigg] Z(\zeta) = 0, \end{aligned} \tag{2.34}$$

in the complex plane and the requirement of convergence at the point $\zeta = 0, 2\pi$ and $\zeta = i\infty$ forces polynomial solutions and determines the energy spectrum $E_n = \omega(2n + 2 \pm k_1 \pm k_2)$, $(n = 0, 1, 2, ...)$.
As a consequence trigonometric and hyperbolic quasi-exactly solvable systems [10] are generated in the form

$$\begin{aligned} \frac{d^2 X}{d\nu^2} + \Bigg[\left(\frac{\alpha^2}{4} + \alpha(2n + 2 \pm k_1 \pm k_2)\right)\cosh^2\nu - \frac{\alpha^2}{4}\cosh^4\nu \\ - \frac{k_1^2 - \frac{1}{4}}{\sinh^2\nu} + \frac{k_2^2 - \frac{1}{4}}{\cosh^2\nu} + \lambda\Bigg] X = 0, \end{aligned}$$

$$\begin{aligned} \frac{d^2 Y}{d\mu^2} - \Bigg[\left(\frac{\alpha^2}{4} + \alpha(2n + 2 \pm k_1 \pm k_2)\right)\cos^2\mu - \frac{\alpha^2}{4}\cos^4\mu \\ + \frac{k_1^2 - \frac{1}{4}}{\cos^2\mu} + \frac{k_2^2 - \frac{1}{4}}{\sin^2\mu} + \lambda\Bigg] Y = 0, \end{aligned}$$

where $\alpha = D^2\omega/2$.

Thus we have established that an integral part of the notion of quasi-exact solvability is the reduction of superintegrable systems to one dimensional problems.

We can express our observation in the form of the following hypothesis: **Quantum mechanical problems which are expressible as one-dimensional quasi-exactly solvable systems can be obtained via separation of variables from N-dimensional Schrödinger equations for superintegrable systems.**

This analogy prompts us to use the term quasi-exactly solvable for equations of type (2.12) or (2.34), defined in the complex plane and which are not exactly-solvable but admit polynomial solutions.

We suggest calling quantum mechanical systems **first-order quasi-exactly solvable** if the polynomial solutions of the one-parametric differential equation obtained fom the N -dimensional Schrödinger equation after separation of variables are defined by a recurrence relation which contain three terms or more and the discrete eigenvalues can be calculated as the solutions of algebraic equations. According to this definition, systems (2.12) and (2.34) are first order quasi-exactly solvable.

We propose also another definition of exact-solvability through the solutions of ordinary differential equation: a *quantum mechanical system is called exactly solvable if the solutions of the Schrödinger equation, can be expressed in terms of hypergeometric functions* $_mF_n$. More precisely this means that the coefficients in power series expansions of the solutions satisfy two-term recurrence relations, rather than recurrence relations of higher order.

It is obvious, that an N-dimensional Schrödinger equation is exactly solvable if it is separable in some coordinate system and each of the separated equations is exactly solvable. Obviously, a superintegrable system is exactly-solvable if is exactly solvable in at least one system of coordinates.

3. Magyari - Uschveridze example. A critical further example is the tenth-order polynomial quasi-exactly solvable problem from an article of Magyari [7] and Ushveridze's book [11].

Consider the Schrödinger equation in the form $H\Psi = E\Psi$ with the Hamiltonian

$$H = \frac{\partial^2}{\partial x^2} + \frac{\partial^2}{\partial y^2} + \frac{\partial^2}{\partial z^2} + 36k_1^2\left[2(x-iy)^2 - 4(x^2+y^2) - z^2\right]$$
$$+48k_1k_2\left[3(x-iy)^2 - (x+iy)\right] - 16(2k_2^2 + 3k_1k_3)(x+iy) - \frac{p(p-1)}{z^2}.$$

This is essentially the Euclidean space superintegrable system with nondegenerate potential

$$V = \alpha\left[z^2 - 2(x-iy)^3 + 4(x^2+y^2)\right) + \beta\left(2(x+iy) - 3(x-iy)2\right]$$
$$+\ \gamma(x+iy) + \frac{\delta}{z^2},$$

where the six basis symmetry operators can be taken in the form

$$H = \partial_x^2 + \partial_y^2 + \partial_z^2 + V,$$

$$L_1 = (\partial_x - i\partial_y)2 + f_1, \quad L_2 = \partial_z^2 + f_2, \quad L_3 = \{\partial_z, J_2 + iJ_1\} + f_3,$$

$$L_4 = \frac{1}{2}\{J_3, \partial_x - i\partial_y\} - \frac{i}{4}(\partial_x + i\partial_y)^2 + f_4, \quad L_5 = (J_2 + iJ_1)^2 + 2i\{\partial_z, J_1\} + f_5,$$

where $\{A, B\} = AB + BA$ and the f_i $(i = 1, 2, ..., 5)$ are the function of potential V. There is a quadratic algebra generated by these symmetries.

One choice of separable coordinates in three dimensions is

$$z = iuvw,$$

$$x + iy = \frac{1}{2}(u^2v^2 + u^2w^2 + v^2w^2) - \frac{1}{4}(u^4 + v^4 + w^4), \quad x - iy = \frac{1}{2}(u^2 + v^2 + w^2).$$

The separation equations in these coordinates have the form

$$\left[\frac{d^2}{d\lambda^2} - 36k_1^2\lambda^{10} - 48k_1k_2\lambda^8 - 8(2k_2^2 + 3k_1k_3)\lambda^6 + \frac{p(1-p)}{\lambda^2} + E\lambda^4 + \ell_2\lambda^2 + \ell_3\right]\Lambda(\lambda) = 0. \tag{3.1}$$

where ℓ_2 and ℓ_3 are the separation constant. This is essentially (for $p = 1, 0$), the Magyari-Ushveridze one-dimensional quasi-exactly solvable problem.

In searching for finite solutions of the Schrödinger equation we write

$$\Psi(u, v, w) = \exp\left[k_1(u^6 + v^6 + w^6) + k_2(u^4 + v^4 + w^4) + k_3(u^2 + v^2 + w^2)\right](uvw)^p\Phi(u, v, w)$$

where the function Φ has polynomial form

$$\Phi(u, v, w) = \prod_{j=1}^{n}(u^2 - \theta_j)(v^2 - \theta_j)(w^2 - \theta_j).$$

The zeros of the polynomials satisfy the relations

$$\frac{2n+1}{2\theta_i} - 12k_1\theta_i^2 - 4k_2\theta_i - k_3 + \sum_{j \neq i}\frac{1}{\theta_i - \theta_j} = 0.$$

Solving these equations we see that the eigenvalues E, ℓ_2, ℓ_3 have the form

$$E = -(30 + 24n + 12p)k_1 - 16k_2k_3,$$

$$\ell_2 = -4k_3^2 - (12 + 16r)k_2 - 24k_1\sum_{j=1}^{r}\theta_j,$$

$$\ell_3 = -(2 + 8r + 4p)k_3 - 16k_2 \sum_{j=1}^{r} \theta_j - 24k_1 \sum_{j=1}^{r} \theta_j^2.$$

The last equations, in fact, give us also the requirements that the one-dimensional differential equation (3.1) with tenth-order anharmonicity admits polynomial solutions, or is quasi-exactly solvable a la Turbiner and Ushveridze.

REFERENCES

[1] S.R. Wojciechowski. *Superintegrability of the Calogero-Moser System.* Phys. Lett., **A 95**: 279, 1983.

[2] N.W. Evans. *Superintegrability in Classical Mechanics*; Phys. Rev., **A 41**: 5666, 1990.

[3] C. Grosche, G.S. Pogosyan, and A.N. Sissakian. *Path Integral Discussion for Smorodinsky - Winternitz Potentials: I. The Two - and Three Dimensional Euclidean Space.* Fortschritte der Physik, **43**: 453–521, 1995.

[4] E.G. Kalnins, W. Miller, Jr., and G.S. Pogosyan. *Superintegrability and associated polynomial solutions. Euclidean space and sphere in two-dimensions.* J. Math. Phys., **37**: 6439–6467, 1996.

[5] E.G. Kalnins, G. Williams, W. Miller, Jr., and G.S. Pogosyan. *Superintegrability in the three-dimensional Euclidean space.* J. Math. Phys., **40**: 708–725, 1999.

[6] E.G. Kalnins, W. Miller, Jr., G.S. Pogosyan, and G.C. Williams. *On superintegrable symmetry-breaking potentials in N-dimensional Euclidean space.* J. Phys. **A 35**(22): 4755–4773, 2002.

[7] E. Magyari. *Exact Quantum-Mechanical Solutions for Anharmonic Oscillators.* Phys. Lett., **A 81**: 116–118, 1981.

[8] A.V. Turbiner and A.G. Ushveridze. *Spectral singularities and quasi-exactly solvable quantum problem.* Phys. Lett., **A 126**: 181–183, 1987.

[9] A.V. Turbiner. *Quasi-Exactly-Solvable Problems and $sl(2)$ Algebra.* Comm. Math. Phys., **118**: 467–474, 1988.

[10] A.G. Ushveridze. *Quasi-exactly solvable models in quantum mechanics.* Sov. J. Part. Nucl., **20**: 504–528, 1989.

[11] A.G. Ushveridze, *Quasi-Exactly solvable models in quantum mechanics.* Institute of Physics, Bristol, 1993.

[12] L.G. Mardoyan, G.S. Pogosyan, A.N. Sissakian, and V.M. Ter-Antonyan. *Hidden symmetry, Separation of Variables and Interbasis Expansions in the Two-Dimensional Hydrogen Atom.* J. Phys., **A 18**: 455–466, 1985.

[13] L.G. Mardoyan, G.S. Pogosyan, A.N. Sissakian, and V.M. Ter-Antonyan. Two-Dimensional Hydrogen Atom: I. Elliptic Bases; *Theor. Math. Phys.*, **61**: 1021, 1984.

[14] E.G. Kalnins, W. Miller, Jr., and G.S. Pogosyan. *Exact and quasi-exact solvability of second order superintegrable quantum systems. I Euclidean space preliminaries.* J. Math. Phys., **47**: 033502, 2006.

A REMARK ON UNITARY CONFORMAL HOLONOMY

FELIPE LEITNER*

Abstract. If the conformal holonomy group $Hol(\mathcal{T})$ of a simply connected space with conformal structure of signature $(2p-1, 2q-1)$ is reduced to $\mathrm{U}(p,q)$ then the conformal holonomy is already contained in the special unitary group $\mathrm{SU}(p,q)$. We present two different proofs of this statement, one using conformal tractor calculus and an alternative proof using Sparling's characterisation of Fefferman metrics.

Key words. Tractor calculus, conformal holonomy, Fefferman construction.

AMS(MOS) subject classifications. 53A30, 53C29, 32V05, 53B30, 53C07.

1. Introduction. Any conformal structure on a space M^n of dimension $n \geq 3$ with signature (r,s) gives rise to a $|1|$-graded parabolic geometry with canonical Cartan connection, which is uniquely determined by a normalisation condition on the curvature. This connection solves the equivalence problem for conformal geometry. The standard representation of the orthogonal group $\mathrm{O}(r+1, s+1)$, which double covers the Möbius group $\mathrm{PO}(r+1, s+1)$, gives rise to a so-called standard tractor bundle $\mathcal{T}(M)$ on M and the canonical Cartan connection induces a linear connection on this vector bundle. This construction allows the definition of the conformal holonomy group $Hol(\mathcal{T})$ with algebra $\mathfrak{hol}(\mathcal{T})$, which represent important conformal invariants.

An interesting task in the frame work of conformal geometry is the classification of possible conformal holonomy groups and the realisation of such holonomy groups by concrete geometric constructions. Some progress in this direction was made in the works [Arm05, Lei04a, Lei04b, Lei05a, Lei05b]. The probably best known case so far is that of conformally Einstein spaces. In this situation the conformal holonomy group acts trivially on a certain standard tractor and via a cone construction a classification of conformal holonomy can be achieved. Obviously, those holonomy groups do not have an irreducible action on standard tractors. A classical case of an irreducible subgroup of $\mathrm{O}(2p, 2q)$ is the special unitary group $\mathrm{SU}(p,q)$. Conformal structures with holonomy reduction to (a subgroup of) $\mathrm{SU}(p,q)$ are obtained by the classical Fefferman construction, which assigns to any integrable CR-space of hypersurface type an invariant conformal structure on some canonical circle bundle.

We aim to prove here (in two different ways) that if the conformal holonomy group $Hol(\mathcal{T})$ of some space M with conformal structure of signature $(2p-1, 2q-1)$ is a subgroup of the unitary group $\mathrm{U}(p,q)$ then the conformal holonomy algebra $\mathfrak{hol}(\mathcal{T})$ is automatically contained in the

*Institut für Geometrie und Topologie, Universität Stuttgart, Pfaffenwaldring 57, D-70569 Stuttgart, GERMANY (leitner@mathematik.uni-stuttgart.de).

special unitary algebra $\mathfrak{su}(p,q)$ (if M is simply connected $Hol(\mathcal{T})$ is automatically reduced to $\mathrm{SU}(p,q)$). So in the first moment it seems that there is a gap in the list of possible conformal holonomy groups. However, as one of our proofs shows the reason for this behaviour is directly implicated by the normalisation condition for the curvature of the canonical Cartan connection, and therefore, the gap should be considered as natural and immediate consequence in the theory of conformal holonomy.

We will proceed as follows. In Sections 2 and 3 we introduce the canonical Cartan connection of conformal geometry and deduce the notion of conformal (tractor) holonomy. In Section 4 we make some necessary considerations about conformal Killing vector fields. In Section 5 we prove our reduction result in terms of tractor calculus (cf. Theorem 5.1). In the final section we give an alternative proof by using Sparlings's characterisation of Fefferman metrics. This reasoning allows us to present a local geometric description of spaces with conformal holonomy reduction to $\mathrm{U}(p,q)$ (resp. $\mathrm{SU}(p,q)$) (cf. Corollary 6.1).

2. Conformal Cartan geometry. We briefly introduce in this section the canonical Cartan geometry, which belongs to any space (M,c) with conformal structure (cf. e.g. [Kob72] or [CSS97a, CSS97b]).

Let $\mathbb{R}^{r,s}$ denote the Euclidean space of dimension $n = r+s \geq 3$ and signature (r,s), where the scalar product is given by the matrix

$$\mathbb{J} = \begin{pmatrix} -I_r & 0 \\ 0 & I_s \end{pmatrix} .$$

We denote by $\mathfrak{g}$ the Lie algebra $\mathfrak{so}(r+1,s+1)$ of the orthogonal group $\mathrm{O}(r+1,s+1)$, which acts by the standard representation on the Euclidean space $\mathbb{R}^{r+1,s+1}$ of dimension $n+2$. With respect to the coordinates $\{x_-, x_1, \ldots, x_n, x_+\}$ we define the invariant scalar product on $\mathbb{R}^{r+1,s+1}$ by

$$\langle x, y \rangle = x_- y_+ + x_+ y_- + (x_1, \ldots, x_n)\ \mathbb{J}\ (y_1, \ldots, y_n)^\top .$$

The Lie algebra $\mathfrak{g} = \mathfrak{so}(r+1,s+1)$ is $|1|$-graded by

$$\mathfrak{g} = \mathfrak{g}_{-1} \oplus \mathfrak{g}_0 \oplus \mathfrak{g}_1 ,$$

where $\mathfrak{g}_0 \cong \mathfrak{co}(r,s)$, $\mathfrak{g}_{-1} \cong \mathbb{R}^n$ and $\mathfrak{g}_1 \cong \mathbb{R}^{n*}$. The 0-part $\mathfrak{g}_0$ decomposes further to the centre $\mathbb{R}$ and a semisimple part $\mathfrak{so}(r,s)$, which is the Lie algebra of the isometry group of the Euclidean space $\mathbb{R}^{r,s}$. We realise the subspaces $\mathfrak{g}_0, \mathfrak{g}_{-1}$ and $\mathfrak{g}_1$ of $\mathfrak{g}$ by matrices of the form

$$\begin{pmatrix} 0 & 0 & 0 \\ m & 0 & 0 \\ 0 & -m^\top \mathbb{J} & 0 \end{pmatrix} \in \mathfrak{g}_{-1} , \quad \begin{pmatrix} -a & 0 & 0 \\ 0 & A & 0 \\ 0 & 0 & a \end{pmatrix} \in \mathfrak{g}_0 ,$$

$$\begin{pmatrix} 0 & l & 0 \\ 0 & 0 & -\mathbb{J}\, l^\top \\ 0 & 0 & 0 \end{pmatrix} \in \mathfrak{g}_1 ,$$

where $m \in \mathbb{R}^n$, $l \in \mathbb{R}^{n*}$, $A \in \mathfrak{so}(r,s)$ and $a \in \mathbb{R}$.

The space

$$\mathfrak{p} := \mathfrak{g}_0 \oplus \mathfrak{g}_1$$

is a parabolic subalgebra of $\mathfrak{g}$. We also denote $\mathfrak{p}_+ := \mathfrak{g}_1$. Whilst the grading of $\mathfrak{g}$ is not $\mathfrak{p}$-invariant, the filtration

$$\mathfrak{g} \supset \mathfrak{p} \supset \mathfrak{p}_+$$

is $\mathfrak{p}$-invariant. The subgroup G_0 of the Möbius group $G := \mathrm{PO}(r+1,s+1)$, which consists of those elements whose adjoint action on $\mathfrak{g}$ preserves the grading, is isomorphic to the reductive group $\mathrm{CO}(r,s) = \mathrm{O}(r,s) \times \mathbb{R}_+$ with Lie algebra $\mathfrak{g}_0 = \mathfrak{co}(r,s)$. The parabolic subgroup P of G with Lie algebra $\mathfrak{p}$, which preserves the filtration of $\mathfrak{g}$, is then isomorphic to the semidirect product of G_0 and the vector subgroup $P_+ := \exp(\mathfrak{p}_+)$ (cf. e.g. [CSS97a]).

The $\mathfrak{g}_0$-homomorphism ∂^* defined by

$$\begin{aligned} \partial^* : Hom(\Lambda^2\mathfrak{g}_{-1},\mathfrak{g}) &\to Hom(\mathfrak{g}_{-1},\mathfrak{g}) , \\ \psi &\mapsto \partial^*\psi = \Big(X \in \mathfrak{g}_{-1} \mapsto \textstyle\sum_{i=1}^n [\eta_i, \psi(\xi_i, X)] \Big) \end{aligned}$$

where $\{\xi_i : i = 1,\dots,n\}$ is some basis of $\mathfrak{g}_{-1}$ and $\{\eta_i : i = 1,\dots,n\}$ is the corresponding dual basis of $\mathfrak{p}_+$ (with respect to the Killing form), is the so-called codifferential, which is adjoint to the differential ∂ computing the Lie algebra cohomology of $\mathfrak{g}_{-1}$ with values in $\mathfrak{g}$ (cf. [CSS97b]).

Now let M^n be a C^∞-manifold of dimension $n \geq 3$. Two smooth metrics g and $\tilde{g}$ on M are called conformally equivalent if there exists a C^∞-function ϕ such that $\tilde{g} = e^{2\phi}g$. A conformal structure c on M^n of signature (r,s), $n = r+s$, is by definition a class of conformally equivalent C^∞-metrics with signature (r,s). Let $GL(M)$ denote the general linear frame bundle on M. The choice of a conformal structure c on M is the same thing as a G_0-reduction $\mathcal{G}_0(M)$ of $GL(M)$. By the process of prolongation we obtain from $\mathcal{G}_0(M)$ the P-principal fibre bundle $\mathcal{P}(M)$, which is a reduction of the second order linear frame bundle $GL^2(M)$ of M (cf. e.g. [Kob72, CSS97b]).

A Cartan connection ω on the P-principal fibre bundle $\mathcal{P}(M)$ is a smooth 1-form with values in $\mathfrak{g}$ such that

1. $\omega(\chi_A) = A$ for all fundamental fields χ_A, $A \in \mathfrak{p}$,
2. $R_g^*\omega = Ad(g^{-1}) \circ \omega$ for all $g \in P$ and
3. $\omega|_{T_u\mathcal{P}(M)} : T_u\mathcal{P}(M) \to \mathfrak{g}$ is a linear isomorphism for all $u \in \mathcal{P}(M)$.

The curvature 2-form Ω of the Cartan connection ω is given by

$$\Omega = d\omega + \frac{1}{2}[\omega,\omega] .$$

It is $\iota_{\chi_A}\Omega = 0$ for all $A \in \mathfrak{p}$, i.e., the curvature Ω is trivial with respect to insertion of vertical vectors on $\mathcal{P}(M)$. The corresponding curvature function

$$\kappa : \mathcal{P}(M) \rightarrow Hom(\Lambda^2\mathfrak{g}_-, \mathfrak{g})$$

is pointwise defined by

$$\kappa(u)(X,Y) := d\omega(\omega_u^{-1}(X), \omega_u^{-1}(Y)) + [X,Y], \quad u \in \mathcal{P}(M), \ \ X,Y \in \mathfrak{g}_- .$$

This function κ decomposes to $\kappa_{-1} + \kappa_0 + \kappa_1$ according to the grading of the space $\mathfrak{g}$. The map $\mathfrak{g}_0 \rightarrow \mathfrak{gl}(\mathfrak{g}_-)$ induced by the adjoint representation is the inclusion of a subalgebra. Therefore, the 0-part κ_0 can be seen as a function on $\mathcal{P}(M)$, which takes values in $\mathfrak{g}_-^* \otimes \mathfrak{g}_-^* \otimes \mathfrak{g}_-^* \otimes \mathfrak{g}_-$.

A basic fact of conformal geometry is the existence of a canonical Cartan connection ω_{nor} on $\mathcal{P}(M)$, which is uniquely determined by the curvature normalisation

$$\partial^* \circ \kappa = 0 .$$

We call ω_{nor} the normal Cartan connection on $\mathcal{P}(M)$ to the space M with conformal structure c. The condition $\partial^* \circ \kappa = 0$ is equivalent to

$$\kappa_{-1} = 0 \qquad \text{and} \qquad tr\kappa_0 := \sum_{i=1}^{n} \kappa_0(\xi_i, \cdot)(\cdot)(\eta_i) = 0 ,$$

i.e., the unique normal Cartan connection ω_{nor} of conformal geometry is torsion-free and the 0-part of the curvature function is traceless (cf. [Kob72, CSS97a]).

3. Some tractor calculus. Using the normal Cartan connection of conformal geometry we define now a linear connection on standard tractors, which allows us to introduce the notion of conformal (tractor) holonomy.

Let (M^n, c), $n \geq 3$, be a space with conformal structure of signature (r,s) and $\mathcal{P}(M)$ the corresponding P-reduction of $GL^2(M)$ equipped with the normal Cartan connection ω_{nor}. The parabolic $P = (\mathrm{O}(r,s) \times \mathbb{R}_+) \ltimes \mathbb{R}^{n*}$ is included in $\mathrm{O}(r+1, s+1)$ by

$$\begin{array}{cccc} \iota : & P & \rightarrow & \mathrm{O}(r+1,s+1) , \\ & (\beta, \alpha, x) & \mapsto & \begin{pmatrix} \alpha^{-1} & x & v \\ 0 & \beta & y \\ 0 & 0 & \alpha \end{pmatrix} \end{array}$$

where $\beta \in \mathrm{O}(r,s)$, $\alpha > 0$ in $\mathbb{R}$, $x \in \mathbb{R}^{n*}$, $y := -\alpha\beta\mathbb{J}x^\top$ and $v := -\frac{\alpha}{2}x\mathbb{J}x^\top$. This inclusion determines the adjoint action of P on $\mathfrak{g} = \mathfrak{so}(r+1, s+1)$ (such that the filtration is P-invariant). Then we define the so-called adjoint tractor bundle $\mathcal{A}(M)$ over (M, c) by

$$\mathcal{A}(M) := \mathcal{P}(M) \times_{Ad(P)} \mathfrak{g} .$$

This associated vector bundle admits a natural filtration, which we denote by

$$\mathcal{A}(M) \supset \mathcal{A}^0(M) \supset \mathcal{A}^1(M) .$$

Thereby, it is $\mathcal{A}^0(M) = \mathcal{P}(M) \times_{Ad(P)} \mathfrak{p}$ and the subbundle $\mathcal{A}^1(M) = \mathcal{P}(M) \times_{Ad(P)} \mathfrak{p}_+$ is isomorphic to the dual T^*M of the tangent bundle of M. Moreover, the quotient $\mathcal{A}(M)/\mathcal{A}^0(M)$ is naturally identified with the tangent bundle TM. For a section $A \in \Gamma(\mathcal{A}(M))$ we denote by $\Pi(A) \in \mathfrak{X}(M)$ the corresponding projection to vector fields on M.

By restriction of the standard representation of $O(r+1, s+1)$ to P via the inclusion ι we obtain a representation of P on the Euclidean space $\mathbb{R}^{r+1,s+1}$, which also preserves the scalar product $\langle\cdot,\cdot\rangle$ (cf. section 2). We use this P-representation to define the standard tractor bundle $\mathcal{T}(M)$ over (M, c) by

$$\mathcal{T}(M) := \mathcal{P}(M) \times_P \mathbb{R}^{r+1,s+1} .$$

The standard tractor bundle is equipped with an invariant scalar product $\langle\cdot,\cdot\rangle$ and it admits a filtration as well, which we denote by

$$\mathcal{T}(M) \supset \mathcal{T}^0(M) \supset \mathcal{T}^1(M) .$$

Thereby, it is $\mathcal{T}^1(M) = \mathcal{P}(M) \times_P \mathbb{R}$, a trivial real line bundle, which is isomorphic to the density bundle $\mathcal{E}[-1]$ of conformal weight -1. (If $\mathcal{R} \subset S^2(T^*M)$ denotes the ray subbundle of metrics g in c with $\mathbb{R}_+$-action $\mathcal{R} \times \mathbb{R}_+ \to \mathcal{R}$ given by $(g, s) \mapsto s^2 \cdot g$ then the densities $\mathcal{E}[w]$ are defined as associated line bundles by $\mathcal{E}[w] := \mathcal{R} \times_{\rho^w} \mathbb{R}$, where $\rho^w : a \in \mathbb{R}_+ \mapsto a^{-w/2} \in Aut(\mathbb{R})$ (cf. e.g. [CG03]).) Moreover, since the standard $\mathfrak{g}$-action on $\mathbb{R}^{r+1,s+1}$ is compatible with the P-action, there exists a natural equivariant action of $\mathcal{A}(M)$ on $\mathcal{T}(M)$, which we denote by

$$\bullet : \mathcal{A}(M) \otimes \mathcal{T}(M) \to \mathcal{T}(M) .$$

In this respect we can understand $\mathcal{A}(M)$ as a subbundle of the endomorphism bundle $End(\mathcal{T}(M))$.

The tractor bundles $\mathcal{T}(M)$ and $\mathcal{A}(M)$ over (M, c) enjoy the existence of naturally defined covariant derivatives. To introduce these we extend the normal Cartan connection ω_{nor} by right translation to a principal fibre bundle connection on the enlarged bundle

$$\tilde{\mathcal{G}}(M) := \mathcal{P}(M) \times_\iota O(r+1, s+1) .$$

The tractor bundles $\mathcal{T}(M)$ and $\mathcal{A}(M)$ are associated to $\tilde{\mathcal{G}}(M)$ by

$$\mathcal{T}(M) = \tilde{\mathcal{G}}(M) \times_{O(r+1,s+1)} \mathbb{R}^{r+1,s+1} \quad \text{and} \quad \mathcal{A}(M) = \tilde{\mathcal{G}}(M) \times_{Ad} \mathfrak{g} .$$

The extended connection on $\tilde{\mathcal{G}}(M)$ induces in the usual manner the linear connections

$$\nabla^{nor} : \Gamma(\mathcal{T}(M)) \to \Gamma(T^*M \otimes \mathcal{T}(M)) \quad \text{and}$$

$$\nabla^{nor} : \Gamma(\mathcal{A}(M)) \to \Gamma(T^*M \otimes \mathcal{A}(M)).$$

The curvature Ω of the normal Cartan connection ω_{nor} on $\mathcal{P}(M)$ is by definition P-equivariant. Therefore, it can be interpreted as a smooth 2-form with values in $\mathcal{A}(M)$. It holds

$$\nabla^{nor}_X \nabla^{nor}_Y t - \nabla^{nor}_Y \nabla^{nor}_X t - \nabla^{nor}_{[X,Y]} t = \Omega(X,Y) \bullet t$$

for all standard tractors $t \in \Gamma(\mathcal{T}(M))$ and $X, Y \in \mathfrak{X}(M)$.

The covariant derivative ∇^{nor} induces a parallel displacement on the standard tractor bundle $\mathcal{T}(M)$ over (M,c). In particular, the parallel displacement along loops in (M,c) generates a Lie subgroup of the structure group $\mathrm{O}(r+1, s+1)$. We denote this Lie group by $Hol(\mathcal{T})$ and call it the conformal holonomy group of (M,c). Moreover, we denote by $\mathfrak{hol}(\mathcal{T})$ the holonomy algebra to $Hol(\mathcal{T})$. Both objects $Hol(\mathcal{T})$ and $\mathfrak{hol}(\mathcal{T})$ are conformal invariants naturally attached to the underlying space (M,c) (cf. [Arm05, Lei04a, Lei04b]).

Now we choose some arbitrary metric g in the conformal class c on M and describe the tractor bundles $\mathcal{T}(M)$ and $\mathcal{A}(M)$ and the tractor curvature Ω with respect to this metric g. First of all, we note that g corresponds in a unique way to a G_0-equivariant section

$$\rho_g : \mathcal{G}_0(M) \to \mathcal{P}(M) .$$

With the help of this lift ρ_g the structure group of the tractor bundles $\mathcal{T}(M)$ and $\mathcal{A}(M)$ can be reduced to $G_0 = \mathrm{CO}(r,s)$ and we obtain identifications

$$\mathcal{T}(M) \cong_g \mathcal{G}_0(M) \times_{\iota(G_0)} \mathbb{R}^{r+1,s+1} \qquad \text{and}$$

$$\mathcal{A}(M) \cong_g \mathcal{G}_0(M) \times_{Ad \circ \iota(G_0)} \mathfrak{g}$$

of vector bundles. We note that these identifications (denoted by $\cong_g$) depend strongly on the choice of the metric g (resp. the lift ρ_g).

The restricted representations of G_0 on $\mathbb{R}^{r+1,s+1}$ and $\mathfrak{g}$ are not any longer indecomposable. In fact, via the lift ρ_g we obtain natural identifications of $\mathcal{T}(M)$ and $\mathcal{A}(M)$ with graded vector bundles:

$$\mathcal{T}(M) \cong_g \mathcal{E}[1] \oplus TM[-1] \oplus \mathcal{E}[-1] \qquad \text{and}$$

$$\mathcal{A}(M) \cong_g TM \oplus \mathfrak{co}(TM) \oplus T^*M,$$

where $TM[-1] := TM \otimes \mathcal{E}[-1]$. With respect to the scalar product $\langle \cdot, \cdot \rangle$ on $\mathcal{T}(M)$ the subbundles $\mathcal{E}[1]$ and $\mathcal{E}[-1]$ are lightlike, i.e., it holds $\langle u, u \rangle = 0$ for any $u \in \mathcal{E}[1]$ resp. $u \in \mathcal{E}[-1]$. The restriction of $\langle \cdot, \cdot \rangle$ to TM in $\mathcal{T}(M)$ reproduces the metric g.

With the choice of a metric g and the corresponding grading on $\mathcal{T}(M)$ we can express any tractor $t \in \Gamma(\mathcal{T}(M))$ by a triple $t = (b, \tau, d)$ resp. a column vector

$$t = \begin{pmatrix} d \\ \tau \\ b \end{pmatrix} .$$

Accordingly, an adjoint tractor $A \in \Gamma(\mathcal{A}(M))$ is represented by a triple (ξ, φ, η), which we will usually write as a matrix of the form

$$A = \begin{pmatrix} -\varphi_c & \eta & 0 \\ \xi & \varphi_{ss} & -\eta^\flat \\ 0 & -g(\xi, \cdot) & \varphi_c \end{pmatrix},$$

where φ_{ss} is the skew-adjoint part of $\varphi \in \Gamma(\mathfrak{co}(TM))$ and $\varphi_c \cdot id|_{TM}$ is the trace part. The vector $\eta^\flat$ is dual to the 1-form η with respect to g. The action $A \bullet t$ of $\mathcal{A}(M)$ on $\mathcal{T}(M)$ computes now as a matrix product.

In particular, the tractor curvature $\Omega \in \Gamma(\mathcal{A}(M))$ decomposes into a triple $(\Omega_{-1}, \Omega_0, \Omega_1)$. Of course, these components Ω_{-1}, Ω_0 and Ω_1 have well known interpretations in terms of g and its curvature expressions. In fact, the 2-form Ω_{-1} with values in TM is the torsion of the Levi-Civita connection ∇^g to g, which is known to be zero. The 2-form Ω_0 with values in $\mathfrak{co}(TM)$ is the Weyl curvature usually denoted by $\mathcal{W}$. Eventually, the 1-part Ω_1 is equal to the Cotton-York tensor $\mathcal{C}$ of g, which is given by

$$\mathcal{C}(X,Y) := (\nabla^g_X \mathcal{K})(Y) - (\nabla^g_Y \mathcal{K})(X),$$

where Ric^g denotes the Ricci curvature of g, $scal^g$ the scalar curvature and

$$\mathcal{K} := \frac{1}{n-2}\left(\frac{scal^g}{2(n-1)} g - Ric^g\right)$$

is the Schouten tensor.

The normalisation condition $\partial^* \circ \kappa = 0$ for the Cartan curvature Ω is reflected in the vanishing of Ω_{-1} and the Ricci-type trace of $\mathcal{W} = \Omega_0$. The Weyl curvature $\mathcal{W}$ of g satisfies also the first Bianchi identity

$$\sum_{cycl} \mathcal{W}(X,Y)Z = 0$$

(which is a consequence of the generalised Bianchi identity for the curvature function κ of ω_{nor} (cf. [CSS97a])).

4. Conformal Killing vector fields as adjoint tractors. In this section we recall a relation of conformal Killing vector fields and adjoint tractors, which satisfy a certain tractor equation. In particular, we will establish Lemma 4.2, which will be crucial in the following sections and which was proved (in a more general setting) in [Cap05] (cf. also [Lei04b]).

Let (M, c) be a space with conformal structure. A vector field $V \in \mathfrak{X}(M)$ is called a conformal Killing vector field if for some metric $g \in c$ and some function λ it holds

$$\mathcal{L}_V g = \lambda \cdot g,$$

where $\mathcal{L}$ denotes the Lie derivative. In fact, if this condition is satisfied by V then it exists for any metric $\tilde{g} \in c$ some function $\tilde{\lambda}$ such that $\mathcal{L}_V \tilde{g} = \tilde{\lambda} \cdot \tilde{g}$. Certain sections in $\mathcal{A}(M)$ give rise to conformal Killing vector fields.

LEMMA 4.1. *(cf. [Cap05]) Let $A \in \Gamma(\mathcal{A}(M))$ be an adjoint tractor. The quotient $V_A := \Pi(A) \in \mathfrak{X}(M)$ is a conformal Killing vector field on (M, c) if and only if*

$$\nabla^{nor}_X A = -\Omega(\Pi(A), X) \qquad \textit{for all} \;\; X \in TM \; .$$

Conversely, it is also true that every conformal Killing vector field V determines uniquely via its 2-jet $j_2(V) \in \mathfrak{X}(\mathcal{P}(M))$ an adjoint tractor A_V, whose quotient $\Pi(A_V)$ reproduces the vector field V and which satisfies the tractor equation of Lemma 4.1. We can express this adjoint tractor A_V explicitly with respect to any metric $g \in c$ by application of a certain second order differential operator. It is

$$A_V \;\simeq_g\; (V, \nabla^g V, \mathcal{D}^g(V)) \;=\; \begin{pmatrix} -\frac{2}{n} div^g V & \mathcal{D}^g(V) & 0 \\ V & asym \nabla^g V & -\mathcal{D}^g(V)^\flat \\ 0 & -g(V,\cdot) & \frac{2}{n} div^g V \end{pmatrix} ,$$

where $div^g V := tr_g(\nabla^g_\cdot V, \cdot)$ is the divergence, the covariant derivative $\nabla^g V$ splits into the anti-symmetric part $asym\nabla^g V$ and the symmetric part $div^g V \cdot id|_{TM}$ and

$$\mathcal{D}^g \;:=\; \frac{1}{n-2} \left(\, tr_g \nabla^2 \;+\; \frac{scal^g}{2(n-1)} \, \right)$$

is the Bochner-Laplacian with a curvature normalisation (cf. [Lei05b]). The applied second order differential operator in this formula is a special instance of a so-called splitting operator known from the construction of BGG-sequences (cf. [CSS01]).

Next we consider the tractor equation

$$\nabla^{nor} A = 0$$

for a section $A \in \Gamma(\mathcal{A}(M))$. We have the following remarkable result of conformal tractor calculus.

LEMMA 4.2. *(cf. [Cap05]) Let $A \in \Gamma(\mathcal{A}(M))$ be a ∇^{nor}-parallel adjoint tractor. Then it holds*

$$\Omega(\Pi(A), \cdot) = 0 \; .$$

In particular, the quotient $\Pi(A)$ is a conformal Killing vector field.

5. Conformal holonomy reduction. In Section 3 we introduced the conformal holonomy group $Hol(\mathcal{T})$ and the holonomy algebra $\mathfrak{hol}(\mathcal{T})$ of a space (M, c) with conformal structure. In this section we assume that there exists a section J of the adjoint tractor bundle $\mathcal{A}(M)$ on (M^n, c) of dimension $n = 2m \geq 4$ with signature $(2p-1, 2q-1)$, which acts as complex structure on the standard tractor bundle $\mathcal{T}(M)$ and which is parallel with respect to the covariant derivative ∇^{nor}, i.e., it holds

$$J^2 := J \bullet J = -id|_{\mathcal{T}(M)} \in \Gamma(End(\mathcal{T}(M)) \qquad \text{and} \qquad \nabla^{nor} J = 0 \,.$$

The existence of such a complex structure J is equivalent (by the very definition of holonomy) to the fact that the holonomy group $Hol(\mathcal{T})$ of (M, c) is contained in the unitary group $\mathrm{U}(p,q)$. However, we aim to show here (in terms of tractor calculus) that the existence of such J implies already that the holonomy algebra $\mathfrak{hol}(\mathcal{T})$ is reduced to the special unitary algebra $\mathfrak{su}(p,q)$! The reason for this reduction is essentially the normalisation condition $\partial^* \circ \kappa = 0$ for the connection ω_{nor}.

We start our reasoning with an observation about complex structures in $\mathfrak{so}(2p, 2q)$.

LEMMA 5.1. *(cf. [Lei05b]) Let*

$$\beta = \begin{pmatrix} -a & l & 0 \\ m & A & -\mathbb{J}l^\top \\ 0 & -m^\top \mathbb{J} & a \end{pmatrix}$$

be a matrix in $\mathfrak{g} = \mathfrak{so}(2p, 2q)$. Then the property $\beta^2 = -id$ is equivalent to the following conditions on $m, -\mathbb{J}l^\top \in \mathfrak{g}_{-1}$ and $A \in \mathfrak{so}(\mathbb{J})$:

1. *m and $-\mathbb{J}l^\top$ are lightlike eigenvectors of A to the eigenvalue a,*
2. *the scalar product of m with $-\mathbb{J}l^\top$ equals $1 + a^2$ and*
3. *A^2 restricted to $(span\{m, -\mathbb{J}l^\top\})^\perp$ in $\mathfrak{g}_{-1}$ is equal to $-id$.*

Proof. It is

$$\beta^2 = \begin{pmatrix} a^2 + lm & -al + lA & -l\mathbb{J}\,{}^tl \\ -am + Am & ml + A^2 + \mathbb{J}\,{}^tl^tm\mathbb{J} & -A\mathbb{J}\,{}^tl - a\mathbb{J}\,{}^tl \\ -{}^tm\mathbb{J}m & -{}^tm\mathbb{J}A - a^tm\mathbb{J} & {}^tm^tl + a^2 \end{pmatrix} .$$

From this matrix square the statement of Lemma 5.1 becomes obvious. □

Let $J \in \mathcal{A}_x(M)$ be a complex structure of $\mathcal{T}_x(M)$ at some point $x \in M$, i.e., $J^2 = J \bullet J = -id|_{\mathcal{T}_x(M)}$, and let $g \in c$ be an arbitrary metric on M. With respect to the grading of $\mathcal{A}_x(M)$ induced by g we can write the complex structure J according to Lemma 5.1 as a matrix

$$J = \begin{pmatrix} -J_c & J_\eta & 0 \\ j & J_{ss} & -J_\eta^\flat \\ 0 & -g(j,\cdot) & J_c \end{pmatrix},$$

where $j, -J_\eta^\flat \in T_xM$ are lightlike J_c-eigenvectors of J_{ss} with $-g(j, J_\eta^\flat) = 1 + J_c^2$. We define the subspace $W(J,g)$ of T_xM as the g-orthogonal complement to $span\{j, J_\eta^\flat\}$ in T_xM. Then it holds

$$\langle \cdot,\cdot \rangle|_{W(J,g)} = g|_{W(J,g)} \qquad \text{and} \qquad J|_{W(J,g)} = J_{ss}|_{W(J,g)} \,,$$

where the restriction of the endomorphism J_{ss} acts as g-orthogonal complex structure on $W(J,g)$. We note that if u_- and u_+ are generating elements of $\mathcal{E}_x[1] \cong_g \mathbb{R}$ resp. $\mathcal{E}_x[-1] \cong_g \mathbb{R}$ with $\langle u_-, u_+\rangle = 1$ then we have

$$W(J,g) := span\{j, J^\flat_\eta\}^{\perp_g} = span\{u_-, Ju_-, u_+, Ju_+\}^{\perp(\mathcal{T},\langle\cdot,\cdot\rangle)}$$

which is a subset in $T_xM \subset \mathcal{T}_x(M)$. In this situation we can choose a complex basis $\{e_\alpha : \alpha = 1,\dots,m-1\}$ of $W(J,g)$ such that $\{e_\alpha, Je_\alpha : \alpha = 1,\dots,m-1\}$ is an orthogonal basis of $W(J,g)$ and

$$\{\ u_-\ ,\ Ju_-\ ,\ u_+\ ,\ Ju_+\ ,\ e_1\ ,\ Je_1\ ,\ \dots\ ,\ e_{m-1}\ ,\ Je_{m-1}\ \}$$

is a basis of the space $\mathcal{T}_x(M)$ of standard tractors at $x \in M$. We call the complex basis of the form

$$\mathcal{B}(J,g) := \{\ u_-\ ,\ u_+\ ,\ e_1\ ,\ \dots\ ,\ e_{m-1}\ \}$$

a (J,g)-adapted basis of $\mathcal{T}_x(M)$.

Now let (M^n, c) be a space of dimension $n = 2m \geq 4$ with conformal structure of signature $(2p-1, 2q-1)$ and a ∇^{nor}-parallel complex structure $J \in \Gamma(\mathcal{A}(M))$. Let us denote by

$$\mathcal{T}^{\mathbb{C}}(M) = \mathcal{T}(M) \otimes \mathbb{C}$$

the complexified standard tractor bundle. We extend the complex structure J on $\mathcal{T}(M)$ to a $\mathbb{C}$-linear complex structure on the complexification $\mathcal{T}^{\mathbb{C}}(M)$, which we denote again by J. The bundle $\mathcal{T}^{\mathbb{C}}(M)$ decomposes into the direct sum $\mathcal{T}_{10} \oplus \mathcal{T}_{01}$, where $\mathcal{T}_{10}$ denotes the i-eigenspace of J and $\mathcal{T}_{01}$ is the complex conjugate. The determinant bundle

$$\mathcal{S} := \Lambda^{m+1*}(\mathcal{T}_{10})$$

is a complex line bundle on M. We call $\mathcal{S}$ the canonical complex line tractor bundle of $(\mathcal{T}(M), J)$. (If we denote by $\tilde{\mathcal{U}}(M)$ the $\mathrm{U}(p,q)$-reduction of the $\mathrm{O}(2p,2q)$-principal fibre bundle $\tilde{\mathcal{G}}(M)$ induced by J then the canonical complex line tractor bundle is given by $\mathcal{S} = \tilde{\mathcal{U}}(M) \times_{det_{\mathbb{C}}^{-1}} \mathbb{C}$.)

The connection ω_{nor} induces on $\mathcal{S}$ a linear connection ∇^{nor}. We denote by $\Omega_{\mathcal{S}}$ the (conformal) curvature of this connection.

LEMMA 5.2. *Let $\mathcal{S}$ be the canonical complex line tractor bundle of $\mathcal{T}(M)$ with ∇^{nor}-parallel complex structure J. Then the curvature $\Omega_{\mathcal{S}}$ vanishes identically on $\mathcal{S}$.*

Proof. We aim to compute the curvature $\Omega_{\mathcal{S}}$ on $\mathcal{S}$. First of all, we remark that with Lemma 4.2 and the assumptions we know that

$$\Omega(\Pi(J),\ \cdot\ ) = 0\ .$$

We set $j := \Pi(J)$ and with respect to an arbitrary metric $g \in c$ we can conclude that $\iota_j\mathcal{W} = 0$ and $\iota_j\mathcal{C} = 0$.

Now let $\{E_\alpha : \alpha = 1, \ldots, m+1\}$ be a local complex frame of $(\mathcal{T}(M), J)$ such that $\{E_\alpha, JE_\alpha : \alpha = 1, \ldots, m+1\}$ is a local orthonormal frame of $(\mathcal{T}(M), \langle\cdot,\cdot\rangle)$. Then we denote

$$\varrho_\alpha := \frac{1}{\sqrt{2}}\langle\ \cdot\ ,\ E_\alpha + iJE_\alpha\ \rangle$$

and the $(m+1)$-form

$$\varrho := \varrho_1 \wedge \ldots \wedge \varrho_{m+1}$$

is a local complex tractor volume form on (M, c), i.e., a local section in $\mathcal{S}$. It holds

$$\Omega_{\mathcal{S}}(X,Y) \circ \varrho = -i \sum_{\alpha=1}^{m+1} \langle E_\alpha, E_\alpha\rangle \cdot \langle \Omega(X,Y) \bullet E_\alpha, JE_\alpha\rangle \cdot \varrho$$

for all $X, Y \in TM$. In particular, this expression proves that

$$\Omega_{\mathcal{S}}(j,\ \cdot\ ) = 0\,.$$

Next we reformulate above expression for $\Omega_{\mathcal{S}}$ with respect to an arbitrary metric $g \in c$ on M and a local (J,g)-adapted frame $\mathcal{B}(J,g) = \{u_-, u_+, e_1, \ldots, e_{m-1}\}$ with $\varepsilon_\alpha := g(e_\alpha, e_\alpha)$. It holds

$$\Omega_{\mathcal{S}}(X,Y) = -2i \cdot \langle \Omega(X,Y) \bullet u_-, Ju_+\rangle - i \sum_{\alpha=1}^{m-1} \varepsilon_\alpha \cdot \langle \Omega(X,Y) \bullet e_\alpha, Je_\alpha\rangle,$$

which is a purely imaginary number for $X, Y \in TM$. We remember that the curvature Ω has with respect to g the matrix form

$$\begin{pmatrix} 0 & \Omega_1 & 0 \\ \Omega_{-1} & \Omega_0 & -\Omega_1^\flat \\ 0 & -g(\Omega_{-1},\cdot) & 0 \end{pmatrix} = \begin{pmatrix} 0 & \mathcal{C} & 0 \\ 0 & \mathcal{W} & -\mathcal{C}^\flat \\ 0 & 0 & 0 \end{pmatrix}.$$

We obtain

$$\begin{aligned} \Omega_{\mathcal{S}}(X,Y) &= -i \sum_{\alpha=1}^{m-1} \varepsilon_\alpha \cdot \langle \Omega(X,Y) \bullet e_\alpha, Je_\alpha\rangle \\ &= -i \sum_{\alpha=1}^{m-1} \varepsilon_\alpha \cdot \langle \Omega_0(X,Y) \bullet e_\alpha, Je_\alpha\rangle \\ &= -i \sum_{\alpha=1}^{m-1} \varepsilon_\alpha \cdot \mathcal{W}(X,Y,e_\alpha, J_{ss}e_\alpha) \\ &= -i \sum_{\alpha=1}^{m-1} \varepsilon_\alpha \cdot \big(\ \mathcal{W}(X, J_{ss}e_\alpha, e_\alpha, Y) - \mathcal{W}(X, e_\alpha, J_{ss}e_\alpha, Y)\ \big) \end{aligned}$$

$$= -i \sum_{\alpha=1}^{m-1} \varepsilon_\alpha \cdot \big(\langle \Omega_0(X, J_{ss}e_\alpha) \bullet e_\alpha, Y \rangle - \langle \Omega_0(X, e_\alpha) \bullet J e_\alpha, Y \rangle \big)$$

$$= -i \sum_{\alpha=1}^{m-1} \varepsilon_\alpha \cdot \big(\langle \Omega_0(X, J_{ss}e_\alpha) \bullet J e_\alpha, JY \rangle + \langle \Omega_0(X, e_\alpha) \bullet e_\alpha, JY \rangle \big)$$

for all $X, Y \in TM$.

Let us assume now that one of the vectors X and Y is an element of $W(J,g)$. In fact, we can assume without loss of generality that $Y \in W(J,g)$ and $X \in TM$ is arbitrary. We set $u_{2\alpha-1} := e_\alpha$ and $u_{2\alpha} := J_{ss}e_\alpha$ for $\alpha = 1, \dots, m-1$. With $\iota_j \mathcal{W} = 0$ and $\iota_j \mathcal{C} = 0$ we obtain

$$\begin{aligned}
\Omega_{\mathcal{S}}(X,Y) = & \; -i \sum_{\alpha=1}^{m-1} \varepsilon_\alpha \cdot \big(\mathcal{W}(X, J_{ss}e_\alpha, J_{ss}e_\alpha, J_{ss}Y) + \mathcal{W}(X, e_\alpha, e_\alpha, J_{ss}Y) \big) \\
= & \; -i \sum_{i=1}^{n-2} \varepsilon_\alpha \cdot \mathcal{W}(X, u_i, u_i, J_{ss}Y) \\
& + \frac{i}{1 + J_c^2} \cdot \Big(\mathcal{W}(X, j, J_\eta^\flat, J_{ss}Y) + \mathcal{W}(X, J_\eta^\flat, j, J_{ss}Y) \Big) \\
= & \; -i \cdot tr_g^{23} \mathcal{W}(X, \cdot\, , \cdot\, , J_{ss}Y) \\
= & \; 0 \, .
\end{aligned}$$

This shows that $\iota_Y \Omega_{\mathcal{S}} = 0$ for all $Y \in W(J,g)$. Since the orthogonal complement of $W(J,g)$ in TM has dimension 2 and spans together with $W(J,g)$ the tangent space TM, we can conclude that the only possible non-vanishing component of the curvature on $\mathcal{S}$ is $\Omega_{\mathcal{S}}(j, J_\eta^\flat)$. However, we know already that $\iota_j \Omega_{\mathcal{S}} = 0$, i.e., the latter component of $\Omega_{\mathcal{S}}$ vanishes as well. □

The proof of Lemma 5.2 uses strongly the symmetries of the Weyl curvature $\mathcal{W}$, which are direct consequences of the normalisation condition $\partial^* \circ \kappa = 0$ for ω_{nor} and the generalised Bianchi identity. We note further that Lemma 5.2 shows that the local complex tractor volume form ϱ is parallel with respect to ∇^{nor}. This property also implies the local existence of a conformal Killing spinor on (M, c) (cf. [Bau99]). Lemma 5.2 is the main ingredient for the proof of our reduction claim on the conformal holonomy.

THEOREM 5.1. *Let (M, c) be a space of dimension $n = 2m \geq 4$ with conformal structure c of signature $(2p-1, 2q-1)$ such that the conformal holonomy group $Hol(\mathcal{T})$ is contained in the unitary group $\mathrm{U}(p,q)$. Then*

1. *the holonomy algebra $\mathfrak{hol}(\mathcal{T})$ is a subalgebra of the special unitary algebra $\mathfrak{su}(p,q)$.*

2. *If, in addition, the space M is simply connected then the holonomy group $Hol(\mathcal{T})$ is contained in the special unitary group* $\mathrm{SU}(p,q)$.

Proof. (1) The assumption of Theorem 5.1 about the holonomy group $Hol(\mathcal{T})$ implies the existence of a ∇^{nor}-parallel complex structure $J \in \Gamma(\mathcal{A}(M))$ on the standard tractor bundle $\mathcal{T}(M)$ of (M,c). The statement of Lemma 5.2 shows that the values $\Omega(X,Y) \in \mathcal{A}(M)$ of the tractor curvature have vanishing complex trace for all $X,Y \in TM$. It follows that the curvature form Ω on $\tilde{\mathcal{G}}(M)$ takes values only in $\mathfrak{su}(p,q) \subset \mathfrak{g}$. Since the special unitary algebra $\mathfrak{su}(p,q)$ is an ideal in $\mathfrak{u}(p,q)$, the Ambrose-Singer Theorem proves that the holonomy algebra $\mathfrak{hol}(\mathcal{T})$ is contained in $\mathfrak{su}(p,q)$.

(2) In general, the holonomy group $Hol(\mathcal{S},\nabla^{nor})$ of the canonical complex line tractor bundle is a closed Lie subgroup of $\mathrm{U}(1)$. Here, since $\mathcal{S}$ is locally flat by Lemma 5.2, $Hol(\mathcal{S},\nabla^{nor})$ is a discrete subgroup of $\mathrm{U}(1)$. With the assumption that M is simply connected it follows that $Hol(\mathcal{S},\nabla^{nor})$ is trivial, i.e., the complex line bundle $\mathcal{S}$ is globally flat and gives rise to a parallel complex tractor volume form on M. This proves that the conformal holonomy $Hol(\mathcal{T})$ is contained in $\mathrm{SU}(p,q)$. □

6. The argument using Sparling's characterisation. An integrable CR-structure of hypersurface type on a manifold N of odd dimension $2m-1 \geq 3$ with signature $(p-1,q-1)$, $p+q=m+1$, of the Levi form gives rise to a $|2|$-graded parabolic geometry with structure group $\mathrm{PSU}(p,q)$. The classical Fefferman construction assigns to any such CR-structure a S^1-principal fibre bundle M over N of total dimension $2m$, which is in a natural way equipped with a conformal structure c of signature $(2p-1,2q-1)$. To be more concrete, the explicit construction of c is usually achieved by choosing a pseudo-Hermitian structure θ on N, which then gives rise to a particular metric g on M. This metric is called the Fefferman metric (corresponding to θ) and it turns out that the conformal class of g does not depend on the choice of pseudo-Hermitian form, but only on the CR-structure of N. We do not aim either to introduce CR-geometry nor do we explain the Fefferman construction, in detail (cf. e.g. [Fef76, Lee86] or [Cap02]). Instead, what is important for us is a characterisation of Fefferman metrics with help of a certain Killing vector field. This is the so-called Sparling's characterisation of Fefferman metrics.

THEOREM 6.1. *(Sparling's characterisation) (cf. [Spa85, Gra87]) Let (M^n,g) be a pseudo-Riemannian space of dimension $n \geq 4$ and signature $(2p-1,2q-1)$. Suppose that g admits a Killing vector j (i.e., $\mathcal{L}_j g = 0$) such that*

1. *$g(j,j)=0$, i.e., j is lightlike,*
2. *$\iota_j \mathcal{W} = 0$ and $\iota_j \mathcal{C} = 0$,*
3. *$Ric(j,j) > 0$ on M.*

Then g is locally isometric to the Fefferman metric of some integrable CR-space N of hypersurface type with signature (p,q) and dimension $n-1$.

On the other side, any Fefferman metric of an integrable CR-space N of hypersurface type admits a Killing vector field j satisfying (1) to (3).

Sparling's characterisation is suitable to reprove Theorem 5.1.

Alternative Proof of Theorem 5.1. Under the assumptions of Theorem 5.1 there exists a ∇^{nor}-parallel complex structure $J \in \Gamma(\mathcal{A}(M))$. From Lemma 4.2, it follows that $j := \Pi(J)$ is a conformal Killing vector on (M, c). We choose an arbitrary metric g in c. It holds $\mathcal{L}_j g = \frac{2}{n} div^g j \cdot g$. Since J is a complex structure, it follows from Lemma 5.1 that the vector field j is lightlike and admits no zeros on M. This implies that the partial differential equation

$$j(\phi) \;=\; \frac{1}{n} \cdot div^g j$$

admits locally always a solution for ϕ.

We choose such a solution ϕ and proceed for the moment with local considerations. We set $f := e^{2\phi} g$ and then the vector field j is a Killing vector with respect to the metric f, i.e., $\mathcal{L}_j f = 0$. From Lemma 4.2, it follows also that j annihilates the conformal curvature Ω. With respect to the metric f this means $\iota_j \mathcal{W} = 0$ and $\iota_j \mathcal{C} = 0$. Moreover, it holds

$$div^f j = 0 \quad \text{and} \quad \mathcal{D}^f(j) \;=\; \frac{1}{n-2}\left(-Ric^f(j,\cdot) \;+\; \frac{scal^f}{2(n-1)} \cdot f(j,\cdot)\right).$$

Since $f(j, \mathcal{D}^f(j)) = -1 - (div^f j)^2$, these identities prove that

$$Ric^f(j,j) \;=\; n-2 \;>\; 0\,.$$

The results so far show that we can apply Theorem 6.1. We conclude that f is (locally) isometric to the Fefferman metric of some integrable CR-space of hypersurface type.

Similar as in conformal geometry it is the case that a CR-structure on some space N is uniquely determined by a normal Cartan connection on some principal fibre bundle. The curvature of this unique normal Cartan connection takes values in $\mathfrak{su}(p,q)$, which is the Lie algebra of the structure group of CR-geometry. Moreover, it is known in the classical Fefferman construction for integrable CR-spaces that the lift along the S^1-fibration of the unique normal Cartan connection of CR-geometry gives rise to the unique normal Cartan connection of the Fefferman conformal class. In our situation this argument says that the conformal curvature Ω of $c = [f]$ takes values only in $\mathfrak{su}(p,q)$.

The latter argument is rather vague, since we use some construction from CR-tractor calculus, which we do not develop here (cf. [Cap02]). However, we can replace this argument by the following one. It is known that any Fefferman metric admits at least locally a certain conformal Killing spinor (cf. [Bau99]). This fact implies again that the conformal curvature Ω of $c = [f]$ has values solely in $\mathfrak{su}(p,q)$.

The latter statement about the curvature is locally true everywhere on (M,c) and we conclude that the conformal curvature Ω has no complex trace on (M,c). Again, as in the proof of Theorem 5.1, we can apply the Ambrose-Singer Theorem to reason that the conformal holonomy algebra $\mathfrak{hol}(\mathcal{T})$ of (M,c) is contained in $\mathfrak{su}(p,q)$. This proves the first statement of Theorem 5.1, which implies as before with further assumption the second statement on the holonomy group $Hol(\mathcal{T})$. □

Sparling's characterisation is not only useful to prove the result of Theorem 5.1. In addition, it gives rise to a geometric characterisation and construction principle for spaces whose conformal holonomy sits in $\mathrm{U}(p,q)$.

COROLLARY 6.1. *Let (M,c) be a space with conformal structure of signature $(2p-1,2q-1)$ and conformal holonomy group contained in $\mathrm{U}(p,q)$. Then the conformal class c is locally the Fefferman conformal class of some integrable CR-space of hypersurface type of signature (p,q).*

REFERENCES

[Arm05] S. ARMSTRONG, *Conformal Holonomy: A Classification*, e-print: arXiv: math.DG/0503388 (2005).

[BEG94] T.N. BAILEY, M.G. EASTWOOD, AND A.R. GOVER, *Thomas's structure bundle for conformal, projective and related structures.* Rocky Mountain J. Math. (1994), **24**(4): 1191–1217.

[Bau99] H. BAUM, *Lorentzian twistor spinors and CR-geometry*, Differential Geom. Appl. (1999), **11**(1): 69–96.

[Cap02] A. CAP, *Parabolic geometries, CR-tractors, and the Fefferman construction*, Differential Geom. Appl. (2002), **17**: 123–138.

[Cap05] A. CAP, *Infinitesimal Automorphisms and Deformations of Parabolic Geometries*, ESI e-Preprint 1684, Vienna, 2005.

[CG03] A. CAP AND A.R. GOVER, *Standard tractors and the conformal ambient metric construction*, Ann. Global Anal. Geom. (2003), **24**(3): 231–259.

[CSS97a] A. CAP, J. SLOVAK, AND V. SOUCEK, *Invariant Operators on Manifolds with Almost Hermitian Symmetric Structures, I. Invariant Differentiation*, Acta Math. Univ. Comenian. (1997), **66**(1): 33–69.

[CSS97b] A. CAP, J. SLOVAK, AND V. SOUCEK, *Invariant Operators on Manifolds with Almost Hermitian Symmetric Structures, II. Normal Cartan Connections*, Acta Math. Univ. Comenian. (1997), **6**(2): 203–220.

[CSS01] A. CAP, J. SLOVAK, AND V. SOUCEK, *Bernstein-Gelfand-Gelfand sequences.* Ann. of Math. (2) (2001), **15**(1): 97–113.

[Fef76] CH. FEFFERMAN, *Monge-Ampere equations, the Bergman kernel, and geometry of pseudoconvex domains*, Ann. Math. (1976), **10**: 395–416.

[Gra87] C.R. GRAHAM, *On Sparling's characterization of Fefferman metrics*, Amer. J. Math. (1987), **10**(5): 853–874.

[Kob72] SH. KOBAYASHI, *Transformation groups in differential geometry*, Ergebnisse der Mathematik und ihrer Grenzgebiete, Band 70. Springer-Verlag, New York-Heidelberg, 1972.

[Lee86] J.M. LEE, *The Fefferman metric and pseudo-Hermitian invariants*, Trans. Amer. Math. Soc. (1986), **296**(1): 411–429.

[Lei04a] F. LEITNER, *Conformal holonomy of bi-invariant metrics*, e-print: arXiv: math.DG/0406299 (2004).

[Lei04b] F. LEITNER, *Conformal Killing forms with normalisation condition*, Rend. Circ. Mat. Palermo Suppl. ser. II. (2005), **75**: 279–292; electronically available: arXiv:math.DG/0406316 (2004).

[Lei05a] F. LEITNER, *On transversally symmetric pseudo-Einstein and Fefferman-Einstein spaces*, to appear in Math. Z., electronically available: arXiv: math.DG/0502287 (2005).

[Lei05b] F. LEITNER, *About complex structures in conformal tractor calculus*, e-print: arXiv:math.DG/0510637 (2005).

[Spa85] G.A.J. SPARLING, *Twistor theory and the characterisation of Fefferman's conformal structures*, Preprint Univ. Pittsburg, 1985.

POLYNOMIAL ASSOCIATIVE ALGEBRAS FOR QUANTUM SUPERINTEGRABLE SYSTEMS WITH A THIRD ORDER INTEGRAL OF MOTION

IAN MARQUETTE*

Abstract. We consider a superintegrable Hamiltonian system in a two-dimensional space with a scalar potential that allows one quadratic and one cubic integral of motion. We construct the most general associative cubic algebra and we present specific realizations. We use them to calculate the energy spectrum. All classical and quantum superintegrable potentials separable in cartesian coordinates with a third order integral are known. The general formalism is applied to one of the quantum potentials.

1. Introduction. The purpose of this article is to study the algebra of integrals of motion of a certain class of quantum superintegrable systems allowing a second and a third order integral of motion. We will consider a cubic associative algebra and we will study its algebraic realization. A systematic search for superintegrable systems in classical and quantum mechanics was started some time ago [8, 10, 18]. The study of superintegrable system with a third order integral is more recent. All classical and quantum superintegrable potentials in $E(2)$ that separate in cartesian coordinate and allow a third order integral were found by S. Gravel [13]. In this article we will be interested in particular in one new potentials found in Ref. [13] that was studied by the dressing chain method in Ref. [14].

It is well known that in quantum mechanics the operators commuting with the Hamiltonian, form an $o(4)$ algebra for the hydrogen atom [1] and a $u(3)$ algebra for the harmonic oscillator [15]. This type of symmetry is called a dynamical or hidden symmetry and can be used give a complete description of the quantum mechanical system. The symmetry determines all the quantum numbers, the degeneracy of the energy levels and the energy spectrum [9, 15].

Many examples of polynomial algebras have been used in different branch of physics. C.Daskaloyannis studied the quadratic Poisson algebras of two-dimensional classical superintegrable systems and quadratic associative algebras of quantum superintegrable systems [2–4]. He shows how the quadratic associative algebras provide a method to obtain the energy spectrum. He uses realizations in terms of a deformed oscillator algebra [5]. We will follow an analogous approach for the study of cases with third order integrals.

In an earlier article [17] we considered cubic Poisson algebras for classical potentials and applied the theory to the 8 potentials separating in cartesian coordinates and allowing a third order integral. The purpose of

*Département de physique et Centre de recherche mathématique, Université de Montréal, C.P.6128, Succursale Centre-Ville, Montréal, Québec H3C 3J7, Canada (ian.marquette@umontreal.ca).

this article is to study cubic associative algebras. We find the realization of these polynomial associative algebras in terms of a parafermionic algebra. From this we find Fock type representations and the energy spectrum. We reduce this problem to the problem of solving two algebraic equations. This article provides another example of the fact that it is very useful to consider not only Lie algebras in the study of quantum systems but also polynomial algebras.

2. Cubic associative algebras and their algebraic realizations. We consider a quantum superintegrable system with a quadratic Hamiltonian and one second order and one third order integral of motion

$$\begin{aligned} H &= a(q_1,q_2)P_1^2 + 2b(q_1,q_2)P_1P_2 + c(q_1,q_2)P_2^2 + V(q_1,q_2), \\ A &= A(q_1,q_2,P_1,P_2) \\ &= d(q_1,q_2)P_1^2 + 2e(q_1,q_2)P_1P_2 + f(q_1,q_2)P_2^2 \\ &\quad + g(q_1,q_2)P_1 + h(q_1,q_2)P_2 + Q(q_1,q_2), \\ B &= B(q_1,q_2,P_1,P_2) \\ &= u(q_1,q_2)P_1^3 + 3v(q_1,q_2)P_1^2P_2 + 3w(q_1,q_2)P_1P_2^2 \\ &\quad + x(q_1,q_2)P_2^3 + j(q_1,q_2)P_1^2 + 2k(q_1,q_2)P_1P_2 \\ &\quad + l(q_1,q_2)P_2^2 + m(q_1,q_2)P_1 + n(q_1,q_2)P_2 + S(q_1,q_2) \end{aligned} \tag{2.1}$$

with

$$P_1 = -i\hbar\partial_1, \qquad P_2 = -i\hbar\partial_2, \tag{2.2}$$

$$[H,A] = [H,B] = 0. \tag{2.3}$$

We assume that our integrals close in a polynomial algebra:

$$\begin{aligned} &[A,B] = C \\ &[A,C] = \alpha A^2 + \beta\{A,B\} + \gamma A + \delta B + \epsilon \\ &[B,C] = \mu A^3 + \nu A^2 + \rho B^2 + \sigma\{A,B\} + \xi A + \eta B + \zeta \end{aligned} \tag{2.4}$$

where $\{\}$ denotes an anticommutator.

The Jacobi identity $[A,[B,C]] = [B,[A,C]]$ implies $\rho = -\beta$, $\sigma = -\alpha$ and $\eta = -\gamma$.

$$\begin{aligned} &[A,B] = C \\ &[A,C] = \alpha A^2 + \beta\{A,B\} + \gamma A + \delta B + \epsilon \\ &[B,C] = \mu A^3 + \nu A^2 - \beta B^2 - \alpha\{A,B\} + \xi A - \gamma B + \zeta. \end{aligned} \tag{2.5}$$

The Casimir operator of a polynomial algebra is an operator that commutes with all elements of the associative algebra. The Casimir operator satisfies:

$$[K, A] = [K, B] = [K, C] = 0 \tag{2.6}$$

and this implies

$$\begin{aligned} K = {} & C^2 - \alpha\{A^2, B\} - \beta\{A, B^2\} + (\alpha\beta - \gamma)\{A, B\} + (\beta^2 - \delta)B^2 \\ & + \Big(\beta\gamma - 2\epsilon\Big)B + \frac{\mu}{2}A^4 + \frac{2}{3}\Big(\nu + \mu\beta\Big)A^3 \\ & + \Big(- \frac{1}{6}\mu\beta^2 + \frac{\beta\nu}{3} + \frac{\delta\mu}{2} + \alpha^2 + \xi\Big)A^2 \\ & + \Big(- \frac{1}{6}\mu\beta\delta + \frac{\delta\nu}{3} + \alpha\gamma + 2\zeta\Big)A. \end{aligned} \tag{2.7}$$

We construct a realization of the cubic associative algebra by means of the deformed oscillator technique. We use a deformed oscillator algebra $\{b^t, b, N\}$ which satisfies the relation

$$[N, b^t] = b^t, [N, b] = -b, b^t b = \Phi(N), bb^t = \Phi(N + 1). \tag{2.8}$$

We request that the "structure function" $\Phi(x)$ should be a real function that satifies the boundary condition $\Phi(0) = 0$, with $\Phi(x) > 0$ for $x > 0$. These constraints imply the existence of a Fock type representation of the deformed oscillator algebra [4, 5]. There is a Fock basis $|n>$, n=0,1,2... satisfying

$$N|n >= n|n >, \quad b^t|n >= \sqrt{\Phi(N + 1)}|n + 1 >, \tag{2.9}$$

$$b|0 >= 0, \quad b|n >= \sqrt{\Phi(N)}|n - 1 > . \tag{2.10}$$

We consider the case of a nilpotent deformed oscillator algebra, i.e., there should be an a integer p such that,

$$b^{p+1} = 0, (b^t)^{p+1} = 0. \tag{2.11}$$

These relations imply that we have

$$\Phi(p + 1) = 0. \tag{2.12}$$

In this case we have a finite-dimensional representation of dimension p+1.

Let us show that there is a realization of the form:

$$A = A(N), \qquad B = b(N) + b^t\rho(N) + \rho(N)b. \tag{2.13}$$

The functions A(N) , b(N) et $\rho(N)$ will be determined by the cubic associative algebra, in particular the first and second relation. We use the commutation relation of the cubic associative algebra to obtain

$$\begin{aligned} [A,B] &= C, \\ [A,B] &= b^t \bigtriangleup A(N)\rho(N) - \rho(N) \bigtriangleup A(N) b, \\ \triangle A(N) &= A(N+1) - A(N), \end{aligned} \tag{2.14}$$

$$\begin{aligned} [A,C] &= \alpha A^2 + \beta\{A,B\} + \gamma A + \delta B + \epsilon \\ &= b^t(\gamma(A(N+1)+A(N)) + \delta)\rho(N) \\ &\quad + \rho(N)(\gamma(A(N+1)+A(N)) + \delta)b + \alpha A^2(N) \\ &\quad + 2\beta A(N)b(N) + \gamma A(N) + \delta b(N) + \epsilon \end{aligned} \tag{2.15}$$

using

$$\begin{aligned} [A,C] &= b^t \bigtriangleup A(N)\rho(N) \bigtriangleup A(N) + \triangle A(N)\rho(N) \bigtriangleup A(N) b \\ &= b^t \bigtriangleup A(N)^2 \rho(N) + \rho(N) \bigtriangleup A(N)^2 b \end{aligned} \tag{2.16}$$

we obtain two equations that allow us to determine A(N) and b(N).

$$\begin{aligned} \bigtriangleup A(N)^2 &= \gamma(A(N+1)+A(N)) + \epsilon, \\ \alpha A(N)^2 + 2\gamma A(N) + b(N) &+ \delta A(N) + \epsilon b(N) + \xi = 0. \end{aligned} \tag{2.17}$$

We shall distinguish two cases.
Case 1. $\beta \neq 0$

$$\begin{aligned} A(N) &= \frac{\beta}{2}\left((N+u)^2 - \frac{1}{4} - \frac{\delta}{\beta^2}\right), \\ b(N) &= \frac{\alpha}{4}\left((N+u)^2 - \frac{1}{4}\right) + \frac{\alpha\delta - \gamma\beta}{2\beta^2} \\ &\quad - \frac{\alpha\delta^2 - 2\gamma\delta\beta + 4\beta^2\epsilon}{4\beta^4}\frac{1}{(N+u)^2 - \frac{1}{4}}. \end{aligned} \tag{2.18}$$

The constant u will be determined below using the fact that we require that the deformed oscillator algebras should be nilpotent. The last equation of the associative cubic algebra contains the cubic term and is

$$[B,C] = \mu A^3 + \nu A^2 - \beta B^2 - \alpha\{A,B\} + \xi A - \gamma B + \zeta \tag{2.19}$$

We obtain the equation,

$$\begin{aligned} &2\Phi(N+1)(\triangle A(N) + \frac{\gamma}{2})\rho(N) - 2\Phi(N)(\triangle A(N-1) - \frac{\gamma}{2})\rho(N-1) \\ &= \mu A(N)^3 + \nu A(N)^2 - \beta b(N)^2 \\ &\quad - 2\alpha A(N)b(N) + \xi A(N) - \gamma b(N) + \zeta \end{aligned} \tag{2.20}$$

The Casimir operator is now realized as

$$\begin{aligned}K = {} & \Phi(N+1)(\beta^2 - \delta - 2\beta A(N) - \triangle A(N)^2)\rho(N) \\ & + \Phi(N)(\beta^2 - \delta - 2\beta A(N) - \triangle A(N-1)^2)\rho(N-1) \\ & - 2\alpha A(N)^2 b(N) + (\beta^2 - \delta - 2\beta A(N))b(N)^2 \\ & + 2(\alpha\beta - \gamma)A(N)b(N) + (\beta\gamma - 2\epsilon)b(N) \\ & + \frac{\mu}{2}A(N)^4 + \frac{2}{3}(\nu + \mu\beta)A(N)^3 \\ & + \left(-\frac{1}{6}\mu\beta^2 + \frac{\beta\nu}{3} + \frac{\delta\mu}{2} + \alpha^2 + \epsilon\right)A(N)^2 \\ & + \left(-\frac{1}{6}\mu\beta\delta + \frac{\delta a}{3} + \alpha\gamma + 2\zeta\right)A(N)\end{aligned} \tag{2.21}$$

and finally the structure function is

$$\begin{aligned}\Phi(N) = {} & \\ & \frac{\rho(N-1)^{-1}}{(\triangle A(N-1) - \frac{\beta}{2})(\beta^2 - \epsilon - 2\beta A(N) - \triangle A(N)^2) + (\triangle A(N) + \frac{\beta}{2})(\beta^2 - \delta - 2\beta A(N) - \triangle A(N-1)^2)} \\ & ((\triangle A(N) + \tfrac{\beta}{2})(K + 2\alpha A(N)^2 b(N) - (\beta^2 - \delta - 2\beta A(N))b(N)^2 \\ & -2(\alpha\beta - \gamma)A(N)b(N) - (\beta\gamma - 2\epsilon)b(N) - \tfrac{\mu}{2}A(N)^4 - \tfrac{2}{3}(\nu + \mu\beta)A(N)^3 \\ & -(-\tfrac{1}{6}\mu\beta^2 + \tfrac{\beta\nu}{3} + \alpha^2 + \xi)A(N)^2 - (-\tfrac{1}{6}\mu\beta\delta + \tfrac{\delta\nu}{3} + \alpha\gamma + 2\zeta)A(N)) \\ & -\tfrac{1}{2}\Big(\beta^2 - \delta - 2\beta A(N) - \triangle A(N)^2\Big)\Big(gA(N)^3 + \nu A(N)^2 - \beta b(N)^2 \\ & \qquad - 2\alpha A(N)b(N) + \xi A(N) - \gamma b(N) + \zeta\Big).\end{aligned} \tag{2.22}$$

Thus the structure function depends only on the function ρ. This function can be arbitrarily chosen and does not influence the spectrum. In Case 1 we choose

$$\rho(N) = \frac{1}{3 * 2^{12}\beta^8(N+u)(1+N+u)(1+2(N+u))^2}\,. \tag{2.23}$$

From our expressions for A(N) , b(N) and $\rho(N)$, the third relation of the cubic associative algebra and the expression of the Casimir operator we find the structure function $\Phi(N)$. For the Case 1 the structure function is a polynomial of order 10 in N. The coefficients of this polynomial are functions of α, β, μ, γ, δ, ϵ, ν, ξ and ζ.

Case 2. $\beta = 0$ et $\delta \neq 0$

$$A(N) = \sqrt{\delta}(N+u), b(N) = -\alpha(N+u)^2 - \frac{\gamma}{\sqrt{\delta}}(N+u) - \frac{\epsilon}{\delta}. \tag{2.24}$$

In Case 2 we choose a trivial expression $\rho(N) = 1$. The explicit expression of the structure function for this case is

$$\Phi(N) = \left(\frac{K}{-4\delta} - \frac{\gamma\epsilon}{4\delta^{\frac{3}{2}}} - \frac{\zeta}{4\sqrt{\delta}} + \frac{\epsilon^2}{4\delta^2}\right)$$
$$+\left(\frac{-\alpha\epsilon}{2\delta} - \frac{d}{4} - \frac{\gamma^2}{4\delta} + \frac{\gamma\epsilon}{2\delta^{\frac{3}{2}}} + \frac{\alpha\gamma}{4\sqrt{\delta}} + \frac{\zeta}{2\sqrt{\delta}} + \frac{\nu\sqrt{\delta}}{12}\right)(N+u)$$
$$+\left(\frac{-\nu\sqrt{\delta}}{4} - \frac{3\alpha\gamma}{4\sqrt{\delta}} + \frac{\gamma^2}{4\delta} + \frac{\epsilon\alpha}{2\delta} + \frac{\alpha^2}{4} + \frac{\xi}{4} + \frac{\mu\delta}{8}\right)(N+u)^2 \tag{2.25}$$
$$+\left(\frac{-\alpha^2}{2} + \frac{\gamma\alpha}{2\delta^{\frac{1}{2}}} + \frac{\nu\sqrt{\delta}}{6} - \frac{\mu\delta}{4}\right)(N+u)^3 + \left(\frac{\alpha^2}{4} + \frac{\mu\delta}{8}\right)(N+u)^4.$$

We will consider a representation of the cubic associative algebra in which the generator A and the Casimir operator K are diagonal. We use a parafermionic realization in which the parafermionic number operator N and the Casimir operator K are diagonal. The basis of this representation is the Fock basis for the parafermionic oscillator. The vector $|k,n>, n = 0,1,2...$ satisfies the following relations:

$$N|k,n>= n|k,n>, \quad K|k,n>= k|k,n> . \tag{2.26}$$

The vectors $|k,n>$ are also eigenvectors of the generator A.

$$A|k,n>= A(k,n)|k,n> ,$$

$$A(k,n) = \frac{\beta}{2}((n+u)^2 - \frac{1}{4} - \frac{\delta}{\beta^2}), \qquad \beta \neq 0, \tag{2.27}$$

$$A(k,n) = \sqrt{\delta}(n+u), \qquad \beta = 0, \delta \neq 0.$$

We have the following constraints for the structure function,

$$\Phi(0,u,k) = 0, \qquad \Phi(p+1,u,k) = 0. \tag{2.28}$$

With these two relations we can find the energy spectrum. Many solutions for the system exist. Unitary representations of the deformed parafermionic oscillator obey the following constraint $\Phi(x) > 0$ for $x = 1,2,...,p$.

3. Examples. There exist 21 quantum potentials separable in cartesian coordinates with a third order integral, we will consider one interesting case in which the cubic algebra allows us to calculate the energy spectrum.

Case Q5.

$$H = \frac{P_x^2}{2} + \frac{P_y^2}{2} + \hbar^2\left(\frac{x^2+y^2}{8a^4} + \frac{1}{(x-a)^2} + \frac{1}{(x+a)^2}\right). \tag{3.1}$$

$$A = \frac{P_x^2}{2} - \frac{P_y^2}{2} + \hbar^2\left(\frac{x^2-y^2}{8a^4} + \frac{1}{(x-a)^2} + \frac{1}{(x+a)^2}\right), \tag{3.2}$$

$$\begin{aligned} B = X_2 = \{L, P_x^2\} &+ \hbar^2\left\{y\left(\frac{4a^2-x^2}{4a^4} - \frac{6(x^2+a^2)}{(x^2-a^2)^2}\right), P_x\right\} \\ &+ \hbar^2\left\{x\left(\frac{(x^2-4a^2)}{4a^4} - \frac{2}{x^2-a^2} + \frac{4(x^2+a^2)}{(x^2-a^2)^2}\right), P_y\right\}. \end{aligned} \tag{3.3}$$

The integrals A, B and H give rise to the algebra

$$\begin{aligned} [A,B] &= C, \\ [A,C] &= \frac{h^4}{a^4}B, \\ [B,C] &= -32\hbar^2A^3 - 48\hbar^2A^2H + 16\hbar^2H^3 + 48\frac{\hbar^4}{a^2}A^2 \\ &\quad + 32\frac{\hbar^4}{a^2}HA - 16\frac{\hbar^4}{a^2}H^2 + 8\frac{\hbar^6}{a^4}A - 4\frac{\hbar^6}{a^4}H - 12\frac{\hbar^8}{a^6}\ . \end{aligned} \tag{3.4}$$

The Casimir operator is

$$K = -16\hbar^2H^4 + 32\frac{\hbar^4}{a^2}H^3 + 16\frac{\hbar^6}{a^4}H^2 - 40\frac{\hbar^8}{a^6}H - 3\frac{\hbar^{10}}{a^8}, \tag{3.5}$$

and we have

$$\begin{aligned} \Phi(x) = &\left(4\frac{a^4}{\hbar^2}H^4 - 12a^2H^3 + 11\frac{\hbar^4}{a^2}H - \frac{15}{4}\frac{h^6}{a^4}\right) \\ &+ \left(8a^2H^3 - 8\hbar^2H^2 - 14\frac{\hbar^4}{a^2}H - 4\frac{\hbar^6}{a^4}\right)(x+u) \\ &+ \left(20\frac{\hbar^4}{a^2}H - 14\frac{\hbar^6}{a^4}\right)(x+u)^2 \\ &+ \left(-8\frac{\hbar^4}{a^2}H + 16\frac{\hbar^6}{a^4}\right)(x+u)^3 - 4\frac{\hbar^8}{a^4}(x+u)^4 \end{aligned} \tag{3.6}$$

$$\begin{aligned} \Phi(x) = &\left(\frac{-4\hbar^8}{a^4}\right)\left(x+u-\left(\frac{-a^2E}{\hbar^2}-\frac{1}{2}\right)\right)\left(x+u-\left(\frac{a^2E}{\hbar^2}+\frac{1}{2}\right)\right) \\ &\left(x+u-\left(\frac{-a^2E}{\hbar^2}+\frac{3}{2}\right)\right)\left(x+u-\left(\frac{-a^2E}{\hbar^2}+\frac{5}{2}\right)\right). \end{aligned} \tag{3.7}$$

We find u with

$$\Phi(0,u,k)=0,$$

$$u_1=\frac{-a^2E}{\hbar^2}-\frac{1}{2},\qquad u_2=\frac{a^2E}{\hbar^2}+\frac{1}{2},\tag{3.8}$$

$$u_3=\frac{-a^2E}{\hbar^2}+\frac{3}{2},\qquad u_4=\frac{-a^2E}{\hbar^2}+\frac{5}{2}.$$

We have four cases:
Case 1. $u=u_1$

$$E=\frac{\hbar^2 p}{2a^2},\quad \Phi(x)=\left(\frac{4\hbar^8}{a^4}\right)x(p+1-x)(x-2)(x-3)\ .\tag{3.9}$$

Case 2. $u=u_2$

$$E=\frac{-\hbar^2(p+2)}{2a^2},\quad \Phi(x)=\left(\frac{4\hbar^8}{a^4}\right)x(p+1-x)(p+3-x)(p+4-x),\tag{3.10}$$

$$E=\frac{-\hbar^2 p}{2a^2},\quad \Phi(x)=\left(\frac{4\hbar^8}{a^4}\right)x(p+1-x)(p-1-x)(p-2-x),\tag{3.11}$$

$$E=\frac{-\hbar^2(p-1)}{2a^2},\quad \Phi(x)=\left(\frac{4\hbar^8}{a^4}\right)x(p+1-x)(p-2-x)(p-x).\tag{3.12}$$

Case 3. $u=u_3$

$$E=\frac{\hbar^2(p+2)}{2a^2},\quad \Phi(x)=\left(\frac{4\hbar^8}{a^4}\right)x(p+1-x)(x-1)(x+2)\ .\tag{3.13}$$

Case 4. $u=u_4$

$$E=\frac{\hbar^2(p+3)}{2a^2},\quad \Phi(x)=\left(\frac{4\hbar^8}{a^4}\right)x(p+1-x)(x+1)(x+3)\ .\tag{3.14}$$

The only case that correspond to unitary representations are (3.10) and (3.14).

4. Conclusion. The main results of this article are that we have constructed the associative algebras for superintegrable potential with a second order integral and a third order integral. We find realizations in terms of deformed oscillator algebras for the cubic associative algebras. We apply our result to a speficic potential [13, 14]. We leave the other quantum cases to a future article.

We see that many systems in classical and quantum physics are described by a nonlinear symmetry that provides information about the energy spectrum.

We note that our polynomial algebras and their realizations are independant of the choice of coordinate system. We could apply our results in the future to systems with a third order integral that are separable in polar, elliptic or parabolic coordinates. The method is independant of the metric and we could apply our polynomial algebras to other cases than superintegrable potentials in E(2).

Acknowledgments. The author thanks Pavel Winternitz for his very helpful comments and suggestions, and also Frederick Tremblay for useful discussions. The author benefits from FQRNT fellowship.

REFERENCES

[1] V. Bargmann, Zur Theorie des Wasserstoffsatoms, Z. Phys. **99**: 576–582 (1936).

[2] D. Bonatsos, C. Daskaloyannis, and K. Kokkotas, Deformed oscillator algebras for two-dimensional quantum superintegrable systems. Phys. Rev. A **50**: 3700–3709 (1994).

[3] D. Bonatsos, C.Daskaloyannis, and K.Kokkotas, Quantum-algebraic description of quantum superintegrable systems in two dimensions, Phys. Rev. A **48**: R3407-R3410 (1993).

[4] C. Daskaloyannis, Quadratic poisson algebras of two-dimensional classical superintegrable systems and quadratic associative algebras of quantum superintegrable systems, J. Math. Phys. **42**: 1100–1119 (2001).

[5] C. Daskaloyannis, Generalized deformed oscillator and nonlinear algebras, J. Phys. A: Math.Gen **24**: L789–L794 (1991).

[6] J. Drach, Sur l'intégration logique des équations de la dynamique à deux variables: Forces conservatrices. Intégrales cubiques. Mouvements dans le plan, C.R. Acad. Sci III **200**: 22–26 (1935).

[7] J. Drach, Sur l'intégration logique et sur la transformation des équations de la dynamique à deux variables: Forces conservatrices. Intégrales. C.R. Acad. Sci III, 599-602 (1935).

[8] N.W. Evans, Group theory of the Smorodinsky-Winternitz system. J. Math. Phys. **32**: 3369–3375 (1991).

[9] V. Fock, Zur Theorie des Wasserstoffsatoms, Z. Phys. **98**: 145–154 (1935).

[10] J. Fris, V. Mandrosov, Ya.A. Smorodinsky, M. Uhlir, and P. Winternitz, On higher symmetries in quantum mechanics, Phys. Lett. **16**: 354–356 (1965).

[11] Ya.I. Granovskii, A.S. Zhedanov, and I.M. Lutzenko, Quadratic Algebra as a Hidden Symmetry of the Hartmann Potential, J. Phys. A **24**: 3887–3894 (1991).

[12] S. Gravel and P. Winternitz, Superintegrability with third-order integrals in quantum and classical mechanics J. Math. Phys. **43**: 5902–5912 (2002).

[13] S. Gravel, Hamiltonians separable in Cartesian coordinates and thirdorder integrals of motion, J. Math. Phys. **45**: 1003–1019 (2004).

[14] S. Gravel, Superintegrability, isochronicity, and quantum harmonic behavior, ArXiv:math-ph/0310004 (2004).

[15] J.M Jauch and E.L Hill, On the problem of degeneracy in quantum mechanics, Phys Rev **57**: 641–645 (1940).

[16] P. Létourneau and L. Vinet, Superintegrable systems: Polynomial algebras and quasi-exactly solvable hamiltonians, Ann. Physics **243**: 144–168 (1995).

[17] I. Marquette and P. Winternitz, Polynomial Poissons algebras for superintegrable systems with a third order integral of motion, ArXiv:math-ph/0608021 (2006).

[18] P. Winternitz, Y.A. Smorodinsky, and M. Uhlir et I.Fris, Symmetry groups in classical and quantum mechanics, Yad. Fiz. **4**: 625–635 (1966). (English translation: Sov. J. Nucl. Phys. **4**: 444–450 (1967).)

SEPARATION OF VARIABLES FOR SYSTEMS OF FIRST-ORDER PARTIAL DIFFERENTIAL EQUATIONS AND THE DIRAC EQUATION IN TWO-DIMENSIONAL MANIFOLDS

RAYMOND G. MCLENAGHAN* AND GIOVANNI RASTELLI†

Abstract. The problem of solving the Dirac equation on two-dimensional manifolds is approached from the point of separation of variables, with the aim of creating a foundation for analysis in higher dimensions. Beginning from a sound definition of multiplicative separation for systems of two first order linear partial differential equations of Dirac type and the characterization of those systems admitting multiplicatively separated solutions in some arbitrarily given coordinate system, more structure is step by step added to the problem by requiring the separation constants are associated with commuting differential operators. Finally, the requirement that the original system coincides with the Dirac equation on a two-dimensional manifold allows the characterization of the orthonormal frames and metrics admitting separation of variables for the equation and of the symmetries associated with the separated coordinates.

1. Introduction.

Exact solutions to Dirac's relativistic wave equation by means of the method of separation of variables have been studied since the equation was postulated in 1928. Indeed, the solution for the electron in hydrogen atom may be obtained by this method. While there is a well developed theory of separation of variables for the Hamilton-Jacobi equation, and the second order linear partial differential equations of mathematical physics (see [1, 2]), an analogous theory for the Dirac equation is still in it's early stages. The complications arise from the fact that one is dealing with a system of first order partial differential equations whose derivation from the invariant Dirac equation depend not only on the choice of coordinate system but also on the choice of an orthonormal moving frame and representation for the Dirac matrices with respect to which the components of the unknown spinor are defined. Further complications arise if the background space-time is assumed to be curved. Much of the progress in the theory has been stimulated by developments in Einstein's general theory of relativity where one wishes to study first quantized relativistic electrons on curved background space-times of physical interest such as the Schwarzschild and Kerr black hole space-times. This required the preliminary development of a theory of spinors on general pseudo-Riemmanian manifolds (see [3]). The solution of the Dirac equation in the Reissner-Nordstrom solution was apparently first obtained by Brill and Wheeler in 1957 [4] who separated the equations for the spinor components in standard orthogonal Schwarzschild coordinates with respect to a moving frame

*Department of Applied Mathematics, University of Waterloo, Waterloo, Ontario, N2L 3G1, CANADA. The work of the first author was supported in part by a Natural Sciences and Engineering Research Council of Canada Discovery Grant.

†Dipartimento di Matematica, Università di Torino, via Carlo Alberto 10, 10123 Torino, ITALY.

adapted to the coordinate curves. A comparable separable solution in the Kerr solution was found by Chandrasekhar in 1976 [5] by use of an ingenious separation ansatz involving Boyer-Lindquist coordinates and the Kinnersley tetrad. The separability property was characterized invariantly by Carter and McLenaghan [6] in terms of a first order differential operator constructed from the valence two Yano-Killing tensor that exists in the Kerr solution, that commutes with the Dirac operator and that admits the separable solutions as eigenfunctions with the separation constant as eigenvalue. Study of this example led Miller [7] to propose the theory of a factorizable system for first order systems of Dirac type in the context of which the separabilty property may be characterized by the existence of a certain system of commuting first order symmetry operators. While this theory includes the Dirac equation on the Kerr solution and its generalizations [8] it is apparently not complete since as is shown by Fels and Kamran [8] there exist systems of the Dirac type whose separablity is characterized by second order symmetry operators. The purpose of this paper is to study the product separabilty of equations of Dirac type in the simplest possible setting namely on two-dimensional Riemannian spaces. The motivation for working in the lowest possible dimension is that it is possible to examine in detail the different possible scenarios that arise from the separation ansatz and the imposition of the separation paradigm that the separation be characterized by a symmetry operator admitting the separable solutions as eigenfunctions. First steps in this direction already been taken by Fels and Kamran [8] and Fatibene, McLenaghan and Smith [9]. The insight obtained from this approach may help suggest an approach to take for the construction of a general separability theory for Dirac type equations. The paper is organized as follows. In Section 2 we write the Dirac equation on a general 2-dimensional Riemannian manifold with respect to a general orthonormal moving frame expressed in a general system of local coordinates. In this way we obtain an explicit system of first order pdes (of Dirac type) the separability properties of which are studied in the sequel. Section 3 contains a discussion of the Levi-Civita separability conditions for the associated Hamilton-Jacobi (HJ) equation for the geodesics of the Riemannian metric. These conditions are used to relate the separability properties of the Dirac equation to that of the HJ and Laplace equation on the manifold. In Section 4 a definition of product separability for equations of Dirac type is given. This is followed in Section 5 by a discussion of symmetry operators and the detailed assumptions used in our anaylsis. The three types of separation studied are also defined in this section. Separation of Type I for equations of Dirac type are studied in Section 6. These results are applied to the Dirac equation in Section 7. Type II separation is studied in Section 8 and applied to the Dirac equation in Section 9. Section 10 contains the conclusion. In this paper we employ the notation of Fatibene, McLenaghan and Smith [9].

2. The Dirac equation in two-dimensional Riemannian manifolds. Let (M, g) denote a two-dimensional manifold. The Dirac equation is given by

$$D\psi - \lambda\psi = i\tilde{\gamma}^\mu \nabla_\mu \psi - \lambda\psi = 0 \tag{2.1}$$

where $\tilde{\gamma}^\mu = \gamma^a e^\mu{}_a$,

$$\gamma^1 = \begin{pmatrix} 1 & 0 \\ 0 & -1 \end{pmatrix}, \quad \gamma^2 = \begin{pmatrix} 0 & -i \\ i & 0 \end{pmatrix}, \tag{2.2}$$

satisfying $\gamma^a\gamma^b + \gamma^b\gamma^a = 2\eta_{ab}I$, is the Dirac representation, and $(e^\mu{}_a)$ an orthonormal frame field (spin frame). The Riemannian metric g is given by

$$g_{\mu\nu} = e_\mu{}^a \eta_{ab} e_\nu{}^b \tag{2.3}$$

where $\eta_{ab} = \delta_{ab}$ and the spinor covariant derivative by $\nabla_\mu\psi = \partial_\mu\psi + \Gamma_\mu\psi$. Because $\Gamma_\mu = \frac{1}{8}[\gamma_a, \gamma_b]\Gamma_\mu^{ab} = \frac{1}{4}\gamma_a\gamma_b\Gamma_\mu^{ab}$, where $\Gamma_\mu^{ab} = e_\alpha{}^a(\Gamma_{\beta\mu}^\alpha e^{\beta b} + \partial_\mu e^{\alpha b})$, it follows that $\Gamma_\mu^{ab} = -\Gamma_\mu^{ba}$, and we have

$$\tilde{\gamma}^1 = \begin{pmatrix} e^1{}_1 & -ie^1{}_2 \\ ie^1{}_2 & -e^1{}_1 \end{pmatrix}, \quad \tilde{\gamma}^2 = \begin{pmatrix} e^2{}_1 & -ie^2{}_2 \\ ie^2{}_2 & -e^2{}_1 \end{pmatrix}$$

and

$$\Gamma_\mu = \frac{1}{2}\begin{pmatrix} 0 & -i\Gamma_\mu^{12} \\ -i\Gamma_\mu^{12} & 0 \end{pmatrix}.$$

Moreover, by putting

$$\begin{cases} X = -2e^\mu{}_2\Gamma_\mu^{12} \\ Y = -2e^\mu{}_1\Gamma_\mu^{12} \end{cases} \tag{2.4}$$

we have

$$i\tilde{\gamma}^\mu\Gamma_\mu = \frac{1}{4}\begin{pmatrix} iX & -Y \\ Y & -iX \end{pmatrix}. \tag{2.5}$$

Finally, the Dirac equation in generic coordinates $(q^1, q^2) = (x, y)$, becomes

$$\mathbf{D}\psi - \lambda\psi = \tilde{\mathbf{A}}\partial_x\psi + \tilde{\mathbf{B}}\partial_y\psi + \tilde{\mathbf{C}}\psi - \lambda\psi = 0 \tag{2.6}$$

where

$$\tilde{\mathbf{A}} = \begin{pmatrix} ie^1{}_1 & e^1{}_2 \\ -e^1{}_2 & -ie^1{}_1 \end{pmatrix} \tag{2.7}$$

$$\tilde{\mathbf{B}} = \begin{pmatrix} ie^2{}_1 & e^2{}_2 \\ -e^2{}_2 & -ie^2{}_1 \end{pmatrix} \tag{2.8}$$

$$\tilde{\mathbf{C}} = i\tilde{\gamma}^{\mu}\Gamma_{\mu} = \frac{1}{4}\begin{pmatrix} iX & -Y \\ Y & -iX \end{pmatrix} \tag{2.9}$$

and

$$\begin{cases} X = -2e^{\mu}{}_{2}\Gamma^{12}_{\mu} = -2\Gamma_{212} \\ Y = -2e^{\mu}{}_{1}\Gamma^{12}_{\mu} = -2\Gamma_{112}. \end{cases} \tag{2.10}$$

It is remarkable that the coefficients Γ_{abc} can be derived from the $(e_{a}{}^{\mu})$ only. Indeed, by introducing the orthonormal moving frame

$$\begin{cases} E_a = e^{\mu}{}_{a}\dfrac{\partial}{\partial x^{\mu}} \\ E^a = \hat{e}_{\mu}{}^{a} dx^{\mu} \end{cases} \tag{2.11}$$

the structure functions C^{a}_{bc} are defined by

$$\begin{cases} dE^a = \dfrac{1}{2}C^{a}_{bc}E^b \wedge E^c \\ [E_a, E_b] = C^{c}_{ab}E_c \end{cases}$$

and connection coefficients by

$$\begin{cases} \nabla_{E_a}E_b = \Gamma^{c}_{ba}E_c \\ \nabla_{E_a}E^b = -\Gamma^{b}_{ca}E^c. \end{cases}$$

It follows that

$$\Gamma_{abc} = \frac{1}{2}\left(C_{cab} - C_{abc} - C_{bca}\right). \tag{2.12}$$

Thus, we can enunciate the following statement that holds for any two-dimensional Riemannian or pseudo-Riemannian manifold:

PROPOSITION 1. *Given a representation of the Clifford algebra and a coordinate system, the Dirac equation (2.1) is uniquely determined by the spin frame components and their derivatives.*

Moreover, thanks (2.3) to (2.7)–(2.9) and (2.12) we are able to write the components of the metric tensor in some undefined separated coordinates as functions of the spin frame components only. Then, by putting conditions on the Riemann tensor, vanishing for the Euclidean plane, we are able to explicitly determine the separated coordinates as functions of Cartesian coordinates. This can be done or by direct integration of the tensorial equations of the metric components, or by applying the method described in [10].

3. Separation of variables for Dirac equation. Coordinates separating the Dirac equation are associated with separation of the Hamilton-Jacobi and Schrödinger equations via the Klein-Gordon equation [11], [12]:

a necessary condition for the separation of the Dirac equation in some coordinate system is the separation of the associated Klein-Gordon equation in the same coordinates. This last equation separates in the classical separable coordinate systems for the Hamilton-Jacobi and Schrödinger equations (for example: ellipsoidal, paraboloidal, sferoidal in three-dimensional Euclidean space).

Coordinate systems allowing (additive) separation of variables for the Hamilton-Jacobi equation with Hamiltonian H are characterized by the Levi-Civita equations [13]

$$\partial_i H \partial_j H \partial^{ij} H + \partial^i H \partial^j H \partial_{ij} H - \partial_i H \partial^j H \partial_j^i H - \partial^i H \partial_j H \partial_i^j H = 0 \quad (3.1)$$

where the repeated indices are not summed, $i \neq j$ and $\partial_i = \partial/\partial q^i$, $\partial^i = \partial/\partial p_i$. A separable coordinate q^i which satisfies

$$\frac{\partial_i H}{\partial^i H} = Q^j(q^k)p_j \quad (3.2)$$

where the j index is summed, is said of first class and if $\partial_i H = 0$ it is said to be ignorable. Ignorable coordinates are associated with Killing vectors (isometries). In separation of variables theory for the Hamilton-Jacobi equation, first class coordinates always can be transformed into ignorable without changing the separability structure [14]. It is unknown if the same holds for Dirac equation.

As described in the previous section, we are able to determine the components of the metric tensor once the matrices in (2.7)–(2.8) are known. If the coordinates separate (2.1) we can build the geodesic Hamiltonian $H = \frac{1}{2} g^{ij} p_i p_j$ and then check whether the Levi-Civita equations and (3.2) are satisfied. It follows that for all separable cases of Dirac equation that are found, we can easily check whether Hamilton-Jacobi separation holds or not, and if coordinates are of first-class, without making use of the Klein-Gordon equation.

4. Multiplicative separation of eigenvalue type PDE systems. Equation (2.6) shows that the Dirac equation in two-dimensional manifolds is nothing but a system of first order PDEs in (ψ_1, ψ_2) of eigenvalue type. Consequently, we can study in general the separability of this kind of systems and, step by step, add more restrictions in order to obtain finally systems associated with (2.1).

Let (x, y) be a coordinate system on the two dimensional manifold and

$$\psi = \begin{pmatrix} \psi_1(x,y) \\ \psi_2(x,y) \end{pmatrix}. \quad (4.1)$$

Let $\mathbf{D}$ be the operator defined by

$$\mathbf{D} = \begin{pmatrix} A_1 & A_2 \\ A_3 & A_4 \end{pmatrix} \partial_x + \begin{pmatrix} B_1 & B_2 \\ B_3 & B_4 \end{pmatrix} \partial_y + \begin{pmatrix} C_1 & C_2 \\ C_3 & C_4 \end{pmatrix}, \quad (4.2)$$

where A_i, B_i and C_i are functions of (x,y), such that

$$\mathbf{D}\,\psi = \begin{pmatrix} A_1\partial_x\psi_1 + A_2\partial_x\psi_2 + B_1\partial_y\psi_1 + B_2\partial_y\psi_2 + C_1\psi_1 + C_2\psi_2 \\ A_3\partial_x\psi_1 + A_4\partial_x\psi_2 + B_3\partial_y\psi_1 + B_4\partial_y\psi_2 + C_3\psi_1 + C_4\psi_2 \end{pmatrix}. \tag{4.3}$$

Let λ be a real or complex number, and consider the eigenvalue equation

$$\mathbf{D}\psi - \lambda\psi = 0. \tag{4.4}$$

We assume here the ansatz of multiplicative separation for ψ, i.e.

$$\psi_i = a_i(x)b_i(y) \tag{4.5}$$

since it is the most appropriate ansatz for a linear equation of the form (2.1). Then, the equation (4.4) becomes

$$\begin{cases} A_1\dot a_1 b_1 + A_2\dot a_2 b_2 + B_1 a_1\dot b_1 + B_2 a_2\dot b_2 + (C_1-\lambda)a_1b_1 + C_2a_2b_2 = 0 \\ A_3\dot a_1 b_1 + A_4\dot a_2 b_2 + B_3 a_1\dot b_1 + B_4 a_2\dot b_2 + C_3a_1b_1 + (C_4-\lambda)a_2b_2 = 0 \end{cases} \tag{4.6}$$

Hence, we are led to the following definition for the separability of $\mathbf{D}$:

DEFINITION 1. *The equations (4.6) (and the operator* $\mathbf{D}$*) are separated in* (x,y) *if there exist nonzero functions* $R_i(x,y)$ *such that the (4.6) can be written as:*

$$\begin{cases} R_1 a_r b_s (E_1^x + E_1^y) = 0 \\ R_2 a_t b_u (E_2^x + E_2^y) = 0 \end{cases} \tag{4.7}$$

for suitable indices r,s,t,u, *where* $E_i^x(x,a_j,\dot a_j)$, $E_i^y(y,b_j,\dot b_j)$. *Moreover, the equations*

$$E_i^x = \mu_i = -E_i^y \tag{4.8}$$

define the separation constants μ_i.

The above definition refers to the so-called "naive" separation of variables that is not always the most general. For example, ellipsoidal coordinates in $n > 2$ dimensional Euclidean spaces cannot be included, although they separate the geodesic Hamilton-Jacobi equation. However, in dimension two all Hamilton-Jacobi separable systems on Riemannian or pseudo-Riemannian manifolds are also "naively" separable.

5. Symmetry operators and separated equations. In the Miller-Shapovalov theory [7], [15] first-order symmetry operators were employed to characterize separation and the associated separated equations are of the form $\partial_i\psi = S_i(q^i,\lambda,\mu_j)\psi$ (factorizable systems). However, some examples of separation were found [8] associated only to second-order symmetry operators. In both cases the operators were strictly related to separated equations for the Dirac operator. Following these suggestions, we adopt for our analysis the following assumptions:

0. We assume a given coordinate system (x, y) and impose separation of (4.6) in (x, y) according to Definition 1. This can be done in several different ways and each way generates separated equations of the form (4.8).
1. We build eigenvalue-type operators $\mathbf{L}\psi = \mu\psi$ with eigenvalues $\mu(\mu_i)$ making use of the terms E_i^x and E_i^y in (4.8) only. When searching for first-order operators, we can have $\mu_1 = \mu_2 = \mu$, in the separated equations.
2. We require that the operators $\mathbf{L}$ are independent of λ (otherwise we are considering some kind of "fixed energy separation"). This imposes the use, for the construction of $\mathbf{L}$, of only the terms among E_i^x or E_i^y independent of λ. In this way, the coordinates (x, y) are associated with $\mathbf{L}$ for any (nonzero) value of the parameter λ and separation is "independent of the energy". The situation is thus similar to that of classical separation for the Hamilton-Jacobi equation.
3. We require $[\mathbf{L}, \mathbf{D}]\psi = (\mathbf{LD} - \mathbf{DL})\psi = 0$ for all ψ. The operators L satisfying the previous equation are called *symmetry operators.*
4. We assume $\lambda \neq 0$. This is a purely technical assumption, the case $\lambda = 0$ will be considered elsewhere.

In this way, the symmetry operators are directly generated by the separated equations and having them enables one to immediately write down the same separated equations. The (widely accepted) assumption that separation constants are associated with eigenvalues of symmetry operators enables one to give quantum mechanical meaning to the same constants. In some papers, for example [8], the Dirac operator $\hat{D}$ is assumed to be, in our notation, $\hat{D} = D - \lambda I$. In this case, all commutation conditions are "energy" (or "mass") dependent if one is not explicitly assuming "$\forall\lambda$". When working with $\hat{D}$ it seems more natural to assume the conformal form of commutation $[\hat{D}, L] = f\hat{D}$ (R-commutation), for arbitrary functions $f(q^i, ..)$ depending on configuration coordinates and possibly on spinorial quantities. For this form of the Dirac operator it only seems natural to assume dependence from λ of operators $\mathbf{L}$. We remark that, if separable coordinates are somehow determined by symmetry operators, in this case the coordinates will depend on λ. However, this extension of separation theory can require a deep revision of many aspects of the theory (see for example [16] relative to Hamilton-Jacobi separation theory) and will be considered elsewhere. By assuming $\psi_i = a_i(x)b_i(y)$, the mechanism of separation of (2.1) in the form (4.7) naturally generates the following three types of separation:

I. $a_1 \neq a_2$ and $b_1 \neq b_2$.
II. $a_1 = a_2 = a$ and $b_1 \neq b_2$ (or vice-versa).
III. $a_1 = a_2 = a$ and $b_1 = cb_2 = b$ (c constant).

We do not consider here "reduced separation" [17] such as assuming from the beginning that $a_1 = e^x$, $a_1 = x^2 - 1$, etc. In this paper we consider Types I and II only, Type III will be considered elsewhere.

6. Type I. We begin by assuming $R_i = 1$. Recall that here and in the sequel $\lambda \neq 0$. Because of the term $\lambda\psi_i$, at least one of the indices l, s in $a_l b_s$ factorizing the $i-th$ equation must be equal to i. Then, all possible schemes of separation arise by multiplying the system (4.6) by the matrix

$$\mathbf{T} = \begin{pmatrix} \frac{1}{a_l b_m} & 0 \\ 0 & \frac{1}{a_p b_q} \end{pmatrix}$$

where one of l, m must be 1 and one of p, q must be 2. According to the allowed 4-tuples of indices (l, m, p, q), we have the following nine separation schemes:

$$\begin{cases} (1,1,1,2)_1 & (1,1,2,1)_2 & (1,1,2,2)_3 \\ (1,2,1,2)_4 & (1,2,2,1)_5 & (1,2,2,2)_6 \\ (2,1,1,2)_7 & (2,1,2,1)_8 & (2,1,2,2)_9 \end{cases} \tag{6.1}$$

The separation conditions $(\mathbf{D} - \lambda)_i \psi = a_l b_m (E_i^x + E_i^y)$ impose restrictions on the functions A_i, B_i and C_i that, corresponding to the different cases, determine respectively the separate forms of $\mathbf{D}$:

$$\mathbf{D}_1 : \begin{pmatrix} A_1(x) & 0 \\ 0 & A_4(x) \end{pmatrix} \partial_x + \begin{pmatrix} B_1(y) & 0 \\ B_3(y) & 0 \end{pmatrix} \partial_y + \begin{pmatrix} C_{11}(x) + C_{12}(y) & 0 \\ C_3(y) & C_4(x) \end{pmatrix},$$

$$\mathbf{D}_2 : \begin{pmatrix} A_1(x) & 0 \\ A_3(x) & 0 \end{pmatrix} \partial_x + \begin{pmatrix} B_1(y) & 0 \\ 0 & B_4(y) \end{pmatrix} \partial_y + \begin{pmatrix} C_{11}(x) + C_{12}(y) & 0 \\ C_3(x) & C_4(y) \end{pmatrix},$$

$$\mathbf{D}_3 : \begin{pmatrix} A_1(x) & 0 \\ 0 & A_4(x) \end{pmatrix} \partial_x + \begin{pmatrix} B_1(y) & 0 \\ 0 & B_4(y) \end{pmatrix} \partial_y + \begin{pmatrix} C_{11}(x) + C_{12}(y) & 0 \\ 0 & C_{41}(x) + C_{42}(y) \end{pmatrix},$$

$$\mathbf{D}_4 : \begin{pmatrix} 0 & A_2(x) \\ 0 & A_4(x) \end{pmatrix} \partial_x + \begin{pmatrix} B_1(y) & 0 \\ B_3(y) & 0 \end{pmatrix} \partial_y + \begin{pmatrix} C_1(y) & C_2(x) \\ C_3(y) & C_4(x) \end{pmatrix},$$

$$\mathbf{D}_5 : \begin{pmatrix} 0 & A_2(x) \\ A_3(x) & 0 \end{pmatrix} \partial_x + \begin{pmatrix} B_1(y) & 0 \\ 0 & B_4(y) \end{pmatrix} \partial_y + \begin{pmatrix} C_1(y) & C_2(x) \\ C_3(x) & C_4(y) \end{pmatrix},$$

$$\mathbf{D}_6 : \begin{pmatrix} 0 & A_2(x) \\ 0 & A_4(x) \end{pmatrix} \partial_x + \begin{pmatrix} B_1(y) & 0 \\ 0 & B_4(y) \end{pmatrix} \partial_y + \begin{pmatrix} C_1(y) & C_2(x) \\ 0 & C_{41}(x) + C_{42}(y) \end{pmatrix},$$

$$\mathbf{D}_7 : \begin{pmatrix} A_1(x) & 0 \\ 0 & A_4(x) \end{pmatrix} \partial_x + \begin{pmatrix} 0 & B_2(y) \\ B_3(y) & 0 \end{pmatrix} \partial_y + \begin{pmatrix} C_1(x) & C_2(y) \\ C_3(y) & C_4(x) \end{pmatrix},$$

$$\mathbf{D}_8 : \begin{pmatrix} A_1(x) & 0 \\ A_3(x) & 0 \end{pmatrix} \partial_x + \begin{pmatrix} 0 & B_2(y) \\ 0 & B_4(y) \end{pmatrix} \partial_y + \begin{pmatrix} C_1(x) & C_2(y) \\ C_3(x) & C_4(y) \end{pmatrix},$$

$$\mathbf{D}_9 : \begin{pmatrix} A_1(x) & 0 \\ 0 & A_4(x) \end{pmatrix} \partial_x + \begin{pmatrix} 0 & B_2(y) \\ 0 & B_4(y) \end{pmatrix} \partial_y + \begin{pmatrix} C_1(x) & C_2(y) \\ 0 & C_{41}(x) + C_{42}(y) \end{pmatrix}.$$

After exchanging x and y some of the previous operators coincide, namely $\mathbf{D}_1 \equiv \mathbf{D}_2$, $\mathbf{D}_4 \equiv \mathbf{D}_8$, $\mathbf{D}_5 \equiv \mathbf{D}_7$, $\mathbf{D}_6 \equiv \mathbf{D}_9$ and $\mathbf{D}_3 \equiv \mathbf{D}_3$. Moreover, by introducing the operator

$$\mathbf{J} = \begin{pmatrix} 0 & 1 \\ 1 & 0 \end{pmatrix}$$

such that

$$\mathbf{J} \begin{pmatrix} \psi_1 \\ \psi_2 \end{pmatrix} = \begin{pmatrix} \psi_2 \\ \psi_1 \end{pmatrix}$$

we observe that $\mathbf{D}_1\psi$ is of the same form as $\mathbf{J}\mathbf{D}_9\mathbf{J}\psi$; thus we can consider $\mathbf{D}_1$ and $\mathbf{D}_9$ as equivalent in the following discussion. Thus we have four distinct classes represented by $\mathbf{D}_1$, $\mathbf{D}_3$, $\mathbf{D}_4$ and $\mathbf{D}_5$. We remark that the equivalences just discussed, apart from the exchange of coordinates, hold unless one is not considering the commutation of the $\mathbf{D}_i$ with the operators built from separated equations; it is then recommended to consider separately the different cases. By writing

$$\mathbf{T}(\mathbf{D}_i\psi - \lambda\psi) = \begin{pmatrix} E^1_x + E^1_y \\ E^2_x + E^2_y \end{pmatrix}$$

and separating it as $E^x_i = \mu_i = -E^y_i$, we obtain the corresponding separated equations.

I) Consider $\mathbf{D}_1\psi - \lambda\psi = 0$: the first component can be separated as

$$\begin{cases} A_1(x)\dot{a}_1 + C_{11}(x)a_1 - \lambda a_1 = \mu_1 a_1 \\ B_1(y)\dot{b}_1 + C_{12}(y)b_1 = -\mu_1 b_1 \end{cases} \tag{6.2}$$

or

$$\begin{cases} A_1(x)\dot{a}_1 + C_{11}(x)a_1 = \mu_1 a_1 \\ B_1(y)\dot{b}_1 + C_{12}(y)b_1 - \lambda b_1 = -\mu_1 b_1 \end{cases} \tag{6.3}$$

according to the alternative grouping of λ with terms in x or y. The second component reads

$$\begin{cases} A_4(x)\dot{a}_2 + C_4(x)a_2 - \lambda a_2 = \mu_2 a_1 \\ B_3(y)\dot{b}_1 + C_3(y)b_1 = -\mu_2 b_2 \end{cases} \tag{6.4}$$

where μ_1 and μ_2 are the separation constants. These separated equations can be decoupled by integrating a_1 and b_1 from the first two and substituting the results in the last. The solutions for a_i, b_i are in all cases obtained by solving by first-order ODE's. By using only the terms independent of λ and assuming $\mu_1 = \mu_2$ one can try to build first-order eigenvalue-type operators $\mathbf{L}$. However, this is impossible here under the assumption of independence among a_i and b_i. The same result holds for higher-order operators, when we can assume $\mu_1 \neq \mu_2$. It follows that no eigenvalue-type operator is associated with the $\mathbf{D}_1$ separation scheme.

II) By considering equation $\mathbf{D}_3\psi - \lambda\psi = 0$ we obtain the following four systems of separated equations, according to the possible different groupings of λ:

$$\begin{cases} A_1(x)\dot{a}_1 + C_{11}(x)a_1 - \lambda a_1 = \mu_1 a_1 \\ B_1(y)\dot{b}_1 + C_{12}(y)b_1 = -\mu_1 b_1 \end{cases} \tag{6.5}$$

$$\begin{cases} A_1(x)\dot{a}_1 + C_{11}(x)a_1 = \mu_1 a_1 \\ B_1(y)\dot{b}_1 + C_{12}(y)b_1 - \lambda b_1 = -\mu_1 b_1 \end{cases} \tag{6.6}$$

for the first component, and

$$\begin{cases} A_4(x)\dot{a}_2 + C_{41}(x)a_2 - \lambda a_2 = \mu_2 a_2 \\ B_4(y)\dot{b}_2 + C_{42}(y)b_2 = -\mu_2 b_2 \end{cases} \tag{6.7}$$

$$\begin{cases} A_4(x)\dot{a}_2 + C_{41}(x)a_2 = \mu_2 a_2 \\ B_4(y)\dot{b}_2 + C_{42}(y)b_2 - \lambda b_2 = -\mu_2 b_2 \end{cases} \tag{6.8}$$

for the second. All equations are already decoupled in a_i, b_i and solutions are always given by solving first-order ODE's. By putting $\mu_1 = \mu_2 = \mu$ we can obtain from the previous systems the following couples of equations, suitable for the construction of first-order operators. From(6.5) and (6.7):

$$\begin{cases} (B_1\partial_y + C_{12})b_1 = -\mu b_1 \\ (B_4\partial_y + C_{42})b_2 = -\mu b_2 \end{cases} \tag{6.9}$$

from(6.5) and (6.8):

$$\begin{cases} (B_1\partial_y + C_{12})b_1 = -\mu b_1 \\ (A_4\partial_x + C_{41})a_2 = \mu a_2 \end{cases} \tag{6.10}$$

from(6.6) and (6.7):

$$\begin{cases} (A_1\partial_x + C_{11})a_1 = \mu a_1 \\ (B_4\partial_y + C_{42})b_2 = -\mu b_2 \end{cases} \tag{6.11}$$

from(6.6) and (6.8):

$$\begin{cases} (A_1\partial_x + C_{11})a_1 = \mu a_1 \\ (A_4\partial_x + C_{41})a_2 = \mu a_2. \end{cases} \tag{6.12}$$

The corresponding operators of the form $\mathbf{L}\psi = \mu\psi$ are then respectively given by

$$\mathbf{L}_1 = -\begin{pmatrix} B_1\partial_y + C_{12} & 0 \\ 0 & B_4\partial_y + C_{42} \end{pmatrix}$$

$$\mathbf{L}_2 = \begin{pmatrix} -B_1\partial_y - C_{12} & 0 \\ 0 & A_4\partial_x + C_{41} \end{pmatrix}$$

$$\mathbf{L}_3 = \begin{pmatrix} A_1\partial_x + C_{11} & 0 \\ 0 & -B_4\partial_y - C_{42} \end{pmatrix}$$

$$\mathbf{L}_4 = \begin{pmatrix} A_1\partial_x + C_{11} & 0 \\ 0 & A_4\partial_x + C_{41} \end{pmatrix}.$$

An easy computation shows that all these operators commute with $\mathbf{D}_3$ and between themselves when applied to some generic ψ. The same holds for their powers.

III) We have for $\mathbf{D}_4\psi - \lambda\psi = 0$ the following separated equations:

$$\begin{cases} A_2(x)\dot{a}_2 + C_2(x)a_2 = \mu_1 a_1 \\ B_1(y)\dot{b}_1 + C_1(y)b_1 - \lambda b_1 = -\mu_1 b_2 \end{cases} \tag{6.13}$$

$$\begin{cases} A_4(x)\dot{a}_2 + C_4(x)a_2 - \lambda a_2 = \mu_2 a_1 \\ B_3(y)\dot{b}_1 + C_3(y)b_1 = -\mu_2 b_2 \end{cases} \tag{6.14}$$

The equations can be easily decoupled by substitution. It follows that

$$\mathbf{L} = \begin{pmatrix} 0 & A_2\partial_x + C_2 \\ -B_3\partial_y - C_3 & 0 \end{pmatrix}. \tag{6.15}$$

However, the requirement $[\mathbf{L}, \mathbf{D}_4]\psi = 0$ for every ψ implies $B_1 = B_3 = A_2 = A_4 = 0$. Thus, no differential operator of this kind can exist provided a_i and b_i are independent. Moreover, no higer-order operator can be constructed. In conclusion, no operator can be associated with $\mathbf{D}_4$ for type I separation.

IV) The separation of $\mathbf{D}_5\psi - \lambda\psi = 0$ is given by

$$\begin{cases} A_2(x)\dot{a}_2 + C_2(x)a_2 = \mu_1 a_1 \\ B_1(y)\dot{b}_1 + C_1(y)b_1 - \lambda b_1 = -\mu_1 b_2 \end{cases} \tag{6.16}$$

$$\begin{cases} A_3(x)\dot{a}_1 + C_3(x)a_1 = \mu_2 a_2 \\ B_4(y)\dot{b}_2 + C_4(y)b_2 - \lambda b_2 = -\mu_2 b_1. \end{cases} \tag{6.17}$$

The equations in a_i can be decoupled by substituting, for instance, a_1 and $\dot{a}_1$ given by the first equation of the first system and its derivative with respect to x into the left-hand term of the remaining equation yielding a second-order ODE in a_2 only. Solutions for a_i are therefore obtained by solving second-order ODE's. The same procedure allows us to decouple the equations for b_i. It is easy to verify that no first-order operator independent of λ can be built by using the previous equations. For higher order operators we have

$$\begin{cases} (A_2\partial_x + C_2)(A_3\partial_x + C_3)a_1 = \mu_1\mu_2 a_1 \\ (A_3\partial_x + C_3)(A_2\partial_x + C_2)a_2 = \mu_1\mu_2 a_2. \end{cases} \tag{6.18}$$

Then, for the operator

$$\mathbf{L}_5 = \begin{pmatrix} (A_2\partial_x + C_2)(A_3\partial_x + C_3) & 0 \\ 0 & (A_3\partial_x + C_3)(A_2\partial_x + C_2) \end{pmatrix}$$

we have

$$\mathbf{L}_5\psi = \mu_1\mu_2\psi$$

and for its powers

$$\mathbf{L}_5^h\psi = (\mu_1\mu_2)^h\psi$$

for any positive integer h. We have $[\mathbf{D}_5, \mathbf{L}_5] = 0$, as do the powers $\mathbf{L}_5^h$.

REMARK. if we call μ the eigenvalue of $\mathbf{L}$ associated with ψ, from $\mu = \mu_1\mu_2$ we obtain as separation constants, for example, μ_1 and μ/μ_1.

We now assume $R_i \neq 1$. The separated equations obtained from $\mathbf{D}$ are of the same form as seen above, except for the terms including λ which now become now λ/R_i. However, this does not change the previous analysis. Because $\lambda \neq 0$, the functions R_i can depend on only one of the coordinates, otherwise the terms containing λ are not separated. Then, the operators $\mathbf{L}$ built as above from separated equations do not contain the functions R_i. Moreover, the terms in $\mathbf{L}$ arising from the i-th separated equation must depend on the coordinate different from that appearing in the factor R_i, being independent of λ. Thus, the operators $\mathbf{L}$ are the same as those obtained in the case $R_i = 1$, while the operators $\mathbf{D}$ consequently take the form

$$\mathbf{D}_k = \begin{pmatrix} R_1 & 0 \\ 0 & R_2 \end{pmatrix} \mathbf{D}'_k \tag{6.19}$$

where the $\mathbf{D}'_k$ are the operators $\mathbf{D}_i$ seen in the previous section.

We analyze now only the cases corresponding to the existence of symmetry operators associated with separation. Let $\mathbf{D} = \begin{pmatrix} R_1 & 0 \\ 0 & R_2 \end{pmatrix} \mathbf{D}_3$. According to the different operators $\mathbf{L}_1, \ldots, \mathbf{L}_4$ we have respectively: $(R_1(x), R_2(x))$, $(R_1(x), R_2(y))$, $(R_1(y), R_2(x))$ and $(R_1(y), R_2(y))$. Since

$\mathbf{D}$ and $\mathbf{L}_i$ are simultaneously diagonalized, it is easy to verify that $\mathbf{L}_i\mathbf{D} = \begin{pmatrix} R_1 & 0 \\ 0 & R_2 \end{pmatrix} \mathbf{L}_i\mathbf{D}_3$ and $\mathbf{D}\mathbf{L}_i = \begin{pmatrix} R_1 & 0 \\ 0 & R_2 \end{pmatrix} \mathbf{D}_3\mathbf{L}_i$. Then the operators $\mathbf{L}_i$ still commute with $\mathbf{D}$. For $\mathbf{D} = \begin{pmatrix} R_1 & 0 \\ 0 & R_2 \end{pmatrix} \mathbf{D}_4$ we must have $R_1(y),\ R_2(x)$. By imposing $[\mathbf{L}, \mathbf{D}] = 0$, we again have the result that there is no differential operator $\mathbf{L}$ of the required form satisfying the previous equation, whatever R_i may be. In a similar way we can see for $\mathbf{D} = \begin{pmatrix} R_1 & 0 \\ 0 & R_2 \end{pmatrix} D_5$, that for the associated operator $\mathbf{L}_5$ described above we must have $(R_1(y), R_2(y))$ and again $[\mathbf{L}_5, \mathbf{D}] = 0$. We collect these results in the following proposition:

PROPOSITION 2. *The only cases of type I separation associated with symmetry operators for (6.19), occur for* $\mathbf{D}_3$, *with four independent first-order symmetry operators and their powers, and for* $\mathbf{D}_5$, *where there is one second order symmetry operator and its powers.*

We remark that $\mathbf{L}_5$ is essentially a generalization of the nonfactorizable symmetry operator found in [8] for their two-dimensional example.

7. The Dirac case, Type I. Up to this point, we have not considered any connection between operators (6.19) and the Dirac equation. Recalling the results of the Section 1 we obtain the following results by considering all possible commuting cases for separation of Type I:

1) $\mathbf{D}_3$: This separation scheme is never associated with the separation of the Dirac equation in the Dirac representation, since it is always associated with singular spin frames: from (2.7),(2.8) we have $e_2^1 = e_2^2 = 0$, then the matrix (e_a^μ) is singular everywhere.

2)$\mathbf{D}_5$: Proceeding as above, recalling that $R_1(y)$, $R_2(y)$, and applying the conditions for the existence of the symmetry operator $\mathbf{L}_5$ we obtain from (2.7)–(2.9) the following separability conditions:

$$\begin{cases} e_1^1 = e_2^2 = 0 \\ ie_1^2 = R_1(y)B_1(y) \\ e_2^1 = R_1(y)A_2(x) \\ -e_2^1 = R_2(y)A_3(x) \\ -ie_1^2 = R_2(y)B_4(y) \\ iX = 4R_1(y)C_1(y) \\ -Y = 4R_1(y)C_2(x) \\ Y = 4R_2(y)C_3(x) \\ -iX = 4R_2(y)C_4(y) \end{cases}$$

from which it follows that $R_2 = cR_1$, where c is a constant, and $A_2 = -cA_3$, $B_1 = -cB_4$, $C_1 = -cC_4$, $C_2 = -cC_3$. The system is therefore equivalent to

$$\begin{cases} e_1^1 = e_2^2 = 0 \\ ie_1^2 = R_1(y)B_1(y) \\ e_2^1 = R_1(y)A_2(x) \\ iX = 4R_1(y)C_1(y) \\ -Y = 4R_1(y)C_2(x). \end{cases} \tag{7.1}$$

Thus the only allowed symmetry operator has the following form:

$$\mathbf{L}_5 = -\frac{1}{c}\begin{pmatrix} (A_2\partial_x + C_2(x))^2 & 0 \\ 0 & (A_2\partial_x + C_2(x))^2 \end{pmatrix}. \tag{7.2}$$

Given the previous expression for the components of the spin frame in these coordinates, the quantities $X = -2\Gamma_{122}$ and $Y = -2\Gamma_{121}$ can be computed from the $e_a{}^\mu$. Moreover, since the $e_a{}^\mu$ are assumed to be real functions one has $B_1(y) = i\bar{B}_1(y)$ where $\bar{B}_1$ is a real function. Thus we have

$$\begin{cases} X = 4A_2(x)(R_1(y)\bar{B}_1(y))^{-1}\dfrac{d(R_1(y)\bar{B}_1(y))}{dy} \\ Y = 4R_1(y)\bar{B}_1(y)A_2(x)^{-1}\dfrac{dA_2(x)}{dx} \end{cases}$$

which is consistent with the above conditions iff

$$\begin{cases} A_2 = k \\ C_2(x) = 0 \\ C_1(y) = (R_1(y)\bar{B}_1(y))^{-1}\dfrac{d(R_1(y)\bar{B}_1(y))}{dy} \end{cases} \tag{7.3}$$

where k is constant and $\bar{C}_1(y)$ and $\bar{B}_1$ are real functions, or

$$\begin{cases} R_1\bar{B}_1 = k \\ C_1 = 0 \\ C_2 = -\dot{A}_2/A_2 \\ \bar{B}_1 = k/R_1. \end{cases} \tag{7.4}$$

In both cases we can write the components of the metric tensor as

$$\begin{cases} g_{11} = (A_2R_1)^{-2} \\ g_{22} = (R_1\bar{B}_1)^{-2} \end{cases} \tag{7.5}$$

that, after rescaling of (x, y) and the requirement that Riemann tensor is zero, correspond to polar or Cartesian coordinates on the Euclidean plane. By choosing $A_2 = 1$, we can write the symmetry operator as

$$\mathbf{L}_5 = -\frac{1}{c}\begin{pmatrix} \partial_x^2 & 0 \\ 0 & \partial_x^2 \end{pmatrix}.$$

REMARK. The operator

$$\mathbf{L} = \begin{pmatrix} \partial_x & 0 \\ 0 & \partial_x \end{pmatrix}$$

that commutes with $\mathbf{D}_5$, it is not a good symmetry operator in this case. Indeed, it is not built from the separated equations and it implies $a_1 = a_2$ (up to a constant factor that can be absorbed by b_i), a case corresponding to Type II separation which is excluded a priori. The differential equations (6.16) and (6.17) for the a_i are in this particular case

$$\begin{cases} \dot{a}_2 = \mu_1 a_1 \\ -\dot{a}_1 = \mu_2 a_2. \end{cases}$$

Solving for a_i we obtain

$$\begin{cases} a_1 = c_1^1 e^{-\sqrt{-\mu_1\mu_2}x} + c_1^2 e^{\sqrt{-\mu_1\mu_2}x} \\ a_2 = \sqrt{-\mu_1/\mu_2}\left(-c_1^1 e^{-\sqrt{-\mu_1\mu_2}x} + c_1^2 e^{\sqrt{-\mu_1\mu_2}x}\right). \end{cases} \tag{7.6}$$

For b_i the separated equations are

$$\begin{cases} i\bar{B}_1(y)\dot{b}_1 + i\bar{C}_1(y)b_1 - R_1(y)^{-1}\lambda b_1 = -\mu_1 b_2 \\ -\frac{i}{c}\bar{B}_1(y)\dot{b}_2 - \frac{i}{c}\bar{C}_1(y)b_2 - R_1(y)^{-1}\lambda b_2 = -\mu_2 b_1 \end{cases}$$

with $\bar{C}_1(y) = (R_1(y)\bar{B}_1(y))^{-1}\dfrac{d(R_1(y)\bar{B}_1(y))}{dy}$. The system can be decoupled as we have seen for the a_i. If also $\bar{B}_1 = -c\bar{B}_4 = 1$, $R_1 = 1$, (Cartesian coordinates), the separated equations can be solved as follows:

$$\begin{cases} b_1 = d_1^1 e^{r_1 y} + d_1^2 e^{r_2 y} \\ b_2 = d_2^1 e^{r_1 y} + d_2^2 e^{r_2 y} \end{cases} \tag{7.7}$$

with $d_2^1 = \dfrac{\lambda - r_1}{\mu_1} d_1^1$, $d_2^2 = \dfrac{\lambda - r_2}{\mu_1} d_1^2$ where r_i are roots of $r^2 - c\lambda(1-c)r - c(\lambda^2 + \mu_1\mu_2) = 0$. We thus have the following proposition:

PROPOSITION 3. *The only symmetry operator for type I separation associated with the nonsingular Dirac equation is* $\mathbf{L}_5$ *with (7.3) or (7.4). The separable spin frames are given by (7.1). The corresponding coordinates separate the geodesic Hamilton-Jacobi equation,* $\mathbf{L}_5$ *is in diagonal form and is associated with first-class coordinate. If the Riemannian manifold is the Euclidean plane, the coordinates, up to a rescaling, coincide with polar or Cartesian coordinates.*

We remark that the case $\mathbf{D}_7$, corresponding to the exchange of x and y, is not different from this one, since the choice of coordinates is arbitrarily.

REMARK. From (7.6) and (7.7) it appears that although a_2 and b_2 depend not only on μ but also on μ_1 and μ_2 individually, the whole eigenfunction $\psi_2 = a_2b_2$ depends on μ only, the eigenvalue of the symmetry operator. The eigenfunctions of the Dirac operator are then parametrized by λ and μ only, as required by the rules of quantum mechanics. This also reflects the completeness of the solution.

8. Type II. Assume $a_1 = a_2$ and $b_1 \neq b_2$. We have

$$\mathbf{D}\psi - \lambda\psi = \begin{pmatrix} A_1\dot{a}b_1 + A_2\dot{a}b_2 + B_1a\dot{b}_1 + B_2a\dot{b}_2 + (C_1-\lambda)ab_1 + C_2ab_2 \\ A_3\dot{a}b_1 + A_4\dot{a}b_2 + B_3a\dot{b}_1 + B_4a\dot{b}_2 + C_3ab_1 + (C_4-\lambda)ab_2 \end{pmatrix} = 0.$$

We now divide each row of by the two functions ab_i, $i = 1, 2$ only, obtaining the following four separated operators

$$\mathbf{D}_1 : \begin{pmatrix} A_1(x) & 0 \\ A_3(x) & 0 \end{pmatrix} \partial_x + \begin{pmatrix} B_1(y) & B_2(y) \\ B_3(y) & B_4(y) \end{pmatrix} \partial_y + \begin{pmatrix} C_{11}(x)+C_{12}(y) & C_2(y) \\ C_{31}(x)+C_{32}(y) & C_4(y) \end{pmatrix},$$

$$\mathbf{D}_2 : \begin{pmatrix} A_1(x) & 0 \\ 0 & A_4(x) \end{pmatrix} \partial_x + \begin{pmatrix} B_1(y) & B_2(y) \\ B_3(y) & B_4(y) \end{pmatrix} \partial_y + \begin{pmatrix} C_{11}(x)+C_{12}(y) & C_2(y) \\ C_3(y) & C_{41}(x)+C_{42}(y) \end{pmatrix},$$

$$\mathbf{D}_3 : \begin{pmatrix} 0 & A_2(x) \\ A_3(x) & 0 \end{pmatrix} \partial_x + \begin{pmatrix} B_1(y) & B_2(y) \\ B_3(y) & B_4(y) \end{pmatrix} \partial_y + \begin{pmatrix} C_1(y) & C_{21}(x)+C_{22}(y) \\ C_{31}(x)+C_{32}(y) & C_4(x) \end{pmatrix},$$

$$\mathbf{D}_4 : \begin{pmatrix} 0 & A_3(x) \\ 0 & A_4(x) \end{pmatrix} \partial_x + \begin{pmatrix} B_1(y) & B_2(y) \\ B_3(y) & B_4(y) \end{pmatrix} \partial_y + \begin{pmatrix} C_1(y) & C_{21}(x)+C_{22}(y) \\ C_3(y) & C_{41}(x)+C_{42}(y) \end{pmatrix}.$$

In order to obtain the most general separated equations we must consider each of the D_i multipliyed by the matrix

$$\begin{pmatrix} R_1 & 0 \\ 0 & R_2 \end{pmatrix}$$

where R_1 and R_2 will be determined in their dependence on x or y by the separated equations employed to construct the symmetry operators.

For D_1 we have the following systems of separated ODEs,

$$\begin{cases} A_1\dot{a} + (C_{11} - \lambda/R_1)a = \mu_1 a \\ B_1\dot{b}_1 + B_2\dot{b}_2 + C_2(y)b_2 + C_{12}b_1 = -\mu b_1 \end{cases} \tag{8.1}$$

or

$$\begin{cases} A_1\dot{a} + C_{11}a = \mu_1 a \\ B_1\dot{b}_1 + B_2\dot{b}_2 + C_2(y)b_2 + (C_{12} - \lambda/R_1)b_1 = -\mu_1 b_1 \end{cases} \tag{8.2}$$

for the first component, corresponding to the possible groupings of the parameter λ; for the second component we have

$$\begin{cases} A_3\dot{a} + C_{31}a = \mu_2 a \\ B_3\dot{b}_1 + B_4\dot{b}_2 + (C_4(y) - \lambda/R_2)b_2 + C_{32}b_1 = -\mu_2 b_1. \end{cases} \tag{8.3}$$

For the case D_2 all the possible separated equations are given by

$$\begin{cases} A_1\dot{a} + (C_1^1 - \lambda/R_1)a = \mu_1 a \\ B_1\dot{b}_1 + B_2\dot{b}_2 + C_2(y)b_2 + C_{12}b_1 = -\mu_1 b_1 \end{cases} \tag{8.4}$$

or

$$\begin{cases} A_1\dot{a} + C_{11}a = \mu_1 a \\ B_1\dot{b}_1 + B_2\dot{b}_2 + C_2(y)b_2 + (C_{12} - \lambda/R_1)b_1 = -\mu_1 b_1 \end{cases} \tag{8.5}$$

for the separation of the first component, and

$$\begin{cases} A_4\dot{a} + (C_{41} - \lambda/R_1)a = \mu_2 a \\ B_3\dot{b}_1 + B_4\dot{b}_2 + C_3 b_1 + C_{42}b_2 = -\mu_2 b_2 \end{cases} \tag{8.6}$$

or

$$\begin{cases} A_4\dot{a} + C_{41}a = \mu_2 a \\ B_3\dot{b}_1 + B_4\dot{b}_2 + C_3(y)b_1 + (C_{42} - \lambda/R_2)b_2 = -\mu_2 b_2 \end{cases} \tag{8.7}$$

for the second component, according to the two possible groupings of λ.

For the operator D_3, the separated equations are

$$\begin{cases} A_2\dot{a} + C_{21}a = \mu_1 a \\ B_1\dot{b}_1 + B_2\dot{b}_2 + (C_1(y) - \lambda/R_1)b_1 + C_{22} = -\mu_1 b_2 \end{cases} \tag{8.8}$$

for the first component and

$$\begin{cases} A_3\dot{a} + C_{31}a = \mu_2 a \\ B_3\dot{b}_1 + B_4\dot{b}_2 + (C_4(y) - \lambda/R_2)b_2 + C_{32}b_1 = -\mu_2 b_1 \end{cases} \tag{8.9}$$

for the second.

Finally, for the operator D_4, the separated equations are

$$\begin{cases} A_2\dot{a} + C_{21}a = \mu_1 a \\ B_1\dot{b}_1 + B_2\dot{b}_2 + (C_1(y) - \lambda/R_1)b_1 + C_{22} = -\mu_1 b_2 \end{cases} \tag{8.10}$$

for the separation of the first component, and

$$\begin{cases} A_4\dot{a} + (C_{41} - \lambda/R_2)a = \mu_2 a \\ B_3\dot{b}_1 + B_4\dot{b}_2 + C_3(y)b_1 + C_{42}b_2 = -\mu_2 b_2 \end{cases} \tag{8.11}$$

or

$$\begin{cases} A_4\dot{a} + C_{41}a = \mu_2 a \\ B_3\dot{b}_1 + B_4\dot{b}_2 + C_3(y)b_1 + (C_{42} - \lambda/R_2)b_2 = -\mu_2 b_2 \end{cases} \tag{8.12}$$

for the second.

9. The Dirac case, Type II. In order to shorten our analysis, we consider in the following only the operators $\mathbf{D}_i$ that can be associated with Dirac equations whose spin frame is nonsingular. The association with a Dirac equation implies the following relations among the components of the matrices $\mathbf{A}$, $\mathbf{B}$ and $\mathbf{C}$:

$$\begin{cases} A_4 = -A_1R_1/R_2 \\ A_3 = -A_2R_1/R_2 \\ B_4 = -B_1R_1/R_2 \\ B_3 = -B_2R_1/R_2 \\ C_4 = -C_1R_1/R_2 \\ C_3 = -C_2R_1/R_2 \end{cases} \tag{9.1}$$

where each of the R_i must depend on only one coordinate because $\lambda \neq 0$. Then, after (2.7)–(2.9) the determinant of the spin frame matrix is

$$|e_a{}^\mu| = iR_1^2(B_1A_2 - A_1B_2). \tag{9.2}$$

We observe that to $\mathbf{D}_1$ always correspond to a singular spin frame, since from the expression of $\mathbf{D}_1$ we have $e_1^1 = e_2^1 = 0$. Similarly for the case $\mathbf{D}_4$ we have immediately $e_1^1 = e_2^1 = 0$, which implies that all related spin frames are necessarily singular. Thus, no Dirac equation can correspond to the separation schemes $\mathbf{D}_1$ and $\mathbf{D}_4$. Moreover, we consider here only the operators commuting with each $\mathbf{D}_i$, i.e. the symmetry operators. However, as in the case of Type I separation, for each index i there also exist separated operators not commuting with $\mathbf{D}_i$.

I) The operator $\mathbf{D}_2$: in this case we have $e_2^1 = 0$. All the possible operators $\mathbf{L}$ that can be built from the separated equations of $\mathbf{D}_2$ and such that $[\mathbf{D}_2, \mathbf{L}] = 0$ have the following form:

$$\mathbf{L}_I = \begin{pmatrix} A_1 & 0 \\ 0 & A_4 \end{pmatrix} \partial_x + \begin{pmatrix} C_{11} & 0 \\ 0 & C_{14} \end{pmatrix}$$

$$\mathbf{L}_{II} = \begin{pmatrix} A_4 & 0 \\ 0 & A_1 \end{pmatrix} \partial_x + \begin{pmatrix} C_{14} & 0 \\ 0 & C_{11} \end{pmatrix}$$

and for which $R_1(y)$, $R_2(y)$, $R_2 = cR_1$ for some constant c. This result follows from $A_i(x)$ and from the requirement of nonsingularity of the spin frame. The calculation shows that the operators commute with $\mathbf{D}_2$ for nonsingular spin frames if and only if $c = -1$ and $C_{14} = C_{11}$. It follows that $A_4 = A_1$ and $\mathbf{L}_I = \mathbf{L}_{II}$.

For second order operators we have

$$\mathbf{L}_{III} = \begin{pmatrix} L_a L_b & 0 \\ 0 & L_c L_d \end{pmatrix}$$

where $a, b, c, d = 1, 2$, $a \neq b$, $c \neq d$ and $L_1 = A_1\partial_x + C_{11}$, $L_2 = A_4\partial_x + C_{41}$. Again we have $R_1(y)$, $R_2(y)$ and $R_2 = cR_1$ for some constant c. All these operators commute with $\mathbf{D}_2$ with nonsingular spin frames iff $C_{11} = -cC_{14} + k$ for some constant k.

Let us consider now $\mathbf{L}_I$, we have from (2.7)–(2.9) the conditions

$$\begin{cases} e^1_1 = -iR_1A_1 \\ e^1_2 = 0 \\ e^2_1 = -iR_1B_1 \\ e^2_2 = R_1B_2 \\ X = -i4R_1(C_{11} + C_{12}) \\ Y = -4R_1C_2(y) \end{cases} \tag{9.3}$$

It follows that $A_1(x) = i\bar{A}_1(x)$, $B_1(y) = i\bar{B}_1(y)$ and, from the calculation of Γ_{abc}:

$$\begin{cases} X = 0 \\ Y = -4\left(B_2\dot{\bar{R}}_1 + \bar{B}_1\dfrac{\dot{\bar{A}}_1}{\bar{A}_1}\right) \end{cases}$$

Therefore, we have either

$$\begin{cases} C_1 = 0 \\ \bar{A}_1 = k_1e^{k_2x} \\ C_2(y) = k_2\bar{B}_1 + B_2\dot{R}_1/R_1 \end{cases} \tag{9.4}$$

where k_i are constants, or

$$\begin{cases} C_1 = 0 \\ \bar{B}_1 = 0 \\ C_2(y) = B_2\dot{R}_1/R_1. \end{cases} \tag{9.5}$$

In both cases the covariant components of the metric tensor are given by

$$\begin{cases} g_{11} = \dfrac{B_2^2 + \bar{B}_1^2}{R_1^2\bar{A}_1^2B_2^2} \\ g_{12} = -\dfrac{\bar{B}_1}{R_1^2\bar{A}_1B_2^2} \\ g_{22} = \dfrac{1}{R_1^2B_2^2}. \end{cases} \tag{9.6}$$

It can be shown that the Levi-Civita equations are always satisfied and x is of first-class. By putting, for example, $R_1 = 1$, $\bar{A}_1 = 1$, $\bar{B}_1 = c_1$, $B_2 = -(c_2 y + c_3)^{-1}$, c_i constants, the Riemann tensor is zero and the manifold is the Euclidean plane.

For $\mathbf{L}_{III}$ we have the conditions

$$\begin{cases} e_1^1 = -iR_1 A_1 \\ e_2^1 = 0 \\ e_1^2 = -iR_1 B_1 \\ e_2^2 = -R_1 B_2 \\ X = -i4R_1(C_{11} + C_{12}) \\ Y = -4R_1 C_2(y) \end{cases} \tag{9.7}$$

Proceeding as before, we obtain

$$\begin{cases} A_1(x) = i\bar{A}_1(x) \\ B_1(y) = i\bar{B}_1(y) \\ X = 0 \\ Y = -4(B_2\dot{\bar{R}}_1 + \bar{B}_1\dfrac{\dot{\bar{A}}_1}{\bar{A}_1}) \end{cases}$$

Therefore, again we have either

$$\begin{cases} C_1 = 0 \\ \bar{A}_1 = k_1 e^{k_2 x} \\ C_2(y) = k_2\bar{B}_1 + B_2\dot{R}_1/R_1 \end{cases}$$

where k_i are constants, or

$$\begin{cases} C_1 = 0 \\ \bar{B}_1 = 0 \\ C_2(y) = B_2\dot{R}_1/R_1. \end{cases}$$

Thus all $\mathbf{L}_{III}$ operators are proportional to the square of $\mathbf{L}_I$. Consequently, the components of the metric tensor coincide with those associated to $\mathbf{L}_I$.

The separated equations for $\mathbf{L}_I$ are

$$\begin{cases} A_1\dot{a} = \mu_1 a \\ A_1\dot{a} = \mu_2 a \\ B_1\dot{b}_1 + B_2\dot{b}_2 + C_2 b_2 - \lambda/R_1 b_1 = -\mu_1 b_1 \\ B_2\dot{b}_1 + B_1\dot{b}_2 + C_2 b_1 + \lambda/R_1 b_2 = -\mu_2 b_2 \end{cases}$$

which imply $\mu_1 = \mu_2 = \mu$ as required. It can easily be shown that the equations for b_i may be decoupled and they are linear combinations of solutions of a second-order ODE.

For $\mathbf{L}_{III}$ we have the separated equations

$$\begin{cases} A_1\dot{a} = \mu_1 a \\ A_1\dot{a} = -c\mu_2 a \\ B_1\dot{b}_1 + B_2\dot{b}_2 + C_2 b_2 - \lambda/R_1 b_1 = -\mu_1 b_1 \\ B_2\dot{b}_1 + B_1\dot{b}_2 + C_2 b_1 + \lambda/R_1 b_2 = c\mu_2 b_2 \end{cases}$$

which imply $\mu_2 = -\mu_1/c$. The solutions coincide with those of $\mathbf{L}_I$ as expected, since $\mathbf{L}_{III}$ is proportional to $\mathbf{L}_I^2$.

II) The operator $\mathbf{D}_3$: in this case we have $e_1^1 = 0$. From the separated equations, the possible first-order symmetry operators associated with nonsingular spin frames are

$$\mathbf{L}_{IV} = \begin{pmatrix} A_2 & 0 \\ 0 & A_3 \end{pmatrix} \partial_x + \begin{pmatrix} C_{12} & 0 \\ 0 & C_{13} \end{pmatrix}$$

and

$$\mathbf{L}_V = \begin{pmatrix} A_3 & 0 \\ 0 & A_2 \end{pmatrix} \partial_x + \begin{pmatrix} C_{13} & 0 \\ 0 & C_{12} \end{pmatrix}.$$

Under our hypothesis we have in both cases $R_2(y) = cR_1(y)$. It follows that $\mathbf{L}_{IV}$ and $\mathbf{L}_V$ commute with $\mathbf{D}_3$ when $c = -1$ and $C_{13} = C_{12}$. Then, $\mathbf{L}_{IV} = \mathbf{L}_V$. From equations (2.7)–(2.9) we have

$$\begin{cases} e_1^1 = 0 \\ e_2^1 = R_1 A_2 \\ e_1^2 = -iR_1 B_1 \\ e_2^2 = R_1 B_1 \\ X = -4iR_1 C_1(y) \\ Y = -4R_1(C_{12} + C_{22}). \end{cases} \tag{9.8}$$

Thus, $B_1 = i\bar{B}_1$, where $\bar{B}_1$ is a real function. From the results of Section I it follows that

$$\begin{cases} X = 4A_2(\dot{R}_1\bar{B}_1 + \dot{\bar{B}}_1 R_1)/\bar{B}_1 \\ Y = -4(B_2\dot{R}_1 - \dot{B}_2 R_1 + B_2\dot{\bar{B}}_1/\bar{B}_1 R_1 + \dot{A}_2/A_2 R_1 \bar{B}_1) \end{cases}$$

which yields

$$\begin{cases} C_1(y) = iA_2(\dot{R}_1/R_1 + \dot{\bar{B}}_1/\bar{B}_1) \\ C_{12} + C_{22} = (B_2\dot{R}_1/R_1 - \dot{B}_2 + B_2\dot{\bar{B}}_1/\bar{B}_1 + \dot{A}_2/A_2\bar{B}_1). \end{cases}$$

These equations are fulfilled either for

$$\begin{cases} A_2 = k \\ C_{12} = 0 \end{cases} \tag{9.9}$$

or

$$\begin{cases} \dot{R}_1/R_1 + \dot{\bar{B}}_1/\bar{B}_1 = 0 \\ C_{12} = 0 \\ A_2 = k_2 e^{k_1 x} \end{cases} \tag{9.10}$$

or

$$\begin{cases} \dot{R}_1/R_1 = 0 \\ \bar{B}_1 = h \\ C_{12} = \dot{\bar{A}}_2 h/\bar{A}_2 \ . \end{cases} \tag{9.11}$$

In all cases

$$\begin{cases} g_{11} = \dfrac{B_2^2 + \bar{B}_1^2}{R_1^2 \bar{B}_1^2 A_2^2} \\ g_{12} = -\dfrac{B_2}{R_1^2 \bar{B}_1^2 A_2} \\ g_{22} = \dfrac{1}{R_1^2 \bar{B}_1^2} \ . \end{cases} \tag{9.12}$$

The Levi-Civita equations for these metrics are satisfied which implies that x is a first-class coordinate. The separated equations are

$$\begin{cases} A_2\dot{a} + C_{12}a = \mu_1 a \\ A_2\dot{a} + C_{12}a = \mu_2 a \\ B_1\dot{b}_1 + B_2\dot{b}_2 + (C_1 - \lambda/R_1)b_1 + C_{22}b_2 = -\mu_1 b_2 \\ B_2\dot{b}_1 + B_1\dot{b}_2 + C_{22}b_1 + (C_1 + \lambda/R_1)b_2 = -\mu_2 b_1 \end{cases}$$

from which it follows $\mu_1 = \mu_2$. This system may be solved in the same manner as the case $\mathbf{D}_2$.

For second-order operators we have

$$\mathbf{L}_{VI} = \begin{pmatrix} L_a L_b & 0 \\ 0 & L_c L_d \end{pmatrix}$$

where $a, b, c, d = 1, 2$, $a \neq b$, $c \neq d$ and $\mathbf{L}_1 = A_2\partial_x + C_{12}$, $\mathbf{L}_2 = A_3\partial_x + C_{13}$. The operators commute and the related spin frames are nonsingular only in the following cases: *i)* $C_{12} = -cC_{13} + k$, *ii)* $B_2 = 0$, $C_{22} = C_{23} = 0$ for the case $a, b, c, d = 1, 2, 2, 1$. Again, we have $R_2(y) = cR_1(y)$, and $B_1 = i\bar{B}_1$ where $\bar{B}_1$ is a real function. For the case *i)* we obtain the same results as those obtained for $\mathbf{L}_{IV}$ and $\mathbf{L}_V$. In the case *ii)* we have ${e_2}^2 = 0$ and

$$\begin{cases} X = 4A_2(\dot{R}_1\bar{B}_1 + \dot{\bar{B}}_1 R_1)/\bar{B}_1 \\ Y = -4\dot{A}_2/A_2 R_1 \bar{B}_1 \end{cases}$$

which imply

$$\begin{cases} C_1(y) = iA_2(\dot{R}_1/R_1 + \dot{\bar{B}}_1/\bar{B}_1) \\ C_{12}(x) = \bar{B}_1\dot{A}_2. \end{cases}$$

These equations are consistent when $A_2 = k$, $C_{12} = 0$ or $\dot{R}_1/R_1 + \dot{\bar{B}}_1/\bar{B}_1 = 0$ etc. as in the previous cases. In all cases

$$\begin{cases} g_{11} = \dfrac{1}{R_1^2 A_2^2} \\ g_{12} = 0 \\ g_{22} = \dfrac{1}{R_1^2 \bar{B}_1^2} \end{cases} \tag{9.13}$$

coincide with the components given in $i)$ by putting $B_2 = 0$. Again, the coordinates can be rescaled, when the Riemann tensor is zero, to obtain polar or Cartesian coordinates. The Levi-Civita equations are satisfied and x is consequently a first class coordinate. Note that the Riemann tensor is zero for suitable values of the parameters.

In the case $i)$ the separated equations are:

$$\begin{cases} A_2\dot{a} + (-cC_{13} + k)a = \mu_1 a \\ A_2\dot{a} - cC_{13}a = -c\mu_2 a \\ B_1\dot{b}_1 + B_2\dot{b}_2 + (C_1 - \lambda/R_1)b_1 + C_{22}b_2 = -\mu_1 b_2 \\ B_2\dot{b}_1 + B_1\dot{b}_2 + (C_1 + \lambda/R_1)b_2 + C_{22}b_2 = c\mu_2 b_1 \end{cases} \tag{9.14}$$

implying $k = 0$ and $\mu_2 = -\mu_1/c$. These equations can be solved as seen above. In the case $ii)$ the separated equations are

$$\begin{cases} A_2\dot{a} + C_{12}a = \mu_1 a \\ A_2\dot{a} + C_{12}a = -c\mu_2 a \\ B_1\dot{b}_1 + (C_1 - \lambda/R_1)b_1 = -\mu_1 b_2 \\ B_1\dot{b}_2 + (C_1 + \lambda/R_1)b_2 = c\mu_2 b_1 \end{cases} \tag{9.15}$$

implying again $\mu_2 = -\mu_1/c$. The system for the b_i can be easily decoupled. It follows that $\mathbf{L}_{VI}$ is proportional to the square of $\mathbf{L}_{IV}$. Again, suitable choices of the parameters A_i, B_i, R_i allow the manifold to be the Euclidean plane. These results are collected in the following proposition:

PROPOSITION 4. *The only symmetry operators for type II separation associated with the nonsingular Dirac equation are* $\mathbf{L}_I$, *with (9.4) or (9.5), and* $\mathbf{L}_{IV}$ *with (9.9), (9.10) or (9.11). The separable spin frames are consequently given by (9.3) and (9.8) respectively. The corresponding coordinates, orthogonal or nonorthogonal, separate the geodesic Hamilton-Jacobi equation, the symmetry operators are associated with first-class coordinates and in diagonal form.*

We remark that, although apparently equivalent, the metric tensor components corresponding to the two symmetry operators are not equivalent. Indeed, while for $\mathbf{L}_I$ a free dependence on x, through A_1, is possible only for orthogonal coordinates, for $\mathbf{L}_{IV}$ there is no such a restriction. Once again, we recall that in Hamilton-Jacobi separation theory all non-orthogonal separable coordinates can be transformed into orthogonal coordinates without changing the separation structure. Here, it remains an open problem to see if similar transformations exist. Moreover, the separable spin frames corresponding to the two symmetry operators of above are different.

10. Example: Constant curvature manifolds. We assume that the sectional curvature K of the manifold is constant; consequently, we have the following equation involving the nonnull component of the Riemann tensor and the determinant of the covariant components of the metric tensor

$$\mathbf{R}_{1212}\, det(g_{ij})^{-1} + K = 0. \tag{10.1}$$

By computing (10.1) for all the separable cases of the Dirac equation shown in the previous sections we obtain differential equations in A, B, R that have to be satisfied in order for the Dirac equation to be well defined on the constant curvature manifold. In the case (7.3), after setting $\bar{B}_1(y) = 1$, by rescaling the coordinates, we obtain

$$R\ddot{R} - \dot{R}^2 + K = 0,$$

a solution of which is

$$R(y) = \frac{1}{2}(e^y + Ke^{-y}).$$

For $K = -1, 0, 1$ we have respectively the hyperbolic plane (pseudosphere), the Euclidean plane and the sphere S_2. By setting $k = 1$ in (7.3) we find isothermal coordinates in all these manifolds, with metric components given by $g^{11} = g^{22} = cosh^2(y), \frac{1}{4}e^{2y}, sinh^2(y)$ respectively, corresponding to reparametrizations of polar coordinates in the first two cases, and geographic coordinates in the last. For $K = 0$ a solution of (10.1) is $R(y) = constant$ which leads to the case of Cartesian coordinates, as shown in Section 2. The non-null components of the spin frame are $e_{\bar{2}}^1 = e_{\bar{1}}^2 = R(y)$, the symmetry operator is $\mathbf{L}_5$ and the Dirac equation follows from (2.6)–(2.9). From the other possible solution for $\mathbf{L}_5$ (7.4) we obtain instead $\bar{B}_1 = R(y)^{-1}$, $k = 1$ and (10.1) which leads to

$$R(y) = 2\sqrt{K}e^{-y\sqrt{K}}.$$

This solution is inconsistent if $K \leq 0$, or $R(y) = constant$ for $K = 0$. The allowed solutions correspond to different parametrizations for the coordinates obtained in the previous case. In the case of the symmetry operator

$\mathbf{L}_I$, for $\bar{B}_1 = 0$ we obtain the same orthogonal coordinates previously seen. If we put $\bar{B}_1 = B_2 = 2^{-\frac{1}{2}}$ then (10.1) and $R(y)$ are the same as the first case of $\mathbf{L}_5$ and for the different values of K the resulting coordinates are non-orthogonal. As consequence of (9.3) the non-null spin frame components are now

$$e_1^1 = R(y)k_1e^{k_2x}, \qquad e_1^2 = e_2^2 = \frac{1}{\sqrt{2}}R(y)$$

and

$$\mathbf{L}_I = \begin{pmatrix} ik_1e^{k_2x} & 0 \\ 0 & ik_1e^{k_2x} \end{pmatrix} \partial_x.$$

The same procedure can be applied to the remaining cases.

11. Conclusion. The operators of above, even if not the most general possible, provide a means for investigating the separability properties of the Dirac equation. Although the physical Dirac equation is in dimension four, separation in two dimension can occur after reduction by symmetries. Such a situation arises in the Kerr solution which admits two commuting Killing vectors. Separable coordinates for the Dirac equation are always Hamilton-Jacobi separable, in both the orthogonal or non-orthogonal cases and always include at least one first class coordinate. The symmetry operators are always in diagonal form, which are always associated with first-order coordinates. In type II separation, the second-order symmetry operators are always powers of the first-order ones. The most general expression for separable spin frame components are associated with symmetry operators $\mathbf{L}_5$, $\mathbf{L}_I$, and $\mathbf{L}_{IV}$,equations (7.1), (9.3) and (9.8). Work is continuing on a number of questions studied in this paper including a comparision with the symmetry operators obtained from geometry expressed in manifestly covariant form in terms of Killing tensors of suitable valence, and the Lorentzian case.

Acknowledgements. The authors wish to thank their reciprocal institutions, the Department of Mathematics, University of Turin and the Department of Applied Mathematics, University of Waterloo, where parts of this paper were written. They also wish to express their appreciation for helpful discussions with Claudia Chanu and Lorenzo Fatibene.

REFERENCES

[1] W. Miller, Jr., *Symmetry and separation of variables*, Addison-Wesley, Reading, 1977.

[2] E.G. Kalnins, *Separation of variables for Riemannian spaces of constant curvature*, Longman, Harlow, 1986.

[3] L. Fatibene and M. Francaviglia, *Natural and gauge natural formalism for classical field theories. A geometric perspective including spinors and gauge theories*, Kluwer, Dordrecht, 2003.

[4] D.R. Brill and J.A. Wheeler, *Interaction of neutrinos and gravitational fields*, Rev. Mod. Phys. **29**: 465–479, 1957.

[5] S. Chandrasekhar, *The mathematical theory of black holes*, Clarendon, Oxford, 1983, p. 531.

[6] B. Carter and R.G. McLenaghan, *Generalized total angular momentum for the Dirac operator in curved space-time*, Phys. Rev. D **19**: 1093–1097, 1979.

[7] W. Miller, Jr., *Mechanism for variable separation in partial differential equations and their relationship to group theory. In Symmetries and Nonlinear Phenomena*, World Scientific, Singapore, 1988, pp. 188–221.

[8] M. Fels and N. Kamran, *Non-factorizable separable systems and higher-order symmetries of the Dirac operator*, Proc. Roy. Soc. London A **428**: 229–249, 1990.

[9] L. Fatibene, R.G. McLenaghan, and S. Smith, *Separation of variables for the Dirac equation on low dimensional spaces. In Advances in general relativity and cosmology*, Pitagora, Bologna, 2003, 109–127.

[10] J.T. Horwood and R.G. McLenaghan, *Transformation to pseudo-Cartesian coordinates in locally flat pseudo-Riemannian spaces*, preprint, University of Waterloo, 2006.

[11] E.G. Kalnins, W. Miller, *Separation of variables methods for systems of differential equations in mathematical physics*, Proceedings of the Annual Seminar of the Canadian Mathematical Society, Lie Theory, Differential Equations and Representation Theory, pp. 283–300, 1989.

[12] S. Smith, *Symmetry operators and separation of variables for the Dirac equation on curved space-times*, PhD thesis, University of Waterloo, 2002.

[13] T. Levi-Civita, *Sulla integrazione della equazione di Hamilton-Jacobi per separazione di variabili*, Math.Ann. **59**: 383–397, 1904.

[14] S. Benenti, Lect. Notes Math. **863**: 512, 1980.

[15] V.N. Shapovalov and G.G. Ekle, *Complete sets and integration of a first-order linear system I and II*, Izv. vyssh. ucheb. zaved. Fiz, (2), 83–92, 1974.

[16] S. Benenti, C. Chanu, and G. Rastelli, *Variable separation theory for the null Hamilton-Jacobi equation*, J.Math. Phys. **46**: 042901/29, 2005.

[17] S. Benenti, C. Chanu, and G. Rastelli, *Remarks on the connection between the additive separation of the Hamilton-Jacobi equation and the multiplicative separation of the Schr¨odinger equations. I. The completeness and Robertson condition*, J. Math. Phys. **43**: 5183–5222, 2002.

DIFFERENTIAL SYSTEMS ASSOCIATED WITH TABLEAUX OVER LIE ALGEBRAS

EMILIO MUSSO* AND LORENZO NICOLODI†

Abstract. We give an account of the construction of exterior differential systems based on the notion of tableaux over Lie algebras as developed in [33]. The definition of a tableau over a Lie algebra is revisited and extended in the light of the formalism of the Spencer cohomology; the question of involutiveness for the associated systems and their prolongations is addressed; examples are discussed.

Key words. Exterior differential systems, Pfaffian differential systems, involutiveness, tableaux, tableaux over Lie algebras.

AMS(MOS) subject classifications. 58A17, 58A15.

1. Introduction. The search for a common structure to various exterior differential systems (EDSs) of geometric and analytic origin led to the algebraic notion of a *tableaux over a Lie algebra* [33]. This notion builds on that of involutive tableau in the theory of EDSs and can be seen as a non-commutative generalization of it. Interestingly enough, from a tableau over a Lie algebra we can canonically construct a linear Pfaffian differential system (PDS) which is in involution and whose Cartan characters coincide with the characters of the tableau.

Particular cases of this scheme lead to differential systems describing well-known integrable systems such as the Grassmannian systems of Terng [37, 4], the curved flat system of Ferus and Pedit [16], and many integrable surfaces arising in projective differential geometry [1, 14, 17]. The tableaux corresponding to Grassmannian and curved flat systems, the so-called Cartan tableaux, are obtained from the Cartan decompositions of semisimple Lie algebras. The tableaux corresponding to the various classes of integrable surfaces in projective 3-space are given by sub-tableaux of a special tableau over $\mathfrak{sl}(4,\mathbb{R})$. This is constructed by the method of moving frames and amounts to the construction of a canonical adapted frame along a generic surface in projective space (e.g., the Wilczynski–Cartan frame [11, 17, 2, 14, 15]). The involutiveness of these examples follows from the involutiveness of the corresponding tableaux. The result that the Grassmannian system and the curved flat system are in involution was first

*Dipartimento di Matematica Pura ed Applicata, Università degli Studi dell'Aquila, Via Vetoio, I-67010 Coppito (L'Aquila), ITALY (musso@univaq.it). Partially supported by MIUR project *Metriche riemanniane e varietà differenziali*, and by the GNSAGA of INDAM.

†Dipartimento di Matematica, Università degli Studi di Parma, Viale G.P. Usberti 53/A - Campus universitario, I-43100 Parma, ITALY (lorenzo.nicolodi@unipr.it). Partially supported by MIUR project *Proprietà geometriche delle varietà reali e complesse*, and by the GNSAGA of INDAM.

proved by Bryant in a cycle of seminars at MSRI [6] and was then taken up and further elaborated by Terng and Wang [38].

Other examples of linear Pfaffian system in involution which fit into the above scheme, and in fact motivated our work, include: the differential systems of isothermic surfaces in Möbius and Laguerre geometry [25, 26, 28, 30]; the differential systems of Möbius-minimal (M-minimal or Willmore) surfaces and of Laguerre-minimal (L-minimal) surfaces [27, 29, 25]; the differential systems associated to the deformation problem in projective geometry and in Lie sphere geometry [11, 10, 3, 22, 32, 13]. The tableaux associated with all these examples are constructed by the method of moving frames on the Lie algebras of the corresponding symmetry groups (cf. Section 5.3). If, on the one hand, the presented approach may be seen as a possibility to discuss various involutive systems from a unified viewpoint, on the other hand, it can be viewed as a possibility to find new classes of involutive systems. In this respect, we mention the class of projective surfaces introduced in [33], which generalizes asymptotically-isothermic surfaces and surfaces with constant curvature of Fubini's quadratic form. An analogous class of surfaces in the context of conformal geometry is discussed in Section 5.3. For an application of the above construction to the study of the Cauchy problem for the associated systems, we refer the reader to [31, 34].

In this article we revisit the definition of tableau over a Lie algebra using the formalism of the Spencer cohomology and extend it to include 2-acyclic tableaux. This allows also non-involutive systems into the scheme, reducing the question of involutiveness of their prolonged systems to that of the prolongations of the associated tableaux (cf. Section 4).

Section 2 contains the basic material about tableaux. Section 3 introduces the notion of tableaux over Lie algebras. Section 4 presents the construction of EDSs from tableaux over Lie algebras and discusses some properties of such systems. Section 5 discusses some examples as an illustration of the theory developed in the previous sections. Further developments are indicated in Section 6. The appendix collects some facts about the Spencer complex of a tableau and the torsion of a PDS.

2. Tableaux. In this section we provide a summary of the results in the algebraic theory of tableaux. As basic reference, we use the book by Bryant, *et al.* [7]. See also [21].

A **tableau** is a linear subspace $\mathbf{A} \subset \mathrm{Hom}(\mathfrak{a}, \mathfrak{b})$, where $\mathfrak{a}$, $\mathfrak{b}$ are (real or complex) finite dimensional vector spaces.

An h-dimensional subspace $\mathfrak{a}_h \subset \mathfrak{a}$ is called *generic* w.r.t. $\mathbf{A}$ if the dimension of

$$\mathrm{Ker}\,(\mathbf{A}, \mathfrak{a}_h) := \{Q \in \mathbf{A} \,|\, Q_{|\mathfrak{a}_h} = 0\}$$

is a minimum. The set of h-dimensional generic subspaces is a Zariski open of the Grassmannian of h-dimensional subspaces of $\mathfrak{a}$.

A flag $(0) \subset \mathfrak{a}_1 \subset \cdots \subset \mathfrak{a}_n = \mathfrak{a}$ of $\mathfrak{a}$ is said *generic* if $\mathfrak{a}_h$ is generic, for all $h = 1, \ldots, n$.

The **characters** of $\mathbf{A}$ are the non-negative integers $s_j(\mathbf{A})$, $j = 1, \ldots, n$, defined inductively by

$$s_1(\mathbf{A}) + \cdots + s_j(\mathbf{A}) = \operatorname{codim} \operatorname{Ker}(\mathbf{A}, \mathfrak{a}_j),$$

for any generic flag $(0) \subset \mathfrak{a}_1 \subset \cdots \subset \mathfrak{a}_n = \mathfrak{a}$.

From the definition, it is clear that

$$\dim \mathfrak{b} \geq s_1 \geq s_2 \geq \cdots \geq s_n, \quad \dim \mathbf{A} = s_1 + \cdots + s_n.$$

If $s_\nu \neq 0$, but $s_{\nu+1} = 0$, we say that $\mathbf{A}$ has *principal character* s_ν and call ν the *Cartan integer* of $\mathbf{A}$.

The **first prolongation** $\mathbf{A}^{(1)}$ of $\mathbf{A}$ is the kernel of the linear map (Spencer coboundary operator, cf. Appendix A)

$$\delta^{1,1} : \operatorname{Hom}(\mathfrak{a}, \mathbf{A}) \cong \mathbf{A} \otimes \mathfrak{a}^* \to \mathfrak{b} \otimes \Lambda^2(\mathfrak{a}^*)$$
$$\delta^{1,1}(F)(A_1, A_2) := \frac{1}{2}\left(F(A_1)(A_2) - F(A_2)(A_1)\right),$$

for $F \in \operatorname{Hom}(\mathfrak{a}, \mathbf{A})$, and $A_1, A_2 \in \mathfrak{a}$.

The **h-th prolongation** of $\mathbf{A}$ is defined inductively by setting

$$\mathbf{A}^{(h)} = \mathbf{A}^{(h-1)^{(1)}},$$

for $h \geq 1$ (by convention $\mathbf{A}^{(0)} = \mathbf{A}$ and $\mathbf{A}^{(-1)} = \mathfrak{b}$). $\mathbf{A}^{(h)}$ identifies with

$$\mathbf{A}^{(h)} = \left(\mathbf{A} \otimes S^h(\mathfrak{a}^*)\right) \cap \left(\mathfrak{b} \otimes S^{h+1}(\mathfrak{a}^*)\right).$$

An element $Q_{(h)} \in \mathfrak{b} \otimes S^{h+1}(\mathfrak{a}^*)$ belongs to $\mathbf{A}^{(h)}$ if and only if $i(X)Q_{(h)} \in \mathbf{A}^{(h-1)}$, for all $X \in \mathfrak{a}$.

THEOREM 2.1. $\dim \mathbf{A}^{(1)} \leq s_1 + 2s_2 + \cdots + ns_n$.

DEFINITION 2.1. $\mathbf{A}$ *is said* ***involutive*** *(or* ***in involution****) if equality holds in the previous inequality.*

THEOREM 2.2. *For any tableau* $\mathbf{A}$ *there exists an integer* h_0 *such that* $\mathbf{A}^{(h)}$ *is involutive, for all* $h \geq h_0$.

THEOREM 2.3. *If* $\mathbf{A}$ *is involutive, then* $\mathbf{A}^{(1)}$ *is involutive and*

$$s_j^{(1)} := s_j(\mathbf{A}^{(1)}) = s_n(\mathbf{A}) + \cdots + s_j(\mathbf{A}),$$

$j = 1, \ldots, n$.

Thus *every prolongation of an involutive tableau is involutive.* Moreover, *the principal character and the Cartan integer are invariant under prolongation of an involutive tableau.*

It is well-known that $\mathbf{A}$ is involutive if and only if

$$H^{q,p}(\mathbf{A}) = (0), \quad \text{for all} \quad q \geq 1, p \geq 0$$

(Guillemin-Sternberg, Serre [20]). See Appendix A for the definition of the Spencer groups $H^{q,p}(\mathbf{A})$.

A weaker notion is the following.

DEFINITION 2.2. *A tableau* $\mathbf{A}$ *is said* ***2-acyclic*** *if*

$$H^{q,2}(\mathbf{A}) = (0), \quad \textit{for all} \quad q \geq 1.$$

This notion plays an essential role in the prolongation procedure of a non-involutive linear Pfaffian system (cf. Kuranishi, Goldschmidt [18, 19]).

As shown in the following examples, tableaux and their prolongations arise naturally in the context of PDE systems and of exterior differential systems.

EXAMPLE 1. Let V and W be vector spaces with coordinates $x^1, \dots, x^n$ and $y^1, \dots, y^s$ dual to bases $v_1, \dots, v_n$ for V and $w_1, \dots, w_s$ for W. Consider the first-order constant coefficient system of PDEs for maps $f : V \to W$ given in coordinates by

$$B_a^{\lambda i} \frac{\partial y^a}{\partial x^i}(x) = 0 \quad (\lambda = 1, \dots, r). \tag{2.1}$$

The linear solutions $y^a(x) = A^a_j x^j$ to this system give rise to a tableau $\mathbf{A} \subset \mathrm{Hom}(V, W)$. Conversely, any tableau $\mathbf{A} \subset \mathrm{Hom}(V, W)$ determines a PDE system of this type.

REMARK 2.1. $\mathbf{A}^{(q)}$ is the set of homogeneous polynomial solutions of degree $q + 1$ to (2.1).

THEOREM 2.4. *The PDE system associated to* $\mathbf{A}$ *is involutive* $\Longleftrightarrow$ $\mathbf{A}$ *is involutive.*

The *symbol* of (2.1) is the annihilator $\mathbf{B} := \mathbf{A}^{\perp} \subset V \otimes W^*$ of $\mathbf{A}$.

EXAMPLE 2. Let $(\mathcal{I}, \omega)$ be a Pfaffian differential system (PDS) on a manifold M with independence condition $\omega \neq 0$, where

$$\mathcal{I} = \{\theta^1, \dots, \theta^s, d\theta^1, \dots, d\theta^s\} \quad \text{(algebraic ideal)}$$

and $\omega = \omega^1 \wedge \dots \wedge \omega^n$. Let $\pi^1, \dots, \pi^t$ be 1-forms such that

$$\theta^1, \dots, \theta^s; \omega^1, \dots, \omega^n; \pi^1, \dots, \pi^t$$

be a local adapted coframe of M.

The Pfaffian differential system $(\mathcal{I}, \omega)$ is called **linear**[1] if and only if

$$d\theta^a \equiv 0 \mod \{\theta^1, \dots, \theta^s, \omega^1, \dots, \omega^n\} \quad (0 \leq a \leq s).$$

[1] In the literature, other names are also used to indicate linear systems: quasi-linear systems, systems in good form, or systems in normal form.

The meaning of this condition is that the variety $V_n(\mathcal{I},\omega) \subset G_n(TM,\omega)$ of integral elements of $(\mathcal{I},\omega)$ is described by a system of inhomogeneous linear equations (cf. Appendix B). A linear PDS is described locally by

$$\begin{cases} \theta^a = 0 \\ d\theta^a \equiv A^a_{\epsilon i}\pi^\epsilon \wedge \omega^i + \frac{1}{2}c^a_{ij}\omega^i \wedge \omega^j \quad \mod \{\theta^1, \dots, \theta^s\} \\ \omega = \omega^1 \wedge \cdots \wedge \omega^n \neq 0, \end{cases}$$

where $c^a_{ij} = -c^a_{ji}$; $1 \le a \le s$; $1 \le i,j \le n$; $1 \le \epsilon \le t$.

Once we fix independent variables and take a point of M, we can associate a tableau to the Pfaffian system as follows. At $x \in M$, let $V^* =$ span $\{\omega^i\}$ and $\{\frac{\partial}{\partial\omega^i}\}$ be the basis of its dual V. Further, let $W^* =$ span $\{\theta^a\}$ and $\{\frac{\partial}{\partial\theta^a}\}$ be the basis of its dual W. We define a tableau $\mathbf{A} \subset W \otimes V^*$ by

$$\mathbf{A} := \text{span}\,\{A^a_{\epsilon i}\frac{\partial}{\partial\theta^a} \otimes \omega^i \;:\; \epsilon = 1, \dots, t\}.$$

The involutiveness of $(\mathcal{I},\omega)$ at x is equivalent to the involutiveness of $\mathbf{A}$ (algebraic condition) together with the integrability condition $V(\mathcal{I},\omega)_{|x} \neq \emptyset$, which in turn is equivalent to the condition $c^a_{ij}(x) = 0$, for each a,i,j ("torsion vanishes at x"). See Appendix B for more on the notion of torsion.

THEOREM 2.5. *The linear PDS $(\mathcal{I},\omega)$ is involutive at x $\iff$ 1) $\mathbf{A}$ is involutive and 2) $V_n(\mathcal{I},\omega)_{|x} \neq \emptyset$.*

The *symbol* of $(\mathcal{I},\omega)$ is the annihilator $\mathbf{B} := \mathbf{A}^\perp \subset V \otimes W^*$ of $\mathbf{A}$:

$$B = \text{span}\,\{B^\lambda = B^{\lambda i}_a \frac{\partial}{\partial\omega^i} \otimes \theta^a : B^{\lambda i}_a A^a_{\epsilon i} = 0, \forall\lambda, \epsilon\}.$$

3. Tableaux over Lie algebras. Let $(\mathfrak{g}, [\,,])$ be a finite dimensional Lie algebra, $\mathfrak{a}, \mathfrak{b}$ vector subspaces of $\mathfrak{g}$ such that $\mathfrak{a} \oplus \mathfrak{b} = \mathfrak{g}$, and $\mathbf{A} \subset \text{Hom}(\mathfrak{a}, \mathfrak{b})$ a tableau. Define the polynomial map $\tau : \mathbf{A} \to \mathfrak{b} \otimes \Lambda^2(\mathfrak{a}^*)$ by

$$\tau(Q)(A_1, A_2) := [A_1 + Q(A_1), A_2 + Q(A_2)]_{\mathfrak{b}} \\ - Q\left([A_1 + Q(A_1), A_2 + Q(A_2)]_{\mathfrak{a}}\right),$$

where $X_{\mathfrak{a}}$ (resp. $X_{\mathfrak{b}}$) denotes the $\mathfrak{a}$ (resp. $\mathfrak{b}$) component of X.

DEFINITION 3.1 ([33]). *A **tableau over** $\mathfrak{g}$ is a tableau $\mathbf{A} \subset \text{Hom}(\mathfrak{a}, \mathfrak{b})$ such that:*

1. *$\mathbf{A}$ is involutive;*
2. *$\tau(Q) \in Im\,\delta^{1,1} \subset \mathfrak{b} \otimes \Lambda^2(\mathfrak{a}^*)$, for each $Q \in \mathbf{A}$.*

REMARK 3.1. A detailed analysis of the examples at our disposal and considerations about the problem of prolongation (cf. Remark 4.1) suggest that *condition* (1) in the above definition *should be replaced by the condition that* $\mathbf{A}$ *is 2-acyclic.* As for condition (2) in the definition, it amounts to the vanishing of a cohomology class in the group

$$H^{0,2}(\mathbf{A}) = \frac{\mathfrak{b} \otimes \Lambda^2(\mathfrak{a}^*)}{\delta^{1,1}(\mathbf{A} \otimes \mathfrak{a}^*)}$$

(cf. Remark 4.1 and Appendix B for the the notion of torsion of a linear PDS).

EXAMPLE 3. If $\mathbf{A} \subset \mathrm{Hom}(\mathfrak{a}, \mathfrak{b})$ is involutive (or 2-acyclic) and $\mathfrak{g}$ is the abelian Lie algebra $\mathfrak{g} = \mathfrak{a} \oplus \mathfrak{b}$, then $\tau(Q) = 0$, for each $Q \in \mathbf{A}$, and $\mathbf{A}$ can be considered as a tableau over $\mathfrak{g}$. Therefore, the concept of tableau over a Lie algebra is a natural (non-commutative) generalization of the classical notion of involutive (or 2-acyclic) tableau.

EXAMPLE 4. Let $\mathfrak{g}$ be a semisimple Lie algebra with Killing form $\langle\,,\rangle$. Let $\mathfrak{g} = \mathfrak{g}_0 \oplus \mathfrak{g}_1$ be a Cartan decomposition. Then

$$[\mathfrak{g}_0, \mathfrak{g}_0] \subset \mathfrak{g}_0, \quad [\mathfrak{g}_0, \mathfrak{g}_1] \subset \mathfrak{g}_1, \quad [\mathfrak{g}_1, \mathfrak{g}_1] \subset \mathfrak{g}_0.$$

Assume that $\mathrm{rank}\, \mathfrak{g}/\mathfrak{g}_0 = k$ and that $\mathfrak{a}$ be a maximal (k-dimensional) abelian subspace of $\mathfrak{g}_1$. Then $\mathfrak{g}_1 = \mathfrak{a} \oplus \mathfrak{m}$, where

$$\mathfrak{m} = \mathfrak{a}^{\perp} \cap \mathfrak{g}_1$$

Further, let

$$\begin{aligned} (\mathfrak{g}_0)_{\mathfrak{a}} &= \{X \in \mathfrak{g}_0 : [X, \mathfrak{a}] = 0\}, \\ (\mathfrak{g}_0)_{\mathfrak{a}}^{\perp} &= \{X \in \mathfrak{g} : \langle X, Y \rangle = 0, \forall\, Y \in (\mathfrak{g}_0)_{\mathfrak{a}}\}, \\ \mathfrak{b} &:= \mathfrak{g}_0 \cap (\mathfrak{g}_0)_{\mathfrak{a}}^{\perp}. \end{aligned}$$

Then, for any regular element $A \in \mathfrak{a}$, the maps

$$\mathrm{ad}_A : \mathfrak{m} \to \mathfrak{b}, \quad \mathrm{ad}_A : \mathfrak{b} \to \mathfrak{m}$$

are vector space isomorphisms and

$$X \in \mathfrak{m} \mapsto -\mathrm{ad}_X \in \mathrm{Hom}(\mathfrak{a}, \mathfrak{b})$$

is injective, hence $\mathfrak{m}$ *can be identified with a linear subspace of* $\mathrm{Hom}(\mathfrak{a}, \mathfrak{b})$.

PROPOSITION 3.1 ([33]). *If* $\mathfrak{g}$ *is a semisimple Lie algebra and* $\mathfrak{a}$, $\mathfrak{b}$, *and* $\mathfrak{m}$ *are defined as above, then* $\mathfrak{m}$, *regarded as a subspace of* $\mathrm{Hom}(\mathfrak{a}, \mathfrak{b})$, *is a tableau over* $\mathfrak{g}$.

DEFINITION 3.2. *The tableau* $\mathfrak{m}$ *is called a* ***Cartan tableau*** *over* $\mathfrak{g}$.

REMARK 3.2. As already indicated in the introduction, the idea of a tableau over a Lie algebra has its origin in the method of moving frames and is related to the existence of canonical adapted frames along generic submanifolds in homogeneous spaces. The tableaux corresponding to systems of submanifold geometry are constructed on the Lie algebras of the transitive groups of transformations of the ambient spaces, e.g., the Wilczynski–Cartan frame in projective differential geometry (cf. [33]), or the canonical Möbius frame in conformal theory of surfaces (cf. Section 5.3).

4. Differential systems associated with tableaux over Lie algebras. Let $\mathbf{A} \subset \mathrm{Hom}(\mathfrak{a}, \mathfrak{b})$ be a tableau over $\mathfrak{g}$ and let G be a connected Lie group with Lie algebra $\mathfrak{g}$. We set $Y := G \times \mathbf{A}$ and refer to it as the *configuration space.*

DEFINITION 4.1. *A basis $(A_1, \ldots, A_k, B_1, \ldots, B_h, C_1, \ldots, C_s)$ of $\mathfrak{g}$ is said adapted to $\mathbf{A}$ if*

1. $\mathfrak{a} = span\{A_1, \ldots, A_k\}$,
2. $\mathrm{Im}\,\mathbf{A} := \sum_{Q\in\mathbf{A}} \mathrm{Im}\,Q = span\{B_1, \ldots, B_h\}$,
3. $\mathfrak{b} = span\{B_1, \ldots, B_h, C_1, \ldots, C_s\}$.

An adapted basis is generic if the flag

$$(0) \subset span\{A_1\} \subset \cdots \subset span\{A_1, \ldots, A_k\} = \mathfrak{a}$$

is generic with respect to $\mathbf{A}$.

For a generic adapted basis, let

$$(\alpha^1, \ldots, \alpha^k, \beta^1, \ldots, \beta^h, \gamma^1 \ldots, \gamma^s)$$

denote the dual coframe on G. Given a basis

$$Q_\epsilon = Q^j_{\epsilon i} B_j \otimes \alpha^i \quad (\epsilon = 1, \ldots m)$$

of the tableau $\mathbf{A}$, Y identifies with $G \times \mathbb{R}^m$ by

$$(g, p^\epsilon Q_\epsilon) \in Y \mapsto (g; p^1, \ldots, p^m) \in G \times \mathbb{R}^m.$$

DEFINITION 4.2 ([33]). *The EDS* ***associated with*** $\mathbf{A}$ *is the Pfaffian system $(\mathcal{I}, \omega)$ on Y generated (as a differential ideal) by the linearly independent 1-forms*

$$\begin{cases} \eta^j := \beta^j - p^\epsilon Q^j_{\epsilon i}\alpha^i, & (j = 1, \ldots, h), \\ \gamma^1, \ldots, \gamma^s, & \end{cases}$$

with independent condition $\omega = \alpha^1 \wedge \cdots \wedge \alpha^k \neq 0$.

An immersed submanifold

$$\Phi = (g; p^1, \ldots, p^m) : N^k \to G \times \mathbf{A} \cong G \times \mathbb{R}^m.$$

is an **integral manifold** of $(\mathcal{I}, \omega)$ if and only if

1. $(\alpha^1 \wedge \cdots \wedge \alpha^k)_{|N} \neq 0$;
2. $\beta^j = p^\epsilon Q^j_{\epsilon i}\alpha^i$, $j = 1, \ldots, h$;
3. $\gamma^1 = \cdots = \gamma^s = 0$.

The main result in the construction is the following.

THEOREM 4.1 ([33]). *Let $\mathbf{A}$ be a tableau over a Lie algebra $\mathfrak{g}$. Then, the EDS $(\mathcal{I}, \omega)$ associated with $\mathbf{A}$ is a linear PDS in involution. In particular, the characters of $\mathbf{A}$ coincide with the Cartan characters of $(\mathcal{I}, \omega)$.*

REMARK 4.1. Condition (2) in the definition of a tableau over a Lie algebra (cf. Definition 3.1) tells us that the PDS associated with a tableau over a Lie algebra is linear and with **vanishing torsion** (cf. Appendix B). This together with the condition that the tableau is 2-acyclic guarantee the existence of a prolongation tower for the associated PDS which can be constructed algebraically from the tableau and its prolongations. The construction of the prolonged systems is a direct consequence of the property of the tableau being 2-acyclic and is entirely based on the Spencer cohomology of the tableau. The vanishing of the torsion is needed only at the first step of the construction (cf. Remark B.2). Therefore, the result stated in Theorem 4.1 can be generalized to the following.

THEOREM 4.2 ([35]). *Let* $\mathbf{A}$ *be a 2-acyclic tableau over a Lie algebra* $\mathfrak{g}$. *Then, the PDS* $(\mathcal{I}, \omega)$ *on* Y *associated with* $\mathbf{A}$ *admits regular prolongations of any order. Moreover, the construction of prolongations is purely algebraic. The configuration space of the h-prolonged system* $(\mathcal{I}^{(h)}, \omega)$ *is*

$$Y^{(h)} := G \times (\mathbf{A} \oplus \mathbf{A}^{(1)} \oplus \cdots \oplus \mathbf{A}^{(h)}).$$

If k *is the least integer such that* $\mathbf{A}^{(k)}$ *is involutive, then the k-prolongation* $(\mathcal{I}^{(k)}, \omega)$ *is in involution and its Cartan characters coincide with that of* $\mathbf{A}^{(k)}$.

5. Examples. In this section, we illustrate the construction developed in Section 4 by discussing some examples.

5.1. The PDS associated with an abelian tableau. Let $\mathbf{A} \subset \mathrm{Hom}(\mathbb{R}^k, \mathbb{R}^h)$ be an m-dimensional involutive tableau over the (abelian) Lie algebra $\mathfrak{g} = \mathbb{R}^k \oplus \mathbb{R}^h$, spanned by the linearly independent $h \times k$ matrices $Q_\epsilon = (Q^j_{\epsilon i})$.

We call $\mathbf{A}^{(1)}$**-system** the linear, homogeneous, constant coefficient PDE system for maps $P = (p^1, \ldots, p^m) : \mathbb{R}^k \to \mathbf{A} \cong \mathbb{R}^m$ defined by the differential inclusion $dP_{|x} \in \mathbf{A}^{(1)}$, for all $x \in \mathbb{R}^k$, where $\mathbf{A}^{(1)}$ is the first prolongation of $\mathbf{A}$. This system can be written

$$\delta^{1,1}(dP) = 0,$$

where $\delta^{1,1}$ is the Spencer coboundary of the tableau $\mathbf{A}$ (recall that $\delta^{1,1} : C^{1,1} = \mathbf{A} \otimes (\mathbb{R}^k)^* \to C^{0,2}$).

LEMMA 5.1. *A map* $P : \mathbb{R}^k \to \mathbf{A} \cong \mathbb{R}^m$ *is a solution to the* $\mathbf{A}^{(1)}$-*system if and only if the* $\mathbb{R}^h$*-valued 1-form*

$$\theta = (\theta^1, \ldots, \theta^h) \in \Omega^1(\mathbb{R}^k) \otimes \mathbb{R}^h, \quad \theta^j = p^\epsilon Q^j_{\epsilon a} dx^a,$$

is closed.

As a consequence, we have

COROLLARY 5.1. *Let* $P : \mathbb{R}^k \to \mathbf{A} \cong \mathbb{R}^m$ *be a solution to the* $\mathbf{A}^{(1)}$-*system and let* $y = (y^1, \ldots, y^h)$ *be a primitive of* θ *(i.e.,* $\theta = dy$*). Then*

$$\mathbb{R}^k \ni x \mapsto (x, y(x), P(x)) \in \mathbb{R}^k \oplus \mathbb{R}^h \oplus \mathbb{R}^m$$

is an integral manifold of the PDS $(\mathcal{I},\omega)$ associated with $\mathbf{A}$. *Moreover, every integral manifold of $(\mathcal{I},\omega)$ arises in this way.*

We can conclude that the integral manifolds of $(\mathcal{I},\omega)$ correspond to the solutions of the $\mathbf{A}^{(1)}$-system. Moreover, $(\mathcal{I},\omega)$ is in involution (as a differential system) and the Cartan characters coincide with those of the tableau $\mathbf{A}$.

5.2. The PDS associated with a Cartan tableau and the G/G_0-system. Let G/G_0 be a semisimple symmetric space of rank k and $\mathfrak{g} = \mathfrak{g}_0 \oplus \mathfrak{g}_1$ a Cartan decomposition of $\mathfrak{g}$. Let $(A_1,\dots,A_k)$ be a regular basis for the maximal abelian subalgebra $\mathfrak{a} \subset \mathfrak{g}_1$. According to Terng [37], the G/G_0**-system** (or the k-dimensional system associated to G/G_0) is the system of PDEs for maps $F : U \subset \mathfrak{a} \to \mathfrak{m} := \mathfrak{g}_1 \cap \mathfrak{a}^\perp$ defined by

$$\left[A_i, \frac{\partial F}{\partial x^j}\right] - \left[A_j, \frac{\partial F}{\partial x^i}\right] = [[A_i, F], [A_j, F]],$$

$1 \le i < j \le k$, where x^i are the coordinates with respect to $(A_1,\dots,A_k)$.

LEMMA 5.2. *A map $F : \mathfrak{a} \to \mathfrak{m}$ is a solution of the G/G_0-system if and only if the $\mathfrak{g}$-valued 1-form*

$$\theta = \alpha + [\alpha, F] \in \Omega^1(\mathfrak{a}) \otimes \mathfrak{g},$$

satisfies the Maurer–Cartan equation $d\theta + \frac{1}{2}[\theta \wedge \theta] = 0$, where $\alpha = \alpha^i \otimes A_i$ is the tautological 1-form on $\mathfrak{a}$.

COROLLARY 5.2. *Let $F : \mathfrak{a} \to \mathfrak{m}$ be a solution of the G/G_0-system and let $g : \mathfrak{a} \to G$ be a primitive of θ (i.e. a solution to $g^{-1}dg = \theta$). Then*

$$\mathfrak{a} \ni x \mapsto (g(x), F(x)) \in G \times \mathfrak{m}$$

is an integral manifold of the PDS $(\mathcal{I},\omega)$ on $Y = G \times \mathfrak{m}$ associated with the Cartan tableau $\mathfrak{m} \subset \mathrm{Hom}(\mathfrak{a},\mathfrak{b})$. Conversely, any integral manifold of $(\mathcal{I},\omega)$ arises in this way.

In conclusion, the integral manifolds of the PDS $(\mathcal{I},\omega)$ associated with the Cartan tableau $\mathfrak{m} \subset \mathrm{Hom}(\mathfrak{a},\mathfrak{b})$ are given by the solutions of the corresponding G/G_0-system. Moreover, $(\mathcal{I},\omega)$ is in involution and its Cartan characters coincide with those of the tableau $\mathfrak{m}$ (i.e., $s_1 = n$, $s_j = 0$, $j = 2,\dots,n$). In particular, the general solution depends on n functions in one variable.

REMARK 5.1. If $(A_1,\dots,A_n)$ is a basis of $\mathfrak{a}$, $(x^1,\dots,x^n)$ the corresponding coordinates, and $F = (F^1,\dots,F^{m-n}) : U \subset \mathfrak{a} \to \mathfrak{b}$, then the G/G_0-system can be written

$$B^a_{\alpha,i}\frac{\partial F^\alpha}{\partial x^j} - B^a_{\alpha,j}\frac{\partial F^\alpha}{\partial x^i} = \Phi^a_{ij},$$

where the coefficients $B^a_{\alpha i}$ are constant and Φ_{ij} are analytic functions.

In general, the PDS associated to a tableau over a Lie algebra corresponds to the nonlinear system of equations

$$B^a_{\alpha,i}\frac{\partial F^\alpha}{\partial x^j} - B^a_{\alpha,j}\frac{\partial F^\alpha}{\partial x^i} + B^a_{\alpha,\beta}\left(\frac{\partial F^\alpha}{\partial x^i}\frac{\partial F^\beta}{\partial x^j} - \frac{\partial F^\alpha}{\partial x^j}\frac{\partial F^\beta}{\partial x^i}\right) = \Phi^a_{ij}.$$

5.3. Old and new involutive systems in conformal surface theory. In this section we discuss some old and new involutive systems/tableaux arising in conformal geometry of surfaces. We start by recalling some preliminary material. Consider Minkowski 5-space $\mathbb{R}^{4,1}$ with linear coordinates $x^0, \dots, x^4$ and Lorentz scalar product given by the quadratic form

$$\langle x, x\rangle = -x^0x^4 + (x^1)^2 + (x^2)^2 + (x^3)^2 = \eta_{ij}x^ix^j. \tag{5.1}$$

Classically, the Möbius space S^3 (conformal 3-sphere) is realized as the projective quadric $\{[x] \in \mathbb{RP}^4 : \langle x, x\rangle = 0\}$. Accordingly, S^3 inherits a natural conformal structure and the identity component $G \cong \mathrm{SO}_0(4,1)$ of the pseudo-orthogonal group of (5.1) acts transitively on S^3 as group of orientation-preserving, conformal transformations (see [5]). The Maurer–Cartan form of G will be denoted by $\omega = (\omega^i_j)$.

Let $f : U \subset \mathbb{R}^2 \to S^3$ be an umbilic free, conformal immersion. A *Möbius frame field* along f is a map $g = (g_0, \dots, g_4) : U \to G$ such that $f(p) = [g_0(p)]$, for all $p \in U$. According to [5], there exists a canonical Möbius frame field[2] $g : U \to G$ along f such that its Maurer–Cartan form $\beta = (\beta^i_j) = g^*\omega$ takes the form

$$\begin{pmatrix} -2q_2\beta^1_0 + 2q_1\beta^2_0 & p_1\beta^1_0 + p_2\beta^2_0 & -p_2\beta^1_0 + p_3\beta^2_0 & 0 & 0 \\ \beta^1_0 & 0 & -q_1\beta^1_0 - q_2\beta^2_0 & -\beta^1_0 & p_1\beta^1_0 + p_2\beta^2_0 \\ \beta^2_0 & q_1\beta^1_0 + q_2\beta^2_0 & 0 & \beta^2_0 & -p_2\beta^1_0 + p_3\beta^2_0 \\ 0 & \beta^1_0 & -\beta^2_0 & 0 & 0 \\ 0 & \beta^1_0 & \beta^2_0 & 0 & 2q_2\beta^1_0 - 2q_1\beta^2_0 \end{pmatrix}$$

with $\beta^1_0 \wedge \beta^2_0 > 0$. The smooth functions q_1, q_2, p_1, p_2, p_3 form a complete system of conformal invariants for f and satisfy the following structure equations

$$d\beta^1_0 = -q_1\beta^1_0 \wedge \beta^2_0, \quad d\beta^2_0 = -q_2\beta^1_0 \wedge \beta^2_0, \tag{5.2}$$
$$dq_1 \wedge \beta^1_0 + dq_2 \wedge \beta^2_0 = (1 + p_1 + p_3 + {q_1}^2 + {q_2}^2)\beta^1_0 \wedge \beta^2_0, \tag{5.3}$$
$$dq_2 \wedge \beta^1_0 - dq_1 \wedge \beta^2_0 = -p_2\beta^1_0 \wedge \beta^2_0, \tag{5.4}$$
$$dp_1 \wedge \beta^1_0 + dp_2 \wedge \beta^2_0 = (4q_2p_2 + q_1(3p_1 + p_3))\beta^1_0 \wedge \beta^2_0, \tag{5.5}$$
$$dp_2 \wedge \beta^1_0 - dp_3 \wedge \beta^2_0 = (4q_1p_2 - q_2(p_1 + 3p_3))\beta^1_0 \wedge \beta^2_0. \tag{5.6}$$

[2] If g is a canonical frame, any other canonical frame is given by $(g_0, -g_1, -g_2, g_3, g_4)$.

5.3.1. The Möbius tableau. The existence of a canonical Möbius frame field suggests the following construction. Let $(\alpha^1, \alpha^2, \beta^1, \ldots, \beta^4, \gamma^1, \ldots, \gamma^4)$ be the basis of $\mathfrak{g}^*$ defined by

$$\begin{cases} \alpha^1 = \omega_0^1, & \alpha^2 = \omega_0^2, \quad \beta^1 = \omega_0^0, \quad \beta^2 = \omega_1^0, \quad \beta^3 = \omega_2^0, \quad \beta^4 = \omega_1^2, \\ \gamma^1 = \omega_3^0, & \gamma^2 = \omega_0^3, \quad \gamma^3 = \omega_0^1 - \omega_1^3, \quad \gamma^4 = \omega_0^2 + \omega_2^3. \end{cases}$$

Next, let $(A_1, A_2, B_1, \ldots, B_4, C_1, \ldots, C_4)$ be its dual basis and set

$$\mathfrak{a} = \text{span}\,\{A_1, A_2\}, \quad \mathfrak{b} = \text{span}\,\{B_1, \ldots, B_4\}.$$

Consider the 5-dimensional subspace $\mathbf{M} \subset \text{Hom}(\mathfrak{a}, \mathfrak{b})$ consisting of all elements $Q(q,p)$ of the form

$$\begin{aligned} Q(q,p) = q_1(B_4 \otimes \alpha^1 + 2B_1 \otimes \alpha^2) + q_2(-2B_1 \otimes \alpha^1 + B_4 \otimes \alpha^2) \\ + p_1 B_2 \otimes \alpha^1 + p_2(-B_3 \otimes \alpha^1 + B_2 \otimes \alpha^2) + p_3 B_3 \otimes \alpha^2, \end{aligned}$$

where $q = (q_1, q_2) \in \mathbb{R}^2$, $p = (p_1 . p_2, p_3) \in \mathbb{R}^3$. A direct computation yields the following.

LEMMA 5.3. *The subspace* $\mathbf{M}$ *is a tableau over* $\mathfrak{g} \cong \mathfrak{so}(4,1)$.

The PDS associated with the tableau $\mathbf{M}$, referred to as the Möbius system, is the PDS on $Y = G \times \mathbf{M} \cong G \times \mathbb{R}^5$ generated by the 1-forms $\gamma^1, \ldots, \gamma^4, \eta^1, \ldots, \eta^4$, where

$$\begin{cases} \eta^1 = \beta^1 + 2q_2\omega_0^1 - 2q_1\omega_0^2, & \eta^2 = \beta^2 - p_1\omega_0^1 - p_2\omega_0^2, \\ \eta^3 = \beta^3 + p_2\omega_0^1 - p_3\omega_0^2, & \eta^4 = \beta^4 - q_1\omega_0^1 - q_2\omega_0^2, \end{cases}$$

with independence condition $\omega_0^1 \wedge \omega_0^2 \neq 0$. The integral manifolds of the Möbius system are the 2-dimensional submanifolds

$$(g; q, p) : M^2 \to G \times \mathbf{M} \cong G \times \mathbb{R}^5$$

such that:

- $f = [g_0] \to S^3$ is an umbilic free, conformal immersion;
- $g : M^2 \to G$ is a canonical Möbius frame along f;
- $q_1, q_2, p_1, p_2, p_3 : M^2 \to \mathbb{R}$ are the conformal invariants of f.

5.3.2. Willmore surfaces. *Willmore immersions* are defined as extremals for the Willmore functional $\int (H^2 - K)dA$ (H the mean curvature, K the Gauss curvature). They are characterized by the Euler–Lagrange equation

$$\Delta H + 2H(H^2 - K) = 0,$$

which expressed in terms of the conformal invariants is equivalent to the equation $p_1 = p_3$ (cf. [5, 30, 39]). Willmore surfaces can be seen as integral manifolds of the Möbius system restricted to the submanifold of Y given by

$$Y_W = \{Q(q,p) \in Y \,|\, p_1 = p_3\}.$$

Now, it is easy to check that the subspace

$$\mathbf{M}_W := \{Q(q,p) \in \mathbf{M} \,|\, p_1 = p_3\}$$

defines a 4-dimensional tableau over $\mathfrak{g} \cong \mathfrak{so}(4,1)$ with characters $s_0 = 8$, $s_1 = 4$, $s_2 = 0$, and that Y_W is the configuration space of $\mathbf{M}_W$. Observe also that the restriction to Y_W of the Möbius system is exactly the PDS associated with $\mathbf{M}_W$.

5.3.3. Other classes of surfaces. More generally, one could consider the class of surfaces whose invariant functions p_1 and p_3 satisfy a linear relation, that is, are expressed by $p_1(t) = t\cos a + b_1$, $p_2(t) = t\sin a + b_2$, for real constants a, b_1, b_2. This class includes Willmore surfaces as special examples and corresponds to the 4-dimensional affine tableau

$$\mathbf{M}_{(a,b_1,b_2)} = \{Q(q, p_1(t), p_2, p_3(t)) \,|\, t, p_2 \in \mathbb{R}, q \in \mathbb{R}^2\}.$$

Also in this case, an algebraic, direct computation shows that $\mathbf{M}_{(a,b_1,b_2)}$ is involutive. Therefore, by the construction developed in the previous section, the associated PDS is in involution. Its Cartan characters are $s_0 = 8$, $s_1 = 4$, $s_2 = 0$.

REMARK 5.2. Similar arguments have been used in connection with the study of surfaces in projective differential geometry [33]. The same approach can also be used to discuss surface theory in the framework of Laguerre geometry [27] and other classical geometries.

6. Further developments.

6.1. Continue the program, initiated with the study of several classes of integrable surfaces in projective differential geometry [33], of identifying the geometry associated to a given tableau/system, i.e., find submanifolds in some homogeneous space whose integrability conditions are given by the PDS associated with the given tableau.

6.2. Study the algebraic structure of tableaux over Lie algebras to understand when a tableau generates an integrable geometry.

6.3. Study the Cauchy problem for the associated systems (cf. [31, 34]).

6.4. Analyze the characteristic cohomology of a tableau over a Lie algebra, its geometric interpretations, and its relations with the characteristic cohomology of Bryant–Griffiths [8, 9].

APPENDIX

A. The Spencer complex. [cf. [7]] Retaining the notation of Section 2, identify the symmetric product $S^q(\mathfrak{a}^*)$ with the space of homogeneous polynomials of degree q on $\mathfrak{a}$. For each $v \in \mathfrak{a}$, let δ_v be the map of $\mathfrak{b} \otimes S^q(\mathfrak{a}^*) \to \mathfrak{b} \otimes S^{q-1}(\mathfrak{a}^*)$ given by partial differentiation w.r.t. v. Let $v_1, \ldots, v_n$ be a basis of $\mathfrak{a}$, and $v^1, \ldots, v^n$ its dual basis.

The operator

$$\mathfrak{b} \otimes S^q(\mathfrak{a}^*) \otimes \Lambda^p(\mathfrak{a}^*) \xrightarrow{\delta^{q,p}} \mathfrak{b} \otimes S^{q-1}(\mathfrak{a}^*) \otimes \Lambda^{p+1}(\mathfrak{a}^*)$$

given by

$$\delta^{q,p}\xi := \sum \delta_{v_i}\xi \wedge v^i$$

($\delta^{0,p} = 0$, for $p \geq 0$) is independent of the basis, $\delta^2 = 0$, and the sequence of the corresponding bigraded complex is exact except when $q = 0$ and $p = 0$.

Let $\mathbf{A} \subset \mathrm{Hom}(\mathfrak{a}, \mathfrak{b})$ be a tableau with prolongations $\mathbf{A}^{(h)}$, $h \geq 0$. Consider the sequence of spaces

$$C^{q,p}(\mathbf{A}) := \mathbf{A}^{(q-1)} \otimes \Lambda^p(\mathfrak{a}^*),$$

for integers $q \geq 0$ and $0 \leq p \leq n$. Note that since $\mathbf{A}^{(q-1)} \subset \mathfrak{b} \otimes S^q(\mathfrak{a}^*)$, the space $C^{q,p}(\mathbf{A})$ is a subspace of $\mathfrak{b} \otimes S^q(\mathfrak{a}^*) \otimes \Lambda^p(\mathfrak{a}^*)$. We have

$$\delta C^{q,p}(\mathbf{A}) \subset C^{q-1,p+1}(\mathbf{A}),$$

but the sequence

$$C^{q+1,p-1}(\mathbf{A}) \xrightarrow{\delta^{q+1,p-1}} C^{q,p}(\mathbf{A}) \xrightarrow{\delta^{q,p}} C^{q-1,p+1}(\mathbf{A})$$

is no longer exact for all p and q. The associated cohomology groups

$$H^{q,p}(\mathbf{A}) := Z^{q,p}(\mathbf{A})/B^{q,p}(\mathbf{A})$$

are called the *Spencer groups* of $\mathbf{A}$, where $B^{q,p}(\mathbf{A}) = \mathrm{Im}(\delta^{q+1,p-1})$ and $Z^{q,p}(\mathbf{A}) = \mathrm{Ker}\,(\delta^{q,p})$. Notice that $Z^{0,p}(\mathbf{A}) = \mathfrak{b} \otimes \Lambda^p(\mathfrak{a}^*)$, for all $p \geq 0$, and $Z^{q,1}(\mathbf{A}) = \mathbf{A}^{(q)}$, for all $q \geq 1$.

A significant result in the subject is that the vanishing of the $H^{q,p}$ is equivalent to involutiveness.

THEOREM A.1 ([20]). *A tableau $\mathbf{A}$ is involutive if and only if $H^{q,p}(\mathbf{A})$ is zero, for all $q \geq 1$ and $p \geq 0$.*

A weaker condition than involutiveness is the following.

DEFINITION A.1. *A tableau $\mathbf{A}$ is called 2-acyclic if $H^{q,2}(\mathbf{A}) = (0)$, for all $q \geq 1$.*

Another way of formulating the condition

$$H^{q,p}(\mathbf{A}) = (0), \quad \text{for all} \quad q \geq 1, p \geq 0$$

is that the sequences

$$0 \to \mathbf{A}^{(k)} \xrightarrow{\delta} \mathbf{A}^{(k-1)} \otimes \mathfrak{a}^* \to \cdots \xrightarrow{\delta} \mathbf{A} \otimes \Lambda^{k-1}(\mathfrak{a}^*) \to$$

$$\cdots \xrightarrow{\delta} \mathfrak{b} \otimes \Lambda^k(\mathfrak{a}^*) \to \frac{\mathfrak{b} \otimes \Lambda^k(\mathfrak{a}^*)}{\delta\left(\mathbf{A} \otimes \Lambda^{k-1}(\mathfrak{a}^*)\right)} \to 0$$

are exact for all k. In particular, we have

$$H^{0,k}(\mathbf{A}) = \frac{\operatorname{Ker}(\delta^{0,k})}{\operatorname{Im}(\delta^{1,k-1})} = \frac{\mathfrak{b} \otimes \Lambda^k(\mathfrak{a}^*)}{\delta\left(\mathbf{A} \otimes \Lambda^{k-1}(\mathfrak{a}^*)\right)}.$$

B. Torsion of a Pfaffian systems. [cf. [7]] Retaining the notation of Example 2, let $(\mathcal{I}, \omega)$ be a Pfaffian system. An admissible integral element $E \in V_n(\mathcal{I}, \omega)_{|x}$ is given by

$$\theta^a = 0, \quad \pi^\epsilon = p_i^\epsilon \omega^i,$$

where the fiber coordinates p_i^ϵ satisfy

$$A^a_{\epsilon\, j}(x) p_i^\epsilon - A^a_{\epsilon\, i}(x) p_j^\epsilon + c^a_{i\, j}(x) = 0.$$

Under a change of coframe

$$\tilde{\theta}^a = \theta^a, \quad \tilde{\omega}^i = \omega^i, \quad \tilde{\pi}^\epsilon = \pi^\epsilon - p_i^\epsilon \omega^i \tag{B.1}$$

the numbers $c^a_{i\, j}(x)$ transform to

$$\tilde{c}^a_{ij}(x) = A^a_{\epsilon\, j}(x) p_i^\epsilon - A^a_{\epsilon\, i}(x) p_j^\epsilon + c^a_{i\, j}(x).$$

This defines an equivalence relation

$$\tilde{c}^a_{ij}(x) \sim c^a_{ij}(x).$$

DEFINITION B.1. *The equivalence class $[c^a_{ij}(x)]$ is called the **torsion** of $(\mathcal{I}, \omega)$.*

LEMMA B.1. *The torsion of $(\mathcal{I}, \omega)$ lives in*

$$H^{0,2}(\mathbf{A}) = \frac{W \otimes \Lambda^2(V^*)}{\delta^{1,1}(\mathbf{A} \otimes V^*)} = \frac{\operatorname{Ker}(\delta^{0,2})}{\operatorname{Im}(\delta^{1,1})}.$$

Proof. If $Q = p_j^\epsilon A^a_{\epsilon\, i} \frac{\partial}{\partial \theta^a} \otimes \omega^i \otimes \omega^j \in \mathbf{A} \otimes V^*$, then

$$\delta^{1,1}(Q) = \sum_{i<j} \left(p_j^\epsilon A^a_{\epsilon\, i} - p_i^\epsilon A^a_{\epsilon\, j}\right) \frac{\partial}{\partial \theta^a} \otimes \omega^i \wedge \omega^j.$$

According to the transformation rule (B.1) of the c^a_{ij} under a coframe change, the cocycle

$$\frac{1}{2}c^a_{i\,j}(x)\,\frac{\partial}{\partial\theta^a}\otimes\omega^i\wedge\omega^j\in C^{0,2}(\mathbf{A})$$

gives a class in $H^{0,2}(\mathbf{A})$. □

REMARK B.1. The vanishing of the torsion is a necessary and sufficient condition for the existence of an integral element over x.

LEMMA B.2. *The torsion of* $(\mathcal{I}^{(1)},\omega)$ *lives in the vector spaces*

$$H^{0,2}(\mathbf{A}^{(1)})\cong H^{1,2}(\mathbf{A}).$$

We also recall the following.

LEMMA B.3.

$$H^{q,p}(\mathbf{A}^{(1)})\cong H^{q+1,p}(\mathbf{A}),\quad q\geq 1.$$

REMARK B.2. Thus, the involutiveness of the tableau A associated to $(\mathcal{I},\omega)$ implies both the involutiveness of the tableau $A^{(1)}$ and the vanishing of torsion of the prolonged system $(\mathcal{I}^{(1)},\omega)$.

REFERENCES

[1] M.A. AKIVIS AND V.V. GOLDBERG, *Projective differential geometry of submanifolds*, North-Holland Mathematical Library, **49**, North-Holland Publishing Co., Amsterdam, 1993.

[2] G. BOL, *Projektive Differentialgeometrie, 2. Teil*, Studia mathematica, B. IX, Vandenhoeck & Ruprecht, Göttingen, 1954.

[3] W. BLASCHKE, *Vorlesungen über Differentialgeometrie und geometrische Grundlagen von Einsteins Relativitätstheorie*, B. **3**, bearbeitet von G. Thomsen, J. Springer, Berlin, 1929.

[4] M. BRÜCK, X. DU, J. PARK, AND C.-L. TERNG, Submanifold geometry of real Grassmannian systems, The Memoirs, Vol. 155, AMS, **735** (2002), 1–95.

[5] R. BRYANT, A duality theorem for Willmore surfaces, *J. Differential Geom.* **20** (1984), no. 1, 23–53.

[6] R.L. BRYANT, Lectures given at MSRI "Integrable system seminar", 2003, unpublished notes.

[7] R.L. BRYANT, S.-S. CHERN, R.B. GARDNER, H.L. GOLDSCHMIDT, AND P.A. GRIFFITHS, *Exterior differential systems*, Mathematical Sciences Research Institute Publications, 18, Springer-Verlag, New York, 1991.

[8] R.L. BRYANT AND P.A. GRIFFITHS, Characteristic cohomology of differential systems, I: General theory, *J. Amer. Math. Soc.* **8** (1995), 507–596.

[9] ———, Characteristic cohomology of differential systems, II: Conservations laws for a class of parabolic equations, *Duke Math. J.* **78** (1995), 531–676.

[10] É. CARTAN, *Sur le problème général de la déformation*, C. R. Congrés Strasbourg (1920), 397–406; or *Oeuvres Complètes*, III 1, 539–548.

[11] É. CARTAN, Sur la déformation projective des surfaces, *Ann. Scient. Éc. Norm. Sup. (3)* **37** (1920), 259–356; or *Oeuvres Complètes*, III 1, 441–538.

[12] É. CARTAN, *Les systèmes différentielles extérieurs et leurs applications géométriques*, Hermann, Paris, 1945.
[13] E. FERAPONTOV, Lie sphere geometry and integrable systems, *Tohoku Math. J.* **52** (2000), 199–233.
[14] ———, Integrable systems in projective differential geometry, *Kyushu J. Math.* **54** (2000), 183–215.
[15] ———, The analogue of Wilczynski's projective frame in Lie sphere geometry: Lie-applicable surfaces and commuting Schrödinger operators with magnetic fields, *Internat. J. Math.* **13** (2002), 956–986.
[16] D. FERUS AND F. PEDIT, Curved flats in symmetric spaces, *Manuscripta Math.* **91** (1996), 445–454.
[17] S.P. FINIKOV, *Projective Differential Geometry*, Moscow, Leningrad, 1937.
[18] H. GOLDSCHMIDT, Existence theorems for analytic linear partial differential equations, *Ann. of Math. (2)* **86** (1967), 246–270.
[19] H. GOLDSCHMIDT, Integrability criteria for systems of nonlinear partial differential equations, *J. Differential Geom.* **1** (1967), 269–307.
[20] V.W. GUILLEMIN AND S. STERNBERG, An algebraic model of transitive differential geometry, *Bull. Amer. Math. Soc.* **70** (1964), 16–47.
[21] T.A. IVEY AND J. M. LANDSBERG, *Cartan for beginners: differential geometry via moving frames and exterior differential systems*, Graduate Studies in Mathematics, **61**, American Mathematical Society, Providence, RI, 2003.
[22] G.R. JENSEN, Deformation of submanifolds of homogeneous spaces, *J. Differential Geom.* **16** (1981), 213–246.
[23] ———, Higher order contact of submanifolds of homogeneous spaces. Lecture Notes in Mathematics, Vol. **610**. Springer-Verlag, Berlin-New York, 1977.
[24] M. KURANISHI, On E. Cartan's prolongation theorem of exterior differential systems, *Amer. J. Math.* **79** (1957), 1–47.
[25] E. MUSSO, Deformazione di superfici nello spazio di Möbius, *Rend. Istit. Mat. Univ. Trieste* **27** (1995), 25–45.
[26] E. MUSSO AND L. NICOLODI, *On the equation defining isothermic surfaces in Laguerre geometry*, New developments in Differential geometry, Budapest 1996, Kluver Academic Publishers, 285–294.
[27] ———A variational problem for surfaces in Laguerre geometry, *Trans. Amer. Math. Soc.* **348** (1996), 4321–4337.
[28] ———, Isothermal surfaces in Laguerre geometry, *Boll. Un. Mat. Ital. (7) II-B, Suppl. fasc. 2,* **11** (1997), 125–144.
[29] ———, Willmore canal surfaces in Euclidean space, *Rend. Istit. Mat. Univ. Trieste* **31** (1999), 1–26.
[30] ———, Darboux transforms of Dupin surfaces, *Banach Center Publ.* **57** (2002), 135–154.
[31] ———, On the Cauchy problem for the integrable system of Lie minimal surfaces, *J. Math. Phys.* **46** (2005), no. 11, 3509-3523.
[32] ———, Deformation and applicability of surfaces in Lie sphere geometry, *Tohoku Math. J.* **58** (2006), no. 2, 161-187; preprint available as `math.DG/0408009`.
[33] ———, Tableaux over Lie algebras, integrable systems, and classical surface theory, *Comm. Anal. Geom.* **14** (2006), no. 3, 475-496; preprint available as `math.DG/0412169`.
[34] ———, A class of overdetermined systems defined by tableaux: Involutiveness and Cauchy problem, Phys. D (2007), to appear; preprint available as `math.DG/0602676`.
[35] ———, in preparation.
[36] P.J. OLVER, *Equivalence, invariants, and symmetry*, Cambridge University Press, Cambridge, 1995.
[37] C.-L. TERNG, Soliton equations and differential geometry, *J. Differential Geometry* **45** (1997), 407–445.

[38] C.-L. Terng and E. Wang, Curved Flats, exterior differential systems and conservation laws, in: Complex, Contact and Symmetric Manifolds (in honor of L. Vanhecke), O. Kowalski, E. Musso, and D. Perrone (Eds.), Progress in Mathematics, Vol. **234**, Birkhäuser, 2005, 235–254.

[39] T.J. Willmore, *Riemannian geometry*, Oxford Science Publications, The Clarendon Press, Oxford University Press, New York, 1993.

CONFORMAL STRUCTURES WITH EXPLICIT AMBIENT METRICS AND CONFORMAL G_2 HOLONOMY

PAWEL NUROWSKI*

Abstract. Given a generic 2-plane field on a 5-dimensional manifold we consider its (3, 2)-signature conformal metric $[g]$ as defined in [7]. Every conformal class $[g]$ obtained in this way has very special conformal holonomy: it must be contained in the split-real-form of the exceptional group G_2. In this note we show that for special 2-plane fields on 5-manifolds the conformal classes $[g]$ have the Fefferman-Graham ambient metrics which, contrary to the general Fefferman-Graham metrics given as a formal power series [2], can be written in an explicit form. We propose to study the relations between the conformal G_2-holonomy of metrics $[g]$ and the possible pseudo-Riemannian G_2-holonomy of the corresponding ambient metrics.

1. The (3, 2)-signature conformal metrics. Consider an equation

$$z' = F(x, y, y', y'', z) \qquad \text{with} \qquad F_{y''y''} \neq 0, \tag{1.1}$$

for two real functions $y = y(x)$, $z = z(x)$ of one real variable x. To simplify notation introduce new symbols $p = y'$ and $q = y''$. Equation (1.1) is totally encoded in the system of three 1-forms:

$$\begin{aligned} \omega^1 &= \mathrm{d}z - F(x, y, p, q, z)\mathrm{d}x \\ \omega^2 &= \mathrm{d}y - p\mathrm{d}x \\ \omega^3 &= \mathrm{d}p - q\mathrm{d}x, \end{aligned} \tag{1.2}$$

living on a 5-dimensional manifold J parametrized by (x, y, p, q, z). In particular, every solution to (1.1) is a curve $\gamma(t) = (x(t), y(t), p(t), q(t), z(t)) \subset J$ on which all the forms $\omega^1, \omega^2, \omega^3$ identically vanish.

We introduce an equivalence relation between equations (1.1) which identifies the equations having the same set of solutions. This leads to the following definition:

DEFINITION 1.1. *Two equations $z' = F(x, y, y', y'', z)$ and $\bar{z}' = \bar{F}(\bar{x}, \bar{y}, \bar{y}', \bar{y}'', \bar{z})$, defined on spaces J and $\bar{J}$ parametrized, respectively, by $(x, y, p = y', q = y'', z)$ and $(\bar{x}, \bar{y}, \bar{p} = \bar{y}', \bar{q} = \bar{y}'', \bar{z})$, are said to be* (locally) equivalent, *if and only if there exists a (local) diffeomorphism $\phi : J \to \bar{J}$ transforming the corresponding forms*

$$\begin{aligned} \omega^1 &= \mathrm{d}z - F(x, y, p, q, z)\mathrm{d}x \\ \omega^2 &= \mathrm{d}y - p\mathrm{d}x \\ \omega^3 &= \mathrm{d}p - q\mathrm{d}x \end{aligned} \qquad \textit{and} \qquad \begin{aligned} \bar{\omega}^1 &= \mathrm{d}\bar{z} - \bar{F}(\bar{x}, \bar{y}, \bar{p}, \bar{q}, \bar{z})\mathrm{d}\bar{x} \\ \bar{\omega}^2 &= \mathrm{d}\bar{y} - \bar{p}\mathrm{d}\bar{x} \\ \bar{\omega}^3 &= \mathrm{d}\bar{p} - \bar{q}\mathrm{d}\bar{x} \end{aligned}$$

*Instytut Fizyki Teoretycznej, Uniwersytet Warszawski, ul. Hoża 69, Warszawa, POLAND (nurowski@fuw.edu.pl). This work was supported in part by the Polish Ministerstwo Nauki i Informatyzacji grant nr: 1 P03B 07529 and the US Institute for Mathematics and Its Applications in Minneapolis.

via:

$$\begin{aligned}\phi^*(\bar{\omega}^1) &= \alpha\omega^1 + \beta\omega^2 + \gamma\omega^3\\ \phi^*(\bar{\omega}^2) &= \delta\omega^1 + \epsilon\omega^2 + \lambda\omega^3,\\ \phi^*(\bar{\omega}^3) &= \kappa\omega^1 + \mu\omega^2 + \nu\omega^3\end{aligned}$$

with functions $\alpha, \beta, \gamma, \delta, \epsilon, \lambda, \kappa, \nu$ on J such that

$$\det\begin{pmatrix}\alpha & \beta & \gamma\\ \delta & \epsilon & \lambda\\ \kappa & \mu & \nu\end{pmatrix} \neq 0.$$

It follows that equation (1.1) considered modulo equivalence relation of Definition 1.1 uniquely defines a conformal class of (3, 2)-signature metrics $[g_F]$ on the space J. In coordinates (x, y, p, q, z) this class may be described as follows. Let

$$D = \partial_x + p\partial_y + q\partial_p + F\partial_z$$

be a total differential associated with equation (1.1) on J. Then a representative g_F of the conformal class $[g_F]$ may be written as

$$\begin{aligned}g_F = \Big[\ & DF^2_{qq}F^2_{qq} + 6DF_qDF_{qqq}F^2_{qq} - 6DF_{qqq}F_pF^2_{qq}\\
&-3DDF_{qq}F^3_{qq} + 9DF_{qp}F^3_{qq} - 9F_{pp}F^3_{qq}\\
&+9DF_{qz}F_qF^3_{qq} - 18F_{pz}F_qF^3_{qq} + 3DF_zF^4_{qq}\\
&-6DF_qF^2_{qq}F_{qqp} + 6F_pF^2_{qq}F_{qqp} - 8DF_qDF_{qq}F_{qq}F_{qqq}\\
&+8DF_{qq}F_pF_{qq}F_{qqq} + 3DDF_qF^2_{qq}F_{qqq} - 3DF_pF^2_{qq}F_{qqq}\\
&-3DF_zF_qF^2_{qq}F_{qqq} + 4(DF_q)^2F^2_{qqq} - 8DF_qF_pF^2_{qqq}\\
&-3(DF_q)^2F_{qq}F_{qqqq} + 4F^2_pF^2_{qqq} + 6DF_qF_pF_{qq}F_{qqqq}\\
&-3F^2_pF_{qq}F_{qqqq} - 6DF_qF_qF^2_{qq}F_{qqz} + 6F_pF_qF^2_{qq}F_{qqz}\\
&-3DF_qF^3_{qq}F_{qz} + 12F_pF^3_{qq}F_{qz} + 3F^2_{qq}F_{qqq}F_y\\
&-6DF_{qqq}F_qF^2_{qq}F_z + 4DF_{qq}F^3_{qq}F_z + 6F_qF^2_{qq}F_{qqp}F_z\\
&+8DF_{qq}F_qF_{qq}F_{qqq}F_z - 4DF_qF^2_{qq}F_{qqq}F_z\\
&-9F_{qp}F^3_{qq}F_z + F_pF^2_{qq}F_{qqq}F_z - 8DF_qF_qF^2_{qqq}F_z\\
&+8F_pF_qF^2_{qqq}F_z + 6DF_qF_qF_{qq}F_{qqqq}F_z - 6F_pF_qF_{qq}F_{qqqq}F_z \qquad (1.3)\\
&+18F^3_{qq}F_{qy} + 6F^2_qF^2_{qq}F_{qqz}F_z + 3F_qF^3_{qq}F_{qz}F_z\\
&-2F^4_{qq}F^2_z + F_qF^2_{qq}F_{qqq}F^2_z + 4F^2_qF^2_{qqq}F^2_z\\
&-3F^2_qF_{qq}F_{qqqq}F^2_z - 9F^2_qF^3_{qq}F_{zz}\ \Big]\ (\tilde{\omega}^1)^2\\
&+\Big[\ 6DF_{qqq}F^2_{qq} - 6F^2_{qq}F_{qqp} - 8DF_{qq}F_{qq}F_{qqq}\\
&+8DF_qF^2_{qqq} - 8F_pF^2_{qqq} - 6DF_qF_{qq}F_{qqqq}\end{aligned}$$

$$+6F_pF_{qq}F_{qqqq} - 6F_qF_{qq}^2F_{qqz} + 6F_{qq}^3F_{qz}$$
$$+2F_{qq}^2F_{qqq}F_z - 8F_qF_{qqq}^2F_z + 6F_qF_{qq}F_{qqqq}F_z \Big]\, \tilde\omega^1\tilde\omega^2$$
$$+\Big[\, 10DF_{qq}F_{qq}^3 - 10DF_qF_{qq}^2F_{qqq} + 10F_pF_{qq}^2F_{qqq}$$
$$-10F_{qq}^4F_z + 10F_qF_{qq}^2F_{qqq}F_z \,\Big]\, \tilde\omega^1\tilde\omega^3$$
$$+30F_{qq}^4\, \tilde\omega^1\tilde\omega^4 + \Big[\, 30DF_qF_{qq}^3 - 30F_pF_{qq}^3 - 30F_qF_{qq}^3F_z \,\Big]\, \tilde\omega^1\tilde\omega^5$$
$$+\Big[\, 4F_{qqq}^2 - 3F_{qq}F_{qqqq} \,\Big]\, (\tilde\omega^2)^2 - 10F_{qq}^2F_{qqq}\, \tilde\omega^2\tilde\omega^3$$
$$+30F_{qq}^3\, \tilde\omega^2\tilde\omega^5 - 20F_{qq}^4\, (\tilde\omega^3)^2$$

where[1]

$$\begin{aligned}
\tilde\omega^1 &= \mathrm{d}y - p\mathrm{d}x \\
\tilde\omega^2 &= \mathrm{d}z - F\mathrm{d}x - F_q(\mathrm{d}p - q\mathrm{d}x) \\
\tilde\omega^3 &= \mathrm{d}p - q\mathrm{d}x \\
\tilde\omega^4 &= \mathrm{d}q \\
\tilde\omega^5 &= \mathrm{d}x.
\end{aligned} \tag{1.4}$$

It follows from the construction described in Ref. [7] that when the equation (1.1) undergoes a diffeomorphism ϕ of Definition 1.1, the above metric g_F transforms conformally.

The conformal class of metrics $[g_F]$ is very special among all the $(3,2)$-signature conformal metrics in dimension 5: the Cartan normal conformal connection for this class, instead of having values in full $\mathfrak{so}(4,3)$ Lie algebra, has values in its certain 14-dimensional subalgebra. This subalgebra turns out to be isomorphic to the split real form of the exceptional Lie algebra $\mathfrak{g}_2 \subset \mathfrak{so}(4,3)$. Thus, conformal metrics $[g_F]$ provide an abundance of examples of metrics with an *exceptional* conformal *holonomy.* This holonomy is always a subgroup of the noncompact form of the exceptional Lie group G_2. We strongly believe that randomly chosen function F, such that $F_{qq} \neq 0$, give rise to conformal metrics $[g_F]$ with conformal holonomy *equal* to G_2.

It is interesting to study the conformal classes $[g_F]$ from the point of view of the Fefferman-Graham ambient metric construction [2]. Since for each F defining equation (1.1) we have a metric g_F in dimension five, then since five is *odd*, Fefferman-Graham guarantees [2] that there is a *unique* formal power series of a *Ricci-flat metric* of signature $(4,3)$ corresponding to g_F. Moreover, since given F the metric g_F is explicitly determined by

[1] Note that formula for g_F differs from the one given in Ref. [7] by tilde signs over the all omegas. In Ref. [7], when copying the calculated metric g_F, by mistake, we forgot to put these tilde signs over the omegas. Hence, in Ref. [7], formula for g_F is true, provided that one puts the tilde signs over the omegas and supplements it by the definitions (1.4) of the tilded omegas.

formula (1.3), we see that starting with *real analytic* F, the metric g_F is *real analytic*. Thus, every analytic F of (1.1) leads to analytic g_F and then, in turn, via Fefferman-Graham, leads to a unique *real analytic* ambient metric $\tilde{g}_F$ of signature $(4,3)$. Since both the Levi-Civita connection for $\tilde{g}_F$ and the Cartan normal conformal connection for the corresponding 5-dimensional metric g_F have values in (possibly subalgebras of) the same Lie algebra $\mathfrak{so}(4,3)$, it is interesting to ask about the relations between them. We discuss these relations on examples.

2. The strategy for constructing explicit examples of ambient metrics. We start with the Fefferman-Graham result [2] adapted to the 5-dimensional situation of conformal metrics $[g_F]$.

Let g_F be a representative of the conformal class $[g_F]$ defined on J by (1.3). Consider a manifold $J \times \mathbb{R}_+ \times \mathbb{R}$. Introduce coordinates $(0 < t, u)$ on $\mathbb{R}_+ \times \mathbb{R}$ in $J \times \mathbb{R}_+ \times \mathbb{R}$. We have a natural projection $\pi : J \times \mathbb{R}_+ \times \mathbb{R} \to J$, which enables us to pullback forms from J to $J \times \mathbb{R}_+ \times \mathbb{R}$. Omitting the pullback sign in the expressions like $\pi^*(g_F)$ we define a formal power series

$$\breve{g}_F = -2\mathrm{d}t\mathrm{d}u + t^2 g_F - ut\alpha + u^2\beta + u^3 t^{-1}\gamma + \sum_{k=4}^{\infty} u^k t^{2-k}\mu_k. \tag{2.1}$$

Here $\alpha, \beta, \gamma, \mu_k$, $k = 4, 5, 6, \ldots.$, are pullbacks of symmetric bilinear forms $\alpha, \beta, \gamma, \mu_k$ from J to $J \times \mathbb{R}_+ \times \mathbb{R}$. Thus $\breve{g}_F$ is a formal *bilinear form* on $J \times \mathbb{R}_+ \times \mathbb{R}$. This formal bilinear form has signature $(4,3)$ in some neighbourhood of $u = 0$. The following theorem is due to Fefferman and Graham [2].

THEOREM 2.1. *Among all the bilinear forms $\breve{g}_F$ which, via (2.1), are associated with metric g_F of (1.3) there is* precisely one, *say $\tilde{g}_F$, satisfying the Ricci flatness condition $Ric(\tilde{g}_F) = 0$.*

Given g_F, all the bilinear forms $\alpha, \beta, \gamma, \mu_k$ in $\tilde{g}_F$ are totally determined. Another issue is to calculate them explicitly. For example, it is quite difficult to find the general formulas for the higher order forms μ_k. Nevertheless the explicit expressions for the forms α, β, γ are known [4, 5]. We write them below in the form obtained by C R Graham. We define the coefficients α_{ij}, β_{ij} and γ_{ij} by $\alpha = \alpha_{ij}\mathrm{d}x^i\mathrm{d}x^j$, $\beta = \beta_{ij}\mathrm{d}x^i\mathrm{d}x^j$, $\gamma = \gamma_{ij}\mathrm{d}x^i\mathrm{d}x^j$, where $(x^i) = (x, y, p, q, z)$ are coordinates on J. Then Graham's expressions for α_{ij}, β_{ij} and γ_{ij} are [4]:

$$\begin{aligned}
\alpha_{ij} &= 2\mathsf{P}_{ij},\\
\beta_{ij} &= -B_{ij} + \mathsf{P}_i{}^k\mathsf{P}_{jk},\\
3\gamma_{ij} &= B_{ij;k}{}^k - 2W_{kijl}B^{kl} + 4\mathsf{P}_{k(i}B_{j)}{}^k - 4\mathsf{P}_k{}^k B_{ij} + 4\mathsf{P}^{kl}C_{(ij)k;l}\\
&\quad -2C^k{}_i{}^l C_{ljk} + C_i{}^{kl}C_{jkl} + 2\mathsf{P}^k{}_{k;l}C_{(ij)}{}^l - 2W_{kijl}\mathsf{P}^k{}_m\mathsf{P}^{ml},
\end{aligned} \tag{2.2}$$

where

$$\mathsf{P}_{ij} = \frac{1}{3}(R_{ij} - \frac{1}{8}Rg_{Fij}),$$

is the Schouten tensor for the metric $g_F = g_{Fij}\mathrm{d}x^i\mathrm{d}x^j$,

$$W_{ijkl} = R_{ijkl} - 2(\mathsf{P}_{i[k}g_{Fl]j} - \mathsf{P}_{j[k}g_{Fl]i})$$

is its Weyl tensor,

$$C_{ijk} = \mathsf{P}_{ij;k} - \mathsf{P}_{ik;j}$$

is the Cotton tensor, and

$$B_{ij} = C_{ijk;}{}^{k} - \mathsf{P}^{kl}W_{kijl}$$

is the Bach tensor.

Of course all the above quantities can be explicitly calculated once F, and in turn the metric g_F, is chosen.

In the rest of the paper we will chose particular functions $F = F(x,y,p,q,z)$, and we will calculate the corresponding forms α,β,γ for them. We will give examples of F's for which the bilinear form γ is identically vanishing,

$$\gamma \equiv 0. \tag{2.3}$$

Given such F's we will consider

$$\bar{g}_F = -2\mathrm{d}t\mathrm{d}u + t^2 g_F - ut\alpha + u^2\beta.$$

Note that $\bar{g}_F$ coincides with the ambient metric $\tilde{g}_F$ up to the terms *quadratic* in the ambient coordinates t,u. If by *chance* the bilinear form $\bar{g}_F$ satisfies the Ricci flatness condition

$$Ric(\bar{g}_F) \equiv 0,$$

then by the *uniqueness* of the ambient metric $\tilde{g}_F$ stated in Theorem 2.1, it will *coincide* with the ambient metric $\tilde{g}_F$:

$$\bar{g}_F \equiv \tilde{g}_F.$$

The uniqueness result of Theorem 2.1, together with the Ricci flatness of $\bar{g}_F$, is powerful enough to guarantee that not only the coefficient γ in the ambient metric $\tilde{g}_F$ identically vanishes, but that *all* the coefficients μ_k, $k = 4,5,6,....$, vanish too!

Thus the strategy of finding explicit ambient metrics $\tilde{g}_F$ for g_F is as follows:

- find $F = F(x,y,p,q,z)$ for which the corresponding metric g_F has identically vanishing form γ of (2.2);
- calculate the approximate ambient metric $\bar{g}_F$ for such F;
- check if the Ricci tensor $Ric(\bar{g}_F)$ of $\bar{g}_F$ is identically vanishing;
- if you have F with the above properties then the approximate metric $\bar{g}_F$ is the ambient metric $\tilde{g}_F$ for g_F.

3. Conformally Einstein example. As the first example, following Ref. [7], we calculate g_F and its approximate ambient metric $\tilde{g}_F$ for a very simple equation:

$$z' = F(y''), \qquad \text{with} \qquad F_{y''y''} \neq 0.$$

It was shown in Ref. [7] that the conformal class $[g_F]$ may be represented by[2]

$$\begin{aligned}
-15(F'')^{10/3} g_F = \\
& 30(F'')^4 \Big[\mathrm{d}q\mathrm{d}y - p\mathrm{d}q\mathrm{d}x \Big] + \Big[4F^{(3)2} - 3F''F^{(4)} \Big] \mathrm{d}z^2 \\
& +2 \Big[-5(F'')^2F^{(3)} - 4F'F^{(3)2} + 3F'F''F^{(4)} \Big] \mathrm{d}p\mathrm{d}z \\
& +2 \Big[15(F'')^3 + 5q(F'')^2F^{(3)} - 4FF^{(3)2} + 4qF'F^{(3)2} + 3FF''F^{(4)} \\
& -3qF'F''F^{(4)} \Big] \mathrm{d}x\mathrm{d}z \\
& + \Big[-20(F'')^4 + 10F'(F'')^2F^{(3)} + 4(F')^2F^{(3)2} - 3(F')^2F''F^{(4)} \Big] \mathrm{d}p^2 \\
& +2 \Big[-15F'(F'')^3 + 20q(F'')^4 + 5F(F'')^2F^{(3)} - 10qF'(F'')^2F^{(3)} \\
& +4FF'F^{(3)2} - 4q(F')^2F^{(3)2} - 3FF'F''F^{(4)} + 3q(F')^2F''F^{(4)} \Big] \mathrm{d}p\mathrm{d}x \\
& + \Big[-30F(F'')^3 + 30qF'(F'')^3 - 20q^2(F'')^4 \\
& -10qF(F'')^2F^{(3)} + 10q^2F'(F'')^2F^{(3)} + 4F^2F^{(3)2} \\
& -8qFF'F^{(3)2} + 4q^2(F')^2F^{(3)2} - 3F^2F''F^{(4)} \\
& +6qFF'F''F^{(4)} - 3q^2(F')^2F''F^{(4)} \Big] \mathrm{d}x^2.
\end{aligned} \tag{3.1}$$

As noted in Ref. [7] this metric is conformal to a Ricci flat metric $\hat{g}_F = \mathrm{e}^{2\Upsilon(q)} g_F$ with a conformal scale $\Upsilon = \Upsilon(q)$ satisfying second order ODE:

$$90F''^2(\Upsilon'' - \Upsilon'^2) - 60F''F^{(3)}\Upsilon' + 3F''F^{(4)} - 4F^{(3)2} = 0.$$

Thus, since for each $F = F(q)$ the conformal class $[g_F]$ contains a Ricci flat metric, its conformal holonomy must be a proper subgroup of the noncompact form of G_2. An interesting feature of this conformal class is that it is very special among all the conformal classes associated with equation (1.1). Not only has g_F very special conformal holonomy, making it very similar to the Lorentzian 4-dimensional Brinkmann metrics; moreover, since its Weyl tensor has essentially only one nonvanishing component (see Ref. [7] for details) it is *not* weakly generic (see Ref. [3] for definition). This makes

[2] The metric presented here differs from this of [7] by a convenient conformal factor equal to $-15(F'')^{10/3}$.

$[g_F]$ analogous to the Lorentzian type N metrics in 4-dimensions, such as for example, Fefferman metrics.

Having g_F of (3.1) we used the symbolic computer calculation program Mathematica to calculate its associated form γ of (2.2). We checked that this form *identically vanishes.* We further used Mathematica to calculate the corresponding approximate ambient metric $\bar{g}_F$. On doing that we observed that, surprisingly, the bilinear form β is also *identically vanishing.* The explicit formula for the approximate ambient metric is given below:

$$\bar{g}_F = t^2 g_F - 2\,\mathrm{d}t\mathrm{d}u \;-\; 2tuF''^{4/3}P\mathrm{d}q^2, \tag{3.2}$$

with

$$P = \frac{4F^{(3)2} - 3F''F^{(4)}}{90(F'')^{10/3}},$$

and g_F given by (3.1). The metric $\bar{g}_F$ is defined locally on $J \times \mathbb{R}_+ \times \mathbb{R}$ with coordinates (x, y, p, q, z, t, u). It obviously has signature $(4, 3)$. We also checked, again using Mathematica, that $Ric(\bar{g}_F) \equiv 0$. Thus, we fulfilled the strategy outlined in Section 2. This enables us to conclude that $\bar{g}_F$ of (3.2) coincides with the ambient metric $\tilde{g}_F$ for g_F. To give expressions for the Cartan normal conformal connection for g_F and the Levi-Civita connection for $\tilde{g}_F = \bar{g}_F$ we first introduce a nonholonomic coframe $(\theta^1, \theta^2, \theta^3, \theta^4, \theta^5)$ on J given by

$$\begin{aligned}
\theta^1 &= \mathrm{d}y - p\mathrm{d}x\\
\theta^2 &= \mathrm{d}z - F\mathrm{d}x - F'(\mathrm{d}p - q\mathrm{d}x)\\
\theta^3 &= -\frac{2}{\sqrt{3}}(F'')^{1/3}(\mathrm{d}p - q\mathrm{d}x)\\
30(F'')^{10/3}\theta^4 &= \left(3F'F''F^{(4)} - 4F'F^{(3)2} - 10(F'')^2F^{(3)}\right)(\mathrm{d}p - q\mathrm{d}x)\\
&\quad + \left(4F^{(3)2} - 3F''F^{(4)}\right)(\mathrm{d}z - F\mathrm{d}x) + 30(F'')^3\mathrm{d}x\\
\theta^5 &= -(F'')^{2/3}\mathrm{d}q.
\end{aligned}$$

In this coframe the metric g_F is simply:

$$g_F = 2\theta^1\theta^5 - 2\theta^2\theta^4 + (\theta^3)^2.$$

By means of the canonical projection

$$\pi(x, y, p, q, z, t, u) = (x, y, p, q, z)$$

the coframe $(\theta^1, \theta^2, \theta^3, \theta^4, \theta^5)$ can be pulled back to five linearly independent forms $(\theta^1, \theta^2, \theta^3, \theta^4, \theta^5)$ on $J \times \mathbb{R}_+ \times \mathbb{R}$. They can be supplemented by

$$\theta^0 = \mathrm{d}t \qquad \text{and} \qquad \theta^6 = \mathrm{d}u$$

to form a coframe $(\theta^0, \theta^1, \theta^2, \theta^3, \theta^4, \theta^5, \theta^6)$ on the ambient space $J \times \mathbb{R}_+ \times \mathbb{R}$.

The Cartan normal conformal connection, when written on J in the coframe $(\theta^1,\theta^2,\theta^3,\theta^4,\theta^5)$ reads:

$$\omega_{G_2}=\begin{pmatrix} 0 & 0 & 0 & 0 & 0 & -P\theta^5 & 0\\ \theta^1 & 0 & Q\theta^2+\frac{9}{2\sqrt{3}}P\theta^3 & \frac{1}{\sqrt{3}}\theta^4 & -\frac{1}{2\sqrt{3}}\theta^3 & 0 & -P\theta^5\\ \theta^2 & 0 & 0 & \frac{1}{\sqrt{3}}\theta^5 & 0 & -\frac{1}{2\sqrt{3}}\theta^3 & 0\\ \theta^3 & 0 & -2\sqrt{3}P\theta^5 & 0 & \frac{1}{\sqrt{3}}\theta^5 & -\frac{1}{\sqrt{3}}\theta^4 & 0\\ \theta^4 & 0 & 0 & -2\sqrt{3}P\theta^5 & 0 & Q\theta^2+\frac{9}{2\sqrt{3}}P\theta^3 & 0\\ \theta^5 & 0 & 0 & 0 & 0 & 0 & 0\\ 0 & \theta^5 & -\theta^4 & \theta^3 & -\theta^2 & \theta^1 & 0 \end{pmatrix}.$$

Here:

$$Q=\frac{40F^{(3)3}-45F''F^{(3)}F^{(4)}+9F''^2F^{(5)}}{90F''^5}.$$

Now we use coframe $(\theta^0,\theta^1,\theta^2,\theta^3,\theta^4,\theta^5,\theta^6)$ to write down the Levi-Civita connection for $\tilde{g}_F$. We have

$$\tilde{g}_F=g_{ij}\theta^i\theta^j,$$

with the indices range: $i,j=0,1,2,...6$, and the matrix g_{ij} given by

$$g_{ij}=\begin{pmatrix} 0 & 0 & 0 & 0 & 0 & 0 & -1\\ 0 & 0 & 0 & 0 & 0 & t^2 & 0\\ 0 & 0 & 0 & 0 & -t^2 & 0 & 0\\ 0 & 0 & 0 & t^2 & 0 & 0 & 0\\ 0 & 0 & -t^2 & 0 & 0 & 0 & 0\\ 0 & t^2 & 0 & 0 & 0 & -2tuP & 0\\ -1 & 0 & 0 & 0 & 0 & 0 & 0 \end{pmatrix}.$$

The Levi-Civita connection for $\tilde{g}_F$ on $J\times\mathbb{R}_+\times\mathbb{R}$, when written in the coframe $(\theta^0,\theta^1,\theta^2,\theta^3,\theta^4,\theta^5,\theta^6)$ reads:

$$\omega_{LC} =$$
$$\begin{pmatrix} 0 & 0 & 0 & 0 & 0 & -tP\theta^5 & 0 \\ \frac{1}{t}\theta^1 + \frac{u}{t^2}P\theta^5 & \frac{1}{t}\theta^0 & Q\theta^2 + \frac{9}{2\sqrt{3}}P\theta^3 & \frac{1}{\sqrt{3}}\theta^4 & -\frac{1}{2\sqrt{3}}\theta^3 & \frac{u}{t^2}P\theta^0 - \frac{u}{3t}Q\theta^5 - \frac{1}{t}P\theta^6 & -\frac{1}{t}P\theta^5 \\ \frac{1}{t}\theta^2 & 0 & \frac{1}{t}\theta^0 & \frac{1}{\sqrt{3}}\theta^5 & 0 & -\frac{1}{2\sqrt{3}}\theta^3 & 0 \\ \frac{1}{t}\theta^3 & 0 & -2\sqrt{3}P\theta^5 & \frac{1}{t}\theta^0 & \frac{1}{\sqrt{3}}\theta^5 & -\frac{1}{\sqrt{3}}\theta^4 & 0 \\ \frac{1}{t}\theta^4 & 0 & 0 & -2\sqrt{3}P\theta^5 & \frac{1}{t}\theta^0 & Q\theta^2 + \frac{9}{2\sqrt{3}}P\theta^3 & 0 \\ \frac{1}{t}\theta^5 & 0 & 0 & 0 & 0 & \frac{1}{t}\theta^0 & 0 \\ 0 & t\theta^5 & -t\theta^4 & t\theta^3 & -t\theta^2 & t\theta^1 - uP\theta^5 & 0 \end{pmatrix}.$$

Note that on $\Sigma = \{(x, y, p, q, z, t, u) : u = 0,\ t = 1\}$ we trivially have $\theta^0 \equiv 0 \equiv \theta^6$. Thus, restricting the formula for ω_{LC} to Σ, we see that $\omega_{G_2} \equiv \omega_{LC|\Sigma}$. Off this set the two connections: ω_{LC} and the pullbacked-by-π-connection ω_{G_2}, differ significantly. To see this it is enough to observe that contrary to ω_{LC}, the connection $\pi^*(\omega_{G_2})$ has *torsion*. Indeed writing the first Cartan structure equations for the $\pi^*(\omega_{G_2})$ in the coframe $(\theta^0, \theta^1, \theta^2, \theta^3, \theta^4, \theta^5, \theta^6)$ we find that the torsion is:

$$d\theta^i + \pi^*(\omega_{G_2})^i{}_j \wedge \theta^j = \begin{pmatrix} 0 \\ -\theta^0 \wedge \theta^1 - P\theta^5 \wedge \theta^6 \\ -\theta^0 \wedge \theta^2 \\ -\theta^0 \wedge \theta^3 \\ -\theta^0 \wedge \theta^4 \\ -\theta^0 \wedge \theta^5 \\ 0 \end{pmatrix}.$$

The vanishing of this torsion on the initial hypersurface Σ confirms our earlier statement that the two connections ω_{G_2} and ω_{LC} coincide there.

It is interesting to note that the curvature $d\omega_{LC} + \omega_{LC} \wedge \omega_{LC}$ does not depend on t, u and is annihilated by ∂_t and ∂_u. Thus it can be considered to be a 2-form on Σ. As such it is precisely equal to the curvature $d\omega_{G_2} + \omega_{G_2} \wedge \omega_{G_2}$ of the connection ω_{G_2}:

$$d\omega_{G_2} + \omega_{G_2} \wedge \omega_{G_2} = d\omega_{LC} + \omega_{LC} \wedge \omega_{LC} = \begin{pmatrix} 0 & 0 & 0 & 0 & 0 & 0 & 0 \\ 0 & 0 & A_5 & 0 & 0 & 0 & 0 \\ 0 & 0 & 0 & 0 & 0 & 0 & 0 \\ 0 & 0 & 0 & 0 & 0 & 0 & 0 \\ 0 & 0 & 0 & 0 & 0 & A_5 & 0 \\ 0 & 0 & 0 & 0 & 0 & 0 & 0 \\ 0 & 0 & 0 & 0 & 0 & 0 & 0 \end{pmatrix} \theta^2 \wedge \theta^5,$$

where[3]

$$A_5 = \frac{-224F^{(3)4} + 336F''F^{(3)2}F^{(4)} - 51F''^2F^{(4)2} - 80F''^2F^{(3)}F^{(5)} + 10F''^3F^{(6)}}{100F''^{20/3}}.$$

[3]We use the letter A_5 to denote the nonvanishing component of the curvature to be in accordance with [7] and Cartan's paper [1]. Note however that in order to avoid collision of notations between the present and the next sections we use capital A_5 instead of a_5 of paper [7].

4. Non-conformally Einstein example. To get quite different example of $[g_F]$ we consider equation (1.1) in the form:

$$z' = y''^2 + a_6 y'^6 + a_5 y'^5 + a_4 y'^4 + a_3 y'^3 + a_2 y'^2 + a_1 y' + a_0 + bz,$$

where $a_i, i = 0, 1, ..., 6$, and b are real constants. This equation has the defining function

$$F = q^2 + a_6 p^6 + a_5 p^5 + a_4 p^4 + a_3 p^3 + a_2 p^2 + a_1 p + a_0 + bz$$

and, via (1.3), leads to a conformal class $[g_F]$ represented by a metric

$$\begin{aligned} 15(2)^{-2/3} g_F = &\Big[9a_2 + 2b^2 + 27a_3 p + 54a_4 p^2 + 90 a_5 p^3 + 135 a_6 p^4\Big] \mathrm{d}y^2 \\ &+ \Big[15a_0 + 2(b^2 - 3a_2)p^2 - 3a_3 p^3 + 9a_4 p^4 + 30 a_5 p^5 \\ &+ 60 a_6 p^6 - 20bpq + 5q^2 + 15bz\Big] \mathrm{d}x^2 \qquad (4.1) \\ &+ \Big[15a_1 + 4(3a_2 - b^2)p - 9a_3 p^2 - 48 a_4 p^3 - 105 a_5 p^4 \\ &- 180 a_6 p^5 + 20bq\Big] \mathrm{d}x\mathrm{d}y + 20\mathrm{d}p^2 - 10(bp + q)\mathrm{d}p\mathrm{d}x \\ &+ 10b\mathrm{d}p\mathrm{d}y - 30\mathrm{d}q\mathrm{d}y - 15\mathrm{d}x\mathrm{d}z + 30p\mathrm{d}q\mathrm{d}x. \end{aligned}$$

This metric is *not* conformal to an Einstein metric. The quickest way to check this is the calculation of the Cotton, C_{ijk}, and the Weyl, W_{ijkl}, tensors for g_F. Once these tensors are calculated, it is easy to observe that they do not admit a vector field K^i such that $C_{ijk} + K^l W_{lijk} = 0$. As a consequence the metric is *not* a *conformal C-space* metric. This proves our statement since every conformally Einstein metric is necessarily a conformal C-space metric (see e.g. Ref. [3]).

Recall that g_F of (4.1), as a member of the family of metrics (1.3), defines a conformal class $[g_F]$ with *conformal* holonomy H *reduced* to the noncompact group G_2 or to one of its subgroups. But since the metric (4.1) is not conformal to an Einstein metric, we do not have an immediate reason to conclude that $H \neq G_2$. We *conjecture* that $H = G_2$ here and try to prove it in a subsequent paper [6].

It is remarkable that the ambient metric $\tilde{g}_F$ for g_F of (4.1) assumes a very compact form:

$$\begin{aligned} \tilde{g}_F = &t^2 g_F \;\; - 2\,\mathrm{d}t\mathrm{d}u \\ &-2\,tu\,\Big[\frac{1}{20}(-2a_2 + 4b^2 + 3a_3 p + 6a_4 p^2 - 20 a_5 p^3 - 120 a_6 p^4)\mathrm{d}x^2 \\ &-\frac{9}{20}(a_3 - 10a_5 p^2 - 40 a_6 p^3)\mathrm{d}x\mathrm{d}y - \frac{9}{10}(a_4 + 5a_5 p + 15 a_6 p^2)\mathrm{d}y^2\Big] \end{aligned}$$

$$+u^2\left[\frac{3}{20(2)^{2/3}}(a_4-10a_5p+60a_6p^2)\mathrm{d}x^2 + \frac{9}{4(2)^{2/3}}(a_5-12a_6p)\mathrm{d}x\mathrm{d}y+\frac{81}{4(2)^{2/3}}a_6\mathrm{d}y^2\right].$$

This is checked by applying our strategy described in Section 2 to the metric (4.1). As in the previous example, using Mathematica, we calculated the bilinear form γ for (4.1). It turned out to be equal to *zero*, $\gamma\equiv 0$. Then we calculated $\bar{g}_F$, and checked that it is *Ricci flat*. Thus we concluded that $\bar{g}_F$ coincides with the ambient metric for $\tilde{g}_F$. The above given formula for $\tilde{g}_F$ is therefore just $\bar{g}_F$, which we calculated using (2.2).

We find this example as a sort of miracle. A priori there is no reason for g_F to have the ambient metric *truncated* at the *second* order in terms of the ambient parameters t and u. We are intrigued by this fact.

Now, following the general procedure outlined in [7], we introduce a special coframe for g_F given by:

$$\begin{aligned}
\theta^1 &= \mathrm{d}y-p\mathrm{d}x\\
\theta^2 &= \mathrm{d}z-F\mathrm{d}x-2q(\mathrm{d}p-q\mathrm{d}x)\\
\theta^3 &= -\frac{2^{4/3}}{\sqrt{3}}(\mathrm{d}p-q\mathrm{d}x)\\
\theta^4 &= 2^{-1/3}\mathrm{d}x\\
15(2)^{1/3}\theta^5 &= (9a_2+2b^2+27a_3p+54a_4p^2+90a_5p^3+135a_6p^4)(\mathrm{d}y-p\mathrm{d}x)\\
&\quad +10b(\mathrm{d}p-q\mathrm{d}x)-30\mathrm{d}q\\
&\quad +15(a_1+2a_2p+3a_3p^2+4a_4p^3+5a_5p^4+6a_6p^5+2bq)\mathrm{d}x.
\end{aligned}$$

In this coframe the metric g_F is:

$$g_F=2\theta^1\theta^5-2\theta^2\theta^4+(\theta^3)^2.$$

As in the previous section, we use the canonical projection

$$\pi(x,y,p,q,z,t,u)=(x,y,p,q,z)$$

to pullback the coframe $(\theta^1,\theta^2,\theta^3,\theta^4,\theta^5)$ to five linearly independent forms $(\theta^1,\theta^2,\theta^3,\theta^4,\theta^5)$ on $J\times\mathbb{R}_+\times\mathbb{R}$, which are further supplemented by

$$\theta^0=\mathrm{d}t \qquad \text{and} \qquad \theta^6=\mathrm{d}u$$

to form a coframe $(\theta^0,\theta^1,\theta^2,\theta^3,\theta^4,\theta^5,\theta^6)$ on the ambient space $J\times\mathbb{R}_+\times\mathbb{R}$.

It turns out that if $b=0$ the coframes on J and $J\times\mathbb{R}_+\times\mathbb{R}$ defined in this way are suitable to analyze the relations between the Cartan normal conformal connection ω_{G_2} for $[g_F]$ and the Levi-Civita connection ω_{LC} for $\tilde{g}_F$. If $b\neq 0$ the connection ω_{G_2} in the coframe $(\theta^1,\theta^2,\theta^3,\theta^4,\theta^5)$ and the connection ω_{LC} in the coframe $(\theta^0,\theta^1,\theta^2,\theta^3,\theta^4,\theta^5,\theta^6)$ do not coincide on $t=1$, $u=0$. We will not analyze this case here.

Restricting to the

$$b = 0$$

case we find the following:

- the connections ω_{G_2} in the coframe $(\theta^1, \theta^2, \theta^3, \theta^4, \theta^5)$ and the connection ω_{LC} in the coframe $(\theta^0, \theta^1, \theta^2, \theta^3, \theta^4, \theta^5, \theta^6)$ coincide on $t = 1$, $u = 0$.
- the torsion of $\pi^*(\omega_{G_2})$ in the coframe $(\theta^0, \theta^1, \theta^2, \theta^3, \theta^4, \theta^5, \theta^6)$ is nonvanishing off the set $t = 1$, $u = 0$
- unlike the example of the previous section the curvature $d\omega_{LC} + \omega_{LC} \wedge \omega_{LC}$ significantly depends on t and u.
- even on $t = 1$, $u = 0$, the curvature $d\omega_{G_2} + \omega_{G_2} \wedge \omega_{G_2}$ and the restriction of $d\omega_{LC} + \omega_{LC} \wedge \omega_{LC}$ do not coincide.

Acknowledgements. I am very grateful to T.P. Branson, M. Eastwood, W. Miller, Jr., and the organizers of the 2006 IMA Summer Program "Symmetries and Overdetermined Systems of Partial Differential Equations", for inviting me to Minneapolis to participate in this very fruitful event. The topic covered by this note is inspired by the talk of C.R. Graham which I heard in Minneapolis during the program. In particular, I am very obliged to C.R. Graham for sending me the formulas (2.2), which I used to prepare the examples included in this note.

REFERENCES

[1] Cartan E., "Les systemes de Pfaff a cinq variables et les equations aux derivees partielles du seconde ordre" *Ann. Sc. Norm. Sup.* **27**: 109–192 (1910).

[2] Fefferman C. and Graham C.R., "Conformal invariants", in *Elie Cartan et mathematiques d'aujourd'hui*, Asterisque, hors serie (Societe Mathematique de France, Paris), 95–116 (1985).

[3] Gover A.R. and Nurowski P., "Obstructions to conformally Einstein metrics in n dimensions" *Journ. Geom. Phys.* **56**: 450–484 (2006).

[4] Graham C.R., Private communications, unpublished.

[5] de Haro S., Skenderis K., and Solodukhin S.N., "Holographic reconstruction of spacetime and renormalization in the AdS/CFT correspondence", *Comm. Math. Phys.* **217**: 594–622 (2001), hep-th/0002230.

[6] Leistner Th. and Nurowski P., in preparation.

[7] Nurowski P., "Differential equations and conformal structures" *Journ. Geom. Phys.* **55**: 19–49 (2005).

DEFORMATIONS OF QUADRATIC ALGEBRAS, THE JOSEPH IDEAL FOR CLASSICAL LIE ALGEBRAS, AND SPECIAL TENSORS

PETR SOMBERG*

Abstract. The Joseph ideal is a unique ideal in the universal enveloping algebra of a simple Lie algebra attached to the minimal coadjoint orbit. Its construction using deformation theory was described by Braverman and Joseph. The same ideal appeared recently in connection with algebra of symmetries of differential operators. The paper presents a review of the corresponding circle of ideas and it discusses a role of special tensors in these questions.

1. Introduction and motivation. The orbit method in the representation theory of simple Lie algebras allows to relate certain irreducible representations of a given simple Lie algebra $\mathfrak{g}$ to coadjoint orbits in its dual $\mathfrak{g}^\star$. A prominent role is played by the minimal (nilpotent) coadjoint orbit, whose annihilator ideal in the universal enveloping algebra of $\mathfrak{g}$ is the Joseph ideal ([8]). A description of the Joseph ideal using deformation theory was discussed in [3]. The authors used there effectively results on quadratic algebras of Koszul type ([2]).

Recently, the Joseph ideal appeared again in connection with the algebra of symmetry of differential operators ([5, 6]). In particular, a role of special tensors in one-parametric family of deformations of homogeneous quadratic algebras is discussed in [6].

Our present article has several sections. Section 2 contains a geometric motivation. It introduces the algebra of symmetries of the Laplace operator and the family of ideals connected with it. This is a special case of a more general situation (corresponding to the orthogonal Lie algebras). A natural generalization to suitable families of nonhomogenous deformations of homogeneous quadratic ideals in the tensor algebra $T(\mathfrak{g})$ of a general simple Lie algebra $\mathfrak{g}$ are described and studied in Section 3. A role of special tensors, which were constructed in [6], is illustrated in the motivating case of the orthogonal Lie algebra. Section 4 contains a discussion of deformation theory for quadratic Koszul algebras and gives an interpretation of computations presented in Section 3 from the point of view of deformation theory and the Hochschild cohomology. The last Section 5 introduces the Joseph ideal in greater detail and explains its relation to the symmetry algebras of differential operators.

2. Symmetry algebras. Symmetries of a differential operator D play a substantial role in a study of properties of the kernel of D. To be more precise, let D be a differential operator mapping the space of sections

*Karlova Universita (Charles University), Czech Republic (somberg@karlin.mff.cuni.cz).

$\Gamma(E)$ of a vector bundle E over a manifold M to the space of sections $\Gamma(F)$ of a vector bundle F over M. A differential operator $\mathcal{D}$ from $\Gamma(E)$ to $\Gamma(E)$ is a symmetry of D, if $D\mathcal{D} = \delta D$ for a linear differential operator δ from $\Gamma(F)$ to itself. The space of symmetries is an algebra with a product given by composition of symmetries.

It is clear that a symmetry $\mathcal{D}$ of D preserves the kernel of D. If $\mathcal{D} = \mathcal{P}\, D$, then $\mathcal{D}$ is trivial on the kernel of D. We shall consider such symmetries to be trivial and we shall consider the quotient of the algebra of symmetries by the ideal of trivial symmetries.

A lot of information is available on symmetries given by first order differential operators (see [9, 4] and references therein). On the other hand, the structure of the full algebra of symmetries is known only in exceptional cases. The basic example is the symmetry of the Laplace operator. Let us write down explicitly definition of the symmetry algebra in this case.

Let (M, g) be a Riemannian manifold, $dim M = n$ $(n \geq 3)$ and let $\triangle$ be the corresponding Laplace operator acting on the space of smooth functions on M.

DEFINITION 2.1. *A symmetry of the Laplace operator $\triangle$ on M is a linear differential operator $\mathcal{D}$ acting on functions with the property that $\triangle\mathcal{D} = \delta\triangle$ for some linear differential operator δ. Two symmetries $\mathcal{D}_1, \mathcal{D}_2$ of $\triangle$ are equivalent if and only if $\mathcal{D}_1 - \mathcal{D}_2 = \mathcal{P}\triangle$ for some differential operator $\mathcal{P}$. The composition of differential operators preserves this equivalence relation, hence the quotient defines the* **symmetry algebra** *$\mathcal{A}_n$ of the Laplace operator.*

An example of a symmetry of the Laplace operator is a conformal Killing vector field on M. They need not exist in general, their appearance is closely related to geometry of M. In the homogeneous case $M = S^n$, the space of all symmetries of the first order is formed by conformal Killing vectors. It can be identified, as a Lie algebra, with the Lie algebra $\mathfrak{so}(n+1, 1)$.

The algebra of all symmetries of the Laplace operator in the flat case $M = \mathbb{R}^n$ has been computed in [5] (see also [11]). It looks as follows.

THEOREM 2.1. *In the flat case $M = \mathbb{R}^n$, the algebra $\mathcal{A}_n$ is isomorphic to the tensor algebra*

$$\bigoplus_{s=0}^{\infty} \bigotimes^{s} \mathfrak{so}(n+1, 1) \tag{2.1}$$

modulo two-sided ideal J generated by the elements

$$V \otimes W - V \circledcirc W - \frac{1}{2}[V, W] + \frac{n-2}{4(n+1)}\langle V, W\rangle \tag{2.2}$$

for $V, W \in \mathfrak{so}(n+1, 1)$. Here $\circledcirc$ denotes the projection on Cartan component in the tensor product, $[-,-]$ denotes the Lie bracket and κ is the Killing form in $\mathfrak{g}$.

3. Deformations of quadratic ideals in $T(\mathfrak{g})$. The two-sided ideal J in the theorem above can be considered as a deformation (depending on a parameter λ) of a simpler quadratic ideal I. We shall describe this idea in a more general setting, following papers [3, 6].

Let $\mathfrak{g}$ be a complex simple Lie algebra. Let $\mathfrak{g} \circledcirc \mathfrak{g} \subset \mathfrak{g} \otimes \mathfrak{g}$ denotes the Cartan part of the tensor product. If β is the highest weight of the adjoint representation, then $\mathfrak{g} \circledcirc \mathfrak{g}$ has the highest weight 2β and it appears in $\mathfrak{g} \otimes \mathfrak{g}$ with multiplicity one. Denote by R the invariant complement of $\mathfrak{g} \circledcirc \mathfrak{g}$, hence

$$\mathfrak{g} \otimes \mathfrak{g} = (\mathfrak{g} \circledcirc \mathfrak{g}) \oplus R.$$

The Lie algebra $\mathfrak{g}$ appears in the decomposition of the wedge product $\mathfrak{g} \otimes \mathfrak{g}$ with multiplicity one. Hence there is a unique (up to a multiple) equivariant map from $\Lambda^2(\mathfrak{g})$ to $\mathfrak{g}$. It is a multiple of the Lie bracket. Let us denote the Lie bracket (as a bilinear map from $\mathfrak{g} \times \mathfrak{g}$ to $\mathfrak{g}$) by α.

The trivial representation $\mathbb{C}$ appears with multiplicity one in the product $\mathfrak{g} \otimes \mathfrak{g}$. Hence there is a unique (up to a multiple) equivariant map $\beta : \mathfrak{g} \otimes \mathfrak{g}$ to $\mathbb{C}$. It is a multiple of the Killing form κ :

$$\beta : \mathfrak{g} \otimes \mathfrak{g} \mapsto \mathbb{C}, \beta = \beta_\lambda = \lambda \cdot \kappa.$$

Let $T = T(\mathfrak{g})$ be the tensor algebra of the vector space $\mathfrak{g}$ with its natural grading $T = \oplus_{n=0}^{\infty} T_k$ and the associated filtration

$$F^k(T) = \oplus_{n=0}^{k} T_n.$$

In these terms, we can write the ideal $J = J(P)$ from the theorem above as the two-sided ideal generated by the set

$$P_\lambda = \{u \in F^2(T) | u = x - \alpha(x) - \beta_\lambda(x);\ x \in R\},$$

for a special value $\lambda_0 = -\frac{n-2}{4(n+1)}$. For a general value $\lambda \in \mathbb{C}$, the two-sided ideals J_λ generated by P_λ can be considered as a deformation of the homogeneous ideal I generated by $R \subset \mathfrak{g} \otimes \mathfrak{g}$.

It is surprising how big the ideals J_λ are for a generic value of λ. For classical simple complex Lie algebras $\mathfrak{g}$, the codimension of J_λ in $T(\mathfrak{g})$ is less or equal to one for all but one value of λ.

We are now going to show why it is so exceptional to find λ, for which J_λ has an infinite codimension. Let us start our discussion with a tensor $V \in \otimes^3 \mathfrak{g}$, which we suppose to belong to the subspace

$$F := (R \otimes \mathfrak{g}) \cap (\mathfrak{g} \otimes R).$$

We can reduce the tensor V in two different ways. We shall use the symbol $\simeq$ to indicate that two tensors differ by an element from the ideal J_λ.

Firstly, V belongs to $R \otimes \mathfrak{g}$. Hence $V = \sum_i S_i \otimes X_i$, with $S_i \in R$ and $X_i \in \mathfrak{g}$. Then

$$V \simeq V_l := \sum_i \alpha(S_i) \otimes X_i + \sum_i \beta(S_i) \otimes X_i \in F^1(\mathfrak{g}) \otimes \mathfrak{g}.$$

Similarly, V is in $\mathfrak{g} \otimes R$, and we get that $V = Y_j \otimes R_j$ (with $R_j \in R, Y_j \in \mathfrak{g}$) is equivalent to

$$V \simeq V_r := \sum_j Y_j \otimes \alpha(R_j) + \sum_j Y_j \otimes \beta(R_j) \in \mathfrak{g} \otimes F^1(\mathfrak{g}).$$

Consequently, the difference $V_r - V_l$ belongs to the ideal $J_\lambda \cap F^2(\mathfrak{g})$.

It can be shown ([3], 3.3) that the two-homogeneous part

$$\sum_i \alpha(S_i) \otimes X_i - \sum_j Y_j \otimes \alpha(R_j)$$

of the difference $V_l - V_r$ belongs always to R and is hence equivalent to

$$\alpha(\sum_i \alpha(S_i) \otimes X_i - \sum_j Y_j \otimes \alpha(R_j)) - \beta(\sum_i \alpha(S_i) \otimes X_i - \sum_j Y_j \otimes \alpha(R_j)) \in F^1(\mathfrak{g}).$$

As shown in ([3], 3.3), the zero homogeneous part

$$\beta(\sum_i \alpha(S_i) \otimes X_i - \sum_j Y_j \otimes \alpha(R_j))$$

vanishes and we are left with an element

$$Z = Z(V) := \alpha(\sum_i \alpha(S_i) \otimes X_i - \sum_j Y_j \otimes \alpha(R_j)) + \sum_i \beta(S_i) \otimes X_i - \sum_j Y_j \otimes \beta(R_j),$$

which belongs to $J_\lambda \cap \mathfrak{g}$.

It is shown in [3] that the element Z vanishes identically (i.e., for any $V \in F$) only for a unique, exceptional value of λ_0. For all other values, we have constructed a nontrivial element in $J_\lambda \cap \mathfrak{g}$.

To illustrate the procedure described above better, let us return back to the specific example of the orthogonal case. Let again $\mathfrak{g} = \mathfrak{so}(n+2, \mathbb{C})$. Let us represent elements in $\mathfrak{g}$ by antisymmetric tensors $U^{ab} = -U^{ba}$. In the following computation, the Penrose abstract index notation is used (for details, see [10], Vol. 1). For each $U \in \mathfrak{g}$, let us define a tensor $V \in \otimes^3(\mathfrak{g})$ by

$$\begin{aligned} V^{abcdef} = {} & 2g^{af}g^{be}U^{cd} - 2g^{ae}g^{bf}U^{cd} - 2g^{cf}g^{de}U^{ab} + 2g^{ce}g^{df}U^{ab} \\ & + g^{ac}g^{be}U^{df} - g^{bc}g^{ae}U^{df} - g^{ad}g^{be}U^{cf} + g^{bd}g^{ae}U^{cf} \\ & - g^{ac}g^{bf}U^{de} + g^{bc}g^{af}U^{de} + g^{ad}g^{bf}U^{ce} - g^{bd}g^{af}U^{ce} \\ & - g^{ac}g^{de}U^{bf} + g^{ad}g^{ce}U^{bf} + g^{bc}g^{de}U^{af} - g^{bd}g^{ce}U^{af} \\ & + g^{ac}g^{df}U^{be} - g^{ad}g^{cf}U^{be} - g^{bc}g^{df}U^{ae} + g^{bd}g^{cf}U^{ae}. \end{aligned} \tag{3.1}$$

The map $U \mapsto V(U)$ is clearly $\mathfrak{g}$-equivariant. It is immediate from the definition of V that it belongs to $\Lambda^2(\mathfrak{g})\otimes\mathfrak{g} \subset R\otimes\mathfrak{g}$. Moreover, the projection of V to $\mathfrak{g}\otimes(\mathfrak{g}\circledcirc\mathfrak{g})$ vanishes (for details, see [6]). Hence we have defined an equivariant map

$$U \in \mathfrak{g} \mapsto V = V(U) \in F = (R\otimes\mathfrak{g}) \cap (\mathfrak{g}\otimes R).$$

We may now reduce V^{abcdef} in two different ways with respect to the given ideal, as indicated above. The Lie bracket (denoted by α) is given in abstract index notation by

$$(\alpha(S^{abcd}))^{ad} = \frac{1}{2}(S^{a}{}_{b}{}^{bd} - S^{d}{}_{b}{}^{ba}).$$

The Killing form κ is given by

$$\kappa(S^{abcd}) = (n-2)S^{ab}{}_{ab}.$$

A short calculation gives,

$$V^{a}{}_{b}{}^{bdef} = (n-4)[g^{af}U^{de} - g^{ae}U^{df} + g^{de}U^{af} - g^{df}U^{ae}],$$

which is skew in ad. Therefore $V^{ab}{}_{ab}{}^{cd} = 0$. In terms of maps α and β, we have

$$([\alpha\otimes Id](V))^{adef} = (n-4)[g^{af}U^{de} - g^{ae}U^{df} + g^{de}U^{af} - g^{df}U^{ae}]$$

and $([\beta\otimes Id](V))^{af} = 0$. The tensor $([\alpha\otimes Id](V))^{adef}$ belongs clearly to $\Lambda^2(\mathfrak{g})$, hence also to R. Applying α once more, we get

$$\alpha([\alpha\otimes Id](V)) = (n-2)U^{af}.$$

Altogether, we have got

$$V^{abcdef} \simeq (n-2)(n-4)U^{af}. \tag{3.2}$$

On the other hand, the right reduction gives

$$([Id\otimes\alpha](V))^{abcf} = -(n-4)[g^{ac}U^{bf} - g^{bc}U^{af} - g^{af}U^{bc} + g^{bf}U^{ac}]$$

and

$$([Id\otimes\beta](V))^{ab} = 2(n-1)(n-2)U^{ab}.$$

Therefore,

$$V^{abcdef} \simeq -\frac{n-4}{2}[g^{ac}U^{bf} - g^{bc}U^{af} - g^{af}U^{bc} + g^{bf}U^{ac}] - 2\lambda(n-1)(n-2)^2U^{af}.$$

But tracing over bc gives

$$g^{ac}U^{bf} - g^{bc}U^{af} - g^{af}U^{bc} + g^{bf}U^{ac} \simeq -(n-2)U^{af}$$

and so

$$V^{abcdef} \simeq (n-2)[\tfrac{n-4}{2} - 2\lambda(n-1)(n-2)]U^{af}. \tag{3.3}$$

Comparing (3.2) with (3.3), we conclude that U^{ab} must be (for generic value of n) in the ideal J_λ unless we have $\lambda = \lambda_0 := -\frac{n-4}{4(n-1)(n-2)}$. Hence for all $\lambda \neq \lambda_0$, the ideal J_λ contains $\mathfrak{g}$ and its codimension in $T(\mathfrak{g})$ is at most one.

In the case under discussion ($\mathfrak{g} = \mathfrak{so}(n+2, \mathbb{C})$), we have hence a complete information on the size of J_λ. For $\lambda \neq \lambda_0$, the codimension of J_λ is at most one, while for $\lambda = \lambda_0$, the quotient $T(\mathfrak{g})/J_\lambda$ is the infinite dimensional algebra $\mathcal{A}_n$ described in the theorem above.

Note that the information on the size of J_λ was obtained by testing different possible reductions on a special tensor $V = V(U)$. It is not easy to find such special tensors, because it is difficult to write down explicit form of tensors in $(R \otimes \mathfrak{g}) \cap (\mathfrak{g} \otimes R)$. The tensors $V(U)$ were constructed in [6] for all classical Lie algebras. Using them, the special values of λ_0 are computed for each classical simple Lie algebra $\mathfrak{g}$.

There is no explicit construction of such special tensors available for exceptional simple Lie algebras. The proof that there is a special value of λ with the property that $J_\lambda \cap \mathfrak{g}$ is trivial can be found in [3]. The proof uses the same ideas but it is based on general arguments without an explicit construction of the corresponding tensors.

We have seen how to get negative results (J_λ is too big for λ different from an exceptional value). On the other hand, the difficult question remains - what is happening for the special value $\lambda = \lambda_0$. We have seen that the right and the left reductions of the special tensors coincide in this case. But there could be similar reductions leading to different results on higher levels. It is a very difficult problem to treat these special cases in a systematic way and it needs much better tools. We are going to describe these tools in the next section.

Before doing that, let us add one more remark. We have seen how to treat the case of orthogonal Lie algebras by an explicit construction. The quotient $T(\mathfrak{g})/J_{\lambda_0}$ was clearly infinite dimensional, because it was identified with the algebras of symmetries of the Laplace operator. The graded version of this algebra was described explicitly, so that we have a full information on its size. This is an alternative to a general abstract approach (discussed in the next Section) - to construct explicitly a geometric realization of the quotient for each simple Lie algebra $\mathfrak{g}$. In [6], it is done for series A, B and D of simple complex Lie algebras.

4. Deformation theory for quadratic algebras. In this section, we are going to show, following papers [3, 2], how the deformations discussed above fit into a general scheme of deformation theory for quadratic algebras.

Let V be a finite dimensional vector space, $T(V) := \oplus_{i=0}^{\infty} T^i(V)$ its tensor algebra. Let $R \subset T^2(V) = V \otimes V$ be the linear subspace of relations and $J(R) \subset T(V)$ a two-sided ideal generated by R. Homogeneous quadratic algebra $Q(V, R)$ corresponding to R is defined as $Q(V, R) := T(V)/J(R)$.

The filtered analog of the previous graded algebra is based on the filtration $F^i(T(V)) = \{\oplus_j T^j(V) | j \leq i\}$ of $T(V)$. Fix a subspace $P \subset F^2(T(V)) = \mathbb{R} \oplus V \oplus (V \otimes V)$ and let $J(P) \subset T(V)$ be an ideal generated by P. Non-homogeneous quadratic algebra $Q(V, P)$ is defined by $Q(V, P) := T(V)/J(P)$.

The algebra $Q(V, P)$ inherits an increasing filtration from $F(T(V))$. Let us define the projection $p : F^2(T(V)) \to V \otimes V$ and denote by R the subspace $R = p(P)$. There is a natural surjection $Q(V, R) \to gr(Q(V, P))$, so we can ask whether this map is a monomorphism. If so, we have a good control on the size of the quotient algebra $Q(V, P)$. Following the standard terminology, we say that a non-homogeneous quadratic algebra $Q(V, P)$ is of Poincare-Birkhoff-Witt type if this surjection $Q(V, R) \to gr(Q(V, P))$ is an isomorphism of graded algebras.

The central task of deformation theory of associative quadratic algebras is to decide when there exists a deformation of $Q(V, R)$ to $Q(V, P)$. This amounts to construct a family $U \to Spec(\mathbb{R}[t])$ of associative quadratic filtered $\mathbb{R}[t]$-algebras over an affine line $Spec(\mathbb{R}[t])$ such that

1. The fiber over $t = 0$ is isomorphic to $gr(Q(V, P))$,
2. The fiber over $t \neq 0$ is isomorphic to $Q(V, P)$.

In other words, $U := \{\sum_i u_i t^i | u_i \in gr_i(Q(V, P))\}$, $deg(t) = 1$.

The structure of deformations of a graded associative algebra $A = Q(V, R)$ is controlled by the Hochschild cohomology groups, where $H^2_{-i-1}(A, A)$ parameterizes isomorphism classes of ith order deformations and $H^3_{-i-1}(A, A)$ parameterizes the space of obstructions of an extension from ith to $(i+1)$th level of deformations. Here the subscript by cohomology groups denotes grading of corresponding cochains, induced by grading of A.

The deformation theory simplifies considerably in the case of quadratic algebras of Koszul type. By definition, a quadratic algebra A is of Koszul type, if $H^i_j(A, M) = 0$ for all $i < -j$ and for all graded bimodules M. For such algebras, it is sufficient to control the deformation only up to the third order. In this situation, we have the following theorem.

THEOREM 4.1. *([2]) Let $A = Q(V, R)$ be a quadratic Koszul algebra and $\alpha : R \to V, \beta : R \to \mathbb{R}$ be linear maps. Let us define $F := (R \otimes V) \cap (V \otimes R) \subset V \otimes V \otimes V$.*

If the following three conditions

1. $Im(\alpha \otimes 1 - 1 \otimes \alpha)(F) \subset R$,
2. $\alpha \circ (\alpha \otimes 1 - 1 \otimes \alpha)(F) = -(\beta \otimes 1 - 1 \otimes \beta)(F)$,
3. $\beta \circ (1 \otimes \alpha - \alpha \otimes 1)(F) = 0$

are satisfied, then there exists a graded deformation of A s.t. the fiber at $t = 1$ is isomorphic to $Q(V, P)$ for $P = \{x - \alpha(x) - \beta(x) | x \in R\}$.

This Theorem is related to the discussion of deformation theory of quadratic Koszul algebras in the following way:

- 1. $\Longrightarrow \alpha$ defines (an isomorphism class of) first order deformation of A,
- 2. $\Longrightarrow \beta$ defines (an isomorphism class of) second order deformation of A,
- 3. $\Longrightarrow \beta$ can be lifted to the 3rd order deformation of A, i.e. there is no obstruction to the prolongation of a deformation.

The Koszulness property then implies that this 3rd order deformation can be prolonged to an (actual) graded deformation of A. Let us indicate the proof of Theorem 4.1.

Sketch of proof: Let $B^i(A)$ be the bar resolution of an associative algebra $A = Q(V, R)$, and

$$K^i(A) := \cap_{j=0}^{i-2}(V^{\otimes j} \otimes R \otimes V^{\otimes i-j-2}),\ \tilde{K}^i(A) := A \otimes K^i \otimes A. \quad (4.1)$$

We have canonical embedding $K^i(A) \hookrightarrow T^i(A)$ and so there is an embedding of complex of $A \otimes A^0$-modules $(\tilde{K}(A) \hookrightarrow B(A)) := \{\tilde{K}^i(A) \hookrightarrow B^i(A)\}_i$. Then a suitable criterion of Koszulness of A, equivalent with the previously mentioned one, is that A is Koszul iff the map $(\tilde{K}(A) \hookrightarrow B(A))$ is a quasi-isomorphism (i.e. it yields an isomorphism of Hochschild cohomology groups.)

Note that both α, β are defined on R, which is equal to K^2, and so we can regard them with the same notation as maps $\tilde{K}^2 \to A$. The first condition in Theorem 4.1 means that α is a 2-cocycle in the complex $Hom_{A\otimes A^0}(\tilde{K}, A)$, i.e. it defines a cohomology class in $H^2(A, A)$ and we can find a Hochschild 2-cocycle f_1 of A such that $f_1|_{K^2} = \alpha$ and f_1 is homogeneous of degree -1. This element defines first order deformation of A.

By the second condition in Theorem 4.1, we have $d\beta = \alpha \circ (\alpha \otimes 1 - 1 \otimes \alpha)$ in the Koszul complex. It is easy to find a Hochschild 2-cochain f_2 of degree -2 satisfying $df_2 = f_1 \circ (f_1 \otimes 1 - 1 \otimes f_1)$ and $f_2|_{K^2} = \beta$. The element f_2 defines an extension of first order deformation to the second order deformation of A.

Having the element f_2 of second order deformation of A, the obstruction of lifting the second order deformation to a third order deformation is given by the Hochschild 3-cocycle $\mathcal{O} = f_1 \circ (f_2 \otimes 1 - 1 \otimes f_2) - f_2 \circ (f_1 \otimes 1 - 1 \otimes f_1)$. Direct check shows that the third condition in Theorem 4.1 is equivalent to the triviality of $\mathcal{O}$ in $H^3(A, A)$, which means that f_2 can be continued to the third level deformation. □

5. The relation to the Joseph ideal. The discussion above is very much related to the Joseph ideal. It is a special ideal in the enveloping algebra $\mathcal{U}(\mathfrak{g})$ of a simple Lie algebra $\mathfrak{g}$. The motivating example can be reformulated to see the connection better.

As a quotient of tensor algebra, $\mathcal{A}_n$ is associative algebra generated by $\mathfrak{so}(n+1,1)$ and subject to two relations

$$\begin{aligned} V \otimes W - W \otimes V &= [V, W], \\ V \otimes W + W \otimes V &= 2V \circledcirc W - \frac{n-2}{2(n+1)} \langle V, W \rangle. \end{aligned} \tag{5.1}$$

This in turn makes $\mathcal{A}_n$ isomorphic to the universal enveloping algebra $\mathfrak{U}(\mathfrak{so}(n+1,1))$ modulo the two-sided ideal J generated by

$$V \otimes W + W \otimes V - 2V \circledcirc W + \frac{n-2}{2(n+1)} \langle V, W \rangle$$

for $V, W \in \mathfrak{so}(n+1,1)$. The ideal J is called the Joseph ideal for $\mathfrak{g} = \mathfrak{so}(n+2, \mathbb{C})$. To describe this ideal properly, we have to add some remarks on relations between representations of simple Lie algebras and co-adjoint orbits in $\mathfrak{g}^*$. We shall follow in this section the paper [1], where more details can be found.

For a simple Lie algebra $\mathfrak{g}$, let $Pol(\mathfrak{g}^*)$ denote the polynomial algebra on the dual $\mathfrak{g}^*$. For a two-sided ideal J' of $\mathfrak{U}(\mathfrak{g})$, we shall consider $I' := gr(J')$ as an ideal in $Pol(\mathfrak{g}^*)$ via the identification

$$gr(\mathfrak{U}(\mathfrak{g})) \xrightarrow{\sim} Sym(\mathfrak{g}) \xrightarrow{\sim} Pol(\mathfrak{g}^*).$$

The associated variety $Ass(J')$ is the zero locus of I' i.e. an algebraic subvariety of $\mathfrak{g}^*$. The important point is that for the ideal J constructed above, the associated variety $Ass(J)$ is (the closure of) the minimal coadjoint orbit in $\mathfrak{g}^*$.

Let $\mathfrak{g}$ be a complex simple Lie algebra different from type A_n. Then it was shown in [8] that there is a unique completely prime primitive ideal $\bar{J}$ with the property that its associated variety is (the closure of) the unique coadjoint orbit of a minimal nonzero dimension. The ideal $\bar{J}$ is called the Joseph ideal. It was proved in [3] that the ideal J is the Joseph ideal, i.e. it coincides with $\bar{J}$. An alternative proof of this fact can be found in [1]. Following the standard philosophy of the orbit method for construction and classification of representations of $\mathfrak{g}$, there should be a representation of $\mathfrak{g}$ attached to the minimal orbit. A geometric realization of the representation is described in [1] using the ambient construction for the conformally invariant Laplace operator.

Let us consider the ambient space $\mathbb{R}^{p,q}$, $p + q = n + 2$, with a nondegenerate quadratic form Q of signature (p, q) and denote by $\mathcal{N}$ the corresponding null cone in $\mathbb{R}^{p,q}$. For a properly chosen real number d, the Laplace operator Δ on the ambient space $\mathbb{R}^{p,q}$ induces a conformally invariant operator Δ' acting on the space of smooth functions on $\mathcal{N}$, which are homogeneous of degree d. The kernel of the operator Δ' is then an infinite dimensional representation of $\mathfrak{so}(p,q)$. Using a characterization of the Joseph ideal from [7], it was proved in [1] that the annihilator ideal $Ker(\mathfrak{U}(\mathfrak{so}(p,q)) \to End(\mathbb{V}_{\triangle'}))$ is the Joseph ideal J. But the ideal defining symmetry algebra of Laplace operator on the flat space $\mathbb{R}^{p-1,q-1}$ is contained in the annihilator ideal of the representation $Ker(\triangle')$ and Theorem 2.1 shows that both ideals coincide.

Acknowledgements. The article is based on the results [6] of joint cooperation with M. Eastwood and V. Soucek. The author is grateful to IMA, Minneapolis, for excellent conditions and working atmosphere during the IMA program "Symmetries of PDE's and prolongation of overdetermined systems." The author thanks the research program MSM 0021620839 and the grants GACR 201/05/2117, GAUK 447/2004 for the suppport.

REFERENCES

[1] B. Binegar and R. Zierau, *Unitarization of a singular representation of* SO(p,q), Comm. Math. Phys. (1991), **138**(2).

[2] A. Braverman and D. Gaitsgory, *Poincaré-Birkhoff-Witt theorem for quadratic algebras of Koszul type*, J. of Algebra (1996), **181**: 315–328.

[3] A. Braverman and A. Joseph, *The minimal realization from deformation theory*, J. of Algebra (1998), **205**: 13–36.

[4] I.M. Benn and J.M. Kress, First-order Dirac symmetry operators, Class. quantum gravity, 2004, **21**(2): 427–431.

[5] M. Eastwood, *Higher symmetries of the Laplacian*, Ann. Math. (2005), **161**: 1645–1665.

[6] M. Eastwood, P. Somberg, and V. Soucek, *The Uniqueness of the Joseph Ideal for the Classical Groups*, math.RT/0512296.

[7] D. Garfinkle, *A new construction of the Joseph ideal*, PhD thesis, MIT.

[8] A. Joseph, *The minimal orbit in a simple Lie algebra and its associated maximal ideal*, Ann. Sci. Ecole Norm. Sup. (1976), **9**: 1–30.

[9] B. Kostant, Verma modules and the existence of quasi-invariant differential operators, Lecture Notes in Mathematics, 466, Springer-Verlag, 1974, pp. 101–129.

[10] R. Penrose and W. Rindler, *Spinors and Space-time*, Vol. **1**, Cambridge University Press, 1984.

[11] M.A. Vasiliev, Higher spin superalgebras in any dimension and their representations, preprint, hep-th 0404124.

ANALOGUES OF THE DOLBEAULT COMPLEX AND THE SEPARATION OF VARIABLES

VLADIMÍR SOUČEK*

Abstract. The Dirac equation is an analogue of the Cauchy-Riemann equations in higher dimensions. An analogues of the $\bar{\partial}$ operator in the theory of several complex variables in higher dimensions is the Dirac operator D in several Clifford variables. It is possible to construct a resolution starting with the operator D, which is an analogue of the Dolbeault complex. A suitable tool for studying properties of this resolution is the separation of variables for spinor valued fields in several vector variables and the corresponding Howe dual pair, which is of independent interest.

Key words. Generalized Dolbeault complex, several Clifford variables, separation of variables, Howe pairs.

AMS(MOS) subject classifications. Primary 30G35, 32W99, 58J10.

1. Introduction. Generalizations of complex analysis to higher dimensions were studied carefully during last seventy years. There are three equivalent definitions of holomorphicity for a complex valued function $f : \mathbb{R}^2 \to \mathbb{C}$ on the complex plane - (complex) differentiability, the Cauchy-Riemann equations $\bar{\partial} f = 0$ and having a (complex) Taylor series. A suitable generalization of the second condition was formulated first in dimension 4, where the Euclidean space can be identified with the space $\mathbb{H}$ of quaternions. An analogue of Cauchy-Riemann equations was studied in 40's by Fueter, a summary of properties of solutions of the Fueter equation and the corresponding quaternionic analysis can be found in [24].

It is now widely accepted that a proper generalization of complex function theory is described by Clifford analysis, which means the function theory for solutions of the Dirac operator [12, 5, 14, 13]. In higher dimension, functions on R^n with values in the Clifford algebra (resp. in the corresponding spinor spaces) are considered instead of complex valued functions.

During the second half of the last century, the theory of several complex variables was highly developed for functions $f = f(z_1, \ldots, z_n) : (\mathbb{R}^2)^k \to \mathbb{C}$ satisfying equations $\bar{\partial}_{z_1} f = \ldots = \bar{\partial}_{z_k} f = 0$. It was soon found that their properties are quite different from the case of one variable. The Hartogs type theorems show that holomorphic functions of several complex variables cannot have compact singularities. It was realized later that it is a consequence of the fact that the corresponding Cauchy-Riemann equations for several variables form an overdetermined system of first or-

*Mathematical Institute, Charles University, Sokolovská 83, Praha, Czech Republic.
The work is a part of the research project MSM 0021620839 financed by MSMT and partly supported by the grant GA UK 447/2004. The author wants to thank the referee for helpful suggestions.

der PDE's. This fact is encoded in the resolution of the Cauchy-Riemann system, which is formed by the Dolbeault complex.

Similar questions were studied also in higher dimensions for maps $f = f(\underline{x}_1, \ldots, \underline{x}_k) : (\mathbb{R}^n)^k \to \mathbb{S}$ depending on several Clifford variables and satisfying system of PDE's of the form $D_1 f = \ldots = D_k f = 0$, where D_j denotes the Dirac operator in variable $\underline{x}_j$. Solutions of the system are usually called monogenic functions (in several variables). The first results were found again in the quaternionic case, in dimension 4. The resolution of the system corresponding to the Fueter operator in several quaternionic variables was first described in [20, 2]. Similar questions were later studied in general dimension using algebraic methods and methods coming from Clifford analysis and constructions of an analogue of the Dolbeault complex corresponding to the Dirac operator in several variables has substantially advanced. The corresponding results can be found in the book by F. Colombo, I. Sabadini, F. Sommen and D. Struppa ([9]).

An important information on systems of PDE's involved in the resolutions above is their symmetry. A study of differential operators invariant with respect to a certain symmetry has a long tradition. Typical examples are invariant differential operators on Riemannian manifolds. A lot of attention was paid recently to invariant differential operators on manifolds with a given conformal, quaternionic or CR structures, which are special cases of invariant differential operators on manifolds with a given parabolic structure. A lot of results were accumulated concerning such operators and they can be effectively used for our present purposes. An important feature of these operators is that they form discrete families, which are in many cases completely classified on homogeneous models for a given parabolic geometry. This fact help a lot in attempts to construct resolutions considered above.

A symmetry of the Fueter (resp. the Dirac) operator is well known, these operators are invariant with respect to the group of conformal motions. It was possible to describe also a symmetry appropriate for the resolution of the Fueter operator in several quaternionic variables ([3, 6, 10, 8]). In this case, the appropriate symmetry group is the symmetry group of the quaternionic projective space $\mathbb{HP}^n$. In higher dimensions, an appropriate symmetry is the symmetry group of isotropic Grassmanians (for more details see [15]). In higher dimensions, there is still a lot of open questions. Neither algebraic methods, nor methods of Clifford analysis and methods coming from a study of symmetries mentioned above are sufficient at present to describe the form of the resolution in the general case.

In the present paper, we shall discuss another very useful tool for studying analogues of the Dolbeault complexes in several Clifford variables. This is the theory of dual pairs created by R. Howe and theorems on separation of variables connected with dual pairs. The aim of the paper is to describe the relevant facts and to explain their use in the study of the (generalized) Dolbeault complexes. We shall review needed facts about the Howe pairs for the orthogonal group, the relation to theorems on separation of variables and their extension to the case of Spin group. Then we shall explain the construction of the Dolbeault complex for two and three variables to illustrate the role of the symmetry in the construction. A construction of the (generalized) Dolbeault complex in general case is not known at present. The full treatment of the complex in the stable range (i.e., in the case that the number of variables is less or equal to the half of the dimension of the variable) is not yet finished and there are very few results in other cases.

2. Separation of variables – one variable case. The separation of variables in the case of scalar valued functions on $\mathbb{R}^n$ leads immediately to a simple description of the classical theory of spherical harmonics. The separation of variables for spinor valued fields is very similar. It explains nicely the structure of homogeneous solutions of the Dirac equation (so called spherical monogenics), which was developed in the last decades in the framework of the Clifford analysis.

2.1. Spherical harmonics. Let us consider the Euclidean space $\mathbb{R}^n$ as the defining representation of the group $SO(n)$. Then the space $\mathcal{P}$ of all polynomials on $\mathbb{R}^n$ with the induced action of the group $SO(n)$ can be decomposed to irreducible (finite dimensional) components as indicated in the diagram below, where the columns collect functions of a particular homogenity. Note that each component appears with an infinite multiplicity (the repetition of the modules in each row of the diagram).

$$
\begin{array}{ccccccc}
\mathcal{P}_0 & \mathcal{P}_1 & \mathcal{P}_2 & \mathcal{P}_3 & \mathcal{P}_4 & \mathcal{P}_5 & \mathcal{P}_6 \\
\mathcal{H}_0 = \langle 1 \rangle & \stackrel{\Delta}{\leftarrow} & r^2\mathcal{H}_0 & \stackrel{\Delta}{\leftarrow} & r^4\mathcal{H}_0 & \stackrel{\Delta}{\leftarrow} & r^6\mathcal{H}_0 \\
 & \mathcal{H}_1 & \stackrel{\Delta}{\leftarrow} & r^2\mathcal{H}_1 & \stackrel{\Delta}{\leftarrow} & r^4\mathcal{H}_1 & \\
 & & \mathcal{H}_2 & \stackrel{\Delta}{\leftarrow} & r^2\mathcal{H}_2 & \stackrel{\Delta}{\leftarrow} & r^4\mathcal{H}_2 \\
 & & & \mathcal{H}_3 & \stackrel{\Delta}{\leftarrow} & r^2\mathcal{H}_3 & \\
 & & & & \mathcal{H}_4 & \stackrel{\Delta}{\leftarrow} & r^2\mathcal{H}_4
\end{array}
$$

There is a hidden symmetry behind the decomposition. We can define an $sl(2) \simeq sp(2)$ action on the space $\mathcal{P}$, which will remove the degeneration. It is an analogue of the use of the $sl(2)$ action on forms in the Kähler geometry. All that is a consequence of the fact that $SO(n)$ and $sp(2)$ form a dual pair. The Lie algebra $sl(2) \simeq sp(2)$ is defined as a subalgebra in the Weyl algebra of all differential operators on $\mathbb{R}^n$ with polynomial coefficients. Its generators are Δ, the multiplication by r^2 and their commutator (a shift of the Euler operator). Note that the spaces $\mathcal{H}_k$ of harmonic polynomials of homogeneity k are irreducible representations of $SO(n)$ with the highest weight $\lambda_k = (k, 0, \dots, 0)$. Their restriction to the sphere are spaces of spherical harmonic of order k. It is often useful to have such a nice and explicit realization of the abstract irreducible module with the highest weight of type $k = (k, 0, \dots, 0)$.

The following simplest version of the theorem on separation of variables proves facts illustrated in the diagram above (see [17]).

THEOREM 2.1. *Let $\mathcal{H}$ be the space of all harmonic functions and let $\mathcal{I}$ be the space of all $SO(n)$ invariants in $\mathcal{P}$. Then*

$$\mathcal{P} = \mathcal{H} \otimes \mathcal{I}.$$

Moreover, the space of invariants $\mathcal{I}$ is the vector space generated by invariants $\{1, r^2, r^4, \dots\}$ and the space $\mathcal{H}$ of harmonic polynomials decomposes as

$$\mathcal{H} = \oplus_{k=0}^{\infty} \mathcal{H}_k.$$

2.2. Spherical monogenics. All facts on spherical harmonics described above are very well known and often used. Consider now the Clifford algebra Cl_n for the Euclidean space $\mathbb{R}^n$ and its basic (irreducible) spinor module $\mathbb{S}$. The action of an element from the Clifford algebra on spinors will be denoted by the left multiplication. Let $\mathcal{P}^S$ denote the space of all polynomials on $\mathbb{R}^n$ with values in $\mathbb{S}$ with the corresponding action of the group $Spin(n)$, given by

$$[s \cdot f](x) = s\, f(s^{-1}\, x\, s);\ x \in \mathbb{R}^n,\ s \in Spin(n).$$

Spinor valued functions in the kernel of Dirac operator $D = \sum_{i=1}^{n} e^i \partial_i$ are usually called monogenic functions.

Polynomial solutions of the Dirac equation for spinor valued fields of a given homogeneity were systematically studied in Clifford analysis in the last decades. The space of all such solutions of homogeneity k will be denoted by $\mathcal{M}_k$. It was shown that the spaces $\mathcal{M}_k$ form again an irreducible module with the highest weight $\mu_k = (\frac{2k+1}{2}, \frac{1}{2}, \dots, \frac{1}{2})$ in the case that n is odd. The spaces $\mathcal{M}_k$ split into two irreducible components with weights $\mu_k^{\pm} = (\frac{2k+1}{2}, \frac{1}{2}, \dots, \frac{1}{2}, \pm\frac{1}{2})$ for n even.

As for spherical monogenics, it is often useful to realize the abstract module with the highest weight of such type by the spaces $\mathcal{M}_k$ (they are called the spaces of spherical monogenics).

The decomposition of the space $\mathcal{P}^S$ into irreducible parts is described by the following diagram, which is very similar to the one given before for the scalar case.

$$
\begin{array}{ccccccccc}
\mathcal{P}_0^S & & \mathcal{P}_1^S & & \mathcal{P}_2^S & & \mathcal{P}_3^S & & \\
 & & & & & & & & \\
\mathcal{M}_0 = \langle 1 \rangle & \stackrel{D}{\leftarrow} & \underline{x}\mathcal{M}_0 & \stackrel{D}{\leftarrow} & |x|^2\mathcal{M}_0 & \stackrel{D}{\leftarrow} & \underline{x}^3\mathcal{M}_0 & \dots \\
 & & \oplus & & \oplus & & \oplus & & \\
 & & \mathcal{M}_1 & \stackrel{D}{\leftarrow} & \underline{x}\mathcal{M}_1 & \stackrel{D}{\leftarrow} & |x|^2\mathcal{M}_1 & \dots \\
 & & & & \oplus & & \oplus & & \\
 & & & & \mathcal{M}_2 & \stackrel{D}{\leftarrow} & \underline{x}\mathcal{M}_2 & \dots \\
 & & & & & & \oplus & & \\
 & & & & & & \mathcal{M}_3 & \dots
\end{array}
$$

The vector space $\mathbb{R}^n$ is, as usually, embedded into the Clifford algebra Cl_n. Hence the element $\underline{x} \in \mathbb{R}^n$ and its powers $\underline{x}^n$ are polynomials with values in Cl_n. In local coordinates $\underline{x} = \sum_i e_i x_i$. There is a natural action of $Spin(n)$ on the space $\mathcal{P}(Cl_n)$ of polynomials with values in Cl_n given by

$$[s \cdot f](x) = sf(s^{-1}xs)s^{-1}.$$

Note that the powers $\underline{x}^n$ are invariants for such action. The proof of these facts can be found in standard texts on Clifford analysis ([12, 5, 14, 13]).

It can be summarized in the following theorem on separation of variables for spinor valued fields.

THEOREM 2.2. *Let $\mathcal{M} = \oplus_0^\infty \mathcal{M}_k$ be the space of all polynomial solutions of the Dirac equation. Let $\mathcal{I}(Cl_n)$ be the space of $Spin(n)$ invariants in the space $\mathcal{P}(Cl_n)$)of all Cl_n valued polynomials. Then $\mathcal{I}(Cl_n)$ is the vector space generated by invariants $\{1, \underline{x}, |\underline{x}|^2, \underline{x}^3 \dots\}$ and*

$$\mathcal{P} = \mathcal{I}(Cl_n) \otimes \mathcal{M}.$$

Moreover, $\mathcal{M}$ decomposes as $\oplus_{k=0}^\infty \mathcal{M}_k$.

There is a hidden symmetry and the dual pair present. Consider the Lie super algebra generated by $\underline{x}$ and D. To close it, we need to add just three more generators (Δ, r^2 and their commutator) corresponding to the scalar case. The 5-dimensional Lie superalgebra is isomorphic to $osp(1, 2)$ (see [16]).

3. Separation of variables – two variables case. Situation in more variables is more complicated (and more interesting). The dual pair action and its consequences are much more important and useful. With increasing number of variables, the decomposition of polynomials depending on several variables is really complicated. But it still possible to understand its structure for the so called stable range (it means cases, where number of variables is less or equal to the half of the dimension n). We shall illustrate the situation in the case of two variables.

3.1. Scalar valued functions. Let us consider now the case of polynomials depending on two variables from $\mathbb{R}^n$. Let $\mathcal{P}^{(2)}$ be the space of polynomials on

$$Mat_{2\times n} = \left\{ \begin{pmatrix} x_1 \\ x_2 \end{pmatrix} \Big| x_1, x_2 \in \mathbb{R}^n \right\}.$$

The action of $SO(n)$ on rows of the matrix induces the action on $\mathcal{P}^{(2)}$. It is not possible to draw a diagram showing the decomposition of the space of polynomials into ireducible parts. The corresponding facts are encoded in the following theorem on separation of variables.

DEFINITION 3.1. *Denote by ∇_1, ∇_2 gradients with respect to individual variables and by $\langle .,. \rangle$ the Euclidean scalar product. The space $\mathcal{H}^{(2)}$ of harmonic polynomials of two variables is defined as the common kernel of three operators*

$$\Delta_{11} = \langle \nabla_1, \nabla_1 \rangle; \; \Delta_{12} = \langle \nabla_1, \nabla_2 \rangle; \; \Delta_{22} = \langle \nabla_2, \nabla_2 \rangle.$$

The group $GL(2) \times SO(n)$ acts naturally on $Mat_{2\times n}$, hence also on $\mathcal{P}^{(2)}$. Note that both actions commute clearly with each other. The space of all $SO(n)$-invariant polynomials will be denoted by $\mathcal{I}^{(2)}$, it is a $GL(2)$-module.

The theorem on separation variables has a standard form. It is easy to guess from the case $k = 2$ its form for more variables (but note that it holds only in the stable range, see [17]).

THEOREM 3.1. *The space $\mathcal{I}^{(2)}$ of all $SO(n)$ invariant polynomials is the polynomial algebra in variables r_{11}, r_{12}, r_{22}, where $r_{ij} = \langle x_i, x_j \rangle$.*

Moreover, if $n \geq 4$,

$$\mathcal{P}^{(2)} = \mathcal{H}^{(2)} \otimes \mathcal{I}^{(2)}.$$

It is more difficult to describe the decomposition of harmonic polynomials. There are now higher multiplicities in the $SO(n)$-module decomposition even in the space of harmonic polynomials. But as a $GL(2) \times SO(n)$ module, the space of harmonic polynomials has a multiplicity one decomposition. Note that we can as well consider the action of $SL(2) \times SO(n)$ only. Representations of the commutative factor of the reductive group $GL(2)$ are one-dimensional and play no role in the decomposition. Irreducible representations of $SL(2)$ can be classified by their highest weight ℓ, where ℓ is a nonnegative integer.

THEOREM 3.2. *Let $n \geq 4$. Let E_{jk}, $j, k \in \mathbb{Z}$, $j \geq k \geq 0$ denote an irreducible $SO(n)$ module with the highest weight $(j, k, 0, \ldots, 0)$.*

Let F_{jk} denote the irreducible $SL(2)$-module with highest weight $(j-k)$ (there is a redundancy coming from the restriction to $SL(2)$).

Then

$$\mathcal{H}^{(2)} \simeq \oplus_{j,k\in\mathbb{Z},\, j\geq k\geq 0} E_{jk} \otimes F_{jk}$$

and there are simple formulae for highest weight vectors in each irreducible piece.

The action of $GL(2)$ on $\mathcal{P}^{(2)}$ is quite natural but it can be extended to a bigger, hidden symmetry. Let $\mathcal{W}^{(2)}$ be the Weyl algebra of differential operators with polynomial coefficients in variables x_1, x_2. The Lie algebra $\mathfrak{g}_0 = gl(2)$ of $GL(2)$ acts by elements of $\mathcal{W}^{(2)}$, generated by

$$E_{11} = \langle x_1, \nabla_1 \rangle + \frac{n}{2}, \qquad E_{22} = \langle x_2, \nabla_2 \rangle + \frac{n}{2},$$

$$E_{12} = \langle x_1, \nabla_2 \rangle, \qquad E_{21} = \langle x_2, \nabla_1 \rangle.$$

Let us define $\mathfrak{g}_1 = L(\{\Delta_{11}, \Delta_{12}, \Delta_{22}\})$ and $\mathfrak{g}_{-1} = L(\{r_{11}, r_{12}, r_{22}\})$, where L indicates the span of the corresponding elements. Then the (graded) Lie algebra $\mathfrak{g} = \mathfrak{g}_{-1} \oplus \mathfrak{g}_0 \oplus \mathfrak{g}_1$ in $\mathcal{W}^{(2)}$ is isomorphic to $sp(4)$.

The groups $SO(n)$ and $sp(4)$ give another example of the Howe dual pair. It means that they generate two subalgebras in $\mathcal{W}^{(2)}$, which are commutants of each other. Consequently, the decomposition of the whole space of polynomials into irreducibles with respect to this dual pair is multiplicity free.

3.2. Spinor valued functions. We shall now treat the case of spinor valued fields depending on two variables. Recall that D_1, D_2 denote the Dirac operators in respective variables.

DEFINITION 3.2. *Let $\mathcal{P}^{(2)}(\mathbb{S})$ be the space of spinor valued polynomials on*

$$Mat_{2\times n} = \left\{ \begin{pmatrix} x_1 \\ x_2 \end{pmatrix} \middle| \, x_1, x_2 \in \mathbb{R}^n \right\}.$$

The space $\mathcal{M}^{(2)}$ of monogenic polynomials of two variables is defined as the common kernel of operators D_1, D_2.

The group $GL(2) \times Spin(n)$ acts naturally (from left and right) on Mat_{2xn}, hence it induces also an action on $\mathcal{P}^{(2)}(\mathbb{S})$, or $\mathcal{P}^{(2)}(Cl_n)$. The space of all $Spin(n)$-invariant Cl_n-valued polynomials will be denoted by $\mathcal{I}^{(2)}(\mathbb{S})$.

As before, we can use methods of Clifford analysis and the previous result on scalar valued case to prove the corresponding theorem on separation of variables for spinor valued polynomials in two Clifford variables ([7]).

THEOREM 3.3. *Let $\mathcal{W}(\mathbb{S}) = \mathcal{W} \otimes Cl_n$ be the Weyl algebra of operators with coefficients in Cl_n. The space $\mathcal{I}^{(2)}(S)$ is a subalgebra of $\mathcal{W}(\mathbb{S})$ generated by $\{1, \underline{x}_1, \underline{x}_2\}$. It can be written as*

$$\mathcal{I}^{(2)}(S) = \mathcal{I}^{(2)} \otimes L\{1, \underline{x}_1, \underline{x}_2, \underline{x}_1\underline{x}_2\}.$$

Moreover, if $n \geq 4$,

$$\mathcal{P}^{(2)}(\mathbb{S}) = \mathcal{I}^{(2)}(\mathbb{S}) \otimes \mathcal{M}^{(2)}.$$

It is possible to describe also the decomposition of the space of all polynomial solutions of the Dirac operator in two variables.

THEOREM 3.4. *Let $n \geq 4$. Let E_{jk}, $j,k \in \mathbb{Z}$, $j \geq k \geq 0$ denote an irreducible $Spin(n)$ module with highest weight $(\frac{2j+1}{2}, \frac{2k+1}{2}, \frac{1}{2}, \ldots, \frac{1}{2}, \frac{1}{2})$ (in even dimensions the sum of two with highest weight differing by the sign of the last component).*

Let F_{jk} denote an irreducible $SL(2)$-module with highest weight $(j-k)$. Then

$$\mathcal{M}^{(2)}(\mathbb{S}) \simeq \oplus_{j,k\in\mathbb{Z},\, j\geq k\geq 0} E_{jk} \otimes F_{jk}.$$

There are simple explicit formulae for highest weight vectors in each irreducible $Spin(n)$ component of the decomposition.

The action of the Lie algebra $\mathfrak{g}_0 = gl(2)$ of $GL(2)$ can be again extended in a quite interesting way. Recall that its generators are

$$E_{11} = \langle x_1, \nabla_1 \rangle + \frac{n}{2}, \qquad E_{22} = \langle x_2, \nabla_2 \rangle + \frac{n}{2},$$

$$E_{12} = \langle x_1, \nabla_2 \rangle, \qquad E_{21} = \langle x_2, \nabla_1 \rangle.$$

Let us define $\mathfrak{g}_1 = \langle D_1, D_2 \rangle$, $\mathfrak{g}_2 = \langle \Delta_{11}, \Delta_{12}, \Delta_{22} \rangle$ and $\mathfrak{g}_{-1} = \langle \underline{x}_1, \underline{x}_2 \rangle$, $\mathfrak{g}_{-2} = \langle r_{11}, r_{12}, r_{22} \rangle$.

Then the vector space

$$\mathfrak{g} = \mathfrak{g}_{-2} \oplus \mathfrak{g}_{-1} \oplus \mathfrak{g}_0 \oplus \mathfrak{g}_1 \oplus \mathfrak{g}_2$$

in $\mathcal{W}^{(2)}(\mathbb{S})$ is closed under the graded commutator! It is another example of a simple Lie superalgebra, isomorphic to $osp(1,4)$ in the standard notation. The group $Spin(n)$ and the algebra $osp(1,4)$ give another example of the Howe dual pair. The decomposition of the space of all spinor valued polynomials with respect to the dual pair is multiplicity free.

4. The generalized Dolbeault sequence. Now we shall return back to the original problem of the Dirac operator D in k Clifford variables in dimension n and an analogue of the Dolbeault complex extending it to a resolution. The Dolbeault complex in several complex variables is a resolution starting with the operator $\bar{\partial} = (\partial_{\bar{z}_1}, \ldots, \partial_{\bar{z}_n})$. The operator $D = (D_1, \ldots, D_k)$ in k Clifford variables should also be the first operator in a complex. We want to construct it.

It is important to note that there is an abstract proof showing that there exists a (finite) minimal resolution starting with the Dirac operator (unique up to an isomorphism). The real problem is, however, to find the form of the resolution and to describe explicitly corresponding differential operators.

In more details, the scheme of the abstract existence proof is going as follows (for details, see [9]). By taking the Fourier transform, we get a matrix P_0, which is the symbol of the operator D. Entries of the matrix are (linear) polynomials in $k \cdot n$ real variables. We shall work now in the (commutative) ring of polynomials in these variables. By the Hilbert syzygy theorem, there is a (minimal) resolution of the transpose matrix P_0^t given by a (finite) sequence of matrices $P_i^t, i = 1, \ldots, \ell$.

We can now to go back (to take the transpose matrices and to define the corresponding differential operators $D_i, i = 1, \ldots, \ell$, with constant coefficients by the Fourier transform). A very important theorem (proved in various versions by L. Ehrenpreis, B. Malgrange and H. Komatsu) now says that the sequence of the operators $D_i, i = 1, \ldots, \ell$ acting on the spaces of smooth sections of corresponding vector bundles form a complex, which is exact on the sheaf level. Instead of sheafs of smooth sections $\mathcal{C}^\infty$, the theorem is true also for sheafs $\mathcal{A}$ of real analytic sections, the sheafs $\mathcal{D}'$ of distributional sections and for sheafs $\mathcal{B}$ of hyperfunction sections (for more details and for references, see [9], Theorems 2.1.1 and 2.1.3, p. 102).

The problem how to describe the generalized Dolbeault complex explicitly is fully solved in dimension 4 (see [20, 1, 2, 3, 8, 9, 10, 23]). It is remarkable that the resolution is completely understood for any number variables.

In higher dimensions, there is an important difference between two possible cases. In the stable range (recall that it means that the number of variables is at most equal to the half of the dimension), the structure of the resolution is mostly understood. For the case of two, resp. three, variables a full description of the resolution and its operators is known. For higher number of variables (still in the stable range), there are precise conjectures for a length and for a form of the resolution. It is still an open problem how to prove these conjectures and, in particular, to find explicit description of the operators of the second order in the corresponding complexes. Out of the stable range, only some examples exists.

We shall now describe the construction of the complex in the stable range for the case of two and three variables and we shall show that the symmetry of the operator D has strong consequences, which makes the description of the complex easier.

4.1. The analogue of the Dolbeault complex in two variables. For two variables, we can construct the complex in an elementary way. If we try to imitate the construction of the de Rham complex, the procedure will fail due to noncommutativity of the operators D_1 and D_2. Indeed, if want to find compatibility condition for the two components $g_1 = D_1 f$ and $g_2 = D_2 f$ of the image, we cannot use the relation $D_2 g_1 = D_1 g_2$. On the other hand, $D_2^2 = \Delta_2$ commutes with D_1 and implies

$$D_2^2 g_1 = D_2^2 D_1 f = D_1 D_2^2 f = D_1 D_2 g_2.$$

Exchanging indices 1 and 2, we get the other similar necessary condition. It leads directly to a complex

$$f \overset{D_0}{\to} \begin{pmatrix} g_1 = D_1 f \\ g_2 = D_2 f \end{pmatrix} \qquad \begin{pmatrix} G_1 \\ G_2 \end{pmatrix} \overset{D_2}{\to} h = D_1 G_1 + D_2 G_2.$$

$$\begin{pmatrix} g_1 \\ g_2 \end{pmatrix} \overset{D_1}{\to} \begin{pmatrix} G_1 = D_2^2 g_1 - D_1 D_2 g_2 \\ G_2 = D_1^2 g_2 - D_2 D_1 g_1 \end{pmatrix}.$$

Note that, unlike the case of the Dolbeault complex, there are second order operators involved in the complex! The complex above is an analogue of the Dolbeault sequence we were looking for.

THEOREM 4.1. *Let $\mathcal{C}(S)$ is the space of smooth spinor valued maps of two variables $x_1, x_2 \in \mathbb{R}^n$, $n \geq 4$. Then the complex*

$$\mathcal{C}(S) \overset{D_0}{\to} \mathcal{C}(S) \oplus \mathcal{C}(S) \overset{D_1}{\to} \mathcal{C}(S) \oplus \mathcal{C}(S) \overset{D_2}{\to} \mathcal{C}(S)$$

defined above is a (minimal) resolution of the sheaf of monogenic functions in two variables.

By construction, the sequence above is a complex. To show that it is a resolution, we need to work more. As described above, we have to translate the complex of differential operators in the theorem (by the Fourier transform and by transposition) to the case, where the Hilbert syzygy theorem is usually applied. The corresponding sequence of matrices P_i^t is a complex and we would like to prove that it is a resolution. The proof is not difficult, if we are able to prove the following implication:

$$|x_2|^2 g \in \underline{x}_1 \mathcal{P}^{(2)}(\mathbb{S}) \Rightarrow g \in \underline{x}_1 \mathcal{P}^{(2)}(\mathbb{S}).$$

The proof of this implication was first given by D. Constales in [11] and was quite difficult. It can be shown that it is an easy consequence of the theorem on separation of variables.

4.2. The case of three variables. We shall first quote the result obtained by method of Clifford analysis ([9]).

THEOREM 4.2. *Let $D = (D_1, D_2, D_3)$ and let $n \geq 6$. Then there exists a complex of operators, which is a (minimal) resolution of the sheaf of monogenic functions in three variables. The shape of the complex is indicated in the diagram below. Exponents indicate how many spinorial components individual maps have.*

$$\mathbb{S} \to \mathbb{S}^3 \to \mathbb{S}^8 \to \begin{matrix} \mathbb{S}^6 \\ \oplus \\ \mathbb{S}^6 \end{matrix} \to \mathbb{S}^8 \to \mathbb{S}^3 \to \mathbb{S}.$$

There is an explicit description of all maps and all operators given involved in the resolution given in [9]. To get a conceptual description of the individual chain spaces in the complex and to see a possible generalizations to a bigger number of variables, it is important to write the same complex in another way using a symmetry of the first operator in the complex. A visible symmetry of the operator D in three variables is the product $GL(3) \times Spin(n)$ acting on the space $\mathbb{R}^{3n}$, hence also on spinor valued functions in three variables.

Let $\mathbb{F}_{ijk}$ denote an irreducible $GL(3)$-module with the highest weight (i, j, k). Values of fields in the complex in the theorem above can be identified with modules defined by $C_{ijk} = \mathbb{F}_{ijk} \otimes \mathbb{S}$ (for suitable choices of i, j, k) and the whole complex has the form

$$C_{000} \to C_{100} \to C_{210} \to \begin{matrix} C_{200} \\ \oplus \\ C_{311} \end{matrix} \to C_{321} \to C_{332} \to C_{333}.$$

The structure of the complex is better visible, if we describe values of maps in individual nodes using the Young diagrams for the corresponding $GL(3)$-modules $\mathbb{F}_{ijk}$. The duality between the first four nodes and the last four nodes is clearly visible, the union of the corresponding Young diagrams fills up exactly the square with 9 boxes. In the next diagram, the orders of individual differential operators are indicated. All operators in the complex are of the first or the second order, as indicated by ordinary (resp. double) arrows in the diagram.

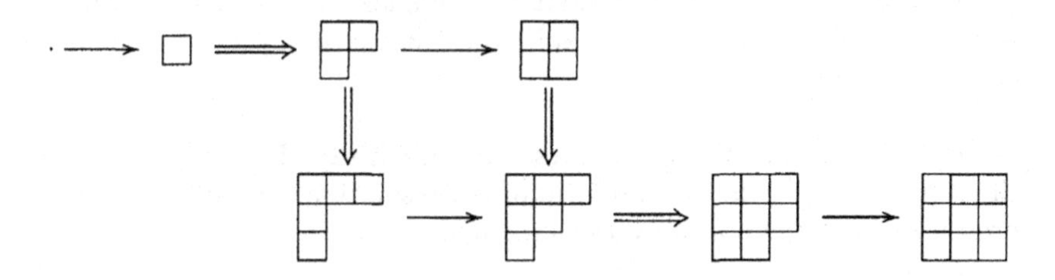

It is easy to check that dimensions of the representations $\mathbb{F}_{ijk}$ coincide with those given in Theorem 4.2. Note that the complex is clearly a union of the two dual versions of the complex for two variables. The order of the operator is the difference between number of boxes in individual components, all Young diagrams are symmetric along the diagonal.

Such a description suggests immediately a conjecture for the complex in four variables saying that it is a union of two dual versions of the complex for three variables, glued together by four operators of the second order (they will join the last four nodes in the first copy with the first four nodes of the second). All Young diagrams are subset of the square with 16 boxes. The dual nodes correspond to complementary Young diagrams.

The conjecture is easily extended to any number of variables in the stable range ($n \geq 2k$). The scheme also indicate an explicit form of the operators involved. They are constructed by the action of individual Dirac operators D_i followed by projections indicated by the corresponding Young digrams. Such a projection is uniquely defined (up to a multiple) for first order operators and can be written in an explicit way. But there is a two dimensional space of projections for the second order operators. A conjecture is saying that for a suitable choice of the projections for second order operators, the resulting sequence is a complex.

As for the case $k = 2$, we can translate the complex to the algebraic situation (by the Fourier transform and the transpose). There is a hope that the separation of variable can be helpful in the proof of the conjectures for $k > 2$, as it was in the case $k = 2$.

4.3. A broader symmetry behind. A general form of the complex above (for any k in the stable range) was guessed from symmetry described above. There is a more systematic way how to deduce it. The Dirac operator in one variable is conformally invariant. There is a parabolic geometry behind the complexes in dimension four. In general dimension, there is also a parabolic geometry with the property that the Dirac operator in several variables is invariant operator for this geometry (under a suitable interpretation). Unlike the quaternionic case, the geometry is now two-graded. To connect fields in such geometry to the studied complexes, we have to consider fields constant along the direction of the second graded component. The (singular) Hasse diagram for such geometry then coincides (in the stable range) with the form of the complex constructed by induction above and, at the same time, it explains the choice of representation in the individual nodes.

4.4. Other available results and methods. In the stable range, the conjectures on the general form of the resolution were formulated and proved in some cases using various methods.

1. The first methods used were algebraic methods Hilbert syzygy theorem, Gröbner basis, computational algebraic analysis. See [1, 2] and a summary in the recent monograph [9]. Fresh information can be found on the web page www.tlc185.com/coala run by A. Damiano.

2. Clifford analysis methods were used in [21, 22]. A recent summary can be found in the monograph by F. Colombo, I. Sabadini, F. Sommen and D. Struppa ([9]).

3. Representation theory methods (parabolic geometry) were used by P. Franek. He used the language of homorphisms of (generalized) Verma modules. In [15], he found corresponding orbits of the affine action of the Weyl group and proved the inductive construction of the orbits. The results coincide with the scheme described above. The resolution was constructed only in the case $k = 2$ (which extends the results obtained in the Clifford analysis setting to the case of a broader symmetry). Some partial information was obtained for $k > 2$.

4. The use of the Penrose transform was the key point in dimension four. R. Baston constructed in [3] the corresponding resolutions for any number of variables together with the many parametric series of more complicated complexes. At the same time, it is the result, which brings a substantial amount of information (in this special case) on the structure of analogues of BGG complexes in singular character. First steps to extend these methods to higher dimensions were made by L. Krump (see [18, 19]).

REFERENCES

[1] W.W. Adams, C.A. Berenstein, P. Loustaunau, I. Sabadini, and D.C. Struppa, *Regular functions of several quaternionic variables and the Cauchy–Fueter complex*, J. Geom. Anal., **9**(1) (1999), 1–15.

[2] W.W. Adams, P. Loustaunau, V.P. Palamodov, and D.C. Struppa, *Hartogs' phenomenon for polyregular functions and projective dimension of related modules over a polynomial ring*, Ann. Inst. Fourier, **47** (1997), 623–640.

[3] R.J. Baston, *Quaternionic Complexes*, J. Geom. Phys. **8**(1–4) (1992), 29–52.

[4] R.J. Baston and M.G. Eastwood, *The Penrose transform - its interaction with representation theory*, Oxford University Press, New York, 1989.

[5] F. Brackx, R. Delanghe, and F. Sommen, *Clifford Analysis.* Research Notes in Mathematics 76, Pittman 1982.

[6] J. Bureš, A. Damiano, and I. Sabadini, *Explicit resolutions for the complex of several Fueter operators*, preprint 1994.

[7] J. Bureš, F. Sommen, V. Souček, and P. Van Lancker, work in progress.

[8] J. Bureš, V. Souček, *Complexes of invariant operators in several quaternionic variables*, Compl. Variables and Elliptic Equations, **51**(5–6) (2006), 463–487.

[9] F. Colombo, I. Sabadini, F. Sommen, and D.C. Struppa, *Analysis of Dirac Systems and Computational Algebra*, Birkhauser, 2004.

[10] F. Colombo, V. Souček, and D.C. Struppa, *Invariant resolutions for several Fueter operators*, preprint, 2004.

[11] D. Constales, *Relative position of L^2-domains in complex and Clifford anaylsis*, PhD. thesis, Ghent, Belgium, 1990.

[12] R. Delanghe, *On regular-analytic functions with values in Clifford algebra*, Math.Ann. **185** (1970), 91–111.

[13] R. Delanghe, *Clifford Analysis : History and Perspective*, Computational methods and Function Theory, Vol. 1 (2001), 17–153.

[14] R. Delanghe, F. Sommen, and V. Souček, *Clifford algebra and spinor valued function: A function theory for the Dirac operator*, Math. and its Appl. 53, Kluwer Acad. Publ., Dordrecht, 1992.

[15] P. Franek, *Dirac operators in more variables from point of view of a parabolic geometry*, PhD thesis, Praha, 2006.

[16] J. Holland and G. Sparling, *Conformally invariant powers of the ambient Dirac operator*, preprint, arXiv math. DG/0112033.

[17] R. Howe, E. Tan, and J. Willenbring, *Reciprocity algebras and branching for classical symmetric pairs*, preprint arXiv RT/0407467.

[18] L. Krump and V. Souček, *Singular BGG sequences for the even orthogonal case*, to appear in the Proceedings of the Winter School 'Geometry and Physics', Srni, 2006.

[19] L. Krump and V. Souček, *The generalized Dolbeault complex in two variables*, to appear in the Proceedings of the Winter School 'Geometry and Physics', Srni, 2007.

[20] D. Pertici, *Funzioni regolari di piú variabili quatrenioniche*, Ann. Mat. Pura Appl. Série IV,CLI (1988), 39–65.

[21] I. Sabadini, F. Sommen, and D.C. Struppa, *The Dirac complex on abstract vector variables: megaforms*, preprint, 2002.

[22] I. Sabadini, F. Sommen, D. C. Struppa, and P. Van Lancker, *Complexes of Dirac operators in Clifford algebras*, Math. Zeit. **239**(2) (2002), 293–320.

[23] P. Somberg, *Quaternionic Complexes in Clifford Analysis and its Applications*, NATO ARW Series, Kluwer Acad. Publ., Dordrecht (F. Brackx, J.S.R. Chisholm, V. Souček eds.) (2001), 293–301.

[24] A. Sudbery, *Quaternionic analysis*, Math. Proc. Camb. Phil. Soc. **85** (1979), 199–225.

LIST OF WORKSHOP PARTICIPANTS

- Stephen C. Anco, Department of Mathematics, Brock University
- Douglas N. Arnold, Institute for Mathematics and its Applications, University of Minnesota
- Donald G. Aronson, Institute for Mathematics and its Applications, University of Minnesota
- Helga Baum, Department of Mathematics, Humboldt University of Berlin
- Gloria Marí Beffa, Department of Mathematics, University of Wisconsin
- Sergio Benenti, Dipartimento di Matematica, Università di Torino
- Melisande Fortin Boisvert, Department of Mathematics and Statistics, McGill University
- Andreas Cap, Fakultät für Mathematik, Universität Wien
- Mark Chanachowicz, Physics Department, University of Waterloo
- Claudia Chanu, Dipartimento di Matematica, Università di Torino
- Jeongoo Cheh, Department of Mathematics, University of St. Thomas
- Peter A. Clarkson, Institute of Mathematics, Statistics, and Actuarial Science, University of Kent at Canterbury
- Michael Cowling, School of Mathematics, University of New South Wales
- Luca Degiovanni, Department of Mathematics, Università di Torino
- Hongjie Dong, Department of Mathematics, University of Chicago
- Boris Doubrov, International Sophus Lie Centre, Belarusian State University
- Michael Eastwood, School of Mathematical Sciences, University of Adelaide
- Mark E. Fels, Department of Mathematics and Statistics, Utah State University
- Eugene Ferapontov, Department of Mathematical Sciences, Loughborough University
- Peter Franek, Department of Mathematics, Karlova Universita (Charles University)
- Michal Godlinski, Institute pf Theoretical Physics, University of Warsaw
- Hubert Goldschmidt, Department of Mathematics, Columbia University
- Sergey Golovin, Lavrentyev Institute of Hydrodynamics SD RAS
- A. Rod Gover, Department of Mathematics, University of Auckland

- C. Robin Graham, Department of Mathematics, University of Washington
- Robert Gulliver, School of Mathematics, University of Minnesota
- Hazem Hamdan, School of Mathematics, University of Minnesota
- Chong-Kyu Han, Department of Mathematics, Seoul National University
- Kengo Hirachi, Graduate School of Mathematical Sciences, University of Tokyo
- Doojin Hong, Algebra and Geometry, Eduard Cech Center
- Evelyne Hubert, Project CAFE, Institut National de Recherche en Informatique Automatique (INRIA)
- Peter Hydon, Department of Mathematics and Statistics, University of Surrey
- Jens Jonasson, Department of Mathematics, Linköping University
- Andreas Juhl, Matematiska institutionen, Uppsala University
- Ernie G. Kalnins, Department of Mathematics, University of Waikato
- Joseph Kenney, Department of Mathematics, University of Minnesota
- Irina Kogan, Department of Mathematics, North Carolina State University
- Jonathan M. Kress, School of Mathematics, The University of New South Wales
- Svatopluk Krysl, Mathematical Institute, Karlova Universita (Charles University)
- Joseph M. Landsberg, Department of Mathematics, Texas A & M University
- Guang-Tsai Lei, GTG Research
- Thomas Leistner, School of Mathematical Sciences, University of Adelaide
- Felipe Leitner, Institut für Geometrie und Topologie, Universitẗ Stuttgart
- Debra Lewis, Institute for Mathematics and its Applications, University of Minnesota
- Xiaolong Liu, Department of Physics and Astronomy, University of Iowa
- Elizabeth L. Mansfield, Institute of Mathematics and Statistics, University of Kent at Canterbury
- Ian Marquette, Département de physique et Centre de recherche mathématique, Université de Montréal
- Jose Kenedy Martins, Department of Mathematics, University of Minnesota
- Vladimir S. Matveev, Mathematisches Institut, Fakultät für Mathematik und Informatik, Friedrich-Schiller-Universität Jena

- Bonnie McAdoo, Department of Mathematical Sciences, Clemson University
- Raymond G. McLenaghan, Department of Applied Mathematics, University of Waterloo
- Willard Miller, Jr., School of Mathematics, University of Minnesota
- Emilio Musso, Department of Mathematics, Università di L'Aquila
- Lorenzo Nicolodi, Department of Mathematics, Università di Parma
- Anatoly Nikitin, Institute of Mathematics, National Academy of Sciences of Ukraine
- Paweł Nurowski, Instytut Fizyki Teoretycznej, Uniwersytet Warszawski
- Luke Oeding, Department of Mathematics, Texas A & M University
- Peter J. Olver, School of Mathematics, University of Minnesota
- Bent Orsted, Institut for Matematiske Fag, Aarhus University
- Saadet S. Ozer, Department of Mathematics, Yeditepe University
- Teoman Ozer, Division of Mechanics, Istanbul Technical University
- Lawrence J. Peterson, Department of Mathematics, University of North Dakota
- George S. Pogosyan, International Center for Advanced Studies, Yerevan State University
- Juha Pohjanpelto, Department of Mathematics, Oregon State University
- Arvind Satya Rao, Department of Mathematics, University of Iowa
- Giovanni Rastelli, Dipartimento di Matematicà, Universita di Torino
- Gregory J. Reid, Department of Applied Mathematics, University of Western Ontario
- Chan Roath, Scientific Research, Ministry of Education, Youth and Sport
- Colleen Robles, Department of Mathematics, University of Rochester
- Siddhartha Sahi, Department of Mathematics, Rutgers University
- Gerd Schmalz, School of Mathematics, Statistics and Computer Science, University of New England
- Neil Seshadri, Graduate School of Mathematics, University of Tokyo
- Astri Sjoberg, Department of Applied Mathematics, University of Johannesburg
- Jan Slovak, Department of Algebra and Geometry, Masaryk University
- Dalibor Smid, Mathematical Institute, Karlova Universita (Charles University)

- Roman G. Smirnov, Department of Mathematics and Statistics, Dalhousie University
- Shane Smith
- Petr Somberg, Mathematical Institute, Karlova Universita (Charles University)
- Vladimír Souček, Mathematical Institute, Karlova Universita (Charles University)
- Dennis The, Department of Mathematics and Statistics, McGill University
- Frédérick Tremblay, Mathematics and Statistics, University of Montreal
- Jukka Tuomela, Department of Mathematics, University of Joensuu
- Francis Valiquette, School of Mathematics, University of Minnesota
- Mikhail Vasiliev, Theory Department, P. N. Lebedev Physics Institute
- Raphael Verge-Rebelo, Centre de Recherches Mathematiques, University of Montreal
- Alfredo Villanueva, Department of Mathematics, University of Iowa
- Ben Warhurst, School of Mathematics, University of New South Wales
- Pavel Winternitz, Centre de Resherches Mathematiques, University of Montreal
- Thomas Wolf, Department of Mathematics, Brock University
- Keizo Yamaguchi, Department of Mathematics, Hokkaido University
- Jin Yue, Department of Mathematics and Statistics, Dalhousie University
- Ismet Yurdusen, Centre de Recherches Mathematiques, University of Montreal
- Igor Zelenko, Sector of Functional Analysis and its Applications, International School for Advanced Studies (SISSA/ISAS)
- Renat Zhdanov, Bio-Key International

1997–1998	Emerging Applications of Dynamical Systems
1998–1999	Mathematics in Biology
1999–2000	Reactive Flows and Transport Phenomena
2000–2001	Mathematics in Multimedia
2001–2002	Mathematics in the Geosciences
2002–2003	Optimization
2003–2004	Probability and Statistics in Complex Systems: Genomics, Networks, and Financial Engineering
2004–2005	Mathematics of Materials and Macromolecules: Multiple Scales, Disorder, and Singularities
2005-2006	Imaging
2006-2007	Applications of Algebraic Geometry
2007-2008	Mathematics of Molecular and Cellular Biology
2008-2009	Mathematics and Chemistry

IMA SUMMER PROGRAMS

1987	Robotics
1988	Signal Processing
1989	Robust Statistics and Diagnostics
1990	Radar and Sonar (June 18–29) New Directions in Time Series Analysis (July 2–27)
1991	Semiconductors
1992	Environmental Studies: Mathematical, Computational, and Statistical Analysis
1993	Modeling, Mesh Generation, and Adaptive Numerical Methods for Partial Differential Equations
1994	Molecular Biology
1995	Large Scale Optimizations with Applications to Inverse Problems, Optimal Control and Design, and Molecular and Structural Optimization
1996	Emerging Applications of Number Theory (July 15–26) Theory of Random Sets (August 22–24)
1997	Statistics in the Health Sciences
1998	Coding and Cryptography (July 6–18) Mathematical Modeling in Industry (July 22–31)
1999	Codes, Systems, and Graphical Models (August 2–13, 1999)
2000	Mathematical Modeling in Industry: A Workshop for Graduate Students (July 19–28)
2001	Geometric Methods in Inverse Problems and PDE Control (July 16–27)
2002	Special Functions in the Digital Age (July 22–August 2)

2003 Probability and Partial Differential Equations in Modern Applied Mathematics (July 21–August 1)
2004 n-Categories: Foundations and Applications (June 7–18)
2005 Wireless Communications (June 22–July 1)
2006 Symmetries and Overdetermined Systems of Partial Differential Equations (July 17–August 4)
2007 Classical and Quantum Approaches in Molecular Modeling (July 23-August 3)

IMA "HOT TOPICS" WORKSHOPS

- Challenges and Opportunities in Genomics: Production, Storage, Mining and Use, April 24–27, 1999
- Decision Making Under Uncertainty: Energy and Environmental Models, July 20–24, 1999
- Analysis and Modeling of Optical Devices, September 9–10, 1999
- Decision Making under Uncertainty: Assessment of the Reliability of Mathematical Models, September 16–17, 1999
- Scaling Phenomena in Communication Networks, October 22–24, 1999
- Text Mining, April 17–18, 2000
- Mathematical Challenges in Global Positioning Systems (GPS), August 16–18, 2000
- Modeling and Analysis of Noise in Integrated Circuits and Systems, August 29–30, 2000
- Mathematics of the Internet: E-Auction and Markets, December 3–5, 2000
- Analysis and Modeling of Industrial Jetting Processes, January 10–13, 2001
- Special Workshop: Mathematical Opportunities in Large-Scale Network Dynamics, August 6–7, 2001
- Wireless Networks, August 8–10 2001
- Numerical Relativity, June 24–29, 2002
- Operational Modeling and Biodefense: Problems, Techniques, and Opportunities, September 28, 2002
- Data-driven Control and Optimization, December 4–6, 2002
- Agent Based Modeling and Simulation, November 3–6, 2003
- Enhancing the Search of Mathematics, April 26-27, 2004
- Compatible Spatial Discretizations for Partial Differential Equations, May 11-15, 2004
- Adaptive Sensing and Multimode Data Inversion, June 27–30, 2004
- Mixed Integer Programming, July 25–29, 2005
- New Directions in Probability Theory, August 5–6, 2005
- Negative Index Materials, October 2-4, 2006
- The Evolution of Mathematical Communication in the Age of Digital Libraries, December 8-9, 2006

- Math is Cool! and Who Wants to Be a Mathematician?, November 3, 2006
- Special Workshop: Blackwell-Tapia Conference, November 3-4, 2006
- Stochastic Models for Intracellular Reaction Networks, May 11-13, 2008

SPRINGER LECTURE NOTES FROM THE IMA:

The Mathematics and Physics of Disordered Media
Editors: Barry Hughes and Barry Ninham
(Lecture Notes in Math., Volume 1035, 1983)

Orienting Polymers
Editor: J.L. Ericksen
(Lecture Notes in Math., Volume 1063, 1984)

New Perspectives in Thermodynamics
Editor: James Serrin
(Springer-Verlag, 1986)

Models of Economic Dynamics
Editor: Hugo Sonnenschein
(Lecture Notes in Econ., Volume 264, 1986)

Current Volumes:

24 **Mathematics in Industrial Problems, Part 2** A. Friedman
25 **Solitons in Physics, Mathematics, and Nonlinear Optics**
P.J. Olver and D.H. Sattinger (eds.)
26 **Two Phase Flows and Waves**
D.D. Joseph and D.G. Schaeffer (eds.)
27 **Nonlinear Evolution Equations that Change Type**
B.L. Keyfitz and M. Shearer (eds.)
28 **Computer Aided Proofs in Analysis**
K. Meyer and D. Schmidt (eds.)
29 **Multidimensional Hyperbolic Problems and Computations**
A. Majda and J. Glimm (eds.)
30 **Microlocal Analysis and Nonlinear Waves**
M. Beals, R. Melrose, and J. Rauch (eds.)
31 **Mathematics in Industrial Problems, Part 3** A. Friedman
32 **Radar and Sonar, Part I**
R. Blahut, W. Miller, Jr., and C. Wilcox
33 **Directions in Robust Statistics and Diagnostics: Part I**
W.A. Stahel and S. Weisberg (eds.)
34 **Directions in Robust Statistics and Diagnostics: Part II**
W.A. Stahel and S. Weisberg (eds.)
35 **Dynamical Issues in Combustion Theory**
P. Fife, A. Liñán, and F.A. Williams (eds.)
36 **Computing and Graphics in Statistics**
A. Buja and P. Tukey (eds.)
37 **Patterns and Dynamics in Reactive Media**
H. Swinney, G. Aris, and D. Aronson (eds.)
38 **Mathematics in Industrial Problems, Part 4** A. Friedman
39 **Radar and Sonar, Part II**
F.A. Grünbaum, M. Bernfeld, and R.E. Blahut (eds.)
40 **Nonlinear Phenomena in Atmospheric and Oceanic Sciences**
G.F. Carnevale and R.T. Pierrehumbert (eds.)
41 **Chaotic Processes in the Geological Sciences** D.A. Yuen (ed.)
42 **Partial Differential Equations with Minimal Smoothness and Applications** B. Dahlberg, E. Fabes, R. Fefferman, D. Jerison, C. Kenig, and J. Pipher (eds.)
43 **On the Evolution of Phase Boundaries**
M.E. Gurtin and G.B. McFadden
44 **Twist Mappings and Their Applications**
R. McGehee and K.R. Meyer (eds.)
45 **New Directions in Time Series Analysis, Part I**
D. Brillinger, P. Caines, J. Geweke, E. Parzen, M. Rosenblatt, and M.S. Taqqu (eds.)
46 **New Directions in Time Series Analysis, Part II**
D. Brillinger, P. Caines, J. Geweke, E. Parzen, M. Rosenblatt, and M.S. Taqqu (eds.)
47 **Degenerate Diffusions**
W.-M. Ni, L.A. Peletier, and J.-L. Vazquez (eds.)
48 **Linear Algebra, Markov Chains, and Queueing Models**
C.D. Meyer and R.J. Plemmons (eds.)

49 **Mathematics in Industrial Problems, Part 5** A. Friedman
50 **Combinatorial and Graph-Theoretic Problems in Linear Algebra**
R.A. Brualdi, S. Friedland, and V. Klee (eds.)
51 **Statistical Thermodynamics and Differential Geometry of Microstructured Materials**
H.T. Davis and J.C.C. Nitsche (eds.)
52 **Shock Induced Transitions and Phase Structures in General Media** J.E. Dunn, R. Fosdick, and M. Slemrod (eds.)
53 **Variational and Free Boundary Problems**
A. Friedman and J. Spruck (eds.)
54 **Microstructure and Phase Transitions**
D. Kinderlehrer, R. James, M. Luskin, and J.L. Ericksen (eds.)
55 **Turbulence in Fluid Flows: A Dynamical Systems Approach**
G.R. Sell, C. Foias, and R. Temam (eds.)
56 **Graph Theory and Sparse Matrix Computation**
A. George, J.R. Gilbert, and J.W.H. Liu (eds.)
57 **Mathematics in Industrial Problems, Part 6** A. Friedman
58 **Semiconductors, Part I**
W.M. Coughran, Jr., J. Cole, P. Lloyd, and J. White (eds.)
59 **Semiconductors, Part II**
W.M. Coughran, Jr., J. Cole, P. Lloyd, and J. White (eds.)
60 **Recent Advances in Iterative Methods**
G. Golub, A. Greenbaum, and M. Luskin (eds.)
61 **Free Boundaries in Viscous Flows**
R.A. Brown and S.H. Davis (eds.)
62 **Linear Algebra for Control Theory**
P. Van Dooren and B. Wyman (eds.)
63 **Hamiltonian Dynamical Systems: History, Theory, and Applications**
H.S. Dumas, K.R. Meyer, and D.S. Schmidt (eds.)
64 **Systems and Control Theory for Power Systems**
J.H. Chow, P.V. Kokotovic, R.J. Thomas (eds.)
65 **Mathematical Finance**
M.H.A. Davis, D. Duffie, W.H. Fleming, and S.E. Shreve (eds.)
66 **Robust Control Theory** B.A. Francis and P.P. Khargonekar (eds.)
67 **Mathematics in Industrial Problems, Part 7** A. Friedman
68 **Flow Control** M.D. Gunzburger (ed.)
69 **Linear Algebra for Signal Processing**
A. Bojanczyk and G. Cybenko (eds.)
70 **Control and Optimal Design of Distributed Parameter Systems**
J.E. Lagnese, D.L. Russell, and L.W. White (eds.)
71 **Stochastic Networks** F.P. Kelly and R.J. Williams (eds.)
72 **Discrete Probability and Algorithms**
D. Aldous, P. Diaconis, J. Spencer, and J.M. Steele (eds.)
73 **Discrete Event Systems, Manufacturing Systems, and Communication Networks**
P.R. Kumar and P.P. Varaiya (eds.)
74 **Adaptive Control, Filtering, and Signal Processing**
K.J. Åström, G.C. Goodwin, and P.R. Kumar (eds.)

75 **Modeling, Mesh Generation, and Adaptive Numerical Methods for Partial Differential Equations** I. Babuska, J.E. Flaherty, W.D. Henshaw, J.E. Hopcroft, J.E. Oliger, and T. Tezduyar (eds.)

76 **Random Discrete Structures** D. Aldous and R. Pemantle (eds.)

77 **Nonlinear Stochastic PDEs: Hydrodynamic Limit and Burgers' Turbulence** T. Funaki and W.A. Woyczynski (eds.)

78 **Nonsmooth Analysis and Geometric Methods in Deterministic Optimal Control** B.S. Mordukhovich and H.J. Sussmann (eds.)

79 **Environmental Studies: Mathematical, Computational, and Statistical Analysis** M.F. Wheeler (ed.)

80 **Image Models (and their Speech Model Cousins)** S.E. Levinson and L. Shepp (eds.)

81 **Genetic Mapping and DNA Sequencing** T. Speed and M.S. Waterman (eds.)

82 **Mathematical Approaches to Biomolecular Structure and Dynamics** J.P. Mesirov, K. Schulten, and D. Sumners (eds.)

83 **Mathematics in Industrial Problems, Part 8** A. Friedman

84 **Classical and Modern Branching Processes** K.B. Athreya and P. Jagers (eds.)

85 **Stochastic Models in Geosystems** S.A. Molchanov and W.A. Woyczynski (eds.)

86 **Computational Wave Propagation** B. Engquist and G.A. Kriegsmann (eds.)

87 **Progress in Population Genetics and Human Evolution** P. Donnelly and S. Tavaré (eds.)

88 **Mathematics in Industrial Problems, Part 9** A. Friedman

89 **Multiparticle Quantum Scattering With Applications to Nuclear, Atomic and Molecular Physics** D.G. Truhlar and B. Simon (eds.)

90 **Inverse Problems in Wave Propagation** G. Chavent, G. Papanicolau, P. Sacks, and W.W. Symes (eds.)

91 **Singularities and Oscillations** J. Rauch and M. Taylor (eds.)

92 **Large-Scale Optimization with Applications, Part I: Optimization in Inverse Problems and Design** L.T. Biegler, T.F. Coleman, A.R. Conn, and F. Santosa (eds.)

93 **Large-Scale Optimization with Applications, Part II: Optimal Design and Control** L.T. Biegler, T.F. Coleman, A.R. Conn, and F. Santosa (eds.)

94 **Large-Scale Optimization with Applications, Part III: Molecular Structure and Optimization** L.T. Biegler, T.F. Coleman, A.R. Conn, and F. Santosa (eds.)

95 **Quasiclassical Methods** J. Rauch and B. Simon (eds.)

96 **Wave Propagation in Complex Media** G. Papanicolaou (ed.)

97 **Random Sets: Theory and Applications** J. Goutsias, R.P.S. Mahler, and H.T. Nguyen (eds.)

98 **Particulate Flows: Processing and Rheology** D.A. Drew, D.D. Joseph, and S.L. Passman (eds.)

99 **Mathematics of Multiscale Materials** K.M. Golden, G.R. Grimmett, R.D. James, G.W. Milton, and P.N. Sen (eds.)

100 **Mathematics in Industrial Problems, Part 10** A. Friedman

101 **Nonlinear Optical Materials** J.V. Moloney (ed.)

102 **Numerical Methods for Polymeric Systems** S.G. Whittington (ed.)

103 **Topology and Geometry in Polymer Science** S.G. Whittington, D. Sumners, and T. Lodge (eds.)

104 **Essays on Mathematical Robotics** J. Baillieul, S.S. Sastry, and H.J. Sussmann (eds.)

105 **Algorithms For Parallel Processing** M.T. Health, A. Ranade, and R.S. Schreiber (eds.)

106 **Parallel Processing of Discrete Problems** P.M. Pardalos (ed.)

107 **The Mathematics of Information Coding, Extraction, and Distribution** G. Cybenko, D.P. O'Leary, and J. Rissanen (eds.)

108 **Rational Drug Design** D.G. Truhlar, W. Howe, A.J. Hopfinger, J. Blaney, and R.A. Dammkoehler (eds.)

109 **Emerging Applications of Number Theory** D.A. Hejhal, J. Friedman, M.C. Gutzwiller, and A.M. Odlyzko (eds.)

110 **Computational Radiology and Imaging: Therapy and Diagnostics** C. Börgers and F. Natterer (eds.)

111 **Evolutionary Algorithms** L.D. Davis, K. De Jong, M.D. Vose, and L.D. Whitley (eds.)

112 **Statistics in Genetics** M.E. Halloran and S. Geisser (eds.)

113 **Grid Generation and Adaptive Algorithms** M.W. Bern, J.E. Flaherty, and M. Luskin (eds.)

114 **Diagnosis and Prediction** S. Geisser (ed.)

115 **Pattern Formation in Continuous and Coupled Systems: A Survey Volume** M. Golubitsky, D. Luss, and S.H. Strogatz (eds.)

116 **Statistical Models in Epidemiology, the Environment, and Clinical Trials** M.E. Halloran and D. Berry (eds.)

117 **Structured Adaptive Mesh Refinement (SAMR) Grid Methods** S.B. Baden, N.P. Chrisochoides, D.B. Gannon, and M.L. Norman (eds.)

118 **Dynamics of Algorithms** R. de la Llave, L.R. Petzold, and J. Lorenz (eds.)

119 **Numerical Methods for Bifurcation Problems and Large-Scale Dynamical Systems** E. Doedel and L.S. Tuckerman (eds.)

120 **Parallel Solution of Partial Differential Equations** P. Bjørstad and M. Luskin (eds.)

121 **Mathematical Models for Biological Pattern Formation** P.K. Maini and H.G. Othmer (eds.)

122 **Multiple-Time-Scale Dynamical Systems** C.K.R.T. Jones and A. Khibnik (eds.)

123 **Codes, Systems, and Graphical Models** B. Marcus and J. Rosenthal (eds.)

124 **Computational Modeling in Biological Fluid Dynamics** L.J. Fauci and S. Gueron (eds.)

125 **Mathematical Approaches for Emerging and Reemerging Infectious Diseases: An Introduction** C. Castillo-Chavez with S. Blower, P. van den Driessche, D. Kirschner, and A.A. Yakubu (eds.)

126 **Mathematical Approaches for Emerging and Reemerging Infectious Diseases: Models, Methods, and Theory** C. Castillo-Chavez with S. Blower, P. van den Driessche, D. Kirschner, and A.A. Yakubu (eds.)

127 **Mathematics of the Internet: E-Auction and Markets**
B. Dietrich and R.V. Vohra (eds.)

128 **Decision Making Under Uncertainty: Energy and Power**
C. Greengard and A. Ruszczynski (eds.)

129 **Membrane Transport and Renal Physiology**
H. Layton and A.M. Weinstein (eds.)

130 **Atmospheric Modeling**
D.P. Chock and G.R. Carmichael (eds.)

131 **Resource Recovery, Confinement, and Remediation of Environmental Hazards**
J. Chadam, A. Cunningham, R.E. Ewing, P. Ortoleva, and M.F. Wheeler (eds.)

132 **Fractals in Multimedia**
M.F. Barnsley, D. Saupe, and E.R. Vrscay (eds.)

133 **Mathematical Methods in Computer Vision**
P.J. Olver and A. Tannenbaum (eds.)

134 **Mathematical Systems Theory in Biology, Communications, Computation, and Finance** J. Rosenthal and D.S. Gilliam (eds.)

135 **Transport in Transition Regimes**
N. Ben Abdallah, A. Arnold, P. Degond, I. Gamba, R. Glassey, C.D. Lawrence, and C. Ringhofer (eds.)

136 **Dispersive Transport Equations and Multiscale Methods**
N. Ben Abdallah, A. Arnold, P. Degond, I. Gamba, R. Glassey, C.D. Lawrence, and C. Ringhofer (eds.)

137 **Geometric Methods in Inverse Problems and PDE Control**
C.B. Cooke, I. Lasiecka, G. Uhlmann, and M.S. Vogelius (eds.)

138 **Mathematical Foundations of Speech and Language Processing**
M. Johnson, S. Khudanpur, M. Ostendorf, and R. Rosenfeld (eds.)

139 **Time Series Analysis and Applications to Geophysical Systems**
D.R. Brillinger, E.A. Robinson, and F.P. Schoenberg (eds.)

140 **Probability and Partial Differential Equations in Modern Applied Mathematics** E.C. Waymire and J. Duan (eds.)

141 **Modeling of Soft Matter**
Maria-Carme T. Calderer and Eugene M. Terentjev (eds.)

142 **Compatible Spatial Discretizations**
Douglas N. Arnold, Pavel B. Bochev, Richard B. Lehoucq, Roy A. Nicolaides, and Mikhail Shashkov (eds.)

143 **Wireless Communications** Prathima Agrawal, Daniel Matthew Andrews, Philip J. Fleming, George Yin, and Lisa Zhang (eds.)

144 **Symmetries and Overdetermined Systems of Partial Differential Equations** Michael Eastwood and Willard Miller, Jr. (eds.)

145 **Topics in Stochastic Analysis and Nonparametric Estimation**
Pao-Liu Chow, Boris Mordukhovich, and George Yin (eds.)

146 **Algorithms in Algebraic Geometry**
Alicia Dickenstein, Frank-Olaf Schreyer, and Andrew J. Sommese (eds.)

147 **Symmetric Functionals on Random Matrices and Random Matchings Problems**
Grzegorz A. Rempała and Jacek Wesołowski